Prestressed
Concrete Structures

Prestressed Concrete Structures

PASALA DAYARATNAM

Formely Professor of Civil Engineering
Indian Institute of Technology, Kanpur

Oxford & IBH Publishing Co. Pvt. Ltd.
New Delhi

CBS

(A Unit of CBS Publishers & Distributors Pvt Ltd)

New Delhi • Bengaluru • Chennai • Kochi • Kolkata • Mumbai
Hyderabad • Jharkhand • Nagpur • Patna • Pune • Uttarakhand

PRESTRESSED CONCRETE STRUCTURES

CBS Publishers & Distributors Pvt Ltd
204 FIE, Patparganj Industrial Area, Delhi-110 092
E-mail: delhi@cbspd.com, cbspubs@airtelmail.in

© 2016, Copyright Reserved

Reprint 2018

ISBN 978-81-204-1791-5

Printed at Chaman Enterprises, New Delhi.

To my parents
Kamalakshmi and Solomon, Pasala

Preface to the Sixth Edition

Prestressed concrete Structures materials that have dominated the construction and industry for more eight decades. Almost all engineers in the construction or design of civil engineering structures have to have basic knowledge of the subject. All civil engineers will have something to do with concrete at sometime or other. A large number of books are also being published for students and practicing engineers. The method of design is being updated based on the new materials, research and practice, Prestressed design is the most popular method of concrete structures during the last four decades throughout the world (IS: 1342-1980) (2004). The sixth revision of the mother code on concrete structures is the most the extensively used structural material that has dominated the construction industry. Almost all engineers in the construction or design of civil engineering structures have to have basic knowledge of the subject. All civil engineers will have something to do with concrete at sometime or other. Codes and specification on materials, design and construction have been developed and are continuously updated. Extensive research and developmental work on cement, concrete, prestressed and construction is continuously carried out in all developed and developing countries. A large number of books are also being published for students and practicing engineers. The method of design is being updated based on the new materials, research and practice. Further, the design is also an art, experience and innovation. The author having been in the field of teaching, research and consultancy for the last forty years has evolved the book to suit the basic needs of the student and the practicing engineer. It is believed that the book will sustain the continuous changes of the modern concrete practice. It is aimed at the training of the mind of a civil engineer.

Design follows simple principles of statics and stability but the overall safety and durability of a structure depends on the art of structural detailing of members and the interconnectivity. A large number of examples with illustrative sketches. The market is flooded by variety of structural analysis computer software. Engineering colleges and the training centers provide training in structural analysis software. Software prints out results based on the input and idealization of the structural skeleton. Sometimes, engineers are misled by long detailed output that may or may not be very relevant to the actual structure. It is important that the user must be an experienced structural engineer who can model the structure to suit the software for reliable results.

Kanpur 1991   PASALA DAYARATNAM

Preface to the First Edition

This book is the natural growth of my several years of teaching and consulting in structural engineering with a particular emphasis on prestressed concrete structures. Economic criteria and a rational approach to design need not necessarily depend on intuition and experience but can be based on sound scientific principles coupled with practical constraints. The book is developed with detailed examples and illustrations to give basic concepts and a feeling for the variety of prestressed concrete structures. It is meant as a complete course in prestressed concrete structures useful for teaching advanced undergraduate and postgraduate students. Considerable literature supplied by various prestressed concrete manufacturers and consultants has been incorporated which gives the book a practical orientation.

The first chapter develops concepts of prestressing and the second chapter presents introductory material for prestressed concrete. These two chapters together form an introduction to the subject. Chapters three to seven discuss the service load design of various discrete structural systems such as simple composite and non-composite beams, tie bars, columns and piers, pipes and ring beams, and continuous beams. There is a greater emphasis on economic criteria of design and the cross-sectional responses of various elements to different parameters. Chapters eight and nine present ultimate flexure and shear-moment failure criteria of design. Rigorous and approximate methods of computation of ultimate capacity of the beams are presented. The ultimate load design of prestressed concrete beams is also illustrated.

Chapters ten and eleven discuss the design of continuum structures such as two-way and flat slabs, folded plates, water tanks, domes, and ring beams, as well as cylindrical shell roofs. These two chapters are written especially for a student or designer who does not have a basic background in the classical plate and shell theories, and still wants to design such structures.

Chapters twelve and thirteen present design of end zones of pretensioned and post-tensioned prestressed concrete structural elements.

A set of tables useful for the design of beams and shells is presented in Appendix A. Relevant references to the subject are given in Appendix B. The references given are not meant as an exhaustive set.

x

I take this opportunity to express my thanks to Mr. S. C. Goel, Indian Institute of Technology, Kanpur, for his constant help at various stages of development of the book. I am grateful to Mrs. Jane G. Merriam, for her kind help in editing the book; the Indian Institute of Technology, Kanpur, for providing encouraging facilities for preparing the manuscript; and the students who have made significant comments for the presentation of the book.

Kanpur 1970   PASALA DAYARATNAM

Contents

List of Symbols

Greek Letters

α = change in slope of cable/tendon; angle of plane of principal stress; angle of deviation in folded plate

β = compatibility factor; bending moment coefficients in slab

γ = unit weight (density) of concrete or water

Δ = shape factor ($= y_b/y_t$)

ϵ = eccentricity ratio ($= e/b$); strain

ϵ_0 = ultimate strain

ϵ_a = strain due to shrinkage

ϵ_b = strain in bottom fibre

ϵ_c = mean elastic strain; strain in concrete

ϵ_{ce} = strain in concrete at working

ϵ_e = strain in prestressed steel

ϵ_{se} = effective strain in prestressed steel

ϵ_t = strain in top fibre; tension strain in concrete

ϵ_u = limiting compressive strain of concrete

η = P_e/P_i

θ = slope of the tendon; rotation of beams etc.

λ = A_{cs}/A

μ = coefficient of friction

ν = Poisson's ratio

ρ = efficiency factor ($= r^2/h^2$)

xviii

Σ_0	=	total parameter of the steel
σ_1	=	maximum principal stress
σ_2	=	minimum principal stress
σ	=	stress
σ_a	=	average stress (= P/A)
σ_{ci}	=	compressive stress in concrete at transfer
σ_s	=	stress in steel
σ_{se}	=	effective stress in steel
σ_{s1}	=	stress in steel at anchorage
σ_{sl}	=	stress in steel at transfer (initial strain)
σ_{sv}	=	allowable stress in vertical stirrup steel
σ_t	=	stress in the folded plate due to shear force T
σ_{te}	=	tensile stress in concrete at working loads
σ_{it}	=	stress at tension fibre at transfer
σ_y	=	stress in y-direction (i.e., transverse stress in beam): bursting stress
σ_{ymax}	=	maximum tensile stress ordinate (bursting stress)
σ_x	=	normal stress on x-plane
τ	=	Shear stress
τ_b	=	bond stress
ϕ	=	ratio of prestressing force at mid-section to that at the end section of the beam; rotation of joints
ω	=	depth factor (= $h\sigma^1_{cu}/\gamma L^2$)

English Letters

Capital Letters

A	=	area of cross section
A_c	=	area of composite section; area of concrete column
A_{cs}	=	area of concrete slab in composite beams ($b_s t_a$)
A_{cse}	=	effective area of slab (= $b_{se} t_s = n_c A_{cs}$)
A_p	=	area of prestressed steel
A_s	=	area of non-tensioned steel
A_{sv}	=	area of shear reinforcement
B	=	breadth of column
C	=	compression
c_c	=	creep coefficient
CA	=	centroidal axis
CGC	=	centre of gravity of gross section

CGS	=	centre of gravity of prestressing steel
D	=	diameter of steel wires; total depth of (column) sections
E_e	=	Young's modulus of concrete
E_s	=	Young's modulus of steel
E_{cs}	=	Young's modulus of material of the slab in composite construction
F	=	axial force (direct tension or compression)
F_{bst}	=	bursting force
F_e	=	axial load at cracking
F_g	=	load factor for dead load
F_1	=	load factor for live load
F_u	=	ultimate axial load
GJ	=	torsional rigidity of section
H	=	horizontal force in edge of shells
I	=	moment of inertia
I_c	=	moment of inertia of the gross composite section about its centroidal axis
I_h	=	moment of inertia about horizontal axis
I_v	=	moment of inertia about vertical axis
L	=	length of span
M	=	bending moment
M_a	=	moment due to external loads ($= M_a + M_s$)
M_{be}	=	bending moment at working load due to unbalanced load
M_{bi}	=	bending moment at transfer due to unbalanced load
M_g	=	bending moment due to self weight
M_{gs}	=	bending moment due to self weight of slab
M_1	=	bending moment due to live load
M_{pb}	=	bending moment due to prestressing balancing force
M_r	=	resisting moment capacity
M_s	=	bending moment due to superimposed load
M_t	=	bending moment due to total load ($= M_a + M_s$)
M_u	=	ultimate bending moment
N	=	normal force
P	=	prestressing force, line load at folds in folded plates
P_a	=	axial force to shrinkage
P_E	=	Euler's buckling load
$P_e = P_W$	=	effective prestressing force after loss of prestress
P_o	=	prestressing force at anchorage (force indicated by jack)
P_i	=	initial prestressing force (at transfer)
P_{ia}	=	prestressing force in long span of slabs

P_{ib} = prestressing force in short span of slabs

P_x = prestressing force at distance

Q = transverse shear in slabs; $M_u/bd^2\sigma'_{cu}$

Q_y = first moment of area

R = moment ratio (= M_d/M_g); reactions of supports; radius of curvature

R_1 = M_d/M_{gs}

R_2 = M_{s1}/M_{gs}

T = age of concrete (in days) at transfer; self balancing shear force at folds in folded plates

V = shear force, vertical force

V_e = effective shear force

V_p = shear force due to prestressing force $\left(= P_e \dfrac{dy}{dx} \right)$

Z = modulus of section

Small Letters

a = distance of neutral axis from top fibre; radius of sphere; depth of anchor plate; span of slab in x-direction, lever arm

b = width; span of slab in y-direction (short span length)

b_b = width of bottom flange

b_f = width of compression flange

b_t = width of top flange

b_s = width of slab

c = constant

c_a = $A/b_t h$

c_b = t_b/h

c_{ce} = compression stress coefficient at effective load

c_{ci} = compression stress coefficient at transfer (initial)

c_f = b_b/b_t

c_g = moment coefficient of self weight

c_t = I/bh^3

c_t = t_t/h

c_{se} = tension stress coefficient at effective load

c_{ti} = tension stress coefficient at transfer

c_w = b_w/b_t

c_{yb} = y_b/h

c_{yt} = y_t/h

d	=	distance from *CGS* to top fibre
d'	=	distance from *CGS* to bottom fibre
e	=	eccentricity
e_x	=	eccentricity at distance x
f	=	strength = ultimate stress
f_{ck}	=	characteristic strength of concrete
	=	strength of 150 mm cube at 28 days of water curing
f_{ci}	=	strength of concrete at the time of transfer of prestress
f_k	=	characteristic strength
f_p	=	characteristic strength of pretensioned steel
f_{pu}	=	allowable strength of pretensioned steel
f_{sy}	=	characteristic strength of stirrup steel
f_y	=	characteristic strength of non-tensioned steel
g	=	maximum sag of cable
h	=	overall depth of the beam; rise of shell
k	=	wobble correction coefficient
k_1 or k_7 =		concrete strength coefficient
k_b	=	bond factor
l_b	=	bond length
l_t	=	transmission length
l_u	=	total bond length
m	=	reinforcement ratio; bending moment in transverse direction in folded plates, modular ratio
m_c	=	Poisson ratio for concrete
m_s	=	Poisson's ratio for steel
p	=	reinforcement ratio
p_o	=	$\dfrac{A_p}{bd} \dfrac{f_{ck}}{f_p}$
p_s	=	ratio shear steel ($= A_m/s_b$)
q	=	intensity of load per unit area; $A_s/b_1 d$
r	=	radius of gyration
r_k	=	radius of gyration of composite section
s_y	=	spacing of stirrups
t	=	thickness, thickness of compression flange
t_b	=	thickness of bottom flange
t_s	=	thickness of slab/plate
t_t	=	thickness of top flange

xxii

$u's$	=	composite section coefficients
v	=	displacement (lateral or horizontal direction)
w	=	vertical deflection; load per unit length
y	=	profile of cable; distance from *CGC*
y_b	=	distance of bottom fibre from *CGC*
y_{bc}	=	distance of bottom fibre of the beam to the centroidal axis of the composite section
y_t	=	distance of top fibre from *CGC*
y_{tc}	=	distance of top fibre of the beam from the centroidal axis of the composite section
z	=	discrete function defining the cable profile both *x*- and *y*-directions

Subscripts

a	=	shrinkage; a span direction in slabs
b	=	bottom flange; bond; beam; balancing: *b* span direction in slabs
c	=	concrete; compression; composite section
d	=	dome
e	=	effective prestressing stage (working load condition)
g	=	self weight
h	=	horizontal
i	=	initial condition
l	=	live load; line load
m	=	moment
o	=	external load; anchorage
s	=	steel, slab; shell
t	=	top flange; transfer condition; total tension
u	=	ultimate
v	=	vertical transverse
w	=	web, working load
ϕ	=	moment; rotation

Superscripts

prime(')	=	ultimate; secondary; maximum
c	=	column strip
m	=	middle strip

CHAPTER 1

Materials Concrete and Prestressed Concrete Structures

1.1. Aggregates and Sand in Concrete

Broken stone material, gravel and sand are called *aggregates* for production of concrete. Unless otherwise specified in this book, aggregate refers to that from stone and used in concrete constructions. *Fine aggregate* is the basic stone material having a particle size less than 5 mm, and 90 to 100 per cent of which pass through 4.75 mm sieve. Natural sand and finely crushed stone are treated as fine aggregate. *Coarse aggregate* is the stone material most of which is retained on 4.75 mm sieve. Gravel or crushed stone are the usual coarse aggregate. There are other types of aggregates, like *broken brick-aggregate*, *cinder-aggregate*, *light-weight*, *slag and heavy-weight aggregate*, etc. that are used in special concrete constructions. Coarse and fine aggregates together are used in concrete whereas only fine aggregate with either cement or lime is used in mortar for plastering and jointing. Particles larger than 4.75 mm are grouped into coarse aggregate, particles between 4.75 mm and 60 microns (μm = micro millimeter) are considered to be fine aggregate, natural particles smaller than 60 microns and are larger than 2 microns are grouped under *silt* and still smaller particle material is called *clay*.

Fine aggregate should originate from rocky material. The fine aggregate is further classified into four sub-groups, as per the Indian Standards. Grade I is the coarsest and grade IV is the finest of the fine aggregate. For grade one, about 90 to 100 per cent of the material must pass through 4.75 mm sieve and about 60 to 80 per cent must pass through the next lower standard sieve, namely 2.36 mm. The first three graded zones are usually acceptable for reinforced or prestressed concrete construction. However, the grade four is too fine and not recommended for concrete construction. It can be used in plastering.

All-in-aggregate: Aggregate containing both coarse and fine aggregates is called all-in-aggregate.

Shape of aggregate: Aggregate shape can be either round, angular or a combination of the both. But it should not be flaky or oblong. It should not contain deleterious materials such as clay lumps, coal particles, soft material and too fine material that pass through 75 micron sieve. The maximum limit of combined deleterious materials in an aggregate is five per cent. *Size of a coarse aggregate* is specified by a sieve size. A 40 mm aggregate normally means that 100 per cent of the aggregate should pass through 63 mm IS sieve (that is next higher of 40 mm IS sieve), and 85 to 100 per cent pass through 40 mm IS sieve. And about 0 to 30 per cent may pass through the next lower size, i.e. 20 mm sieve. Commonly used size terminology of the aggregate is: Very large aggregate 80 mm to 150 mm large aggregate is 40 mm to 80 mm, medium size aggregate is 20 mm to 40 mm, and small size coarse aggregate is 4.75 mm to 20 mm. The Indian Standard (IS) sieves are designated by 63, 40, 20, 16, 12.5, 10, 4.75, 2.38, 1.18, (all in mm), 600, 300, 150, and 75 microns. The size of the sieve is specified by the nominal size of the aperture either in millimeters (mm) or in micrometers (microns).

The important characteristics that one has to look for in aggregate are:

1. **Shape, size and surface texture**, 2. Specific **gravity**, 3. **Void** ratio, 4. **Bulk** density, 5. **Moisture** content, 6. **Porosity and** absorption value, 7. Aggregate **crushing** strength, 8. **Abrasion** value, 9. **Flakiness** index, 10. **Elongation** index, 11. Presence of **deleterious** materials, 12. **Soundness**, 13. **Alkali** Aggregate reaction, and 14. **Fineness Modulus.**

Shape, size and surface texture of aggregate is either acquired from natural erosion or through crushing process. The roundness is indicated by the degree of wear of the faces of the particles. The compactness or compatibility of the aggregate depends on the shape of the aggregate. The angularity number of the spherical aggregate is zero, which means that the percentage of the solid volume of rounded aggregate is 67. The angularity number indicates the percentage of voids in excess of the rounded aggregate. The tensile strength of the mortar or concrete depends on the shape and texture of the aggregate.

Fineness Modulus (FM):   All particles in an aggregate are not exactly of the same size. The aggregate consists of particles of different sizes in a range. The aggregate is specified by the distribution of the particles and it is analyzed by sieving through a set of standard sieves. The *fineness modulus* is defined as the sum of the cumulative percentages of aggregate retained on a set of standard sieves numbered 63, 40, 20, 10, 4.75, 2.36, 1.18 mm, and 600, 300, and 150 micron.

Table 1.1.1 indicates the range of fineness moduli of the aggregate. Smaller the fineness modulus means finer aggregate. The range of *FM* of fine aggregate is 1.0 to 3.5. The fineness modulus of 1.0 indicates the finest sand that can be used for plastering and not for mortar or for concrete. The fineness modulus of coarse aggregate is in the range of 5 to 7.5. Fineness modulus of aggregate is an important parameter in the design of the concrete mix.

TABLE 1.1.1 Ranges of fineness modulus of aggregates

Fine Aggregate		Coarse aggregate			Mixed aggregate		
Very fine	Fine	20	25	40	20	25	40
1.0-2.0	2.0-3.5	5.8-6.4	6.0-7.0	6.5-7.5	4.7-5.2	5.0-5.6	5.3-6.0

Specific gravity of a material is the ratio of its weight to the weight of the equivalent volume of distilled water. Stone may contain pores or capillary voids. The absolute specific gravity is the one that excludes the pores in the solid mass. *Absolute specific gravity* is the ratio of the mass (weight) of the solid in vacuum to the mass of distilled water. The absolute specific gravity is of academic interest only. *Apparent specific gravity* of the material is the ratio of the oven dried mass at 100 to 110 degrees Celsius for 24 hours to the weight of the equal volume distilled water. The volume referred here includes the pores in the aggregate. The apparent specific gravity is the most commonly used term or quantity in the engineering design. The specific gravity of a material in any form either in large or small aggregate is same as that of the base stone from which it is derived from. It varies from 2.6 to 2.85 for most stone based aggregate. Specific weight is the product of the specific gravity and the unit weight of water.

The *bulk density* of aggregate is the weight of unit bulk volume of material. The bulk density is measure in two ways, one loose bulk density and the other compact bulk density. Aggregate is pored in a standard container and levelled to the top in a specified manner. The ratio of weight of the loose aggregate measured from the container to the volume of the container is called as *loose bulk density*. In case of compact bulk density, the container is filled in three layers, and each layer is compacted by tamping with a 16 mm standard rod to a specified number of times. The ratio of the weight of the compacted aggregate to the volume of the container is called *compact bulk density*. The compact bulk density is commonly referred as the bulk density. The bulk density depends on the void ratio and it varies from 0.54 to 0.65 times the specific weight of the material. The range of bulk density is 1500 kg/cum to 1900 kg/cum. High-density aggregate may have a bulk density in the range of 1800 kg/cum to 2100 kg/cum.

Bulk aggregate contains solid particles and void spaces in between the particles. *Void ratio* is the ratio of the volume of the voids to that of the solid particles. The void ratio is determined by the following expression:

$$Void\ ratio = 1 - bulk\ density/specific\ weight \qquad (1.1.1)$$

Void ratio in most single-size aggregate is around 0.42 to 0.46 and in graded aggregates it is around 0.38 to 0.44. Less is the void ratio better is the grading of the aggregate. The void ratio can be determined from the amount of water that is needed to fill the voids of a measured volume of the dry aggregate.

Porosity, Moisture content and absorption value: Aggregate contains pores some visible on the surface and others inside but not visible. If the pores are too small or the capillaries are very fine, the viscosity of water may delay the absorption of the water. *Saturated aggregate* is the one in which the pores are fully saturated with water. The saturation must be free from the surface wetness. In other words the percentage of saturation is measured in surface dry condition.

The saturated aggregate that is allowed to surface dry in room temperature is called *air dry aggregate*. Aggregate dried in oven for a prolonged period is called *oven dry* or *bone-dry aggregate*. Aggregate when exposed to rain and atmosphere collects moisture on its surface and absorbs certain portion of it through the pores in the particles. The total water content in the aggregate is divided into two parts. Moisture content increases the bulk volume of fine aggregate whereas the absorbed water content does not change the volume of the aggregate. The increase in volume of the sand due to surface moisture film pushing the particles apart is called *bulking of sand*. The volume of sand increases with increase in the moisture content upto a certain limit, but further increase in the moisture content reduces the bulk volume of the sand. The maximum bulking of the sand depends on the type of sand and it is about 30 per cent with normal occurrence around 8 per cent of the moisture content. The ratio of the volume of moist sand and that of dry sand is called *bulking factor*. The bulking factor varies from 1.0 to 1.35 depending on the moisture content and the type of sand.

Strength of aggregate: The strength of an aggregate depends on the nature of the basic rock from which it is derived, size distribution, shape and texture of the aggregate and the impurities present. Aggregate *crushing, value and aggregate impact value* are the usual indices used to specify the strength quality of an aggregate. The aggregate crushing value is determined by a simple test. As per the Indian standards, a dry sample of aggregate passing through 12.5 mm sieve and retained on 10 mm sieve is subjected to compression test in a specified manner under 400 kN load. The crushed material that passes through 2.36 mm sieve is measured and the percentage of that material with respect to the original sample weight is called the *aggregate crushing value*. It would have been better to measure the strength by the percentage of the particles retained rather than passing through the 2.36 mm sieve. A crushing value more than 40 reflects a weak aggregate. Crushing is value less than 30 mean a strong aggregate. The following maximum limits of crushing value are recommended:

For lean and plain concrete: 45,
for Plain concrete in dams and hydraulic structures: 40,
for reinforced concrete in buildings: 40,
for concrete roads and high strength concrete or High Performance Concrete: 30.

The toughness of the aggregate is measured by the impact test. An impact value of upto 30 indicates good quality aggregate; however, a value upto 45 is acceptable for plain concrete. Aggregate crushing value or impact value for concrete should not be more than 45 for ordinary concrete works, and for wearing surfaces such as roadways and runways it should not be more than 30.

The strength of aggregate is also determined by Los Angeles test. A specified sample of aggregate is loaded into a rotating drum machine in which steel balls are also loaded. The mixture is subjected to a specified number of revolutions and the percentage loss of the weight of the aggregate indicates the weakness or the strength of the aggregate. The Los Angeles test has some good characteristics that indicate the strength of the aggregate under wear and tear conditions. But this and impact tests make more noise compared to the compression test. A minimum of 80 MPa as core crushing strength of rock is normally acceptable for good concrete aggregate. This test is of academic interest. In general either aggregate crushing or impact or Los Angeles tests are preferred.

Abrasion value of aggregate: Concrete roads and runways are subjected to constant wear. In addition to good compressive strength, the aggregate must have good wear resistance. There are special abrasion apparatus such as Los Angeles; Deval machines designed to test the wear of an aggregate. The percentage of material lost in wear in the abrasion test that is performed on a sample of aggregate in a specified manner is considered as the abrasion value of the aggregate. More material lost, more is the abrasion value, and therefore the wear resistance of the aggregate is inversely proportional to the abrasion value.

Flakiness Index: Thin and flaky pieces aggregate may result if the rock is not crushed properly in the aggregate crusher or if the basic rock has thin layers. Sieving the aggregate through a set of sieves having oblong openings indicates the order of magnitude of the flakiness of an aggregate. *Flakiness index* is the total weight of aggregate passing through a set of special sieves designed exclusively to measure the property, expressed as a percentage of the total weight of the sample. Flaky aggregates shouldn't be used for reinforced or strong concrete.

Elongation Index: Oblong aggregate may result because of poor aggregate crushing mechanism or in sieve separation. Concrete produced with oblong aggregate results in poor strength of concrete. The oblong character of the aggregate is indicated by an index called elongation index. The nomenclature of the *elongation index* is not appropriate term to indicate the oblong property. However it is measured by sieving a sample of the aggregate through a set of sieves having oblong openings, specially designed to test the property. It is the total weight of the material retained on various length gauges, expressed as percentage of the sample.

Bond of aggregate: The cement is the bonding agent in concrete, and the shape, size, grading and texture of the aggregate influence the level of bond. The interlocking of the aggregate is an important property that will result in better bonding and better strength of concrete. Interlocking of the aggregate is inversely proportional to the flakiness index or the elongation index. As already mentioned earlier, the bond of the aggregate is directly proportional to the regular angularity and rough texture of the aggregate.

Soundness of an aggregate: Some aggregates disintegrate into smaller particles when exposed to weather and changing temperature even without load. Soundness is the property of the aggregate to withstand the weathering conditions. It depends on the chemical and physical properties of the aggregate. The soundness of an aggregate is measured by subjecting it to alternate wetting in saturated solution of sodium or magnesium sulphate and drying it in an oven through a set of cycles. The reduction in the particle size of the aggregate, which is obtained through a sieve analysis of the tested sample, indicates the soundness or unsoundness of the aggregate. The percentage absorption of moisture is a good indicator of soundness of the aggregate. Unsoundness is partly indicated by higher absorption of moisture. The soundness is more or less inversely proportional to the absorption.

Aggregate and alkali reaction: Sodium and potassium oxides are called as alkalis (Na_2O and K_2O). Small amounts of alkalis are normally present in cement. The alkalis in cement or in other materials of the concrete react with silica of the aggregate in the presence of moisture resulting into a gel. *The silica alkali gel* formation in the pores or in the planes of weakness aggregate increases the volume thus resulting in cracking of concrete. The gel also destroys the bond between the aggregate and the cement paste. The gel formation is a slow phenomenon and takes years before it is manifested through cracking. The rate of gel formation increases with increase in temperature and moisture content. It is also proportional to the porosity of the aggregate. The absence of moisture or high humidity avoids the gel formation thus the alkali reaction. Hydraulic structures must be designed with sound aggregate and low alkali cements to avoid the alkali reaction. Alkali aggregate reaction is measured by mechanical and chemical tests. Mortar bar with pre-assigned alkali content, which is stored over water at about 380 C and then tested for expansion in time.

1.2. Portland Cement

The calcareous material such as limestone or chalk and argillaceous material such as clay or shale containing silica are mixed thoroughly and burnt at a clinkering temperature of about 13500 C to 15000 C. The clinker is then powdered with small amount of gypsum (about 3.5 per cent of clinker). The two basic methods of manufacturing of Portland cement are Dry process and Wet process. The principles of the methods of manufacturing are mentioned here. Figure 1.1 illustrates the basic principle and components in the manufacture of cement. Water is the medium by which the cement develops combining mechanism, therefore these cement are also referred as *hydraulic cements.*

Dry process: The granules also called pellets are baked by hot air and fed into a rotating kiln where the material undergoes chemical changes at temperature in the range of 13500 C to 15000 C. The clinker formed in the kiln is cooled and ground with gypsum in ball mills. The powder ground to a satisfactory level (about 1012 particles per Newton) is separated as cement.

Wet process: The slurry is fed into a rotary kiln from one end and pulverized coal is fed from the other end of the kiln and burnt. The slurry as it moves down in the kiln gets burnt and finally forms into clinker. The clinker is then let through cooling chambers and afterwards pulverized in ball mills along with gypsum. The powder ground to a satisfactory level is separated as cement.

The manufacture of cement is energy intensive; the wet process consumes more energy when compared to the dry process. More than fifty per cent of the expenditure is in the energy input. Further one kilogram of cement production results in producing one kilogram of carbon dioxide as pollution into the air. The dust pollution from cement industry is controlled to a great extent. The cement is packed in 50-kilogram bags and marketed in most countries. In advanced countries, cement is also packed in drums or even transported in bulk. India is yet to pick up bulk packaging and transportation

1.3. Principal Compounds of Portland Cement

The burning up of calcareous and argillaceous material at high temperature results into a clinker of chemical compounds made up of calcium oxide, silicate, aluminate and aluminoferrite. The main chemical compounds in the cement are Tricalcium silicate, Dicalcium silicate, and Tricalcium aluminate and Tetracalcium aluminoferrite. First tricalcium aluminate and Tetracalcium aluminoferrite are formed in the kiln, and then dicalcium silicate is formed, and after its saturation, tricalcium silicate is formed. Tricalcium silicate is mostly responsible in giving early strength of cement whereas the dicalcium silicate contributes to the strength at later stage. The tricalcium aluminate contributes towards the strength developed in the first 48 hours. The principal chemical compounds and the range of percentage of the quantity of them in ordinary Portland Cement are listed in Table 1.3.1.

The basic chemical compounds in Portland cement are CaO, SiO_2, Al_2O_3, Fe_2O_3, MgO, K_2O, and Na_2O. And they are combined to form the principal compound listed in Table 1.3.1.

TABLE 1.3.1   Chemical compounds in Portland Cement

No.	Compound	Composition	Notation	%Range
1	Tricalcium Silicate	$3CaO.SiO_2$	C3S	45–58
2	Dicalcium Silicate	$2CaO.SiO_2$	C2S	15–32
3	Tricalcium Aluminate	$2CaO.Al_2O_3$	C3A	6–13
4	Tetra Calcium Aluminoferrite	$4CaO.Al_2O_3Fe_2O_3$	C4AF	6–12
5	Minor Compounds	MgO, K_2O, Na_2O		3–6

1.4.   Physical Properties of Portland Cement

Some of the Cement chemical requirements of ordinary, rapid hardening and Portland pozzolana cements are given in table 1.4.1. The important physical properties that are of interest to construction and quality control engineers are briefly described in this section. The minimum requirements that are needed for different types of cements are listed by the codes of practice. The specifications listed by the Indian standard on cement are given in a table 1.4.1 The properties are:

❖ Fineness of cement, Normal consistency of cement, Initial setting time of cement, Final setting time of cement, Strength of cement, Soundness of cement, Heat of Hydration of cement.

TABLE 1.4.1   Chemical composition requirements in OPC, Rapid hardening and Pozzolana cement

No.	Substance	Percentage limits
1	Ratio of Lime to Silica, Alumina & Iron Oxide	$0.66 < p < 1.02$
2	Ratio of Alumina to that of Iron Oxide	Not less than 0.66%
3	Insoluble residue	Not more than 2%
4	Magnesia	Not more than 6%
5	Total sulpher content, sulphuric anhydride (SO_3)	Not more than 2.75%
6	Total loss on ignition	Not more than 5%

Fineness of cement: The cement is ground to such fineness, it is almost impossible to grind it to a single size particle. There are experiments that compute the surface area of unit weight of a powder. The chemical compound formation starts on the surface of the cement particles. Similarly the heat of hydration developed by the chemical reaction of cement with water depends on the surface area. The technology of grinding the cement powder is improving every day. The residue weight of the cement after sieving through 90-micron sieve also indicates the approximate size distribution of cement. The percentage residue on 90-micron sieve should not exceed 10 per cent for ordinary Portland cement. American specification list 45-micron sieve instead of 90-micron sieve. The surface area of cement in square metres per one kilogram of cement is called *specific surface* of cement. The surface area should be in the range of 300 square metres per kilogram of cement. The minimum surface area that is required for ordinary Portland cement is 225 m^2 per kg of cement; however this is on the lower side. The present day cements have a specific surface more than 300 square meters per kilogram of the cement.

The particle size of powder can be determined by measuring the free fall of the particles in an inert medium using the principle of fluid mechanics of free falling bodies. The concentration of turbidity of the cement powder in kerosene can be measured by passing light through the solution and the amount of light passed through gives an indication of the number of particles in the solution. *Wagner turbidimeter* is the common instrument that is used in the determination of the specific surface of cement. This method not only indicates the size but also the type of shape of the particles. *Air permeability* test is another commonly used test for the determination of specific surface. Air is allowed pass through a bed of the particle at constant velocity under constant pressure. A manometer measures the pressure drop of the air that passes through a bed of the powder. The surface area is then computed using the specific gravity of cement, porosity of the bed, velocity of flow, and the pressure drop through the bed. There is another method called *Blaine air permeability* test that uses slightly different principle of the permeability. A known volume of air is allowed pass through at an average pressure through a bed of known porosity. The time taken for the air to pass through is measured from which the specific surface is computed. Specific surface of very fine powder such as fine flyash and silica fume cannot be determined by the permeability method. The gas absorption techniques are used in the determination of the specific surface of very fine powders. The silica fume is about ten to fifty times finer than the cement. There could be a variation in the result of specific surface measured by different methods. The codes of practice will indicate the method and the minimum specific surface requirement.

Normal (standard) Consistency of Cement

Cement reacts with moisture and forms complex compounds. This chemical reaction is termed as hydration and heat that is generated during the reaction of cement with water is called *heat of hydration*. The rate of hydration, the chemical kinetics and the physical behavior of the cement paste depend on the amount of water added to the cement. The percentage of water required to produce a reasonable workable cement paste under prescribed testing is called *the normal consistency or standard consistency* of cement. An instrument called Vicat apparatus is used in the determination of standard consistency. A standard plunger of 10 mm diameter is allowed to penetrate partly through a cement paste. The percentage of water that allows the plunger to penetrate a specified distance through a cement-water paste is taken as the *standard consistency* of the cement. The test is conducted under specified humidity and temperature conditions in a specified duration of time after mixing the cement and water. The range of acceptable normal consistency of Portland cement is 26 to 33 per cent of the mass of the cement.

Initial setting time of Cement: Cement paste changes from semi-fluid state to stiffened state, then to hard state and finally solid mass state. The process of transforming the cement paste from semi-fluid state to stiffened or hard state is called setting of cement. The transformation of states is measured through a prescribed test procedure using Vicat apparatus. The time needed for the standard consistency cement paste to become stiffened in permitting 1.13 mm needle under specified weight, to penetrate through a specified distance in a standard mould is called initial setting of cement. The time that is required for the initial setting is called *initial setting time*. The cement mortar or concrete should give reasonable time for mixing, transporting, placing and finishing of the product. Concrete or mortar must be workable without developing bonding for about 30 minutes after the water is added to the Portland cement. The minimum initial setting time for most Portland cement is 30 minutes. The initial setting time is delayed by addition of some compounds in case the placement of cement takes more than 30 minutes.

False Setting: The premature stiffening or hardening of the cement paste is called false set. The false set of cement takes place much before the initial setting. In normal setting, heat of hydration takes place and no such heat is generated in the false set. The process of hardening of cement paste caused by the reaction of pure calcium aluminate may be the reason for false set. Or it may be due to the non-uniform distribution of gypsum in the cement. Gypsum is the material that delays fast setting of the cement. False set doesn't generate good bonding mechanism and hence the concrete can be re-mixed and finished in position. False set is avoided by addition of gypsum to the cement.

Final setting time: The time taken for initiation of hardening of cement paste from the time it is mixed with water is called final setting time. A standard procedure by Vicat apparatus measures the so-called initiation of hardening of the cement paste. The paste should become hard enough to withstand light pressure. A circular cutting edge needle fitted to the Vicat apparatus is gently lowered on to the surface of the stiffened mortar. The time that is taken to ensure that no indentation is made on the mortar specimen is called the final setting time. The concrete can resist minor disturbances or small pressures immediately after the final setting. However, it should not be disturbed till reasonable curing takes place. The chemical reactions between cement and water are very fast in the initial stages, considerable heat of hydration is generated during the first few hours of mixing of cement with water. As the reactions continue, the cement paste becomes harder and harder and the process may continue for a long period. The water that is added to the cement absorbs the heat of hydration in the initial stages may be upto twelve to twenty four hours depending on the water cement ratio. There after water has to be sprayed over the surface of the concrete to absorb the heat of hydration. The concrete or mortar can resist the loads only after it is cured as specified by the engineer. The final setting time of most Portland cements is about 10 hours.

Initial setting of cement indicates the beginning of noticeable stiffening of the cement paste and the final set indicates the beginning of the hardening of the paste. The normal consistency, initial and final setting times are determined by Vicat needle or by Gillmore needle tests. The limits of initial and final setting times are indicated in Table 1.4.2.

Strength of Cement

Strength of cement is determined by compression test on hardened cement-sand mortar specimen. Cement mortar cube specimens are made in the ratio of 1 : 3 cement and standard sand, and are cast in cubes of 50 cm^2 face. The specimens are cured in water for 28 days and tested under controlled conditions. The amount of water used in making the mortar is selected from the normal consistency of cement. The amount of water selected is equal to $(0.25p + 3)$ percent of the combined weight of cement and sand, in which p is the percentage of the normal consistency of cement. Three cement mortar cubes are tested after one day, three days, seven days and 28 days of curing. The average strength of three cubes tested at a time is taken as a basis for acceptability criteria. The codes of practice specify the minimum average strength of required after one day, three days, seven days and twenty-eight days. The ordinary Portland cements of grade 33, 43 and 53 must give a minimum average strength of 33, 43, and 53 N/mm^2 respectively after 28 days of curing. The curing of the specimens is done in water at 200 C. Standard sand is siliceous, round and natural sand. The sand particles are graded between 80-micron to 1.6 mm. In India, sand available at Ennore in Tamil Nadu is considered to be the standard sand. The absolute minimum strength of mortar cubes of any type of cement must give 33 N/mm^2.

TABLE 1.4.2 Physical properties required of different cements

No.	Type of test	Ordinary grade 33	Rapid Hardening	Low Heat	Pozzolana
1	*Fineness of cement*				
	(a) Residue by weight shouldn't exceed (per cent) after sieving the cement through 90-micron IS sieve,	10	5	–	5
	Or	225	325	320	300
	(b) Specific surface (m^2/kg) by air permeability test method shouldn't be less than				
2	*Setting times*				
	Initial setting time shouldn't be less than	30	30	60	30
	Final setting time shouldn't be more than	600	600	600	600
3	*Minimum average compressive strength of standard mortar cube*				
	(a) 24 ± 0.5 hours	–	16	–	–
	(b) 72 ± 1 hours	–	27.5	10	–
		16	33	16	22
	(c) 7 days ± 2 hours	22		35	33
	(d) 28 days ± 4 hours	33			
	(14 days ± 4 hours for pozzolana)				
4	*Soundness (maximum)*				
	(a) Expansion (in mm) unaerated cement by Le Chatlier method	10	10	10	10
	(b) Expansion (in mm)of aerated sample by being spread out to a depth of 75 mm at relative humidity of 50 to 80% for seven days	5	5	5	5
	(c) Expansion unaerated cement by autoclave test (in per cent)	0.8	0.8	0.8	0.8
5	*Heat of Hydration in calories per gram*				
	(a) After 7 days			65	
	(b) After 28 days			75	

* Cement grades 43 &53 are to give minimum average strength of 43 MPa & 53MPa respectively on the 28 days curing. The 7 days strength is about 67 per cent of 28 day strength.

Soundness of Cement

The word soundness here refers to the tendency of expansion of hardened cement mortar with time. Most of the chemical reactions must take place during the 28 days of curing of the cement or concrete. The micro-cracking of concrete leads to deterioration and failure of concrete. That is why, expansion of cement paste beyond a point is said to be unsound. Le Chatelier apparatus tests the expansion of cement caused by the free lime. The assembly as a whole along with the cement paste is placed at 2000 C and at 98% relative humidity for a prescribed period. The assembly is placed in water and the water is raised to boiling point in thirty minutes. The increase in the opening of the pointers is measured to determine the soundness of the cement. The admissible expansion of the pointers is specified by the codes of practice. A 25.4 mm side square bar of 254 mm length (one inch by 10 inches bar) is cured for 24 hours in humid air. The cured bar is placed in autoclave, the temperature of the water is raised to 2160 C at a pressure of 2 MPa, in 60 minutes and maintained for three hours. The expansion of the bar after cooling back to room temperature should be less than 0.8 mm.

The rise of temperature in the Le Chatlier or the autoclave test accelerates the hydration process and simulates the long term effects. Fine grinding of cement maximizes the free lime mixing with the other compounds and resulting in earlier hydration thus minimizing the unsoundness of the cement.

Heat of Hydration

The basic silicates and aluminates of cement combine with water resulting into insoluble solid compounds. Calcium silicate and tricalcium aluminate hydrates are the main resulting compounds. Calcium hydroxide is one of the other products in the hydration process. The heat developed during the chemical changes is called the heat of hydration.

The reaction of calcium aluminate is almost instantaneous and it may lead to false setting of the cement. The gypsum reacts with aluminate resulting calcium sulfoaluminate. Similarly the gypsum reacts with calcium aluminoferrite resulting calcium sulfoferrite. Most of the exothermic reaction of the cement with water takes place in about two hours. The cement paste becomes stiff during the hydration and then hardens into a solid mass. The calcium hydroxide that is generated from the hydrolysis of the calcium silicates provides an alkaline environment to the concrete.

Specific gravity of ordinary Portland cement varies from 3.12 to 3.16, and for all practical purposes, it is taken as 3.15. The percentage of voids in cement is around 40 and bulk density is about 18 to 20 kN/m3. Usually cement is supplied in bags of 50 kg.

1.5. Types of Portland Cements

Portland cement is produced in different grades and in different types in different countries. The classification of Portland cements varies with the country. The Indian code of practice classifies the Portland cement into the following classes.

Ordinary Portland Cement (grades 33, 43 and 53), **Rapid** Hardening **Portland** Cement, Portland Slag Cement , Portland **Pozzolana** Cement (flyash based), Portland Pozzolana Cement (**Calcined** clay based), **Low** Heat Portland Cement, **Sulphate** Resisting Portland Cement, **Hydrophobic** Cement.

There are other types of cements available in the global market and also in the Indian Market. The number is rather too large, only very commonly used types are mentioned here:

- ❖ Portland Blastfurnace Cement,
- ❖ White Cement,
- ❖ Very Rapid hardening Cement,
- ❖ Supersulphated Cement.

A brief description of some of the commonly used cements is given.

Ordinary Portland Cement (OPC) Grades 33, 43 and 53.

Ordinary Portland cement is known as just cement or simply ordinary cement for laymen. OPC is the most commonly used cement in building construction. A grade notation refers the strength of the cement. Cement whose standard sand mortar cube made with normal consistency water level should give a minimum strength of 33 N/mm^2 at 28 days of curing is assigned **grade 33**. Similarly **grades 43 and 53** are expected to give a minimum strength of cement mortar at 43 MPa and 53 MPa respectively. The strength refers to an average strength of three mortar cubes of face area 50 cm^2. The strength of cement on 7 days curing is about two-thirds of that at 28 days strength. The cubes are made in a specific manner and tested on 28 days of curing in water. The expected strength after 2 days curing of grade 33 and 43 cement mortar cubes are 10 MPa (mega Pascal = one million Newton per square metre) and 20 MPa respectively. At the moment, the most commonly used grades of cement are 43 and 53 primarily because of marketing strategies of the cement companies. Cement **grade 63** is also available in the global market. The fineness of the cement increases with the increase in the grade of the cement and so the rate of heat of hydration increases with the increase of the grade of the cement. Some countries specify the maximum strength of the cement at 28 days to avoid abuse of high strength cements.

Rapid Hardening Cement

Special cement called rapid hardening cement also called as early strength cement is produced to meet the demand of quick setting. The strength developed by rapid hardening cements in about 3 days (72 hours) is equal to that of the ordinary cement developed in 28 days. The factors that influence the early strength are chemical composition, degree of chemical combination of the raw materials, blending, grinding and burning of the raw materials and the fineness of cement. The initial setting time of the rapid hardening cement is still about 30 minutes. The specific surface of the cement is in the **range of 450 to 600 square metres per kilogram** of cement. The rapid hardening cement is used when formwork is to be removed at an early age or the construction is under water. The rate of heat development in this cement is also higher when compared with the ordinary cement. Very rapid hardening cement or ultra early strength cements are also produced with specific surface more than **700 square metres per kilogram**. The percentage of tricalcium silicate in such cements is also high in the range of 60 per cent.

Portland Pozzolana Cement (PPC)

Pozzolana is a material that contains active silica that reacts with lime in the presence of moisture, producing calcium silicate. Pozzolana by itself is not an active binding material, but in the presence of calcium oxide and moisture, develops the binding character. Flyash, rice husk that are fine and rich in silica are the common pozzolanas. Amorphous porous material like rice husk and silica fume have very high specific surfaces as high as ten times or even more than that of cement. There are natural pozzolanas such as diatomaceous earth. Cement produced by grinding Portland cement clinker and pozzolana with addition of gypsum is called Portland Pozzolana cement. A uniform blending of the Portland cement with fine pozzolana can also obtain it. The percentage of pozzolana in such cements varies from 10 to 30 (to 50) percent. There is also some cement with much higher percentage of pozzolanic material. Portland pozzolana cement produces less heat of hydration and offers greater resistance to the attack of aggressive waters. The initial setting time of the cement is same as the ordinary cement. It develops the strength at a slower rate when compared to the ordinary Portland cement with extended period of moist curing. However, the 28 days strength of PPC should be same as that of the OPC. At the moment it is reported that the consumption of Portland pozzolana cement in the world is higher than the ordinary Portland cement. India is yet to catch up with the world in the use of the pozzolana cement. It is particularly useful in marine and hydraulic structures and also for mass concrete construction.

Portland (Blast furnace) Slag Cement

Hot slag from blast furnace is granulated with cold water and then mixed with Portland cement clinker in different proportions. The grinding of the mixture with gypsum will result in Portland Slag cement. Blending of OPC with ground granulated slag can also produce it. The mixture along with some gypsum is ground to the same fineness as that of Portland cement. This cement yields about the same strength as that of the ordinary Portland cement and can be used in place of ordinary cement. Slag is considered to be a waste (better call it a byproduct) product in the production of pig iron. It is a mixture of silica, alumina and lime. The percentage of lime, silica and alumina in the slag are about 45, 35 and 10 respectively. However the chemical and physical quality of the slag differs with the process and the rate of cooling. About 300 kg of slag is produced for one tonne of pig iron. As the slag contains lime and silica, the ground-granulated slag (*ggbs*) can also be used as a cementitious product to a limited extent. The percentage of slag in the slag cement may vary from 35 per cent to 95 per cent. The heat of hydration in the slag cement is less than that of the Ordinary Portland cement as the tricalcium and dicalcium content is less. This may be considered as low heat cement. The slag cements are further classified into two to three groups depending on the percentage of the slag. The seven days strength can be as low as forty per cent of that at 28 days but this shouldn't discourage use of this cement for masonry and similar constructions.

Low Heat Cement

Low heat cement is the one that produces lower heat of hydration when compared to the ordinary Portland cement. The percentage contents of tricalcium and dicalcium silicates are less in the low heat cement. Such cements may also contain pozzolana or ground granulated slag to control the heat of hydration. These cements are useful in mass concreting in which it is difficult to cure the interiors of the concrete. The net calcium oxide content is also less in such cements.

High Alumina Cement

Bauxite and lime stone are ground in appropriate proportions together and the mixture is subjected to high temperature to form into clinker. The kiln used in the Portland cement is not used in this process. Bauxite is the ore for aluminum. The main compounds are calcium aluminate, hydrates and hydrated colloidal alumina. The silica content is small and it may be in the range of 5 to 10 per cent. Bauxite is relatively more expensive when compared with the clay so the cement is more expensive. The alumina hydrates very fast and attains strength at an early stage. About eighty per cent of the strength is developed in the very first one day of curing. The 72 hours curing of the cement mortar will give about the same as that of Portland cement at 28 days curing. Most of the heat of hydration takes place in about first 24 hours, results in high heat of hydration. This cement can be used for underwater construction. Its use in mass concrete and concreting in dry weather is not desirable as it is difficult to control heat of hydration in mass concreting,. The cement also resists sulphate and even acidic attacks. It also has refractory characteristics and withstands high temperatures.

White Cement

White cement is made out of chalk or limestone free from impurities and having very little of oxides of iron as raw material. Oil or gas is used as a fuel in place of coal ash in the kiln to produce the clinker. The clinkering temperature is taken to about 16500 C. Grinding of the clinker is done in special mill so as to avoid contamination by iron oxide. The specific gravity of white cement is slightly less than that of the *OPC*. White cement is used for architectural purposes, in mosaic floors, in fixing glazed tiles, etc. The colour of the white cement may fade away with time so some colour pigments are often mixed to give a more stable colour.

Expansive (or Expanding) Cement

Expansive agents such as gypsum, bauxite, calcite, etc. are burnt to form clinker. The clinker is then ground with Portland cement clinker to form cement. The tendency of shrinking of cement during drying is reduced in expansive cements. In the presence of water, calcium aluminate and calcium react to form calcium sulpho-aluminate hydrate resulting in expansion of the paste. A stabilizing agent such as Blastfurnace slag reacts with excess calcium sulphate and stabilizes the expansion. This cement usually contains 5 to 14 per cent of expansive agent, 10 per cent of stabilizer and the rest being Portland cement clinker.

Natural Cement

Cement rock containing clayey limestone upto 25 per cent is calcified and grounded to obtain natural cement. It is easier to manufacture natural cement and but it is somewhat inferior to Portland cement.

Masonry Cement

Masonry cement is obtained by inter grinding of mixture of Portland cement clinker with inert materials such as limestone, conglomerates, and dolomite, and gypsum and air entraining agent in suitable proportions. This has slow hardening and high workability and water retention properties. It is suitable for masonry construction.

Oil Well Cement

Hydraulic cement that is suitable for resisting high pressure and temperature in sealing water and gas pockets and setting castings during the drilling and repair of oil wells is called oil well cement. This cement contains reduced content of tricalcium aluminate and a retarding additive to suit the requirements.

Hydrophobic Cement

Ordinary Portland cement clinker when ground with an additive that will impart a water repelling property is called hydrophobic cement. The water repelling property is eliminated when the cement is mixed with water. This cement can be stored for long periods in wet and highly humid climates.

1.6. Pozzolanas and Admixtures

Admixtures are the materials that improve some properties of cement such as setting time, workability, colour, and impermeability and resistance aggressive agents when added to the cement. Some of the commonly used admixtures are:

Pozzolana, Flyash (powdered fuel ash), Rice husk ash, Silica fume, Metakoline, Ground granulated blast furnace slag,

Pozzolana

Pozzolana is a natural or artificial material containing large percentage of reactive silica. It is a fine powder of siliceous and aluminous material that reacts with calcium hydroxide in the presence of water to form cementing compounds. The silica in amorphous form reacts with calcium hydroxide to form cementing compound. The volcanic ash, Diatomaceous earth and burnt clay are the common natural pozzolanic material. Flyash and rice husk burnt at regulated temperature of about 6000 C are main artificial pozzolanic materials. One has to be selective in fly ash to ensure sound pozzolanic action. Coal is pulverized and fed into boilers for generation of thermal power. The ash is collected either mechanically or electrostatics in the chimneys. This ash is called flyash. The ash when pulverized is called pulverized fuel ash. The specific surface of the flyash suitable for Pozzolana can vary from 300 to 600 square metres per kilogram. The flyash also contains small amount of unburned coal particles. The coal particle size and quantity must be limited to a minimum value of not more than 3 per cent. The quality of flyash depends on the type of coal that is used in the burning. Flyash derived from Lignite is not considered to be very suitable unless it is processed. It is desirable to process the flyash before using it in the cement. In some sense, the flyash is not really an admixture since more than 30 percent of flyash can be added to the cement.

Burning kaolinitic clay at a temperature of about 7000 C and grinding it to a fine powder of specific surface in the range of 700 square metres obtains the Metakaolin. The Pozzolana is blended with cement either at the grinding stage or at powdered stage in desired proportions. The Pozzolana blended cement is called Pozzolana cement or Pozzolana blended cement. The percentage of Pozzolana in blended cements varies with the desired property. Following are some of the properties of pozzolanic cements when compared with ordinary Portland cements:

(a) Total curing time required is likely to be more may be of the order of 20 per cent,
(b) Heat of hydration developed is less than that of the ordinary Portland cement,
(c) Gains strength slower than that of Portland in the first 15 days but by 28 days it is about the same,
(d) Reacts less with salts and sulphates,
(e) Reduced permeability because of the fine particles of the pozzolana,
(f) Improved workability to some extent,
(g) Reduce bleeding.

Silica Fume

Silica fume is an important pozzolanic material used in high performance concrete. It is obtained as a by-product in ferrosilicon alloy furnaces. It is also called as microsilica or condensed silica fume. It has very high active silica content and having a low bulk density of the order 200 to 300 kg/m^3. The specific gravity of the silica fume is in the range of 2.2 and particle size in the range of atleast ten times finer than the cement particles. The specific surface can be in the order of 3000 to 10,000 square meters per kilogram. Because of its fineness and amorphous nature, it reacts well and fills up the fine voids in the cement paste. Silica fume is used in high performance concrete in strengths of 60 to 150 MPa. The finest particles reduce the permeability of the concrete. Silica fume is being much lighter than water and a fume rather than a powder, it is supplied in compressed pallets and packed well so as to not to get powdered in transportation. Silica fume pallets are fed while mixing with cement in the concrete mixer. Silica fume is mixed in about five to ten percent by weight of cement. It is not really an admixture as it is added as an essential ingredient to achieve very high strengths and low permeability. For lack of an alternate terminology, it is grouped under admixtures.

Admixtures are really additives to cement in small quantities to improve specific character of concrete or mortar. Most of them are artificial and the main purposes of admixtures are:

(a) Improve **workability**,
(b) **Accelerate** setting time,
(c) **Reduce** setting time,
(d) Aid in **curing**,
(e) Decrease **permeability**,
(f) Improve wear **resistance**,
(g) Improve **durability**,
(h) Reduce **shrinkage**,
(i) Reduce **weight**,
(j) Reduce **bleeding**,
(k) Reduce heat of **hydration**.

The admixtures are usually finely ground powders having certain chemical or physical reactions with cement thus producing the desired action. Some admixtures may be obtained in liquid form. The admixtures are added to concrete at the time of actual mixing or sometimes they are premixed with cement just before use. The quantity of admixture is usually limited to one to three percent in most cases. Excessive use of admixture can have secondary effects that are not desirable.

Workability admixtures: Powdered hydrated lime, diatomaceous earth, bentonite and fly ash can be used as admixtures to improve workability. However, use beyond certain proportion and the strength of concrete will decrease, and shrinkage will increase. A number of patented artificial admixtures are available in the market and they are usually referred as plasticizers and super-plasticizers. Calcium chloride, stearate, some oily compounds which function as wetting or dispersing agents are used as workability admixtures. Air in the form of minute bubbles in the concrete also improves the workability. Foaming agent or a chemical that produces gas bubbles on reaction with cement is used as an air entrained agent. Natural resins, sulphonated soaps and oils are the common basic materials for this purpose. Aluminum and zinc powders, hydrogen peroxide are some of the elements used to produce gas in concrete to increase workability. Neutralized vinsol resin when added to cement disperses the cement particles and entraps air. Air entrained concrete is usually lighter and may be less strong when compared with ordinary concrete. Chloride in concrete accelerates corrosion of steel so it is not recommended for reinforced concrete construction. Ready mixed concrete is often pumped through pipeline to reach distances and heights. The slump of such concrete has to be more than 200 mm. Super-plasticizers are the most commonly used workability agent in pumped concrete.

Accelerators: Admixtures that accelerate the setting and hardening are called accelerators. Powdered calcium chloride is one of the commonly used accelerators. Calcium chloride of one to two per cent of cement content is likely to reduce the initial setting time by 10 to 15 minutes. Some soluble carbonates and silicates also help in reducing setting time. The heat of hydration is likely to be more in the first two days. Too much of accelerator can cause too early setting thus resulting in poor workability and consequently low ultimate strength. Chloride admixtures shouldn't be used in the reinforced concrete construction as the chlorine ion will cause corrosion of steel reinforcement. Some of the accelerators such as sodium silicate may result in poor strength and durability.

Retarders: Admixtures that prolong the setting and hardening of cement concrete are called retarders. Retarding agents are used when placement of concrete requires more time because of distance or transportation or other mechanical problems. Gypsum is one of the commonly used retarders. Indiscriminate use of retarders, can, however, affect the ultimate strength. Ready mixed concrete that has to be transported to long distances make use of retarders in combination with plasticizers.

1.7. Water Quality in Cement Concrete Construction

Water is one of the most important ingredients of concrete. Water that is used in mixing and curing of concrete should be free from solids, acids, alkalis, organic materials and salts. Potable and clean bathing water is normally acceptable for concrete mixing but not necessarily without any reservations. The actual solids or salts present in drinking or bathing water may be even higher than that required for durable concrete. The permissible limits of solids in water that can be used in making concrete are given in table 1.7.1. The water that is used in curing should also be of similar standard as that used in mixing the concrete.

TABLE 1.7.1 Permissible limits of solids in water
(pH value not less than 6)

Material	Maximum limit
1. Suspended	2000 mg/litre
2. Organic	200 mg/litre
3. Inorganic	3000 mg/litre
4. Sulfates as SO_3	400 mg/litre
5. Chlorides	2000 mg/litre for plain concrete
	500 mg/litre for reinforced concrete

1.8. Introduction to Concrete

Concrete is a hardened composite matrix consisting of cement, sand and course aggregate mixed with water and cured under moist condition. The composite wet mix ready for placement or placed in position but not hardened is called *green concrete*. Spray of water or supply of moisture to the setting concrete to gain strength and absorb heat of hydration is called *curing of concrete*. The word curing should not be misunderstood in the medical terminology as if the concrete is sick and needs curing. This heat of hydration if not dissipated properly, the solidified concrete mass will not gain full potential and further develops micro-cracks. Civil engineers and builders should be interested in the properties of the green as well as hardened concrete.

    (a) Lean Cement Concrete (LCC) may be called as Lean Concrete (LC), (LC may also refer to lime concrete),

    (b) Plain Cement Concrete (PCC or PC),

    (c) Reinforced Cement Concrete (RCC or simply RC),

    (d) Prestressed Concrete (PSC),

    (e) High strength concrete (HSC),

    (f) Ultra-High Performance Concrete (UHPC),

    (g) High Concrete Performance Concrete (HPC),

    (h) Light Weight Concrete (LWC), and

    (i) Heavy Weight Concrete (HWC).

The Indian Standard code (abbreviated as IS code) of practice on plain and reinforced concrete gives a classification of concretes as: *ordinary, Standard and High strength concrete*. The ordinary concrete has strength in the range of 10 N/mm^2 to 20 N/mm^2 (also called as grades M10 to M20). The range of grades of standard concrete is M25 to M60. The grade of high strength concrete is in the range of M60 to M150. All grades are at an incremental value of 5 N/mm^2. This classification appears to be somewhat arbitrary and not aimed at the common usage concrete. Further it is not necessary that the strength should increase at the rate of 5 N/mm^2 only.

Concrete is generally identified by its strength and referred as grade. A notation in specifying the grade of concrete by the Indian code of practice is prefixed by M. Letter M refers to mix. Grade M10 refers to a concrete having a characteristic strength of 10 N/mm^2. Similarly M15, M20, M25, M30, M35, M40, M45 etc are referred at an incremental value of 5 N/mm^2. The mother code IS: 456-2000(2005) lists grads up to M60, however, concretes upto grade M120 are produced in India at the moment. The days are not very far off when concrete of grade M150 and more will be used in construction. The notation of identifying the concrete may vary with a country. The grade identifies strength and this strength is referred as characteristic strength. The *characteristic strength is the strength of a material below which not more than five percent of the test results are expected to fall*. The strength data of most material follow normal distribution. If f_k is the characteristic strength of a material, then the probability of not more than a specified sample fall below it can be expressed as:

The probability that f (strength) should not fall below fk is = 5 per cent

$$P(f < f_k) = p_f$$

where

   P = probability distribution function,

   f_m = mean strength of the sample,

   p_f = is the accepted probability,

   f = sample strength, and

It is observed that the distribution of strength of most materials follows a normal distribution. The above expression can be rewritten for normal distribution as:

$$\phi\left(\frac{f_k - f_m}{s}\right)$$

where

   ϕ = Normally distributed probability function,

   s = standard deviation of the sample strength.

The inverse of the probability equation can be expressed as:

$$\frac{f_k - f_m}{s} = -k$$

where

   k = is the accepted probability of failure,

A negative sign is assigned to k as the quantity is at the left tail end of the distribution.

For 5 per cent acceptability of failure, the value of k can be obtained from normal distribution table as 1.6.5. Therefore the above equation reduces to:

$$fm = fk + ks$$

The significance of the above expression is to aim the mean strength of the samples higher than the characteristic strength by about the product of the acceptable probability and the standard deviation.

The characteristic strength of concrete (*fck*) can be written as:

$$fck = fm - 1.65s$$

where

   fck = is the characteristic strength of concrete

Lean Cement Concrete

Lean cement concrete (LCC) also called as lean concrete (LC) is made of very small proportion of cement Lean concrete is not a structural concrete and doesn't resist loads. It is used as a base-course in flooring and foundations, and at times it is also used as a filler material. The thickness of the leveling course of the lean concrete varies 75 mm to 150 mm depending on the ground conditions and the strength of the structural concrete that is placed over it. 120 mm thick layer is the most common one. The course aggregate used in the lean mixes is in the range of 40 mm to 63 mm size and depends on the thickness of the concrete. The size of the aggregate should be one third or less than the thickness of the concrete layer. Nominal mixes used in the lean concretes with mix proportions in volume are listed in table 1.8.1. The lean concrete can be broken easily by pickaxe.

TABLE 1.8.1   Nominal mix proportions by volume

Grade (approximate)	Nominal mix
Lean concrete M5	1:5:10
Lean concrete M7.5	1:4:8
Plain concrete M10	1:3:6
M15	1:2:4
M20	1:1.5:3

Phases of Concrete

Concrete goes through six phases of transformation. The approximate duration of the periods of these phases for ordinary Portland cement are:

- Phase 1: *Initial Hydration*, first 15 minutes after mixing with water,
- Phase 2: *Induction period*, from 15 minutes to about 4 hours,
- Phase 3: *Setting process*, 4 hours to 8 hours,
- Phase 4: *Hardening period*, 8 hours to 24 about hours,
- Phase 5: *Curing period*, 24 hours to upto 28 days and
- Phase 6: Service period: about 28 days of curing on wards.

In the first phase, the nucleation of cement hydration products is formed. The hygroscopic particles of cement react with water and chemical kinetics set in at a very fast rate. The surface of the cement is coated with the hydration products. Lot of ionic reactions also takes place in the short duration. Some admixtures may interface with the chemical kinetics of cement even at this stage. The admixture will influence active reaction of the cement. Reaction of the Tricalcium aluminate with water results into hydrates of calcium aluminates during the induction period. In case of inadequate supply of gypsum, flash set of the cement takes place. In case of excess of gypsum, false set of cement takes place during this period. Nucleation of gypsum crystals takes place during the period. The concrete can be re-mixed in flash or false set of cement. The setting of the other compounds of cement also introduced during this period.

The third phase is dominated by the setting of cement. The cement compounds combine with water and result into hydrates of calcium silicates etc. The final setting time for most types of cements is 10 hours. Admixtures can influence the setting time. Retarders that are often used in ready mixed concrete delay especially the initial set. Too much of the retarder can even effect the final setting time of the cement, and consequently the setting of the concrete. Too much delay in the final setting also affects the strength of the concrete. An admixture that reacts with the cement compound can affect the crystallization of the hydrates. The fourth phase is the hardening of concrete. Setting of cement even though leads to some hardening of concrete, the real combination of aggregate with hydrates and concrete solid development takes place during the fourth stage. The period is indicated as 24 hours but the plasticizer and the retarder again influence it. Incompatible two admixtures when used can delay the hardening for longer period. Plasticizer or super-plasticizer along with retarder is added to the ready mixed concrete. Depending on the dosage of the retarder, the hardening of the concrete may be delayed considerably. Sometimes it takes four or even more time to transport and pump the ready mixed concrete because of other reasons such as mechanical, power or formwork problems. That is why more than one dose of retarder is added to the concrete.

The fifth phase of the concrete is curing state. A more detailed discussion on curing is given elsewhere in this chapter. This phase is very important often neglected. Builder is very much interested in laying good concrete but not pay same attention to the curing of the concrete. The final state is the service state. It is often said that the concrete needs no maintenance. This is an over simplification or a poor excuse for bad maintenance of the concrete. Concrete needs care and maintenance to sustain it for the designed life. A scheme of maintenance, quick repairs and protection against aggressive agents is needed for durable concrete.

Plain Concrete (Plain Cement Concrete)

The plain concrete is structural concrete used in foundations, retaining walls, roads, dams and similar applications. There are many structures such as retaining walls, hydraulic dams in which the mass and weight of the structure plays an important role. In such structures the tensile stresses developed are small when compared with the compressive stresses. In roads, foundations and in floors where the plain concrete is used, load is mostly bearing type and doesn't cause much bending moment. The commonly used range of the grades of the concrete is from $M10$ to $M30$. There is no restriction to limit the plain concrete to grade $M30$ however the plain concrete constructions beyond $M35$ are rare. The size of the aggregate used in the plain concrete varies widely. The size of the aggregate can be as much as 150 mm, in mass concrete structures like in hydraulic dams. While in road construction, it can be 40 mm down to 20 mm, and similar size is used in the foundations. The size of the aggregate should be less than one sixth of the thickness of the structure or element, preferably less than one tenth. Curing of mass concrete is not as easy as the curing in thin and slender elements. Use of Portland pozzolana cement in mass concrete is better as the rate of heat of hydration is less.

Reinforced Concrete (Reinforced Cement Concrete, RC or RCC)

Concrete in which reinforcement is embedded is called reinforced concrete. Reinforced concrete is the most extensively used structural material. The reinforcement bars should not slip from the concrete surface when subjected to forces. Further the reinforcement must be protected against any corrosion or rusting. Exposure of steel reinforcement to moisture and oxygen will cause corrosion of the steel; therefore, the concrete must be impermeable. Porous and honey combed concrete is not suitable to reinforced concrete or in fact any type of structural work.

The grade of the concrete that is normally used in the reinforced concrete is in the range of $M20$ to $M40$. However grade $M20$ concrete is not acceptable in many advanced countries even though it is permitted by the Indian code of practice for mild exposure condition. Even though there is no upper limit to the strength of concrete to be used in reinforced work, $M40$ is invariably considered as an upper bound either because of economics or because of other considerations such as deflections etc. The commonly used reinforcement is high yield deformed bars or cold twisted bars. The proof strength of these bars is in the range of 415, 500 and 550 N/mm^2. 20 mm size is the commonly used aggregate; of course the graded aggregate is preferred. The normal weight of the reinforced concrete is 24 kN/m^3 to 25 kN/m^3.

Prestressed Concrete

Concrete in which stresses are induced even before the external loads act on the structure is called *Prestressed Concrete*. Prestressed concrete structures are commonly used for bridges, towers, water tanks, shell structures, folded plates, nuclear reactors, and long span girders. Tensioned high-tension steel wires and cables and anchored to the concrete apply prestressing force on the concrete. The grades of concrete used in the construction start from $M35$ onwards. Till recently, concrete grade $M60$ was about the upper limit in many constructions. However with the introduction of high performance concrete, concrete grades in the range of $M80$ to $M150$ are coming into practice. The size of the aggregate is very similar to that used in reinforced concrete, namely 20 mm aggregate but graded to give maximum density to the concrete. The water cement ratio is limited to the least value and it is normally less than 0.45. The water-cement ratio as low as 0.3, along with super-plasticizers is used in high performance concrete. The tensile proof strength of the high tensile steel wires is of the order 1500 to 2000 N/mm^2. The shrinkage strain of concrete is comparable with the allowable strain in the mild and medium grade steels, so only high tensile steels are used in this construction. The concrete also has to be high strength concrete to match with the strength of the steels and to compensate the losses of prestress. Prestressed concrete is the main solution for long span structures. The normal weight of the prestressed concrete is about 25 to 26 kN/m^3.

High Performance Concrete (HPC)

Very high strength concrete that has the desired strength and workability is often called *high performance* concrete. The technology of high performance concrete is about fifteen years old and becoming popular in prestressed concrete construction. Strong well-graded aggregate with silica fume and low water cement ratio are used in high performance concrete. Silica fume sometimes called as micro-silica is a very fine powder (in fact a fume), and it is ten times finer than the cement. The specific surface of the fume is in the range of 3000 to 10,000 m^2 per kilogram. It is amorphous active silica, reacts with lime forming calcium silica hydrates in the microform. It fills up the finest voids in the concrete, making the concrete impermeable and very strong. Silica fume is more expensive when compared to cement. It is byproduct of Ferro-silica alloys and at the moment it is being imported in to India. The quantity of silica fume used in high performance concrete is about five to ten percent of cement. Use of super-plasticizer is a must to obtain good workability of the concrete. The present range of high performance concrete is M60 to M150. Sooner than expected, the strength of the concrete will go beyond 150 N/mm^2.

Ultra-High Performance Concrete

Ultra-High Performance Concrete (UHPC), also known as Reactive Powder Concrete (RPC), is a high-strength, ductile material formulated by **combining Portland cement, silica fume, quartz flour, fine silica sand, high-range water reducer, water, and steel or organic** fibbers. The material provides compressive strengths up to 150 MPa and flexural strengths up to 25 MPa.

The materials are usually supplied in a **three-component premix**: powders (**Portland cement, silica fume, quartz flour, and fine silica sand**) pre-blended in bulk-bags; super plasticizers; and organic fibbers. The ductile behaviour of this material is a first for concrete, with the capacity to deform and support flexural and tensile loads, even after initial cracking. The use of this material for construction is simplified by the elimination of reinforcing steel and the ability of the material to be virtually self placing or dry cast.

The superior durability characteristics are due to a combination of fine powders selected for their grain size (maximum 600 micrometer) and chemical reactivity. The net effect is a maximum compactness and a small, disconnected pore structure.

The following is an example of the range of material characteristics for UHPC:

Strength

Compressive: 120 to 150 MPa
Flexural: 15 to 25 MPa
Modulus of Elasticity: 45 to 50 GPa

Durability

Freeze/thaw (after 300 cycles): 100%

Salt-scaling (loss of residue): < 60 g/m^2 Abrasion (relative volume loss index): 1.7 Oxygen

permeability: $<10\text{-}20$ m^2 Cl - permeability (total load): < 10 C Carbonation depth: < 0.5 mm

Manufacturing and Installation

The precast canopy components were individually cast and consist of **half-shells, columns, tie beams, struts, and troughs**. The columns and half-shells were injection cast in closed steel forms Troughs were cast through displacement moulding, while struts and tie beams were produced using conventional gravity two-stage castings.

The columns were installed on the concrete platform first. Then, the right and left half-shells, along with the tie beams, were pre-assembled in the plant and transported to the site where they were lifted (by crane) over the railway tracks, for placement on the columns. Upon arrival at the site, the canopies were set on temporary scaffolding, and struts were attached to the shells and previously installed columns with welded connections.

The material's unique combination of superior properties and design flexibility facilitated the architect's ability to create the attractive, off-white, curved canopies. Overall, this material offers solutions with advantages such as speed of construction, improved aesthetics, superior durability, and in permeability against corrosion, abrasion and impact which translates to reduced maintenance and a longer life span for the structure.

Concrete High-Performance American Concrete Institute

"A concrete which meets special performance and uniformity requirements that cannot always be achieved routinely by using only conventional materials and normal mixing, placing and curing practices". The requirements may involve enhancements of characteristics such as placement and compaction without segregation, long-term mechanical properties, and early age strength or service life in severe environments. Concretes possessing many of these characteristics often achieve High Strength, but High Strength concrete may not necessarily be of High Performance .A classification of High Performance Concrete related to strength is shown below.

Compressive strength (MPa)	50	75	100	125	150
High Performance Class	I	II	III	IV	V

Advantages

Reduction in size of the columns, Speed of construction, More economical than steel concrete composite columns, Workability and pump ability, Most economical material in terms of time and money, Increased rentable\useful floor space, Reduced depth of floor system and decrease in overall building height, Higher seismic resistance, lower wind sway and drift, Improved durability in aggressive environment, Wearing resistance, abrasion resistance, Durability against chloride attack, Increased durability in marine environment, Low shrinkage and high strength, Service life more than 100 years, High tensile strength, Reduced maintenance cost.

Light Weight Concrete

Lightweight concrete uses lightweight aggregate and is lighter than the normal concrete. A porous aggregate results in to lightweight material. Further the filling up the voids in the coarse aggregate by fine aggregate is minimized and air content is increased. The aim is to obtain more voids in the concrete and at the same time make it homogeneous. The range of the density of the lightweight concrete is quite wide. The weight of the lightweight concrete varies from 400 kg/m^3 to 1900 kg/m^3. Some lightweight concrete is lighter than water and is used for thermal and sound insulation. Very light concrete is also called as *cellular concrete* and coarse aggregate is not used in such concrete. Further large volume of air bubbles is introduced into the concrete with lightest aggregate. Finally there is the structural lightweight concrete having a unit weight of 1400 to 1900 kg/m^3 and having a minimum compressive strength of 15 N/mm^2. Pumice stone, slag, cinder etc are the common lightweight aggregates. Artificial lightweight aggregate is manufactured using industrial byproducts.

Heavy Weight Concrete

Heavy weight concrete is the one that is made of heavy-density aggregate. Heavy-density aggregate such as hematite, an iron ore material is extensively used in heavy weight concrete. This concrete is used in industrial floors where considerable wear and tear is expected. It is also used in nuclear containment vessel constructions. The unit weight of heavy weight concrete is more than 2600 kg/m^3.

Concrete Plant

A concrete plant, also known as a batch plant or batching plant, is a device that combines various ingredients to form concrete. Some of these inputs include sand, water, aggregate (rocks, gravel, etc.), fly ash, potash, and cement. There are two types of concrete plants: *ready mix* plants and *central mix* plants. A concrete plant can have a variety of parts and accessories, including but not limited to: mixers (*either tilt-up or horizontal* or in some cases both), cement batchers, aggregate batchers, conveyors, radial stackers, aggregate bins, cement bins, heaters, chillers, cement silos, batch plant controls, and dust collectors (to minimize environmental pollution).

The centre of the concrete batching plant is the mixer. There are three types of mixer: Tilt, pan, and twin shaft mixer. The twin shaft mixer can ensure an even mixture of concrete and large output, while the tilt mixer offers a consistent mix with much less maintenance labour and cost.

A *ready mix* plant combines all ingredients except for water at the concrete plant. This mixture is then discharged into a ready mix truck (also known as a concrete transport truck). Water is then added to the mix in the truck and mixed during transport to the job site.

A *central mix* plant combines some or all of the above ingredients (including water) at a central location. The final product is then transported to the job site. Central mix plants differ from ready mix plants in that they offer the end user a much more consistent product, since all the ingredient mixing is done in a central location and is computer-assisted to ensure uniformity of product. A *temporary batch plant* can be constructed on a large job site. A concrete plant becomes central mix with the addition of a concrete mixer.

Concrete batching plants are widely used to produce various kinds of concrete including quaking concrete and hard concrete, suitable for large or medium scale building works, road and bridge works and precast concrete plants, etc.

More recently is the availability of the mobile concrete batch plant. This innovative device was designed for the production of all types of concrete, mixed cements, cold regenerations and inertizations of materials mixed with resin additives. The design includes multiple containers that separately transport all the elements necessary for the production of concrete, or any other mixture, at the specific job site. In this way, the operator can produce exactly what he wants, where he wants and in the quantity he wants through the use of an on-board computer. Once production is started, the various components enter the mixer in the required doses and the finished mixed product comes out continuously ready for final use. It is also suitable for the recovery of materials destined for landfill disposal, such as cement mixtures regenerated from masonry rubble. The mobile batching plant is easy to transport. It can be fixed-mounted on a truck, mounted on a truck with tipping box or mounted on an interchangeable cradle.

Modern concrete batch plants (both ready mix and central mix,) employ computer aided control to assist in fast, accurate measurement of input constituents or ingredients, as well as tie together the various parts and accessories for coordinated and safe operation. With concrete performance so dependent on accurate water measurement, systems will often use moisture probes to measure the amount of water that is part of the aggregate (sand and rock) material while it is being weighed, and then automatically compensate the mix design water target.

A non-profit association brings together all of the main concrete plant manufacturers on common matters related to the industry. According to the CPMB's website, "The National

Ready Mixed Concrete Association (NRMCA) endorses the members of the Concrete Plant Manufacturers Bureau as the preferred providers of concrete plants and associated equipment as providing quality equipment conforming to the standards and specifications of NRMCA's plant certification program and the concrete plant manufacturers' standards.' " "The primary function of the CPMB is to establish minimum standards for rating various components of concrete plants for the protection and assurance to the user that the plated components of the plants conform to these Standards.

1.9. Important Properties of Concrete

A number of characteristics and properties of concrete are of importance to the structural engineer and builder. Many properties are interdependent and should be studied together. Some properties influence in manufacturing the concrete and some in resisting the loads and environmental forces. The important properties relevant to structural and construction engineer are:

- ❖ Workability of concrete,
- ❖ Segregation and Bleeding,
- ❖ Curing of concrete,
- ❖ Shrinkage and Creep,
- ❖ Strength, and
- ❖ Durability of concrete.

As the concrete is a matrix having a number of components, the end product depends on several factors. The main factors that affect the characteristics of the concrete are:

- ❖ Quality of ingredients, such as aggregate, sand, cement, water and admixtures,
- ❖ Relative proportions of the ingredients,
- ❖ Making of green concrete, mixing time, transportation, laying etc.,
- ❖ Methods in making, laying, finishing of concrete,
- ❖ Protection of green concrete and curing,
- ❖ Temperature and weather conditions during laying and curing,
- ❖ Exposure conditions and maintenance.

Some properties are associated with the making and some with the hardened concrete. Different properties of the concrete are briefly discussed in the following sections.

1.10. Workability of Concrete

Workability of concrete is the property that indicates the ease in mixing, placing, compacting and finishing with least segregation of the particles. So the workability involves the following aspects:

- ❖ Ease in mixing the ingredients,
- ❖ Ease in placing the concrete in position,
- ❖ Ease in compacting,

❖ Achieving the homogeneity and no segregation, and

❖ Finishing the surface of the concrete.

Workability of concrete is a function of relative mix proportions, the quality of the aggregate and cement, water cement ratio, mixing time and method of mixing, admixtures, time of transportation and the method of placing. Layman thinks that the concrete is more workability with more water. The laborer who handles the mixing of concrete is happy to add more water to make the concrete easily workable. But water beyond a point weakens and strength and durability of the concrete. The workability requirement depends on the size of the element, amount of reinforcement, location of placing the concrete and the weather conditions at the time of concreting. One needs more workable concrete if the percentage of the reinforcement is higher under similar situations. Therefore the workability specification depends on many factors. The influences of different factors on the workability of concrete are discussed briefly here:

(a) *Water-cement ratio.* The workability increases with increase in the water-cement ratio, however the strength and durability of the concrete decrease if the ratio is beyound a point. It is therefore desirable to limit the water-cement ratio to the minimum possible based on the strength criterion. For high strength and high performance concrete, the water-cement ratio is to be limited to 0.45 or even down to 0.30. The real quantity of water required to complete the chemical reaction of the cement may be about 30 percent by weight of cement. Even the water applied during curing of the concrete assists in the formation of the hydrates. Such a low water-cement ratio will not give workable concrete unless a super-plasticizer is added in making the concrete. It is always advisable to lower the water-cement ratio and use a water reducing agent or super-plasticizer for better quality. Some more discussion will be given on the water-cement ratio on strength of concrete.

(b) *Size of aggregate:* Course aggregate improves the workability of the concrete. Larger the size of the aggregate, better is the workability. Higher the size of the aggregate, higher is the strength of the concrete. However the size of member, percentage of reinforcement, cover and surface finish specifications also control the size of the aggregate. 20 mm or 25 mm size of aggregate is recommended for reinforced concrete. The desired size of the aggregate depends not so much on the workability but on the type of structural elements, and cover to the reinforcement. Fine aggregate decreases the workability of the concrete.

(c) *Grading of aggregate:* A well-graded aggregate will result in better compaction and minimum segregation. The specification on size of the aggregate normally refers to the largest value used in the concrete. But it is desirable to use at least two different sizes of the coarse aggregate for good concrete compaction and strength. For example one can use 20 mm and 10 mm combination in the coarse aggregate selection. Better workability and strength are obtained by the well-graded aggregate.

(d) *Shapes of aggregate:* Rounded and angular shape aggregates are the basic desirable shapes for good concrete. The rounded aggregate demands least amount of water

28

when compared with that for the other shapes. Angular aggregate with almost equal size faces is good for interlocking and strength. The angles between the faces of the aggregate should not be acute or obtuse. Angular aggregate is preferred for strength consideration aspect. Crushed aggregate is usually an angular shaped one.

(e) *Flakiness of the aggregate:* Flaky or oblong aggregate decreases the workability and also the strength of the concrete. Therefore such aggregate should not be used in making of concrete.

(f) *Aggregate-cement ratio:* More cement means more water for the same water-cement ratio. Consequently the increase in the aggregate content with respect to the cement decreases the ratio of the water to the total mass of the concrete. Normally the aggregate-cement ratio is chosen for a given water-cement ratio to obtain a desirable strength of the concrete. Higher aggregate-cement ratio decreases the workability as well as the strength of the concrete.

(g) *Air content in concrete mix:* Air in green concrete acts as a lubricant and increases the fluidity and consequently the workability. Sometimes air entraining agents are added to concrete to increase the workability at a lower water-cement ratio. However the entrapped air must be driven out by proper vibration for good consolidation of concrete.

(h) *Time of transit:* A minimum mixing time is needed for a given water-cement ratio to obtain good workability of concrete. The transit time of concrete from the mixer to the location of placement needs to be as minimum as possible. Green concrete gets stiffened during transit even before the initial setting of the cement. Five minutes of transit time is desirable, at the most ten minutes can be considered as a last resort. A retardant must be added if the transit time is more than ten minutes. Ready mixed concrete needs more transit time. It can be anywhere from 30 minutes to four hours. Addition of retardant in the transit mix is very common. The retardant may have to be added in periodically depending on the transit time. Too much of retardant can effect the hardening and the final setting time and strength. Four hours of retarding the setting time is probably on the higher side. Concretes with four hours of initial setting time will effect the final setting time.

(i) *Admixture to concrete:* Workability admixtures such as air entraining agent or plasticizer or super-plasticizer are often used in high strength concretes to improve the workability at low water-cement ratio. Super-plasticizer is a must in the high performance concrete. Selection of minimum water-cement ratio to achieve good strength at low cost and then addition of super-plasticizer to obtain the desired workability is the best strategy in good concrete practice.

Workability of concrete is quantified and measurable by a number of tests. That is mixing, placing, compacting, finishing and obtaining good homogeneity. The following are the tests that measure the workability of concrete:

(a) Slump test,
(b) Compaction factor,

(c) Vee-bee (or *V-B* or *Vebe test*) consistometer,
(d) Flow test,
(e) Remolding test,
(f) Ball penetration or Kelley ball test, and
(g) Two-point test.

Slump test is the most extensively used test to measure the workability of the concrete. It is probably the simplest to apply in field with least equipment. A frustum of a conical tube of 300 mm high, 200 mm diameter at bottom, 100 mm at top with two handles attached to the surface at about the middle height is used as a mould. Concrete is placed in three layers, each layer being consolidated by 25 tamping of a rod and then the top surface is leveled. The mould is pulled up and the wet molded concrete is allowed to slump or slide down. The magnitude of the slump of the concrete from the 300 mm height is called *slump of the concrete*. It is considered to be reasonably good for low slump or high slump but not very good for the middle range of the slump. The fresh concrete is classified into four broad groups based on slump, and they are listed in table 1.10.1.

TABLE 1.10.1   Classification of fresh concrete based on slump

Class->	Low	Medium	High	Very high
Slump (mm)	0-30	40-75	80-150	160 & above

The concrete gets stiffened with time even in a small duration of 10 minutes transit. The slump of the concrete at time of placing is likely to be less than that measured at mixing time. The phenomenon of reduction of slump with time is called slum loss. Ready mixed concrete is transported for longer distance and duration when compared to the site mixed concrete.

The required slump is always expected to be at the time of placing the concrete. The normally recommended slump for different types of concreting environment in dry weather condition is listed in table 1.10.2. Slump test is the most preferred test in field.

TABLE 1.10.2   Recommended concrete slump for environment

Type of work	Slump
PCC, Shallow slabs & beams with nominal reinforcement	Low
Beams and columns with reasonable reinforcement	Medium
Heavily reinforced concrete members and Prestressed	High
Deep members with heavy reinforcement (pumped concrete)	Very high

Compacting Factor Test

The compacting factor test is based on the degree of compactness achieved in placing (or dropping) the concrete from a specific height. The ratio of the density of the concrete actually achieved to the density that is possible in fully compacted condition is called compacting factor. Two conical hoppers with doors at bottom are placed one over the other at a specified distance. The top hopper is bigger than the lower one. A cylindrical mold of 150 mm by 300 mm is placed below the hoppers. The top hopper is filled with fresh concrete without any compaction. On opening of the bottom door of the top hopper, the concrete falls into the lower hopper. The concrete fills up the lower hopper by gravity and then overflows. Similarly the bottom door of the lower hopper is released and the concrete is allowed to fall and fill the bottom cylinder. The density of the concrete in the cylinder is calculated. The ratio of this density to that of the fully is compacted density of the concrete in the cylinder is called the compacting factor. The workability of the concrete based on the compacting factor can be divided into four classes and they are listed in table 1.10.3.

TABLE 1.10.3 Classification of workability of fresh concrete based on compacting factor

Class	Low	Medium	High	Very high
Compacting factor	0.75	0.85	0.92	0.95

There is some correlation of the classification of the workability of concrete based on the slump and compaction factor but not necessarily having a one to one correspondence.

Vebe Test

The Vebe test equipment consists of vibrating table with slump cone placed in a shallower cylindrical tube. The table is set in light vibration after the slump cone is lifted. The time taken for the freestanding concrete cone to fill the cylinder is measured. The time required for remolding the concrete from the conical shape to that of cylinder is measured in seconds. The time in *Vebe test* vary from 5 seconds to 25 seconds depending on the inverse of the workability of the concrete. Five seconds of time of *Vebe* test represents highly workable concrete and 25 seconds is stiff concrete. The *vebe* test is less commonly used when compared to the two tests mentioned earlier.

1.11. Curing of Concrete

Curing of concrete is a misleading word for a common man. Curing means to cure one from a problem or decease. Curing of concrete has altogether a different meaning. Cement when comes in contact with moisture, reacts chemically on the surface of the cement particles. Bonding with the surrounding particles of cement is developed during the chemical reaction of cement. The process of formation of compounds of cement with water is called hydration. The hydration of cement generates heat and the heat is called heat of hydration. The hydration process is initiated at about eighty percent of humidity environment. Uncontrolled heat of hydration causes micro cracking in concrete. Further, moisture is needed continuously for some period in and around cement particles to continue with the formation of hydrates of cement. Therefore moist environment is a necessity for the hardening concrete. Supply of moisture or water to the hardening concrete to dissipate the heat of hydration and to aid further chemical reaction of cement is called *curing of concrete*. The chemical reaction gets accelerated with increase of the temperature of the water or moisture.

The concrete gains strength with increase in duration of curing. An approximate strength of concrete at 90 days with different periods of moisture curing is given table 1.11.1. Concrete that is allowed to gain strength by exposing to air is called air curing. It is not really any curing but given a notation for no curing with water and for lack of better terminology.

TABLE 1.11.1 Strength of concrete with curing

Days of water curing >	0	3	7	28
Ratio of strength on 90 days to that at 28 days curing >	0.5	0.75	1.0	1.2

Moist curing: The concrete after casting should be allowed to reach almost the final setting time of cement. The concrete is considered to be green till about the final setting time of cement. The final setting time of cement is not necessarily the final set of the concrete. The final setting time of most cements is about 10 hours but the concrete after placing in position under normal temperature of the order 20 degrees Celsius is likely to set in about six hours. Five to ten hours of pre-curing time is adequate in hot weather conditions as in most parts of India. The hardening of concrete in warm weather of about 30+ degrees Celsius is faster so the curing with moisture can start after eight hours of casting. After about 6 to 10 hours of casting, the exposed surfaces of hardening concrete must be kept continuous under damp or wet condition by sprinkling of water. Or the surface can be covered with jute sacks that are kept under wet condition. The concrete can be kept under submerged condition wherever feasible for curing. This is probably done in pre-cast concrete construction. The concrete test samples are always cured in water submerged condition. But such a situation is hardly available for most structures. Commonly accepted practice is to cover the concrete with wet jute bags and sprinkling of water as frequently as needed to keep the bags wet all the time. Most asked question is how frequently the water must be sprinkled? The frequency of spraying water depends on the weather conditions. In rainy season when the humidity is high, the sprinkling of water can be once in four or five hours. But in the dry summer, sprinkling has to be more frequent. The aim is to keep the gunny bags wet all the time. In any case, the surface of the concrete should be wet or under 80 percent humidity condition. Another question that is asked is how long the curing of concrete be continued without interruption? 28 days curing is the best and most desirable to result into a strong and durable concrete. Under no circumstances the period of curing should be less than 7 days. A canvas or polyethylene sheet placed over the wet surface of the concrete helps in reducing loss of moisture from the concrete. Seven days of water curing will atleast give strength of that of 28 days cured value on 90th day. However one must remember the durability of the concrete will be affected even if the strength is achieved if the curing time is cut down.

Curing of many concrete structures built in the unorganized sector in India is given the lowest importance. The sprinkling of water over the concrete is usually handed over to a small boy or a woman worker or a watchman. Further, the construction is in progress while some portion requiring curing. Many parts of the structure are not easily accessible to the helpless boy who is supposed to maintain the wetness of the surface of the concrete. The engineer and the supervisor must pay more attention to the curing of the concrete. Curing of slabs is about the easiest, and curing of the concrete in columns is about the most difficult one. The slabs being relatively thin and are flat, pounding of water on the top face of the slab may be adequate.

However the bottom face of the slab is sprayed with water now and then. The concrete slabs have built-in higher factor of safety because of continuity in two directions when compared to that available to columns. The slabs will collapse only after developing a yield line mechanism. Columns are the most critical elements and are the least compacted and least cured in common practice. Covering the columns with wet jute bags and keeping the bags under wet condition is a must. The side faces of the beams are the other items that are not properly cared/cured.

Curing of mass concrete: Thick concrete members such as raft slabs, pile caps and other thick concrete sections need special attention in curing. Concrete raft slabs of the order 1000 mm to 1500 mm or even more thicker are common in large constructions. Since the bottom face is resting on lean concrete, loss of moisture on that surface is almost none. The lean concrete need to be atleast M7.5 grade concrete. Concrete sections thicker than 1500 mm need special curing arrangements. The temperature of the concrete at the time of laying need to be about 15 degrees Celsius or even less in mass concrete. Ice cold water can be used in mixing the concrete so that the interior of the concrete is at a lower temperature to start with. Surface curing will not reach the interiors of such thick members. If the thickness of the concrete is more than 2000 mm, cooling pipelines have to be laid in the concrete so that cool water can be circulated through such lines for curing. Rigid perforated pipes or tubes can be embedded in mass concrete at the time of laying the concrete. Cold water is pumped through the pipes to provide heat absorption and to maintain 80 per cent humidity inside. Too much of un-dissipated heat of hydration can cause micro cracking and damage to the mass concrete.

Steam curing: Concrete can be cured by controlled steam. The concrete must be allowed to harden for about 3 hours, then it is covered under tarpaulin or put in steam chambers. Pre-cast concrete elements can be placed in steam chambers. Steam at about 800 to 1000 C is allowed over the concrete, gradually raising temperature of the concrete chamber upto 700 C in about 2 hours time. The steam curing cycle consists of four parts. Pre-heating period, heating period, steaming period and post-steam period. The pre-steam period is the period the concrete is allowed to set partly and it may be two to three hours. Too early exposure of concrete to the steam can cook the concrete resulting in bubbling or boiling surface. The concrete shouldn't be subjected to sudden rise in temperature. The heating period is the period in which the steam is allowed into the chamber at atmospheric pressure to heat the chamber. This period depends on the mass of the concrete and the size of the member and the arrangements of steaming such as chamber or tarpaulin. May be couple of hours of heating is adequate. Steam curing for about 8 to 10 hours can be considered as the minimum required obtaining an equivalent 7 days strength of water cured concrete. Steam curing of 18 to 24 hours gives strength equal to that of 28 days water cured one. The entire cycle of steam curing can be about 18 to 24 hours. The steam is cut off slowly in about an hour's time and the concrete is allowed to cool in the chamber for another hour. Improperly done steam curing can cause cooking and spalling of the surface concrete. Steam curing is normally done for prestress concrete or pre-cast concrete members. It is a matter of economics of storage and supply of the product, reuse of prestressing equipment and the bed for quick turn over. The steam curing appears to be economical in producing prestressed concrete pole, railway sleepers, pipes etc.

1.12. Shrinkage and Creep of Concrete

Shrinkage of concrete is the decrease in volume due to evaporation of moisture from concrete and hardening of the concrete. The shrinkage is divided into three types, and they are plastic *shrinkage, drying shrinkage and thermal shrinkage*. There is some settlement or subsidence of concrete even before the concrete gets hardened, and this reduction in volume of fresh concrete is called *plastic shrinkage*. Stiffening of the top layer of the concrete may cause settlement of the inside concrete. The plastic shrinkage may be caused by bleeding, absorption of water by subgrade soil, settlement of formwork, and rapid evaporation of moisture from the surface etc. Plastic shrinkage produces horizontal cracking on the top surface. All precautions should be taken to minimize the loss of moisture from the surface and into the sub-grade to minimize or eliminate the plastic shrinkage.

Fresh concrete when exposed to ambient humidity undergoes volume change due to change in the moisture content in the concrete. The humidity in the fresh concrete is normally more than the ambient humidity. Therefore a reduction in the volume of the concrete takes place. The decrease in volume of the concrete due to change in the moisture caused by ambient condition is called *drying shrinkage*. Thermal shrinkage is the one that is caused by the cooling of the concrete. Decrease in volume of the concrete takes place as it is exposed to the ambient temperature that is lower than that of the concrete. The decrease in the volume of concrete either due to plastic shrinkage or dry shrinkage or thermal shrinkage will result in cracking, sometime visible and or invisible. The drying shrinkage can be reversed to some extent by induction of moisture into the concrete. Curing of the concrete at right time and for the right period minimizes the shrinkage.

The shrinkage of concrete is proportional to the amount of water added at the time of mixing and the amount of cement content. Some aggregate characteristics will also influence the shrinkage. Finer cement reacts faster and develops heat of hydration at faster rate thus resulting in shrinkage of the concrete. Temperature at the time of placement of the concrete influences the thermal shrinkage. Higher the temperature at placement, higher is the shrinkage. Pre-cooling of the concrete either with cold water or cooled aggregate reduces the shrinkage. The shrinkage of concrete takes place for a long period. In some constructions, the elements may be cast at different periods. There may be a reasonable difference between the castings of concretes in different parts of the structure. The concrete cast at different periods wills shrinks at different rates. The difference in the shrinkage of the different parts is called the differential shrinkage. Part of the shrinkage is recoverable by wetting of the concrete. Structural engineer is primarily interested in the total and differential shrinkage. The quality control should look after the different components to minimize each of the shrinkage. The shrinkage is a function of the following factors:

❖ Shrinkage decreases with increase in the aggregate-cement ratio,
❖ Shrinkage increases with increase in the cement content,
❖ Shrinkage increases with increase in the water-cement ratio,
❖ Shrinkage decreases with increase in the curing period.

The plastic shrinkage is irrecoverable, whereas the drying shrinkage is partly recoverable by adding moisture to the dried concrete. Half of the shrinkage takes place in the first one-month, and seventy five per cent in the first six months after commencement of drying of concrete. The plastic shrinkage can be reduced considerably by protecting the surface of the concrete immediately after casting of the concrete. The best principle to reduce the shrinkage is to reduce cement and water-cement ratio, and then add an admixture or plasticizers to improve the workability. The shrinkage of concrete can be eliminated by use of expansive cement, the cement that expands while setting.

Shrinkage is basically a strain without stress, and yet produces cracking in the concrete. Restraint on the shrinkage will cause stresses in the concrete. The shrinkage starts at the surfaces of the concrete and extends into the interior. Approximate shrinkage strain for a water-cement ratio of 0.7 can be as much as 0.0007 while it is about 0.0003 for a water-cement ratio of 0.45.

The total free shrinkage for most reinforced concrete construction that is cured for 28 days is taken as 0.0003.

$$\in sh = 0.0003$$

In practice, the columns are cast on one day and the beam and slab are cast may be few days later. Or the slab including the beams is cast in different stages. The members cast on different days will have different rates of shrinkage. In bridge building, the beams may be pre-cast members and the concrete slab is placed at site over the pre-cast beams. The difference in of shrinkage of the two or more members is called *differential shrinkage*. Differential shrinkage produces interface forces in the members and even cracking.

Creep of Concrete

Concrete when subjected to loads undergoes elastic strain to start with. The stress in the member produces strain and the strain increases if the stress is maintained for a longer duration. The time depended strain under constant stress at 100 percent humid concrete is called *basic creep*. Basic creep of concrete takes place in mass concrete structures. The time dependent strain under constant stress at normal conditions is called *creep strain*. The creep strain per unit stress is called *specific creep*. The ratio of creep strain to the elastic strain is called *creep coefficient*. On unloading, most of the elastic strain is recovered and only part of the creep strain is recovered with time. Creep is a function of relative proportions of aggregate, water, cement, type of aggregate, porosity of concrete, curing period, level of stress and age of loading.

Creep of concrete increases with increase in cement content and water-cement ratio. It also increases with decrease in the age of concrete at loading. Creep also increases with increase in the exposed temperature of the concrete. The creep normally referred is the ultimate creep. Seventy percent of the creep strain takes place in the first one month and the balance may take place in the next three years period time.

Creep strain relation is:

$$\in cc = Cc \in se$$

where

$\in cc$ = creep strain, Cc = creep coefficient, and $\in se$ = elastic strain

The total strain at any given period is equal to the sum of creep strain and elastic strain. The creep coefficients are given in table 1.12.1.

TABLE 1.12.1 Creep coefficients

Age of concrete at loading	Creep coefficient
7 days	2.2
28 days	1.6
1 year	1.1

As the creep strain increase, deflections in the beams and slabs also increase. The deflection caused by the dead load is to be multiplied by the creep coefficient to obtain the creep deflection. However because of the reinforcement doesn't creep at the same rate as that of concrete, the creep deflection of the reinforced concrete members is only a portion of the creep coefficient. Some redistribution of the stresses takes place indirectly due to creep. In case of columns, the concrete will undergo creep and transfer part of the load to the reinforcement. Therefore the reinforcement in columns under direct compression is subjected to higher levels of stress than that estimated by simple compatibility condition. The creep can cause failure of a structure because of excessive strain in the concrete. This can happen when the dead load is much higher than the live load. In such situations, the permanent load dominates the total strain.

1.13. Strength of Concrete

Strength of concrete is its ability to resist load before collapse. The coarse aggregate in concrete is like a bone skeleton in human structure. The bone primarily resists compression, so much so the aggregate in concrete is the prime source to resist the compression. Aggregate can also resist tension to some extent but in the concrete matrix, there are other weak links in tension and hence the aggregate doesn't control tension problem. The cement-sand mortar in concrete is like muscle in human body. The mortar fills the voids in the aggregate and binds it together to form in to a homogeneous mass. Water in the concrete is like the life giving blood of human being. Water by itself may not resist load except hydraulic load, but when combined with cement, it results into compounds of strength and durability. The strength of concrete is measured by its ability to resist compressive stress. 150 mm concrete cube is taken as basic specimen to measure the compressive strength in India and few other parts of the world. 200 mm cube is considered as a standard testing specimen in some parts of Europe. 150 mm by 300 mm cylinder is the compression test specimen in North America. Therefore the strength of concrete is referred with respect to a test specimen and it bears a relation with the concrete prism. The compressive strength of prism in direct compression is different from the under bending compression. The strength of concrete is associated with the hardened concrete and depends on several factors as listed below:

1. *Strength decreases with increase in water-cement ratio.* Water-cement ratio is one of the most **important factors in determining the strength of concrete. This can be compared to the ratio of the blood** content to the weight of a person. There is always an optimal value of this ratio. Too less blood causes anemic and too much of it can cause high blood pressure or other effects. It is already mentioned that the decrease in the water-cement ratio decreases the workability of the concrete. The

advantage of the low water-cement ratio is obtained only if good consolidation of the concrete is achieved. The water-cement ratio shouldn't be indiscriminately decreased leading to poor workability of concrete Workability agents such as super-plasticizers have to be added to achieve required workability and consequently good consolidation and strength. Some of the water that is added into the concrete mix is directly consumed for chemical reactions, and over and the water available beyound that is required for chemical reaction remains as free water in the concrete. This free water in the hardened concrete evaporates leaving micro-voids in the concrete. The voids in concrete decrease the strength of concrete. More water added at mixing of concrete results in more shrinkage and porosity. One has to strike a balance between strength and workability. The workability can be improved by adding admixtures to cement, so it is wiser to use low water cement ratio.

2. *Strength increases with increase in cement content.* Cement acts as a binder in the presence of moisture. More cement means, more binding material. The binding material helps in binding aggregate in to a homogeneous solid mass. Cement content beyond a point is of no use and in fact it may cause bad side effects. Protein is good for human body but if one eats too much of proteins, the body will go sick, therefore an optimal cement content must be added for a given situation. More cement also means more heat of hydration and more shrinkage. Further, cement is the expensive material among the basic components of the concrete. It is therefore necessary to make an optimal use of cement in making concrete.

3. *Strength of concrete increases with increase in the size of aggregate.* Strength of the concrete depends on the strength, shape and size of the aggregate. As already mentioned, the aggregate is the filler and the strength giver. Basic strength of the aggregate is important and reflects in the final strength of the concrete. Every thing else being same, the strength of the concrete increases marginally with increase in the size of the aggregate. The type of construction, size of the members and intensity of reinforcement and cover requirement control the size of the aggregate that can be used in the concrete. In plain and mass concrete, one would like to select larger aggregate. 20 mm aggregate is the most commonly used aggregate in reinforced and prestressed concretes. Aggregate crushing value less than 45 is recommended in ordinary plain and reinforced concrete. A value of 30 is preferred in prestressed concrete and reinforced concrete exposed to very severe environment. 30 or 25 mm size aggregate is recommended in road pavement with aggregate crushing value not less than 30.

4. *Strength of concrete depends on the grading, shape and texture of aggregate.* Well-graded aggregate results into compact homogeneous mass therefore better strength. About equal angles of angular shaped aggregate gives better interlocking therefore gives higher strength. Round aggregate also gives good strength. The texture of the aggregate needs to be rough and not smooth to provide good interlocking. Flaky or oblong aggregate should be avoided in making concrete.

5. *Strength increases with curing period of concrete.* As mentioned earlier, cement generates heat of hydration during its reaction with water. This heat of hydration if not absorbed

by external agency, produces cracking in the concrete thus leading to a poor product. The heat of hydration can be absorbed by water that is sprinkled on the surface of the concrete. The process of dissipating the heat of hydration and also assisting in saturated chemical reaction of hardening concrete is called *curing of concrete*. In the process curing, the heat of hydration is dissipated and calcium silicates are formed and the concrete gains strength with time. Curing of concrete normally starts after about 10 hours of casting of concrete. The methods of curing may differ but the essentiality of curing is to protect the concrete surface from the loss of moisture and maintenance of wet surface around the concrete. The strength of concrete increases with curing. The suggested standard curing period is 28 days. All standard test specimens must be cured for twenty-eight days. Similarly the concrete structures need to be cured for 28 days to gain full potential strength. Concrete with ordinary Portland cement gains about two-thirds of the 28 days strength in seven days of curing. The rate of gain in strength of concrete with curing increases with curing period but its gain is rapid to start with and slows down with time. The gain of strength beyond 28 days of curing is considered to be marginal and may be about 30 per cent in a year period. If the curing is stopped after 7 days, then the concrete will not give its full potential. The gain in strength in such a case will be about eighty to eight-five percent of its full potential.

6. *Concrete gains strength faster with increase in the temperature at curing.* Temperature helps in accelerating in the formation of chemical compounds. It doesn't mean hot water should be poured on concrete. The temperature of the curing water needs to be increased slowly. Hot water is not recommended in curing but faster curing is achieved with warmer water when compared with cold water. Cold water is used in curing the mass concrete. Curing of concrete with steam is also practiced for very quick results. One day of steam curing is considered to be equal to 28 days of water curing. However there is a method of applying the steam curing.

Approximate strength in relation with the age of concrete with respect to that of 28 days curing is given in Table 1.13.1 The designer may give an appropriate correction to the strength of concrete when he is certain about the time of actual application of the load. However the revised code of practice on plain and reinforced concrete doesn't permit correction to the strength of the concrete with age.

TABLE 1.13.1 Strength of concrete with age

Age of concrete	Relative strength with respective to 28 days
7 days	0.65 to 0.7
28 days	1.0
3 months	1.10 to 1.15
6 months	1.15 to 1.20
12 months	1.20 to 1.25

Strength and Stress-strain Behaviour

Strength and stress-strain behaviour of concrete is measured by uni-axial compressive test performed on concrete cube or cylinder specimen. 150 mm concrete cube is the standard specimen accepted in India and some other countries. The mould of the specimen must be either cast iron or steel, machined to high dimensional accuracy. Concrete is placed in the mould and vibrated by a specified method and cured in water for 28 days. After 28 days, the cube surface is dried and tested in a compression-testing machine at a prescribed rate of loading. Typical stress-strain behavior of concrete cube is indicated in Figure 1.3.1.

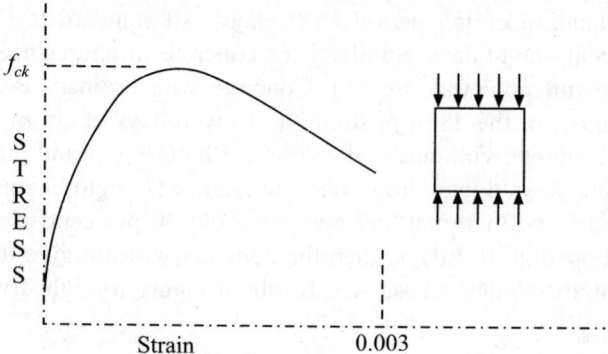

Figure 1.13.1 Stress-strain of concrete.

The testing is under controlled rate of loading rather than the strain controlled. The rate of loading is 2.5 N/mm^2/sec. The stress-strain curve of concrete in compression is linear upto 30 percent of the strength, and then an increase in the rate of strain from 30 to 50 per cent of the strength, and from there to about 85 per cent of the strength, the rate of increase in strain is faster. It is difficult to control the rate of loading after the load reaches 75 per cent of the strength, and the crushing of the concrete takes place suddenly. The maximum bending compressive strain in concrete is idealized to 0.0035. The limiting strain can be higher in case of strain controlled tests. The crushing strain in direct compression is limited to 0.002. The instantaneous loading on concrete indicates an increase in the strength. This loading may be termed as impact loading and the strength reading of the concrete with impact will be higher. Repeated cyclic loading of about 5000 cycles on concrete cube can reduce the strength to 70 percent. Similarly sustained load can decrease the strength to about 80 percent of the normal test load. The cylinder strength of concrete is about 85 percent of the cube strength. It is therefore seen

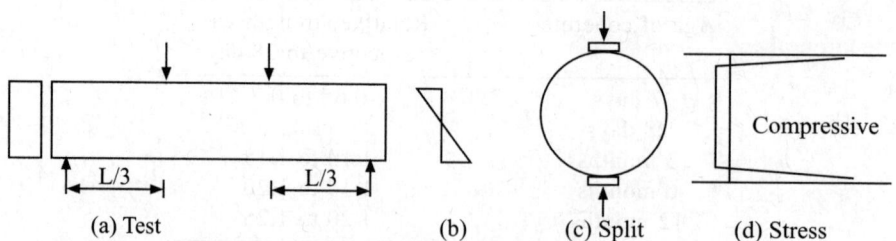

Figure 1.13.2 Flexure and split cylinder tests.

that the procedure in testing of the concrete has to be standardized to aim at consistency in understanding of the strength of the concrete.

There are other strengths such as prism, flexure, tension, split, shear, bond, etc., which are generally inter-related with the cube strength.

The tensile strength in bending, which is called strength in flexure of concrete is obtained by a flexure test of plain concrete beam specimen. The codes of practice specify a standard specimen subjected to two concentrated loads at one-third span points as shown in Figure 1.13.2

The bending tensile strength of concrete from the flexure test is also called as modulus of flexure and it is:

$$f_{cr} = \frac{M}{Z} = \frac{2WL}{bh^2}$$

where

M = bending moment in the pure bending moment zone = $WL/3$
W = concentrated load at one-third span
L = Span of the test beam, b = width of the section, and h = depth of the section

In the absence of a flexure test, one can get the approximate bending tension of the concrete from the compression strength of the concrete and it is given by:

Modulus of rupture = bending tensile strength = fcr = 0.7 √fck

The direct tensile strength of concrete is sometimes needed in design of specific members. Unfortunately it is difficult to design a test to determine direct tensile strength of concrete. There used to be a test called briquette test for the determination of direct tensile strength. A specially designed specimen called briquette with gripping ends was suggested at one time. But it is found that it is almost impossible to apply axial tension load without causing bending on the specimen. Therefore the test was considered to be not good enough to determine the real tensile strength of the concrete. An empirical expression is often used to get an approximate direct tensile strength of concrete and it is:

Direct tensile strength = fct = 0.35√ fck

where

fct = direct tensile strength of the concrete.

Another test called split cylinder test was developed to estimate tensile strength of the concrete. The test is also called a Brazilian test. A 150 mm by 300 mm concrete solid cylinder is laid horizontal between the two platens of a compression-testing machine and subjected to compression load on diagonally opposite faces of the cylinder. Figure 1.13.2(c) illustrates the test and Figure 1.13.2(d) illustrates the stress distribution across the depth of the cylinder. The split tensile strength can be computed as:

$$f_{ct} = \frac{P}{\pi DL}$$

where

P = compressive force, L = length of the cylinder, D = diameter of the cylinder.

The direct tensile strength of the concrete is approximately equal to about sixty percent of the flexure strength of the concrete. The cracking strength of concrete is normally referred as that equal to the split cylinder strength. The tensile strength of the concrete in bending is equal to the modulus of rupture in case of beams subjected to bending. The allowable tensile stress is obtained by dividing the tensile strength by partial factor of safety.

The reinforcement bars are embedded in concrete and the bar should not slip from concrete when it is subjected to pull or push. The bonding capacity of the concrete with the bar is called bond strength of the concrete. The bond strength of the concrete is determined by pullout test. A bar is embedded in a concrete cylinder and pulled from one end. Pulling force is applied at one end of the bar and slip at the other end of the bar is measured. The bond strength is determined in two stages. One that corresponds to no slip condition and the other is slip condition of the bar. The load at which the slip is initiated is called the no slip bond strength. The ultimate bond strength is the bond stress corresponding to load at which 0.25 mm slip takes place. The bond stress is equal to the pulling force divided by the surface area of the embedded bar. Table 1.13.1 gives different strengths of the concrete recommended for design.

TABLE 1.13.1 Modulus of rupture and bond strength (stress in N/mm^2)

Concrete strength = fck =	10	15	20	25	30	35	40	45
Modulus rupture	2.2	2.7	3.1	3.5	3.8	4.1	4.4	4.7
Bond stress for 0.25 mm slip								
Plain bars	–	1.9	2.4	2.9	3.3	3.4	3.5	3.5
Deformed bars	–	3.5	4.4	5.2	5.8	6.3	6.5	6.6
Bond stress for no slip:								
Plain bars	–	1.5	1.7	2.0	2.2	2.5	2.7	2.8
Deformed bars	–	1.9	2.1	2.5	2.8	3.1	3.4	3.5

Cube test: Concrete cube under compression has its top and bottom faces in contact with the compression platens. As the cube is subjected to uni-axial compression, the cube will get shortened and the lateral dimensions tend to expand by the Poisson's effect. However the contact top and bottom faces of the cube are constrained against the lateral expansion. The fracture of the cube is influenced by the frictional force on the top and bottom faces of the cube in addition to the axial compression. The failure is not by uni-axial compression but by combined action of the forces on the cube. The failure or the crushing pattern of the test cube is like a double symmetrical pyramid. Figure 1.13.3 illustrates the failure pattern of a concrete cube. The failure is brittle and is due to the combined effect of compression and friction on the faces. On the other hand when a concrete prism is subjected to uni-axial compression, the surface friction of the platens at the top and bottom faces damp out in short length and the middle portion of the prism experiences direct compression or uniaxial compression. The compressive strength of the concrete prism has a relation with that of the cube.

TABLE 1.13.2 Compressive strength of prism with respect to that of 150 mm cube

Length/width	0.5	1.0	2.0	3.0	4.0	5 and more
Relative strength	1.5	1.0	0.8	0.72	0.68	0.67

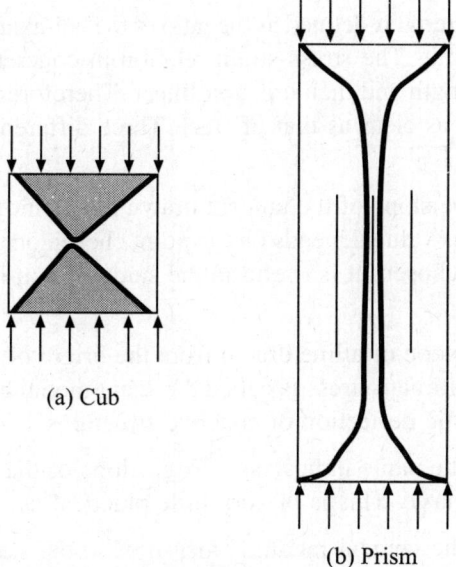

(a) Cub

(b) Prism

Figure 1.13.3 Failure pattern of concrete cube and prism under uni-axial compression.

The strength of a prism specimen decreases with increase in height-to-side ratio and converges to a value when this ratio is 4 or 5. The compressive strength of prismatic element where the length to thickness ratio is five or more, is 0.67 times that of the 150 mm cube. Similarly the 150 by 300 mm cylinder strength is about 0.80 times that of the 150 mm cube.

TABLE 1.13.3 Strength of concrete cubes with respect to that of 150 mm cube

Cube size (in mm) =	100	150	200	300
Relative strength wrt. 150 mm cube	105	1.0	0.95	0.87

The strength that is indicated by a cube specimen is estimated to be more than that by the cylinder specimen. The size of the cube and the diameter of the cylinder are to be same. Its relation is:

$$fcu = (0.75 \text{ to } 0.80) \, fcy$$

where

fcy = cylinder strength having height to diameter ratio equal to two, and

fcu = strength of concrete cube having the same size as that of the diameter of the cylinder.

The lower value applies to smaller sizes.

Moisture content in concrete provides lubrication effect and reduces the strength when compared with a dry sample.

Strength of a dry sample = (1.1 to 1.20) times the strength of the saturated sample.

Modulus of Elasticity of Concrete

Modulus of elasticity of a material is defined as the ratio of the uni-axial stress to the corresponding strain with in the elastic limit. The stress-strain relation in concrete is almost linear upto 30 percent of the crushing strength and then it is non-linear. Therefore the measurement of elastic modulus in concrete is not as clear as that of steel. Three different moduli are mentioned in such non-linear materials.

Tangent modulus: The slope of the tangent drawn at a point to the stress-strain curve is called tangent modulus. This value depends on the point chosen on the curve. It decreases with increase in the stress level chosen. It is useful in the study of non-linear behavior of concrete structure.

Scent modulus: The slope of a line drawn from the origin of the stress-strain curve to a point at 40 per cent of the ultimate stress is called the scent modulus. This probably is realistic in the determination of elastic deflection of concrete structures.

Chord modulus: As the name indicates, it is the slope of the chord drawn between two points on the stress-strain curve. This is of very little practical usefulness.

Dynamic modulus: The tangent modulus measured at the starting stress-strain curve is considered to be the dynamic modulus of elasticity. This can be obtained from the compression test of a cube or a cylinder. This value is likely to be atleast 20 to 40 per cent higher than the scent modulus.

Flexural modulus of elasticity: A simply supported plain concrete beam is subjected to concentrated load at mid-point and the deflection of the beam is measured for incremental loads. Flexural modulus of elasticity is computed from the load deflection relation of a simply supported beam. The test beam has to be slender so that the deflection due to shear is negligible. The flexural modulus of elasticity is given by:

$$E_c = \frac{WL^3}{48Iv}$$

The modulus of elasticity of concrete is essential to compute deflections of structures. The modulus of elasticity of concrete is taken close to the flexural modulus of elasticity. The code of practice suggests the modulus of elasticity in terms of the strength of the concrete. The Indian code of practice on plain and reinforced concrete suggests the following. The modulus of elasticity of the concrete Ec (in N/mm^2) can be taken as

$$Ec = 5000 \ \sqrt{fck}$$

A linear stress distribution across the depth of a beam is assumed for a beam subjected to bending moment. As the bending moment increase, the plain concrete beam fails when the tensile stress concrete at the extreme fibre reaches the flexural strength. All the tensile force is resisted by the reinforcement in reinforced concrete beam under bending. Redistribution of stresses in concrete compression zone takes place as the load tends to failure level. Consequently the strength of concrete in bending is higher than that in direct compression. There is no such redistribution of stress in columns under compression.

1.14. Durability of Concrete Structures

The property of concrete structure to withstand the environmental effects along with loads is the durability property. The durability is measured on qualitative scale rather than a specific quantitative scale. The internal and external factors that influence the durability and life of a structure are listed and explained briefly in this section. The degree of the effect is different for different factors. A list of parameters that influence the durability is given and explained later.

Physical and mechanical factors:

Chemical Factors:

Biological and Environmental factors

1. *Non-structural factors*
    (a) *Physical Internal factors:*
        Compaction of concrete, Porosity and permeability of concrete, Surface finish of concrete, Duration of curing of concrete; Cover to the reinforcement and Surface cracking to start with.
    (b) *External factors:*
        Abrasion and erosion (wear & tear),
        Exposure to wetting and drying (moisture) and Freezing and thawing
2. *Chemical Factors:*
    Chemical composition of Aggregate, Quality of cement and water used, Exposure to acidic environment, Sulphate attack, Aggregate alkaline reaction, Other chemical aggressive actions.
3. *Biological and Environmental factors*
    Biological growth, such a moss, algae etc on the surface of concrete
4. *Non-structural factors*
    Shape and size of structural elements, Drainage from the structure,
    Joints, Inserts, Bearings, Railings, Anchorage and Fixtures.
    There is interdependence between the factors that are listed above.

Physical Factors

The main physical factors that influence the durability of concrete are explained in brief:

- ❖ **Compaction of concrete and Moisture transport:** Production of good concrete should be the aim of any builder. Good workable concrete leading to dense and well-compacted concrete minimizes the transport of moisture within. The porosity or permeability of the concrete is a direct result of poor compaction of concrete. Transportation of moisture induces physical and chemical reactions that are responsible for the deterioration of the concrete. Diffusion of air, carbon dioxide and chloride ions etc. into concrete combined with moisture add to the problem. The carbonation of concrete leads to increase in volume and cracking. Similarly the presence of moisture and air leads to corrosion of reinforcement.

- ❖ **Surface Finish:** Water in contact with the concrete surface is transported through capillary action even in well-compacted concrete. Uneven and rough surface retains moisture and transports. Micro-cracked surfaces of concrete permit the movement of water freely into the concrete. An impermeable concrete surface is the most ideal finish even though not possible, but even such surfaces develop micro cracking in due course because of changes in temperature and chemical actions. Adequate precautions must be taken to maintain an even and un-cracked surface of concrete.

❖ **Wetting and drying:**   Water carries dissolved salts such as sulphates, chlorides, carbonates etc. into the pores of the concrete under wet condition. The moisture in the concrete evaporates during drying leaving dissolved salts in crystallized state. The accumulation of such crystals increases the volume leading to cracking of the concrete. Further the concentrated crystallized chemical act on the aggregates and reinforcement causing carbonation and corrosion as the case may be. Efflorescence of crystallized salts at the surface of concrete is also a result of wetting and drying.

❖ **Freezing and Thawing:**   Water or moisture transported into the pores of the concrete when exposed to freezing and thawing causes volume changes in addition to the acceleration of chemical degradation. This leads to cracking of the concrete and deterioration.

❖ **Chemical Factors**
Chemical reactions take place due to internal concrete structure or external exposure environment. Some of the important chemical attacks on concrete are briefly mentioned. Moisture transportation aids the chemical attacks.

❖ **Carbonation, Acid formation and attack:** Moisture along with environmental gases, liquids and solids lead to formation of acids that react with hydrates of cement. The chemical attack breaks the chemical bonds in the calcium silicates and hydrates thus destroying the strength of the concrete. The extent of deterioration depends on the density and porosity of the concrete, cement and aggregate properties and the environmental conditions.

❖ **Sulphate attack:** The sulphate ions from soil or water or even from cement react with calcium hydroxide in the concrete forming sulphates. The increase in volume due to formation of sulphates within the pores of the concrete result in cracking and further degradation of the concrete. The presence of aluminates in cement has a tendency to expand in volume on reaction with sulphates.

❖ **Alkali and chemical attacks:** Concrete is initially saturated with lime therefore has an alkali environment. The sodium and potassium ions in the alkali solutions attack the silica of the aggregate. The rate of attack depends on the active silica in the aggregate and dust particles, porosity of the concrete and moisture transport. Concrete surface when exposed to acidic environment reacts with aggregates resulting in increase of volume and consequent cracking.

❖ **Biological and Environmental Factors**
Continuous presence of moisture on the surface of concrete promotes biological growth such as moss, algae and small plants. Penetration of plant roots and biological products into concrete results in cracking and deterioration.

❖ **Corrosion of Reinforcement**
Oxides of iron and steel are formed when the reinforcement is exposed to moisture and oxygen. Similarly the chloride ions accelerate the formation of oxides of iron. The alkali environment in the concrete with pH value in the range of 12 provides good

protection against formation of iron oxides. The diffusion of carbon dioxide into concrete reacts with calcium hydroxide forming calcium carbonate. Therefore the pH value in concrete decreases to 9 and the protection to the reinforcement decreases. The dissolution of positively charged iron ions and the combination of the released electrons with moisture and oxygen results into hydroxyl ions. The ion activates the formation of ferric oxide. The formation of ferric oxide on the surface of the reinforcement results into increase in volume of rusted surface. The increase can be as much as 50 to 200 percent of the iron content.

The high alkalinity and relatively high electrical resistivity of Portland cement under moist condition protects the embedded reinforcement in concrete. Factors that are likely to cause corrosion to the reinforcement are:

(a) Inadequate cover to reinforcement,
(b) Cracking of the concrete,
(c) Honey combing of concrete,
(d) Carbonation of hydrates,
(e) Electrolysis.

Good concrete with well-finished surface and having adequate cover to the reinforcement provide good protection to reinforcement against corrosion. Cracks or voids in the concrete provide an access to moisture, air and other environmental chemical agents. Rusted steel will expand in volume thus causing cracking along the reinforcement. The hydrated cement when exposed to carbon dioxide gets carbonated and results in increased shrinkage and cracking. The alkali protection to the reinforcement is lessened. Similarly passage of electricity or development of static potential difference around the reinforcement causes oxidation and corrosion. The following are recommended to protect the reinforcement against corrosion. The main aim is to minimize the permeability.

❖ The most important is to place good concrete with excellent surface finish and cure it well.
❖ Select lowest feasible water-cement ratio.
❖ Minimize the use of admixtures containing calcium chloride, or soluble chlorides. Restrict the entry of salts through the water used either in the concrete mix or in curing.
❖ At least 14 days of uninterrupted curing be done so that good hydration takes place without micro-cracking.
❖ Adequate cover to the reinforcement be provided.
❖ **Stress Corrosion in Pretension Steel**
The oxidation of stressed steel wires results in sudden splitting of the wires. This phenomenon is present in stressed steels of prestressed concrete constructions.
❖ Non-Structural Component Factors.

The discontinuities in construction, construction and expansion joints are liable for problems. Corrosive and even non-corrosive fixtures and inserts exposed to moisture are liable for causing cracking in the concrete. Such non-structural elements must be planned, positioned, placed, protected and maintained carefully.

As per the IS: 456:2000(2005), the code of practice of plain and reinforced concrete, the factors influencing the durability of concrete structures include:

❖ The Environment,
❖ The cover to the embedded steel,
❖ Type and quality of construction material,
❖ Cement content and water-cement ratio of concrete,
❖ Workmanship to obtain full compaction and efficient curing,
❖ The shape size of the member.

1.15. Durability Design Considerations

The exposure conditions of a structure decide the selection of structural and architectural materials, design and detailing of the elements. The basic five exposure conditions for the durability design considerations are:

1. Mild exposure,
2. Moderate exposure,
3. Severe exposure,
4. Very severe exposure, and
5. Extreme exposure.

Mild exposure: The mild exposure condition may be called as protected environment. Interiors of dwellings, offices, commercial complexes and workshops of non-aggressive environment and other structures protected against weathering, wetting and drying are classified in this group.

Moderate exposure: Foundations buried under non-aggressive soils and ground water, structures exposed to rain and not under frequent wetting and drying, outside non-aggressive environment, high humidity halls, running water (canals, dams, weirs etc.) bath rooms and water tanks come under this class.

Severe exposure: Structures exposed to frequent wetting and drying of ordinary water, partially submerged under water, occasional freezing situations, structures in coastal regions, structures subjected to constant vibrations, foundations in aggressive soils are considered to be under this classification.

Very severe exposure: Structures exposed to sea water spray, extreme freezing, chemical fumes, colour dying halls, partially submerged sea water (all harbor structures in contact with sea water), septic tanks come under this classification.

Extreme exposure: Concrete roads; structures under constant abrasion in wet and dry conditions, direct contact with aggressive chemicals, floors of chemical plants are classified in this category.

The important parameters are listed here. Table 1.15.1 lists the desirable specifications for the different materials etc for the five exposure conditions.

(a) **Materials:**
Quality and Quantity of aggregate,
Type and Quantity of cement,

Water Quality, Water-Cement ratio,
Admixtures if any to be used,
Concrete mix design and its workability,
Reinforcement and
Inserts and Fixtures.

(b) **Detailing and Workmanship:**
Surface finish, shape and texture,
Cover to the reinforcement and precautions,
Finishing of joints,
Drainage of water,
Curing of concrete,
Accessibility for maintenance and
Quality assurance scheme.

(c) **Maintenance:**
Inspection regulations,
Systematic maintenance scheme and implementation,
Immediate attention to damages,
Replacement and Repairs,
Renovation and Prevention of accumulation of water.

TABLE 1.15.1 Recommended limits for durability considerations
(unless otherwise stated it is by weight)

| Material / Property | Exposure Classification | | | |
	Mild (1)	Moderate (2)	Severe (3)	Very Severe & extreme (5)
Aggregate	Normal	Normal	Good	Good
Grading	40	40	30	30
Crushing value (Max.)	2000	2000	2200	2400
Aggregate density (kg/cm) (minimum)				
Amount of Silts/clay/fine (less than 0.15 mm) (Maximum %				
In Coarse	2	2	1	1
In Sand	4	3	2	2
Sulphates Admissible (max %)	1	1	0.5–1.0	0.5–1.0
2. **Cement (OPC)** Minimum(kg/cum)				
PCC	220	240	250	260–280
RCC	300	300	320	340–360
PSC	350	350	400	400–400

Maximum content (kg/cum)<----------------------- 450 ------------------------------->
(unless special considerations are given)

3.	Maximum Water Cement ratio				
	PCC	0.60	0.60	0.50	0.45–0.40
	RCC	0.55	0.50	0.45	0.45–0.40
	PSC	0.50	0.50	0.45	0.45–0.35
	(Better avoid PSC in highly aggressive environment)				
4.	Minimum grade on concrete				
	PCC	M10	M15	M20	M20, M25
	RCC	M20	M25	M30	M35, M40
	PSC	M35	M40	M45	M45-150
5.	Water quality for mixing				
	minimum. pH value	4.5	6.0	6.0	6.5
6.	Maximum chloride content Maximum acid soluble Chloride content as % of chloride ion by mass of concrete				
	PCC	1%			
	RCC	0.15%			
	PSC	0.10%			
7.	Minimum cover (in mm) to reinforcement bars in RCC (The cover specification applies to main and secondary reinforcement)				
	Slab	20	20	25	30 to 35
	Walls	20	20	25	30 to 35
	Beams	25	30	45	50 to 60
	Columns	40	40	45	50 to 75
*	(1) A wearing or maintenance protective coat to be provided to the concrete in extreme exposure condition (Reinforcement be coated with protective coating) (2) Cover should not be less than the maximum size of bar				
8.	Inserts projecting beyond the concrete surface should be made of:	Steel	GI	GI	Stainless steel
9.	Minimum clear spacing of reinforcement	Diameter of the bar, or Maximum size of aggregate + 5 to 10 mm,			
10	Minimum cover (in mm) to prestressed concrete wires or cables				
	Slabs	20	25	25	25–30
	Beams	25	30	45	45–75
11.	Recommended maximum crack widths (in mm)	0.3	0.2	0.1	0.1
12.	Minimum clear spacing of cables ducts in PSC	<---------------------- 1.5 to 2 times the diameter of the ducts (preferably 2 times the dia of duct.)---------->			

Structural Concrete: The constituents of concrete, namely aggregate, cement, water and admixtures must satisfy the standard requirements. Further there are limits on qualities and qualities of the basic materials to suit the exposure requirements. Similarly the mix proportions, methods of mixing, laying, consolidation, finishing, formwork and curing are important. Testing methods and quality assurance must be given adequate importance.

Aggregate: Normal aggregate well graded and containing least amount of fine particles (less than 0.15 mm) should be selected. The crushing value of the aggregate reflects the grading and angularity. Some recommended limits on aggregate are given in Table 1.15.1. Some variations in the limits are permitted based on the type of construction such as Plain concrete, Reinforced concrete etc.

Cement and Water: Portland cements of grades 33, 43 and 53 are suitable for structural concrete. Portland composite (Pozzolana) cements can also be used depending on the type of construction conditions. 53-grade cement is faster in hardening when compared with the 33 grade. A minimum quantity of cement content, especially in reinforced concrete is required to provide adequate alkaline environment to protect the steel from corrosion. Similarly maximum limit is suggested to minimize shrinkage and heat of hydration effects. Shrinkage of concrete increase with water contents hence a reduction in water-cement ratio enables higher cement content. Water used in concrete making should confirm to the standards and should not contain oil, organic matter, humic acid etc. The water-cement ratio should be as low as possible not only for strength consideration but also to minimize the porosity of the concrete.

Admixtures: Admixture added to cement or concrete to improve the properties such as workability, setting time or to decrease the permeability should not normally exceed 5 per cent and preferably in the range of 2 per cent or as recommended by the manufacturer. Admixtures or additives should not contain any chlorides in any form in reinforced or prestressed concrete constructions.

Strength of concrete: Strength of concrete need not necessarily indicate the durability, but yet it is considered as a desirable index for durability. Concrete must have a minimum strength for durability considerations. It indirectly reflects the quality of production and porosity. Minimum acceptable grades of concrete under different exposure conditions in the opinion of the author are listed in Table 1.17. The ones suggested by the code is listed separately.

Reinforcement: Some suggestions on cement and water contents were already explained and the limits are indicated in the Table 1.17 to protect the reinforcement from corrosion. The other suggestions and design parameters are:

(a) Uninterrupted curing to avoid micro-cracking and complete hydration,
(b) Blended cements with slag or fly ash or silica fumes can be used,
(c) Limit the total sodium oxide to a maximum of 0.6%,
(d) Use Low water–cement ratio, and
(e) Provide adequate cover to the reinforcement and not less than that mentioned in Table 1.15.1.

Proper placing of reinforcement to ensure the cover to the reinforcement is ensured. Unfortunately cover to the reinforcement is the most neglected factor as the workers trample during construction. Table 1.15.1 suggests the minimum cover to be provided to the reinforcement.

Similarly the clear spacing of the bars and cables should be adequate enough to achieve sound concrete around the bars. A special protective treatment to the concrete surface or to the reinforcement should be provided in aggressive environment as of extreme exposure.

Inserts: Steel or galvanized iron or stainless steel inserts can be embedded in concrete depending on the exposure conditions. Wrongly placed or wrong inserts exposed to moisture or retention of moisture because surface damages, which leads to deterioration of the structure. Table 1.15.1 recommends desirable type of inserts.

The minimum cement contents and the water-cement ratio as suggested by the code are listed in table 1.15.2.

TABLE 1.15.2 Minimum quantity of cement or cementetious material required, and Maximum W/C ratio permitted as per IS: 456-2000(2005)

Exposure	Plain Concrete		Reinforced Concrete		Minimum Grade Concrete	
	Min. Cements (kg/m^3)	Max. free w/c	Min. Cements (kg/m^3)	Max. free w/c	Plain Concrete	Reinforced Concrete
Mild	220	0.6	300	0.55	-	M20
Moderate	250	0.60	300	0.50	M15	M25
Severe	260	0.50	350	0.45	M20	M30
Very Severe	280	0.45	375	0.45	M20	M35
Extreme	300	0.40	375	0.40	M25	M40

Note: Cement content specified is irrespective of grade of cement,

It is inclusive of additions made to concrete such as flyash, blast furnace slag etc., with respect to water cement ratio, Minimum grade of concrete for plain concrete is not specified for mild exposure conditions.

TABLE 1.15.3 Adjustment to Minimum Cement contents for aggregate other than 20mm Nominal size

Nominal size of aggregate (mm)	Adjustment to minimum cement content of Table 1.18 (kg/m^3)
10	+40
20	0
40	−30

Maximum Cement Content: The code on practice of plain and reinforced concrete specifies maximum cement content. Cement paste shrinks on drying, more cement means more shrinkage. Further, heat of hydration is also increases with increase in cement content. Heat of hydration leads to micro cracking. Micro cracking further leads to ingress of water and ultimately deterioration of concrete, corrosion of reinforcement. It is therefore desirable to limit the maximum quantity of cement content. The maximum cement content recommended by the Indian code of practice is 450 kg/m^3 unless otherwise approved by the engineer in-charge.

Chlorides in Concrete: The chloride ion may be present in the components of concrete, such as water, aggregate, sand, admixtures, and cement. Or chloride may be diffusing from the environment into the exposed surface of the concrete. The chloride ion causes corrosion of reinforcement at an accelerated rate.

Maximum limit in sulphate: Sulphates are present in many aggregates, and even in cement. The soil on which the structure stands may also contain sulphates. Even the environment contains some forms of sulpher and sulphates. Sulphates combining with water results in compounds of higher volume. It is necessary to limit the total sulpher trioxide (SO_3) content in the concrete to 4 per cent by mass of the cement. This limit doesn't apply to super-sulphated cement complying to IS: 6909.

Alkali-aggregate reaction: Almost all aggregates contain silica as the main element. Alkalis such as sodium (Na_2O) and potassium (K_2O) react with silica in the aggregate producing expansive compound. This reaction is produced if the concrete is exposed to high moisture. The cement or the aggregate may contain alkali reactive constituents. Increase in the volume of the reacted aggregate will cause cracking and deterioration of the concrete. The following precautions should be taken.

❖ Use non-reactive aggregate,

❖ Use fly ash conforming to IS: 3812 or blast furnace slag conforming to IS: 12089. Or use Portland Pozzolana cement etc.,

❖ Protect the concrete from constant exposure to moisture,

❖ Limit the alkalis to 1.1% in cement.

1.16. Fire Protection Specifications

Domestic facilities such as cooking gas and electric power lines are provided in most homes. A liability of fire hazards to homes exists. Modern buildings has large amount of combustible material such as clothes, wooden furniture, books and paper, plastic items and wall and curtain hangings etc. The permanent buildings are built with brick and concrete. These building materials resist the fire well. The strength of the material deteriorates fast under fire. The increase in vertical heights of the buildings and compactness of the multistory buildings make the fire fighting a complex problem. The buildings must be built to resist fire to desirable safe levels. This section is only an introduction to the fire protection specifications of simple buildings. Fire protection measures to tall buildings are not dealt in this book. The time period for which a building can withstand fire without serious damages is called fire rating. The basic fire ratings and the protection specifications to the structural materials based on the code suggestions are mentioned in this section. Minimum dimensions and minimum cover required to the reinforcement for different periods of fire exposure are listed in tables 1.16.1 and 1.16.2.

TABLE 1.16.1 Minimum required dimensions for members (in mm) for fire rating in hours

	Beam & floor dimensions			Column dimension		Wall thickness	
Rating hours	Beam width	Rib width	Floor thickness	Fully exposed	One face exposed	$0.4\% < p < 1.0\%$	$p > 1\%$
0.5	200	125	75	150	100	100	100
1.5	200	125	110	250	140	140	100
2.0	200	125	125	300	160	160	100
4.0	280	175	170	450	240	240	180

p = percentage of reinforcement

It can be seen that the code has come with precise minimum dimensions and minimum covers to the reinforcement. In the present practice the minimum thickness of ribs of waffle slabs are made in the range of 100 mm as against 125 mm now recommended. The thickness of such waffle floor slabs may even start from 60mm at the moment as against 75 mm thickness suggested for half an hour fire rating.

TABLE 1.16.2 Nominal cover requirement to reinforcement (in mm)

Rating	Beams		Slabs		Ribs		Columns
In hours	Ss	Cont.	Ss	Cont.	Ss	Cont.	
1.0	20	20	20	20	20	20	40
1.5	20	20	25	20	35	20	40
2.0	40	30	35	25	45	35	40(45)*
3.0	60	40	45	35	55	45	40(55)
4.	70	50	55	45	65	55	40(65)

Ss = simply supported, Cont. = continuous, *the values given in the brackets recommended by the author on the assumption that if the ribs need 45 mm cover the columns that are exposed to the same order of heat should have atleast that much cover. Columns at about the middle height of the floors are seen to have had maximum damaged in fire.

The waffle slabs are usually provided in large halls for public gatherings and also in tall buildings. The minimum fire rating of such public halls is 1.5 hours and above. If that is the case, the waffle slab thickness can't be less than 110 mm. The minimum cover to the reinforcement in ribs for 1.5 hours of fire rating is 35 mm in simply supported waffle slabs. The minimum width of the rib will be atleast 160 mm. Further the cover requirement as per the present code applies to main as well as secondary reinforcement.

1.17. Quality Assurance in Concrete Structures

Concrete is a powerful material leading to innovative structures and methods of construction. The innovative structures require innovative and quality controls to give high reliability. The concrete construction uses a wide spectrum of skilled and semi-skilled workers. Experienced engineers and construction companies, build outstanding structures with excellent quality controls; and on the other end of the spectrum, skilled and semi-skilled workers build concrete houses for common man with little or no supervision. The quality assurance of concrete structures varies from reliability level to uncertainty level. There are many non-engineered buildings being built in India. There is a need to educate the semi-skilled and engineers on the quality assurance of concrete structures.

A number of questions do arise on the quality assurance programme. The questions are how and what exactly is to be assured? How a quality of a structure can be assured? In the final analysis, it is the quality of the structure to be assured and not only the quality of the materials as dominantly observed today. Components that control the quality of a structure are:

1. Quality of the basic materials such as aggregate, cement, water, admixtures, etc.
2. Mix proportions of concrete and its production,
3. Construction and fabrication of the structure, formwork,
4. Transportation, placing, compaction and finishing of concrete,
5. Tolerances in quantities and qualities, and detailing,
6. Curing of concrete,
7. Strength and serviceability design of the structure,
8. Durability specifications and maintenance,
9. Life expectance.

At the moment, the quality control is leaning heavily towards items one and two of the above list because of the historical and technological background. The idea that the quality control in the basic materials and the production of the concrete ensures the quality of the final structure is a necessary condition but not sufficient Input of excellent material doesn't guarantee an excellent product. There are a number of inter links between the quality of basic materials and the final structure. Between the cup and the lip, there are a number of slips.

Code Provisions on Quality Control

The clauses on the quality control of concrete structures given in the code are primarily on materials and production of concrete and not on the final structure. Characteristic strength of the concrete is the centre of focus in the design and even construction of concrete structures. The characteristic strength is defined as the strength of the material below which not more than five percent of the test results fall. Strength of concrete, as a matter of fact, strength of any material will follow a normal distribution. Figure 1.17.1 illustrates a normal distribution curve in which the five-percent cumulative value is indicated.

The characteristic strength in terms of probability of failure definition can be expressed as:

$$P(fc < fck) = pf$$

For a normally distributed strength of concrete where pf is the accepted probability of failure. The above probability acceptance can be expressed as:

$$\phi(fck - fm)/s) = pf$$

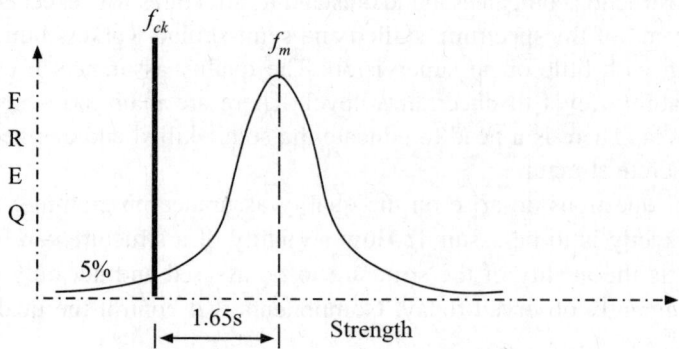

Figure 1.17.1 Normal distribution curve.

where

fm and s are the population mean value and standard deviation of the normal distribution, fck characteristic strength, ϕ is the cumulative normal distribution function.

The expression can be inverted to establish the relation between the mean and characteristic strengths, and it is:

$$\phi-1(pf) = (fck - fm)/s = -k$$

where k is an index that signifies the acceptable probability of failure. The minus sign is assigned as the five per cent is on the negative end of the curve. The value of k is 1.65 for five-per cent acceptability of failure.

The inverse of the equation can be rearranged as:

$$fm = fck + 1.65s$$

The minimum size of the sample needs to be about fifty in general for a normal distribution, however the code accepts the size of thirty. Even that is not a practicable size for day to day quality control of concrete. Three concrete cube specimens form a sample. In most constructions, thirty-sample size is not practicable in day-to-day control, therefore four consecutive non-overlapping samples are considered to be a practicable size to test the acceptability of the concrete. As the size is smaller than the minimum population size, a modification to the acceptability expression is suggested. An expression, which satisfies the five percent acceptability criterion with a sample smaller than thirty, is given by:

$$fm1 = fck + 1.65 \ s \ (1-1/\sqrt{n})$$

in which n is the size of the sample and fm1 is the mean value of the smaller size sample. The equation reduces to the earlier one for n tending to infinity and the mean value comes out to be same as characteristic strength for a single sample.

The above equation for four non-overlapping samples reduces to:

$$fm1 = fck + 0.825 \ s$$

Besides the pre-assigned checks on the input materials, the main clauses on the quality control acceptability of concrete as per IS: 456-2000(2005) is:

(a) The concrete is said to acceptable if the following relation is satisfied

$$fm1 \geq fck + 0.825 \ s \ ; \ \text{or}$$

$$fm1 \geq fck + A, \text{ and}$$

$$fi \geq fck - B$$

where fck = characteristic strength,

$fm \ l$ = mean strength of any four consecutive non-overlapping samples,

fi = strength of a sample,

A = 3 MPa for $M15$ concrete and 4 MPa for $M20$ grade concrete and above

B = 3 MPa for $M15$ concrete and 4 MPa for $M20$ grade concrete and above

The concrete is liable for rejection if it is porous or honeycombed, improper construction joints and tolerances on size of the members. It can also be rejected for improper placement of reinforcement and inadequate cover, and not following the specifications.

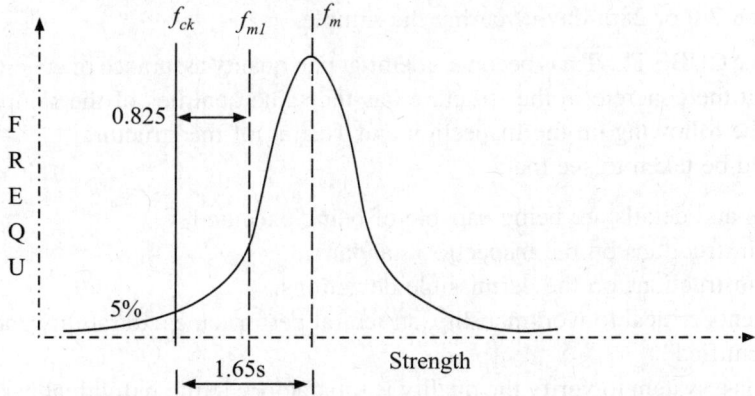

Figure 1.17.2   Mean value for 4 sample acceptance.

The mean strength requirement set up by the code is on the liberal side especially for concretes of M35 and above. The mean strength of four consecutive samples needs to be four Newtons per square mm or 0.825 times the standard deviation is easy to satisfy. The value 0.825 derived from the normal distribution of the statistics is acceptable if the sample size is large. Consider a case of eight sets of four consecutive non-overlapping samples. Let each of the samples satisfies the criterion of the mean value greater than the characteristic strength by 0.825 times standard deviation. But when all the thirty two samples are considered as a large sample, the global quality control expression demands that the mean value should be greater than the characteristic strength by 1.65 time the standard deviation of the total sample. The standard deviation of the total sample will be more than the standard deviation of any one of the eight samples. Further the multiplier of the standard deviation is 1.65. The chances of satisfying the acceptance criterion of the large population of thirty-two samples are not high.

The acceptability criteria are supposed to be valid for all concretes upto 60 MPa. That may mean even if the concrete is of $M60$ grade, the mean value of the order 64 MPa of four consecutive samples is acceptable. In that sense the section is liberal. To say that strength of every sample should have a value greater than characteristic strength minus 4 MPa is more stringent. For example that for $M60$ concrete, no sample strength should be less than 56 MPa. This could cause some problem because the margin for individual sample strength is rather small. For example, the Indian Road Congress (IRC) permits strength of a sample not less than 80 per cent of the characteristic value, and the earlier version also permitted this eighty-per cent.

Concrete Cube Test

As per Indian building code of practice, a concrete sample consists of three 150 mm-cube concrete specimens, cured for 28 days in water and tested under moist surface dry condition. The sample is subject to the following conditions:

1. Sample should to be selected randomly,
2. The mould of the cube is well-finished steel or cast iron,
3. The concrete is placed and vibrated under careful control conditions,
4. Cured under good moist condition,
5. Tested on 7th or 28th day of curing the sample.

The concrete CUBE TEST has become a demi-god in quality assurance of concrete structures. It is a hope that the concrete in the structure has the same qualities of the sample! The code further states the following on the Inspection and Testing of the structure:
Care should be taken to see that:

(a) Design and details are being capable of being executed,
(b) Clear instructions on the inspection standards,
(c) Clear instructions on the permissible deviations,
(d) "Elements critical to workmanship, structural performance, durability and appearance are identified"
(e) There is a system to verify the quality is satisfactory in the individual parts, especially the critical.

Other Deficiencies of Acceptable Criteria in Cube Test

A concrete sample consists of 3 specimens to be tested on 7th day, another 3 specimens to be tested on 28th day after casting and curing. Standard also specifies how the samples to be collected, cast, cured and tested. The small constructions in India are mostly site mixed. Even though concrete is made by weigh batching, the batching plant itself is not necessarily automated in many cases. Consequently, there are some inherent deficiencies in the acceptability cube test criterion. Compressive and flexural strength tests are specified of which the flexural strength test is considered only in special conditions. Some of the weak points in the acceptance criteria are:

(1) The concrete batch from which the samples are collected may be prepared with care even though the samples are supposed to be randomly chosen,
(2) Filling, compaction of samples is different from placing and vibration of the concrete in the structure,

(3) The formwork and shuttering to the concrete in structure is not as grout leak proof like the cube mould,

(4) The time difference in transportation and lying of the concrete in position could be different from that of the preparation of cube specimens,

(5) The congestion of reinforcement that might cause honey combing is not reflected in making of the cubes,

(6) The curing of cubes is done systematically when compared to the curing of the concrete in structure,

(7) Testing of 3 specimens as a sample and taking an average strength of the sample is not dependable. Even if more number of samples is considered but the total number in a batch of construction is not high for a statistically acceptable level.

(8) There is a likelihood of less quality control in real concrete structure, detailing of reinforcement, formwork and curing as the actual concrete is not subjected to any testing.

In case where the tensile strength of the concrete plays an important role, flexure test is recommended by the code.

1.18. Non-destructive Testing of Concrete Structures

Methods of non-destructive testing are developed during the last four decades. Reliable equipment for such non-destructive testing is manufactured only in recent times. Further a number of non-destructive tests are available. Ultra sonic non-destructive testing has been accepted as a reliable method of testing of welding in all-important steel structures. At the moment, rebound hammer test, ultrasonic test, and core-cutting tests are considered to be dependable tests in concrete construction. Of these, ultrasonic test has its own advantages and considered being efficient in locating honeycomb and porous concrete. The core cutting sampling test is adapted to a limited extent because of practical difficulties associated with core drilling, specimen preparation and drilling in thin elements etc. Rebound hammer test that indicates the hardness of concrete surface reflects the strength of the concrete indirectly. This test is used quite extensively in assessing the quality of actual concrete in the structure. The test can be carried out quite extensively and a reliable statistical approach can be applied for acceptance of the test. Some of the strong points in quality control through rebound hammer test are:

(1) A very large number of test readings can be taken in short duration,

(2) Large number of locations on the structure can also be selected,

(3) A statistical approach with high degree of confidence can be applied,

(4) Actual structure is tested, therefore the test result reflects the totality of the final product indicating the concrete mix, its consolidation, surface finish, formwork and curing,

(5) The builder/contractor or even the supervisor will be careful during the real concreting because the actual structure is under testing.

An indiscriminate application of the non-destructive testing without proper correlation factors can lead to misleading results. Number of precautions and correction factors to the results of concrete rebound hammer test are needed. These special aspects are:

(1) Surface texture preparation,

(2) The size or the thickness of the element under testing,

(3) Moisture content in the concrete,

(4) Level of maturity of the concrete with respect to 28 days strength,

(5) Accuracy of the test hammer,

(6) Stability of the supporting base used during testing.

The results may have the marginal variation if the method of testing and speed of impact is not uniform, so while calibrating the equipment, the person actually testing can be involved to reflect the method of testing.

The following are some of the observations regarding the concrete test hammer results:

(1) Moist concrete surface will show a lower strength when compared with a dry one,

(2) Thin elements such as slabs, reflect higher strength when compared with thick elements and mass concrete,

(3) Rough surface will give a lower reading when compared with smooth one,

(4) Grout coated or plastered surface will also show lower strength when compared with the original concrete,

(5) A direct reading on aggregate surface will indicate a higher value.

Since very large sampling can be done even on a single element without much effort, data will reflect the final quality of the concrete. A rationalized statistical approach to obtain the strength of the concrete is dependable when compared to the results of standard cube tests. Many Engineers feel that the results of the concrete rebound hammer gives the strength of the concrete on the surface of the structure but does not reflect the quality of concrete inside. Testing of the surface extensively should give much more confidence when compared to cube strength randomly selected and tested. As per present practice, a cube test consisting of 3 to 6 cubes is expected to represent the quality of the concrete of the structure irrespective of where and from which portion of the structure the concrete cube is supposed to have been taken as a sample. If one can accept such a concrete cube as representative sample, there is no need to doubt the reliability of statistically arrived result based upon concrete rebound hammer test.

Acceptable Criteria of Non-destructive Test

The present section deals primarily with rebound hammer test.

Sampling and Acceptable criteria:

(1) Minimum number of locations should be 10 for an element or for a given pour of concrete,

(2) At least 10-hammer reading should be taken at each location. A location means a spot of about 75 mm square.

(3) All readings beyond 20% of the average value should be rejected. If the number of readings to be rejected exceeds 20% of the readings, then the reliability of the location should be examined. There could be a special problem such as honey combing or porous concrete at such locations.

(4) The total number of readings should be atleast 100 for given element or of a pour of concrete.

(5) The co-efficient of variation of the readings should be within 15%. The characteristic strength of the concrete should be calculated after converting each of the readings to the equivalent strengths with appropriate correction factors.

(6) The characteristic strength of the concrete can be calculated by the formula given below:

$$fck = fm - 1.65s$$

In which fm = mean value of the strength; s = standard deviation

This implies that the probability of the strength falling below the specified value is not more than Five per cent.

(7) In case 20% of the locations have fallen under the rejected, then the quality of concrete is considered not acceptable.

Example

A simple illustration of results of non-destructive test undertaken on roof slab of a building is given in this example.

The example gives statistics and frequency distribution of the data in table 1.18.1. Table 1.18.2 also gives most probable characteristic strength predicted by the computer program design exclusively for NDT test. There is a considerable scope in predicting a reliable strength of the concrete based upon of the statistical analysis of the test results.

TABLE 1.18.1 Statistics of non-destructive test(Data in N/mm^2)

Frequency Distribution: for MIN. CHI-SQ. value

Total number of samples:	80
Max. Value of the data:	36.30
Min. value of the data:	10.70
Number of interval:	7
Class interval width:	3.71

Final class is extended upto +ve infinity

Remaining LEFT side area under normal curve is considered as a separate class interval

Class-Interval	Actual.	Cum.	Expected	Cum. Expected.
.1000E+02 .1371E+02	2	2	5.08	6.86
.1371E+02 .1743E+02	15	17	11.83	18.69
.1743E+02 .2114E+02	20	37	18.55	37.24
.2114E+02 .2486E+02	23	60	19.57	56.81
.2486E+02 .2857E+02	8	68	13.89	70.70
.2857E+02 .3229E+02	8	76	6.63	77.34
.3229E+02 .3600E+02	3	79	2.66	80.00

MEAN: 21.64   MEDIAN: 21.63
MODE: 21.76   Standard Deviation: 5.8009
SKEWNESS Coefficient: .596   KURTOSIS Coefficient: 2.203
CHI-SQUARE: 0.38308E+02   CONFIDENCE Level:0.000
Correlated STD DEV: 4.6407,   Coefficient of Variation:0 .21440

Specified Strength of the Concrete = 20.0
Most probable Characteristic Strength = 14.8

X-axis shows Mid-Values of strength;   y-axis shows Frequency

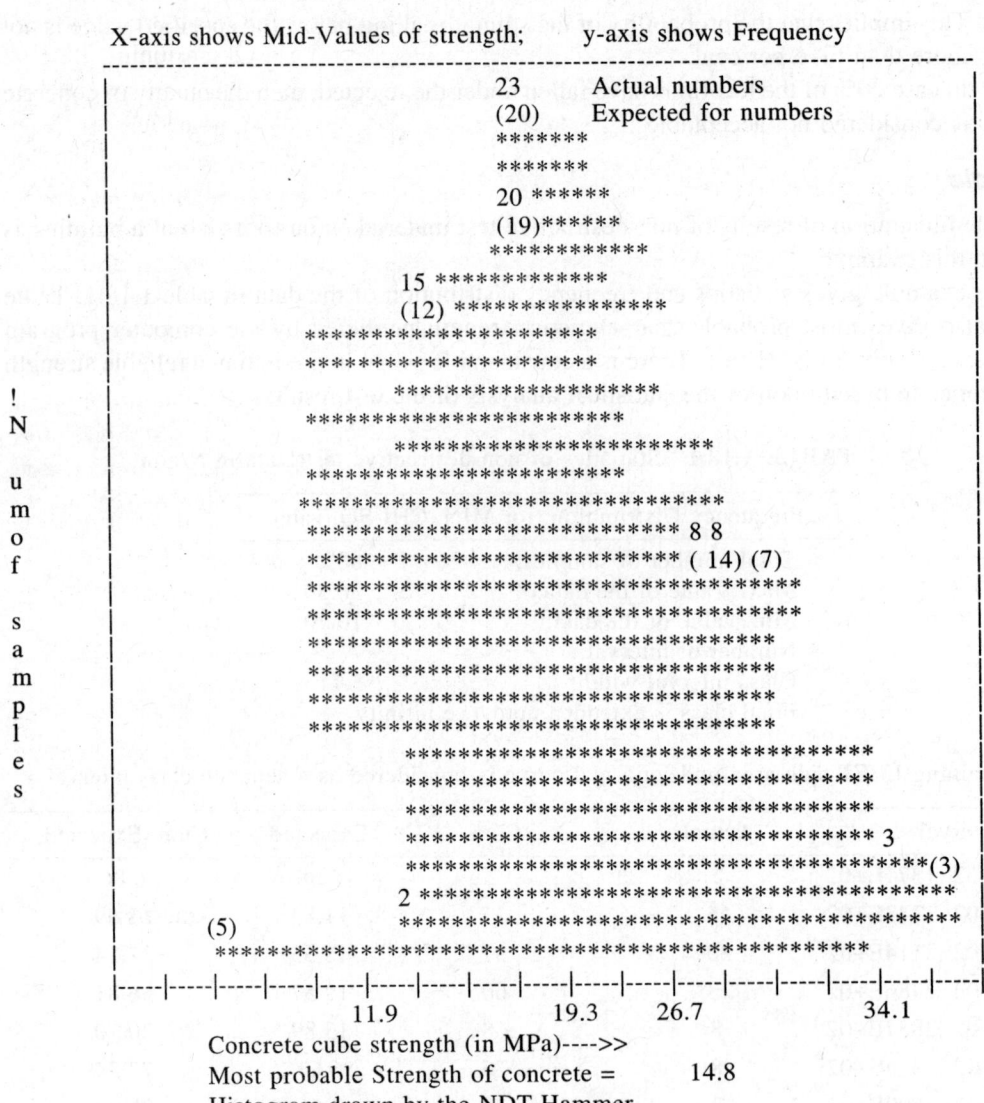

Concrete cube strength (in MPa)---->>
Most probable Strength of concrete =   14.8
Histogram drawn by the NDT Hammer

1.19. Definition

Prestressed concrete may be defined as the concrete in which effective internal stresses are induced artificially with tensioned steel before loading the structure. A simple engineering example which has used the principle of prestressing since several centuries is that of the formation of a wooden barrel. As steel ropes or bands are wound around wooden staves and tightened, compressive stresses are produced in the barrel. Under working load conditions, the liquid in the barrel will cause tensile stresses which are counterbalanced by the previously induced compressive stresses. This type of operation of tightening the steel bands before working operations is called prestressing.

Concrete is poor in tension, so at sections of concrete where tension is expected, compressive stresses are introduced before working loads. There are several methods of tensioning the steel and transferring the steel force to the concrete. The prestressing steel, when used in concrete members, is usually called tendon. The systems and operations of prestressing are discussed later.

An eccentric force acting on a column causes direct thrust and bending moment. The bending moment caused by the eccentric force is equal to the product of the axial force (P) and the eccentricity (e) with respect to the centroid of the section. In case of T-beam or unequal flanged symmetrical sections, the distances of the top and bottom fibre from the centroid of the cross-section are different. The extreme stresses are caused at the top or bottom fibres (also called extreme fibres) due to eccentric load. If the eccentricity is more than the kern distance, the extreme fibre away from the load axis will be subjected to tension. Assuming that no cracking takes place in the cross-section and the material follows a linear stress-strain relation the stresses caused by an eccentric load on slender members can be obtained from simple beam-column theory. They may be expressed using simple column and beam theories.

$$\text{Stress due to axial compression} = -\frac{P}{A} \qquad (1.19.1)$$

$$\text{Stress due to moment} = \frac{My}{I} \qquad (1.19.2)$$

in which

A = area of the cross-section

I = moment of inertia of the section

y = distance of the fibre from centroidal axis (CA)

M = bending moment on the section.

Let σ_t and σ_b are the stresses caused by P and M at the top and bottom extreme fibres which are situated at top & bottom fibre distances y_t and y_b from the centroid. Then

$$\sigma_t = -\frac{P}{A} + \frac{My_t}{I} \qquad (1.19.3)$$

$$\sigma_b = -\frac{P}{A} + \frac{My_b}{I} \qquad (1.19.4)$$

In which the tensile stress is treated as positive and the compressive stress as negative. The negative sign for My_t is assigned assuming the bending moment is such it causes compression on the top fibre. The explanation given here is with reference to a member whose axis is horizontal. This is given as a matter of convenience and can be extended to any element in any orientation. In most cases, tensile stress is treated positive but in case of concrete structures, the concrete is mostly under compression. The notation is only for convenience and will be modified at times for simplicity in the formulation of the expressions. In case of a beam column in which the axial force is acting with an eccentricity 'e', the bending moment caused by the force produces tension at the top fibre. Therefore the moment is

$$M = -Pe \qquad (1.19.5)$$

The substitution of Eq. 1.19.5 in Eqs. 1.19.3 and 1.19.4 gives

$$\sigma_t = -\frac{P}{A} + \frac{Pey_t}{I} \qquad (1.19.6)$$

$$\sigma_b = -\frac{P}{A} - \frac{Pey_b}{I} \qquad (1.19.7)$$

The above two equations can be modified by setting $I = Ar^2$ in which r = radius of gyration :

$$\sigma_t = -\frac{P}{A}\left(1 - \frac{ey_t}{r^2}\right) \qquad (1.19.8)$$

$$\sigma_b = -\frac{P}{A}\left(1 + \frac{ey_b}{r^2}\right) \qquad (1.19.9)$$

The bounds for eccentricity of a thrust for which no tension is caused can be obtained from Eqs. 1.19.8 and 1.19.9, and they are :

$$\frac{P}{A}\left(1 - \frac{ey_t}{r^2}\right) > 0 \quad \text{and} \qquad (1.19.10)$$

$$\frac{P}{A}\left(1 + \frac{ey_b}{r^2}\right) > 0 \qquad (1.19.11)$$

The two inequivalities 1.19.10 and 1.19.11 give

$$e < \frac{r^2}{y_t} \qquad (1.19.12)$$

$$e > -\frac{r^2}{y_b} \qquad (1.19.13)$$

For symmetrical sections $y_t = y_b = \dfrac{h}{2}$

where h = overall depth of the section.

The ineqs. 1.19.11 and 1.19.13 give

$$-\frac{2r^2}{h} < e < \frac{2r^2}{h} \tag{1.19.14}$$

in case of rectangular sections the value of r^2 is given by

$$r^2 = h^2/12.$$

Therefore, the ineq. 1.19.14 for rectangular sections reduces to

$$\frac{-h}{6} < e < \frac{h}{6} \tag{1.19.15}$$

The bounds given by ineq. 1.19.14 are referred as the *kern* distances.

Let there be an external bending moment M acting on the cross-section. The moment is considered positive if it is causing compression on the top fibre of the section. Such a moment causes a compression equal to My_t/I at the extreme top fibre and tension My_b/I at the extreme bottom fibre. The final stresses caused by an eccentric thrust on a cross-section which is subjected to a moment M can be obtained by simple superimposition. They are :

$$\sigma_t = \frac{-P}{A} + \frac{Pey_t}{I} - \frac{My_t}{I} \tag{1.19.16}$$

$$\sigma_b = \frac{-P}{A} - \frac{Pey_b}{I} + \frac{My_b}{I} \tag{1.19.17}$$

The above equations can be rewritten as:

$$\sigma_t = \frac{-P}{A}\left(1 - \frac{ey_t}{r^2}\right) - \frac{My_t}{I} \tag{1.19.18}$$

$$\sigma_b = \frac{-P}{A}\left(1 + \frac{ey_b}{r^2}\right) + \frac{My_b}{I} \tag{1.19.19}$$

Depending on the relative magnitudes of each of the terms in the Eqs. 1.19.18 and 1.19.19, the stresses σ_t and σ_b can either be positive (i.e., tension) or negative.

In most concrete structures, the tensile stresses in concrete are usually avoided or neglected.

It is also desirable to review the difference between the centroidal axis and the neutral axis in association with the set of equations generated so far. The centroidal axis is associated with the property of the section alone and it is the axis passing through the centroid of the section. Stress caused by moment ($\sigma = My/I$) at the centroidal axis is always zero. It can be seen from the fact that as y is equal to zero so the stress is zero. The neutral axis is defined as the axis at which the stress (or more often it is the strain) changes from positive to negative. At this level the stress is equal to zero and the nature of the stresses on either side of the axis, are different. When the direct stress, P/A is superimposed on the stress caused by the bending moment My/I, the net stress is $P/A \pm My/I$. In such a case, the stress at the centroidal axis equal to P/A and the neutral axis does not coincide with the centroidal axis. This is what really happens in all beam columns and prestressed beams. The neutral axis may or may not fall within the section. If the value of P/A is more when compared with the maximum of My/I then the neutral axis falls outside the section.

1.20. Stress Distribution

The distribution of stress and strain on a section in a simply supported beam is shown in Figure 1.20.1 assuming that the plane section remains plane even after bending. Certain amount of tension is produced in the bottom fibre by the external loads and, because concrete is poor in tension, some kind of tension-bearing capacity has to be induced in the tensile zone. The tension capacity is provided in two ways : (i) sufficient steel can be embedded in the beam at the tension zone to resist the tension directly. This method is defined as reinforced concrete construction : (ii) a precompressive stress can be induced at the tension fibre through tensioned steel. A prestressed concrete construction is the one in which the concrete is precompressed by tensioned steel to compensate the tensile stress which is likely to be developed by working load.

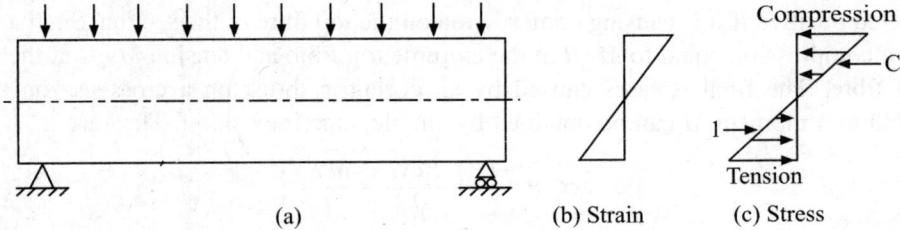

Figure 1.20.1 Simply supported beam (strain and stress distribution due to bending).

If a section is first subjected to prestressing force of P, then the stresses are given by

$$\sigma_t = \frac{-P}{A}\left(1 - \frac{ey_t}{r^2}\right) - \frac{My_t}{I} \qquad (1.20.1)$$

$$\sigma_b = \frac{-P}{A}\left(1 + \frac{ey_b}{r^2}\right) + \frac{My_b}{I} \qquad (1.20.2)$$

where

P = prestressing force introduced through the tendon

A = area of cross-section

I = moment of inertia

r = radius of gyration

e = eccentricity

y_t = distance of the extreme top fibre from CGC

y_b = distance of the extreme bottom fibre from CGC

σ_t = stress at top fibre (= σ_1)

σ_b = stress at bottom fibre (= σ_2)

M = bending moment caused by external load

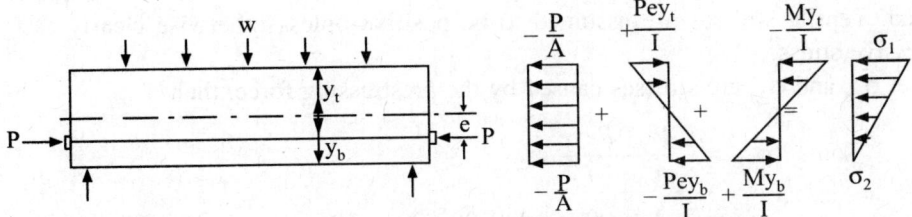

Figure 1.20.2 Simply supported beam with prestress (stress variation).

Eqs. 1.20.1 and 1.20.2 illustrate that some of the stresses caused by prestressing force are of opposite in nature of those caused by the external moment. The prestressing force and the eccentricity of the force can be adjusted to compensate a desired amount of the stress caused by the external loads. Figure 1.20.2 shows the nature of stress distribution across the depth of a beam caused by prestressing force and external load. Example 1.20.1 illustrates the effect of prestressing force on stresses in a beam.

P cannot be increased arbitrarily just to compensate the tensile stress. As P increases, σ_b also increases so the allowable compressive stress of concrete governs. Compensation of tensile force can also be achieved by increasing the eccentricity 'e'; but as the eccentricity increases, σ_t in eq. 1.20.1 might turn out to be a tensile stress beyond allowable limits. The values of P and e have to be carefully selected. The advantage of providing thrust P at the bottom is illustrated in example 1.20.1.

Example 1.20.1 A simply supported beam of span 8 m is loaded with a uniformly distributed load of 5000 N/m. The cross-section of the beam is 30 cm by 80 cm. Determine the stresses with and without a horizontal thrust of 240,000 N acting at 20 cm below the centroidal axis.

Solution

Bending moment at midsection is given by

$$M = \frac{wL^2}{8} = \frac{5000(64)(100)}{8} = 4 \times 10^6 \text{ N cm}$$

$$I = \frac{bh^3}{12} = \frac{30(80)^3}{12} = 1.28 \times 10^6 \text{ cm}^4$$

The stresses caused by the moment at extreme top and bottom fibres are

$$\sigma_{1l} = \frac{My_1}{I} = \frac{4(10^6)(-40)}{1.28(10^6)} = -125 \text{ N / cm}^2$$

$$\sigma_{2l} = \frac{My_2}{I} = \frac{4(10^6)(40)}{1.28(10^6)} = 125 \text{ N / cm}^2$$

where the subscripts 1 and 2 indicate extreme top and bottom fibres. The subscript l indicates live load. Tensile stresses are assumed to be positive unless otherwise clearly assigned as compressive stress.

Let σ_{1p} and σ_{2p} are stresses caused by the prestressing force, then

$$\sigma_{1p} = -\frac{P}{A} - \frac{Pey_1}{I}$$

$$= -\frac{240,000}{80(30)} - \frac{240,000(20)(-40)}{1.28(10^6)}$$

$$= -100 + 150 = 50\,\text{N}/\text{cm}^2$$

$$\sigma_{2p} = -\frac{P}{A} - \frac{Pey_2}{I} = -100 - 150 = -250\,\text{N}/\text{cm}^2$$

where $y_1 = -40$ cm, $y_2 = 40$ cm and Pe is the bending moment causing compression on the bottom fibre which is considered negative.

Prestressing force causes a tension of 50 N/cm² at top fibre whereas the live load produces a compressive stress of 125 N/cm². So the effective stress at extreme top fibre is

$$\sigma_1 = \sigma_{1l} + \sigma_{1p} = -125 + 50 = -75\,\text{N}/\text{cm}^2$$

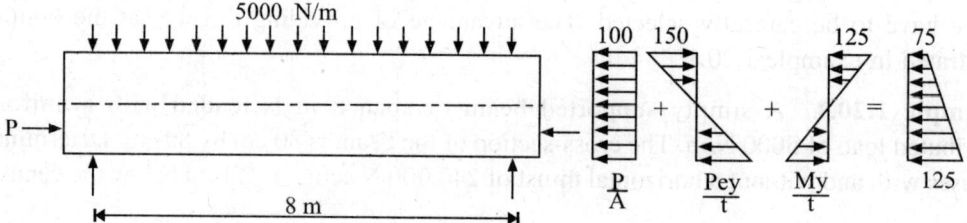

Figure 1.20.3 Stress distribution at mid-section.

Similarly the effective stress at the extreme bottom fibre is

$$\sigma_2 = \sigma_{2l} + \sigma_{2p} = 125 - 250 = -125\,\text{N/cm}^2$$

The stresses at both the extreme fibres are compressive so the entire cross-section is under compression. The prestressing force eliminates the tensile stresses caused by the external loads thus safeguarding concrete against tension. The stresses at the end section where external moment is zero are

$$\sigma_1 = -\frac{P}{A} - \frac{Pey_1}{I} = 50\,\text{N}/\text{cm}^2$$

$$\sigma_2 = -\frac{P}{A} - \frac{Pey_2}{I} = -250\,\text{N}/\text{cm}^2$$

1.21. Profile of Tendons

It may be seen from example 1.20.1 that there are two critical sections, one at mid-span and another at end-span. The stress distributions of these two sections are shown in Figure 1.21.1. The stress distribution at the end section is more dominant than that at middle of the beam. The stress distribution due to external bending moment starts with zero values at the end and reaches a maximum at the centre of the beam, whereas the stress distribution provided by the horizontal thrust stays constant throughout the beam. The critical stress distribution at the end section can easily be avoided by providing zero eccentricity at the end while keeping the 20 cm eccentricity at the middle of the beam. Now it is necessary to discuss the profile of the eccentricity along the span. The stress variation along the span due to the uniformly distributed load is parabolic; so a tendon of parabolic profile will be effective to compensate the tensile stresses due to the external loads. Sometimes the profile is also taken as two straight lines with a bend at the middle of the section. These profiles are shown in Figure 1.21.2. The eccentricity of the tendon at the end of the beam need not necessarily be zero. To provide for a steep curvature, a negative eccentricity can also be given.

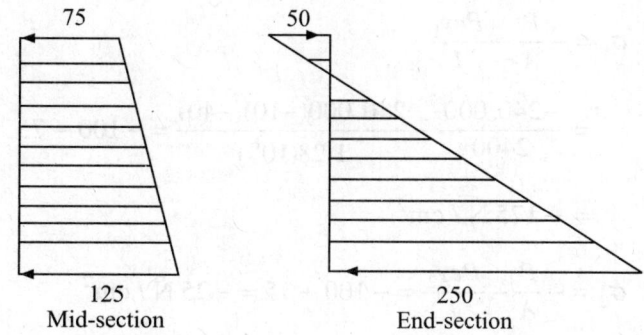

75   50

125   250
Mid-section   End-section

Figure 1.21.1   Stress distribution—Example 1.20.1. (stresses in N/cm^2).

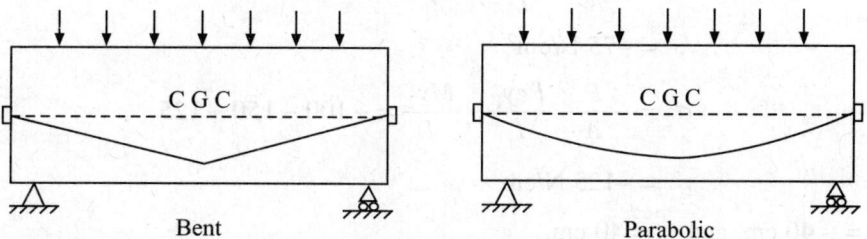

C G C   C G C

Bent   Parabolic

Figure 1.21.2 Possible profiles of tendons.

Example 1.21.1   Same as example 1.20.1 with a modification in the profile of the tendon : The eccentricity at the end is –10 cm and at the mid-section it is 20 cm with straight line variation. Determine the stresses at the end and mid-section (negative sign of e indicates eccentricity above centroidal axis, CA). Figure 1.21.3 illustrates the cable profile.

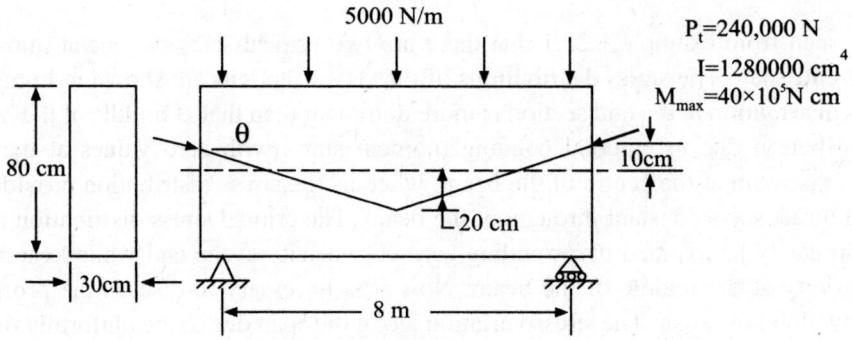

Figure 1.21.3 Example 1.21.1

Solution

Stresses at the end section where $e = -10$ cm are

$$\sigma_1 = -\frac{P}{A} - \frac{Pey_1}{I}$$

$$= \frac{-240,000}{2400} - \frac{240,000(-10)(-40)}{1.28(10^6)} = -100 - 75$$

$$= -175 \text{ N} / \text{cm}^2$$

$$\sigma_2 = -\frac{P}{A} - \frac{Pey_2}{I} = -100 + 75 = -25 \text{ N} / \text{cm}^2$$

Stresses at midsection where $e = 20$ cm are

$$\sigma_1 = -\frac{P}{A} - \frac{Pey_1}{I} + \frac{My_1}{I} = -100 + 150 - 125$$

$$= -75 \text{ N/cm}^2$$

$$\sigma_2 = -\frac{P}{A} - \frac{Pey_1}{I} + \frac{My_2}{I} = -100 - 150 + 125$$

$$= -125 \text{ N/cm}^2$$

where $y_1 = -40$ cm, and $y_2 = 40$ cm.

Figure 1.21.4 shows the stress distribution at the end and middle sections of the beam. It may be seen from the two previous examples that the profile of the cable has considerable effect on the distribution of stress at different sections of the beam. Another two examples are given to illustrate the effect of the cable profile on the stresses in the beam.

(Note: The stress pointing towards the beam is assumed to be compression as given in the figures).

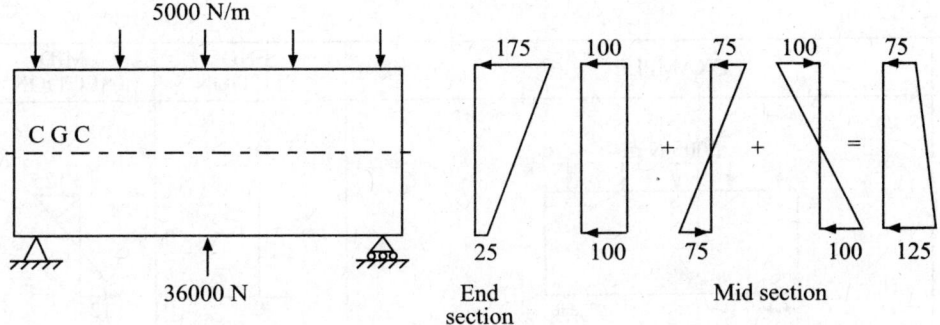

Figure 1.21.4   Stress distribution- Example 1.21.1 (stress in N/cm^2).

Example 1.21.2   Same as example 1.21.1 except that the eccentricity at the ends is 10 cm instead of −10 cm.

Solution

The stress distribution due to the external load and prestressing force computed as done in example 1.21.1 and shown in Figure 1.21.5.

The stresses at the end section are

$$\sigma_1 = -\frac{P}{A} - \frac{Pey_1}{I} = -100 + 75 = -25 \text{ N / cm}^2$$

$$\sigma_2 = -\frac{P}{A} - \frac{Pey_2}{I} = -100 - 75 = -175 \text{ N / cm}^2$$

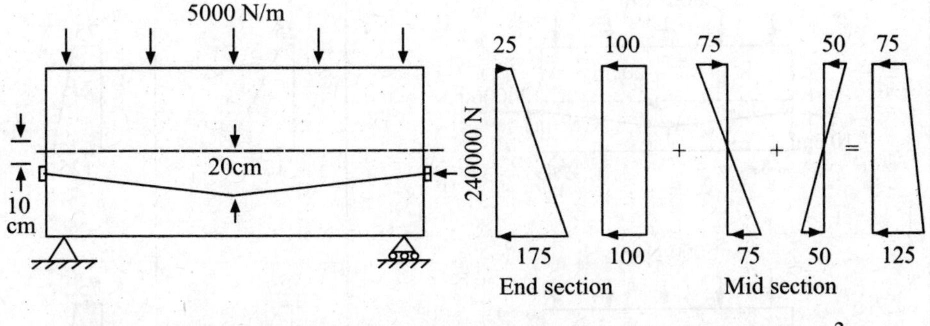

Figure 1.21.5   Stress distribution-Example 1.21.2 (stresses in N/cm^2).

A comparative study of stress distribution of the five examples is given in table 1.21.1. Beams 2 to 5 have used the same amount of material under same external load conditions except for the profile of the tendon. It can be seen that the profile and the eccentricity of the tendon has considerable effect on the stress.

70

TABLE 1.21.1 Stress distribution on beam with tendon (stresses in N/cm²)

S No	EXAMPLE	END SECTION	MID SECTION
1	5000 N/m	0 / 0	+125 / −125
2	5000 N/m; 20 cm; 240000 N	50 / 250	75 / 125
3	5000 N/m; 20 cm; 10 cm; 240000 N	175 / 25	75 / 125
4	5000 N/m; 20cm; 10 cm; 240000 N	25 / 175	75 / 125
5	5000 N/m; 20 cm; 10 cm; 240000 N	25 / 175	75 / 125

1.22. Load Balancing Method

A prestressing cable in concrete exerts a set of forces on concrete so it can be replaced by the set of forces for the purpose of analysis. T. Y. Lin (1.1)* is one of the early investigators who explained the concept and used it in design of prestressed concrete beams. Consider an infinitesimal length of a cable as shown in Figure 1.22.1. The equilibrium of force of the element in the vertical direction for a constant force in the cable is given by

$$P \sin \theta_2 - P \sin \theta_1 + w_b \delta x = 0 \qquad (1.22.1)$$

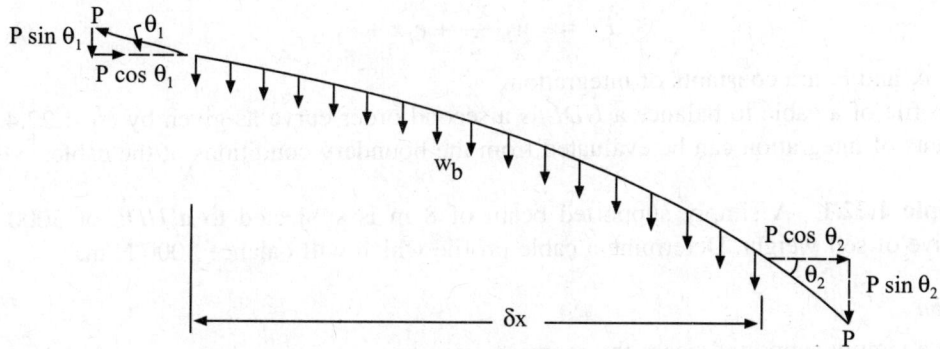

Figure 1.22.1 Equilibrium of a small cable.

where θ_1 and θ_2 are the slopes of the cable, w_b is the balancing force and δx is the length of the element. w_b is considered positive if acting downwards. In case of a shallow curved cable, the slope of the cable is very small as compared to unity, so

$$\sin \theta_2 = \theta_2 \text{ and } \sin \theta_1 = \theta_1$$

Eq. 1.22.1 can be rewritten as

$$P(\theta_2 - \theta_1) = - w_b \delta x \qquad (1.22.2)$$

If the slope of the curve is differentiable θ_1 and θ_2 can be expressed as

$$\theta_1 = \frac{de}{dx}, \theta_2 = \frac{de}{dx} + \frac{d^2 e}{dx^2} \delta x$$

and eq. 1.22.2 reduces to

$$P \frac{d^2 e}{dx^2} = - w_b \qquad (1.22.3)$$

where e is the ordinate of the cable profile measured positive downward from the *CA*.

A cable can be replaced by an equivalent set of forces and the beam can be analysed for such a set. Anchorage force can be replaced by axial, shear and bending forces and the shallow curve by a vertical force proportional to the curvature. The axial force through *CGC* will not

---

* The numerals in the bracket indicate the reference listed at the end.

cause any moment for small deflection theory. The transverse component acting at a support point is directly transferred to the support. Otherwise if it is situated in between the supports, it will cause moment. Moments caused by eccentricities and radial components of the cable force will have to be included in the analysis.

It can be stated from eqs. 1.22.2 and 1.22.3 that the balancing force is given by the product of the cable force and the rate of change of slope of the cable. The cable can be replaced by an equivalent force which is equal to w_b as given in eqs. 1.22.2 or 1.22.3.

If the balancing force is constant, i.e. w_b = constant, integration of eq. 1.22.2 gives

$$Pe = -w_b \frac{x^2}{2} + c_1 x + c_2 \tag{1.22.4}$$

where c_1 and c_2 are constants of integration.

Profile of a cable to balance a *UDL* is a second order curve as given by eq. 1.22.4. The constants of integration can be evaluated from the boundary conditions of the cable.

Example 1.22.1 A simply supported beam of 8 m is subjected to a *UDL* of 5000 N/m inclusive of self weight. Determine a cable profile which will balance 3000 N/m.

Solution

In a simply supported beam the moments at support ends are zero so eccentricity of the cable at the two supported ends must be equal to zero. The boundary conditions are :

$$e = 0 \text{ at } x = 0 \text{ and } x = L$$

Substitution of the above boundary conditions in eq. 1.4.4 gives

$$c_2 = 0 \text{ and } c_1 = w_b L/2$$

Eq. 1.22.4 reduces to

$$Pe = w_b x \, (L - x)/2 \tag{1.22.5}$$

Let g = maximum sag in the cable which is at $x = L/2$, then from eq. 1.22.5,

$$Pg = \frac{w_b L^2}{8} \quad \text{or} \quad w_b = \frac{8Pg}{L^2}$$

If the balancing load is made equal to 3000 N/m

$$w_b = 3000 \text{ N/m}$$

$$Pg = \frac{w_b L^2}{8} = \frac{3000 \times 8 \times 8}{8} = 24,000 \text{ Nm}$$

The sag of the cable is normally constrained by the depth of the beam. The maximum sag that can be permitted allowing an effective cover of 10 cm is

$$g = d/2 - 10 = 30 \text{ cm so}$$

$$P = \frac{24,000}{g} = \frac{24,000}{0.3} = 80,000 \text{ N}$$

Slopes of the cable at the supports are

$$\text{at } x = 0 \quad \theta_1 = \left(\frac{de}{dx}\right) = \frac{w_b L}{2P} = \frac{4g}{L}$$

$$\text{at } x = L \quad \theta_2 = \left(\frac{de}{dx}\right) = -\frac{4g}{L}$$

Figure 1.22.2 illustrates the cable with load on the beam along with the equivalent load of the cable. The effective load acting on the beam is obtained as a difference of the external and cable equivalent loads and it is shown in the Figure 1.22.2.

Effective load acting on a beam can be obtained from the difference of the external load and the cable equivalent forces.

$$w_e = w - w_b \tag{1.22.7}$$

where w and w_e are external and effective loads respectively. In addition to the equivalent distributed load of w_e, the cable produces horizontal thrust equal to $P \cos \theta$ and vertical load equal to $P \sin \theta = P \theta$ at each supports as shown in Figure 1.22.2.

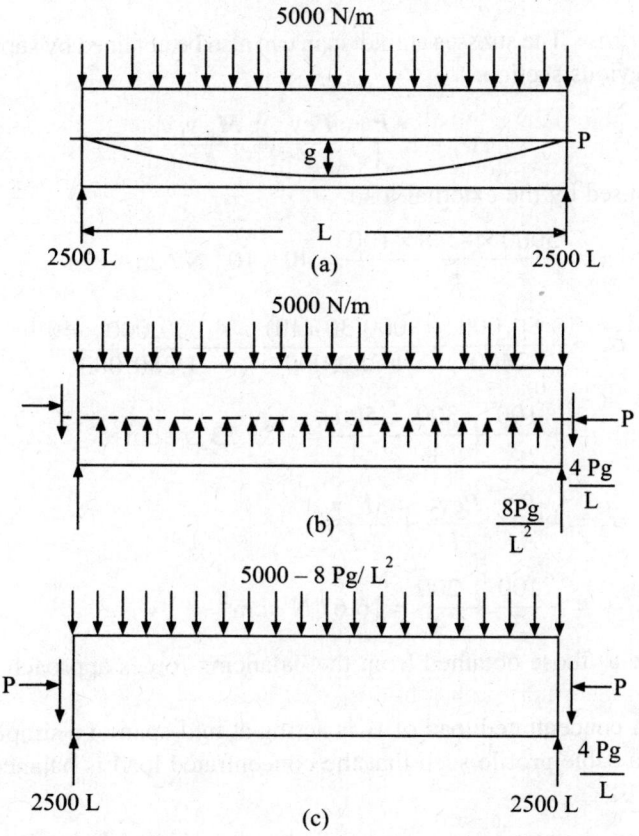

Figure 1.22.2 UDL load balancing cable in a simply supported beam.

Example 1.22.2 The cross-section of the beam of example 1.22.1 is 30 cm by 80 cm. Determine the stresses at mid-span using balancing load concept.

Solution

Effective load = w_e = 5000 – 3000 = 2000 N/m
Bending moment at mid-span is

$$= M = \frac{w_e L^2}{8} = \frac{2000 \times 8 \times 8}{8}$$

$$= 16,000 \text{ Nm} = 160,000 \text{ Ncm}$$

Stresses at mid-span are

$$\sigma_1 = -\frac{P}{A} - \frac{My_1}{I} = -\frac{80,000}{2400} - \frac{1,600,000 \times 4}{1,280,000}$$

$$= -100/3 - 50 = -83.33 \text{ N/cm}^2$$

$$\sigma_2 = -\frac{P}{A} + \frac{My_2}{I} = -\frac{100}{3} + 50 = 16.67 \text{ N / cm}^2$$

Alternate approach. The stresses at mid-span can also be obtained by superposition principle discussed in the previous section.

$$\sigma_1 = -\frac{P}{A} + \frac{Pey_1}{I} - \frac{M_w y_1}{I}$$

where M_w = BM caused by the external load

$$= \frac{5000 \times 8 \times 8 \times 100}{8} = 40 \times 10^5 \text{ N / cm}$$

$$\sigma_1 = -\frac{80,000}{2400} + \frac{8000(30)(40)}{1,280,000} - \frac{4,000,000 \times 40}{1,280,000}$$

$$= -\frac{100}{3} + \frac{300}{4} - \frac{500}{4} = -83.33 \text{ N / cm}^2$$

$$\sigma_2 = -\frac{P}{A} - \frac{Pey_2}{I} + \frac{M_w y_2}{I}$$

$$= -\frac{100}{3} + \frac{200}{4} = 16.67 \text{ N / cm}^2$$

The values are same as those obtained from the balancing forces approach.

Example 1.22.3 A concentrated load of W is acting at mid-span of a simply supported beam of span L. Suggest a cable profile such that the concentrated load is balanced.

Solution

The load to be balanced is not a continuous load so eq. 1.22.2 in which δx tends to a point will hold good in which case $W_b \delta_x$ tends W_b. Since there is no moment at the ends of the beam, eccentricities at the support ends should be zero. The balancing load at all points except at mid-span is identically equal to zero which gives the equation of the profile as

$$e = c_1 x + c_2$$

and at $\qquad\qquad x = 0,\ e = 0,\ \text{so}\ c_2 = 0$

The cable profile is a straight line with kink at mid-span. Let the change in slope at mid-span is 2θ then from symmetry $\theta_2 = \theta_1 = \theta$. Eq. 1.22.2 gives

$$P(2\theta) = W_b \quad \text{or} \quad P\theta = \frac{W_b}{2}$$

If g is the sag at mid-span, then for shallow sags

$$\theta = \frac{g}{L/2} = \frac{2g}{L} \quad \text{and}$$

$$P\theta = \frac{2Pg}{L} = \frac{W_b}{2} \quad \text{or}$$

$$Pg = \frac{W_b L}{4} \quad \text{or} \quad W_b = \frac{4Pg}{L}$$

Figure 1.22.3 shows the cable's and the equivalent forces on the beam.

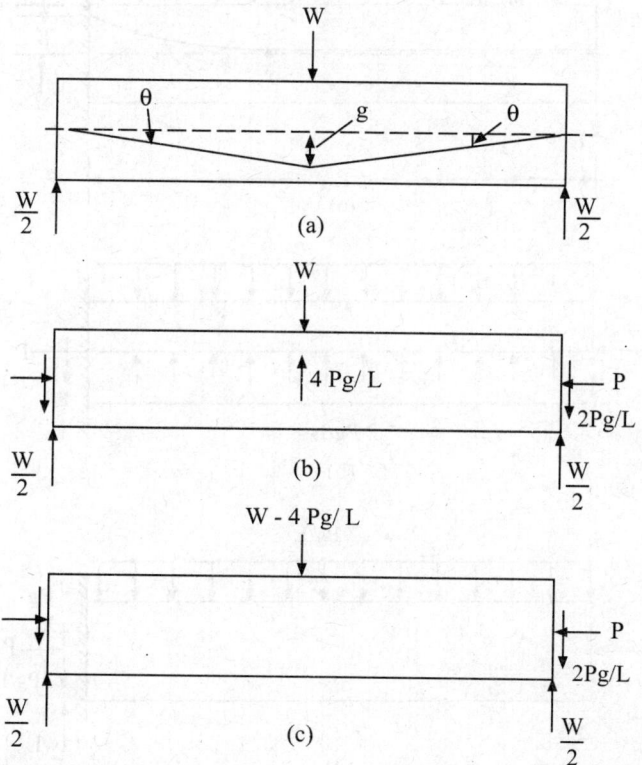

Figure 1.22.3 Balancing cable for concentrated load.

Example 1.22.4 A cantilever beam of span L is subjected to *UDL* of w. Design a cable profile such that a load of w is balanced by the cable.

Solution

As the load is *UDL*, the eq. 1.4.4 can be used

$$Pe = -w_b \frac{x^2}{2} + c_1 x + c_2$$

Let the origin of the above equation be at the free end as shown in Figure 1.22.4. Since there is no end moment or end shear force, the profile of the cable should be such that it will neither cause moment nor shear force which means the eccentricity as well as the slope of the cable at the free end must be zero

$$e = \frac{de}{dx} = 0 \text{ at } x = 0$$

The profile reduces to

$$Pe = \frac{-w_b x^2}{2} \qquad\qquad (1.22.8)$$

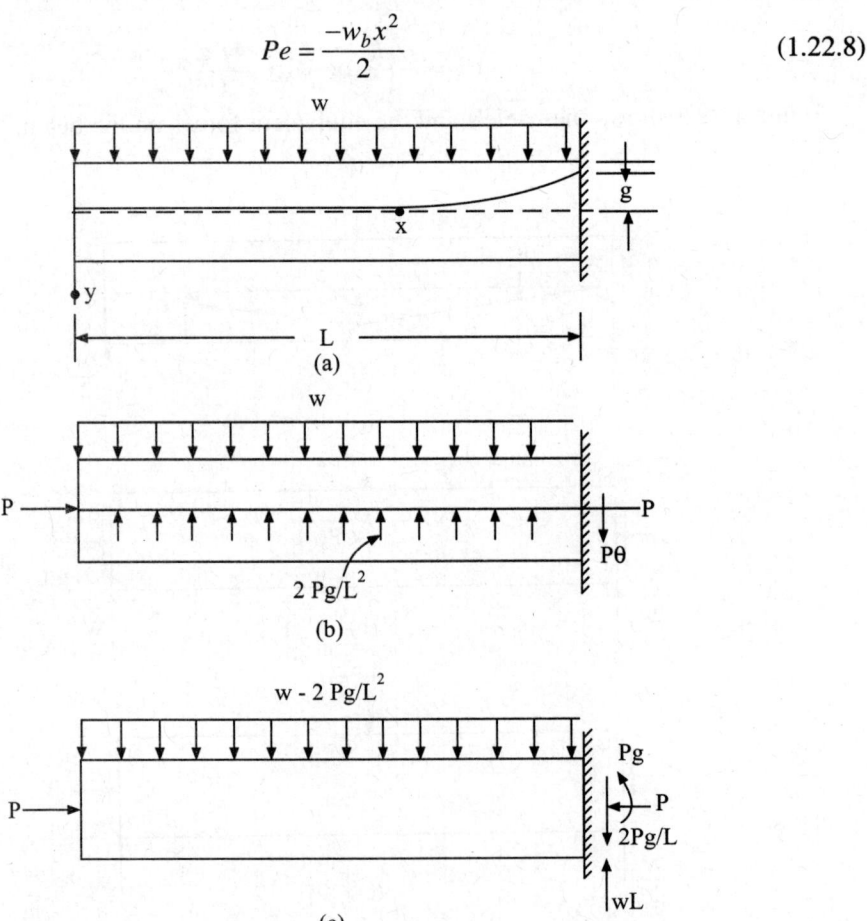

Figure 1.22.4 Load balancing cable in a cantilever beam.

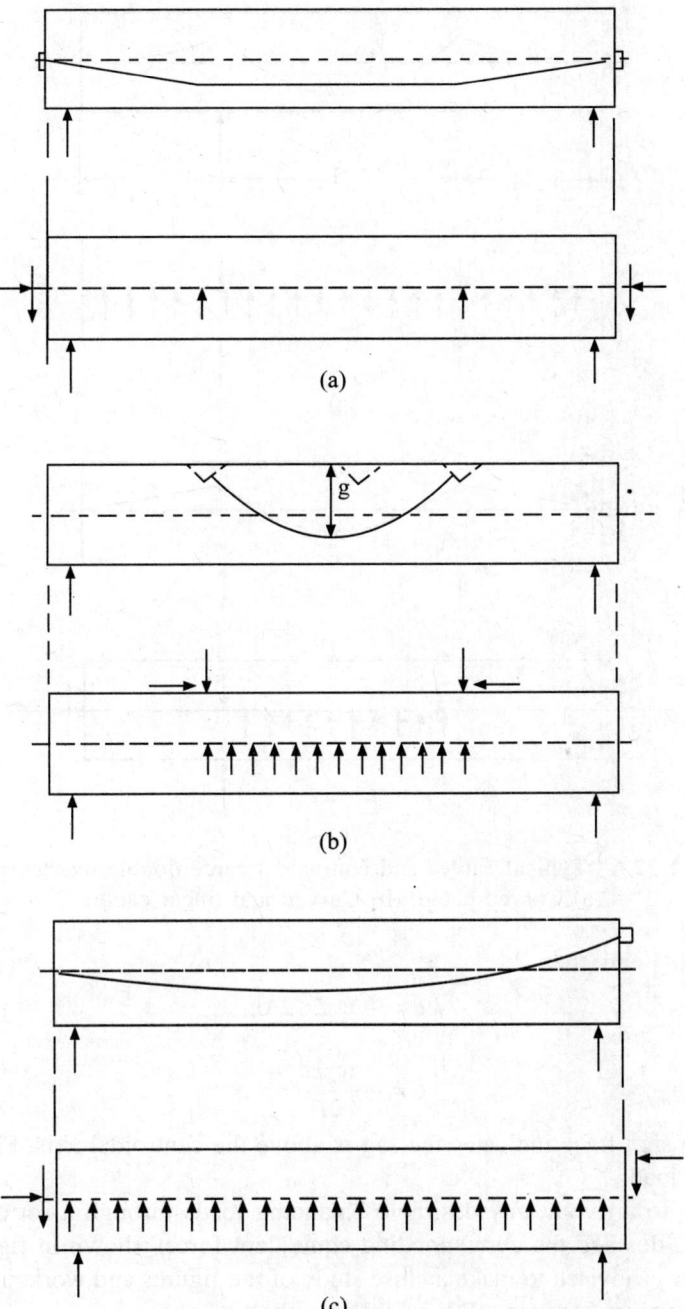

Figure 1.22.5 Typical cables and equivalent forces.
(a) Three straight line segment cable
(b) Cable in a segment of a beam
(c) Eccentrically anchored cable

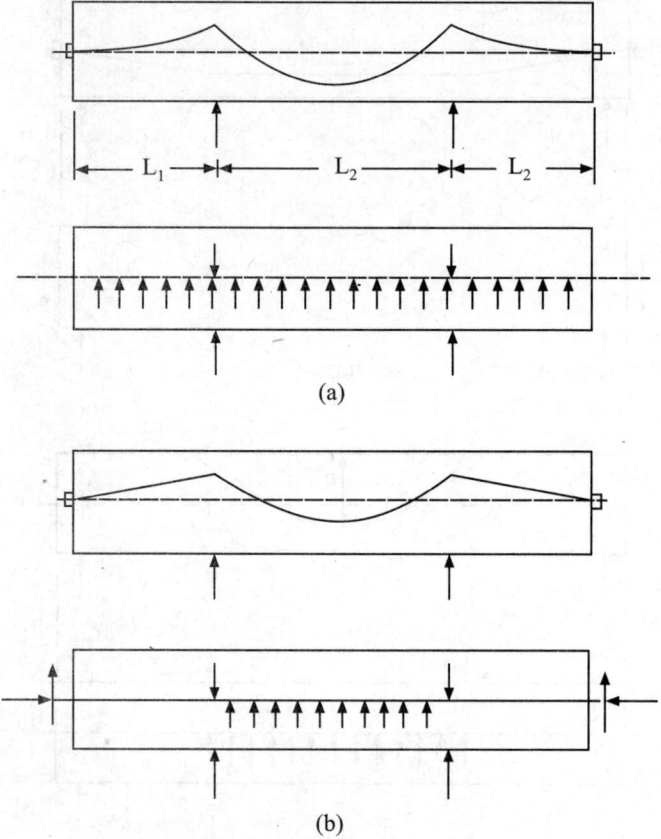

Figure 1.22.6 Typical cables and equivalent force double overhang beam, (a) Curved cable, (b) Curved and linear cable.

Let g = sag at fixed end, then

$$Pg = -w_bL^2/2 \text{ or}$$

$$g = -\frac{w_bL^2}{2P} \tag{1.22.9}$$

The negative sign for g indicates the sag is above the centroidal axis. Figure 1.4.4 gives the details of the loads.

It is possible to arrive at any desirable balancing loads through a set of cable profiles. Typical cable profiles and the corresponding equivalent forces shown in figures. 1.22.5 and 1.22.6. The reader is advised to make a close study of the figures and workout some examples giving numerical values. Assume all the cables as shallow ones.

1.23. Basic Phases of Loading in Prestressed Construction

Prestressed concrete structures are subjected to different load conditions of which only few are critical. The main load conditions for which a structure should be designed are :

1. Transfer of prestress from the jacks or from the anchor abutments to the structure. This condition is usually called transfer or initial condition. In this condition the concrete has the least strength whereas the prestressing force is maximum. The external load acting on the structure is only self weight.
2. Working load condition in which normal working loads such as live as dead loads are placed on the structure so as to cause critical stresses. In this condition the loads are supposed to act for long periods of time so the permissible stresses are taken lower than those in the transfer condition.
3. Handling and transportation load condition is not common to all structures. The structure, if it is to be transported to the site, then the handling procedure should be specified and design be made so as to withstand the handling load. The permissible stresses in this load condition are usually higher than those in the working load.
4. Combined wind or earthquake load condition may not be critical for many simple structures. However, large structures should be designed to resist wind and earthquake loads. The permissible stresses in this load condition are usually higher than those in the working load condition as the duration of peak wind and earthquake actions are of small duration.

1.24. Materials

This section is devoted to basic materials used in the prestressed concrete construction. Because of variety of reasons seen later, it is necessary to have some concept about the material properties used in the construction. The basic two materials used in the prestressed concrete construction are : (i) steel and (ii) concrete.

Steel : Mild and hard-drawn steels used in reinforced concrete have a yield stress of 240 MPa to 360 MPa. If this steel is used in prestressed concrete and tensioned to about 200 MPa at the initial stage, the effective prestressing force available after shrinkage and creep of concrete and creep of steel will be negligible. The approximate loss of prestressing due to shrinkage of concrete, creep of concrete and steel is estimated to be of the order of 175 to 250 MPa. Obviously no prestressing force will be left in the mild steel after the losses. The high tensile steels (HTS) have ultimate strength capacity as high as 1000 to 2100 MPa and the use of such steels will provide considerable amount of effective prestressing force even after losses in prestress.

TABLE 1.24.1 Properties of high tensile steel (HTS) wire

Diameter (in mm)	3	4	5	7	8
(1) Ultimate stress (MPa)					
(Cold-drawn)	1900	1750	1600	1500	1400
(As-draw wires)	1800	1750	1600	–	–
(2) Percentage elongation	2.5	3	4	6	8

High tensile (HT) steel available is mainly in the form of wires which are cold or as-drawn from high tensile steel bars. The process of cold drawing tends to realign the crystals and the strength of the wires is increased by each drawing so that the strength of the wire increases as its diameter decreases. However, the cold drawing process decreases the ductility of the material which is a disadvantage. High tensile steel wires are also obtained as 'as-drawn' wires. These wires have low proportional limit and to increase the proportional limit, the wires are subjected to stress-relieving processes. A typical variation of ultimate stress of HT steel with respect to the diameter of the wire is shown in Figure 1.24.1.

The prestressing steel is also available in the form of strands which are obtained by twisting wires together. By using strands, the number of units to be handled decreases. Small diameter wires of very high tensile strength can be used for strands. Approximate chemical composition of the high tensile steel wires is :

carbon = 0.60 to 0.85%; manganese = 0.7 to 1.0%,

phosphorus = 0.05%, sulphur = 0.055 % and a small amount of silicon.

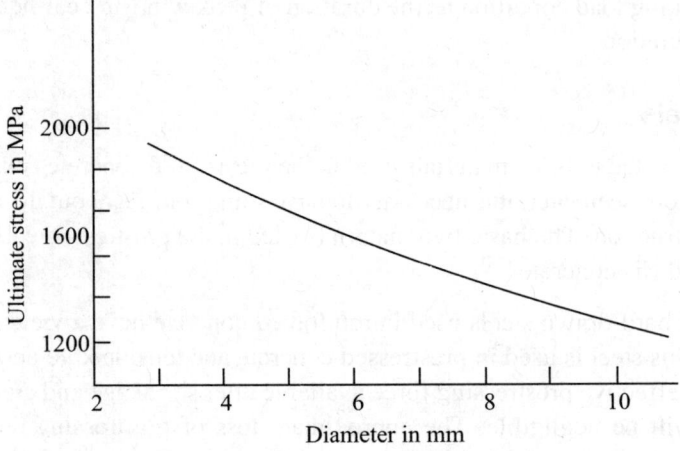

Figure 1.24.1 Typical variation of ultimate strength of HT steel wires.

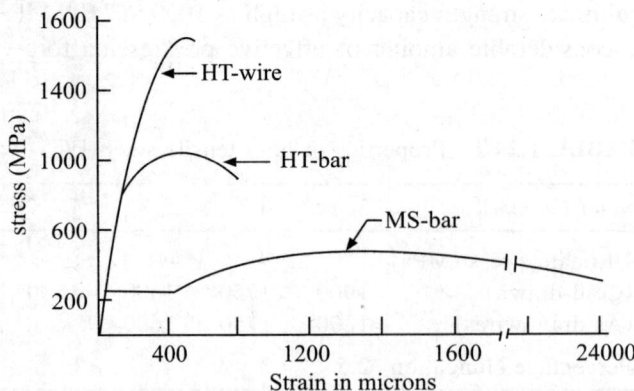

Figure 1.24.2 Typical stress strain variation of steels.

Physical properties : The proportional limit of high tensile steel is rather hard to find as the yield point does not exist. A typical stress strain curve of a high tensile steel is shown in Figure 1.24.2.

From Figure 1.24.2 it can be observed that the proportional limit or yield of high tensile steel is not very clearly seen. The yield stress is replaced by proof stress in HT steel and it is taken to be equal to the stress at which 0.2 per cent permanent strain is obtained. Table 1.24.2 gives approximate values of proof stresses of different steels.

TABLE 1.24.2 Approximate Yield and Proportional Limits

Wire	(Proof stress)/ f_p'	(Proportional stress) f_p'
Wire as drawn	0.75	0.35
Prestretched	0.85	0.55
Temp, treated wires	0.87	0.70
Strands-stress relieved	0.90	0.75

f_p' = ultimate stress of pretension steel

The modulus of elasticity of high tensile steel may be taken as 2.00 GPa.
The creep characteristics of high tensile wires are given in Figure 1.24.3.

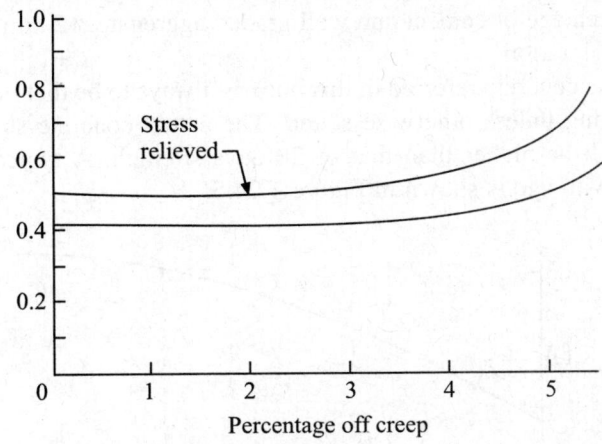

Figure 1.24.3 Creep in steel.

In most of the high tensile wires, the creep is negligible up to 0.45 f_p, and it is about 3% at 0.55 f_p stress level. The steel is subjected to high tension during the first stages of construction, so most of the creep occurs in the early stage of construction. The wires and strands are generally supplied in coils of sufficient big diameter as not to cause any inelastic deformation.

Concrete : Consequent to the use of high tensile steel in prestressed concrete construction, the concrete has to be of high strength character. The Indian code of practice suggests a minimum cube strength of 400 MPa for pre-tensioned system and 300 MPa for post-tensioned system. A typical concrete strength vs. water-cement ratio is shown in Figure 1.24.4.

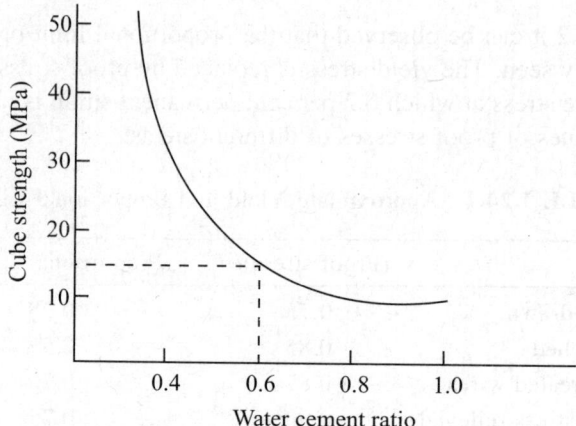

Figure 1.24.4 Effect of water cement ratio on strength of concrete.

As the water-cement ratio increases, the strength of concrete decreases. For most prestressed concrete constructions, a water-cement ratio in the range of 0.35 to 0.45 is used. For water-cement ratio less than 0.4, the workability of concrete decreases, with the result that compact and high density concrete is difficult to obtain. High workability with less water-cement ratio requires a higher percentage of cement and well graded aggregate. 12 mm to 25 mm slump is used with controlled vibration.

The strength of the concrete referred in this book is always to be that of a 15 cm cube after 28 days of water curing unless otherwise stated. The actual concrete strength after several months of casting will be higher than that of 28 days strength. A typical relation between strength of concrete with age is shown in Figure 1.24.5.

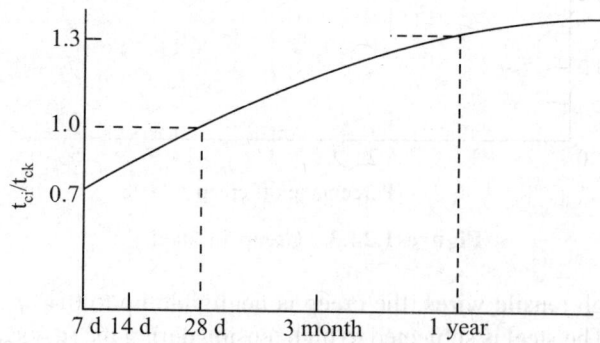

Figure 1.24.5 Relative strength variation of concrete with time.

A typical stress-strain curve of a well graded concrete is shown in Figure 1.24.6 and the stress strain relation (1.3) may be expressed as

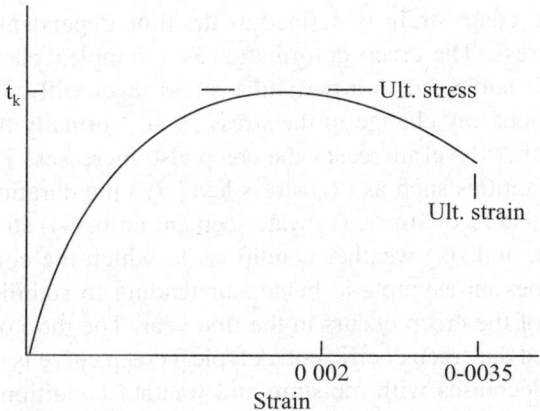

Figure 1.24.6 Stress-strain relation of concrete.

$$\sigma_c = E_c \in \left(1 - \frac{\in}{2\in_0}\right) \qquad (1.24.1)$$

where $\in_0$ = ultimate strain (about 0.004) and and ε = strain.

Determination of Young's modulus for concrete is not as specific as it is for steel. This can be easily observed from the curve. There have been many formulae developed to determine Young's modulus of concrete. It can be taken as

$$E_c = 5700 \sqrt{f_{ck}}$$

where f_{ck} is the 28 days 15 cm cube strength for a portland cement and 7 days cube strength for a rapid-hardening cement.

Young's modulus is considered to be a function of the ultimate strength of the concrete. The modulus of elasticity plays an important role in computations of the deformation aud loss of prestress characteristics. A certain amount of tolerance has to be assumed in the use of Young's modulus. The Poisson's ratio for prestressed concrete varies from 0.10 to 0.22 and an average of 0.15 may be adopted in the calculations.

Time dependent character of concrete shrinkage in concrete : As concrete is cast, cured and exposed to weather conditions, some drying and chemical changes take place. Consequently, the concrete undergoes change in volume depending upon time, aggregate moisture content, initial water-cement ratio, and weather conditions. The change of deformation independent of stress and temperature is called shrinkage. The amount of shrinkage varies from 0 to 0.001. Some amount of the shrinkage is recoverable upon restoring the moisture content in the concrete. The amount of shrinkage is proportional to the water-cement ratio. It is desireable to use a low water-cement ratio to reduce the effect of shrinkage. In some extreme wet conditions with certain aggregate, the concrete is likely to expand. The aggregate properties also affect the shrinkage. Aggregate with low percentage absorption will give less shrinkage. The chemical decomposition of cement or even the chemical reactions in the cement will have some influence on the shrinkage. For purposes of design, the shrinkage may be assumed as 0.0002 to 0.0004.

Creep in concrete : The creep strain is defined as the time dependent deformation resulting from the presence of stress. The creep deformation is a complex phenomenon especially in concrete. Most materials subjected to a particular stress level will continue to deform for a certain period even without any change in the stress level. Normally there is no creep at low stress level, but as the stress level increases the creep also increases. The creep of concrete is a function of several quantities such as : (i) stress level, (ii) the duration of stress, (iii) age of concrete, (iv) previous history of stress, (v) water-cement ratio, (vi) strength of concrete, (vii) aggregate, (viii) cement, and (ix) weather conditions to which the concrete is exposed. The creep of concrete assumes an asymptotic behaviour tending to stabilise the strain in a long period. However, most of the creep occurs in the first year. The ratio of the final strain to the initial of concrete is called the creep coefficient. A typical creep curve is shown in Figure 1.24.7. The creep of concrete decreases with moisture and weather conditions. For the purposes of design the creep coefficient of concrete is taken as 2.0 to 3.0.

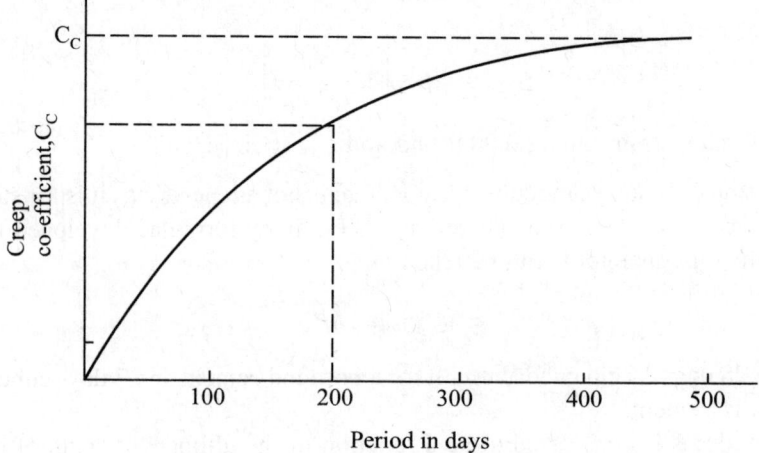

Figure 1.24.7 Creep of concrete.

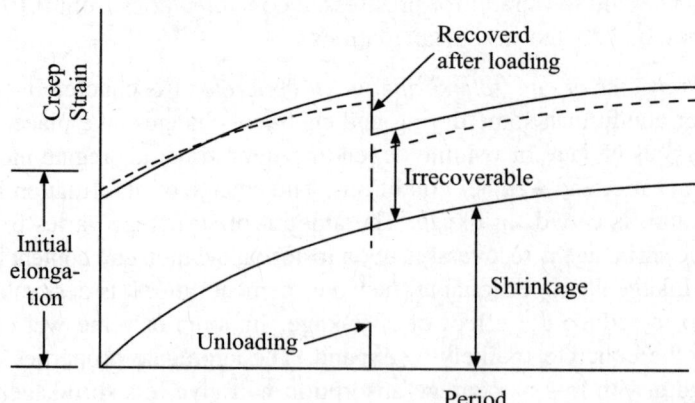

Figure 1.24.8 Creep concrete.

Figure 1.24.8 illustrates the recoverable and irrecoverable creep train along with shrinkage strain. Creep coefficients given by ndian code of practice (1.2) are given in Chapter II.

PROBLEMS

1.1 A simply supported beam of span L and depth h is subjected to self weight w_g and a live load of w_e. Both the loads are distributed uniformly over the span. Find a cable and profile so that a load of $w_g + w_e/2$ is balanced by the cable.

1.2 A simply supported beam of span L and depth h is subjected to self weight w_g uniformly distributed and four concentrated loads of each F, spaced at equal interval. Find load balancing cables for these loads. Adopt two different cables for the two load systems.

1.3 A double overhang beam has the middle span as L and the overhang span on either side as $L/4$. Determine a prestressing cable for balancing a uniformly distributed load on the entire beam.

1.4 A hollow box section has 10 cm by 30 cm inside dimensions and 25 cm by 50 cm outside dimensions. The beam is simply supported on 4.5 m span and subjected to a live load of 2kN/m. The beam is prestressed with an axial force of 300kN. Determine the stresses without the live load.

1.5 A solid rectangular cross-section of 100 cm by 15 cm is provided with an axial force of 1.2 MN with 2 cm eccentricity at mid-span. If the beam is simply supported on a span of 3 m and subjected to a load of 2 kN/m calculate the stresses at working load. Working load includes the self weight and live load.

1.6 Calculate the balancing cable force in problem 2.5 if the maximum sag in the cable is 2 cm at mid-span. Also calculate the effective loads on the beam.

1.7 A simply supported beam of span 8 m is carrying a live load of 10 kN/m. The self weight of the beam is 5 kN/m and its total depth is 40 cm. Design a parabolic cable of sag 10 cm such that it balances full self weight and 50 per cent of live load.

1.8 A 6 m cantilever beam of rectangular cross-section is provided with two sets of cables. One set of cable is balancing the self weight and other set of cables are balancing 12 kN/m live load. The cross-section at free end is 20 cm by 80 cm and at fixed end is 40 cm by 80 cm. Design the cables. (Hint: Provide zero eccentricity and zero slope for both the cables at free end. The self weight is uniformly varying so the profile of the cable will be a third order curve.)

1.9 A double overhang beam with 1 m overhangs on either side and 4 m within the supports is provided with a continuous cable. Assuming the self-weight of the beam as 3 kN/m design a cable to balance the total self weight. The depth of the beam is uniform and it is 20 cm. (Hint: Provide zero eccentricity and zero slope at each free ends of the beam. Calculate bending moments at support and at midspan. Select the eccentricities of the cable at these two locations and a cable force such that the moments are completely balanced. Provide a parabolic cable with negative curvature.)

1.10 If there are two concentrated loads of each 15 kN at each free end of the beam in problem 1.9, design a cable to balance the loads.

CHAPTER 2

Prestressing System and Losses of Prestress

2.1. Classification

Prestressed concrete system may be classified into a number of groups based on design, construction, method of applying prestressing and purpose of the structure. Some of them are discussed here.

2.2. External or Internal Prestressing System

Introduction of a horizontal thrust through the ends of a beam can be achieved either through an external force or through a built up internal thrust. The external prestressing can be obtained by external reactions introduced through different support conditions. For a simply supported beam, the external reaction can be obtained by jacking against abutments as shown in Figure 2.2.1 a continuous beam can be prestressed externally by jacking the appropriate supports to get the desired effect (Figure 2.2.2). Economical external prestressing of statically indeterminate structures is not simple and is often impossible. External prestressing is not common even

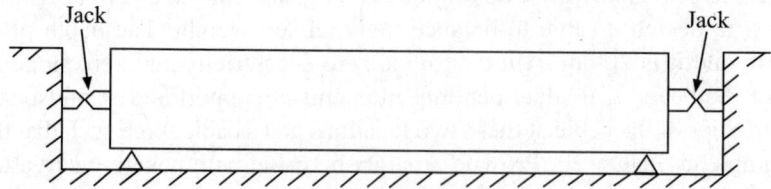

Figure 2.2.1 A Externally prestressed concrete simply supported beam.

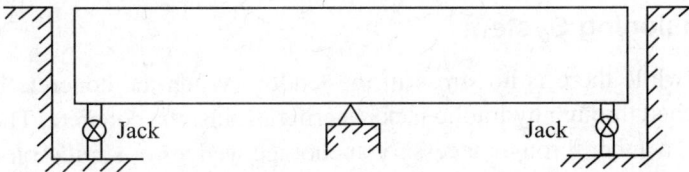

Figure 2.2.2 Externally prestressed concrete continuous beam.

though theoretically possible. It requires greater degree of accuracy in planning, maintenance and execution. Shrinkage of the beam releases certain thrust of the external jack thus reducing the prestressing force on the beam. Similarly a small expansion of the beam will create very high thrust in the beam due to the restraint of the jacks.

2.3. Linear and Circular Prestressing

Linear prestressing is done in straight members such as beams, piles and slabs whereas circular prestressing is applied to circular structures such as tanks, pipes, soils and pressure vessels in which the wires are put in curve going around the structure.

2.4. Pre-tensioning System

Pre-tensioning is obtained by stressing the tendons in position to a predetermined amount and placing the concrete in position while maintaining the stress in the tendons through an external force system. As the concrete hardens, the tensioned tendons are bonded; when the tendons are released from external jacks, the tendons will try to regain the original length. The concrete, as a result, is stressed. The tendons are strained (stressed) through hydraulic jacks and the strain is maintained during curing. A typical pre-tensioning bed is illustrated in Figure 2.4.1. Methods of prestressing, anchoring and other items of construction vary with patent. Since tendon force is transferred from steel to the concrete through bond, it is desirable to use small diameter wires. If the profile of the tendon is curved, then necessary profile fixing arrangements have to be provided.

The tendon is stretched between two bulk heads or rigid supports and then anchored to supports till its force is transferred to the proper members. Some type of anchorage arrangements have to be provided for holding the tendons to the supports.

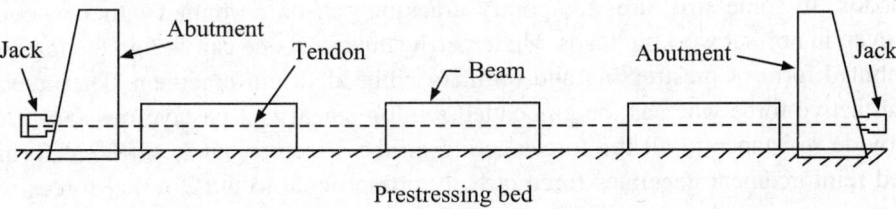

Prestressing bed

Figure 2.4.1 Typical pre-tensioning bed.

2.5. Post-tensioning System

Concrete is cast while there is no stress in the tendon. When the concrete is .hardened, the tendons are stretched through hydraulic jacks bearing against the concrete. The tendon force is transferred to the member through necessary anchorage wedges or similar blocks at the end of the member. A flexible hose or rubber tube may be used to house the tendons so that bond is not developed between the wires and concrete while casting or curing the concrete. The hose gets bonded to the concrete as the concrete hardens. Very often the void between the hose and the wires is filled with a grout mix in which case the wire develops a bond with the concrete. The method is called bonded construction in post-tensioning system. In some construction, an inflated water-filled rubber hose is placed in position. As the concrete hardens, the pressure in the hose is released, and the hose is pulled out leaving a duct in the concrete. The tendon is inserted in the duct and stressed with the help of jacks. Then the tendon force is transferred to the concrete through anchorages which are usually fixed at the ends of the member. In some cases of construction, the cable is coated with grease or bituminous material and then wrapped in paper so that no bond is developed between the concrete and the tendon. This type of construction is called unbonded construction.

Post-tensioning system can be used for either in situ or precast construction. In most cases, the tensioning of the tendons is done in stages either at site or in factory. Post-tensioning construction does not require rigid abutments, so it is more versatile for large spans. The anchorage is obtained through several patent methods using wedges or cones and bearing plates etc. Some typical anchorage arrarigements are shown in Figure 2.7.1.

2.6. Partial Prestressed Concrete

A construction where some amount of tensile stresses are allowed to be taken by mild steel bars provided as reinforcement is called a partial prestressed concrete construction. Ordinary reinforcement is also provided to take some of the stresses caused due to handling.

A fully prestressed concrete structural element is the one in which the prestressing force is provided such that all the stresses caused by working or handling loads etc. are within the permissible limits. Many structures are designed as fully prestressed ones. In some structures, the stresses caused by combined load conditions or high wind or earthquake load conditions may be too high when compared with those under normal working load conditions. Higher prestressing may be required for such load conditions when compared to that needed under normal loads. In some structures temporary cracking can be permitted either in combined loads or even in normal working loads. Under such situations, one can design the tension steel in a combined form of prestressing and ordinary embedded reinforcement. The prestressing steel is an active force whereas the embedded reinforcement is a passive resistant. Both the steels provide resistance to all the forces acting on the structure but in different forms. The embedded reinforcement generates force directly proportional to the external forces whereas the increase in force in the prestressing steel is only marginal. In some light precast elements, it becomes almost essential to provide ordinary steel so as to avoid failure of the elements at the time of jacking.

2.7. Classification Based on Methods of Prestressing

The basic principle of prestressing and the behaviour of prestressed beam is more or less the same in all prestressed construction. However, the methods of application of prestressing force or the anchorage devices used in different constructions are likely to vary. Based on the methods of application and construction, prestressed concrete is divided into several systems. There are hundreds of patents covering a number of prestressing systems. Every patent has its own method of prestressing force and its own details of anchorages.

(a) **Pre-tensioning systems:** As described previously, the pre-tensioning system is very adaptable for factory manufacturing prestressed concrete members. There are several patents for pre-tensioned systems; one of the systems, the long line system has been used in various countries. It consists of stretching the wires between bulk heads placed several metres apart on prestressing bed. The prestressing bed itself is capable of adjusting the bulk heads and span, and the positioning of the tendons in profile. To keep the vertical profile of tendons, some kind of tie downs are provided. The wires are tensioned with hydraulic jacks of long strokes. The anchorage arrangement is simple and can be based on the wedge friction principle. There are several other popular patents in practice which are not discussed in this book.

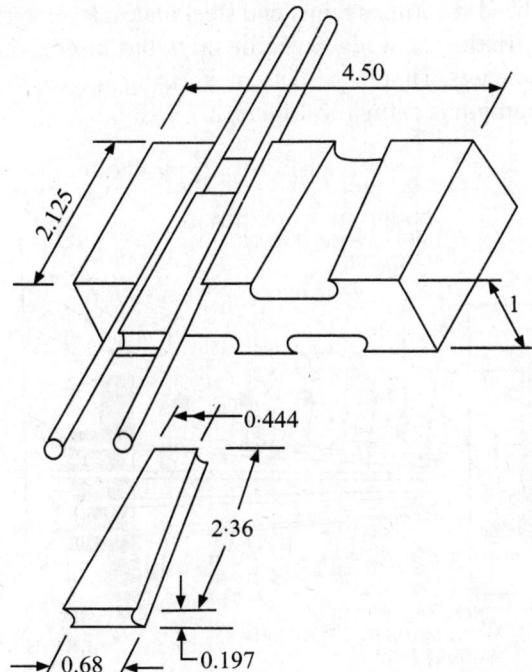

Figure 2.7.1 Anchorage of Magnel system.
(*Courtesy of Stressed Concrete Construction Ltd., Madras*)
(dimensions in inches)

(b) Post-tensioning systems: In post-tensioning systems, the tensioning is normally done with hydraulic jacks developed particularly to suit the patented system'. The anchoring of the wires to the concrete itself need more care than the anchoring to bulk heads done in a pre-tensioning system. Some popular post-tensioning systems are: (i) Freyssinet, (ii) Magnel, (iii) Hobling, (iv) Leon-hardt, (v) Lee-McCall, and (vi) Ryerson BBRV system.

The Freyssinet double acting jack pulls a number of wires at a time using a main ram which reacts against the embedded anchorage. As the required tension of the wire is reached, the inner piston pushes the plug into the anchorage thus causing an anchorage of the wires, without much loss of tension from anchorage take up. Similarly other systems use a specific type of jacking to achieve the required tension.

The principles of anchorages can broadly be listed in three groups: (i) wedge action producing frictional grip on wires, (ii) direct bearing from rivet, or bolt heads, or shims provided at the end of the wires, and (iii) looping of the wires around the concrete. The Freyssinet system uses a concrete cylinder with a conical interior opening and cones reinforced with steel wires. The conical plug has longitudinal grooves to receive the wires. The cylinder is placed flush with the face of the concrete and serves to transmit the reaction of the jack and 'the prestressing force to the member. The Magnel system uses tapered sandwich plate with wedges as shown in Figure 2.7.1. The Lee-McCall system uses steel rod with nut arrangement. Strescon and Preson systems use rivet heads bearing against end steel plates. Ryerson BBRV post-tensioning system does not use the friction or wedge type of grips but inserts flat plate shims after the necessary extension of the wires. There is variety of stressing and anchors used to suit different situations; one typical example is shown in Figure 2.7.3.

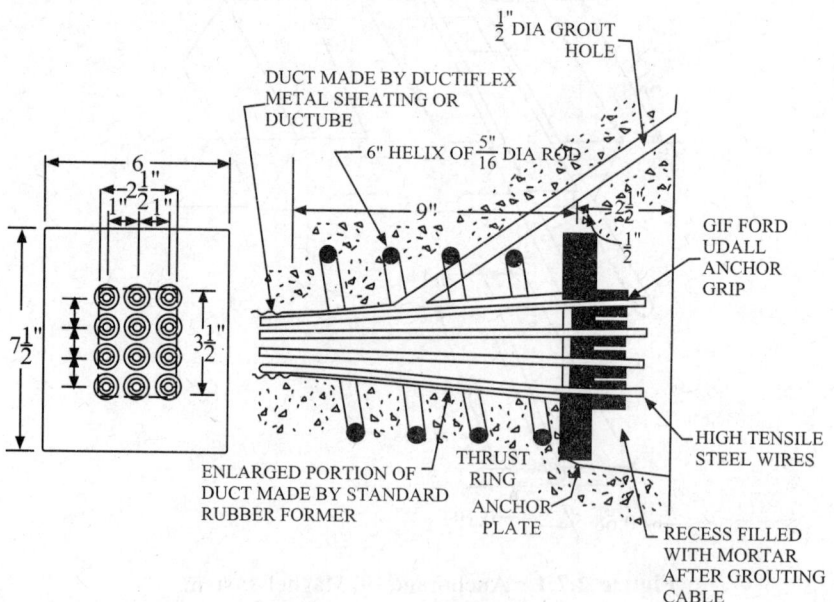

Figure 2.7.2 Gifford-Udall-CCL Anchor Grip for 4, 5 and 7 mm wires used in pre-tensioning and in plate anchorages for pre-tensioning.
(*Courtesy of Killick, Nixon & Co., Ltd., Bombay*)

Anchorage in place

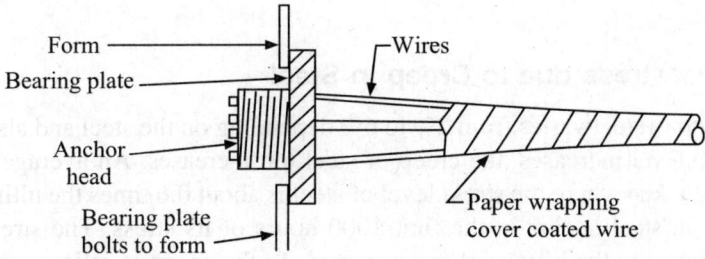

Form ——— ——Wires
Bearing plate ———
Anchor head
Bearing plate bolts to form ———
——Paper wrapping cover coated wire

Anchorage during stressing

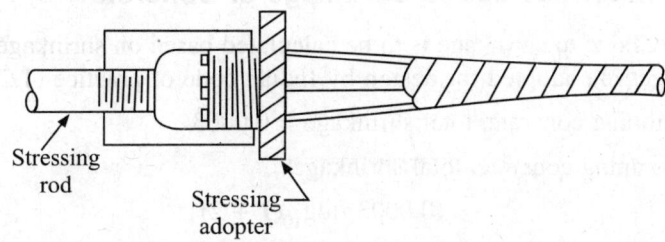

Stressing rod
Stressing adopter

Anchorage in stressed position

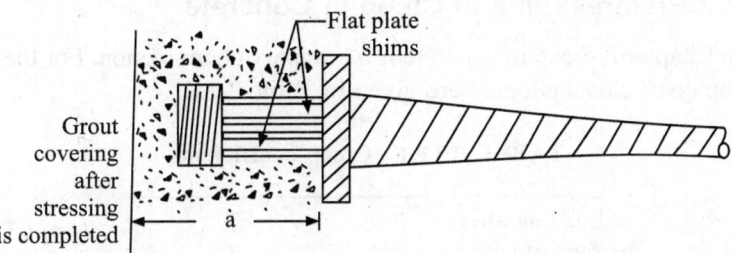

——Flat plate shims
Grout covering after stressing is completed
à

Figure 2.7.3 Type M stressing anchorage Ryerson BURV system.
(Courtesy of Joseph T. Ryerson and Son, Inc., U.S.A.)

2.8. Loss of Prestress

Prestressing force is measured either through the extension of the prestressed wire or by a pressure gauge reading of hydraulic jack at the time of jacking. The force recorded at the time of jacking at the jack position is not necessarily the amount of force available at all sections of the beam and at various working stages of the beam. In addition to the analytical calculations based on reasonable assumptions, a large amount of experimental data based on material behaviour has also been used to predict the loss of prestress. The various losses of prestressing may be due to: (i) creep in steel, (ii) shrinkage of concrete, (iii) creep in concrete, riv) steam curing,

(v) elastic shortening of the member, (vi) anchorage take up, (vii) bending of the member, and (viii) friction.

2.9.   Loss of Prestress due to Creep in Steel

The amount of creep in steel varies from 2% to 6% depending on the steel and also on its stress level. As the stress level increases, the creep of steel too increases. An average of 2% to 3% creep strain may be taken where the stress level of steel is about 0.6 times the ultimate strength. Most of the creep in steel occurs in the first 1000 hours of its stress. The stress on steel or concrete is not constant in the history of the structure. As the concrete relieves some prestress, the stress in steel drops; therefore, the creep in steel could be assumed as relatively low.

2.10.   Loss of Prestress due to Shrinkage of Concrete

The loss of prestress due to shrinkage is to be calculated based on shrinkage of concrete. The following shrinkages are adopted for design by Indian code of practice (1.2).

For pre-tensioning concrete, total shrinkage = 0.0003

For post-tensioning concrete, total shrinkage

$$= (0.0003)/\log_{10}(T + 2)$$

where T = age of concrete in days at transfer.

2.11.   Loss of Prestress due to Creep in Concrete

As explained in Chapter I, creep in concrete is a complex phenomenon. For the purpose of the design, the creep coefficients adopted are given in Table 2.11.1.

TABLE 2.11.1   Creep coefficients

Loading after days of casting	C_c
7 days	3.2
28 days	2.6
365 days	2.1

2.12.   Loss of Prestress due to Steam Curing

Steam curing is often used with a pre-tensioning system. The curing period has to be accelerated so as to release the pretensioning bed and the equipment for further fabrication. Raising the temperature of concrete with steam before hardening heats the prestressed steel and thus relaxes some of the prestressing force. On cooling, the concrete contracts and offsets the strain recovery in the steel. Creep characteristics of steel changes at temperature higher than 100° C, consequently there will be a higher loss of prestress in the member.

2.13. Loss of Prestress due to Elastic Shortening of the Member

Elastic shortening depends upon the modulus of elasticity of concrete, steel and the initial prestressing force. In pre-tensioned construction, the prestressing force is measured while jacking against bulk heads. As soon as the transfer of the prestressing force takes place, the member undergoes an elastic deformation. While computing prestress losses, all the transfer force should be considered towards elastic shortening.

In post-tensioned construction, wires are stretched with reaction against the member itself, so the elastic deformation takes place simultaneously and the jacking or the initial stresses are independent of the elastic shortening. If the tensioning of wires is done in stages, the wires that were anchored will undergo elastic shortening while the jacking is done for the next stage of prestressing. The wires anchored at the very first time will loose a maximum of prestressing force whereas the wires stretched later will not undergo any loss of prestress from elastic shortening. If the prestressing is done at equal intervals, an average prestressing force rould be taken for computing the elastic shortening.

2.14. Loss of Prestress due to Anchorage Take-up

In a post-tensioned system, an allowance should be made for slipping of steel at the time of transferring the tendon force from the jacks to the member In most of the systems, there is likely to be a small amount of adjustment of anchorage cones resulting in a small amount of relaxation of prestress. An average of about 0.25 cm is estimated as anchorage slip in wedge type of grips; for heavy strands the slip may be 0.5 cm.

2.15. Loss of Prestress due to Bending of the Member

Loss of prestress due to bending depends on the bonding character of the construction. In an unbonded construction when the beam bends due to external load, there is bound to be some change in the length of the prestressed steel, and the loss of prestress due to bending is illustrated in Figure 2.15.1. In bonded construction, even though there will be a change in the length of

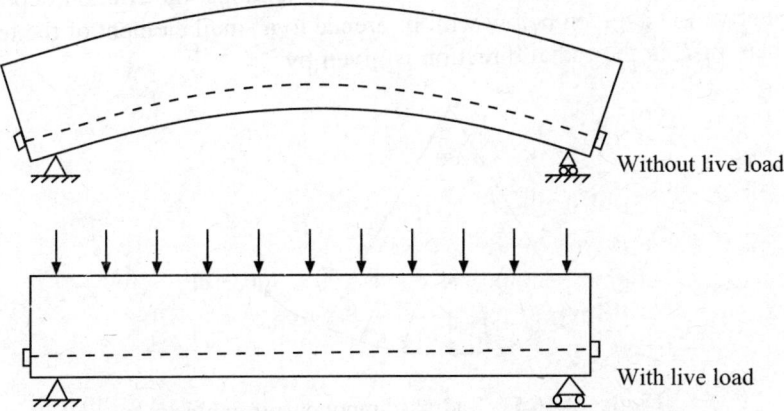

Figure 2.15.1 Effect cf bending on loss of prestress.

the tendon as the live load comes, it would cause only a local effect and no effect on the prestressing force. As soon as the live load is taken away, the loss due to bending will be recovered. For design calculation while working with live loads, the loss due to bending has to be considered for all unbonded construction.

2.16. Loss of Prestress due to Friction

There will be considerable movement of sliding of tendon relative to the surrounding duct during the tensioning operation. Because the tendon is in direct contact with the duct or with spacers provided, the friction will cause a reduction in prestressing force as the distance from the jack increases. There will also be some amount of friction in the jack and in the anchorage system. The loss of prestress due to friction is discussed below:

(a) *Friction in jack and anchorage :*   This friction is directly proportional to the jack pressure and will considerably depend upon the system of stressing. The loss of prestress due to this friction has to be specified by the patent.

(b) *Friction due to alignment variations :*   Whether a duct is straight or curved, there will be a certain amount of variation caused by vibrations or constructional methods in the actual alignment. The variation in the alignment will cause some points of contact of the wire with the duct and also with other wires or spacers; this will cause friction. The prestressing force at any distance x from jack may be expressed as:

$$P_x = P_0 e^{-kx}$$

(2.16.1)

where

P_0 =   prestreising force at jacking end

e =   the base of Napierian logarithm = 2.718

k =   constant depending upon the type of duct (wobble correction factor)

The various values of k are shown in table 2.16.1.

(c) *Friction in duct due to curvature :*   For curved tendons, the friction depends on the curvature. The derivation is given below with reference to a small element of the tendon. The normal component of P in the radial direction is given by

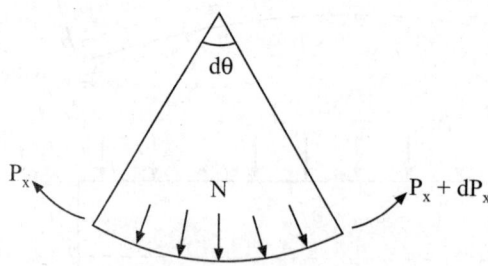

Figure 2.16.1 Radial component of cable.

$$N = (P_x + dP_x)(0 + d\theta) - P_x\theta = P_x d\theta$$
(neglecting second order terms)

$$\text{Frictional loss} = dP_x = \mu N = \mu P_x(d\theta)$$

where μ = coefficient of friction, and

$\quad N$ = normal component of force

$$\frac{dP_x}{P_x} = \mu \, d\theta.$$

On integration from $x = x$ to $x = 0$ we get from the above equation

$$\log \frac{P_o}{P_x} = \mu(\theta_o - \theta_x)$$

or

$$P_o = P_x \, e^{\mu(\theta_0 - \theta_x)} \tag{2.16.2}$$

Let $(0_o - 0_x) = \alpha$ (change in slope between jacking point and the point under consideration) then

Loss of prestress $= P_o - P_x = P_x(e^{\alpha\mu} - 1)$ or

$$\text{Ratio of frictional loss} = \frac{P_o - P_x}{P_x} = (e^{\mu\alpha} - 1) \tag{2.16.3}$$

The values of frictional coefficient μ varies from 0.25 to 0 55 depending upon the duct surface. Approximate values are tabulated in table 2.16.1.

TABLE 2.16.1 Frictional coefficient

Duct	μ	k
Steel moving on smooth concrete	0.30 to 0.45	0.0015
Duct formed by with drawing tubing	0.35 to 0.55	0.0015 to 0.003
Rubber core	0.40 to 0.55	0.0015 to 0.003

The total effect of friction due to alignment and curvature may be written as (adding the frictional force due to alignment variations to that of curvature) :

$$\frac{dP_x}{P_x} = (\mu \, d\theta + k \, dx)$$

On integration

$$Po = P_x e^{(\alpha\mu + kx)} \tag{2.16.4}$$

$$P_o - P_x \, [e^{(\mu\alpha + kx)} - 1] \, P_x \simeq [\mu\alpha + kx] \, P_x \tag{2.16.5}$$

If P_i, is the initial prestress then jacking stress P_o is given as:

$$P_o - P_i + \text{(loss of prestress due to frictional effects)}.$$

The various losses in prestress due to different causes are shown it table 2.16.2.

TABLE 2.16.2 Percentage of loss of prestress (Average values in normal conditions)

Nature of loss due to	Post-tensioning system		Pre-tension ing system	Remarks
	unbonded	bonded		
Creep in steel	2 to 3%	2 to 3%	1.5 to 2%	
Shrinkage in concrete	0.0002	0.0002	0.0003	
Creep in concrete	2.0 to 4.0%	2.0 to 3.0%	2.0 to 3.0%	
Elastic shortening (strain)	$P_t/2AE$	$P_t/2AE$	P_t/AE	if prestressing is done in stages at equal intervals
Anchorage take up	0.25 cm to 0.50 cm	0.25 cm to 0.50 cm		for wires to strand
Bending	as per moment curvature			
Stream curing effect			as per cal- culation	
Frictional effect	$(\mu\alpha + kx)\,P_i$			values of μ and k are given in table 2.16.1

Example 2.16.1 Determination of loss of prestress in a pre-tensioned prestressed concrete beam of 8 m span to carry a load of 5000 N/m. The beam is prestressed with 1.8 MN at transfer and the cable has a parabolic profile with maximum eccentricity of 10 cm at the middle of the span.

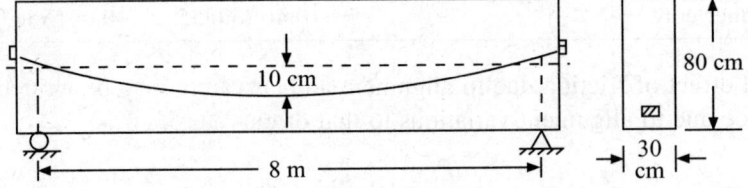

Figure 2.16.2 Example 2.16.1.

Given data:

Cross section = 30 cm by 80 cm

$A_s = 18$ cm^2, $P_i = 1.8$ MN

$E_s = 200$ GPa; $E_c = 30$ GPa.

Solution

Prestress at transferror $= \sigma_{si} = \dfrac{P_i}{A_s} = \dfrac{1.8}{0.0018} = 1000$ MPa

Strain in steel $= \varepsilon_s = \dfrac{\sigma_{si}}{E_s} = 5 \times 10^{-3}$

Losses of prestress due to various causes are computed one after another and finally tabulated at the end.

(i) *Elastic shortening:*   Elastic shortening of beam which will affect the prestressing force is the cumulative compressive strain of the concrete fibre at the steel profile level. It is rather hard to calculate exact shortening of the fibres at steel level if the cables do not have smooth profiles. Calculations could become very tedious if the cables have some complicated profiles. Approximate and fairly accurate methods of determination of loss due to elastic shortening are illustrated in this example. Profile of the cable which is of a parabolic form, can be given by

$$e_x = \frac{4gx\,(L-x)}{L^2} \tag{2.16.6}$$

where

e_x = eccentricity of the cable at distance x from the end of the span g = sag of the cable

$$\left.\begin{array}{l}\text{Then compressive strain at streel}\\ \text{level due to prestressing force}\end{array}\right\} = \frac{1}{E_c}\left(\frac{P_i}{A} + \frac{P_i e^2 x}{I}\right)$$

If $\in_c$ is the mean elastic strain due to prestressing'force, then

$$\in_c = \frac{1}{LE_c}\int_0^L \left(\frac{P_i}{A} + \frac{P_i e^2 x}{I}\right) dx$$

$$= \frac{P_i}{E_c}\left(\frac{1}{A} + \frac{8g^2}{15I}\right)$$

$A = 2400$ cm^2, $I = 1280000$ cm^4, $g = 10$ cm

On substitution of the respective quantities in the above expression, $\in_c$ works out to be

$$\in_c = \frac{825}{E_c}$$

where E_c is in N/cm^2.

An approximate mean strain could be obtained by taking a mean value of strains at maximum and minimum strain points. In this problem minimum compressive strain occurs at support and maximum strain occurs at mid-span.

Strain at end section $= \dfrac{P_i}{AE_c} = \dfrac{750}{E_c}$

Strain at mid section $= \dfrac{1}{E_c}\left(\dfrac{P_i}{A} + \dfrac{P_i e^2 x}{I}\right) = \dfrac{890}{E_c}$

Approximate mean strain $= \epsilon_c = \dfrac{890 + 750}{2E_c} = \dfrac{820}{E_c}$

As seen from the above calculations the difference between the approximate mean strain and the mean strain is not very much. The difference is negligible in this particular problem.

Loss of strain due to elastic shortening $= \dfrac{825}{E_c} = 2.75 \times 10^{-4}$

(ii) *Shrinkage strain :* As per Indian code of practice (1.2) the shrinkage strain may be taken as 0.0003.

(iii) *Creep strain in concrete :* Creep coefficient in average humid weather conditions may be taken as 2.5 per transfer of prestress at 28 days of curing of concrete.

$$\text{Creep strain} = (C_c - 1)\,\epsilon_c = 4.12 \times 10^{-4}$$

(iv) *Creep in steel:* Let creep in steel be about 2% then

$$\text{Creep strain in steel} = 0.02\,\epsilon_s$$

(v) *Other losses:* Loss of prestress due to anchorage take-up and friction of spacers and end block is small in pretensioned system. About 2% of strain loss may be counted due to anchorage take-up and friction of the spacers. The loss of prestress and the corresponding percentage is shown in table 2.16.3.

The example gives an approximate idea of losses of prestress due to various reasons. Percentage loss of prestress in this example is 23.7.

TABLE 2.16.3 Loss of prestress

Loss due to	Loss of strain	Percentage loss
Elastic shortening	2.75×10^{-4}	5.5
Shrinkage	3.0×10^{-4}	6.0
Creep in concrete	4.12×10^{-4}	8.2
Creep in steel	1×10^{-4}	2.0
Anchorage etc.	1×10^{-4}	2.0
Total	11.87×10^{-4}	23.7

The ratio of effective prestress to the initial =

$$\eta = \frac{\epsilon_s - \text{loss}}{\epsilon_s} = \frac{P_e}{P_i} = (1 - 0.237) = 0.763$$

$$P_e = \eta P_i = 0.763 \,(1.8) = 1.3734 \text{ MN}$$

where P_e = effective prestressing force.

Example 2.16.2 A post-tensioned prestress concrete beam of 50 m span is subjected to a transfer prestress force of 9.38 MN. Transfer of force is at 28 day strength. Profile of the cable is parabolic with maximum eccentricity at mid sections as 110 cm.

Determine the loss of prestress and find the jacking force required. The beam is subjected to a live, load of 24 kN/m and jacking is. done from both ends of the beam.

Given data:

P_i = 9.38 MN, A = 9471 cm^2,

A_s = 89.3 cm^2, I = 6224 × 10^4 cm^4,

σ_s = 1050 MPa

E_s = 210 GPa

Ec = 38.2 GPa

Solution

Elastic strain in steel at transfer condition is given by

$$\epsilon_s = \frac{\sigma_s}{E_s} = \frac{1050}{210000} = 50 \times 10^{-4}$$

(i) *Loss due to elastic shortening* :

$$\text{Elastic strain at end section} = \frac{P_i}{AE_c} = 2.60 \times 10^{-4}$$

$$\text{Elastic strain at mid section} = \left(\frac{P_i}{AE_c} + \frac{Me}{IE_c} \right) = 3.66 \times 10^{-4}$$

where

M = bending moment due to prestressing force and self weight of the beam

$$\text{Mean elastic strain} = \left(\frac{3.66 + 2.6}{2} \right) \times 10^{-4} = 3.13 \times 10^{-4}$$

The strands are strained at intervals so half the mean strain is the effective loss in the shortening.

$$\text{Loss of strain} = \frac{3.13 \times 10^{-4}}{2} = 1.56 \times 10^{-4}$$

(ii) *Shrinkage in concrete :*

$$\text{Shrinkage strain} = \frac{3 \times 10^{-4}}{\log (T + 2)} = 2 \times 10^{-4}$$

(iii) *Creep in concrete :*
Assuming a creep coefficient of 2.5
Creep strain $(2.5-1)\epsilon_c = 1.5 \times 3.13 \times 10^{-4} = 4.69 \times 10^{-4}$

(iv) *Creep in steel :*
Assuming a creep in steel as about 2.0%
Creep strain in steel $0.02 \, \epsilon_s = 10 \times 10^{-4}$

(v) *Anchorage take-up :*
Let 0.25 cm be the anchorage take-up at each jacking end,

$$\text{So loss of strain} = \frac{0.25}{L/2} = 1.0 \times 10^{-4}$$

(vi) *Strain due to external force :*
Strain in concrete at the steel level at mid-span due to external load is given by

$$\frac{M_i e}{I E_c} = -3.47 \times 10^{-4}$$

where M_i is bending moment at mid-span due to external loads.

The negative sign is given to indicate that there is a gain in strain instead of loss (local strain at mid-span).

(vii) *Frictional loss :*
Assuming a metallic sheathing, the frictional coefficients may be taken as

$$\mu = 0.25, \, k = 0.0015$$

As the jacking is done from both ends, effective length of 25 m span is taken for calculations. The profile of the cable is given by

$$e_x = \frac{4e}{L^2}(Lx - x^2)$$

$$\left[\frac{de_x}{dx}\right]_{x=0} = 0.088$$

The slope change $= \alpha = 0.088$ radians

$$\frac{P_0 - P_i}{P_i} = \left(\mu\alpha + \frac{kL}{2}\right) = 0.0595$$

TABLE 2.16.4 Loss of prestress

Loss due to	Loss of strain	Percentage loss
Shortening	1.56×10^{-4}	3.13
Shrinkage	2.00×10^{-4}	4.00
Creep in concrete	4.69×10^{-4}	9.38
Creep in steel	1.00×10^{-4}	2.00
Anchorage take-up	1.00×10^{-4}	2.00
Bending	-3.47×10^{-4}	-6.94
Total	6.78×10^{-4}	13.57

$$\text{Strain loss} = \frac{P_0 - P_i}{E_s A_s} = 0.0003$$

The loss of prestress due to various causes are tabulated in table 2.16.4.

$$\eta = 1 - 0.1357 = 0.8643$$

$$P_e = 0.8643 \, P_i = 8.09 \text{ MN}$$

$$P_0 = P_i + \text{frictional loss} = (1 + 0.0595) \, P_i, = 9.94 \text{ MN}$$

Loss of prestress in pre-tensioned construction is usually around 240 MPa and in post-tensioned construction it is 180 MPa.

2.17. Terminology

- Tension in the steel corresponding to the state of the final prestress.

- Initial Prestress – The Prestress in the concrete at transfer.

- Initial Tension – The maximum stress induced in the prestressing tendon at the time of the stressing operation.

- Post-tensioning – A method of prestressing concrete in which prestressing steel is tensioned the hardened concrete.

- Prestressed Concrete – Concrete in which permanent internal stresses are deliberately introduced, usually by tensioned steel, to counteract to the desired degree the stresses caused in the member in service

- Pre-tensioning – A method of prestressing concrete in which the tendons are tensioned before concreting.

    Shrinkage Loss – The loss of stress in the prestressing steel resulting from the shrinkage of the concrete.

    Stress at Transfer – The stress in both the prestressing tendon and the concrete at the stage when the prestressing tendon is released from the prestressing mechanism.

Tendon – A steel element, such as a wire, cable, bar, rod or strand used to impart prestress to concrete when the element is tensioned.

Transfer – The act of transferring the stress in prestressing tendons from the jacks or pre-tensioning bed to the concrete member.

Transmission Length – The distance required at the end of a pretensioned tendon for developing the maximum tendon stress by bond.

Cold-drawn indented wire conforming to IS : 6003-1970 (b) High tensile steel bar conforming to IS : 2090-1962 and (c) Uncoated stress relieved strand conforming to IS : 6006-1970

All prestressing steel shall be free from splits, harmful scratches, surface flaws; rough, jagged and imperfect edges and other defects likely to impair its use in prestressed concrete. Slight rust may be permitted provided there is no surface pitting visible to the naked eye.

Coupling units and other similar fixtures used in conjunction with the wires or bars shall have an ultimate tensile strength of not less than the individual strengths of the wires or bars being joined.

Modulus of Elasticity – The value of the modulus of elasticity of steel used for the design of prestressed concrete members shall preferably be determined by tests on samples of steel to be used for the construction. For the purposes of this clause, a value given by the manufacturer of the prestressing steel shall be considered as fulfilling the necessary requirements.

Losses in prestress – While assessing the stresses in concrete and steel during tensioning operations and later in service, due regard shall be paid to all losses and variations in stress resulting from creep of concrete, shrinkage of concrete, relaxation of steel, the shortening (elastic deformation) of concrete at transfer, and friction and slip of anchorage. Unless otherwise determined by actual tests, allowance for these losses shall be made in accordance with the values specified.

In computing the losses in prestress when untensioned reinforcement is present, the effect of the tensile stresses developed by the untensioned reinforcement due to shrinkage and creep shall be considered.

Loss of prestress due to creep of concrete – The loss of prestress due to creep of concrete under load shall be determined for all the permanently applied loads including the prestress. The creep loss due to live load stresses, erection stresses and other stresses of short duration may be ignored. The loss of prestress due to creep of concrete is obtained as the product of the modulus of elasticity of the prestressing steel and the ultimate creep strain of the concrete fibre integrated along the line of centre of gravity of the prestressing steel over its entire length. The total creep strain during any specific period shall be assumed for all practical purposes, to be the creep strain due to sustained stress equal to the average of the stresses at the beginning and end of the period.

For tendons at higher temperatures or subjected to large lateral loads, greater relaxation losses as specified by the engineer-in-charge shall be allowed for. No reduction in the value of the relaxation losses should be made for a tendon with a load equal to or greater than the relevant jacking force that has been applied for a short time prior to the anchoring of the tendon. **Loss of prestress due to shortening of concrete** – This type of loss occurs when the prestressing tendons upon release from tensioning devices cause the concrete to be compressed. This loss is proportional to the modular ratio and initial prestress in the concrete and shall be calculated as below, assuming that the tendons are located at their centroid:

(a) For pretensioning the loss of prestress in the tendons at transfer shall be calculated on a modular ratio basis using the stress in the adjacent concrete.

(b) For members with post-tensioned tendons which are not stressed simultaneously, there is a progressive loss of prestress during transfer due to the gradual application of the prestressing forces.

This loss of prestress should be calculated on the basis of half the product of the stress in the concrete adjacent to the tendons averaged along their lengths and the modular ratio. Alternatively, the loss of prestress may be exactly computed based on the sequence of tensioning. Loss of prestress due to slip in anchorage – Any loss of prestress which may occur due to slip of wires during anchoring or due to the strain of anchorage shall be allowed for in the design. Loss due to slip in anchorage is of special importance with short members and the necessary additional elongation should be provided for at the time of tensioning to compensate for this loss. Loss of prestress due to friction - The design shall take into consideration all losses in prestress that may occur during tensioning due to friction between the prestressing tendons and the surrounding concrete or any fixture attached to the steel or concrete.

Transmission Zone in Pre-tensioned Members. Transmission length – The considerations affecting the transmission length shall be the following:

(a) The transmission length depends on a number of variables, the most important being the strength of concrete at transfer, the size and type of tendon, the surface deformations of the tendon, and the degree of compactness of the concrete around the tendon. (b) The transmission length may vary depending on the site conditions and therefore should be determined from tests carried out under the most unfavourable conditions. In the absence of values based on all tests, the following values may be used, provided the concrete is well-compacted, and its strength at transfer is not less than 35 N/mm^2 and the tendon is released gradually.

$$0.48 \; f_{cii} \sqrt{\frac{Acr}{Apun}} = \text{ or } 0.8 \, f_{ci} \text{ whichever is smaller, where } f_{ci} \text{ is the cube strength at transfer,}$$

bearing area and is the punching area.

(b) During tensioning, the allowable bearing stress specified in a) may be increased by 25 percent, provided that this temporary value does not exceed f_{ci}.

(c) The bearing stress specified in (a) and (b) for permanent and temporary bearing stress may be increased suitably if adequate hoop reinforcement complying with IS : 456-1978 (2000) (2005) is provided at the anchorages.

(d) When the anchorages are embedded in concrete, the bearing stress shall be investigated after accounting for the surface friction between the anchorage and the concrete.

(e) The effective punching area shall generally be the contact area of the anchorage devices which, if circular in shape, shall be replaced by a square of equivalent area. The bearing area shall be the maximum area of that portion of the member which is geometrically similar and concentric to the effective punching area.

(f) Where a number of anchorages are used, the bearing area A b_r shall not overlap. Where there is already a compressive stress prevailing over the bearing area, as in the case of anchorage placed in the body of a structure, the total stress shall not exceed the limiting values specified in (a), (b) and (c). For stage stressing of cables, the adjacent unstressed anchorages shall be neglected when determining the bearing area.

Bursting tensile forces (a) The bursting tensile forces in the end blocks, or regions of bonded post-tensioned members, should be assessed on the basis of the tendon jacking load. For unbonded members, the bursting tensile forces should be assessed on the basis of the tendon jacking load or the load in the tendon at the limit state of collapse, whichever is greater. The bursting tensile force, Fbst existing in an individual square end block loaded by a symmetrically placed square anchorage or bearing plate, may be derived from the equation below:

When V, the shear force due to the ultimate loads, is less than Vc, the shear force which can be carried by the concrete, minimum shear reinforcement should be provided in the form of stirrups such that:

However, shear reinforcement need not be provided in the following cases: (a) where V is less then 0.5 Vc, and (b) in members of minor importance. When V exceeds Vc, shear reinforcement shall be provided such that:

In rectangular beams, at both corners in the tensile zone, a stirrup should pass around a longitudinal bar, a tendon or a group of tendons t having a diameter not less than the diameter of the stirrup. The depth dt is then taken as the depth from the extreme compression fibre either to the longitudinal bars or to the centroid of the tendons whichever is greater. The spacing of stirrups along a member should not exceed 0.75 dt nor 4 times the web thickness for flanged members. When V exceeds 1.8 Vc, the maximum spacing should be reduced to 0.5 dt. The lateral spacing of the individual legs of the stirrups provided at a cross section should not exceed 0.75 dt.

where Asv = total cross-sectional area of stirrup legs effective in shear, b = breadth of the member which for T, I and L beams should be taken as the breadth of the rib, bw, sv = stirrup spacing along the length of the member, and fv = characteristic strength of the stirrup reinforcement which shall not be taken greater than 415 N/mm².

PROBLEMS

2.1 A simply supported beam of 16 m span is provided with a straight cable profile with a kink at the mid-span. The cable has zero eccentricities at both end sections and e at mid section. If P_i is the prestressing force required at the mid section, determine the prestressing forces at both the end sections of the beam. Wobble and frictional coefficients may be taken as k and n respectively.

2.2 A simply supported post-tensioned concrete beam of 8 m span is rectangular in cross section of 30 cm by 80 cm size. It is provided with a parabolic cable with 200 kN prestressing force and a sag of 20 cm. Determine the loss of prestress and find the jacking force. Use the following coefficients: End anchorages are at CG of the section, $k = -0.0015$ per metre, $\mu = 0.45$, $C_e = 2.0$, $E_s = 200$ GPa, $E_c = 32$ GPa.

2.3 A double cantilever beam of each span of 30 m is provided with straight cable with kink at the support and anchored at the free ends. The cables are pulled from each end. The beam is a T-sectioncd with flange size as 200×15 cm. The web is of 30 cm thick and varies in depth from 50 cm at the free end to 200 cm at the support. The prestressing force at transfer is 1000 kN. Determine the loss of prestress due to different effects.

The concrete is M 35 and the beam is post-tensioned. The cable is anchored at the centroid of the free end section and it is 20 cm from top fibre at the support. The stress in the cable at the time of transfer is 100 kN/cm^2.

2.4 A 20 m long simply supported pre-tensioned prestressed concrete beam is of a box section. The top and bottom flanges are 100×20 cm and 100×15 cm respectively. The thickness of the vertical web is 10 cm each. The overall depth of the section is 80 cm and it is uniform. The beam is prestressed with a set of cables having a total force of 4000 kN at transfer. The stress in the steel at transfer is 105 kN/cm^2. The net sag of the cables at mid span is 30 cm. Determine the loss of prestress ia the beam. $E_s = 200$ GPa, $E_c = 30$ GPa.

CHAPTER 3

Working Stress Design of Simple Beams

3.1. Introduction

In working stress design, the stresses in a member caused by various working loads are calculated assuming elastic and linear behaviour of the material. These stresses are limited to allowable values of fne material. The following basic assumptions are used in working stress design :

  (i) Elastic and linear behaviour of the material.
 (ii) Homogeneous property of the material.
(iii) Plane section remains plane even after bending.
(iv) No cracking at any section.
 (v) Principal stresses not used as criteria of the design.

3.2. Critical Load Conditions

Prestressed concrete beams are subjected to two limiting conditions of loading: (i) transfer of prestress, and (ii) working load.

Prestressing bed and equipment can be released for continuous operation by transferring the prestress before the concrete attains its full strength. Prestressing force decreases with time because of various losses of prestress and reaches a steady state after long period and by this time the concrete attains its full strength. A qualitative behaviour of strength of concrete and the prestressing force with time is shown in Figure 3.2.1. Prestressing force is maximum and concrete strength is minimum at the time of transfer. Therefore transfer condition is a critical condition.

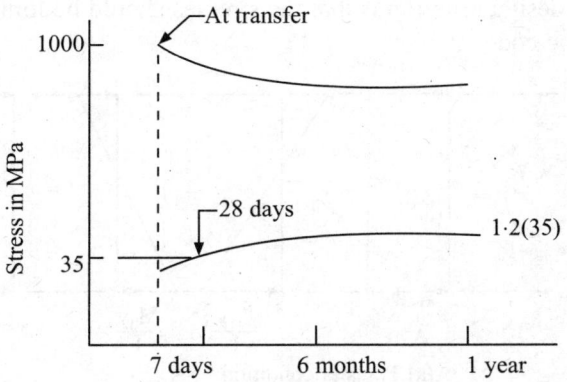

Figure 3.2.1 Variation of strength of concrete and progressing force with time.

The second design criterion is when the concrete and steel reach a steady state with critical working loads. In this condition strength of concrete is maximum and prestressing force is minimum while superimposed and live loads are at maximum.

3.3. Flexural Design Criterion

The main flexural design criterion is that the stress caused by external loads at any point in the beam must be limited to the permissible stress of the material in that condition. Critical stresses occur at extreme fibres at critical sections under critical load conditions. The critical load conditions are: transfer and effective working load conditions. Let and $P_t = P_i$ and $P_w = P_e = \eta P_t$ are the prestressing forces at transfer and effective load conditions. Let M_g and M_i be the bending moments at the critical sections caused by dead weight and live loads respectively. M_g is the self weight bending moment at transfer condition while $M_w = M_g + M_i$ is the external bending moment at working load condition. Figure 3.3.1 illustrates the load conditions.

The stresses at top and bottom fibres in transfer condition are

$$\sigma_{1t} = -\frac{P_t}{A} + \frac{P_t e y_t}{I} - \frac{M_g y_t}{I} \tag{3.3.1}$$

$$\sigma_{2t} = -\frac{P_t}{A} + \frac{P_t e y_b}{I} - \frac{M_g y_b}{I} \tag{3.3.2}$$

where av and σ_{1t} and σ_{2t} are stresses at top and bottom fibres at transfer condition respectively. The tensile stress is treated as positive, y_t and y_b are top and bottom fibre distances from centroid respectively and are measured positive. Similarly stresses at effective working load condition are:

$$\sigma_{1w} = -\frac{P_w}{A} + \frac{P_w y_t}{I} - \frac{M_w y_t}{I} \tag{3.3.3}$$

$$\sigma_{2w} = -\frac{P_w}{A} + \frac{P_w e y_b}{I} - \frac{M_w y_b}{I} \tag{3.3.4}$$

The working load design criterion is that the stresses should be limited to the permissible stresses specified by the code.

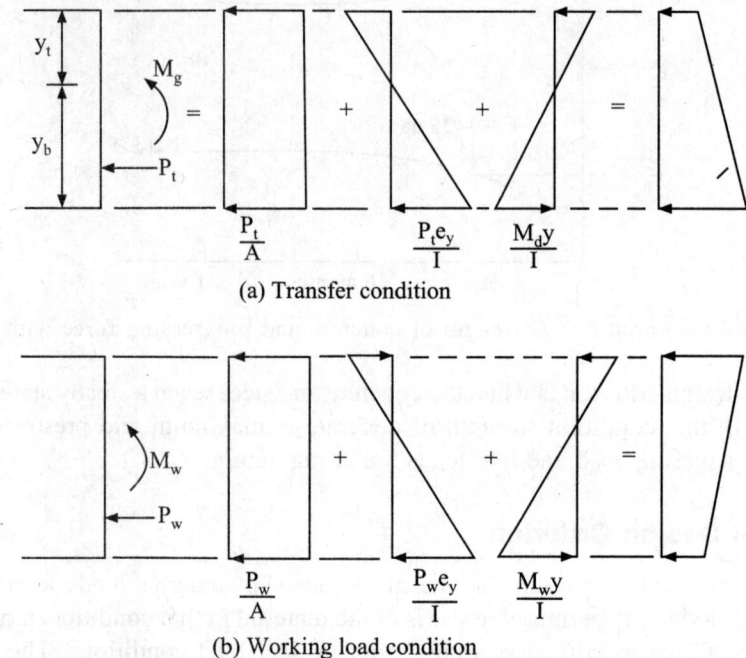

(a) Transfer condition

(b) Working load condition

Figure 3.3.1 Stress at critical load conditions.

Example 3.3.1 A prestressed concrete box section as shown in Figure 3.3.2 is prestressed with 5 numbers of 5 mm straight wires. The beam is 3 m cantilever span and loaded with 3 kN at free end. Transfer prestress in each of the wires is 25 kN. Determine stresses at transfer and at working load.

Properties of the section are calculated in table 3.3.1. The cross section is divided into 3 rectangular sections and the computations are made in the table.

TABLE 3.3.1 Properties of the section

Section	Area A_t (cm^2)	CG of A_i from top d_t (cm)	Moment of area $A_i d_t$ (cm^3)	$\bar{y}_t$ $d_i - y_t$ (cm)	I_{ci} (cm^4)	$A_i \bar{y}_{t_2}$ (cm^4)
(1)	(2)	(3)	(4)	(5)	(6)	(7)
Top flange	77	2	154	−7.3	100	4100
Bottom flange	56	18	1008	8.7	75	4250
Web	96	10	960	0.7	1152	47
Total	229		2122		1327	8397

A deduction for the area of the ducts occupied by the steel has to be accounted in the computations. Area of the steel = 5 (π) $(0.25)/4 = 0.98$ cm^2. This area is small when compared with the area of the concrete. In this example, the actual area of steel is about 0.43 per cent. In post-tensioned beams, the actual area of the ducts is likely to be about two to four times the actual area of steel. In such cases, an approximate deduction of about two per cent towards the ducts is reasonable. In this example, the area of the ducts is neglected.

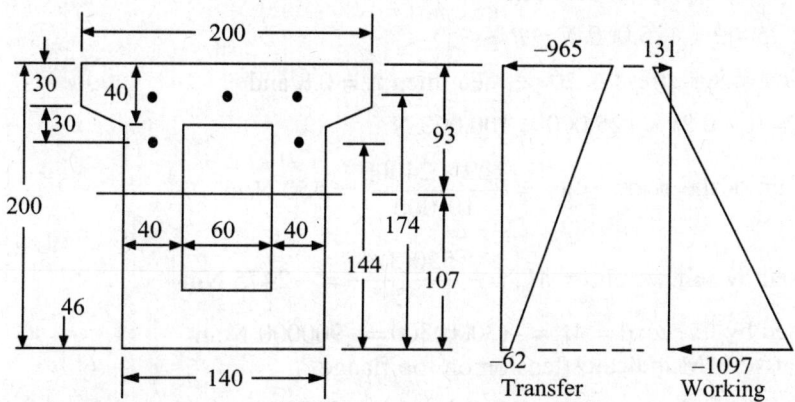

Figure 3.3.2   Example 3.3.1 (dimensions in mm and stresses in N/cm^2).

Columns 2, 3 and 4 of the Table 3.3.1 are computed using the dimensions shown in the figure. Column 4 gives the first moment of the area about the top fibre and column 2 gives the area of the section. The distance of the *CG* of the section from the top fibre is obtained from these two columns. It is

$$y_t = \frac{\Sigma A_i \, d_t}{\Sigma A_i} = \frac{2122}{229} = 9.3 \text{ cm}$$

The moment of inertia of the section is computed from the moment of inertia of each individual segments taken about the centroid. Column 5 of the table gives the distance of the *CG* of the individual segments from the *CG* of the section. Column 6 gives the moment of inertia of each of the segments about its own *CG*. The moment of inertia of the section is therefore obtained as

$$I = \Sigma I_{ci} + \Sigma A_i \bar{y}_i^2$$

in which

I_{ct} = moment of inertia of the ith segment about its own *CG*.

$\bar{y}_i$ = distance of CG of the segment from the *CG* of the section.

I_{ci} and $A_i \bar{y}_{i2}$ are computed in columns 6 and 7 of the table 3.3.1 and the sums are given at the bottom of the table.

$$I = \Sigma I_{ci} + \Sigma \bar{A}_i y_{t2} = 1327 + 8397 = 9724 \text{ cm}^4$$

The eccentricity of the wires can be calculated by taking moment of the steel wires about the tpp fibre. The distance of centroid of steel from top fibre is (d)

$$d = \frac{2(20 - 14.4) + 3(20 - 17.4)}{5} = 3.8 \text{ cm}$$

and the eccentricity is

$e = d - y_t = 3.8 - 9.3 = -5.5$ cm

$P_i = 5 \times 25000 = 125,000 \, N = P_i$

Let the loss of prestress is 20 per cent then $\eta = 0.8$ and

$P_c = P_w = (1 - 0.2) \times 125,000 = 100,000 \, N$

Self weight of the beam $= w_g = \dfrac{229 \times 24000}{10,000} = 550 \text{ N/m}$

BM caused by self weight $= M_g = -\dfrac{550(3)(3)}{2} = -2475 \text{ Nm}$

BM caused by live.load $= M_i = -(3000)300 = -900000$ Ncm
The negative 2?M indicates tension on top flange.

$$M_w = M_s + M_i = -1147500 \text{ Ncm.}$$

Stress at transfer condition in top fibre is

$$\sigma_{1t} = \frac{P_i}{A} + \frac{P_i e y_t}{I} - \frac{M_g y_t}{I}$$

$$= -\frac{125000}{229} + \frac{125000(-5.5)(9.3)}{9724} - \frac{(-247500(9.3))}{9724}$$

$$= -546 - 658 + 237 = -967 \text{ N/cm}^2$$

Bottom fibre stress is

$$\sigma_{2t} = -\frac{P_i}{A} - \frac{P_i e y_b}{I} + \frac{M_g \, y_b}{I}$$

$$= -\frac{125000}{229} - \frac{125000(-5.5)(10.7)}{9724} + \frac{(-247500)(10.7)}{9724}$$

$$= -546 + 757 - 272 = -61 \text{ N/cm}^2$$

The subscripts 1 and 2 refer to top and bottom extreme fibres.
The stresses at working load condition are:

$$\sigma_1 w = -\frac{P_w}{A} - \frac{P_w e y_t}{I} - \frac{M_w \, y_t}{I}$$

$$= -\frac{100000}{227} + \frac{100000(-5.5(9.3))}{9724} \frac{(-1147500) (9.3)}{9727}$$

$$= -440 - 526 + 1097 = 131 \text{ N/cm}^2$$

$$\sigma_{2w} = \frac{P_w}{2} - \frac{P_w e y_t}{I} + \frac{M_w y_b}{I}$$

$$= -440 + 605 - 1262 = -1097 \text{ N/cm}^2$$

The negative sign indicates the compressive stress. In a cantilever beam, the top fibres are usually subjected to tension by the external loads and the bottom fibres are subjected to compression. The prestressing force is designed to produce opposite effect. At transfer prestress condition, the dead load and prestressing forces are acting on the section. It can be seen that at transfer condition that the stress at the top fibre is compressive and it is much more than that at the bottom fibre. This stress distribution is reversed at working load condition. The actual stress distribution at the two load conditions are shown in Figure 3.3.2.

The flexural design criterion in working stress design (WSD) requires that the actual stresses must be limited to permissible stresses of the material. The permissible stresses also called as allowable stresses are chosen as a fraction of ultimate stress of the material. Transfer condition is only a temporary condition of short duration so the permissible stresses at transfer are higher than those permitted at working load. Again permissible stresses in compression of concrete is different from that in tension for obvious reasons. Let the following be denoted as permissible stresses in concrete

σ_{ati} = permissible tensile stress at initial

σ_{aci} = permissible compressive stress at initial

σ_{ate} = permissible tensile stress at working load

σ_{ace} = permissible compressive stress at working load

The second subscript refers to nature of stress and the third subscript refers to condition of loading.

All the above four quantities are treated positive definite which means permissible compressive stress σ_{aci}, gives the magnitude of stress allowed in compression. The governing equations for design can be generated by first identifying maximum tensile and compressive stresses in the beam and limiting them to the permissible stress limits. A beam cross section which is subjected to positive moment, i.e., moment causing compression on the top fibre, maximum tensile stress occurs on top fibre at transfer condition. The corresponding limiting equation can be written from eq. 3.3.1 as

$$\frac{-P_i}{A} + \frac{P_i \, e y_t}{I} - \frac{M_g \, y_t}{I} = \sigma_{it} \leq \sigma_{ati} \qquad (3.3.5)$$

E.q. 3.3.2 which represents critical stress in compression, giving compressive stress as negative should be compared with the limiting compressive permissible stress. For the purpose of comparing the actual stress with the limiting compressive stress, only magnitude of the actual stress should be compared since σ_{cti} is positive. The limiting compressive stress at transfer can be written by just changing the sign of stress σ_{2t} of eq. 3.3.2 into compression as positive and it is

$$\frac{P_i}{A} + \frac{P_i \, e y_b}{I} - \frac{M_g \, y_b}{I} = \sigma_{ci} \leq \sigma_{aci} \qquad (3.3.6)$$

Similarly the limiting stress conditions at critical working load conditions can be obtained from eqs. 3.3.3 and 3.3.4 and they

$$\frac{\eta P_i}{A} - \frac{\eta P_i e y_t}{I} + \frac{M_w y_t}{I} = \sigma_{ci} \leq \sigma_{act} \qquad (3.3.7)$$

$$-\frac{\eta P_i}{A} - \frac{\eta P_i e y_b}{I} + \frac{M_w y_b}{I} = \sigma_{ti} \leq \sigma_{ait} \qquad (3.3.8)$$

Allowable stresses specified by the code are indicated by subscript a. The solution of 3.3.5 to 3.3.8 will give the design para meters.

3.4. Permissible Stresses

Maximum permissible stresses at working load are taken as a friction of the ultimate or proof stresses and are specified by the codes. The permissible stresses are selected as functions of load conditions, ductility and ultimate strength of the materials. Higher stress can be permitted in a load condition in which the load intensity may be high but its duration is small when compared with the working loads. Permissible stresses in concrete at transfer of prestress,

TABLE 3.4.1 Allowable stresses (MPa) in concrete in WSD

Stress	Post-tesioned				Pre-tensioned			
$f_{ck} =$	35	40	45	50	40	45	50	55
General purpose $f_{ci} \geq 0.7 f_{ck}$								
σ_{aci}/f_{ci}	0.50	0.48	0.46	0.44	0.5	0.48	0.45	0.41
σ_{ace}	14.0	15.5	17.0	18.0	16.0	17.5	19.0	20.5
σ^*_{ace}	11.5	12.5	13.5	14.0	13	14.5	15.5	16.5
$\sigma_{ati} = \sigma_{ate}$	1.0	1.0	1.0	1.0	1.4	1.4	1.4	1.4
σ_{ach}/f_{ch}	0.5	0.5	0.5	0.5	0.5	0.5	0.5	0.5
σ_{ath}	1.5	1.5	1.5	1.5	1.5	1.5	1.5	1.5
Bridges $f_{ci} \geq 0.8 f_{ck}$								
σ_{aci}	14.0	15	16	17				
σ_{ace}	11	12	13	14				
σ_{ati}	1.4	1.5	1.6	1.7				
σ_{ate}	0	0	0	0				
σ_{ach}/f_{ch}	0.45	0.45	0.45	0.45				

f_{ci}	= strength of concrete at transfer
σ_{ach}	= allowable stress in handling
σ_{ath}	= allowable tensile stress in handling
f_{ch}	= strength of concrete at handling
σ^*_{ace}	= allowable stress in concrete in case of loads likely to increase.

handling and working load as per Indian code of practice for building and Indian Road Congress (3.1) are given in table 3.4.1. Similarly permissible stresses in high tensile steel are given in table 3.4.2. The actual stress in a structure should be limited to the permissible stress which means the actual stress in the structure should not exceed the limit. Least weight is obtained in the case of statically determinate structures subjected to a fixed critical load if the actual stresses are made equal to the limiting stresses (3.3). However such a situation is often difficult as there are practical limitations on the cross section of the beam, constructional and environmental conditions.

TABLE 3.4.2 Allowable stresses in HTS

Stress	
$\sigma_{asa} = \sigma_{sa} =$	$0.8 f_p$ or f_y
$\sigma_{asi} = \sigma_{si} =$	$0.7 f_p$
$\sigma_{ase} \quad\ =$	$0.6 f_p$

σ_{asa} = allowable stress at jacking (anchorage) and

σ_{asi} = allowable stress at critical section at initial stage

σ_{ase} = allowable effective stress at working loads

Example 3.4.1 A prestressed concrete electric pole carries a 1800 N transverse wind load at 6 m above the ground level. Design the pole with constant width and tapering depth. Given

$$\sigma_{ati} = \sigma_{ate} = 0; \sigma_{aci} = 20 \ MPa, \sigma_{ace} = 18 \ MPa$$

Total length of pole = 7.5 m

Depth of pole in the ground = 1.2 m

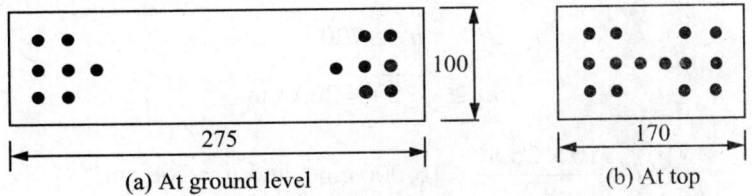

(a) At ground level   (b) At top

Figure 3.4.1 Example 3.4.1

Solution

The wind load can occur in either direction so the pole has to be designed with axial prestressing. Therefore $e = 0$. Indian code of practice for prestressed concrete poles (3.9) is used in this design. The cross section of the pole is assumed to be rectangular with width = b = 10 cm as the minimum based on constructional and handling limitations. Let h_1 and h_2 are the overall depths at top and ground level of the pole respectively. The cross section at the top of the pole is subjected to zero bending but axial prestress. So the condition at transfer is critical at top of the pole. The pole is in vertical position so the *BM* caused by self weight is zero. $M_g = 0$. *BM* caused by external load is

$$M_w = WL_o = 1800 \ (600) = 1080 \ 000 \ N \ cm$$

Eq. 3.3.6 when applied to the cross section at the top of the pole give

$$-\frac{P_i}{bh_1} \leq 0$$

$$\frac{P_t}{bh_1} \leq 2000$$

(All units are in N, cm unless mentioned)

Substituting $b = 10$, the above equation gives

$$P_i \leq 20000h_1 \tag{3.4.1}$$

Cross section at the ground level is critical at working as the bending moment is maximum at that section. Ineqs. 3.3.7 and 3.3.8 give for $\eta = 0.8$

$$\frac{0.8P_i}{A} + \frac{M_w y_t}{I} \leq 1800 \tag{3.4.2}$$

$$-\frac{0.8P_t}{A} + \frac{M_w y_b}{I} \leq 0 \tag{3.4.3}$$

$$y_b = y_t = h_2/2, \quad A = 10h_2 \text{ and } I = 10h_2^3/12$$

The sum of ineqs. 3.4.2 and 3.4.3 gives

$$\frac{M_w h_2}{I} \leq 1800$$

or

$$\frac{I}{h_2} \geq \frac{M_w}{1800} = \frac{1080000}{1800} = 600$$

or

$$h_2^2 \geq 720$$

$$h_2 \geq \sqrt{720} = 26.8 \text{ cm}$$

$$I = \frac{10 \times 26.8^3}{12} = 16,000 \text{ cm}^4 \text{ and } A = 268 \text{ cm}^2$$

Substitution of M_w, I, y_b and A in eq. 3.4.3 gives

$$P_i \geq \frac{1080000 \, (13.4)(268)}{0.8 \times 16,000} = 303000 \text{ N}$$

Let $P_i = 303000$ N

Substituting value of P_i in eq. 3.4.1 one gets

$$h_1 \geq \frac{303000}{20000} = 15.15 \approx 15.2 \text{ cm}$$

Selecting 14 wires the capacity of each wire = (303000/14) = 2I650 N. 5 mm diameter high tensile steel bars can carry 22600 N. Figure 3.4.1 illustrate the cross section of the pole with slightly increased sides. The section is found acceptable to carry (1800/4) N load in the lateral direction.

3.5. Axialiy Prestressed Members

Axial prestressing means that the centroid of the prestressing force coincides with that of the cross section. Wind or earthquake loads are likely to act in any direction. An electric pole is to be designed for wind load acting from any direction. In such cases, the prestressing force has to be axial. Some precast elements are made so thin that it is difficult to ensure any accurate eccentricity. In addition, the handling conditions may be uncertain. So in such cases the prestressing is to be made concentric with the section. The governing stress limitations (Ineqs. 3.3.5 to 3.3.8) for axialiy prestressed sections can be written as (for $e = 0$).

$$-\frac{P_i}{A} - \frac{M_g y_t}{I} \leq \sigma_{ait} \tag{3.5.1}$$

$$\frac{P_i}{A} - \frac{M_g y_t}{I} \leq \sigma_{aci} \tag{3.5.2}$$

$$\frac{\eta P_t}{A} + \frac{M_w y_b}{I} \leq \sigma_{ace} \tag{3.5.3}$$

$$\frac{-\eta P_t}{A} + \frac{M_w y_b}{I} \leq \sigma_{ate} \tag{3.5.4}$$

Ineq. (3.5.1) is a trivial one in most cases. Ineq. (3.5.2) is not likely to govern the same section as much as that will be governed by Ineqs. (3.5.3) and (3.5.4). Addition of Ineqs. 3.5.3 and 3.5.4 gives

$$\frac{M_w (v_t + y_b)}{I} \leq \sigma_{ate} + \sigma_{ace}$$

or

$$I \geq \frac{M_w h}{\sigma_{ate} + \sigma_{ace}} \tag{3.5.5}$$

in which $y_b + y_t = h$.

Similarly addition of Ineqs. 3.5.1 to 3.5.4 gives

$$I \geq \frac{M_1 h}{\sigma_{ati} + \sigma_{act} + \sigma_{ace} + \sigma_{ate}} \tag{3.5.6}$$

where $M_1 = M_w - M_g$.

Ineq. 3.5.5 dominates over Ineq. 3.5.6, therefore the cross section is governed by the Ineq. 3.5.5.

The bounds for prestressing force can also be obtained from Ineq. 3.5.1 to 3.5.4. Addition of Ineq. 3.5.1 and 3.5.4 gives

$$-\frac{P_i}{A} (1 + \eta) + \frac{1}{I} (M_w y_b - M_g y_t) \leq (\sigma_{ati} + \sigma_{ate})$$

Let $I/A = r^2$, where r = radius of gyration. Then the above can be rearranged as

$$P_i \geq \frac{(M_w y_b - M_g y_t)}{r^2 (1 + \eta)} - \frac{A (\sigma_{ati} + \sigma_{ate})}{1 + \eta} \qquad (3.5.7)$$

Axially prestressed members are usually symmetrical, therefore,

$$y_t = y_b = h/2$$

Ineqs. 3.5.7 reduces to

$$P_i \geq \frac{M_1 h}{2(1 + \eta) r^2} - \frac{A (\sigma_{ati} + \sigma_{ate})}{1 + \eta} \qquad (3.5.8)$$

Similarly addition of Ineqs. 3.5.2 and 3.5.3 gives

$$P_i \leq \frac{A (\sigma_{aci} + \sigma_{ace})}{1 + \eta} - \frac{M_1 h}{2(1 + \eta) r^2} \qquad (3.5.9)$$

Ineq. 3.5.8 gives the lower bound and Ineq. 3.5.9 gives an upper bound for prestressing force. The two inequalities can be combined as

$$\frac{M_1 h}{2 (1 + \eta) r^2} - \frac{A(\sigma_{ati} + \sigma_{ate})}{1 + \eta} \leq P_i \leq \frac{A(\sigma_{aci} + \sigma_{ace})}{1 + \eta} - \frac{M_1 h}{2(1 + \eta) r^2} \qquad (3.5.10)$$

It is desirable to select the actual prestressing force corresponding to the lower bound and then check for permissible stresses.

Example 3.5.1   A oneway prestressed concrete roof slab of 5 m span carries a live load of 2000 N/m². The length of the slab is large compared with 5 m. Design the slab for the following.

$$\sigma_{ati} = 0; \ \sigma_{ate} = -100 \text{ N/cm}^2$$
$$\sigma_{aci} = 2000 \text{ N/cm}^2; \quad \sigma_{ace} = 1600 \text{ N/cm}^2$$
$$\sigma_{st} = 112{,}000 \text{ N/cm}^2.$$

Solution

Select and design 1 m width slab as beam. All units are in N or cm units unless mentioned. The area of the section is

$$A = bh = 1000h \text{ cm}^2$$

Let the weight of the concrete be 25000 N/m⁸, then the self weight of the slab is

$$W_g = \frac{25000 \, A}{10000} = 2.5 \, A \text{ N/m} = 0.0025 \text{ N/mm}$$

$$I = \frac{bh^3}{12} = \frac{1000h^3}{12} \text{ cm}$$

$$M_g = \frac{w_g L^2}{8} = \frac{2.5A\,(2500)}{8} = 782\,A = 78200\,h$$

$$M_i = \frac{w_t L^2}{8} = \frac{2000\,(2500)}{8} = 625000 \text{ N cm}$$

From Ineq. 3.5.5 we have

$$\frac{M_w h}{I} = \frac{(782,000\,h + 625,000)\,(12)}{100h^2} \le (\sigma_{ace} + \sigma_{ate})$$

or
$$\frac{9380}{h} + \frac{75000}{h^2} \le (1600 - 100) = 1500$$

or
$$938h + 7500 \le 150h^2$$

or
$$h^2 - 6.25h - 50 \ge 0$$

or
$$h \ge \frac{6.25 + \sqrt{(240)}}{2} > 10.9 \text{ cm}$$

The value of P_i can be obtained from the other inequalities such as:

$$\frac{P_i}{A} \le \sigma_{act} - \frac{M_g h}{2I}$$

$$\le 2000 - \frac{78200h^2(12)}{2(100h^3)}$$

$$\le 2000 - \frac{782(6)}{h} = 1580$$

and
$$\frac{P_i}{A} \ge \frac{M_g h}{2I\eta} = 935$$

or
$$P_i > 935\,A = 935(100)(10.9) = 1002\,000 \text{ N/m}$$

(For $\eta = 0.8$)

Provide 25 numbers of 5 mm wires on each face of the slab, then P_i is given by

$$P_i = (A_s)\,\sigma_{si}$$

$$P_i = 2(25)(0.196)(112000) = 1097\,600 \text{ N}$$

Basic design.variables in prestressed concrete beam are: area of cross section of the beam (A), shape of the cross section which is governed by the location of centroid (y_t and y_b); moment of inertia of the cross section (I), area of prestressing steel (A_s) and cable profile (e). The above six design variables have to be obtained from the four limiting governing equations. Six variables cannot be obtained uniquely from the four equations so at least two design variables have to be assumed. Some of the variables are subjected to practical constraints. For example, the profile of the cable must be limited to within the sectional depth with a necessary provision for cover. Similarly, the thickness of the web must be limited to a minimum practicable thickness. Some of the design constraints may be stated as

$$e \leq (y_b - d')$$
$$b_w \geq \text{minimum thickness of web}$$
$$t_t \geq \text{minimum thickness of top flange}$$
$$t_b \geq \text{minimum thickness of bottom flange}$$
$$b_t \geq \text{minimum width of top flange}$$
$$A_s \geq pA$$

in which b_w and b_t are widths of web and top flanges respectively, t_i and t_b are thicknesses of top and bottom flanges respectively, and d' is the minimum cover for steel cable. The widths of the flanges are constrained in bridges. The percentage of steel is also constrained in most prestressed concrete construction.

3.6. Design of Prestressing Cable for a Given Cross Section

In some cases, the cross section is selected and it is only necessary to find the prestressing force and its profile. In some cases a cross section can be assumed based on experience and the steel cable is designed. Any modifications in the section are carried out by trial method. Ineqs. 3.3.5 to 3.3.8 can be rearranged as

$$-P_i\left(I - \frac{ey_t}{r^2}\right) \leq A\sigma_{ati} + \frac{M_g y_t}{r^2} \tag{3.6.1}$$

$$P_i\left(1 + \frac{ey_b}{r^2}\right) \leq A\sigma_{aci} + \frac{M_g y_b}{r^2} \tag{3.6.2}$$

$$\eta P_i\left(1 - \frac{ey_t}{r^2}\right) \leq A\sigma_{ace} - \frac{M_w y_t}{r^2} \tag{3.6.3}$$

$$-\eta P_i\left(1 + \frac{ey_b}{r^2}\right) \leq A\sigma_{ate} - \frac{M_w y_b}{r^2} \tag{3.6.4}$$

Addition of Ineqs. 3.6.1 and 3.6.2 gives

$$P_i\frac{e(y_b + y_t)}{r^2} \leq A(\sigma_{ati} + \sigma_{act}) + \frac{M_g}{r^2}(y_b + y_t)$$

But $y_b + y_t = h$, so the above equation reduces to

$$P_i e \leq (\sigma_{ati} + \sigma_{aci})\, I/h + M_g \tag{3.6.5}$$

Similarly the Ineqs. 3.6.3 and 3.6.4 give

$$-\eta P_i e \leq (\sigma_{ace} + \sigma_{ate})\, I/h - M_w$$

or

$$\eta P_i e = P_e e \geq M_w - (\sigma_{ace} + \sigma_{ate})\, I/h \tag{3.6.6}$$

All the quantities on the right hand side of Ineqs. 3.6.5 and 3.6.6 are known while P_i and e on the left hand side are the two unknowns. Assuming a limiting eccentricity (e_0) for a critical section, the value of P_i can be determined from the two equations and it is given by

$$[M_w - (\sigma_{ace} + \sigma_{ate})\ I/h]/e_0\eta \leq P_i \leq [M_g + (\sigma_{aci} + \sigma_{act})\ I/h]/e_0 \qquad (3.6.7)$$

Once the value of P_i is determined from Ineq. 3.6.7 the cable profile along the beam can be determined from Ineqs. 3.6.5 and 3.6.6. The two Inequalities can be rearranged as

$$e \leq [M_g + (\sigma_{ati} + \sigma_{act})\ I/h]\ P_i \qquad (3.6.8)$$
$$e \geq [M_w - (\sigma_{ace} + \sigma_{ate})\ I/h]\ \eta P_i \qquad (3.6.9)$$

Example 3.6.1   A simply supported beam of span 4 m is subjected to a UDL of 6000 N/m. Determine the prestressing force and eccentricity of a cable after assuming an approximate section. *M*-42 concrete is used in the beam and f_p = 1600 MPa.

Solution

(i) Permissible stresses (in N/cm^2)

Assuming the beam is pre-tensioned the permissible stresses are:

σ_{act} = 0.5 f_{ck} = 0.5 × 4200 = 2100

σ_{ati} = 0.1 ×2100 = 210 or 140, select 140

σ_{ace} = 0.4 × 4200 = 1680

σ_{ate} = 140

(ii) Selection of a section

Depth of a beam for a simply supported span can be assumed in the range of *L*/20.

Let *h* = *L*/20 = 400/20 = 20 cm

Select a cross section as shown in Figure 3.6.1. Dimensions of the section are same as that of the box section in example 3.3.1 except the section is made as *I*-section.

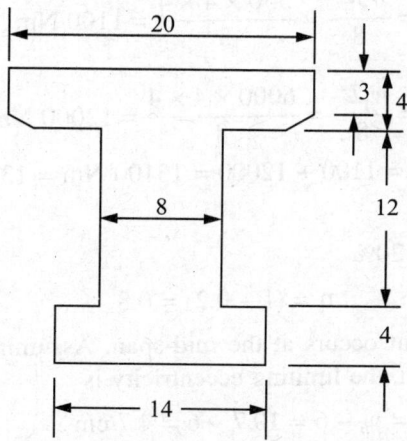

Figure 3.6.1   Cross section of beam–Example 3.6.1
(Dimensions are in cm).

The properties of the section are computed in table 3.6.1 considering flanges and web separately.

$$y_t = \frac{\Sigma A_i d_i}{\Sigma A_i} = \frac{2122}{229} = 9.3 \text{ cm}$$

$$y_b = h - y_t = 10.7 \text{ cm}$$

TABLE 3.6.1 Properties of the section (in cm)

Section	Area A_i	Distance from top	Moment area	y_1	I_{ci}	$A_i y_i^2$
		d_1	$A_i d_i$	$d_i - y_t$		
	(cm²)	(cm)	(cm³)	(cm)	(cm⁴)	(cm⁴)
Top flange	77	2	154	−7.3	100	4100
Bottom flange	56	18	1008	8.7	75	4250
Web	96	10	960	0.7	1152	47
Total	229		2122		1327	8397

$$I = \Sigma I_{ct} + \Sigma A_i y^2 = 1327 + 8397 = 9724 \text{ cm}^3$$
$$I/h = 9724/20 = 486.2 \text{ cm}^3$$

(iii) Moment calculations

$$w_g = A_\gamma = \frac{229 \times 25000}{10,000} = 550 \text{ N/m}$$

where γ = density of the concrete = 25000 N/m³

$$M_g = \frac{w_g L^2}{8} = \frac{550 \times 4 \times 4}{8} = 1100 \text{ Nm}$$

$$M_1 = \frac{w_1 L^2}{8} = \frac{6000 \times 4 \times 4}{8} = 12000 \text{ Nm}$$

$$M_w = M_g + M_1 = 1100 + 12000 = 13100 \text{ Nm} = 1310000 \text{ Ncm}$$

(iv) Prestressing force

Let the loss of prestress = 20%

$$\eta = (1 - 0.2) = 0.8$$

Maximum bending moment occurs at the mid-span. Assuming the centre of steel fiom bottom fibre is situated at 6 cm, the limiting eccentricity is

$$e_0 = y_b - 6 = 10.7 - 6 = 4.7 \text{ cm}$$

$$\frac{M_g}{e_0} = \frac{110000}{4.7} = 23600 \text{ N}$$

$$\frac{M_w}{\eta e_0} = \frac{1310000}{0.8 \times 4.7} = 349000 \text{ N}$$

$$\sigma_{ati} + \sigma_{aci} = 140 + 2100 = 2240 \text{ N/cm}^2$$

$$\sigma_{ace} + \sigma_{ate} = 1680 + 140 = 1820 \text{ N/cm}^2$$

From ineq. 3.6.7

$$P_i \leq \frac{M_g}{e_0} + \frac{(\sigma_{ati} + \sigma_{aci})}{e_0}\left(\frac{I}{h}\right) = 255.600 \text{ N}$$

and

$$P_i \geq \frac{M_w}{\eta e_0} - \frac{(\sigma_{ace} + \sigma_{ate})}{\eta e_0}\left(\frac{I}{h}\right) = 159,000 \text{ N}$$

Select P_i approximately equal to lower limiting value of the governing conditions.

Let $P_i = 160000$ N

Selecting 5 mm high tensile wire, the permissible force at transfer in each wire

$$= \frac{\pi(0.25)}{4}(0.7 f_p) = \frac{\pi}{16}(0.7)(160000) = 22600 \text{ N}$$

Number of wires required $= \dfrac{160000}{22600} \simeq 8$

Select 8 wires and the value of P_i in the wire is

$$P_i = 8(22600) = 180,800 \text{ N}$$

Let $e = 4.5$ cm

(v) Check for permissible stresses (all stresses are in N/cm^2)

$$\frac{P_i}{A} = \frac{180800}{229} = 790$$

$$\frac{P_i e}{I} = \frac{180800 \times 4.7}{9724} = 83$$

$$\frac{M_g}{I} = \frac{110000}{9724} = 11.3$$

$$\frac{M_g}{I} = \frac{1310000}{9724} = 135$$

Substitution of the above quantities in Ineqs. 3.3.5 to 3.3.8 gives

$$\sigma_{ti} = -\frac{P_i}{A} + \frac{P_i e y_t}{I} - \frac{M_g y_t}{I}$$
$$= -790 + 9.3 \times 83 - 11.3 \times 9.3 = -123 < \sigma_{ati}$$

$$\sigma_{ci} = -\frac{P_i}{A} + \frac{P_i e y_b}{I} - \frac{M_g y_b}{I}$$
$$= -790 - 83 \times 10.7 + 11.3 \times 10.7 = 1570 < \sigma_{aci}$$

$$\sigma_{ce} = -\frac{\eta P_i}{A} - \frac{\eta P_i e y_t}{I} + \frac{M_w y_t}{I}$$
$$= 632 - 0.8 \times 83 \times 9.3 + 135 \times 9.3 = 1270 < \sigma_{ace}$$

$$\sigma_{te} = -\frac{\eta P_i}{A} - \frac{\eta P_i e y_b}{I} + \frac{M_w y_b}{I}$$
$$= -632 - 0.8 \times 83 \times 10.7 + 135 \times 10.7$$
$$= 130 \leq \sigma_{ate}$$

All the stresses are within the permissible limits so the design is acceptable. Provide a parabolic cable profile with eccentricities at both ends as zero and 4.5 cm at span.

Comments: Ineq. 3.6.7 is an important one in the design. An ideal section is the one in which the lower and upper bounds of P_t from Ineq. 3.6.7 are close to each other. In case the section selected is too small for the external load, then the lower bound of P_i is likely to be more than the upper bound. In such a case a bigger section must be selected. If the margin between the lower and upper bounds of P_i is large then the section selected is too big, so it must be reduced.

Example 3.6.2 Check whether the section selected in example 3.6.1 is suitable or not, if the external load acting on the beair of example is 8000 N/m instead of 6000 N/m.

Solution (Units are in N and cm unless mentioned)

$$M_i = \frac{w_i L^2}{8} = \frac{8000(4)(400)}{8} = 1600,000 \text{ N cm}$$

$$M_w = 1600,000 + 110,000 = 1710,000 \text{ N cm}$$

$$\frac{M_w}{\eta e_0} = \frac{1710,000}{0.8 \times 4.7} = 455,000 \text{ N}$$

The upper bound remains same as that in example 3.6.1 and it is
$$P_i \leq 255600 \text{ N}$$

The lower bound is

$$P_i \geq \frac{M_w}{\eta e_0} - \frac{(\sigma_{ace} + \sigma_{ate})}{\eta e_0}\left(\frac{I}{h}\right) = 455000 - 190000 = 265000 \text{ N}$$

The lower bound of P_i is 265000 N and upper bound is 255600 N which simply means that there cannot be any value of P_i which will satisfy the governing equation. The section selected is small, so increase the area of the section by about 10%.

Example 3.6.3 Check whether the section selected in example 3.6.1 is suitable in case the live load is only 1000 N/m.

Solution

$$M_i = \frac{w_i L^2}{8} = \frac{1000\,(1600)}{8} = 200,000 \text{ N cm}$$

$$M_w = M_g + M_i = 200,000 + 110,000 = 310,000 \text{ N cm}$$

$$\frac{M_w}{\eta e_0} = \frac{310000}{0.8 \times 4.7} = 48,500 \text{ N}$$

The lower bound of P_i is

$$P_t \geq \frac{M_w}{\eta e_0} - \frac{(\sigma_{ace} + \sigma_{ate})}{\eta e_0}\left(\frac{I}{h}\right) = 48500 - 190000 = -141500 \text{ N}$$

The upper bound is same at the previous example namely 255600 N. Even though many values exist between the two bounds, the lower bound value of P_i works out to be negative which implies the section is too big for the load.

The main problem in the design of a beam is centred around the determination of the cross section. Once the section is either assumed or solved, the evaluation of prestressing force is relatively simple. It was discussed in the previous examples that the prestressing force close to the lower bound is likely to be economical. A possibility can be approached in which the upper and lower bounds of the prestressing force tend to a single value. Such a situation can be obtained by equating the two bounds given in Ineq. 3.6.7. It gives

$$(M_w - (\sigma_{ace} + \sigma_{ace})\,I/h) = (M_g + (\sigma_{ati} + \sigma_{aci})\,I/h)\,\eta$$

Rearrangement of the above equation gives

$$\frac{I}{h}(\sigma_{ace} + \sigma_{ate}) + (\sigma_{ati} + \sigma_{act})\,\eta = M_w - M_g\eta$$

or

$$I = \frac{(M_w - M_g\eta)\,h}{\sigma_{ace} + \sigma_{ate} + \eta\,(\sigma_{ati} + \sigma_{aci})} \tag{3.6.10}$$

A beginner in the design is usually posed with the question of how to select an initial section which is efficient and economical. Unlike in the design of *RCC* and steel sections, higher the area of the initial section, higher would be the prestressed force required in the case of prestressed concrete beams. Therefore, the designer must be more conscious in the selection of the section. Eq. 3.6.10 gives the lower bound value of moment of inertia of prestressed concrete Yearns. The depth of the beam can normally be assumed in the ranges given in Table 3.6.2

In case of *T*or flanged beams, the area and moment of inertia of the section can be expressed as

$$A = c_a b_t h \tag{3.6.11}$$

$$I = c_i b_t h^3$$

124

TABLE 3.6.2 Recommended depths of beams (L = span, h = depth)

Type of beam	L/h
Cantilever	8* to 14
Simply supported	18 to 30
Continuous beams	30 to 40

* The lower value to be chosen in case of longer spans or for heavier loads.

in which c_a and c_t are non-dimensionalised coefficients in the range of 0.3 to 0.6 for c_a and 0.02 to 0.05 for c_i. Eq. 3.6.10 along with the coefficients listed above will be able to establish rationalised sectional properties. The use of the expressions is illustrated in the following solved example of design of a beam.

Example 3.6.4 A simply supported beam of clear span of 20 m is subjected to a uniformly distributed load of 8000 N/m. Design the beam as a post-tensioned prestressed concrete one using M 35 concrete. The beam is T-shaped with flange width as 100 cm.

Let the depth of the beam = span/20 = 1 m

Effective span = 20 + 1 = 21 m

A preliminary guess of the weight of the beam is made for the purpose of computation of bending moment caused by the self weight. Assuming an average thickness of flange and the web as 12 cm, the area of the section is

$$A = 100 \times 12 + 88 \times 12 = 2256 \text{ cm}^2$$

Let $\quad A = 2500 \text{ cm}^2,$

then the self weight

$$w_g = 2500(0.250) = 62.5 \text{ N/cm}$$
$$= 6250 \text{ N/m}$$

$$M_g = \frac{w_g L^2}{8} = \frac{6250(21)(21)}{8} = 344,500 \text{ Nm}$$

$$M = \frac{w_g L^2}{8} = \frac{8000(21)(21)}{8} = 441,000 \text{ Nm}$$

$$M_w = M_g + M_i$$

Let $\quad \eta = 0.8$ then

$$M_w - \eta M_g = 785500 - 275600 = 509\,900 \text{ Nm}$$

The full permissible stress in N/cm^2 are:

$$\sigma_{aci} = 1750 \quad \sigma_{ati} = 100$$

$$\sigma_{ace} = 1400, \sigma_{ate} = 100$$

$$\sigma_{ace} + \sigma_{ate} + \gamma_i (\sigma_{acl} + \sigma_{ail})$$

$$= 1500 + 0.8 \,(1850) = 2980 \text{ N/cm}^2$$

From Eq. 3.6.10

$$I = \frac{(M_w - \eta M_g)\, h}{\sigma_{ace} + \sigma_{ate} + \eta(\sigma_{act} + \sigma_{ati})}$$

$$= \frac{50,990,000(100)}{2980} = 1,711,070 \text{ cm}^4$$

The width of the flange (b_f) is given = 100 cm
The moment of inertia of T-beams is equal to

$$I = c_t \, b_f \, h^3$$

where c_i is in the range of 0.02 to 0.05.

Selecting $c_i = 0.03$

$$h^3 = \frac{I}{c_i \, b_f} = \frac{1711070}{(0.03)\,100} = 570,360$$

$$h = 82.9 \text{ cm}$$

Trial section with the following data:
The dimensions are listed below and the moment of inertia is computed in Table 3.6.3. (All dimensions in cm). Vide Figure 3.6.2.

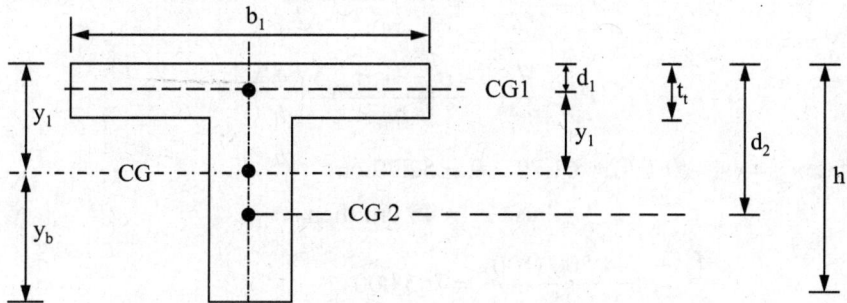

Figure 3.6.2 T-beam section.

$$b_f = 100, \, t_f = 12, \, h = 90$$
$$b_w = 15 \; h_w = 90 - 12 = 78$$
$$A_1 = 100\,(12) = 1200$$
$$A_2 = 15\,(78) = 1170$$

($i = 1, 2$ for flange and web respectively)

126

TABLE 3.6.3 Sectional properties (in cm) trial

Segment	A_i	d_i	$A_i d_i$	y_t	I_i	$A_i y_i^2$
Flange	1200	6	7200	−22.21	14,400	591,940
Web	1170	51	59670	22.79	593,190	607,679
Total	23710		66870		607,590	1199,619

$$y_i = \frac{\Sigma A_i d_i}{\Sigma A_i} = \frac{66870}{2370} = 28.21$$

$$y_h = h - y_t = 90 - 28.21 = 61.79 \text{ cm}$$

$$I = \Sigma I_i + \Sigma A_i y_i^2$$

$$= 607,590 + 1,199,619 = 1,807,209 \text{ cm}^4$$

Use $\qquad\qquad I = 1807\ 200 \text{ cm}^4$

The moment of inertia is adequate, therefore the section appears to be suitable. One must compute the actual stresses developed by the loads on the section. The value of M_g selected is in the range of the actual section.

Prestressing Force: The value of the prestressing force is limited to the following as per the earlier derivation

$$P_i \le \frac{M_g}{e_o} + \frac{(\sigma_{aci} + \sigma_{act})}{e_o} \frac{I}{h}$$

$$P_i \ge \frac{M_g}{\eta e_o} - \frac{(\sigma_{ace} + \sigma_{ate})}{\eta e_o} \left(\frac{I}{h}\right)$$

Approximate $e_o = (y_b - 0.1\ h) = 61.79 - 9 = 52.79$ cm

$$\text{use } e_o = 45 \text{ cm then}$$

$$\frac{M_g}{e_o} = \frac{344\ 500\ (100)}{45} = 765\ 000$$

$$\frac{M_w}{\eta e_o} = \frac{785500(100)}{0.8\ (45)} = 2181900$$

$$\frac{I}{h} = \frac{1807200}{50} = 20080$$

$$P_i \le \frac{M_g}{e_o} + \frac{\sigma_{aci} + \sigma_{act}}{e_o} \frac{I}{h}$$

$$\le 765000 + \frac{1850}{45}(20080) = 1590511 \text{ N}$$

$$P_i \ge \frac{M_w}{\eta e_o} - \frac{\sigma_{ace} + \sigma_{ace}}{\eta e_o}\left(\frac{I}{h}\right)$$

$$\ge 2181900 - \frac{1500}{0.8\,(45)}(20080) = 1345230 \text{ N}$$

Provide P_i on the lower side

Let $P_i = 1,300.000$ N $= 1300\ kN$

Check for stresses (the stresses are in N/cm^2)

$$\frac{P_i}{A} = \frac{1300000}{2370} = 548$$

$$\frac{\eta P_i}{A} = \frac{0.8\,(1300000)}{2370} = 439$$

$$\frac{P_i e}{I} = \frac{1300000\,(45)}{1807200} = 32.37$$

$$\frac{M_g}{I} = \frac{344500\,(100)}{1807200} = 19$$

$$\frac{M_w}{I} = \frac{785500\,(100)}{1807200} = 43$$

$$\sigma_{ti} = -\frac{P_i}{A} + \frac{P_i e y_t}{I} - \frac{M_g y_t}{I}$$
$$= -548 + (32.37 - 19)\,28.21$$
$$= -548 + 377 = -171$$

$$\sigma_{ci} = \frac{P_i}{A} + \left(\frac{P_i e}{I} - \frac{M_g}{I}\right)y_b$$
$$= 548 + (32.37 - 19)\,61.29$$
$$= 548 + 819 = 1367 \text{ N/cm}^2$$

$$\sigma_{ce} = \frac{\eta P_i}{A} - \left(\frac{\eta e}{I} - \frac{M_w}{I}\right)y_t$$
$$= 439 - (0.8(32.37) - 43)\,28.21 = 921 \text{ N/cm}^2$$

$$\sigma_{te} = \frac{-\eta P_i}{A} - \left(\frac{\eta P_i e}{I} - \frac{M_w}{I}\right)y_b$$
$$= -439 - (0.8\,(32.37) - 43)\,61.79$$
$$= -439 + 1057 = 618 \text{ N/cm}^2$$

128

σ_{te} is much higher than the permissible value whereas the other stresses are within the limits. This implies that the minimum prestressing force which primarily governs the transfer load condition is not adequate under working load condition. It may be noted here that the problem selected the moment of inertia as well as the prestressing force on the lower bound level. Even though it is optimum giving rise to least material use but is not exactly satisfying the governing equations. It is desirable to try with the least values and modify the design.

Second Trial: Increase the width of the web to 17 cm from 15 cm. The properties of the section are:

(All dimensions in cm)

$$b_f = 100, \ t_f = 12$$
$$b_w = 17, \ h_w = 88$$

The properties of the section are calculated in table 3.6.4.

TABLE 3.6.4 Sectional properties: second trial

Segment	A_i	d_i	$A_i d_i$	y	t_i	$A_i y_i^2$
Flange	1200	6	7200	−72.74	14,400	923,409
Web	1496	56	82776	22.26	965,418	741,279
Total	2696		89976		979,818	1664,688

$$y_t = d_t = y_t$$

$$y_t = \frac{\Sigma A_i d_i}{\Sigma A} = \frac{89976}{2696} = 33.37$$

$$y_b = h - y_t = 100 - 33.37 = 66.63 \text{ cm}$$

$$I = \Sigma(I_i + A_i y_i^2) = 2624{,}506 \text{ cm}^4$$

Use $\qquad P_i = 1650{,}000 \text{ N}$

$\qquad e = 47 \text{ cm}$

Then $\qquad \dfrac{P_i}{A} = \dfrac{1650000}{2696} = 612 \text{ N/cm}^2$

$$\frac{P_i e}{I} = \frac{1650000(47)}{2624506} = 29.5$$

$$M_g = \frac{2696\,(344500)}{2500} - 371500 \text{ Nm}$$

(Proportioned to the revised area of the cross section)

$$\frac{M_g}{I} = \frac{37150000}{2624506} = 14$$

$$M_w = M_g + M_i = 371400 + 441000 = 812400 \text{ Nm}$$

$$\frac{M_w}{I} = \frac{81240000}{2624506} = 31$$

Check for stresses (the stresses are computed in N/cm^2)

$$\sigma_{ti} = \frac{P_i}{A} + \left(\frac{P_i e}{I} - \frac{M_g}{I}\right) y_t$$

$$= -612 + (29.50 - 14)(33.37)$$

$$= -612 + 15.50(33.37) = -95$$

$$\sigma_{ci} = \frac{P_i}{A} + \left(\frac{P_i e - M_g}{I}\right) y_b$$

$$= 612 + 15.50(66.63) = 1645$$

$$\sigma_{ce} = \frac{\eta P_i}{A} - \left(\frac{\eta P_i e}{I} - \frac{M_w}{I}\right) y_t$$

$$= 490 - (24.16 - 31)(33.37) = 718$$

$$\sigma_t = \frac{-\eta P_t}{A}\left(\frac{\eta P_i e}{I} - \frac{M_w}{I}\right) y_t$$

$$= -490 - (24.16 - 31)(66.63) = -34$$

All stresses are within the permissible values.

3.7. Dimensionless Design Variables

Selection of cross section by intuition or by experience need not necessarily be the most economical section even though with experience one might tend to select a rational section. A method of design which is based on dimensionless variable was developed by Khachaturian (3.10) gives a rational approach to a design problem.

The governing eqs. 3.3.5 to 3.3.8 can be rearranged as

$$\frac{P_i}{A f_{ck}}\left[\frac{e}{h}\left(\frac{h}{r}\right)^2 \frac{1}{(y_b/y_t + 1)} - 1\right]$$

$$-\left(\frac{c_g \gamma L^2}{h f_{ck}}\right)\left(\frac{1}{(y_b/y_t + 1)}\right)\left(\frac{h}{r}\right)^2 = c_{ti} \qquad (3.7.1)$$

$$\frac{P_i}{Af_{ck}}\left[\frac{e}{h}\left(\frac{h}{r}\right)^2 \frac{y_b / y_t}{(y_b / y_t + 1)} + 1\right]$$

$$-\left(\frac{c_g \gamma L^2}{h f_{ck}}\right)\left(\frac{y_b / y_t}{(y_b / y_t + 1)}\right)\left(\frac{h}{r}\right)^2 = c_{ci} \tag{3.7.2}$$

$$-\eta \frac{P_i}{Af_{ck}}\left[\frac{e}{h}\left(\frac{h}{r}\right)^2 \frac{1}{(y_b / y_t + 1)} - 1\right]$$

$$+\left(\frac{c_g \gamma L^2}{h f_{ck}}\right)\left(\frac{(1 + M_a / M_g)}{(y_b / y_t + 1)}\right)\left(\frac{h}{r}\right)^2 = c_{ce} \tag{3.7.3}$$

$$-\eta \frac{P_t}{Af_{ck}}\left[\frac{e}{h}\left(\frac{h}{r}\right)^2 \frac{y_b / y_t}{(y_b / y_t + 1)} + 1\right]$$

$$+\left(\frac{c_g \gamma L^2}{h f_{ck}}\right)\left(\frac{y_b / y_t (1 + M_a / M_g)}{(y_b / y_t + 1)}\right)\left(\frac{h}{r}\right)^2 = c_{te} \tag{3.7.4}$$

In which

$$\sigma_{ti} = c_{ti} f_{ck}; \quad \sigma_{ci} = c_{ci} f_{ck}$$

$$\sigma_{te} = c_{te} f_{ck}; \quad \sigma_{ce} = c_{ce} f_{ck}$$

$$c_g = \frac{M_g}{\gamma A L^3}$$

These four equations can further be simplified by introducing non-dimensionalised parameters as suggested by Khachaturian (3.10) and the equations are given below:

$$m\left[\frac{\in}{\rho(1 + \Delta)} - 1\right] - \frac{c_g}{\omega_\rho(1 + \Delta)} = c_{ti} \tag{3.7.5}$$

$$m\left[\frac{\in \Delta}{\rho(1 + \Delta)} + 1\right] - \frac{\Delta c_g}{\rho \omega(1 + \Delta)} = c_{ci} \tag{3.7.6}$$

$$-\eta m\left[\frac{\in}{\rho(1 + \Delta)} - 1\right] + \frac{(1 + R) c_g}{\omega \rho(1 + \Delta)} = c_{ce} \tag{3.7.7}$$

$$-\eta m\left[\frac{\in \Delta}{\rho(1 + \Delta)} + 1\right] + \frac{(1 + R) \Delta c_g}{\rho \omega(1 + \Delta)} = c_{te} \tag{3.7.8}$$

where

$$hf_{ck} / \gamma L^2 = \omega \quad \ldots \quad \text{depth factor}$$
$$(r/h)^2 = \rho \quad \ldots \quad \text{efficiency factor}$$
$$y_b / y_t = \Delta \quad \ldots \quad \text{shape factor}$$
$$M_a / M_g = R \quad \ldots \quad \text{moment ratio}$$

$$\left. \right\} \text{section properties}$$

$$P_i / Af_{ck} = m \quad \ldots \quad \text{reinforcement ratio}$$
$$e/h = \in \quad \ldots \quad \text{eccentricity ratio}$$

$$\left. \right\} \text{steel properties}$$

Known and unknown quantities of the above four governing equations may be classified as

c_{ci}, c_{ti}, c_{ce} and c_{te} .... known quantities

$\rho, \omega, \Delta, \in, \eta, m$ and R .... unknown quantities

Out of the seven unknowns, η can reasonably be estimated and recomputed if necessary. With this assumption, there are six unknowns in the four equations which means that at least two of the nondimensionalised parameters must be estimated to get a unique solution. Arbitrary estimate of any of the two parameters is done based on practical and economic considerations.

To assume values of any two parameters, a constructive discussion is necessary. Four out of the six parameters (ω, ρ, Δ, R) are associated essentially with the property of the cross section and the other two parameters ($m, \in$) are associated with steel.

(a) *Depth factor* ($\omega = hf_{ck}/\gamma L^2$): The expression contains three known quantities (hf_{ck}, γ, L) and one unknown quantity h. Depth of the beam in many cases is controlled by architectural and site conditions. Even if it is not given, it can reasonably be assumed. The effect of the depth of the beam on economy is discussed later.

(b) *Flexural efficiency factor* ($\rho = r^2/h^2$): For a given depth, the efficiency of the section to carry bending moment increases with the increase of radius of gyration. The possible range of values of ρ are tabulated in table 3.7.1.

TABLE 3.7.1 Approximate limits of ρ

Section	Limits of ρ
All area concentrated at CGC	0
All area concentrated at the extreme fibres	0.25
Rectangular section	0.0833
I-section	0.10 to 0.16
T-section	0.08 to 0.11

Higher the value of ρ, higher is the flexural efficiency of the section.
The effect and mode of selection of efficiency ratio for design is discussed later.

132

(c) *Shape factor* ($\Delta = y_b/y_t$) :   The range of shape factor for I-section is large. For any symmetric sections the value of shape factor is unity. Approximate range of shape factors for practical sections is given in table 3.7.2.

TABLE 3.7.2   Limits of Δ

Section	Limits of Δ
Symmetric sections (symmetry about centroidal axis)	1.0
I-section–top flange heavy	1.0 to 2.0
I-section–bottom flange heavy	0.6 to 10
T-section	1.2 to 2.0
Inverted T	0.6 to 0.9

For heavy live loads and for long spans, the top flange has to be heavier, in which case Δ should be chosen between 1.2 and 2.0. The effect Δ on the limiting equations is considerable, so it is desirable that this value is not assumed to start with the design.

(d) *Moment ratio* $\left(R = \dfrac{M_a}{M_g} = \dfrac{M_g + M_i}{M_g} \right)$:   It is difficult to fix any limit to the range of

the value of R. As the length of the span increases, the ratio R decreases whereas M_g is indirectly a function of M_a. The value of R varies from 6.0 for short spans to 0.3 for long spans. It should not be assumed but should be solved from the equations.

(e) *Reinforcement ratio* ($m = P_i/Af_{ck}$): P_i/A may be defined as average stress. The reinforcement ratio is simply the ratio of average stress to strength of concrete. For all practical purposes, it varies from 0.15 to 0.26. The value of m should be computed from the governing equations for economical section.

(f) *Eccentricity ratio* ($\in = e/h$):   Eccentricity ratio is a measure of the effective utilisation of the prestressing force. As the value of $\in$ increases, the required prestressing force decreases. $\in$ can be taken as high as possible but it has to be bounded by practical considerations such as cover. The two hypothetical limiting cases of $\in$ are: the lower kern point, and the extreme of the lower fibre. These limits of eccentricity ratio vary from 0.25 to 0.60. The extreme value of $\in$ is limited by cover distance. If d' is the minimum cover required then

$$\in_{\text{limit}} = \frac{y_b - d'}{h} = \frac{y_b}{y_b + y_t} - \frac{d'}{h} = \left(\frac{\Delta}{\Delta + 1} - \frac{d'}{h} \right) \tag{3.7.9}$$

If the value of d'/h is around 0.1, then

$$\in_{\text{limit}} = \left(\frac{\Delta}{\Delta + 1} \right) - 0.1.$$

This value varies from 0.25 to 0.40 for heavy bottom flange and 0.40 to 0.6 for heavy top flange I-sections.

Table 3.7.3 gives approximate practical limits of various non-dimensionalised parameters.

TABLE 3.7.3   Limiting values of non-dimensionalised variable

Section	h	ρ	Δ	R	m	ϵ
I-section with	L/15	0.11	1.0	0.3	0.15	0.40
heavier top	to	to	to	to	to	to
flange	L/25	6.16	1.8	2.0	0.25	0.60
				(long span)		
I-section with	L/15	0.10	0.6	1.0	0.15	0.35
heavier bottom	to	to	to	to	to	to
flange	L/25	0.13	1.0	5.0	0.30	0.40
				(short span)		
T-section	L/15	0.08	1.2	0.5	0.15	
	to	to	to	to	to	
	L/25	0.11	1.8	2.5	0.30	

3.8.   Solution of the Equations

In many practical problems, values of ω and ρ may be assumed and then four unknowns with the four equations can be solved for a given set of values of $c's$ subject to the practical constant. The four equations are rearranged below:

Multiply eq. 3.7.6 by η and add to eq. 3.7.8

$$\frac{c_g \Delta (1 + R - \eta)}{\rho \omega (1 + \Delta)} = \eta c_{ci} + c_{te} \tag{3.8.1}$$

Multiply eq. 3.7.5 by η and add to eq. 3.7.7

$$\frac{c_g (1 + R - \eta)}{\rho \omega (1 + \Delta)} = \eta c_{ti} + c_{ce} \tag{3.8.2}$$

Eq. 3.8.1 and eq. 3.8.2 yield

$$\Delta = \frac{\eta c_{ei} + c_{te}}{\eta c_{ti} + c_{ce}} \tag{3.8.3}$$

Addition of eqs. 3.8.1 and 3.8.2 yields

$$\frac{c_g (1 + R - \eta)}{\rho \omega (1 + \Delta)} (1 + \Delta) = \eta (c_{ci} + c_{ti}) + (c_{te} + c_{ce})$$

or $$R = \rho \omega [c_{te} + c_{ce}) + \eta(c_{ci} + c_{ti})] c_g - (1 - \eta) \tag{3.8.4}$$

Dividing eq. 3.7.6 by Δ and subtracting from eq. 3.7.5 gives

$$m \left(-1 - \frac{1}{\Delta} \right) = c_{ti} - \frac{c_{ei}}{\Delta}$$

or
$$m = \frac{c_{et} - \Delta c_{ti}}{1 + \Delta} \tag{3.8.5}$$

Similarly from eq. 3.7.7 and 3.7.8

$$m = \frac{\Delta c_{ce} - c_{te}}{\eta(1 + \Delta)} \tag{3.8.6}$$

Substituting the value of Δ in eq. 3.8.5 yields

$$m = \frac{c_{ei} c_{ce} - c_{ti} c_{te}}{\eta(c_{ti} + c_{ci}) + (c_{te} + c_{ce})} \tag{3.8.7}$$

Adding eqs. 3.7.5 and 3.7.6 gives

$$\in = \left[\rho (c_{ti} + c_{ei}) + \frac{c_g}{\omega} \right] \left[\frac{(c_{te} + c_{ce}) + \eta(c_{ti} + c_{ci})}{(c_{ci} c_{ce} - c_{ti} c_{te})} \right] \tag{3.8.8}$$

Eqs. 3.8.3 and 3.8.4 define the properties of the section and eqs. 3.8.7 and 3.8.8 define the prestressing steel and its location. An interesting aspect about the equations is that eqs. 3.8.3 and 3.8.7 are functions of stress coefficients and effectiveness of prestressing which can always be estimated reasonably well.

Rearrangement of the above equations are presented below. These equations will be used for discussion on the economy of the section.

$$\Delta = \frac{\eta c_{ct} + c_{te}}{\eta c_{ti} + c_{ce}} \tag{3.8.3}$$

$$R = \frac{\rho \omega}{c_g} (c_{te} + \eta c_{ci}) \left(\frac{1}{\Delta} + 1 \right) - (1 - \eta) \tag{3.8.9}$$

$$= \frac{\rho \omega}{c_g} (\eta c_{ti} + c_{ce}) (\Delta + 1) - (1 - \eta) \tag{3.8.10}$$

$$m = \frac{c_{ci} - \Delta c_{ti}}{(1 + \Delta)} = \frac{\Delta c_{ce} - c_{te}}{\eta(1 + \Delta)}$$

$$\in = \frac{1 + \Delta}{c_{ci} - \Delta c_{ti}} \left[\rho(c_{ti} + c_{ci}) + \frac{c_g}{\omega} \right] \tag{3.8.11}$$

Economic design of a beam may be achieved by obtaining minimum concrete and steel with cost factor built in. But in most cases, the minimum weight design is likely to give overall economy. Hence minimum weight design is discussed in this article.

Minimum weight will give minimum self weight moment and consequently maximum value of the moment ratio ($R = M_d/M_g$). R will be maximum for maximum values of the stress coefficients for any given values of ρ and ω i.e., R is maximum for

$$c_{ci} = c_{act}, \, c_{ti} = c_{ati}$$

$$c_{ce} = c_{ace}, \, c_{ate} = c_{ate}$$

If area of steel is minimum, the eccentricity will be maximum and if area of steel is maximum, the eccentricity will be minimum. For long spans, the eccentricity computed from the solution of simultaneous equations is likely to fall beyond the bottom fibre. But the cable has to be kept within the cross section. The shape factor Δ could be so adjusted that the eccentricity of the cable is within the limits of the cross section. A marginal variation of some of the permissible stress coefficients would enable to modify the shape of the cross section. For example, by permitting stress at transfer less than the permissible stress, the shape of the cross section will adjust such that there will be more room available for accommodation of the cable.

The permissible stresses fix the shape of the cross section. The range of the Δ as per the permissible stresses of the Indian code of practice is given in table 1 of appendix A.

3.9. Properties of Idealised Sections

Most shapes of sections in actual construction are of unsymmetrical I-section. Some typical prestressed concrete sections are shown in figures 3.9.1, 3.9.2 and 3.9.3. Present section discusses sectional properties of idealised sections for the sake of simplicity.

Properties of cross section of an idealised unsymmetrical I-section.

$$A = (b_t t_i + b_b t_b + h_w b_w)$$

$$= b_t t_t + b_b t_b + (h - t_i - t_b)\, b_w \tag{3.9.1}$$

$$= (b_t h)[c_t + c_f c_b + (1 - c_t - c_b)\, c_w]$$

$$= c_a (b_t h) \tag{3.9.2}$$

where

$$c_t = t_i/h, \ c_b = t_b/h$$

$$c_w = b_w/b_t, \ c_f = b_b/b_t \tag{3.9.3}$$

$$y_b = (1/A)\left[b_t t_t \left(h - \frac{t_i}{2} \right) + (h - t_i - t_b)\, b_w \left(\frac{h - t_i + t_b}{2} \right) + \frac{b_b t_b^2}{2} \right]$$

$$= \left(\frac{b_t h}{A} \right) h \left[c_t \left(1 - \frac{c_t}{2} \right) + c_w\, (1 - c_t - c_b) \left(\frac{1 - c_t + c_b}{2} \right) + \frac{c_f c_b^2}{2} \right]$$

$$= \frac{h \left[c_t \left(1 - \dfrac{c_t}{2} \right) + c_w\, (1 - c_t - c_b) \left(\dfrac{1 - c_t + c_b}{2} \right) = c_f\, \dfrac{c_b^2}{2} \right]}{[c_t + c_f c_b + (1 - c_t - c_b)\, c_w]}$$

$$= c_{yb} h \tag{3.9.4}$$

$$y_t = h - y_b$$

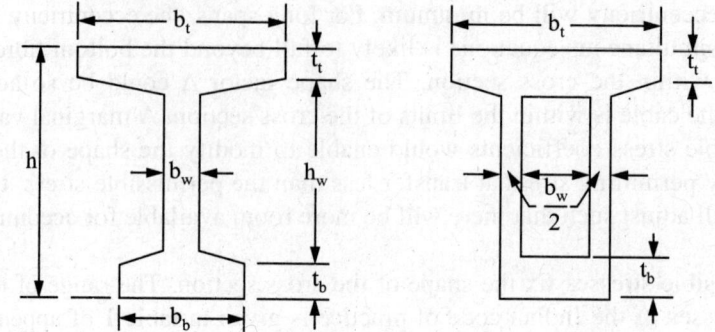

Figure 3.9.1 Typical I-section for long spans.

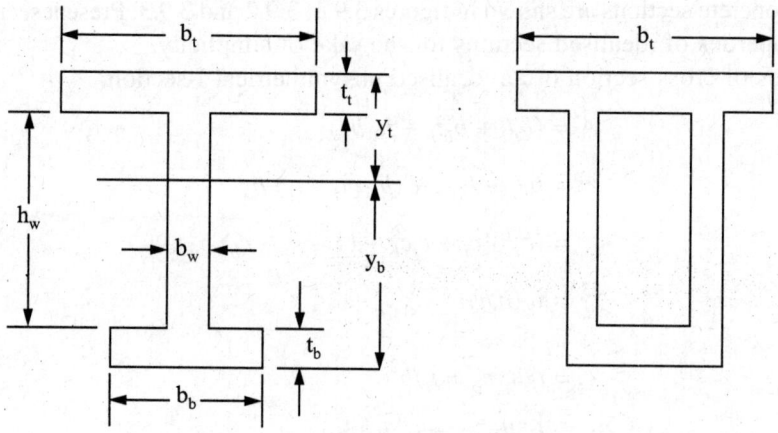

Figure 3.9.2 Idealised form of the I-section.

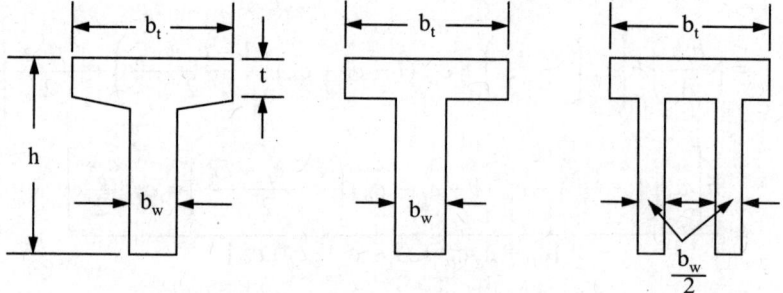

Figure 3.9.3 Typical T-section.

$$= \frac{h\left[c_b\left(1-\frac{c_b}{2}\right)c_f + c_w\left(1-c_t-c_b\right)\left(\frac{1-c_b+c_t}{2}\right)+\frac{c_t^2}{2}\right]}{\left[c_t + c_f c_b + (1-c_t-c_b)\,c_w\right]}$$

$$= c_{yt}h \tag{3.9.5}$$

$$I = b_t h^3 \left[\frac{1}{12}\{c_t^3 + c_i c_{b^3} + c_w\,(1-c_t-c_b)^3\} + c_t\left(c_{yt}-\frac{c_t}{2}\right)^2 \right.$$

$$\left. + c_w\,(1-c_t-c_b)\left(\frac{1+c_t-c_b}{2}-c_{yt}\right)^2 + c_f c_b\left(c_{yb}-\frac{c_b}{2}\right)^2\right]$$

$$= c_t b_t h^3 \tag{3.9.6}$$

where

c_a, c_t, c_{yb} and c_{yt} are non-dimensional coefficients as given in the above expressions.

$$\Delta = \frac{y_b}{y_t} = \frac{c_t\left(1-\frac{c_t}{2}\right)+c_w\,(1-c_t-c_b)\left(\frac{1-c_t+c_b}{2}\right)+c_f\,\frac{c_{b2}}{2}}{c_f c_b\left(1-\frac{c_b}{2}\right)+c_w\,(1-c_t-c_b)\left(\frac{1-c_b+c_t}{2}\right)+\frac{c_t^2}{2}} \tag{3.9.7}$$

$$\rho = \frac{c_t}{c_a} \tag{3.9.8}$$

The expressions listed in eqs. 3.9.1 to 3.9.8 are useful in studying the various relationships of the section. Sectional properties for various values of c_t, c_b, c_w, c_f are given in table 2, appendix A.

3.10. Design Procedure Based on Flexure

A discussion on the analysis of the beam has been made in the previous articles. It is desirable to consolidate the procedure for designing a prestressed concrete beam. A recommended design procedure is:

(i)   Select the materials to be used and list the allowable stresses and stress coefficients as per the code requirements.

(ii) Select ω and ρ in the range as described in the previous articles. Value of ω is generally obtained by other criteria and value of ρ is selected as high as possible.

(iii) After selecting the stress coefficients, solve the eqs. 3.8.3, 3.8.3, 3.8.7 and 3.8.8 for Δ, R, m and $\in$.

(iv) Select the section for the values of ρ and Δ, from table 2 in appendix A.

(v)   Check the eccentricity obtained from the governing equations whether it is within the practical limitations of clear cover.

138

(vi) If the cover is not enough, change the stress coefficients or efficiency factor to meet the requirements.

(vii) Select a set of non-dimensionalised design variables and check whether the stresses are within the limits.

(viii) If all the three requirements are satisfied, proceed to determine the losses of prestress. If the requirements are not satisfied, select another section to suit the requirements.

(ix) Compute the losses in prestress if necessary.

(x) Check the assumed value of η. If the computed value of η does not check with the estimated value, recheck the design with new value of η.

(xi) Fix the profile of the cable.

(**Note:** Design for shear, anchorage, bond and other reauirements are discussed later.)

Design example 3.10.1: Design of a post-tensioned beam for the following data:

$$L = 50 \text{ m}, w_1 = 20000 \text{ N/m}, f_{ck} = 4500 \text{ N/cm}^2,$$

$$f_p = 1500 \text{ MPa} = 150000 \text{ N/cm}^2.$$

Solution

Method of designing a beam is illustrated through series of steps.

(i) *Specifications*

Maximum permissible stresses for loads likely to increase are taken as:

$c_{ati} = 0.02$, $c_{aci} = 0.46$, $c_{ace} = 0.31$ and cate = 0.02

$s_{st} = 0.7 (150000) = 105000 \text{ N/cm}^2$

$s_{se} = 0.6 (150000) = 90000 \text{ N/cm}^2$

$E_c = 38.2 \text{ GPa} = 3820000 \text{ N/cm}^2$

(ii) *Selection of depth of beam and efficiency factor*

For simply supported beams, the depth of the beam can be selected in the range of $L/20$.

$h \approx L/20 = 250 \text{ cm}$

Let $h = 240 \text{ cm}$, $\rho = 0.13$ and $\eta = 0.8$

For simply supported beam with *UDL*, the moment coefficient $c_g = 1/8$

(iii) *Design variables*

$$\Delta' = \frac{\eta c_{aci} + c_{ate}}{\eta c_{ati} + c_{ace}} = \frac{0.8 (0.46) + 0.02}{0.8 (0.02) + 0.31} = \frac{0.388}{0.326} = 1.2$$

$$\omega = \frac{hf_{ck}}{\gamma L^2} = \frac{240\,(4500)}{24000\,(25)} = 1.8$$

$$\epsilon_{teq.} = \frac{1+\Delta'}{c_{act} - \Delta' c_{ati}}\left[\rho(c_{ati} + c_{act}) + \frac{1}{8\omega}\right]$$

$$= \left(\frac{2.2}{0.46 - 0.024}\right)[0.13 \times 0.48 + 0.07] = 0.665$$

$$\epsilon_{\lim} = \frac{y_b = d'}{h}\left(\frac{\Delta'}{1+\Delta'} - 0.1\right) = 0.445$$

The eccentricity required ($\epsilon_{req.}$) is much larger than the eccentricity limit ($\epsilon_{\lim.}$), therefore the shape of the section has to be altered such that the depth of the beam below CA is more and hence Δ must be increased.

Let $c_{ti} = -0.08$ while the remaining stresses are equal to the maximum permissible values.

$$\Delta = \frac{\eta c_{aci} + c_{ate}}{\eta c_{ti} + c_{ace}} = \frac{0.388}{0.236} = 1.64$$

$$\epsilon_{req.} = \frac{1+\Delta}{c_{aci} - \Delta c_{cti}}\left[\rho(c_{ti} + c_{act}) + \frac{1}{8\omega}\right]$$

$$= \frac{2.64}{0.46 + 0.13}[(0.13 \times 0.38) + 0.07] = 0.54$$

$$\epsilon_{\lim.} = \frac{\Delta}{1+\Delta} - 0.1 = \frac{1.64}{2.64} - 0.1 = 0.52$$

(iv) *Selection of section*

It is possible to adjust the eccentricity for $\rho = 0.13$ and $\Delta = 1.64$. So a section whose values of ρ and Δ are nearest to the above values can be selected from table 2 of appendix A. Properties of such a section from table 2 of appendix A are:

$$c_f = 0.5, \qquad c_w = 1.12, \quad c_t = 0.12 \qquad \text{and } c_b = 0.08$$
$$c_a = 0.256, \qquad c_i = 0.032\text{m} \; c_{yb} = 0.626, \qquad c_{yt} = 0.374$$
$$\rho = 0.128 \qquad \text{and } \Delta = 1.68$$

Substitution of the above quantities in eqs. 3.8.10 and 3.8.5 one gets

$$R = 8\rho\omega(c_{ate} + \eta c_{act})\left(\frac{1+\Delta}{\Delta}\right) - (1 - \eta)$$

$$= 8(0.125)\,(1.68)\,(0.02 + 0.368)\,(2.68)/1.68 - 0.2 = 0.95$$

$$m = \frac{c_{aci} + \Delta c_{ti}}{1 + \Delta} = \frac{0.46 + 0.134}{2.68} = 0.2222$$

Select $\epsilon = 0.53$, $R = 0.95$ and $m = 0.22$

(v) *Check for permissible stresses*

Actual stress produced in the cross section can easily be found by using the non-dimensionalised design variables eqs. 3.7.5 to 3.7.8.

$$c_{ti} = m\left[\frac{\in}{\rho(1+\Delta)} - 1\right] - \frac{1}{8\rho\omega\,(1+\Delta)}$$

$$= 0.22\left[\frac{0.53}{0.1280 \times 2.68} - 1\right] - \frac{1}{8 \times 0.128 \times 1.8 \times 2.68}$$

$$= 0.22(1.55 - 1) - 0.204 = -0.118 \le c_{ati} \text{ (permissible)}$$

$$c_{ci} = m\left[\frac{\in \Delta}{\rho(1+\Delta)} + 1\right] - \frac{\Delta}{8\rho\omega\,(1+\Delta)}$$

$$= 0.22\,(1.55 \times 1.68 + 1) - 1.68 \times 0.204$$

$$= 0.22\,(2.6 + 1) - 0.343 = 0.8 - 0.343 = 0.457$$

$$\le c_{aci} = 0.46 \text{ (permissible)}$$

$$c_{ci} = -m\left[\frac{\in \Delta}{\rho(1+\Delta)} + 1\right] - \frac{\Delta(1 + R}{8\rho\omega\,(1+\Delta)}$$

$$= -0.8 \times 0.122 + 0.204 \times 1.95 = 0.3 \le c_{ace} = 0.31 \text{ (permissible)}$$

$$c_{te} = -\eta m\left[\frac{\in \Delta}{\rho(1+\Delta)} + 1\right] + \frac{\Delta(1 + R)}{8\rho\omega(1+\Delta)}$$

$$= -0.64 + 0.66 = 0.02 \le c_{ate} = 0.02 \text{ (permissible)}$$

The actual stresses in the beam are less than the permissible stresses, so the section can be selected based on the non-dimensionalised design variables.

If the actual stresses are on margin, then actual value of η should be computed for the particular section and the stresses be checked with the actual value of η.

Loss of prestress for calculation of η:

(i) *Elastic shortening:* A mean elastic shortening could be obtained by totalling the elastic strains at steel level over the length and dividing the total elongation by the length. Due to symmetry of the beam, only half the length of the beam is considered and the loss of prestress is computed.

Let $\quad \in_c$ = mean elastic strain

An approximate elastic strain may also be obtained without going through the integration.

At $x = L/2$

$$\in_c = \frac{c_{ci}\,f_{ck}}{E_c}\left(\frac{e}{y_b}\right) = \frac{c_{ci}\,f_{ck}\,(1+\Delta)\in}{E_c\,\Delta} = 46 \times 10^{-5}$$

At $x = 0$

$$\in_c = \frac{P_i}{AE_c} = \frac{m\,f_{ck}}{E_c} = 26 \times 10^{-5}$$

Taking a mean value of the above two

$$\in_c = \frac{0.00046 + 0.000260}{2} = 0.00036$$

Using $\in_c = 0.00036$, since the cables are strained in stages, the loss of strain due to elastic shortening is $\in_c/2 = 0.00018$

(ii) Shrinkage strain $= \dfrac{3 \times 10^{-4}}{\log_{10} (T + 2)}$

Taking transfer at 28 days, the loss due to shrinkage is given by

$$= \frac{0.0003}{\log_{10}(30)} = 0.0002$$

(iii) Creep in concrete for a creep coefficient of 2.5 $= (2.5 - 1) \in_c = 0.00054$

(iv) Creep in steel (3%) $= 0.005 \times 0.03 = 0.00015$

(v) Anchorage take-up of 0.25 cm on each end gives loss due 0.25 to anchorage $= \dfrac{0.25}{2500} = 0.0001$

(vi) *Frictional loss* :

Let $k = 0.0015$ (wobble effect per metre length)

$u = 0.25$ (frictional coefficient)

$$e_x = y = \frac{4e}{L^2} (Lx - x^2)$$

$$\left(\frac{dy}{dx}\right)_{x = 0} = \frac{4e}{L} = \frac{4e}{h}\frac{h}{L} = \frac{4\in h}{L}$$

$$= 4(0.53)(1.4)/50 = 0.102$$

$\therefore \alpha = 0.102$ (total change in the slope of the cable)

$$\frac{P_0 - P_t}{P_i} = \left(\mu\alpha + k\frac{L}{2}\right) = (0.25)(0.102) + (0.0015)(25) = 0.063$$

TABLE 3.10.1 Loss of prestress

Loss due to	Loss of strain	Percentage of loss
(i) Elastic shortening	0.00018	3.6
(ii) Shrinkage	0.00020	4.0
(iii) Creep in concrete	0.00054	10.8
(iv) Creep in steel	0.00015	3.0
(v) Anchorage take up	0.00010	2.0
(vi) Bending	−0.00048	−9.6
Total	0.0007	14

There is a gain in prestress due to bending of the beam under working load condition. The gain in strain can be computed approximately from the change in strain at level of steel from transfer condition to the working load condition.

Change in strain at the extreme bottom fibre

$$= \frac{(c_{te} + c_{ct}) f_{ck}}{E_c}$$

The change of strain at the level of the cables can be computed proportionately and it is

$$= \frac{(c_{te} + c_{ci})}{E_c} \left(\frac{e}{y_b} \right)$$

$$= \frac{(c_{te} + c_{ci}) f_{ck}}{E_c} \frac{(1 + \Delta) \in}{\Delta}$$

$$= \frac{(0.02 + 0.457)(4500)(2.68)(0.53)}{3.8(10^6) 1.68} = 4.8 \times 10^{-4}$$

$$\eta = 1 - 0.14 = 0.86$$

The value of η works out to be close to the assumed value and the stress coefficients with new value of η are not altered very much.

(vi) *Selection of actual section*

The area of cross section can be determined from the value of R.

$$R = \frac{M_t}{M_g} = \frac{w_t L^2/8}{w_g L^2/8} = \frac{w_t}{w_g}$$

$$w_g = \frac{w_t}{R} = \frac{20000}{0.95} = 21000 \text{ N/m}$$

The self weight of the beam is

$$w_g = \gamma A = 25000 A$$

$$A = \frac{w_g}{\gamma} = \frac{21000}{25000} = 0.840 \ m^2$$

Use $A = 0.875$ m^2 = 8750 cm^2

Various details of the section are computed using the non-dimensionalised coefficient of the section

$$A = c_a h b_t = 0.256 (240) b_t = 8750 \text{ cm}^2$$

so,

$$b_t = 8750/(240)(0.256) = 137 \text{ cm and}$$

$$b_b = c_f b_t = 0.5 (137) = 68.5 \text{ cm}$$

$$t_t = c_t A = (0.12)(240) = 28.8 \text{ cm}$$

$$t_b = c_b h = (0.08)\ (240) = 19.2 \text{ cm}$$

$$b_w = c_w b_t = (0.12)\ (137) = 16.5 \text{ cm (250 desirable)}$$

$$y_b = c_{yb} h = (0.626)\ (240) = 151 \text{ cm}$$

$$y_t = c_{yt} h = (0.374)\ (240) = 89 \text{ cm}$$

$$e = \varepsilon h = (0.53)\ (240) = 127 \text{ cm}$$

$$P_i = m A f_{ck} = (0.22)\ (8750)\ (4500) = 8660000 \text{ N}$$

$$A_p = \frac{P_i}{0.7 f_p} = \frac{8660000}{105000} = 82.5 \text{ cm}^2$$

Assuming 7 mm wires are used in the beam

$$\text{Number of wires} = \frac{82.5}{\pi(0.49)/4} = 222$$

The wires are grouped into a set cables and placed as per the specifications. Two to four cables can be grouped together with a minimum spacing of the groups equal to 40 mm or 6 mm in excess of the largest size of the aggregate whichever is greater. As many cables as possible should be draped with the maximum eccentricity at the mid-span and zero eccentricity at end section. The end section should be made into a solid block so as to take care of the stress concentrations at end section. Figure 3.10.1 illustrates the idealised and practicable sections. Figure 3.10.2 gives the longitudinal' profile of the beam. The actual area of cross section is at least 3 per cent more than the minimum so as to compensate for the area occupied by the cables.

Intermediate diaphrams. Diaphrams along the length of the beam at 600 cm centre to centre be provided to safeguard against lateral buckling of the beam.

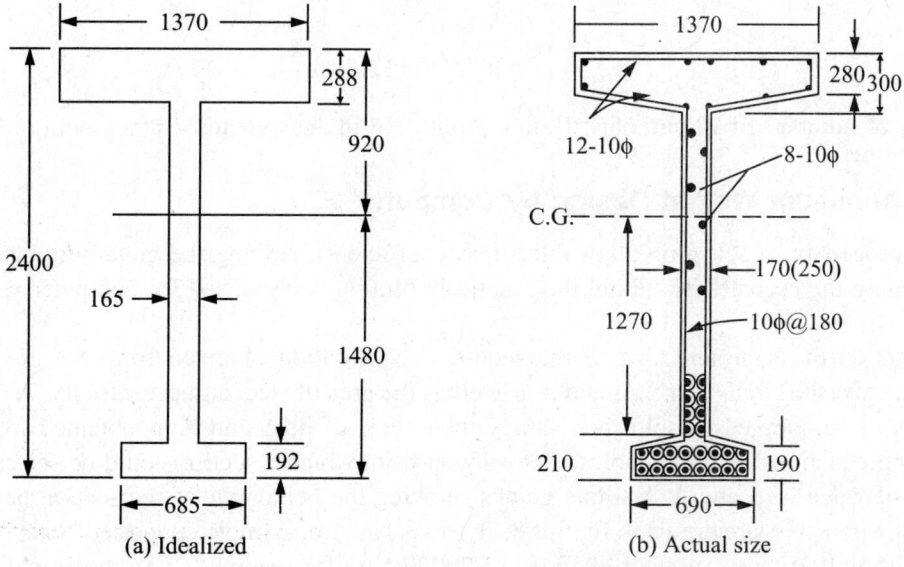

Figure 3.10.1 Cross section at mid-span.

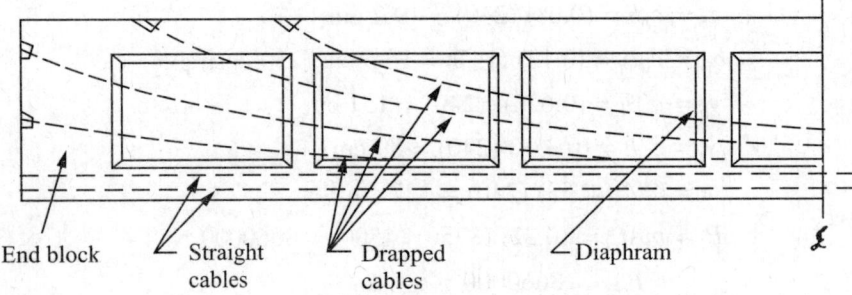

Figure 3.10.2 Half of the longitudinal profile of the cables and beam.

(vii) *Minimum requirements as per IRC specifications*

Minimum clear cover measured from the outside of the sheathing shall be at least 30 mm. In case of location of the bridge in coastal areas, the minimum cover should be 40 mm. A minimum clear spacing of the cable or group of cable should be 40 mm or 6 mm plus the largest size of the aggregate size. The vertical clear spacing between the groups shall be 50 min. The minimum thickness of the web shall be 100 mm plus 1/40 of the overall depth of the beam.

A minimum web reinforcement of 0.25 per cent of the web area should be provided

$$A_{sw} = \frac{0.25}{100}\,(17)\,(100) = 4.25 \text{ cm}^2/\text{m}$$

Use 10 mm *MS* bars at 18 cm apart.

A minimum mild steel reinforcement of 0.2 percent in the cross section is to be provided evenly.

$$A_{sm} = \frac{0.2}{100}\,(8750) = 17.5 \text{ cm}^2$$

Use 24 numbers of 12 mm bats, 10 in top flange, 8 in the web and 4 in the bottom flange.

3.11. Minimum Weight Design by Computer

Design procedure of selecting allowable stress coefficients, solving the governing equations and keeping the eccentricity within the practicable limits is organised for computer working (3.11).

Selection of shape and size of the section: The solution obtained from the governing equations gives the values of ρ, Δ and A as well as the area of steel and eccentricity. A section which is to be adopted, should then satisfy the values of ρ, Δ and A as obtained from the governing equations. A set of tables is developed from which a section could be selected for any set of value of ρ and Δ. Various graphs showing the behaviour of the section based on different parameters are drawn in figures 3.11.1 to 3.11.3 for a simply supported beam of span 50 m with uniformly distributed live load of 20000 N/m for loads likely to increase.

(a) *Effect of* ρ (efficiency factor):   Figure 3.11.1 illustrates that the area of cross section decreases with an increase in the efficiency factor, so it is desirable to select a maximum efficiency factor. The maximum efficiency for practical sections is about 0.16 (see table 2 in appendix A). The design can always be started assuming a maximum efficiency. However, there are some practical limitations for maximum efficiency factor. The limitations are:

(i) As the efficiency factor increases, the eccentricity also increases; and beyond a certain efficiency factor, it will not be possible to keep the eccentricity within the limits of the section.

(ii) For a very high efficiency factor, the web thickness might become so thin that it may not accommodate the cables and may not be able to resist shear force.

While working on an exaniple, it is desirable to start with high efficiency factor and check for the two limitations.

(b) *Effect of depth of the beam*:   Figure 3.11.2 gives the effect of depth on the area of cross section for a given set of efficiency factor and cube strengths. As the depth increases (from $L/25$ to $L/15$) the area of cross section decreases. It is desirable to take a large depth but again the depth is subjected to practical limitation. In case of

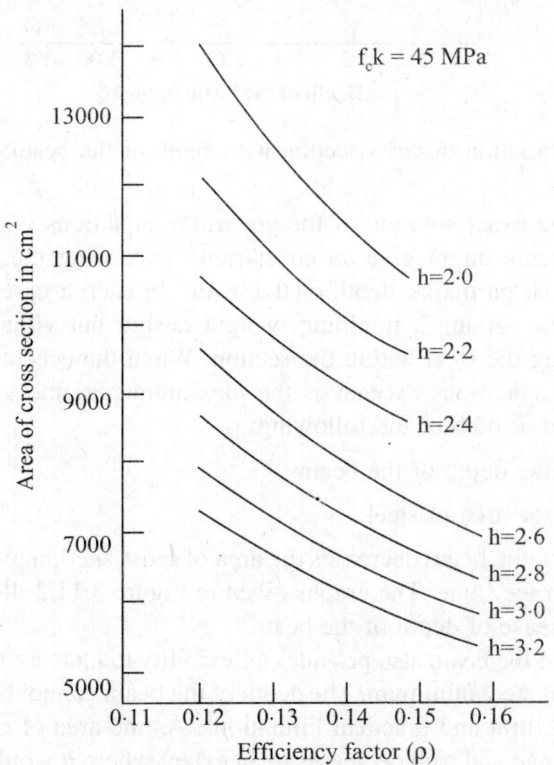

Figure 3.11.1   Variation of cross section with efficiency ratio for f_{ck} = 45 MPa.

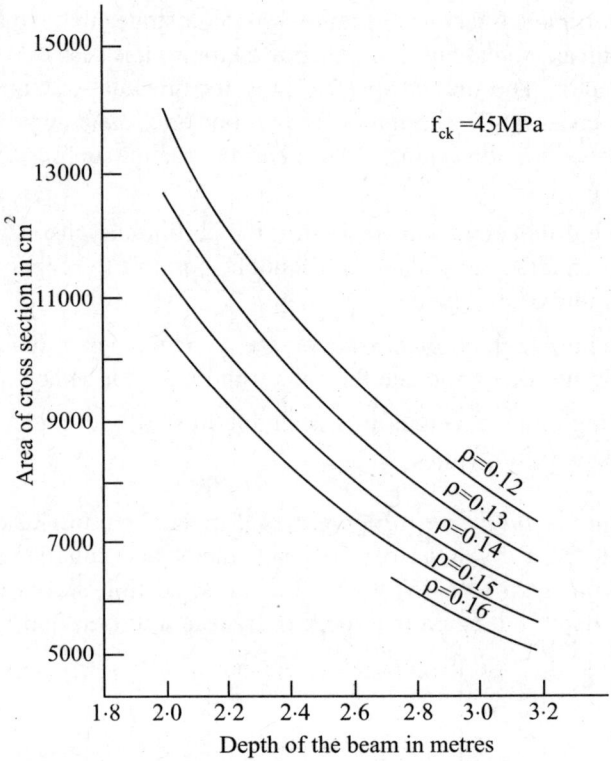

Figure 3.11.2 Variation of cross section with depth of the beam are f_{ck} = 45 MPa.

long spans, the exact solution of the governing equations with maximum allowable stress coefficients might give an eccentricity exceeding the maximum eccentricity possible for that particular depth of the beam. In such a case, the solution becomes absolutely ideal, giving a minimum weight design but violating the basic need of accommodating the steel within the section. When the eccentricity, as calculated by the governing equations exceeds is the maximum possible value, it is necessary to adjust any one or both of the following:

(i) Increase the depth of the beam.

(ii) Increase the area of steel.

Increase in depth of the beam decreases the area of cross section required, provided all the other quantities remain the same. The graphs given in Figure 3.11.2 illustrate the reduction of cross section with increase of depth of the beam.

Increase of depth of the beam also provides a flexibility to increase the limiting eccentricity thus keeping the area of steel minimum. The depth of the beam cannot be increased indefinitely because of the architectural and practical limitations. As the area of cross section decreases, the thickness of the flange and web decreases to an extent where it would be almost impossible to accommodate the cables. On the other hand fot a given large depth, if the thickness of the flange and web are kept within the practicable limits, the weight of the member is likely to be more than that of a smaller depth.

In case, it is not possible to increase the depth of the beam, it is possible to adjust the allowable stress coefficients so as to keep the eccentricity within the allowable limits. The eccentricity could be reduced by decreasing the value of c_{ti}. In other words the stress coefficients should be selected as

$$c_{ci} = c_{aci}, c_{ti} < c_{ati}, c_{ce} = c_{aci}, c_{te} = c_{ace}$$

(c) *Effect of strength of concrete*: Variation in the area of cross section with strength of concrete is shown in Figure 3.11.3. As the strength of concrete increases, the area of cross section decreases.

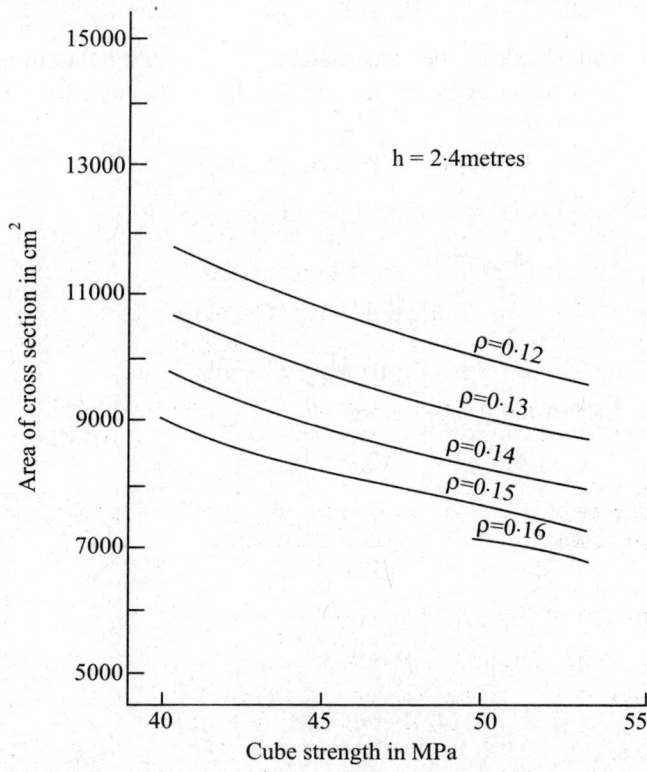

Figure 3.11.3 Variation of cross section with cube strength.

The straight forward conclusion could be made even without graphs. The graphs not only -illustrate the qualitative nature, but also give quantitative values which are helpful in selecting the concrete strength. The decrease in the area of cross section with that of depth or efficiency factor. It is is desirable to use a high strength concrete. However, the rate of cost of high strength concrcce is more, so a proper selection of quality in the concrete is to be made such that the total cost of the structure is minimum. Other limitations in selecting a high strength concrete are the workability, availability of proper equipment and skilled man power for good and uniform compaction.

3.12. Design by Load Balancing Method

The main quantities given in a design problem are the external loads and the span; one is expected to select some quantities and solve for the other quantities. The properties of the section and the prestressing force are the two main design variables. Design of sectional details present some problems for most of the beginners and also to the experienced ones as the economic design is not that easy. Instead of starting with sectional properties, one can start with prestressing force based on load balancing method and then determine the section. The self weight or the bending moment caused by the self weight has to be assumed to start with. It is easier and often more convenient to work with balancing moment rather than the balancing load. The bending moment caused by the prestressing force can be selected in the range of:

$$P_i e = M_g + jM_i \tag{3.12.1}$$

It is assumed that M_g and M_t are of the same nature, where j can be assumed in range of 0.5 to 0.7 depending on the relation between M_g and M_i. The unbalanced moment at transfer of prestress is:

$$M_{ei} = M_g - P_i e = -jM_i \tag{3.12.2}$$

Similarly, the k unbalanced bending moment at working load is

$$M_{ew} = M_w - \eta P_i e$$

$$= M_g + M_i - \eta(M_g + jM_i)$$

$$= (1 - \eta) M_g + (1 - j\eta) \tag{3.12.3}$$

Substitution of Eq. 3.12.1 in ineq. 3.6.5 gives

$$M_g + jM_i \le (\sigma_{ati} + \sigma_{aci}) I/h + M_g$$

or

$$I \ge \frac{jM_t h}{\sigma_{ati} + \sigma_{aci}} \tag{3.12.4}$$

Similarly, the substitution of Eq. 3.12.1 in eq. 3.6.6 gives

$$\eta P_i e = \eta(M_g + jM_t) \ge M_w - (\sigma_{ace} + \sigma_{ate}) I/h$$

or

$$I \ge \frac{(M_g(1 - \eta) + M_t (I - \eta)) h}{\sigma_{ace} + \sigma_{ate}} \tag{3.12.5}$$

Higher of the two values of I from ineqs. 3.12.4 and 3.12.5 can be selected. One can optimise the value of I by selecting such a value of i which will give same lower bound from ineqs. 3.12.5 and 3.12.4.

The method is illustrated through an example.

Example **3.12.1** A simply supported beam of clear span of 20 m is subjected to a uniformly distributed load of 8000 N/m. Design the beam as post-tensioned' prestressed concrete one using M 35 concrete. The beam is T-shaped with flange width as 100 cm. (same as example 3.6.4 with some change in the balancing moment). Actual prestressing force introduced at transfer is only 70 per cent of P_i. It is a multistage prestress. Let the effective length $= L = 20$

+ 1 = 21 m. Let the area of the section be = 2500 cm^2 for the purpose of computation of self weight moment.

$$w_g = 2500 \ (0.025) = 60 \ \text{N/cm} = 6250 \ \text{N/m}$$

$$M_g = \frac{w_g L^2}{8} = 344500 \ \text{Nm}$$

$$M_i = \frac{w_i L^2}{8} = 441000 \ \text{Nm}$$

$$M_w = M_g + M_t = 785500 \ \text{Nm}$$

The permissible stresses in N/cm^2 are:

$$\sigma_{aci} = 1750, \ \sigma_{ati} = 100$$

$$\sigma_{ace} = 1400, \ \sigma_{ate} = 100$$

Let the moment to be balanced at transfer is:

$$P_i e = M_g + 0.6 \ M_t$$

That is, $\quad\quad\quad i = 0.6$

The value of / from Ineq. 3.12.4 is

$$I \geq \frac{0.6(441000) \ (100) \ (100)}{1850}$$

$$\geq 1,430,270 \ \text{cm}^4$$

The approximate depth of the beam can be either taken from $L/25$ to $L/15$ or a more rational value from the following relation.

$$I = c_t b_f h^3$$

where c_t can be taken for T-beams as 0.03.
Therefore,

$$h^3 = \frac{I}{c_i b_f} = \frac{1,430,270}{(0.03) \ (100)} = 476,757$$

$$h = 78.2 \ \text{cm}$$

Select the overall depth of the beam (h) = 85 cm. Typical T-beam section is shown in Figure 3.6.2 and its properties are computed in systematic manner in Table 3.12.1 using h = 85 cm

$$t_f = \text{thickness of flange} \ = 12 \ \text{cm}$$

$$b_f = \text{width of flange} \ \ \ \ = 100 \ \text{cm}$$

$$o_w = \text{width of web} \ \ \ \ \ \ \ = 15 \ \text{cm}$$

$$A_1 = b_f b_f = 100(12) \qquad = 1200 \text{ cm}^2$$

$$A_2 = b_w t_w = 12(85 - 12) = 1095 \text{ cm}^2$$

TABLE 3.12.1 Section properties in cm for T-beam

Segment	A_i	d_i	$A_i d_i$	y_i	I_i	$A_i y_i^2$
Flange	1200	6.0	7200	−20.28	14,400	493,534
Web	1095	48.5	53107	22.22	486,271	540,632
Total	2295		60307		500,671	1,034,166

$$y_t = d_i - y_t$$

$$y_t = \frac{\Sigma A_i d_i}{\Sigma A_i} = \frac{60307}{2245} = 26.28$$

$$I = \Sigma I_i + \Sigma A_i\, y_i^2$$
$$= 500671 + 1{,}034{,}166 = 1{,}534{,}837 \text{ cm}^4$$

Use

$$I = 1534800 \text{ cm}^4$$

$$y_b = h - y_t = 85 - 26.28 = 58.72 \text{ cm}$$

The limit of eccentricity is about

$$y_b - 15 = 43.72 \text{ cm}$$

use

$$e = 43 \text{ cm}$$

$$P_i = \frac{M_g + 0.6\, M_i}{e}$$

$$= \frac{344500 + 0.6\,(4410001)}{0.43}$$

$$= 1416500 \text{ N}$$

use

$$P_i = 1400000 \text{ N}$$

Ckeck for Stresses: Only 70 per cent of the cables are tensioned at transfer. Therefore, the prestressing force at transter eqal to

$$0.7\, P_i = kP_i$$

$$\frac{kP_i}{A} = \frac{0.7(1400000)}{2295} = 427$$

$$\frac{\eta P_i}{A} = 0.8 \frac{(1400000)}{2295} = 488$$

$$\frac{P_i e}{I} = \frac{1400000}{1534800} = 39.2$$

$$\frac{M_g}{I} = \frac{344500(100)}{1534800} = 22.5$$

$$\frac{M_w}{I} = \frac{7855000}{1534800} = 51.2$$

$$\sigma_{ti} = \frac{-kP_i}{A} + \left(\frac{kP_i e - M_g}{I} \right) y_t$$

$$= -427 + (0.7\,(39.2) - 22.5)\,(26.28)$$

$$= -427 + 130 = -297 \ \text{N/cm}^2$$

$$\sigma_{ci} = \frac{k\,P_i}{A} + \left(\frac{kP_i e - M_g}{I} \right) y_b$$

$$= 427 + 290 = 717 \ \text{N/cm}^2$$

$$\sigma_{ce} = \frac{\eta P_i}{A} - \frac{(\eta P_i e - M_w) y_t}{I}$$

$$= 488 - (31.4 - 51.2)\,(26.28) = 1008 \ \text{N/cm}^2$$

$$\sigma_{te} = \frac{-\eta\,P_i}{A} - \left(\frac{\eta P_i e - M_w}{I} \right) y_b$$

$$= -488 + 1160 = 672 \ \text{N/cm}^2$$

The tensile stress at working load is far higher than the allowable value. Therefore'modifications one or more of the following are needed.

(1) Increase the prestressing force
(2) Increase the eccentricity
(3) Increase the section

The eccentricity is already taken to the maximum therefore one has to increase the section and the prestressing force. Increase of the section automatically calls for increase the prestress, hence the prestress is increased and then checked for stresses.

Use $\qquad\qquad P_i = 1800,000 \ \text{N}$

$$e = 44 \ \text{cm}$$

$$\frac{kP_i}{A} = \frac{0.7\,(1800000)}{2295} = 548$$

152

$$\frac{\eta P_i}{A} = \frac{0.8\,(1800000)}{2295} = 626$$

$$\frac{kP_ie}{I} = \frac{0.7\,(1800000)\,(44)}{1534800} = 36.1$$

$$\frac{\eta P_ie}{I} = \frac{0.8\,(1800000)\,(44)}{1534800} = 41.3$$

$$\sigma_{ti} = -548 + (36.1 - 22.5)\,26.28$$

$$= -548 + 358 = 190 \text{ N/cm}^2$$

$$\sigma_{te} = -548 + (36.1 - 22.5)(58.72) = 1347 \text{ N/cm}^2$$

$$\sigma_{te} = -626 - (41.3 - 51.2)(58.72) = 44 \text{ N/cm}^2$$

The stresses are less than the permissible values therefore the section is adequate. The prestressing force can be reduced to 1750000 N.

Provide the final design as follows:

(*all dimensions in mm*)

$$b_f = 1000$$

$$t_f = 120$$

$$h = 850$$

$$b_w = 150$$

$$e = 440$$

$$P_i = 1750 \ kN$$

Prestressting force at transfer

$$= 0.7 \ P_i = 1225 \ kN$$

Provide 44 numbers of 7 mm wires of which 31 wires are prestressed at the time of transfer condition of the beam.

3.13. Multiple Stage Prestressing

In post-tensioned concrete construction, it is possible to prestress the wires or cable at different stages. There are two main reasons based on which the beams are prestressed at different time intervals. It may be desirable to prestress the beam even before the concrete gained full strength so that the beam can be transported and erected in position at an early stage. In case the self weight of the member is too small when compared with the external load, the prestressing force needed for the working load is large enough to cause damage to the beam at transfer. Bridge girders and long span beams are prestressed at different stages. Let kP_i = the prestressing force at the first stage, where k is less than 1 and may be equal to 0.6 to 1.0 depending upon the problem. The critical transfer condition can be treated as the first stage prestressing whereas the other stages may not be so critical yet check for stresses must be made in each of the stage. The governing equations for working stress design can be written similar to Ineqs. (3.3.5) to (3.3.8)

$$\frac{-kP_i}{A} + \frac{kP_i e_i y_t}{I} - \frac{M_g y_t}{I} = \sigma_{ti} \leq \sigma_{ati} \tag{3.13.1}$$

$$\frac{kP_i}{A} + \frac{kP_i e_t y_b}{I} - \frac{M_g y_b}{I} = \sigma_{ct} \leq \sigma_{act} \tag{3.13.2}$$

where e_t = eccentricity of kP_t at transfer condition. Similarly e_w = eccentricity at working load.

$$\frac{\eta P_i}{A} - \frac{\eta P_i e_w y_t}{I} + \frac{M_w y_t}{I} = \sigma_{ce} \leq \sigma_{ace} \tag{3.13.3}$$

$$\frac{-\eta P_t}{A} - \frac{\eta P_i e_w y_b}{I} + \frac{M_w y_b}{I} = \sigma_{te} \leq \sigma_{ate} \tag{3.13.4}$$

Addition of Ineqs. (3.13.1) and (3.13.2) gives

$$\frac{kP_i e_i h}{I} - \frac{M_g h}{I} \leq \sigma_{ati} + \sigma_{aci}$$

or

$$I \geq \frac{(kP_i e_i h - M_g)h}{\sigma_{ati} + \sigma_{aci}} \tag{3.13.5}$$

or

$$kP_i e_i \leq (\sigma_{ati} + \sigma_{aci}) \, I/h + M_g \tag{3.13.6}$$

Similarly Ineqs. (3.13.3) and (3.13.4) give

$$I \geq \frac{(M_w - \eta P_i e_w)}{\sigma_{ace} + \sigma_{ate}} \tag{3.13.7}$$

or

$$\eta P_i e_w \geq M_w - (\sigma_{ace} + \sigma_{ate}) \, I/h \tag{3.13.8}$$

For a given value of I (section if assumed), the upper and lower bounds of the prestressing force can be obtained from Ineqs. (3.13.6) and (3.13.8). On the other hand if the balancing prestressing force is given, one can obtain the section from Ineqs. (3.13.5) and (3.13.6).

It may be observed from example 3.6.2 that the lower bound of P_i is more than the upper bound and therefore it was recommended that the section be increased. Such problems can easily be solved by multiple stage prestressing. An example illustrates the advantages of the multistage prestressing.

Example 3.13.1 A post-tensioned concrete beam of effective span 16 m is subjected to a load of 20,000 N/m. Design a rectangular beam using M35 concrete. Let the first stage post-tensioning be done at a time the concrete has gained strength of 30 N/mm². (That is $f_{ck} = 35$ MPa, $f_{ci} = 30$ MPa).

The permissible stresses are:

$$\sigma_{aci} = 0.5 \, f_{ci} = 0.5 \, (30) = 15 \text{ M/mm}^2$$

$$= 1500 \text{ N/cm}^2$$

$$\sigma_{ati} = 0, \ \sigma_{ate} = 0$$

$$\sigma_{\alpha ce} = 0.4 \ (3500) = 1400 \ \text{N/cm}^2$$

Let

$$h = 16/16 = 1.0 \ \text{m} = 100 \ \text{cm}$$

Maximum eccentricity admissible is about $= 50 - 10 = 40$ cm. Let the eccentricities of the net prestressing forces at transfer and working loads be

$$e_i = 35 \ \text{cm and}$$

and

$$e_w = 40 \ \text{cm}$$

Let $\eta = 0.8$ and the prestressing is done in two stages. The first stage prestressing is about eighty per cent of total force. So $k = 0.8$. For the purpose of computation of self weight moment, let the width of the beam be equal to 40 cm.

$$w_g = (40) \ (100) \ (0.025) = 100 \ \text{N/cm}$$

$$= 10000 \ \text{N/m}$$

$$M_g = \frac{w_g L^2}{8} = \frac{10000 \ (256)}{8} = 320000 \ \text{Nm}$$

$$M_1 = \frac{w_e L^2}{1} = \frac{20,000 \ (256)}{8} = 640000 \ \text{Nm}$$

The balancing moment caused by the effective prestressing force be

$$\eta P_i e_w = M_g + 0.6 \ M_t$$

$$= (320000 + 384000) = 704000 \ \text{Nm}$$

$$P_i = \frac{70400000}{0.8(40)} = 220000 \ \text{N}$$

$$(kP_i e_i - M_g) = 0.8 \ (2200000) \ (35) - 32,000,000$$

$$= 30880,000 \ \text{N cm}$$

$$M_w - \eta P_i e_w = M_g + M_t - \eta P_i e_w$$

$$= 0.4 \ M_1 = 25600000 \ \text{N cm}$$

From Ineqs. (3.13.5) and (3.13.7)

$$I \geq \frac{30880000 \ (100)}{140} = 2206,000 \ \text{cm}^4$$

use

$$I = 2000000 \ \text{cm}^4$$

$$= bh^3/12$$

or

$$b = \frac{12I}{h^3} = \frac{12 \ (2000000)}{1000000}$$

$$= 12(2.00) = 24 \ \text{cm}$$

use $\qquad b = 24$ cm (against 40 cm assumed)

and $\qquad h = 100$ cm

self weight $= w_g = 24\,(100)\,(0.025)\,(100) = 60$

and
$$M_g = \frac{24\,(250)\,(256)\,(100)}{8} = 19200000 \text{ Ncm}$$

$$M_i = 64{,}000{,}000$$

$$M_w = M_g + M_t = 83200{,}000 \text{ Ncm}$$

Since the self weight is reduced, the balancing prestressing force also reduced, the balancing prestressing force with modified M_g is obtained as

$$\eta P_i e_w = M_g + 0.6\,M_t \text{ or}$$

$$P_i = \frac{19200 + 0.6\,(64000)}{0.8\,(40)} = 1800 \; kN$$

and $\qquad kP_i = 0.8\,(1800) = 1800 \; kN$

Check for Stresses (which are in N/cm^2)

$$A = 2400 \text{ cm}^2, \; y_t = y_b = 50 \text{ cm}$$

$$I = 2 \times 10^6 \text{ cm}^4$$

$$\frac{kP_i}{A} = \frac{1440000}{2400} = 600$$

$$\frac{kP_i y_t}{I} = \frac{1440000\,(50)}{2 \times 10^6} = 36$$

$$\frac{M_g y_t}{I} = \frac{19200000\,(50)}{2\,(10^6)} = 480$$

$$\frac{M_w y_t}{I} = \frac{83200000\,(50)}{2\,(10^6)} = 2080$$

The substitution of different values in Ineqs. (4.13.1) to (3.13.4) gives

$$\sigma_{ti} = -600 + 36\,(35) - 480 = 180$$

$$\sigma_{ci} = 600 + 36\,(35) - 480 = 1380$$

$$\sigma_{ce} = 600 - 36\,(40) - 2080 = 1240$$

$$\sigma_{te} = -600 - 36\,(40) + 2080 = 40$$

It can be observed that the actual stress in tension exceeds the permissible value. Therefore decrease e_i

Let

$$P_i = 1800 \; kN$$

$$kP_i = 1440 \; kN$$

$$e_i = 28 \; cm$$

$$e_w = 41 \; cm$$

$$\frac{kP_i}{A} = \frac{1440000}{2400} = 600$$

$$\frac{kP_i y_t}{I} = \frac{1440000 \, (50)}{2(10^6)} = 36$$

The actual stresses in N/cm² are

$$\sigma_{ti} = -600 + 36 \, (28) - 480 = 72$$

$$\sigma_{ci} = 600 + 36 \, (28) - 480 = 1128$$

$$\sigma_{ce} = 600 - 36 \, (41) + 2080 = 1204$$

$$\sigma_{te} = -600 - 36 \, (41) + 2080 \doteq 4$$

Provide 45 numbers of 7 mm HTS steel wires of which 36 are to be pretensioned at the time of transfer with an effective eccentricity of 28 cm. The eccentricity of the 45 wires to be 41 cm.

PROBLEMS

3.1 Design a pre-tensioned prestressed concrete simply supported beam of span 24 m for a live load of 24000 N/m. Design by non-dimensionalised parameters and load balancing methods keeping the depth of the beam not exceeding 120 cm, use $f_{ck} = 45$ MPa, $f_p = 1500$ MPa.

3.2 Design a post-tensioned prestressed concrete balanced cantilever bridge of spans as shown in Figure 3.1. The bridge is used for two lane traffic with amoving load of 150 kN/m. The girders are arranged such that the top flanges of the girders form the main slab of the bridge. A 5 cm thick wearing coat is to be provided on the top of the flanges. Use $f_{ck} = 40$ MPa.

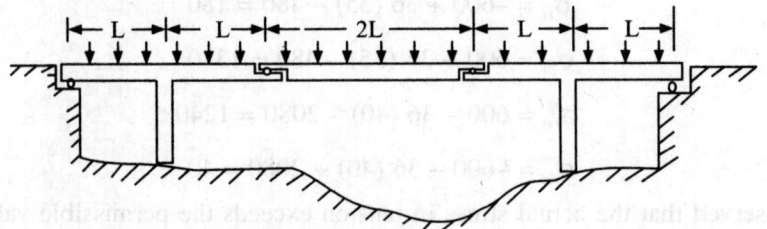

Figure 3.1   Problem 3.2

3.3 A slab is simply supported with an effective span of 6 m and carries a load of 5 kN/m^2. Design a solid slab with axial prestressing force. Use f_{ck} = 35 MPa and f_p = 1600 MPa. (Assume the depth of the slab in the range of L/20.)

3.4 Design the beam mentioned in problem 3.3 if a straight cable with an eccentricity of 3 cm is provided.

3.5 A prestressed concrete pole of 9 m high is to carry a horizontal load of 1800 N at 6.7 m from ground. 1.5 m of the pole is embedded into the ground. Design the pole with M 40 concrete and 5 mm HT wires. Assume the width of the pole as 10 cm and the depth is linearly varying.

3.6 A 10 m cantilever beam of hollow box section has 100 cm outside width. The outside depth of the beam at free end is 50 cm and at fixed end is 100 cm. The inside dimensions at free and fixed ends respectively are 60 cm by 20 cm and 60 cm by 70 cm. Design a prestressing cable if the beam is to carry 10 kN/m.

3.7 A double overhang beam has 1 m overhangs on either ends and 6 m in between the supports. The beam is made of M 35 concrete with 100 cm width. Design an axial prestressing force and depth of the beam if it is to carry 12 kN/m.

3.8 A simply supported beam of span 40 m is carrying load of 12 kN/m. Design a post-tensioned beam with M 35 concrete using the permissible stresses of IRC.

3.9 A two lane highway bridge of 20 m simply supported span and 6.8 m width is provided with 5 numbers of I-section beams adjacent to each other. Design the beams 40 MPa concrete. Use IRC loading and specifications.

3.10 A two lane highway bridge of 16 m cantilever span is provided with one large box section. Design the section with M 35 concrete for IRC loading.

3.11 A one way slab is made of hollow rectangular box strips of 50 cm width. The slab is carrying a live load of 4 kN/m^2 and simply, supported over a span of 4 m. Design the slab strip with inside opening of 30 cm by 6 cm. Use M 420 concrete.

Prestressed Concreate Composite Beams

4.1. Introduction

A prestressed concrete composite beam consists of two essential parts : (i) a prestressed concrete beam either pre-tensioned or post-tensioned, and (ii) concrete slab reinforced, or prestressed or even of plain concrete. Usual construction procedure consists of two stages :

   (i) A prestressed concrete beam which is either fully or partly prestressed is placed in position.

   (ii) A plain or reinforced concrete slab is cast in-situ over the top of the prestressed concrete beam.

    If the beam is partly prestressed before casting the top slab, the remaining prestressing is done after casting the slab. Placing of the top slab is done in two ways. The slab is cast directly on the beam with or without overhang so that the total shuttering is wholly supported by the prestressed concrete beam itself. If the slab is fairly wide and the beam is not to be overstrained with the load of shuttering and wet concrete of the slab, the slab is cast partly supported by a shuttering not directly resting on the beam.

    The concrete slab hardens and acts together with the girder to form a composite section. The precast beams are cast with controlled concrete of high strength, the slab with less strong concrete. If the slab has to act together with the precast beam, a strong bond has to be generated between the beam and the slab. When the composite section bends, the top fibre of the beam and the bottom fibre of the slab should have the strain compatibility of a single member. Some kind of connectors joining the two parts have to be provided to achieve single unit action. These connectors are essentially under shear force so they are called "shear connectors". The integral connection is achieved by providing shear keys at the top fibre of the beam along the span, or leaving the top of the beam unfinished and roughened for casting the slab over the beam. Vertical ties are also provided along the span to prevent separation of the slab from the beam. The deformation of composite section with and without shear connectors is shown in figure 4.1.1.

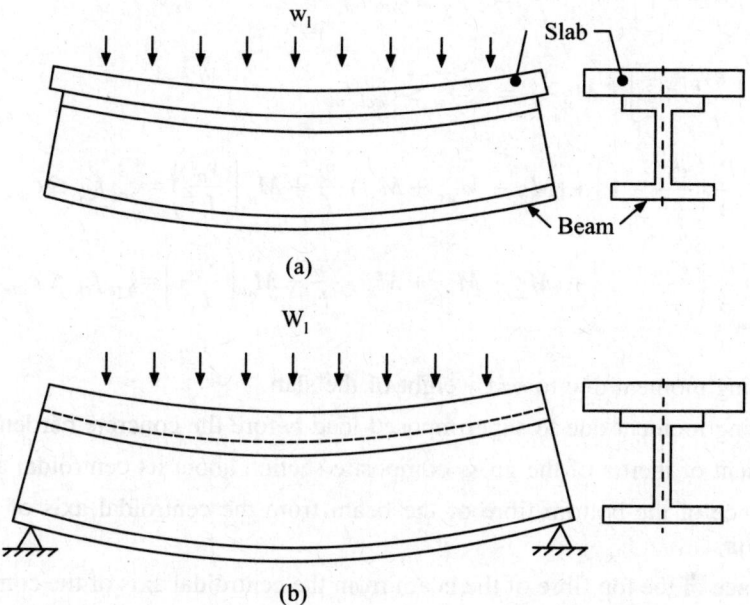

Figure 4.1.1   Effect of shear connectors in composite construction:
(a) Composite section without shear connectors.
(b) Composite section with shear connectors.

4.2. Allowable Stress Considerations

There are three critical conditions of loading in composite construction which may be considered for design purposes :

- (i) *Transfer condition* : The beam is subjected to the transfer prestressing force and self weight of the beam.

- (ii) *Casting of slab condition* : The beam is subjected to transfer prestress, self weight of the beam, self weight of the wet slab and other superimposed loads due to shuttering.

- (iii) *Working load condition* : The composite beam is subjected to effective prestressing force, self weight of the beam and the slab, other superimposed and working loads.

In general, the third condition dominates the second condition so the first and the last conditions are real for design considerations. The stresses due to live loads are to be superimposed on the stresses caused due to second condition. This is a reasonably good and safe assumption because the slab hardens in the elastic deformed shape of the beam at second stage. The differential shrinkage of the slab and the beam affects the actual stress distribution. For the sake of simplicity of design, the shrinkage stresses are treated separately and superimposed on working load stresses. The expressions for stresses the extreme fibre of the beam for these two critical conditions can be written as

$$\frac{P_i}{A}\left(\frac{ey_t}{r^2}-1\right)-M_g\frac{y_t}{I}=c_{ti}f_{ci}\le c_{ati}f_{ci} \tag{4.2.1}$$

$$\frac{P_i}{A}\left(\frac{ey_b}{r^2}+1\right)-M_g\frac{y_b}{I}=c_{ci}f_{ci}\le c_{aci}f_{ci} \tag{4.2.2}$$

$$-\eta\frac{P_i}{A}\left(\frac{ey_t}{r^2}-1\right)+(M_g+M_{gs}+M_{si})\frac{y_t}{I}+M_a\left(\frac{y_{te}}{I_c}\right)=c_{ce}f_{ck}\le c_{ace}f_{ck} \tag{4.2.3}$$

$$-\eta\frac{P_i}{A}\left(\frac{ey_b}{r^2}-1\right)+(M_g+M_{gs}+M_{si})\frac{y_b}{I}+M_a\left(\frac{y_{be}}{I_a}\right)=c_{te}f_{ck}\le c_{ate}f_{ck} \tag{4.2.4}$$

where

M_{gs} = bending moment due to self weight of the slab

M_{sl} = bending moment due to superimposed load before the concrete hardens

I_c = moment of inertia of the gross composite section about its centroidal axis

y_{bc} = distance of the bottom fibre of the beam from the centroidal axis of the composite section

y_{tc} = distance of the top fibre of the beam from the centroidal axis of the composite section

The stress distribution at three conditions are shown in Figure 4.2.1

The moment of inertia of the composite section is to be calculated by considering the relative properties of the materials of the slab and the beam. Since the strength of the slab is different from that of the beam, the stresses in the slab and the beam will differ for any given strain. It is desirable to transform the area of the slab based on relative strength. The transformed area of the slab is to be considered while working with the sectional properties of the composite section. Areas of the beam and the slab have to be transformed proportional to the modulus of elasticity of the two materials.

The total width of the slab will not be effective in acting together with the beam if the width of the slab exceeds certain limits, so the width of the slab should also be restricted. The width of the slab should not be greater than

(i) one-fourth of the span of the beam,

(ii) centre to centre distance of the adjacent beams, and

(iii) twelve times the least thickness of the slab plus web thickness

(iv) $L_0/6 + 6t + b_w$

where L_0 = distance between the two points consecutive points of contraflexure

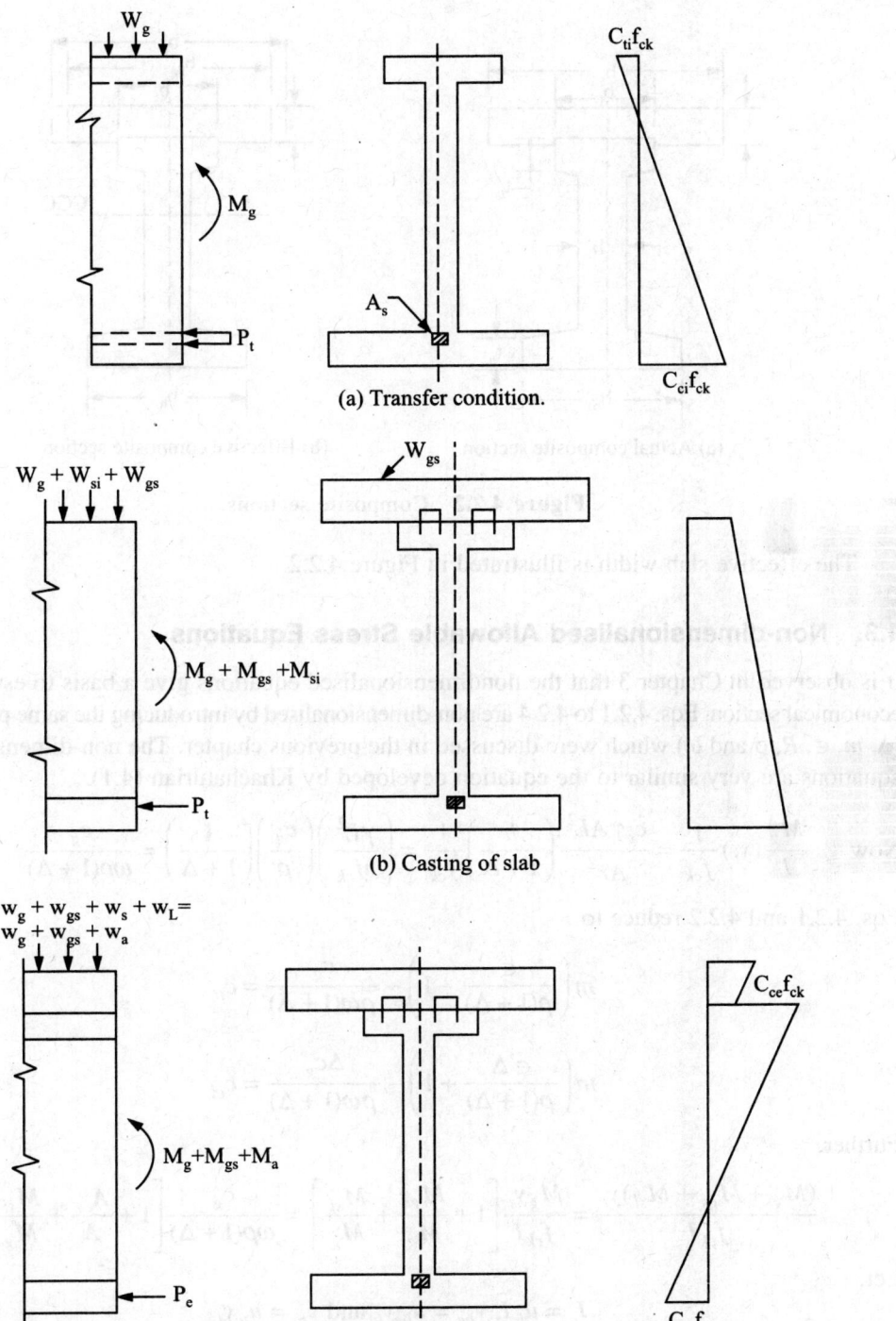

(a) Transfer condition.

(b) Casting of slab

(c) Working load condition

Figure 4.2.1 Load Conditions.

162

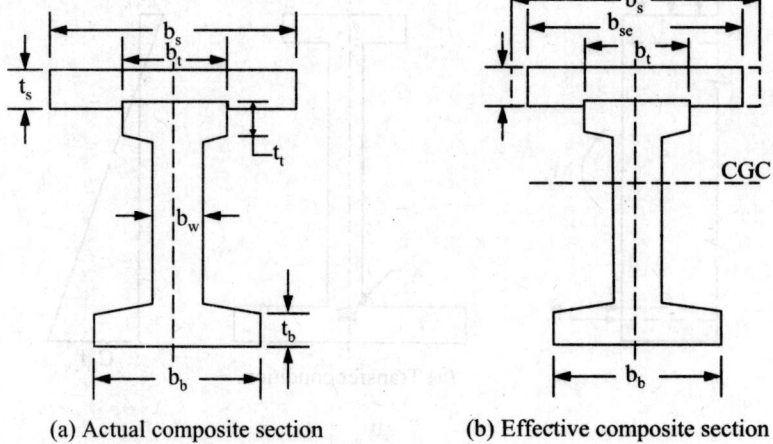

(a) Actual composite section   (b) Effective composite section

Figure 4.2.2 Composite sections.

The effective slab width is illustrated in Figure 4.2.2.

4.3. Non-dimensionalised Allowable Stress Equations

It is observed in Chapter 3 that the nondimensionalised equations give a basis to establish an economical section. Eqs. 4.2.1 to 4.2.4 are non-dimensionalised by introducing the same parameters (Δ, m, $\in$, R, ρ and ω) which were discussed in the previous chapter. The non-dimensionalised equations are very similar to the equation developed by Khachaturian (4.1).

Now
$$\frac{M_g}{I}(y_t)\frac{1}{f_{ck}} = \frac{c_g \gamma AL^2}{Ar^2}\left(\frac{h}{1+\Delta}\right)\frac{1}{f_{ck}} = \left(\frac{\gamma L^2}{hf_{ck}}\right)\left(\frac{c_g}{\rho}\right)\left(\frac{1}{1+\Delta}\right) = \frac{c_g}{\omega\rho(1+\Delta)}$$

Eqs. 4.2.1 and 4.2.2 reduce to

$$m\left(\frac{\in}{\rho(1+\Delta)}-1\right)-\frac{c_g}{\rho\omega(1+\Delta)} = c_{ti} \qquad (4.3.1)$$

$$m\left(\frac{\in\Delta}{\rho(1+\Delta)}+1\right)-\frac{\Delta c_g}{\rho\omega(1+\Delta)} = c_{ci} \qquad (4.3.2)$$

Further,

$$\frac{(M_g + M_{gs} + M_{sl})y_t}{f_{ck}I} = \frac{M_g y_t}{f_{ck}I}\left[1+\frac{M_{gs}}{M_g}+\frac{M_{sl}}{M_g}\right] = \frac{c_g}{\omega\rho(1+\Delta)}\left[1+\frac{A_{cs}}{A}+\frac{M_{sl}}{M_g}\right]$$

Let,

$$I_c = u_{ic}I, \ y_{be} = u_{be}y_b \text{ and } y_{tc} = u_{tc}y_t$$

Now
$$\frac{M_a y_{tc}}{f_{ck}I_c} = \frac{M_g R y_{tc}}{f_{ck}I_c}$$

$$= \left(\frac{M_g y_t}{f_{ck} I} \right) \left(\frac{R u_{tc}}{u_{ic}} \right)$$

$$= \left(\frac{c_g}{\omega \rho \, (1 + \Delta)} \right) R \left(\frac{u_{tc}}{u_{ic}} \right)$$

Let $\dfrac{M_{sl}}{M_g} = R_{sl}$ and $\dfrac{A_{cs}}{A} = \lambda$, then on substitution of the above quantities, the eqs. 4.2.3 and 4.2.4 reduce to

$$-\eta m \left(\frac{\in}{\rho(1 + \Delta)} - 1 \right) + \frac{c_g}{\omega\rho(1 + \Delta)} \left(1 + \lambda + R_{sl} + R \frac{u_{tc}}{u_{tc}} \right) = c_{ce} \qquad (4.3.3)$$

$$-\eta m \left(\frac{\Delta \in}{\rho(1 + \Delta)} + 1 \right) + \frac{\Delta c_g}{\omega\rho(1 + \Delta)} \left(1 + \lambda + R_{sl} + R \frac{u_{bc}}{u_{tc}} \right) = c_{te} \qquad (4.3.4)$$

For ready reference, all the governing equations are given here

$$c_{ti} = m \left(\frac{\in}{\rho(1 + \Delta)} - 1 \right) - \frac{c_g}{\omega\rho(1 + \Delta)} \qquad (4.3.1)$$

$$c_{ci} = m \left(\frac{\in \Delta}{\rho(1 + \Delta)} + 1 \right) - \frac{\Delta c_g}{\omega\rho(1 + \Delta)} \qquad (4.3.2)$$

$$c_{ce} = -\eta m \left(\frac{\in}{\rho(1 + \Delta)} - 1 \right) + \frac{c_g}{\rho\omega(1 + \Delta)} \left(1 + \lambda + R_{sl} + \frac{R u_{te}}{u_{tc}} \right) \qquad (4.3.3)$$

$$c_{ie} = -\eta m \left(\frac{\in \Delta}{\rho(1 + \Delta)} + 1 \right) + \frac{\Delta c_g}{\rho\omega(1 + \Delta)} \left(1 + \lambda + R_{sl} + \frac{R u_{oc}}{u_{ic}} \right) \qquad (4.3.4)$$

Eqs. 4.3.1 and 4.3.3 give

$$\frac{c_g \left[-\eta + 1 + \lambda + R_{sl} + R \dfrac{u_{tc}}{u_{ic}} \right]}{\rho\omega(1 + \Delta)} = \eta c_{ti} + c_{ce}$$

and eqs. 4.3.2 and 4.3.4 give

$$\frac{c_g \Delta \left[1 - \eta + 1 + \lambda + R_{sl} + R \dfrac{u_{bc}}{u_{ic}} \right]}{\rho\omega(1 + \Delta)} = \eta c_{ci} + c_{ce}$$

A combination of the above two expressions yields

$$\Delta \frac{\left[(1-\eta) + \lambda + R_{sl} + R\dfrac{u_{bc}}{u_{ic}}\right]}{\left[(1-\eta) + \lambda + R_{sl} + R\dfrac{u_{tc}}{u_{ic}}\right]} = \frac{\eta c_{ci} + c_{te}}{\eta c_{ti} + c_{ce}}$$

or

$$\Delta = \frac{\eta c_{ci} + c_{te}}{\eta c_{ti} + c_{ce}} \left(\frac{1-\eta + \lambda + R_{sl} + R\dfrac{u_{tc}}{u_{ic}}}{1-\eta + \lambda + R_{sl} + R\dfrac{u_{bc}}{u_{ic}}} \right) = \frac{\eta c_{ci} + c_{te}}{\eta c_{ti} + c_{ce}} \lambda_c \qquad (4.3.5)$$

where

$$\lambda_c = \frac{1 - \eta + \lambda + R_{sl} + R\dfrac{u_{tc}}{u_{ic}}}{1 - \eta + \lambda + R_{sl} + R\dfrac{u_{bc}}{u_{ic}}} \qquad (4.3.6)$$

λ_c is in the range of 0.9 to
Form eqs. 4.3.1 and 4.3.2

$$m = \frac{c_{ci} - \Delta c_{ti}}{1 + \Delta} \qquad (4.3.7)$$

$$\frac{m \in}{\rho(1+\Delta)} - m = c_{ti} + \frac{c_g}{\rho\omega(1+\Delta)}$$

$$\frac{m \in \Delta}{\rho(1+\Delta)} + m = c_{ci} + \frac{(\Delta)c_g}{\rho\omega(1+\Delta)}$$

A combination of the above two expressions yields

$$\frac{m \in (1+\Delta)}{\rho(1+\Delta)} = (c_{ti} + c_{ci}) + \frac{(1+\Delta)c_g}{\rho\omega(1+\Delta)}$$

or

$$m \in = (c_{ti} + c_{ci})\rho + \frac{c_g}{\omega}$$

Therefore,

$$\in = \frac{1}{m}\left[\rho(c_{ti} + c_{ci}) + \frac{c_g}{\omega}\right] \qquad (4.3.8)$$

From eqs. 4.3,3 and 4.3.4

$$\left(1 + \lambda + R_{sl} + R\frac{u_{tc}}{u_{ic}}\right) = \frac{\omega\rho(1+\Delta)}{c_g}\left[c_{ce} + \eta m\left(\frac{\in}{1(1+\Delta)} - 1\right)\right]$$

$$\left(1 + \lambda + R_{sl} + R\frac{u_{bc}}{t_{ic}}\right) = \frac{\omega\rho(1+\Delta)}{\Delta c_g}\left[c_{te} + \eta m\left(\frac{\Delta \in}{\rho(1+\Delta)} + 1\right)\right]$$

or

$$\frac{R}{u_{tc}}(u_{tc} - u_{bc}) = \frac{\omega\rho(1+\Delta)}{c_g}\left[c_{ce} - \frac{c_{te}}{\Delta} + \eta m\left(-1 - \frac{1}{\Delta}\right)\right]$$

$$= \omega\rho\frac{(1+\Delta)}{\Delta c_g}\left[\Delta c_{ce} - c_{te} - \lambda m(1+\Delta)\right]$$

Therefore,

$$R = \frac{\rho\omega u_{ic}(1+\Delta)}{c_g(u_{tc} - u_{bc})\Delta}\left[\Delta c_{ce} - c_{te} - \eta m(1+\Delta)\right] \qquad (4.3.9)$$

4.4. Solution of the Governing Equations

The four eqs. 4.3.1 to 4.3.4 give a criteria for design of a section. The unknown values in the four equations are η, Δ, m, $\in$, R, ω, λ, R_{si}, u_{bc}, u_{tc}, and u_{ic}. Out of these 12 unknowns, some of them are indirectly known. In general, the slab dimensions are known and, therefore, all the quantities associated with concrete slab are defined. Some quantities like η, σ, ω could be assumed reasonably well. The four equations may now be solved with the following assumptions :

Assume quantities η, ρ, ω, λ, R_{si}, u_{bc}, u_{tc}, and u_{ic}, and then solve for Δ, $\in$, m, and R which are the basic sectional details. On the above basis, the solution of the equations may now be written as :

$$\Delta = \lambda_c \frac{\eta c_{ci} + c_{tw}}{\eta c_{tt} + c_{cw}} \qquad (4.3.5)$$

$$m = \frac{c_{ci} - \Delta c_{ti}}{1 + \Delta} \qquad (4.3.7)$$

$$\in = \frac{1}{m}\left[\rho(c_{ti} + c_{ci}) + \frac{c_g}{\omega}\right] \qquad (4.3.8)$$

$$R = \frac{u_{ic}\omega\rho(1+\Delta)}{c_g(u_{tc} - u_{bc})\Delta}\left[\Delta c_{ce} - c_{te} - \eta m(1+\Delta)\right] \qquad (4.3.9)$$

Let y_0 be the distance between the centroids of the composite and non-composite sections, then

$$y_{tc} = y_t - y_0 \quad \text{and} \quad y_{bc} = y_b + y_0$$

$$u_{tc} = \frac{y_{tc}}{y_t} = \left(1 + \frac{y_0}{y_b}\right) \quad \text{or} \quad y_0 = (1 - u_{tc})y_t$$

$$u_{bc} = \frac{y_b + y_0}{y_b} = 1 + \frac{y_0}{y_b} = 1 + \left[\frac{y_o}{y_t}\right]\frac{1}{\Delta} = 1 + \frac{1 - u_{tc}}{\Delta} \qquad (4.3.10)$$

$$I_c = I + A(y_0)^2 + A_{es}(y_{tc} + t_s/2)^2$$

$$u_{ic} = \frac{I_c}{I} = 1 + \frac{A}{I} y_0^2 + \frac{A_{cs}}{I} \left(y_{tc} + t_s/2 \right)^2$$

$$= 1 + \left[\frac{y_0}{r} \right]^2 + \frac{A_{cs}}{Ar^2} \left[y_{tc} + t_s/2 \right]^2$$

$$= 1 + \frac{1}{\rho} \left[\frac{1 - u_{tc}}{1 + \Delta} \right]^2 + \frac{\lambda}{\rho} \left[\frac{u_{tc}}{1 + \Delta} + \frac{t_s}{2h} \right]^2 \qquad (4.3.11)$$

Once u_{tc} is assumed, u_{bc} and u_{ic} can be computed from eqs. 4.3.10 and 4.3.11.

4.5. Ranges of Non-dimensionalised Parameters

(i) Depth factor ($\omega = hf_{ck}/\gamma L^2$): efficiency factor ($\rho = r^2/h^2$); shape factor ($D = y_b/y_t$); eccentricity ratio ($\in = e/h$); reinforcement factor ($m = P_i/Af_{ck}$); and moment ratio ($R = M_d/M_g$) which have already been discussed in the previous chapter.

(ii) Composite section factor (λ_c): It can be observed from eq. 4.3.6 that the value of λ_c tends to unity for non-composite section.

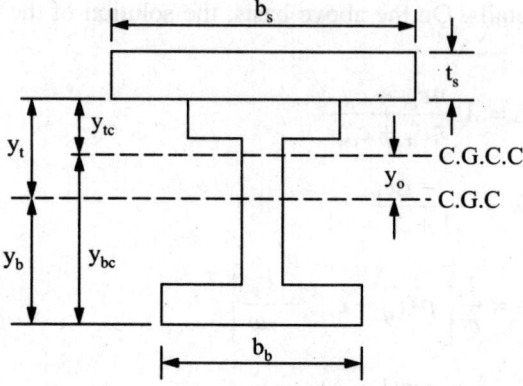

Figure 4.5.1 Composite section.

It could be seen from Figure 4.5.1 that $u_{tc} \leq 1$ and $u_{bc} \geq 1$. The value of u_{tc} is in the range of 0.80 to 1, u_{bc} in the range of 1 to 1.20 and the value of R generally varies in the range of 0.5 to 6. Hence, the range of λ_c will be from 0.90 to 1. For many practical purposes, λ_c may be assumed as 0.95 and Δ be solved from eq. 4.3.5. The procedure for design is same as given in chapter 3. All these equations reduce to a non-composite section by setting A_{cs} equal to zero.

4.6. Shrinkage Stresses

The composite construction may consist of concrete of two different mixes cast at two different time intervals. The shrinkage in the beam is more or less completed by the time the slab starts shrinking. The differential shrinkage causes stresses in composite construction. Assuming a certain shrinkage in the slab, strain compatibility as to be satisfied from which the stresses due to shrinkage are computed. Let $\in_a$ = shrinkage strain in slab and if the slab is completely restrained, the force developed due to the restraint is given by

$$F_a = \in_a E_{cs} A_{cs} \tag{4.6.1}$$

where F_a is the force caused by shrinkage.

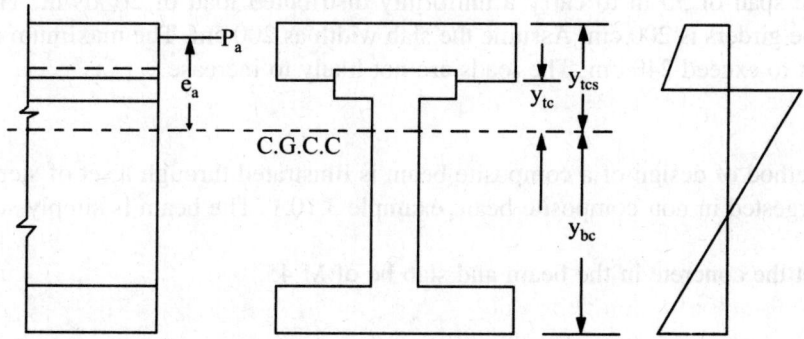

Figure 4.6.1 Differential shrinkage stress.

The stresses caused by shrinkage force are now computed using a simple beam theory. The extreme fibre stresses caused by shrinkage are

$$\sigma_{bb} = -\left(\frac{F_a}{A_c} + \frac{F_a e_a y_{bc}}{I_c} \right) \tag{4.6.2}$$

$$\sigma_{tb} = -\left(\frac{F_a}{A_c} + \frac{F_a e_a y_{tc}}{I} \right) \tag{4.6.3}$$

$$\sigma_{bs} = -\left(\frac{F_a}{A_c} + \frac{F_a e_a y_{tc}}{I_c} \right) + \sigma_a \tag{4.6.4}$$

$$\sigma_{ts} = -\left(\frac{F_a}{A_c} + \frac{F_a e_a y_{tc}}{I_c} \right) + \sigma_a \tag{4.6.5}$$

where

σ_{bb} = stress at bottom fibre of the beam

σ_{tb} = stress at the top fire of the beam

σ_{bs} = stress at bottom fibre of the slab

σ_{ts} = stress at top fibre of the slab

y_{tcs} = stress at top fibre of the slab from the *CGC* of the composite section

$\sigma_a = \epsilon_a E_{cs}$

E_{cs} = Young's modulus of concrete of the slab

The stresses that are computed are approximate but are good for practical purposes.

The shrinkage stresses are in addition to the stresses at working load conditions. It is necessary to superimpose the stresses at working load on the shrinkage stress and then check for the allowable stresses.

Design Example 4.6.1   Design of a post-tensioned prestressed concrete composite beam for an effective span of 50 m to carry a uniformly distributed load of 20 *k*N/m. The centre to centre of the girders is 200 cm. Assume the slab width as 200 cm. The maximum depth of the beam is not to exceed 240 cm. The loads are not likely to increase.

Solution

The method of design of a composite beam is illustrated through a set of steps similar to the one suggested in non-composite beam example 3.10.1. The beam is simply supported, so $c_g = \dfrac{1}{8}$. Let the concrete in the beam and slab be of M 45.

(i)   *Code specifications, assumptions and other data*

Maximum permissible stresses in N/cm^2 are :

$$\sigma_{aci} = 0.46(4500) = 2080$$

$$\sigma_{ati} = 100$$

$$\sigma_{ace} = 0.38(4500) = 1740$$

$$\sigma_{ate} = 100$$

It is already known from example 3.10.1 that the minimum eccentricity required will fall out of the section for full permissible stresses, so assume $c_{ti} = -0.08$ and the rest full stresses.

Let $\quad\quad\quad\quad h = 220$ cm, $\rho = 0.121$, $\eta = 0.82$, $\lambda_c = 0.95$, $u_{tc} = 0.85$ and $\lambda = 0.2$

(ii)   *Design variables*

$$\Delta = (\lambda_c)\frac{\eta c_{ci} + c_{tc}}{\eta c_{ti} + c_{ce}} = 0.95\left[\frac{(0.82)(0.46)+(0.02)}{-(0.82)(0.08)+0.38}\right] = 1.20$$

$$\omega = \frac{hf_{ck}}{\gamma L^2} = \frac{(220)(4500)}{(0.024)(5000)^2} = 1.65$$

Select a section from the table 2 of appendix A such that $\rho = 0.121$ and $\Delta = 1.378$.

The minimum eccentricity requirement is

$$\epsilon_{req.} = \frac{1+\Delta}{c_{ci} - \Delta c_{ti}} \left[\rho\left(c_{ci} + c_{ci}\right) + \frac{c_g}{\omega} \right]$$

$$= \frac{2.378}{0.572} \left[0.121(0.38) + 0.0756 \right] = 0.5$$

$$\epsilon_{lim.} = \frac{\Delta}{1+\Delta} - 0.1 = 0.49$$

It is possible to accommodate the cable within the section since $\epsilon_{req.}$ is approximately equal to $\epsilon_{lim.}$

From eqs. 4.3.10 and 4.3.11 one gets

$$u_{bc} = 1 + \frac{1 - u_{tc}}{\Delta} = 1 + \frac{0.15}{1.378} = 1.108$$

$$u_{ic} = 1 + \frac{1}{\rho}\left[\frac{1 - u_{tc}}{1 + \Delta}\right]^2 + \frac{\lambda}{\rho}\left[\frac{u_{tc}}{1 + \Delta} + \frac{t_s}{2h}\right]^2$$

$$= 1 + \frac{1}{0.121}\left[\frac{0.15}{2.378}\right]^2 + \frac{0.2}{0.121}\left[\frac{0.85}{2.378} + \frac{8}{440}\right]^2$$

$$= 1 + 0.033 + 0.23 = 1.26$$

From eq. 4.3.7

$$m = \frac{c_{ct} - \Delta c_{tt}}{1 + \Delta} = \frac{0.46 + 0.11}{2.378} = 0.24$$

From eq. 4.3.9

$$R = \frac{\rho\omega u_{ic}\left(1 + \Delta\right)}{c_g\left(u_{bc} - u_{tc}\right)\Delta} \left[\eta m\left(1 + \Delta\right) + c_{te} - \Delta c_{ce} \right]$$

$$= \frac{8(0.121)(1.65)(1.26)(2.378)}{(0.258)(1.378)} \left[(0.82)(0.240)(2.378) + 0.02 - 1.378(0.38) \right] = 1.072$$

It should be noted from the above calculations that the value of R gets affected very much by improper choice in u_{tc} or Δ etc. It is very often difficult to select right matching values of u_{tc} and Δ so reader is advised to compute R from the simply supported beam formula (Eq. 3.8.10) and increase it by about $1 + \lambda$ times. Using this approximation and increasing the value of R by $(1 + 0.15)$ times, one gets

$$R = (1.15)\left[(8\rho\omega)\left(c_{tc} + \eta_{ci}\right)\left(\frac{1 + \Delta}{\Delta}\right) - (1 - \eta) \right]$$

$$= (1.15)\left[(8)(0.121)(1.65)(0.396)\left(\frac{2.378}{1.378}\right) - 0.18\right] = 1.08$$

The value of assumed λ is 0.2 so an approximate increase of 15 per cent is allowed and checked with that obtained from eq. 4.3.9. Since the two values of R are very close to each other, there is no need for further trial.

$$w_g = \frac{w_a}{R} = \frac{20000}{1.07} = 18700 \text{ N/m}$$

$$A = \frac{18700}{24000} = 0.78 \text{ m}^2 = 7800 \text{ cm}^2$$

(iii) *Calculation of properties of composite section*

The design coefficients from table 2 of appendix A for

$\rho = 0.121$ and $\Delta = 1.378$ are

$c_f = 0.5$, $c_t = 0.12$, $c_b = 0.12$, $c_w = 0.2$

$c_a = 0.332$, $c_t = 0.04$

$c_{yb} = 0.58$ and $c_{yt} = 0.42$

$y_b = 0.58\,(220) = 127.6$ cm, $y_t = 92.4$ cm

$A_c = 1600 + 7800 = 9400 \text{ cm}^2$; $\lambda = A_{cs}/A = 1600/7800 = 0.206$

Taking moments of areas of the beam and the slab about the centroid of the beam, the value of y_0 can be obtained.

$A_c y_0 = (A_{cs})(t_s/2 + y_t)$

$y_0 = 1600\,(4 + 92.4)\,/9400 = 16.4$ cm

$y_{tc} = 92.4 - 16.4 = 76.0$ cm; $u_{tc} = 76\,/92.4 = 0.825$

$y_{bc} = 127.6 + 16.4 = 144.0$ cm; $u_{bc} = 144\,/127.6 = 1.13$

$I_c = Ah^2\rho + Ay_0^2 + A_{cs}\,(t_s/2 + y_{tc})^2$

$\quad = 7800\,((220)^2\,(0.121) + (16.4)^2) + 1600\,(80)^4$

$\quad = 7800\,(5850 + 269 + 1350) = 7800\,(7469)\text{ cm}^2$

$u_{tc} = I_c/I = 7800\,(7469)\,/7800\,(5850) = 1.27$

The assumed values of λ, u_{tc}, u_{bc} and u_{ic} are 0.2, 0.85, 1.108 and 1.26 respectively, while the actual corresponding values are 0.206, 0.825, 1.13 and 1.27. The quantities compare closely so a check for actual stresses can be made before finalising the section.

(iv) Check for stresses

$m = 0.24$, $\epsilon = 0.5$, $\rho = 0.121$, $\Delta = 1.378$, $\omega = 1.65$, $\eta = 0.82$, $\lambda = 0.206$, $R = 1.07$, $u_{tc} = 0.825$, $u_{bc} = 1.13$, $u_{ic} = 1.27$. Substitution of the above quantities in eqs. 4.3.1 to 4.3.4 gives the actual stresses and they are

$$c_{ti} = m\left[\frac{\epsilon}{\rho(1+\Delta)} - 1\right] - \frac{1}{8\rho\omega(1+\Delta)}$$

$$= 0.24\left[\frac{0.5}{(0.121)(2.378)} - 1\right] - \frac{1}{8(0.121)(1.65)(2.378)}$$

$$= 0.24\,(1.73 - 1) - 0.264 = 0.175 - 0.264$$

$$= -0.089 < c_{ati}$$

$$c_{ci} = m\left[\frac{\epsilon\Delta}{\rho(1+\Delta)} + 1\right] - \frac{\Delta}{8\rho\omega(1+\Delta)}$$

$$= 0.24\,[1.73\,(1.378) + 1] - 1.378\,(0.264) = 0.815 - 0.366$$

$$= 0.449 < c_{aci}$$

$$c_{ce} = -\eta m\left[\frac{\epsilon}{\rho(1+\Delta)} - 1\right] + \frac{1}{8\rho\omega(1+\Delta)}\left[1 + \lambda + R\frac{u_{tc}}{u_{ic}}\right]$$

$$= -0.82\,(0.175) + 0.264\,[1.206 + 1.07\,(0.825)/1.27]$$

$$= 0.36 < c_{ace}$$

$$c_{te} = -\eta m\left[\frac{\epsilon\Delta}{\rho(1+\Delta)} + 1\right] + \frac{\Delta}{8\rho\omega(1+\Delta)}\left[1 + \lambda + R\frac{u_{bc}}{u_{ic}}\right]$$

$$= -0.82\,(0.85) + 0.366\,(1.206 + 0.95)$$

$$= -0.667 + 0.78 = 0.113 \not< c_{ate}\ (\text{not permissible})$$

(v) Modification of the section

The tensile stress at working load exceeds the permissible stress marginally while the remaining stresses are within the permissible limits. A marginal revision in the sectional details is necessary to ensure the stresses even after superposition of shrinkage stresses are within the permissible limits. Any of the following modifications or a combination of them can be tried :

(a) Increase the depth of the beam.

(b) Increase the efficiency factor.

(c) Increase the areas of cross section which means decrease in the value of R.

(d) Increase the prestressing force.

The depth and area of cross section selected for this example are less than those of example 3.10.1 which is justified as in this case the loads are not likely to increase. Keeping the depth and the area of cross section same, increase the flexural efficiency of the cross section while keeping the shape factor Δ around 1.378.

Select the following section from table 2 of appendix A.

$$c_f = 0.6, \qquad c_w = 0.16, \qquad c_t = 0.1, \qquad c_b = 0.08, \qquad c_a = 0.279,$$

$$c_i = 0.036, \qquad c_{yb} = 0.577, \qquad c_{yt} = 0.423, \qquad \rho = 0.129, \qquad \Delta = 1.366$$

The least dimension in this section is that of the thickness of the bottom flange and it is $0.08(220) = 17.6$ cm which is within the limits.

(vi) *Calculation of properties of the composite section*

$$y_b = 0.577 \ (220) = 126.9 \text{ cm}, \ y_t = 93.1 \text{ cm}$$

The following computations are made using many quantities already computed in subsection (iii).

$$y_0 = 1600 \ (4 + 93.1) \ /9400 = 16.5 \text{ cm}$$

$$y_{tc} = 93.1 - 16.5 = 76.6 \text{ cm}; \ u_{tc} = 76.6 \ /93.1 = 0.825$$

$$y_{bc} = 126.9 + 16.5 = 143.4 \text{ cm}; \ u_{bc} = 143.4 \ /126.9 = 1.13$$

Similarly $u_{ic} = 1.27$

(vii) *Recheck for stresses*

The values of u's have remained same. Therefore, the stresses can be calculated using the steps of subsection (iv).

$$c_{ti} = 0.24 \left[\frac{0.5}{(0.129)(2.366)} - 1 \right] - \frac{1}{8(0.129)(2.366)(1.65)}$$

$$= 0.24 \ (1.64 - 1) - 0.249 = 0.154 - 0.249 = -0.095 < c_{ati}$$

$$c_{ci} = 0.24 \ [(1.64)(1.366) + 1] - 1.366 \ (0.249) = 0.776 - 0.34 = 0.436 < c_{cai}$$

$$c_{ce} = -0.82 \ (0.154) + 0.249 \ [1.206 + 1.07 \ (0.825) \ /1. \ 27]$$

$$= -0.126 + 0.473 = 0.347 < c_{ace}$$

$$c_{te} = -0.82 \ (0.776) + 0.34 \ (2.156) = -0.635 + 0.73 = 0.095 \ngtr c_{ate}$$

(viii) *Final modification of the section*

Since the tensile stress at working is still more than the permissible, increase the prestressing force and the depth of the beam.

Let $m = 0.25$ instead of 0.24 and $h = 240$ cm instead of 220 then the values of u's remain same and the c_{ti}, c_{ci} and c_{ce} are reduced which are on the safer side. c_{te} can be calculated by proportioning the quantities of the previous calculations

$$c_{te} = -0.635\,(0.25)\,/0.24 + 0.73\,(220)\,/240$$

$$= -0.66 + 0.67 = 0.01 < c_{ate}$$

(ix) *Details of the section*

The details of the beam can be calculated by using the non-dimensionalised parameters which are already finalised.

$$c_f = 0.6,\ c_w = 0.16,\ c_t = 0.1,\ c_b = 0.08,\ c_a = 0.279$$

$$c_i = 0.036,\ c_{yb} = 0.577,\ c_{yt} = 0.423,\ \rho = 0.129,\ \Delta = 1.366$$

$$m = 0.25,\ h = 240\ \text{cm},\ A_{cs} = 1600\ \text{cm}^2,\ A = 7800\ \text{cm}^2$$

$$y_b = 0.577\,(240) = 138.4\ \text{cm},\ y_t = 240 - y_b = 101.6\ \text{cm}$$

$$c_a b_t h = 7800\ \text{or}\ b_t = \frac{7800}{(0.279)(240)} = 117\ \text{cm}$$

$$b_b = c_f b_t = 0.6\,(117) = 70.2\ \text{cm},\ b_w = 0.16\,(117) = 18.7\ \text{cm}$$

$$t_t = 0.1\,(240) = 24\ \text{cm},\ t_b = 0.08\,(240) = 19.2\ \text{cm}$$

$$P_i = mAf_{ck} = 0.25\,(7800)(4500) = 880\,0000\ \text{N}$$

$$A_s = \frac{P_i}{0.7\sigma_{st}} = \frac{880\,0000}{10\,5000} = 83.8\ \text{cm}^2$$

$$e = 0\,5h = 120\ \text{cm}$$

$$I = A\rho h^2 = 7800\,(0.129)\,(240)^2 = 56 \times 10^6\ \text{cm}^4$$

$$I_c = 1.27\,(56 \times 10^6) = (75)\,10^6\ \text{cm}^4$$

(x) *Shrinkage stresses*

Let the differential shrinkage of the slab and beam be 0.0002 cm/cm $= \epsilon_a$
From eq. 4.6.1

$$F_a = \epsilon_a E_{cs} A_{cs} = (0.0002)\,(3.82 \times 10^6)\,(1600) = 122\,0000\ \text{N}$$

$$\frac{F_a}{A_c} = \frac{122\,0000}{9400} = 130\ \text{N/cm}^2$$

$$e_a = (t_s/2 + y_{tc}) = (4 + u_{tc}y_t) = [4 + 0.825\,(101.6)] = 88\ \text{cm}$$

$$y_{bc} = 1.13\,(138.4) = 156\ \text{cm}$$

From eqs. 4.6.2 to 4.6.5 one gets the stresses caused by shrinkage. All stresses are in N/cm^2.

Direct shrinkage stress in the slab

$$= \varepsilon_a E_c = 0.0002 \, (3.82)(10^6) = 764 \text{ (tension)}$$

$$\sigma_{tb} = -\frac{F_a}{A_c} - \frac{F_a e_a}{I_c} y_{tc}$$

$$= -130 - \frac{(122\,0000)(88)(84)}{75 \times 10^6} = -130 - 120 = -250 \text{(comp)}$$

$$\sigma_{bb} = -\frac{F_a}{A_c} + \frac{F_a e_a}{I_c} y_{bc}$$

$$= -130 + \frac{(122\,0000)(88)(156)}{75 \times 10^6} = -130 + 220 = 90$$

$$\sigma_{ts} = -130 - \frac{(122\,0000)(88)(92)}{75 \times 10^6} + 764 = 500$$

$$\sigma_{bs} = -130 - 120 + (0.0002)(3.\,82 \times 10^6) = -250 + 764 = 514$$

(xi) *Final stresses in the section*

The stresses in the beam at transfer are unaffected by the shrinkage so only the stresses at working load which are affected by shrinkage are given here. Stress at the top fibre of the beam is

$$\sigma_{1w} = \sigma_{ce} + \sigma_{tb} = -c_{ce} f_{ck} + (-250) = -0.347 \,(4500) - 250 = 1810$$

Stress at the bottom fibre of the beam is

$$\sigma_{2w} = \sigma_{te} + \sigma_{bb} = c_{te} f_{ck} + \sigma_{bb} = 0.01 \,(4500) + 90 = 135$$

whereas permissible tension is 100 N/cm^2.

The stresses in the beam are marginally excess and can be accounted by increasing the area of the actual section. Stresses in Slab.

$$\sigma_{1s} = \frac{M_a y_{tcs}}{I_c} + \sigma_{ts} = \frac{(20000)(250\,000)(-92)}{75 \times 10^6} + 500 = -(1.13)$$

$$\sigma_{2s} = \frac{M_a y_{tc}}{I_c} + \sigma_{bs} = \frac{(20000)(250\,000)(-84)}{75 \times 10^6} + 514 = -50$$

The stresses in slab at working load are within the permissible limits, however, the stress caused by shrinkage are much in excess of the permissible stress. So longitudinal mild steel is to be provided in the slab. Provide 8 mm bars at 15 cm spacing.

The actual cross section is shown in Figure 4.6.2. The idealised sizes of the flange and web are modified to suit the construction requirements. Minimum nominal *MS* reinforcement of 0.2 per cent in the cross section and 0.25 per cent in the web has to be provided.

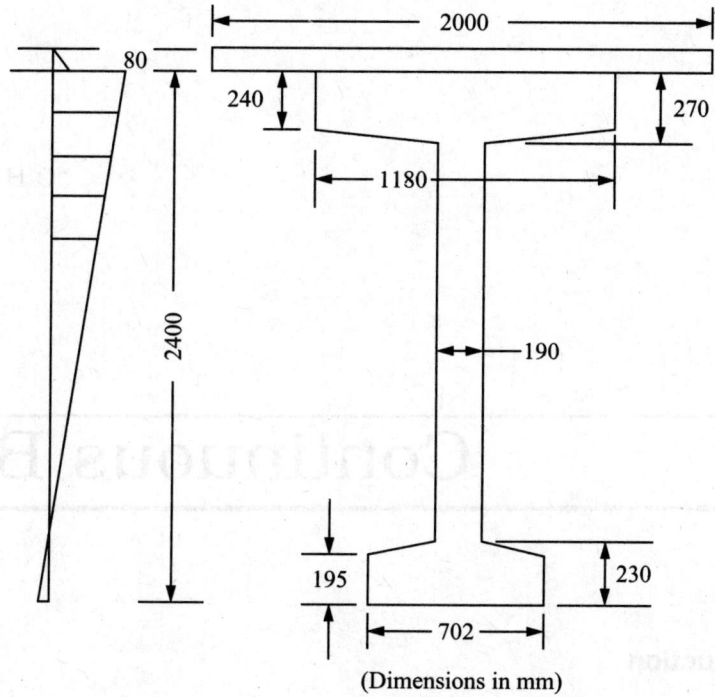

Figure 4.6.2 Cross section of example 4.6.1

PROBLEMS

4.1 A simply supported beam of span 20 m is carrying a load of 24 kN/m. Design a composite beam with top slab of 15 cm thickness and 150 cm width. Use f_{ck} = 40 MPa and f_p = 1500 MPa for beam and f_{ck} = 20 MPa for slab concrete. Make a comparative study of this design with problem 4.1.

4.2 The bridge described in problem 4.2 is provided with a top slab of 15 cm thickness using M20 concrete. Design a composite beam bridge, using the data provided in the problem 4.2.

4.3 Design a rectangular prestressed concrete beam to support a slab of 4 m width and made of M20 concrete. The beam is simply supported and made of M40 concrete with pretensioned HTS wires. The span of the beam is 8 m and the slab is cast over the beam as a composite section. The thickness of the slab is 100 mm and it is carrying a uniformly distributed load of 4 kN/m².

CHAPTER 5

Continuous Beams

5.1. Introduction

Prestressing force causes deformation which can take place freely in a statically determinate structure without causing any change in the reactions. In a statically indeterminate structure, the deformation due to prestressing force will affect the compatibility conditions resulting in some secondary stresses. This effect is explained through a simple example of two span continuous beam shown in Figure 5.1.1. Let the beam rest freely on three supports at same level and let it be prestressed with a straight tendon. The tendency of the tendon will be to lift the beam from the middle support. Hence, there should be a force at the middle support to keep the compatibility of the beam resting on all the three supports. This downward force 'R_b' has to be balanced by upward reactions at the extreme supports. The stresses caused due to these reactions are called secondary stresses.

Figure 5.1.1 gives a complete picture of the problem. [(a) Two span continuous beam (no external force including self-weight); (b) Prestressing force acting and beam free to deform; (c) Constraint due to the continuous support; (d) Secondary shear force; (e) Bending moment due to eccentric prestressing force; (f) Bending moment due to secondary reactions; (g) Actual bending moments]. The shear force and bending moment caused by the prestressing force are also shown in the figure. The secondary stresses depend mainly upon the profile of the prestressing tendon and, if a proper profile is not chosen, the secondary stresses can be very large. On the other hand, if an appropriate profile of the tendon is chosen, the secondary stresses could be reduced to zero. A tendon of this profile is called "concordant cable". A concordant cable may be defined as the cable which will not produce any statically indeterminate reactions. The concordant cable profile is not unique in any given statically indeterminate structure but it is located within a certain narrow zone. The restriction to the concordant profile zone has sometimes the disadvantage of not utilising the depth of the beam effectively for a suitable profile in working load conditions. There are two possible important practical considerations: (i) the

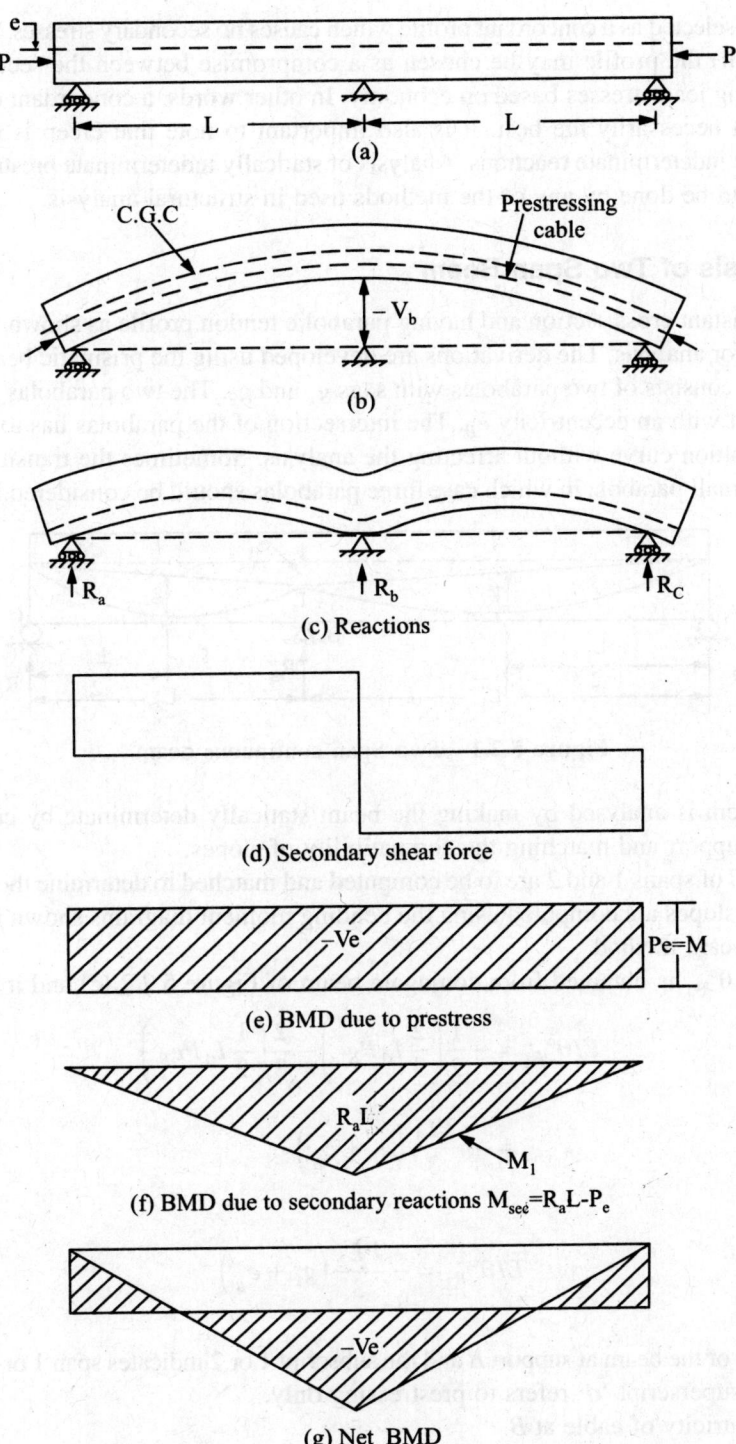

Figure 5.1.1 Two span continuous prestressed beam.

profile may be selected as a concordant profile which causes no secondary stresses, not considering economy, or (ii) the profile may be chosen as a compromise between the secondary stresses and the working load stresses based on economy. In other words, a concordant cable is a good choice but not necessarily the best. It is also important to note that creep is likely to cause some statically indeterminate reactions. Analysis of statically indeterminate prestressed concrete structures could be done by any of the methods used in structural analysis.

5.2. Analysis of Two Span Beam

A beam of constant cross section and having parabolic tendon profile as shown in Figure 5.2.1 is considered for analysis. The derivations are developed using the prismatic beam theory. The tendon profile consists of two parabolas with sags g_1 and g_2. The two parabolas intersect at the middle support with an eccentricity e_B. The intersection of the parabolas has to be rounded to a smooth transition curve without affecting the analysis. Sometimes the transition curve may be taken as a small parabola in which case three parabolas should be considered in the analysis.

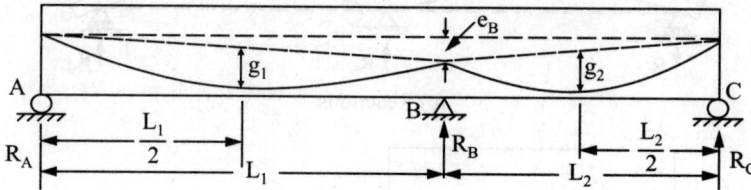

Figure 5.2.1 Two Span continuous beam.

The problem is analysed by making the beam statically determinate by cutting it at the intermediate support and matching the compatibility of slopes.

Slopes at B of spans 1 and 2 are to be computed and matched to determine the indeterminate reactions. The slopes are computed using the bending moment diagrams shown in Figure 5.2.2 by conjugate beam method.

The slope $\theta^\circ{}_{B_1}$ is obtained from conjugate beam of Figure 5.2.2 (c) and it is

$$EI\theta^\circ{}_{B_1} = -\frac{1}{2}\left(\frac{2}{3}L_1 P g_1\right) - \frac{2}{3}\left(\frac{1}{2}L_1 P e_B\right)$$

$$= -\frac{PL_1}{3}\left(g_1 + e_B\right) \tag{5.2.1}$$

Similarly

$$EI\theta^\circ{}_{B_2} = -\frac{PL_2}{3}\left(g_1 + e_B\right) \tag{5.2.2}$$

where

θ_B = slope of the beam at support B and the subscript 1 or 2 indicates span 1 or 2 respectively. The superscript 'o' refers to prestressing only.

e_B = eccentricity of cable at B.

Let θ'_B = slope at B caused by statically indeterminate force of second order.

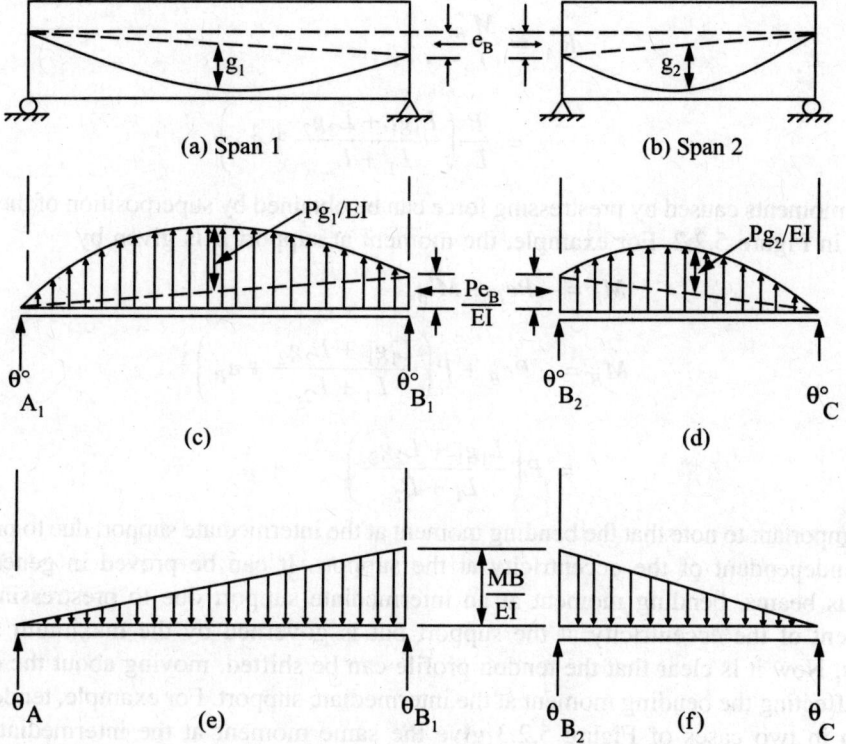

(a) Span 1   (b) Span 2

(c)   (d)

(e)   (f)

Figure 5.2.2   Bm of separated sections due to prestressing only.

Then from Figure 5.2.2 (d) and 5.2.2 (e) we get

$$\left.\begin{array}{l} EI\theta'_{B_1} = \dfrac{1}{3} L_1 M'_B \\[2mm] EI\theta'_{B_2} = \dfrac{1}{3} L_2 M'_B \end{array}\right\}$$ (5.2.3)

where M'_B = secondary BM at B.

The prime (') indicates the effect of the secondary force.

The condition of continuity is that the change in the slope between the right and left of the intermediate support should be zero i.e.,

$$\theta^°_{B_1} + \theta^°_{B_2} + \theta'_{B_1} + \theta'_{B_2} = 0$$ (5.2.4)

$$\therefore \frac{M'_B}{3}\left(L_1 + L_2\right) - \frac{P}{3}\left[L_1\left(e_B + g_1\right) + L_2\left(e_B + g_2\right)\right] = 0$$

or

$$M'_B = P\left(\frac{L_1 g_1 + L_2 g_2}{L_1 + L_2} + e_B\right)$$ (5.2.5)

Now,
$$R'_A = \frac{M'_B}{L_1}$$

$$= \frac{P}{L_1}\left(\frac{L_1 g_1 + L_2 g_2}{L_1 + L_2} + e_B\right) \tag{5.2.6}$$

Final moments caused by prestressing force can be obtained by superposition of the moments as shown in Figure 5.2.2. For example, the moment at support B is given by

$$M_B = -Pe_B + M'_B$$

$$M_B = -Pe_B + P\left(\frac{L_1 g_1 + L_2 g_2}{L_1 + L_2} + e_B\right)$$

$$= P\left(\frac{L_1 g_1 + L_2 g_2}{L_1 + L_2}\right) \tag{5.2.7}$$

It is important to note that the bending moment at the intermediate support due to prestressing force is independent of the eccentricity at the support. It can be proved in general that in continuous beams, bending moment at an intermediate support due to prestressing force is independent of the eccentricity at the support but is governed by the maximum sag of the parabolas. Now it is clear that the tendon profile can be shifted, moving about the end points without affecting the bending moment at the intermediate support. For example, tendon profiles as shown in two cases of Figure 5.2.3 give the same moment at the intermediate support. Hence particular attention should be paid while fixing the sag.

For a concordant cable

$R'_A = 0$, then from eq. 5.2.6,

$$e_B = -\left(\frac{L_1 g_1 + L_2 g_2}{L_1 + L_2}\right) \tag{5.2.8}$$

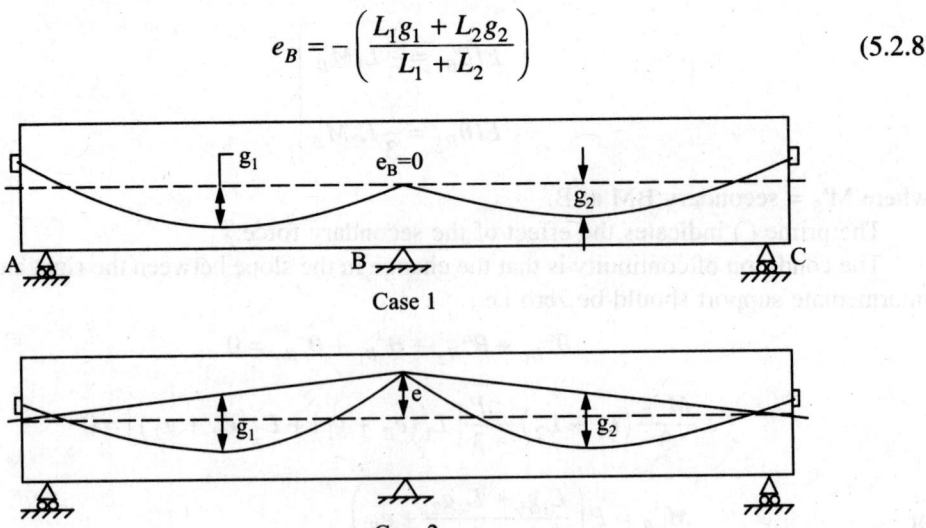

Figure 5.2.3 Two tendon profiles shown in the above two figures give the same moment at B.

Assuming a parabolic cable profile, the equation of the profile of the tendon can be expressed as

$$ex = a + bx + cx^2 \qquad (5.2.9)$$

where ex is the ordinate of the eccentricity at distance x from support, and a, b and c are constants. As per the boundary conditions

$$ex = 0 \text{ at } x = 0, \text{ or } ex = g_1 + e_B/2 \text{ at } x = L_1/2 \text{ and}$$

$$ex = + e_B \text{ at } x = L_1$$

Substitution of the boundary conditions of the cable in eq. 2.5.9 gives

$$a = 0$$

$$bL_1 + cL_1{}^2 = + e_B$$

$$\frac{bL_1}{2} + \frac{cL_1{}^2}{4} = g_1 + \frac{e_B}{2}$$

$$cL_1{}^2\left(1 - \frac{1}{2}\right) = -2g_1$$

$$c = -4g_1/L_1{}^2$$

$$b = (e_B + 4g_1)/L_1$$

$$e_x = x[(4g_1 + e_B)L_1 - 4g_1x]/L_1{}^2 \qquad (5.2.10)$$

Determinations of shear forces : Shear force in the span AB caused by prestress is

$$V_x = R'_A + P\sin\alpha = R'_A - P\left(\frac{de_x}{dx}\right)$$

$$= \frac{P}{L_1}\left[\frac{L_1g_1 + L_2g_2}{L_1 + L_2} + e_B\right] - P\left[\frac{4g_1 + e_B}{L_1} - \frac{8g_1x}{L_1{}^2}\right]$$

$$= \frac{P}{L_1}\left[\frac{L_1g_1 + L_2g_2}{L_1 + L_2} + e_B\right] - \left[4g_1 + e_B - \frac{8g_1x}{L_1}\right]$$

$$= \frac{P}{L_1}\left[L_1g_1\left(8x - 3L_1\right) + L_2\left(L_1g_2 - 4g_1L_2 + 8g_1x\right)\right]\frac{1}{L_1\left(L_1 + L_2\right)}$$

$$= \frac{P}{L_1}\left[\frac{g_1\left(8x - 3L_1\right) + L_2\left(g_2 - 4g_1\right)}{L_1 + L_2} + \frac{8g_1xL_2}{L_1\left(L_1 + L_2\right)}\right] \qquad (5.2.11)$$

182

Example 5.3.1 A two span continuous beam of 16 m each span is provided with a prestressing cable. The prestressing force is 2000 kN and the sag of the cable at mid-span is 40 cm. The cable has an eccentricity of –30 cm at the intermediate support. Determine the secondary bending moments and reactions in the beam. Find a cable profile if the cable is to give maximum effectiveness with or without secondary bending moments. The cross section of the beam is symmetric and depth is 80 cm.

Solution

$L_1 = L_2 = L = 16$ m, $g_1 = g_2 = g = 40$ cm, $e_B = -30$ cm, $P = 2000$ kN.

The secondary bending moment can be obtained from eq. 5.2.5 and it is

$$M'_B = P\left(\frac{L_1 g_1 + L_2 g_2}{L_1 + L_2} + e_B\right) = P\left[\frac{2Lg}{2L} + e_B\right]$$
$$= 2000\ (40 - 30) = 20000\ kN\ cm$$

M'_B is positive so it causes compression on the top fibre

$$R'_A = \frac{P}{L_1}\left(\frac{L_1 g_1 + L_2 g_2}{L_1 + L_2} + e_B\right) = \frac{2000}{1600}(40 - 30) = 12.5\ kN$$

The maximum effectiveness of the cable for ultimate strength of the sections is obtained by keeping the cable closest to the tension fibre. In a two span continuous beam the top fibre of the section at intermediate support and bottom fibre of the section at mid-span are subjected to tension at working load so keep the cable closest to these fibres. Let the minimum cover distance from the cable be 10 cm, then maximum possible eccentricities at mid support and mid-span are –30 and 30 cm respectively.

$$e_B = -30,\ g = 30 + \frac{30}{2} = 45\ cm$$

Secondary bending moment caused by such a cable is

$$M'_B = 2000\ (45 - 30) = 30000\ kN\ cm$$

However, for a concordant cable profile, the eccentricity relations are given by eq. 5.2.8.

$$e_B = -\frac{L_1 g_1 + L_2 g_2}{L_1 + L_2} = -g$$

Providing maximum possible eccentricity at mid support, i.e.,

$$e_B = -30\ cm$$

$$g = -e_B = 30\ cm$$

5.3. Analysis of Two Span Continuous Beam with Eccentricities at Outer Supports

For a cable profile having eccentricities at the end supports, the problem can be divided into two parts: (i) moment caused by simple parabolic cable with zero eccentricities at outer supports plus, (ii) moment caused by straight cables with eccentricities at the end supports. The superposition of the two moments will give the resultant moment due to the cable profile. The slope compatibility can easily be obtained by considering the two parts separately. The moments are shown in Figure 5.3.1.

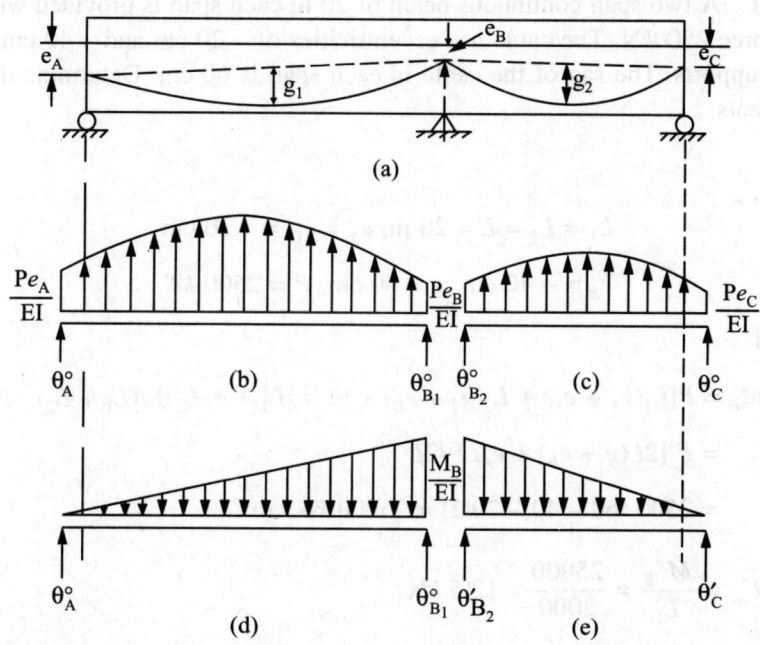

Figure 5.3.1 Two span continuous beam with end eccentricities.

Summing up the slope at B

$$(\theta^\circ_{B_1})_P + (\theta^\circ_{B_2})_P + (\theta^\circ_{B_1})_A + (\theta^\circ_{B_2})_C + (\theta^\circ_{B_1}) + (\theta^\circ_{B_2}) = 0$$

$$-\frac{P}{3}\left[L_1(g_1 + e_B) + L_2(g_2 + e_B) + \frac{1}{2}(e_A L_1 + e_C L_2)\right] + \frac{M'_B}{3}(L_1 + L_2) = 0$$

where e_A, e_B and e_c are eccentricities of the cable at A, B and C respectively

$$M'_B = P[L_1(g_1 + e_B) + L_2(g_2 + e_B) + \frac{1}{2}(e_A L_1 + e_C L_2)]/(L_1 + L_2) \qquad (5.3.1)$$

$$M_B = M'_B + Pe_B$$

$$= P\frac{L_1 g_1 + L_2 g_2 + \frac{1}{2}(e_A L_1 + e_C L_2)}{L_1 + L_2} \qquad (5.3.2)$$

For concordant cable, set $M'_B = 0$. So from eq. 5.3.1

$$e_B = -[L_1 g_1 + L_2 g_2 + \frac{1}{2}(e_A L_1 + e_C L_2)]/(L_1 + L_2) \qquad (5.3.3)$$

Same procedure could be extended to three or more spans. However, relaxation at the supports leads to tedious calculations in solving simultaneous equations. Moment distribution procedure could be used to solve problems of higher indeterminancy.

Example 5.3.1 A two span continuous beam of 20 m each span is provided with a cable of prestressing force 2500kN. The cable has eccentricities of – 20 cm and – 40 cm at outer and intermediate supports. The sag of the cable in each span is 60 cm. Determine the secondary bending moments.

Solution

$$L_1 = L_2 = L = 20 \text{ m}, \ e_A = e_C = -20 \text{ cm}$$

$$e_B = -40 \text{ cm}, \ g = 60 \text{ cm}, \ P = 2500 \ kN$$

From eq. 5.3.1

$$M_B = P[L_1(g_1 + e_B) + L_2(g_2 + e_B) + \tfrac{1}{2}(e_A L_1 + e_C L_2)] / (L_1 + L_2)$$

$$= P\,[2L(g + e_B) + e_A L]\,/2L$$

$$= 2500\,(60 - 40 - 20/2) = 25000 \ kN \text{ cm}$$

$$R'_A = \frac{M'_B}{L_1} = \frac{25000}{2000} = 12.5 \ kN$$

5.4. Fixed End Bending Moments

The moment distribution procedure starts by fixing the section of the beam at the supports against rotation. It is essential to calculate the fixed end moments due to prestressing force and distribute them in the same way as done for fixed end moments caused by external loads. Figure 5.4.1 shows a fixed end beam with prestressing cable.

Figure 5.4.1 (b) gives the conjugate beam which is subjected to moments caused by the prestressing cable. The figure illustrates the cable with positive eccentricities which cause compression on the bottom fibre. Therefore, the M/EI load is acting upwards. The slopes of the beam caused by the cable force are given by the reactions of the conjugate beam and they are

$$EI\theta^\circ{}_A = -\frac{2}{3}\left(\frac{Pe_A L}{2}\right) - \frac{1}{3}\left(\frac{Pe_B L}{2}\right) - \frac{1}{2}\left(\frac{2PgL}{3}\right)$$

$$= -\frac{PL}{6}\left[2e_A + e_B + 2g\right]$$

Similarly, the slope at B caused by the cable is

$$EI\theta^\circ{}_B = -\frac{PL}{6}\left[e_A + 2e_B + 2g\right]$$

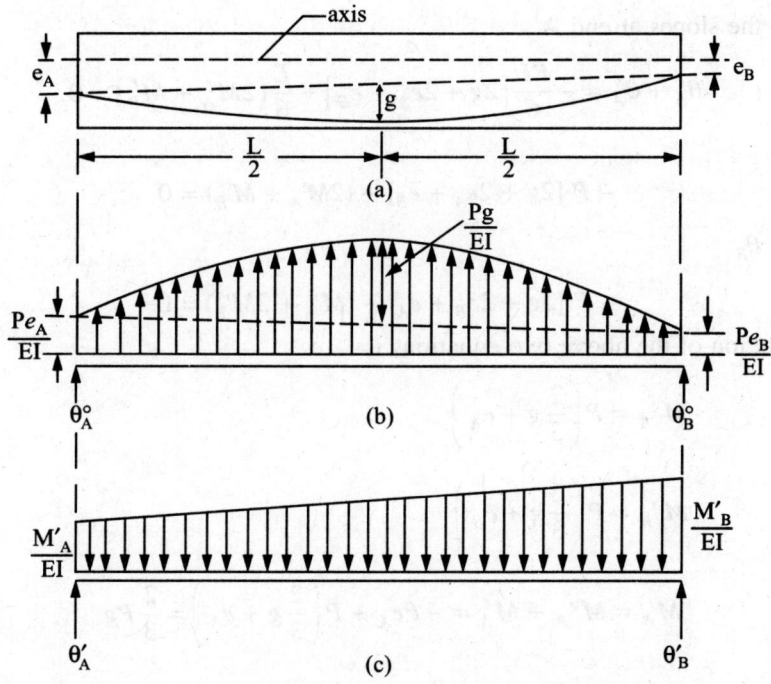

Figure 5.4.1 Fixed end beam.

Figure 5.4.2 (b) gives the conjugate beam with secondary moment as the load on the beam and the slopes caused by the secondary moments can be obtained from it. The slopes are

$$EI\theta'_A = \frac{2}{3}\left(\frac{M'_A L}{2}\right) + \frac{1}{3}\left(\frac{M'_B L}{2}\right) = \frac{L}{6}\left[2M'_A + M'_B\right]$$

$$EI\theta'_B = \frac{L}{6}\left[M'_A + 2M'_B\right]$$

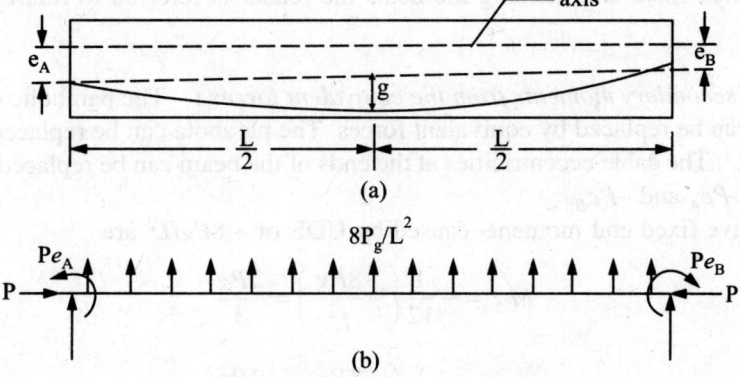

Figure 5.4.2 Fixed end beam.

186

Summing up the slopes at end A

$$\theta'_A + \theta'_A = -\frac{PL}{6}[2g + 2e_A + e_B] + \frac{L}{6}(2M'_A + M'_B) = 0$$

or

$$- P [2g + 2e_A + e_B] + (2M'_A + M'_B) = 0 \tag{5.4.1}$$

Similarly for θ_B

$$- P [2g + 2e_B + e_A] + (M'_A + 2M'_B) = 0 \tag{5.4.2}$$

The solution of the above two equations is

$$M'_A = P\left(\frac{2}{3}g + e_A\right) \tag{5.4.3}$$

$$M'_B = P\left(\frac{2}{3}g + e_B\right) \tag{5.4.4}$$

$$M_A = M^\circ_A + M'_A = - Pe_A + P\left(\frac{2}{3}g + e_A\right) = \frac{2}{3}Pg$$

$$M_B = M^\circ_B + M'_B = \frac{2}{3}Pg \tag{5.4.5}$$

$$M_C = - Pg + \frac{2}{3}Pg = -\frac{1}{3}Pg \tag{5.4.6}$$

It is interesting to note that the end moments are independent of the end eccentricities. For a concordant profile, $M'_A = M'_B = 0$ that leads to

$$e_B = e_A = -\frac{2}{3}g \tag{5.4.7}$$

The fixed end moments due to various tendon profiles are given in table 3 (appendix A). For more detailed fixed end bending moments the reader is referred to references (5.1) and (5.2).

Calculation of secondary moments from the equivalent forces :   The parabolic cable with end eccentricities can be replaced by equivalent forces. The parabola can be replaced by a UDL of intensity $8Pg/L^2$. The cable eccentricities at the ends of the beam can be replaced by equivalent end moments $-Pe_A$ and $-Pe_B$.

The effective fixed end moments caused by UDL of $- 8Pg/L^2$ are

$$M_A = -\frac{1}{12}\left(-\frac{8Pg}{L^2}\right) = \frac{2Pg}{3} \tag{5.4.8}$$

$$M_B = -\frac{1}{12}\left(-\frac{8Pg}{L^2}\right) = \frac{2Pg}{3} \tag{5.4.9}$$

The end moments $-Pe_A$ and $-Pe_B$ are directly transformed to the fixed supports. The secondary moment at section A is

$$M'_A = M_A - M°_A = \frac{2Pg}{3} + Pe_A = P\left(e_A + \frac{2g}{3}\right)$$

$$M'_B = M_B - M°_B = P\left(e_B + \frac{2g}{3}\right)$$

The above quantities match with those computed earlier.

Example 5.4.1 A two ends fixed beam of span 12 m is provided with a cable of 960 kN prestressing force. The eccentricity of the cable at each support is -20 cm and sag at mid-span is 45 cm. Determine the secondary bending moments. If the depth of the beam is 60 cm find a concordant cable.

Solution

$$L = 12 \text{ m}, e_A = e_B = -20 \text{ cm}, g = 45 \text{ cm}, h = 60 \text{ and}$$

$$P = 960 \text{ } k\text{N}$$

From eq. 5.4.3

$$M'_A = M'_B = P (2g/3 + e_A) + 96{,}000 (30 - 20)$$

$$= 9600 \text{ } k\text{N cm}$$

For a concordant cable profile, $e_A = e_B = -2g/3$

Assuming $e_A = e_B = -(60/2 - 10) = -20$ cm

$$g = -\frac{3}{2} \, e_A = 30 \text{ cm}$$

5.5. Application of Moment Distribution Procedure

Continuous beams are usually prestressed with a continuous wire. Sometimes the magnitude of prestressing force may vary within the span by anchoring the cable at intermediate points. The cable can be replaced by an equivalent set of forces and the beam can be analysed by moment distribution. Anchorage force can be replaced by axial, shear and bending forces and the shallow curve by a radial force proportional to the curvature. The axial force at CGC will not cause any moments for small deflection theory. The transverse component acting at a support point is directly transferred to the support. Otherwise, if it is situated in between the supports, it will cause fixed end moments which have to be accounted in the moment distribution. Moments caused by eccentricities and radial components of the cable will cause moments which have to be included in the analysis. The procedure of analysis of prestressed concrete beams by moment distribution is illustrated through an example.

Example 5.5.1 A three span continuous prismatic beam of each span L is fixed at both ends. The beam is provided with a cable of constant prestressing force P and eccentricities e_1 and e_2

at the outer and inner supports respectively. The sag of the cable in the outer span is g_1 and interior span is g_2. The middle span is loaded by a *UDL* of *w*. Determine the moments caused by the cable, external load, and the cable and load together.

Solution

The beam is shown in Figure 5.5.1 and it is designated as *ABCD* where *A*, *B*, *C* and *D* are the supports. The stiffnesses of the spans are :

$$K_{AB} = \frac{4EL}{L} = K_{BA} = K_{BC} = K_{CB} = K_{CD} = K_{DC}$$

(a) *Moments caused by prestressing cable*

The fixed end bending moments for each span are calculated from eq. 5.4.5 and they are :

$$\left. \begin{aligned} M_{fAB} &= M_{fBA} = 2Pg_1/3 \\ M_{fBC} &= M_{fCB} = 2Pg_2/3 \\ M_{fCD} &= M_{fDC} = 2Pg_1/3 \end{aligned} \right\}$$

(5.5.1)

where the subscript *f* refers to the fixed amd bending moment. The anticlockwise moments on the elements are considered positive in moment distribution. Table 5.5.1 gives the moment distribution using the relative stiffnesses and the fixed end bending moments caused by the cable. It is assumed that student is familiar with the moment distribution procedure so no additional explanation is given in this book.

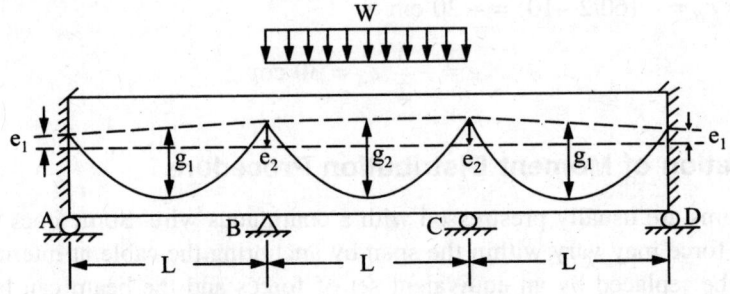

Figure 5.5.1 Examples 5.5.1

The balancing of the moments is carried for only three times by which time the carryover moments have become small as compared to the fixed end moments. The final moments caused by the prestressing cable from table 5.5.1 are

$$\left. \begin{aligned} M_{ABp} &= -P(19g_1 - 3g_2)/24 = -M_{DCp} \\ M_{BAp} &= P(11g_1 + 21g_2)/48 = -M_{CDp} \\ M_{BCp} &= -P(11g_1 + 21g_2)/48 = -M_{CBp} \end{aligned} \right\}$$

(5.5.2)

The subscript *p* refers to prestressing force.

TABLE 5.5.1 Moment distribution due to prestressing

Joint	A	B		C		D
Member	AB	BA	BC	CB	CD	DC
Stiffness (relative)	1	1	1	1	1	1
Distribution factor	0	0.5	0.5	0.5	0.5	0
Fixed end moment due to prestressing	$-\frac{2}{3}Pg_1$	$\frac{2}{3}Pg_1$	$-\frac{2}{3}Pg_2$	$\frac{2}{3}Pg_2$	$-\frac{2}{3}Pg_1$	$\frac{2}{3}Pg_1$
Balance	0	$-\frac{P}{3}(g_1-g_2)$	$-\frac{P}{3}(g_1-g_2)$	$-\frac{P}{3}(g_2-g_1)$	$-\frac{P}{3}(g_2-g_1)$	0
Carryover	$-\frac{P}{6}(g_1-g_2)$	0	$-\frac{P}{6}(g_2-g_1)$	$-\frac{P}{6}(g_2-g_1)$	0	$-\frac{P}{6}(g_2-g_1)$
Balance	0	$\frac{P}{12}(g_2-g_1)$	$\frac{P}{12}(g_2-g_1)$	$\frac{P}{12}(g_1-g_2)$	$\frac{P}{12}(g_1-g_2)$	0
Carryover	$-\frac{P}{24}(g_2-g_1)$	0	$\frac{P}{24}(g_2-g_1)$	$\frac{P}{24}(g_2-g_1)$	0	$-\frac{P}{24}(g_2-g_1)$
Balance	0	$-\frac{P}{48}(g_2-g_1)$	$-\frac{P}{48}(g_1-g_2)$	$-\frac{P}{48}(g_2-g_1)$	$-\frac{P}{48}(g_2-g_1)$	0
Total	$\frac{P}{24}(3g_2-19g_1)$	$\frac{P}{48}(11g_1+21g_2)$	$-\frac{P}{48}(11g_1+21g_2)$	$\frac{P}{48}(11g_1+21g_2)$	$-\frac{P}{48}(11g_1+21g_2)$	$-\frac{P}{24}(3g_2-19g_1)$

(b) *Moments caused by the external loud w acting on the mid-span*

The fixed bending moments caused by w are

$$M_{fAB} = M_{fBA} = M_{fCD} = M_{fDC} = 0$$

$$M_{fBC} = -M_{fCB} = wL^2/12$$

Table 5.5.2 gives the distribution of the moment for the above fixed end moments. The balancing of the moments is applied thrice by which lime the carryover moments are less than two per cent of the fixed end moments. The final moments from the table 5.5.2 are

$$\left. \begin{array}{l} M_{ABw} = -5wL^2/192 = -M_{DCw} \\ M_{BAw} = -21wL^2/384 = -M_{CDw} \\ M_{BCw} = 21wL^2/384 = -M_{CBw} \end{array} \right\} \qquad (5.5.3)$$

TABLE 5.5.2 Moment distribution due to external load

Joint	A	B		C		D
Member	AB	BA	BC	CB	CD	DC
Fixed end moment due to external load	0	0	$\dfrac{wL^2}{12}$	$-\dfrac{wL^2}{12}$	0	0
Balance	0	$-\dfrac{wL^2}{24}$	$-\dfrac{wL^2}{24}$	$\dfrac{wL^2}{24}$	$\dfrac{wL^2}{24}$	0
Carryover	$-\dfrac{wL^2}{48}$	0	$\dfrac{wL^2}{48}$	$-\dfrac{wL^2}{48}$	0	$\dfrac{wL^2}{48}$
Balance	0	$-\dfrac{wL^2}{96}$	$-\dfrac{wL^2}{96}$	$\dfrac{wL^2}{96}$	$\dfrac{wL^2}{96}$	0
Carryover	$-\dfrac{wL^2}{192}$	0	$\dfrac{wL^2}{192}$	$-\dfrac{wL^2}{192}$	0	$\dfrac{wL^2}{192}$
Balance	0	$-\dfrac{wL^2}{384}$	$-\dfrac{wL^2}{384}$	$\dfrac{wL^2}{384}$	$\dfrac{wL^2}{384}$	0
Total	$-\dfrac{5wL^2}{192}$	$-\dfrac{21wL^2}{384}$	$\dfrac{21wL^2}{384}$	$-\dfrac{21wL^2}{384}$	$\dfrac{21wL^2}{384}$	$\dfrac{5wL^2}{192}$

where the subscript w refers that of the load w. The final moments for which the section has to be designed are obtained by superposition of the eqs. 5.5.2 and 5.5.3

$$M_{AB} = M_{ABp} + M_{ABw}$$

$$M_{AB} = -P(19g_1 - 3g_2)/24 - 5wL^2/192 = -M_{DC}$$

Similarly

$$M_{BA} = P(11g_1 + 21g)/48 - 21wL^2/384 = -M_{CD} \qquad (5.5.4)$$

$$M_{BC} = -P(11g_1 + 21g_2)/48 + 21wL^2/384 = -M_{CB}$$

Example 5.5.2 Calculate the bending moments of example 5.5.2 if $L = 8$m, $w = 24$ kN/m, $g = 20$cm, $g = 30$ cm and $P = 400$ kN

Solution

The bending moments caused by the prestressing force are obtained from eq. 5.5.2. (use $g_1 = g_2 = 20$ cm)

$$M_{ABp} = -P(19g_1 - 3g_2)/24 = -400(380 + 90)/24$$

$$= -48.3 \ kNm$$

$$M_{BAp} = -M_{CDp} = -400(2.20 + 6.30)/48 = 70.83 \ kNm$$

$$M_{BCp} = -M_{CBp} = -70.83 \ kNm$$

Bending moments caused by the external load are obtained from eq. 5.5.3.

$$M_{ABp} = -M_{DCw} = -5wL^2/192 = -5(24)(64)/192$$

$$= -40 \ kN \ m$$

$$M_{BAw} = -M_{DCw} = -21(24)(64)/384 = -84 \ kN \ m$$

$$M_{BCw} = -M_{CBw} = 84 \ kN \ m$$

The final moments are

$$M_{AB} = M_{ABP} + M_{ABw} = -48.3 - 40 = -88.3 \ kN \ m$$

$$M_{BA} = -M_{CD} = -M_{BC} = M_{CB}$$

$$= 70.83 - 84 = -13.17 \ kN \ m$$

Example 5.5.3 Calculate approximate relation of eccentricities of a fixed ended three equal span continuous beam such that the secondary moments are zero.

Solution

The spans are equal so the prestressing cable should be symmetric. Let e_1 be the eccentricity at outer ends and e_2 the eccentricity at each of the intermediate supports. Let g_1 and g_2 be the sags in the exterior and interior spans respectively. The effective bending moments for such a beam are computed by a moment distribution in example 5.5.1. The secondary moment relation is

$$M_{ABp} = M^{\circ}_{AB} + M'_{AB}$$

192

where M°_{AB} = primary moment which is equal to Pe_1 which is positive if causing anticlockwise moments as per sign conversion of the moment distribution.

$$M'_{AB} = M_{ABp} - M^\circ AB$$

The moment at each of the supports must be equal to zero for a cable to be concordant, so

$$M'_{AB} = M_{ABp} - M^\circ_{AB} = 0$$

$$M'_{AB} = M_{ABp} - Pe_1 = 0$$

$$- P(19g_1 - 3g_2)/24 - Pe_1 = 0$$

or $$e_1 = - (19g_1 - 3g_2)/24 \qquad (5.5.5)$$

Similarly $M'_{BC} = 0$ gives

$$P(11g_1 + 21g_2)/48 + Pe_2 = 0$$

$$e_2 = - (11g_1 + 2lg_2)/48 \qquad (5.5.6)$$

Let the depth of the beam be equal to h then the approximate limiting sags are

$$g_1 = 0.6h \text{ and } g_2 = 0.75h$$

The eccentricities are

$$e_1 = -(19g_1 - 3g_2)/24 = - h[19(0.6) - 3(0.75)]/24$$

$$= - 0.38h$$

$$e_2 = - h [11(0.6) + 21(0.75)]/48 = - 0.46h$$

e_2 works out approximately equal to half the depth which will not give any cover for the cable so g_2 must be reduced to say 0.6h. Then the value of e_2 works out to be

$$e_2 = - h[11(0.6) + 21 (0.6)]/48 = - 0.40h$$

5.6. Continuous Beams with Variable Section

In continuous beams, the magnitude of moment is comparatively large at intermediate supports, so efficiency of the beam can be improved by increasing the thickness of the beam at the intermediate supports. The variable moment of inertia affects the distribution of moments in a continuous beam. The cable profile is measured from the centroidal axis and if it is to be a smooth curve, should follow the profile of the centroidal axis. As shown in Figure 5.6.1, the centroidal axis has a kink at a distance $\xi_1 L_1$ so the cable profile should also have the same amount of kink.

Example 5.6.1 A two span continuous beam is made of haunched sections. The depth of the beam is $2h$ at the intermediate support and decreases linearly to h on either side at a distance of 0.2L. Determine a concordant cable for such a beam (Figure 5.6.2 a).

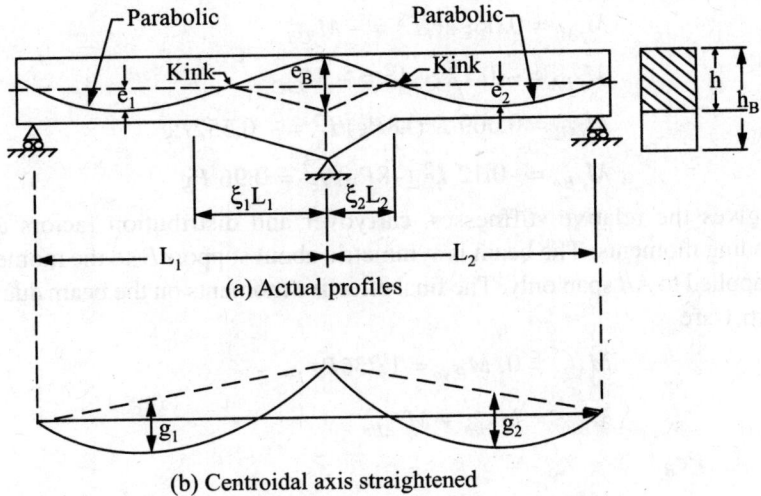

(a) Actual profiles

(b) Centroidal axis straightened

Figure 5.6.1 Beam with variable thickness.

Solution

The carryover relative stiffness and factors for a haunched beam are (5.3)

$$C_{AB} = 0.7, \; C_{BA} = 0.5$$
$$K_{AB} = 4.4, \; K_{BA} = 6.67$$

A smooth parabolic cable gives a uniformly distributed equivalent load. Since the centroidal axis of the beam takes a sudden deviation at $0.8\,L$ distance from the end support, the cable also should take a deviation from the parabolic profile at this point. The centroid of the section at mid support has moved downwards by $h/2$ so the actual eccentricity of the cable at mid support should move downwards by $h/2$ while maintaining the same constant curvature. The equivalent cable load on the beam is $-8Pg/L_2 = w_b$. The fixed end bending moments for a haunched beam are

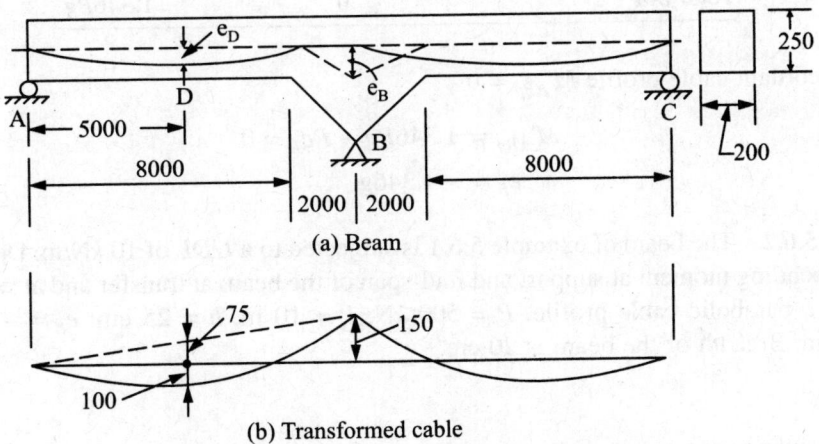

(a) Beam

(b) Transformed cable

Figure 5.6.2 Examples 5.6.1 and 5.6.2

$$M_{fAB} = 0.069 \ w_b L^2 = -M_{fCB} \qquad (5.6.1)$$

$$M_{fBA} = -0.12 \ w_b L^2 = -M_{fBC} \qquad (5.6.2)$$

$$M_{fAB} = 0.069 \ L^2(-8Pg)/L^2 = -0.552Pg$$

$$M_{fBA} = -0.12 \ L^2 \ (-8Pg)/L^2 = 0.96 \ Pg$$

Table 5.6.1 gives the relative stiffnesses, carryover and distribution factors along with the fixed end bending moments. The beam is symmetric about support B so the moment distribution procedure is applied to AB span only. The final effective moments on the beam due to prestressing from table 5.6.1 are

$$M_{ABp} = 0, \ M_{BAp} = 1.236Pg$$

$$M_{ABp} = M_{ABp} + M^{\circ}{}_{ABp}$$

where $M^{\circ}{}_{ABp} = -Pe_B$

$$M'_{ABp} = M_{ABp} - M^{\circ}{}_{ABp} = 1.236Pg + Pe_B \qquad (5.6.3)$$

TABLE 5.6.1 Moment distribution–Example 5.6.1

Joint	A	B
Member	AB	BA
Carryover factor	0.7	0.5
Relative stiffness	4.4	6.67
Distribution factor	1	0.5
Fixed end moment due to P	$-0.552Pg$	$0.96Pg$
Balance	$0.552Pg$	0.0
Carryover	0	$0.386Pg$
Total BM	0	$1.346Pg$

For a concordant cable profile $M'_{ABp} = 0$

$$M'_{ABp} = 1.346Pg + Pe_B = 0$$

$$e_B = -1.346g \qquad (5.6.4)$$

Example 5.6.2 The beam of example 5.6.1 is subjected to a *UDL* of 10 *k*N/m. Determine the effective bending moment at support and mid-span of the beam at transfer and at working load assuming a parabolic cable profile. $P = 500 \ kN$, $L = 10 \ m$, $h = 25 \ cm$, $e_B = -15 \ cm$ and $e_D = 10 \ cm$. Breadth of the beam is 20 cm.

Solution

The bending moment at mid support B and mid-span D be denoted by M_B and M_D. The secondary reactions are zero as a concordant cable profile is selected.

(a)   *Transfer condition*

Total self weight of the beam is computed from the total volume of the concrete and it is

$$w_g = [8(0.2)(0.25) + 2(0.2)(0.25) + (0.2)(0.5)/2] \, 24000 = 13200 \text{ N}$$

For simplicity sake the total load is assumed as *UDL* since it is much smaller than the working load. The bending moment at intermediate support caused by the *UDL* is obtained from the previous example.

$$M_B = 0.12wL^2 - 0.068wL^2/(0.7) = - 0.1676wL^2 \qquad (5.6.5)$$

where $w = UDL$ acting downwards

$M_B = BM$ at B in counterclockwise direction and it is also causing compression on top fibre

The reaction at A called R_A acting upwards can be obtained by taking moments about B

$$R_A L - (wL)L/2 = M_B = - 0.1676 \, wL^2$$
$$R_A = (0.5 - 0.1676)wL = 0.3324 \, wL \qquad (5.6.6)$$

Bending moment at mid-span is given by

$$M_D = R_A L/2 - wL^2/8 = 0.0412wL^2 \qquad (5.6.7)$$

Let the bending moment caused by self weight be denoted by subscript g, that of prestressing force by P and that of live load by L. The bending moments are computed using eqs. 5.6.5 and 5.6.6

$$M_{Bg} = - 0.1676w_g L^2 = 0.1676 \, w_g L = - (0.1676)(13200)(10)$$
$$= - 22100 \text{ Nm}$$
$$M_{Dg} = - 0.0412w_g L^2 = (0.0412)(13200)(10) = 5430 \text{ Nm}$$

The sag of the cable as seen from Figure 6.6.2 is

$$g = e_D - e_B/2 = 10 + 7.5 = 17.5 \text{ cm} = 0.175 \text{ m}$$

Balancing force of the cable at transfer is

$$w_{bt} = - 8P_i g/L^2 = - 8(500000)(0.175)/100 = - 7000 \text{ N/m}$$
$$M_{Bp} = - 0.1676w_{bt}L^2 = - (0.1676)(-7000)(100) = 117300 \text{ Nm}$$
$$M_{Dp} = 0.0412(-7000)(100) = - 28800 \text{ Nm}$$

Effective bending moments at transfer are

$$M_{Bt} = M_{Bg} + M_{Bp} = - 22100 + 117300 = 95200 \text{ Nm}$$
$$M_{Dt} = M_{Dg} + M_{Dp} = 5430 - 28800 = - 23370 \text{ Nm}$$

(b)   *Working load condition*

The bending moments caused by the live load are

$$M_{Bl} = - 0.1676 \, w_l L^2 = - 0.1676 \, (10000)(100) = -167600 \text{ Nm}$$
$$M_{Dl} = 0.0412w_l L^2 = 41200 \text{ Nm}$$

Let the loss of prestress at B is 30 per cent. The bending moments caused by the prestress are

$$M_{Bp} = -0.1676\,(0.7w_{bt})L^2 = 82000 \text{ Nm}$$

$$M_{Dp} = 0.0412\,(0.7w_{bt})L^2 = -20000 \text{ Nm}$$

The effective bending moments at working load are

$$M_{Bw} = M_{Bg} + M_{B1} + M_{Bp} = 22100 - 167600 + 82000 = -63500 \text{ Nm}$$

$$M_{Dw} = M_{Dg} + M_{D1} + M_{Dp} = 5430 + 41200 - 20000 = 26630 \text{ Nm}$$

The stresses at different load conditions can be obtained from the beam formula. The stresses at intermediate support at working load are

$$\sigma_{1w} = -\frac{Pw}{A} - \frac{M_{Bw}y_t}{I} = -\frac{500000}{(20)(50)} - \frac{6350000\,(12)(25)}{(20)(125000)}$$

$$= -500 - 765 = -1265 \text{ N/cm}^2$$

$$\sigma_{2w} = -\frac{Pw}{A} + \frac{M_{Bw}y_t}{I} = -500 + 765 = 265 \text{ N/cm}^2$$

5.7. Design of Continuous Beams

The continuity of the continuous beams gives less bending moments when compared with simple beams. Prestressing cable, unequal settlement of foundations; thermal, shrinkage and creep forces cause secondary bending moments in continuous beams. Therefore, continuous beams are usually designed for small spans in which the unequal settlements of the foundations is not expected. The design is illustrated through simple examples.

Example 5.7.1   A two span continuous culvert beam of 8 m each span is subjected to *UDL* of 8000 N/m. The beam is prismatic and rectangular cross section with a width of 60 cm. Design a prestressing cable assuming the strength of the concrete to be 4000 N/cm².

Solution

The permissible stresses (in N/cm²) are

$$\sigma_{aci} = 1920$$

$$\sigma_{ati} = \sigma_{ate} = 100$$

$$\sigma_{ace} = 1560$$

Assume $h = L/20 = 40$ cm, then

$$w_g = (0.6)(0.4)(24000) = 4760 \text{ N/m}$$

$$w_1 = 8000 \text{ N/m}$$

Maximum bending moment in a two span continuous beam for *UDL* occurs at middle support and it is

$$M = - wL^2/8$$

$$M_g = - w_g L^2/8 = - 4760(64)/8 = - 38080 \text{ Nm}$$

$$M_1 = - w_1 L^2/8 = - 8000(64)/8 = - 64000 \text{ Nm}$$

$$M_w = M_g + M_1 = - 102080 \text{ Nm}$$

The beam is a wide shallow beam with tension occurring both on top and bottom fibres, of course at different locations. A set of straight cables with effective eccentricity equal to zero can be provided. The critical stresses occur at middle support so the cable is first designed to suit this section; later a check on stresses at mid-span is made.

$$A = 60 \times 40 = 2400 \text{ cm}^2$$

Section modulus = $Z = bh^2/6 = 60(1600)/6 = 16,000 \text{ cm}^3$

The limiting stress equation at top fibre in transfer condition is

$$\frac{P_i}{A} - \frac{M_g}{I}\left(\frac{h}{2}\right) \le \sigma_{aci} \quad \text{(since } M_g \text{ is negative)}$$

$$P_i \le A\left(\sigma_{aci} - \frac{M_g}{Z}\right) = \left(1920 - \frac{380\,8000}{16000}\right)2400 = 402\,0000 \text{ N} \tag{5.7.1}$$

Similarly the limiting stress in the bottom fibre at transfer condition is

$$-\frac{P_i}{A} - \frac{M_g}{Z} \le \sigma_{ati}$$

$$P_i \ge \left(\frac{380\,8000}{16,000} - 100\right)2400 = 330,000 \text{ N} \tag{5.7.2}$$

Stress in the bottom fibre at working load condition gives

$$\frac{\eta P_i}{A} - \frac{M_w}{Z} \le \sigma_{ace}$$

$$P_i \ge \frac{A}{\eta}\left(\sigma_{ace} + \frac{M_w}{Z}\right) = \frac{2400}{0.85}\left(1560 - \frac{10208000}{16,000}\right) = 260\,0000 \text{ N} \tag{5.7.3}$$

Stress in the top fibre at working load condition is

$$-\frac{\eta P_i}{A} - \frac{M_w}{Z} \le \sigma_{ate} = 100$$

$$P_i \ge \frac{2400}{0.85}\left(\frac{1,020,8000}{16,000} - 100\right) = 15150,000 \text{ N} \tag{5.7.4}$$

Considering the above four constraints the value of P_i is governed by eqs. 5.7.3 and 5.7.4 and it is

$$1515\,000 \le P_i \le 2600\,000 \text{ N}$$

Select 10 numbers of cables, each having five wires of 7 mm diameter, then the prestressing force provided is

$$P_i = (10)(5)(0.384)(0.7)(150000) = 1980\,000 \text{ N}$$

Provide five cables on either face of the section with 4 cm cover at shown in Figure 5.7.1.

Check for stresses (in N/cm²)

$$P_i/A = 198\,0000 / 2400 = 825$$

$$M_g/Z = -237$$

$$M_w/Z = -\frac{-10208\,000}{16\,000} = -639$$

$$\frac{P_i}{A} - \frac{M_g}{Z} = 825 + 237 = 1062 < \sigma_{aci}$$

$$\frac{P_i}{A} - \frac{M_g}{Z} = -825 + 237 = -588 < \sigma_{ati}$$

$$\frac{\eta P_i}{A} - \frac{M_w}{Z} = 600 + 639 = 1299 < \sigma_{ace}$$

$$-\frac{\eta P_i}{A} - \frac{M_w}{Z} = -660 + 639 = -21 < \sigma_{ate}$$

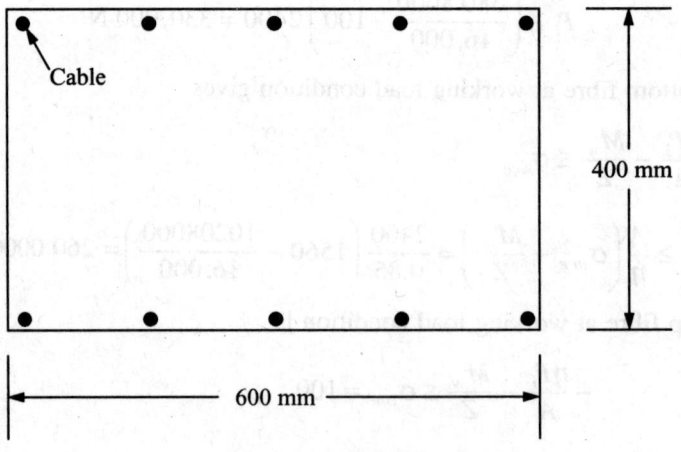

Figure 5.7.1 Example 5.7.1

All the stresses are within the permissible limits. This design is not a unique one as the depth is assumed and there are many possible depths which will satisfy the criterion. Next example illustrates a more rational approach where minimum depth of the beam is obtained.

Example 5.7.2   Design the beam of example 5.7.1 without assuming the depth of the beam, while keeping the width as 6 cm.

Solution

From the previous example have

$$\sigma_{aci} = 1920 \text{ N/cm}^2; \ \sigma_{ati} = 100 \text{ N/cm}^2 = \sigma_{ate}$$

$$\sigma_{ace} = 1560 \text{ N/cm}^2$$

Let depth of the beam be h cm then

$$w_g = (0.60) \ (h/100) \ (24000) = 144 \ h \ kN\backslash$$

$$M_g = - w_g L^2/8 = - 144h \ (64)/8 = - 1152h \text{ Nm}$$

$$= - 115200h \text{ Ncm}$$

$$A = 60h \text{ cm}^2 \text{ and } Z = 60h^2/6 = 10h^2 \text{ cm}^3$$

The governing stress conditions at transfer condition as seen from the previous example are not critical as compared to those of the working load condition. Taking only working load condition, the stresses at bottom and top fibres are (in N/cm^2)

$$\frac{\eta P_i}{A} - \frac{M_w}{Z} \leq \sigma_{ace} = 1560 \tag{5.7.5}$$

$$-\frac{\eta P_i}{A} - \frac{M_w}{Z} \leq \sigma_{ate} = 100 \tag{5.7.6}$$

Where $M_w = -(6400000 + 115200 \ h)$

$$-\frac{2M_w}{S} \leq 1600 \text{ or}$$

$$\frac{2(640\ 0000 + 11\ 5200h)}{10h^2} \leq 1660$$

$$1660 \ h^2 - 23{,}040h - 640{,}000 \geq 0$$

$$h^2 - 13.9 \ h + 386 \leq 0$$

$$h \geq \frac{13.9 \pm 51.8}{2}$$

Neglecting the negative sign as it gives negative depth, the value of h is given by

$$h > 32.9 \text{ cm}$$

Let $h = 33$ cm

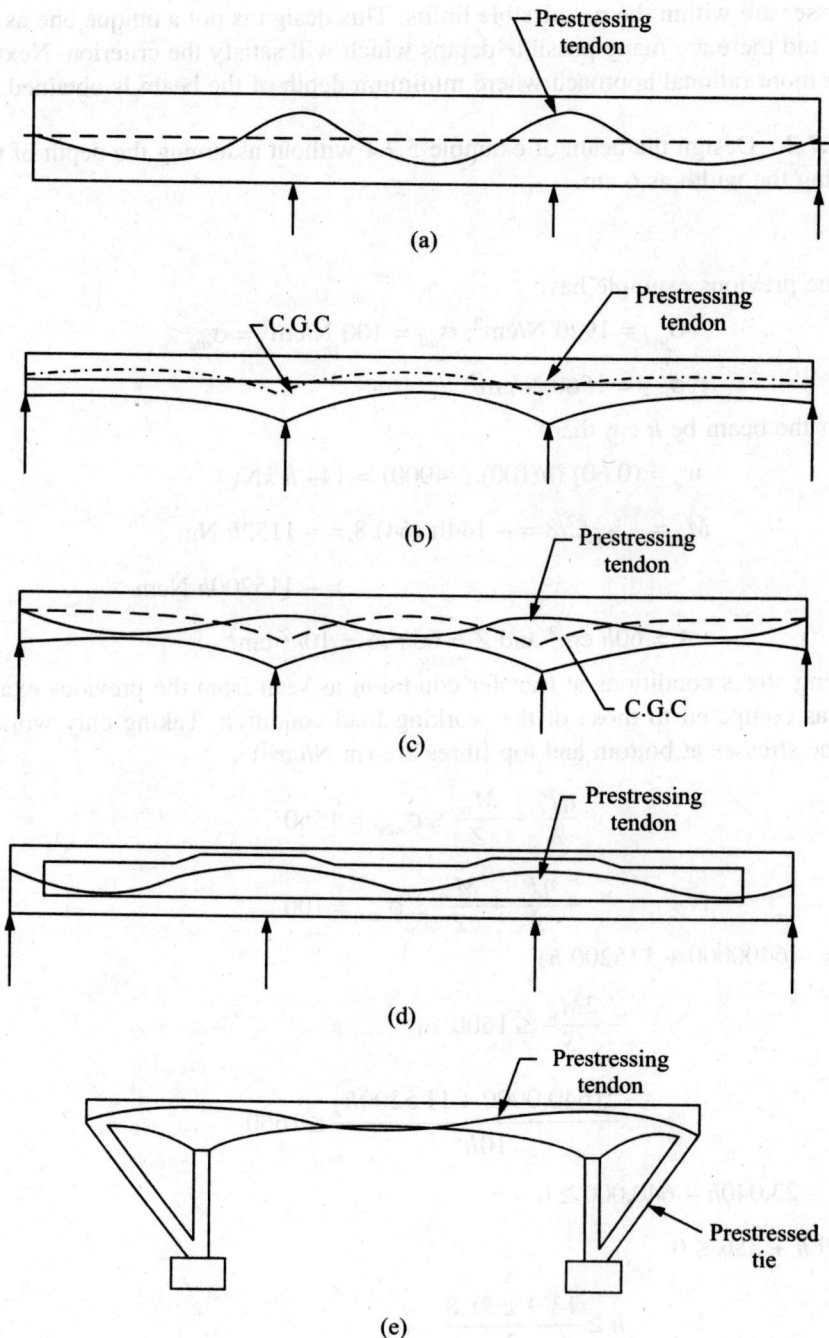

Figure 5.7.2 Typical profiles in Continuous beams.

201

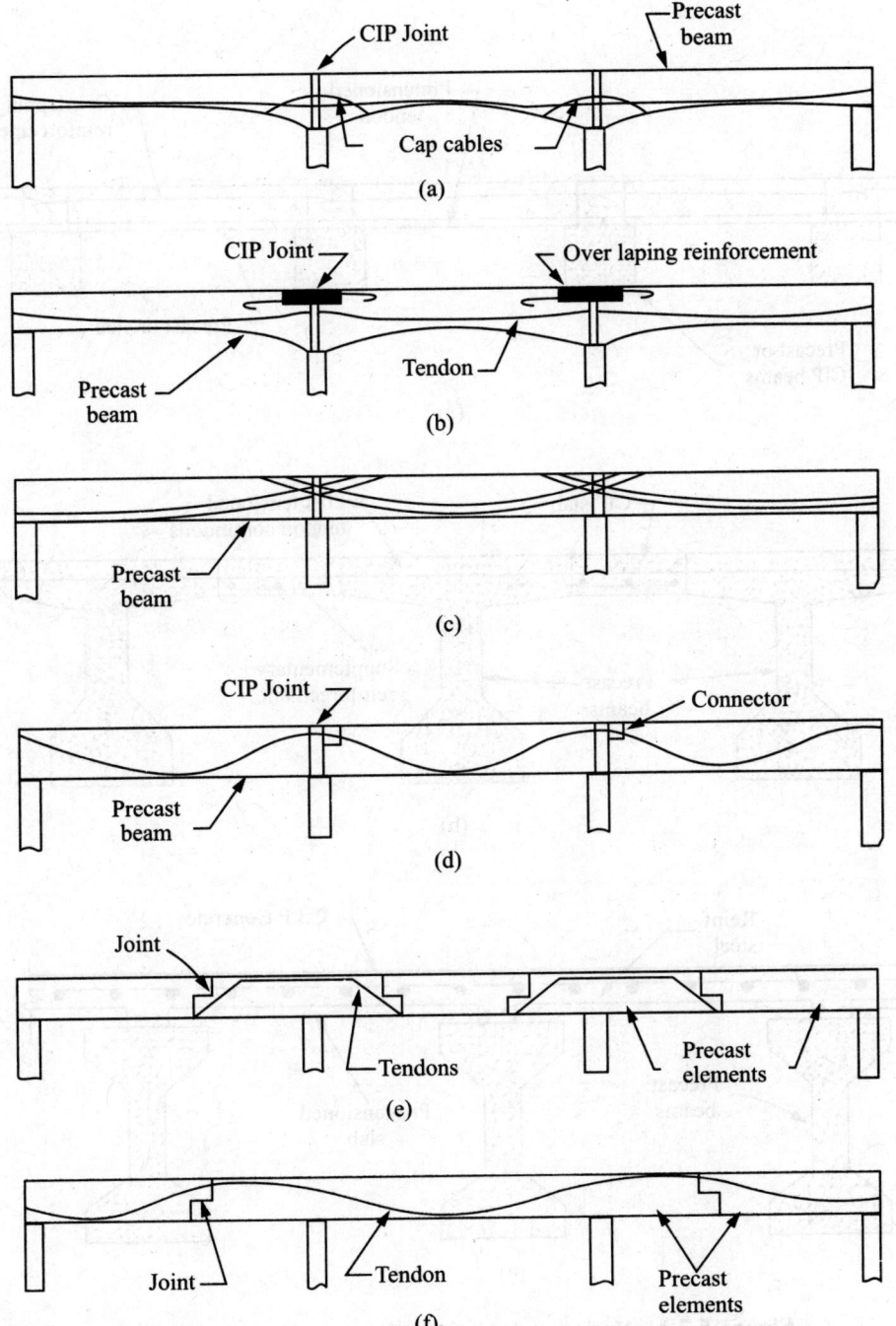

Figure 5.7.3 Continuous beams of precast elements.

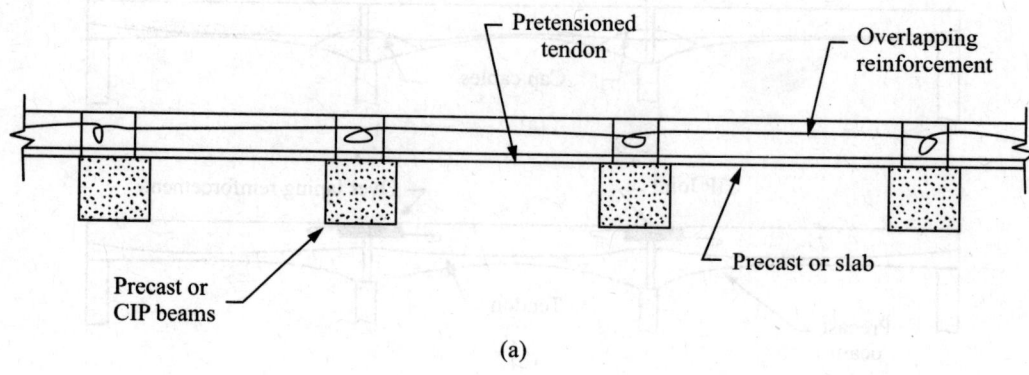

(a)

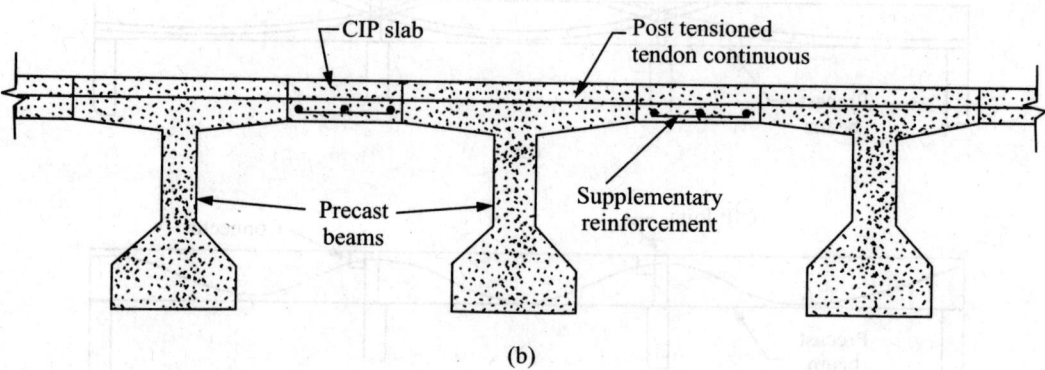

(b)

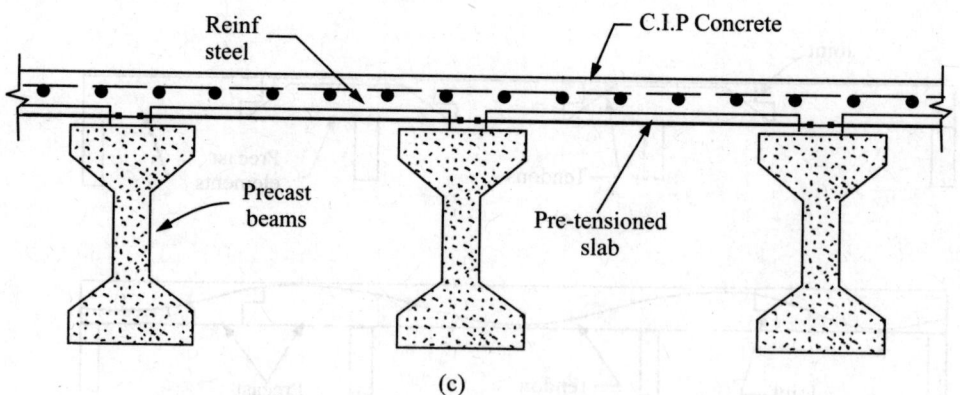

(c)

Figure 5.7.4 Various types of continuous prestressed slabs.

Subtraction of eq. 5.7.6 from eq. 5.7.5 gives

$$\frac{2P_i}{A} = 1460$$

$$P_i = 1460 \ (60) \ (33)/2 = 1445 \ 400 \ \text{N}$$

It can be observed from this example that the depth of the beam as well as the prestressing are less than those in the previous example. The stresses at transfer condition are within the permissible stresses.

If one assumes too much of area of cross section, then the prestressing force required also increases thus making the design very uneconomical.

Typical cable profiles for statically indeterminate structures are shown in Figures 5.7.2. to 5.7.4. For more detailed continuity connections, the reader is referred to references (5.3) and (5.4).

5.8. Load Balancing Method

Analysis and design of continuous beams is very much simplified by load balancing method. If the external load on each span is balanced by the prestressing force of the span, then the stress at any point in the beam is P/A. However, the external load is variable in most of the problems, so the complete balancing of external load for all conditions of loading is not possible. A partial balancing has to be done deriving the maximum benefit out of this method. The procedure is best illustrated by the following example :

Example 5.8.1 Design a post-tensioned two span continuous beam of 30 m each span and subjected to external load of external load of 20 *kN/m*.

Solution

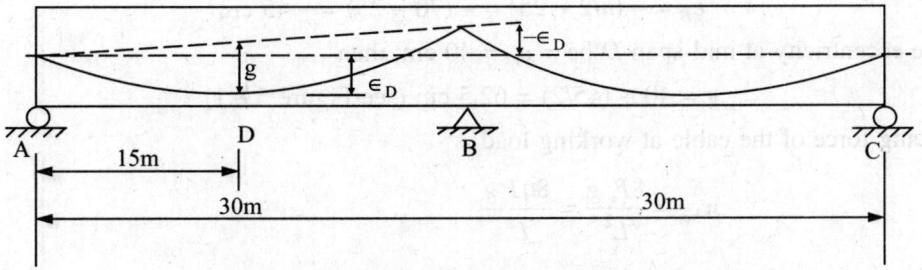

Figure 5.8.1 Example 5.8.1

Let $$f_{ck} = 50 \ MPa = 5000 \ \text{N/cm}^2$$

$$f_p = 1500 \ \text{MPa} = 150000 \ \text{N/cm}^2$$

The permissible stresses could be taken as (all in N/cm^2)

$$\sigma_{aci} = 1720$$

$$\sigma_{ati} = 0 = \sigma_{ate}$$

$$\sigma_{ace} = 1260$$

For a continuous beam of two spans the maximum bending moment occurs at support and it is $-wL^2/8$. The span to depth ratio in two span continuous beams can be taken as 20 to 25.

Let $h \simeq L/21 \simeq 140$ cm and $\eta = 0.8$.

It is desirable to assume a symmetric section as the bending moment along the span varies from positive to negative. So the shape factor Δ is 1. Select $\rho = 0.136$ from table 2 of appendix A for which $\Delta = 1$.

(ii) *Approximate selection of balancing load*

An approximate self weight of the beam can be assumed to start with and checked with the final result. The beam is carrying 20 kN/m and its span is 30 m so an approximate self weight for -30 m span is 0.6 times the external load.

Let
$$w_g = 0.6(20000) = 12000 \text{ N/m}$$

$$w_w = w_g + w_1 = 12000 + 20000 = 32000 \text{ N/m}$$

Let $\qquad w_{bw}$ = balancing load from cable at working load condition

Let the cable be designed such that full self weight and 50 per cent of the live load is balanced.

$$w_{bw} = 12000 + 0.5(20000) = 22000 \text{ N/m}$$

(iii) *Preliminary selection of cable and cross sectional area*

A parabolic cable profile is selected with negative eccentricity at middle support. Let the centroid of steel at middle support be 25 cm from top fibre, then

$$e_B = -(h/2 - 25) = -(70 - 25) = -45 \text{ cm}$$

Let the eccentricity at mid-span D be $= e_D = 40$ cm, then

$$g = 40 + (45/2) = 62.5 \text{ cm (see Figure 5.8.1)}$$

Balancing force of the cable at working load is

$$w_{be} = \frac{8P_w g}{L^2} = \frac{8\eta P_i g}{L^2}$$

$$P_i = \frac{w_{bw} L^2}{8\eta g} = \frac{22000(900)}{8(0.8)(0.625)} = 505\,0000 \text{ N}$$

$$P_e = 0.8 P_i = 404,0000 \text{ N}$$

Let
$$\sigma_{ca} = \frac{P_i}{A} = 900 \text{ N/cm}^2$$

$$A \simeq P_i/900 = 5610 \text{ cm}^2$$

Let $\qquad A = 5600 \text{ cm}^2$

$$w_g = A\gamma = 0.560\,(24000) = 13440 \text{ N/m}$$

which is very close to the assumed value; so there is no need to revise the section at this stage.

Let $\qquad w_{be} = 13000 + 0.5\,(20000) = 23000 \text{ N/m}$

(iv) Check for stresses

The prestressing force and area of cross section are known but actual size and shape of the section are not known. Before going into detailed calculations it is desirable to check for the actual stresses by using non-dimensionalised parameters.

(a) Stresses at transfer condition

$$w_{bi} = w_{be}/\eta = \frac{23000}{0.8} = 28750 \text{ N/cm}^2$$

Unbalanced load = effective load at transfer

$$= w_g - w_{bi} = 13440 - 28750 = -15310 \text{ N/m}$$

Moment at support B during transfer of prestress is

$$M_{Bi} = -\frac{\left(w_g - w_{bi}\right)L^2}{8} = \frac{15310\,(90000)}{8}$$

$$= 172{,}500{,}000 \text{ N cm}$$

All stresses are computed in N/cm^2 tensile stress as positive for convenience. The stress on the top fibre of the concrete is

$$\sigma_{it} = -\frac{P_i}{A} - \frac{M_{Bi}y_t}{I} = -904 - \frac{M_{Bi}y_t}{I}$$

$$= -904 - \frac{M_{Bt}}{Ah}\frac{yt}{h}$$

$$= -904 - \frac{M_{Bt}}{Ah}\left(\frac{1}{1+\Delta}\right)$$

$$= -904 - \frac{17250\,0000\,(1)}{(5600)(140)(0.136)(2)}$$

$$= -904 - 810 = -1714 \text{ (permissible)}$$

The stress at bottom

$$\sigma_{2t} = -\frac{P_i}{A} + \frac{M_{Bt}y_b}{L} = -904 - \frac{M_{Bt}}{Ah}\left(\frac{\Delta}{1+\Delta}\right)$$

$$= -904 + 810 = -94 \text{ (permissible)}$$

(b) Stresses at working load condition

$$w_{be} = w_w - w_{be} = (13440 + 20000) - 23000 = 10440 \text{ N/m}$$

BM at *B* during working load condition is

$$M_{Be} = -(w_w - w_{be})\frac{L^2}{8} = -10440\frac{(90000)}{8}$$
$$= -117,300,000 \text{ N cm}$$

The stress at the top fibre of the concrete is

$$\sigma_{1e} = -\frac{P_e}{A} + \frac{M_{Be}y_t}{I} = -\frac{4040\,000}{5600} - \frac{M_{Be}}{Ah(1+\Delta)\rho}$$

$$= -720 + \frac{117\,300\,000}{5600\,(0.136)(14)(2)}$$

$$= -720 + 540 = -180$$

The stress at the bottom fibre is

$$\sigma_{2e} = -\frac{P_e}{A} + \frac{M_{Be}y_b}{I} = -720 - 540$$
$$= -1260 \text{ N/cm}^2$$

All stresses are within the permissible values.

(v) *Details of the section*

The sectional parameter for $\rho = 0.136$ and $\Delta = 1$ from table 2 of appendix *A* are

$$c_f = 1.0,\ c_w = 0.20,\ c_t = c_b = 0.1,\ c_a = 0.36$$

$$c_i = 0.046,\ c_{yb} = c_{yt} = 0.5$$

$$A = c_a h b_t = 0.36(140)\ b_t = 5600$$

$$b_t = \frac{A}{(0.36)(140)} = \frac{5600}{(0.36)(140)} = 111 \text{ cm}$$

$$b_f = b_t = 111 \text{ cm}$$

$$t_t = 0.1\ h = 14 \text{ cm},\ t_b = 0.1\ h = 14 \text{ cm}$$

$$b_w = 0.26 b_t = 22.2 \text{ cm}$$

The permissible force at transfer in 7 mm wire is

$$0.384(0.7)(15,0000) = 40000 \text{ N}$$

$$\text{Number of 7 mm Wires} = \frac{P_i}{40000} = \frac{505,0000}{40000} = 125$$

Use 19 cables of 7 numbered 7 mm wires such that

$$e_B = -45 \text{ cm}, \quad e_D = 40 \text{ cm}$$

The sectional details given above are modified to a practicable section and shown in Figure 5.8.2.

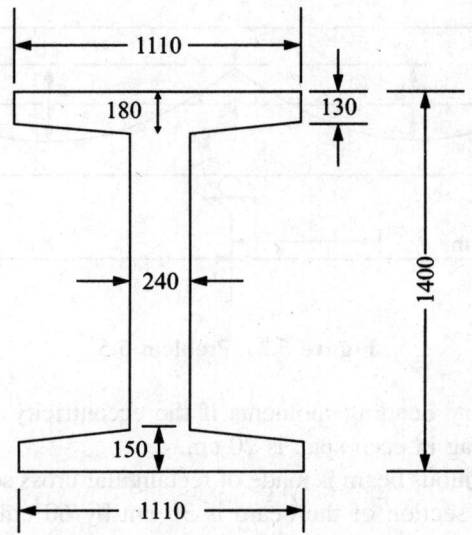

Figure 5.8.2   Example 5.8.1 (dimensions in mm).

PROBLEMS

5.1 Determine the secondary moments developed in the prestressed concrete beam shown in Figure 5.1 of square section.

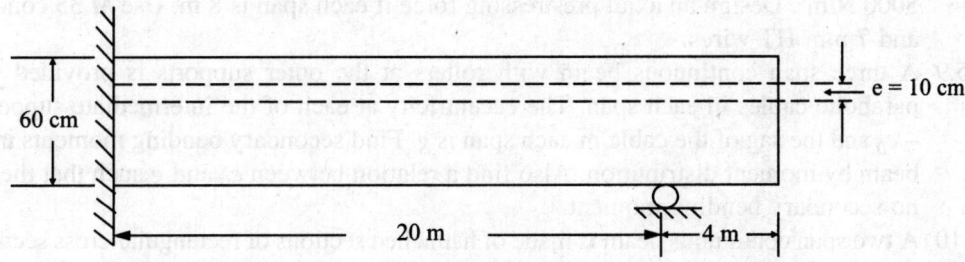

Figure 5.1.

5.2 A three span continuous beam of equal spans is subjected to a uniformly distributed load. Determine a cable profile which will not cause any secondary moments.

5.3 A two span continuous beam of 20 m each span, fixed at one end and simply supported at the other end is subjected to a uniformly distributed load of 16kN/m. Design the beam as a pre-tensioned concrete structure with 45 MPa concrete.

5.4 A two span continuous beam of each span 20 m has a symmetric cross section. The depth of the beam is 100 cm. The cable has an eccentricity of – 30 cm at mid support. Find a sag such that the cable will not produce any secondary bending moment.

5.5 A two span continuous beam with over hangs is provided with a parabolic cable profile as shown in Figure 5.2. Find a concordant cable profile.

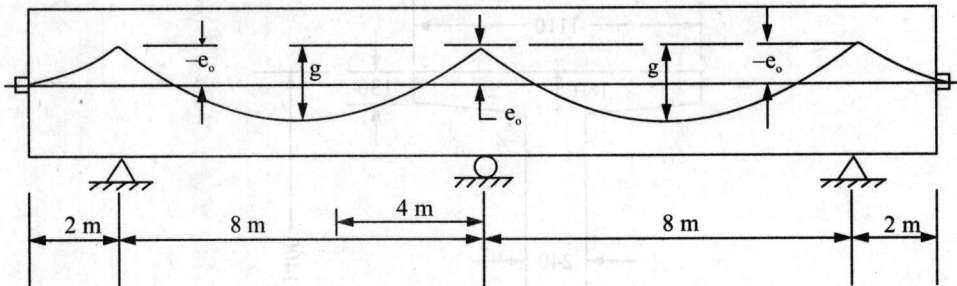

Figure 5.2 Problem 5.5.

Find the secondary bending moments if the eccentricity at each of the supports is –10 cm and the sag in each span is 20 cm.

5.6 A two span continuous beam is made of rectangular cross section with 10 m and 12 m spans. The cross section of the beam is 30 cm by 60 cm. Find a concordant cable profile if a cable of parabolic profile is provided in each span.

5.7 The beam mentioned in problem 5.6 is provided with a straight cable with – 5 cm eccentricity. Find the secondary moments. Also find a prestressing force of the cable such that the beam can carry a load of 10 kN/m.

$$f_{ck} = 40 \text{ MPa}, f_p = 1500 \text{ MPa}$$

5.8 A two span culvert slab is made of a hollow box section of outside dimensions 200 cm by 50 cm and inside dimensions 160 cm by 20 cm. The beam carries a load of 5000 N/m^2. Design an axial prestressing force if each span is 8 m. Use M 35 concrete and 7 mm HT wires.

5.9 A three span continuous beam with rollers at the outer supports is provided with parabolic cables in each span. The eccentricity at each of the intermediate support is – e_0 and the sag of the cable in each span is g. Find secondary bending moments in the beam by moment distribution. Also find a relation between e_0 and g such that there is no secondary bending moment.

5.10 A two span continuous beam is made of haunched sections of rectangular cross sections. Each span is 8 m and width of the beam is 60 cm throughout. The depth of the beam is 40 cm from either ends up to 5 m and linearly increasing to 60 cm at the middle support. Find secondary bending moments assuming a parabolic cable with – 10 cm eccentricity at mid support and 15 cm sag. Design a cable force to support an external load of 8000 N/m.

CHAPTER 6

Miscellaneous Structural Members

6.1. Introduction

Several structural elements or simple members such as columns, piles, tie rods, tension elements, retaining walls, portal and building frame members are being designed and constructed in prestressed concrete. Recent trends in prefabricated construction have opened several areas of structural elements being made by prestressed concrete (6.1 to 6.6). This chapter discusses simple structures and structural members of prestressed concrete construction. Members are classified into two major groups:

(i) members essentially subjected to compressive forces, and
(ii) members subjected to tensile forces.

6.2. Compression Members

The prestressing concept was originally developed to take care of the low tensile capacity of concrete. So prestressing a concrete compression member appears to be redundant and superfluous. However, some of the compressive members like columns and piles are subjected to handling and combined axial and bending force. The prefabricated columns are subjected to bending stresses during transportation and erection. For these reasons, it has been found necessary to use reinforcement to take care of secondary stresses. Prestressing a compressive member subjected to handling or bending moments results in using less steel with higher efficiency.

(1) **Prestressed concrete columns subjected to combined bending and axial force:** Consider a column subjected to axial and bending moment as shown in

Figure 6.2.1. There are two conditions

Let F = axial force

M = bending moment (assumed to be small)

Assuming the same notations as discussed in the previous chapters, the stresses at the two load conditions may be obtained as follows:

(i) *At transfer condition with axial prestress force:*

$$\sigma_{ci} = \frac{P_i}{A_c}$$

$$P_i = \sigma_{si} A_p \qquad (6.2.1)$$

where A_c = area of the column.

(ii) *At working load condition:* In addition to the losses of prestress with time, there will be a further loss of prestress due to elastic shortening caused by the external loads (in case

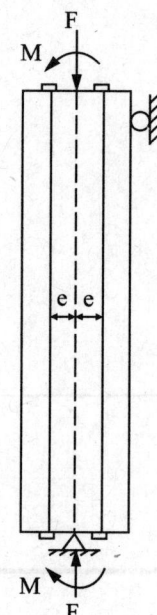

Figure 6.2.1 Column under external forces.

of beams, there is a gain in prestress due to external loads). Considering the failure condition, the compressive strain in the column at the time of failure will be in the order of 0.003. If the prestrain in steel is about 0.004, then the effective prestrain at ultimate load on the member is about 0.001. This illustrates that most of the prestressing force is released as the ultimate load on the column is reached, thus the ultimate strength of the column is not much affected by the pre-compression.

Let the compressive force dominate at working load condition: then the approximate strain in steel is given by

$$\in_s = \in_{se} - \frac{F}{A_c\, El_c} \quad \text{and the prestressing force is}$$

$$P = (\in_s E_s)\, A_p$$

$$= A_p\, E_s\left(\in_{se} - \frac{F}{A_c E_c}\right)$$

$$= Pe - pmF = \eta P_i - pmF \qquad (6.2.2)$$

where $\qquad p = A_p/A_c$ (reinforcement ratio)

$$m = E_s/E_c \text{ (modular ratio)}$$

Compressive stress is given by

$$\sigma_{ce} = \frac{P}{A_c} + \frac{F}{A_c} + \frac{M_y}{I}$$

$$= \frac{P}{A_c} + \frac{F}{A_c} + \frac{MD}{2I} \tag{6.2.3}$$

where $\qquad D$ = total depth of the section.

On substituting the value of P in eq. 6.2.3

$$\sigma_{ce} = \eta \frac{P_i}{A_c} - \frac{pmF}{A_c} + \frac{F}{A_c} + \frac{MD}{2I}$$

$$= p\eta\sigma_{si} - \frac{pmF}{A_c} + \frac{F}{A_c} + \frac{MD}{2I}$$

$$= p\eta\sigma_{si} + \frac{F}{A_c}(1 - pm) + \frac{MD}{2I} \tag{6.2.4}$$

Equations 6.2.1 and 6.2.3 are the governing equations from which the column design is to be determined. Here the known quantities are F and M, while stresses may be selected either from code or economical considerations. Value of η is to be assumed which, if necessary, could be checked after the completion of the design. So only three quantities (A_p, A_c, and shape and size of the section) are to be determined.

From eq. 6.2.1, the reinforcement ratio is given by

$$p = (\sigma_{ci}/\sigma_{si}) = \sigma_{aci}/\sigma_{si} \tag{6.2.5}$$

Equation 6.2.4 may be modified for rectangular section as

$$\sigma_{ce} = p\eta\sigma_{si} + \frac{F}{A_c}(1 - pm) + \frac{6M}{A_c D}$$

$$= p\eta\sigma_{si} + \frac{F}{A_c}\left(1 - pm + \frac{6M}{FD}\right) \leq \sigma_{ace} \tag{6.2.6}$$

Assuming the value of depth D is given by other considerations, the area of the section is given by

$$A_c = \frac{F\left(1 - pm + \dfrac{6M}{FD}\right)}{(\sigma_{ce} - p\eta\sigma_{si})} \tag{6.2.7}$$

Also note that for convenience, the following are used:

$$s_{ce} = \sigma_{ace}, \ \sigma_{ci} = \sigma_{aci}$$

All the quantities on the right hand side of eq. 6.2.7 are known, so the area of the concrete section could be obtained.

Allowable stress in steel at transfer is usually taken as $0.7 f_p$. Precompression is mainly for the secondary stresses so it may be taken in the range of $0.02 f_{ck}$ to $0.2 f_{ck}$. Allowable compressive force under working load may be taken in the range of $0.25 f_{ck}$ to $0.35 f_{ck}$ depending upon the utility of the member.

Example 6.2.1 Design of a prestressed concrete column of 5 metres high for a combined, axial force of 500 kN and bending moment of 1000 kN cm.

Solution

(a) Selection of stresses (all stresses in N/cm^2)

Let

$$f_{ck} = 4500$$

$$f_p = 150000$$

$$\sigma_{si} = 105000$$

$$\sigma_{ci} = 210$$

$$\sigma_{ce} = 0.4\, f_{ck} = 1800$$

(b) Selection of section:

Let a rectangular section of depth 25 cm be assumed, i.e.,

$$D = 25 \text{ cm}$$

(c) Calculation of areas of steel and concrete:

$$p = \frac{\sigma_{ci}}{\sigma_{si}} = \frac{210}{105000} = 0.002$$

$$m = \frac{E_s}{E_c} = \frac{2.1\,(10^6)}{3.8\,(10^6)} = 5.5$$

Let

$$\eta = 0.8, \text{ then from Eq. 6.2.7}$$

$$A_c = 500000 \frac{\left(1 - \dfrac{5.5}{500} + \dfrac{6(1000000)}{500000\,(25)}\right)}{1800 - \dfrac{0.8\,(105000)}{500}}$$

$$= 500000 \left(\frac{1.469}{1632}\right) = 449 \text{ cm}^2$$

Let

$$B = \text{breadth of the section, then it is}$$

$$B = \frac{A_c}{D} = \frac{449}{25} = 18 \text{ cm}$$

The area of the pretensioned steel is

$$A_p = pA_c = 0.002(449) = 0.89 \text{ cm}^2$$

Select 6 numbers of 5 mm wires.

$$A_p \text{ (provided)} = 1.176 \text{ cm}^2$$

The sectional details are shown in Figure 6.2.2.

Example 6.2.2 Design a column subjected to an eccentric load of 500 kN acting at 2 cm eccentric with the column axis and determine the factor of safety. Vide Figure 6.2.2 for column size.

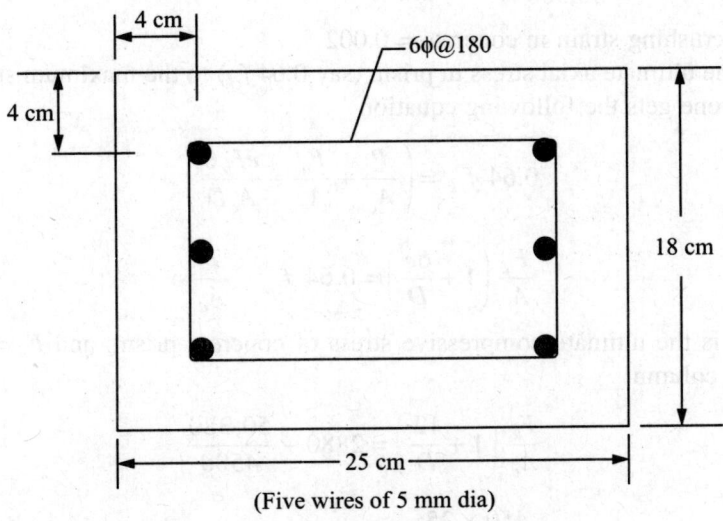

4 cm

6φ@180

4 cm

18 cm

25 cm

(Five wires of 5 mm dia)

Figure 6.2.2 Section of the column example 6.2.1.

Solution

The eccentric loading could be separated into axial force and bending moment as

$$F = 500,000 \text{ N} \quad \text{and} \quad M = 1000,000 \text{ N cm.}$$

Adopt the same section as designed in the example 6.2.1,

$$\text{Let } f_{ck} = 4500 \text{ N/cm}^2 \quad \text{and} \quad \epsilon_{cu} = 0.0035$$

Determination of the ultimate load capacity: As the stress-strain curve up to the crushing of concrete is not linear, eq. 6.2.7 cannot be used directly. Equation 6.2.3 is general and the same can be used here.

The initial strain in steel

$$\epsilon_{si} = \frac{105000}{21 (10^6)} = 0.005$$

The effective prestress strain is

$$\sigma_{se} = \eta \epsilon_{si} = 0.8 (0.005) = 0.004$$

The effective strain in concrete is

$$\epsilon_{ce} = \frac{P_e}{A_c E_c} = \frac{0.8 (0.7 f_p A_p)}{A_c E_c} = 0.00004$$

The net prestress at crushing of concrete is

$$P = A_p \sigma_s = A_p E_s \, (\epsilon_{se} - \epsilon_c + \epsilon_{ce})$$

$$= 1.176(21)(10^6)(0.004 - 0.002 + 0.00004)$$

$$= 50{,}380 \text{ N}$$

in which ϵ_c = crushing strain in concrete = 0.002

Equating the ultimate axial stress in prism (say $0.64 f_{ck}$) to the maximum stress developed in the column, one gets the following equation

$$0.64 \, f_{ck} = \left(\frac{P}{A_c} + \frac{F_u}{A_c} + \frac{e F_u 6}{A_c D} \right) \tag{6.2.8}$$

or

$$\frac{F_u}{A_c} \left(1 + \frac{6e}{D} \right) = 0.64 \, f_{ck} - \frac{P}{A_c}$$

where $0.64 f_{ck}$ is the ultimate compressive stress of concrete prism, and F_u = ultimate load capacity of the column.

$$\frac{F_u}{A_c} \left(1 + \frac{12}{D} \right) = 2880 - \frac{50{,}380}{4500}$$

$$\therefore \qquad F_u = \frac{450 \times 25}{37} (2880 - 11) = 872330 \text{ N}$$

Factor of safety $= \dfrac{F_u}{F} = \dfrac{872330}{500000} = 1.74$

(*Note:* It could be observed from this example that effective compression due to prestressing force at the ultimate load is 18 N/cm^2.)

Ultimate load for a column without prestress can be obtained from eq. 6.2.8 with $P = 0$ and it is

$$F_u = 0.64 \, f_{ck} \, A_c = 872000 \text{ N}$$

$$\frac{F_u}{F} = \frac{872000}{500000} = 1.744$$

The factot of safety of a non-preeompressed column is 1.744 against 1.74 of a pre-compressed column. It shows that the factor of safety is not affected by pre-compressing the column. This is more or less true in most cases. Therefore it is desirable to use prestressed columns when it is expected that the column will be subjected to bending moments either during working or handling conditions.

(2) **Prestressed concrete piles:** Piles are usually cast and cured in the factory and then transported to the site. Such piles are subjected to four load conditions: (i) transfer of prestress, (ii) transportation, (iii) driving, and (iv) service. Out of these four conditions, the transfer condition is not a dominant condition. The piles are likely to be subjected to tensile stresses during transportation and driving. Any small tilting of the pile during the driving operation will result in undesirable stresses. Special precautions have to be

taken at the top of the pile where pile cap and driving force come into contact. Sheet piles are also made of prestressed concrete and pre-compressed axially. If post-tensioning is done, the ducts must be grouted before driving (6.7, 6.8).

Example 6.2.3   A foundation pile is to carry an axial load of 600 kN with a possible moment of 800 kNcm. Design the pile with M350 concrete.

Solution

$$F = 600\ kN = 600000\ N$$

$$M = 800000\ N\ cm$$

$$f_{ck} = 3500\ N/cm^2$$

Let

$$\sigma_{ci} = 0.05\ f_{ck} = 175\ N/cm^2$$

$$\sigma_{ce} = 0.4\ f_{ck} = 1400\ N/cm^2$$

$$\sigma_{si} = 105000\ N/cm^2;\ \eta = 0.75$$

$$B = 25\ cm;\ E_c = 3.32\ (10^6)\ N/cm^2$$

then

$$p = \frac{\sigma_{ci}}{\sigma_{si}} = \frac{175}{105000} = 0.0017$$

$$m = \frac{E_s}{F_c} = \frac{210}{33.2} = 6.35$$

From Eq. 6.27, we have

$$A_c = \frac{600000\ [1 - 6.35(0.0017) + 6(800000)/600000(25)]}{1460 - 0.75\ (105000)\ (03.0017)} = 628\ cm^2$$

then

$$D = \frac{A_c}{B} = \frac{628}{25} \approx 25\ cm$$

The area of tension steel is

$$A_p = pA_c = 0.0017(625) = 1.06\ cm^2.$$

Use 25 cm by 25 cm cross section with 5 numbers of 5 mm wires.

(3) **Prestressed concrete poles and towers:**   Prestressed concrete poles are gaining their use as transmission poles and towers. The transmission poles are generally cast at factory and supplied to the site. The design of poles is simple but the main problem is their transportation. Since the length of poles that could be transported is restricted, the long poles can be cast in small lengths and assembled at site. The connections have to be designed using simple slices, extension pieces or additional reinforcements. Long prestressed concrete poles are used for radio and other communication towers (6.9). Poles are primarily flexural elements so they are presented in Chapters 3,7, 8 and 9.

(4) **Prestressed concrete piers and abutments:** Piers and abutments are prestressed for two reasons. Prestressing is used as a means of anchoring the pier to bed rock, and in some cases, if the bending moment due to water or soil pressure is large, the piers are prestressed against any possible tension. During the worst cases of pending moments, if the piers are subjected to vibrations due to moving loads, the piers are likely to crack. The prestresing force not only stops cracking but also reduces deformations. In most cases of prestressed piers, the prestressing is relatively nominal.

6.3. Tension Members

If a member is subjected to tensile forces only, it appears superfluous to use any concrete construction because the tensile resistance of concrete is negligible. Even though the steel takes all the tension force, a prestressed concrete construction for tension member has distinct advantage of resulting in small deformations. Except for a small variety of tension members, most members requires rigidity. For example, tie rods of arches and ring beams of water tanks require rigidity in addition to strength.

(1) *Tie members*:   Analysis and design of direct tension member is very simple and the governing equations for design can be written by considering the transfer and service load conditions.

Let F = direct tension force to be carried by the member,

then

$$\sigma_{ci} = \frac{P_i}{A_c} = \frac{A_p \sigma_{si}}{A_c} = p\sigma_{si} \qquad (6.3.1))$$

and

$$\sigma_{ce} = \frac{P}{A_c} - \frac{F}{A_c} = \frac{(P-F)}{A_c} \qquad (6.3.2)$$

where

$$p = \eta P_i + \left(\frac{FE_s}{A_c E_c} \right) A_p$$

$$= \left(\eta\sigma_{si} + \frac{mF}{A_c} \right) A_p = (P_e + pmF) \qquad (6.3.3)$$

and

$$\sigma_s = \left(\eta\sigma_{si} + \frac{Fm}{A_c} \right) \qquad (6.3.4)$$

Equations 6.3. 3 and 6.2.2 are similar, except for change in sign

$$\sigma_{ce} = \frac{1}{A_c} (\eta P_i + pmF - F)$$

$$= \frac{1}{A_c} [\eta P_i - F(1 - mp)] \qquad (6.3.5)$$

$$= \eta p\sigma_{si} = \frac{F}{A_c}(1 - mp) \qquad (6.3.6)$$

Allowable stresses : σ_{ci} could be taken as high as $0.5\,f_{ck}$ as the member will be relieved of the compression as soon as the external loads act. In service load conditions, allowable stress of concrete may be taken as practically equal to zero. To have a good serviceability, the member may be designed to have small compression even at working load condition. The value of σ_{cc} may be taken in the range of $0.1\,f_{ck}$ to 0. The stress in steel at working load condition should also be checked.

Example 5.3.1 Design a tie member of 20 m length subjected to 400 kN to axial tension and determine the factor of safety.

Solution

The data and the allowable stresses are (stresses are in N/cm^2)

$$f_{ck} = 4500; \qquad\qquad f_p = 150000$$

$$\sigma_{ci} = 0.48\,f_{ck}; \qquad\qquad \sigma_{si} = 0.6\,f_p = 90000$$

$$\sigma_{ce} = 0.1\,f_{ck}; \qquad\qquad \eta = 0.85 \text{ and } m = 5.5$$

The ratio of the reinforcement is

$$p = \frac{\sigma_{ci}}{\sigma_{si}} = 0.024$$

From Eq. 6.3.6, we have

$$\sigma_{ce} = \eta p \sigma_{si} - \frac{F}{A_c}(1 - mp)$$

or

$$A_c = \frac{F(1 - mp)}{\eta p \sigma_{si} - \sigma_{ce}} = \frac{400000(1 - 5.5(0.024))}{0.85(0.024)(90000) - 450} = 251 \text{ cm}^2$$

Use 16 by 16 cm tie section, then $A_c = 256$ cm^2 the area of pre-tensioned Steel is

$$A_p = pA_c = 0.024(256) = 6.14 \text{ cm}^2$$

Provide 16 numbers of 7 mm wires

$$A_p \text{ (provided)} = 6.15 \text{ cm}^2$$

$$P_i = 6.15(90000) = 553500 \text{ N}$$

Calculation of cracking load and factor of safety

Let ultimate tensile stress of concrete = 400 N/cm^2 and ultimate tensile strain = 0.00015. At cracking load, the strain in steel is

$$\epsilon_s = \epsilon_{se} + \epsilon_{ce} + \epsilon_t$$

$$= \left(\frac{0.85 \times 90000}{21 \times 10^6} + \frac{\eta P_i}{A_c E_c} + 0.00015 \right)$$

$$= (0.00364 + 0.0057 + 0.00015) = 0.00436$$

The corresponding stress in steel is

$$\sigma_s = 0.00436 \times 21 \times 10^6 = 91560 \text{ N/cm}^2$$

If F_c = load at cracking, then equilibrium condition at cracking gives

$$-400 = \frac{1}{256} (91560 \, (6.15) - F_c)$$

or

$$F_c = 102\,400 + 562\,180 = 664\,580 \text{ N}$$

Factor of safety at cracking $= \dfrac{F_c}{F} = \dfrac{664580}{400000} = 1.66$

Total elongation at cracking load $= (0.00057 + 0.00015) \times 2000 = 1.44$ cm

If only high tensile steel is used instead of prestressed concrete tie, the elongation at the cracking load is

$$\frac{F_c L}{A_p E_s} = \frac{664580 \times 2000}{6.15 \times 21 \times 10^6} = 10.3 \text{ cm}$$

Deformation of steel tie-bar at a load of 664580 N is as high as 10.3 cm and will affect the arch, whereas the deformation of a prestressed concrete tie is 1.44 cm for the same load.

The effect of the concrete is lost as soon as the cracking takes place and then all the load is to be taken by steel.

$$F_u = f_p A_p = (150000) \, 6.15 = 921000 \text{ N}$$

Factor of safety $= \dfrac{921000}{400000} = 2.303$

The factor of safety is not affected by providing the concrete.

(2) *Prestressed concrete ring beams and circular tanks*: Ring beams are used in domes, water tanks and in similar constructions. These ring beams are subjected to tensile forces caused by radial pressure. Similarly circular tanks are also subjected to radial pressures which cause hoop tensions in the tank walls. If P is the prestressing force, then the radial component is given by

$$P \frac{d^2 y}{dx^2} = \frac{P}{R} \text{ , where } R \text{ is the radius.}$$

Most of the load is taken by the steel and the concrete acts like a stiffener with impervious medium. The relation between the hoop force and the radial pressure could be obtained from simple statics as shown in Figure 6.3.1. There are two conditions of loading and the governing equation can be written directly for the two conditions.

The governing equations are:
  At service condition

$$\sigma_{ci} = \frac{P_i}{A_c} = \frac{A_p \sigma_{si}}{A_c} = p\sigma_{si} \text{ and} \qquad (6.3.7)$$

$$p = \left(\frac{\sigma_{ci}}{\sigma_{si}} \right) \qquad (6.3.8)$$

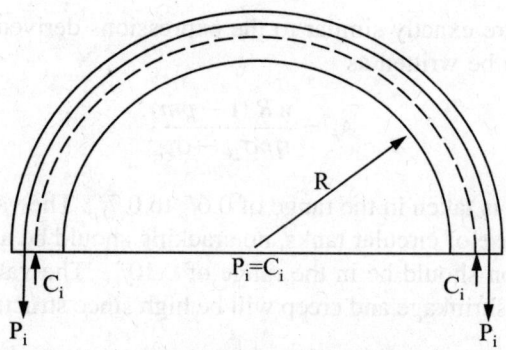

(a) Transfer condition: Prostress could be replaced by equivalent
radial component P_i/R

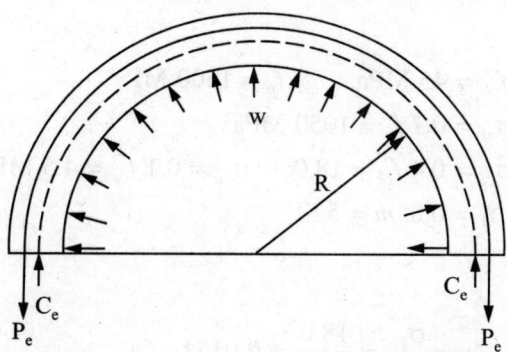

(b) Service load condition where w = intensity of radial pressure

Figure 6.3.1 Free body diagrams at two load conditions.

At service condition,

$$\sigma_{ce} = \frac{C_e}{A_c} = \frac{P - wR}{A_c} \tag{6.3.9}$$

where C_e = compressive force caused by the effective force.

Now

$$\epsilon_s = \epsilon_{se} + \frac{wR}{A_c E_c}$$

$$P = P_e + pmwR \tag{6.3.10}$$

$$\sigma_{ce} = \frac{(\eta P_i + pmwR) - wR}{A_c}$$

$$= \frac{\eta P_i - wR(1 - pm)}{A_c}$$

or

$$A_c = \frac{\eta P_i - wR(1 - pm)}{\sigma_{ce}} \tag{6.3.11}$$

The above expressions are exactly similar to the expressions derived in the previous section. Equation 6.3.11 can also be written as

$$A_c = \frac{wR\,(1 - pm)}{\eta p \sigma_{is} - \sigma_{ce}} \tag{6.3.12}$$

Allowable stresses: σ_{si} is taken in the range of $0.6f_p$ to $0.7f_p$. The value of σ_{ci} could be taken from $0.4f_{ck}$ to $0.5f_{ck}$. Incase of circular tanks, no cracking should be allowed. Hence, the stress at working load condition should be in the range of $0.10f_{ck}$. The value of η will also be low because of losses due to shrinkage and creep will be high since structure will be under wet and dry conditions.

Example 6.3.2 Design of a free edge water tank of 20 metre radius to store 5 metre head of water.

Assume

$$f_{ck} = 45 \text{ MPa}, \qquad f_p = 1500 \text{ MPa}$$

$$\sigma_{si} = 0.7\,f_p = 1050 \text{ MPa}$$

$$\sigma_{ci} = 0.4\,f_{ck} = 18.0; \qquad \sigma_{ce} = 0.1\,f_{ck} = 4.5 \text{ MPA}$$

$$\eta = 0.8, \; m = 5.5$$

Solution

$$p = \frac{\sigma_{ci}}{\sigma_{si}} = \frac{18.0}{1050} = 0.0171$$

$$A_c = \frac{wR\,(1 - pm)}{\eta p \sigma_{si} - \sigma_{ce}}$$

$$w = \gamma h = 0.01\,(500) = 5 \text{ N/cm}^2$$

$$A_c = \frac{5(2000)\,(1 - 0.095)}{0.8\,(0.0171)\,(105000) - 450} = 10.2 \text{ cm}^2$$

Considering 1 cm height, the thickness of the wall is

$$t_c = A_c/1 = 9.2 \text{ cm}$$

$$A_p = 0.021 \times 9.2 = 0.151 \text{ cm}^2$$

The pressure of the water reduces to zero at the top of the tank. Keeping the thickness of the wall almost constant, the percentage of steel could be reduced practically to zero at the ,top of the wall. A nominal steel of 0.1% is kept at the top. Use 5 mm wires at the top of the wall.

Area of steel provided (5 mm wires at 12 mm at bottom)

$$A_p = \frac{0.196}{1.2} = 0.163 \text{ cm}^2/\text{cm}$$

The prestressing force is

$$P_i = A_p\,\sigma_{si} = 0.163\,(105000) = 17150 \text{ N/cm}$$

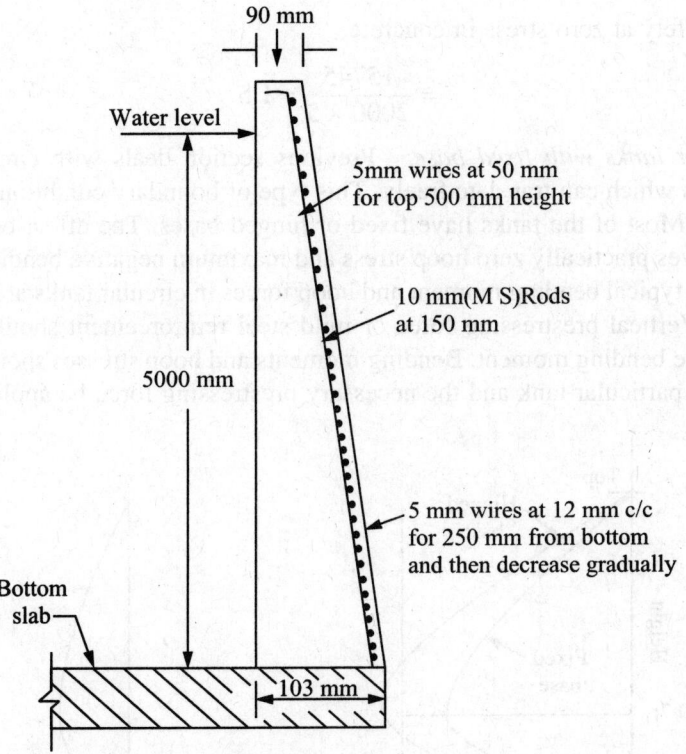

Figure 6.3.2 Section of vertical wall of the water tank.

Cracking load and factor of safety at cracking:

Let tension crack occur at 0.00015 strain and 400 N/cm² stress of concrete.

At cracking load the strain in steel is

$$\epsilon_s = \epsilon_{se} + \epsilon_{ce} + \epsilon_t$$

$$= \frac{0.8 \times 10.5000}{2.1 \times 10^6} + \frac{17150}{10.2 \times 3.82 \times 10^6} + 0.00015$$

$$= 0.004 + 0.00045 + 0.00015 = 0.00460$$

$$P = E_s \epsilon_s A_p = 0.00460 \times 0.163 \times 21 \times 10^6$$

$$= 15745 \text{ N}$$

Now
$$\sigma_{ce} = (P - w_e R)/A_c$$

or
$$-400 = \frac{(15745 - 2000 w_e)}{10.0}$$

or
$$w_e = 9.9 \text{ N/cm}^2$$

Factor of safety at cracking $= \dfrac{9.9}{5.0} = 2.0$

Factor of safety at zero stress in concrete

$$= \frac{15745}{2000 \times 5} = 1.5$$

(3) *Circular tanks with fixed base:* Previous section deals with circular tank having bottoms which can translate freely. This type of boundary condition does not exist in reality. Most of the tanks have fixed or hinged bases. The effect of clamping at the base gives practically zero hoop stress and maximum negative bending moment at the base. A typical bending moment and hoop forces in circular tanks are shown in Figure 6.3.3. Vertical prestressing force or mild steel reinforcement should be provided to resist the bending moment. Bending moments and hoop stresses should be worked out for any particular tank and the necessary prestressing force be applied.

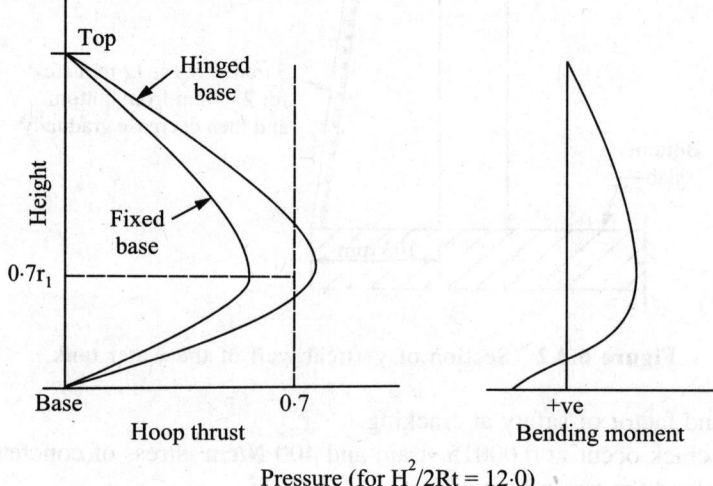

Figure 6.3.3 Hoop force in fixed and hinged bases.

(4) *Prestressed concrete pipes:* Prestressed concrete pipes have become very popular because of the built in factor of safety during handling as well as service conditions. There is also an ease in manufacturing the pipes. Construction and transportation problems are major one's when compared to design of the pipes themselves (6.10).

Example 6.3.3 A high pressure conduit of 2 m inside radius is transporting water at 100 N/cm^2. Design a prestressed conduit with M 35 concrete.

Solution

$$R = 200 \text{ cm}; f_{ck} = 35 \text{ MPa} = 3500 \text{ N/cm}^2$$

$$\sigma_{si} = 90000 \text{ N/cm}^2, w = 100 \text{ N/cm}^2$$

$$\sigma_{ce} = 0.06 f_{ck} = 210 \text{ N/cm}^2$$

$$\sigma_{ci} = 0.5\, f_{ck} = 1750 \text{ N/cm}^2$$

$$\eta = 0.8,\ m = 6.35$$

$$p = \frac{\sigma_{ci}}{\sigma_{si}} = \frac{1750}{90000} = 0.0195$$

From eq. 6. 3. 12

$$A_c = \frac{wR\,(1-pm)}{\eta p\sigma_{si} - \sigma_{ce}} = \frac{100(200)\,[1-0.0195(6.35)]}{0.8(0.0195)\,(90000)-21} = 11.7 \text{ cm}^2$$

$$t = A_c/B = 11.7/1 - 11.7 \text{ cm}$$

$$A_s = A_c p = 11.7\ (0.0195) = 0.228 \text{ cm}^2/\text{cm} = 22.8 \text{ cm}^2/\text{m}$$

Use 12 cm thick pipe with 175 windings of 4 mm wire/m and provide 2 cm thick gunniting over the wires.

6.4. Prestressed Concrete Pavements

(1) *Roads and runways:* Roads and runways have thin slabs which are generally treated as plates on elastic foundation. The thickness of the slab is very small as compared to the length. Secondary effects due to temperature and shrinkage are considerable. The prestressing cables are generally put at the bottom of the slab with sufficient cover and the prestressing is done in the longitudinal direction with ordinary or prestressing steel in the transverse directions. Special precautions have to be taken at the ends of the slab to prevent warping. The expansion and contraction of the pavements is very much reduced by prestressing force.

Prestressed concrete construction is also used for revetments of canals. Precast units to suit the profile of canal are manufactured in factory, transported and then placed in position. Cast in site concrete is more expensive because of the form-work involve'd along the canal whereas small precast units could very weil be used with sufficient care at the joints. Construction and expansion joints call for attention in this problem (6.11 to 6.15).

(2) *Prestressed concrete sleepers:* Prestressed concrete sleepers are made in mass scale and are used on the rail roads. As the train loads are fixed, the design is done once for a particular track. The great advantage of prestressed sleepers is their high resilience with small deformation.

PROBLEMS

6.1 A bridge pier is to carry an axial load of 4000 kN with a horizontal pressure of 10000 N/m for a height of 4 m from the base of the pier. The height of the pier is 8 m above foundation. Design a prestressed concrete pier with 40 MPa concrete.

6.2 A cylindrical barrel pipe line of 2 m diameter is to carry water at a pressure of 300 N/cm^2. Design a prestressed concrete pipe with a factor of safety of 3. At working load condition, the pipe should be at 200 N/cm^2 hoop compressive stress. Use M 45 concrete.

6.3 Design a tie-bar member of a tied arch, in which the tie is subjected to an axial force 1000 kN. Adopt 45 MPa concrete without cracking at working load and 20 MPa compressive stress at transfer condition. Also determine the factor of safety for this member.

6.4 A foundation pile carries a 1003 A.N load. The pile is precast in 8 m length and handled with two point hold. Pick up points are 2.5 m on either side from the centre of the pile. Design a M 40 concrete pile. (Hint: Assume a self weight of the pile as 2000 N/m and calculate the maximum bending moment while handling the pile. Design the pile against the 1000 kN axial force and BM caused by self weight.)

6.5 A column is to carry an axial load of 600 kN and BM of 400 kN. Design the column with M 45 concrete.

6.6 A 150 cm diameter pipe line is carrying water at 80 N/cm^2 pressure. Design the pipe such that it is under compressive stress of 100 N/cm^2 at working condition. Use M 45 concrete and 3 mm HT wires.

6.7 A cylindrical containment vessel of 10 m diameter is to be designed for 140 N/cm^2 pressure. Design the walls and reinforcement of the vessel by using simple thin tube theory. M 40 concrete and 5 mm wires are to be used.

CHAPTER 7

Limit States Design of Beams

7.1. Introduction

A structure should be designed so as to give a satisfactory and sate performance when exposed to natural and environmental forces. The resistance of a structure and the forces acting on the structure are usually assumed as fixed quantities even though they are of randomly varying types. The strength of a structure is a function of the strength of the materials, the geometric configuration of the structure, dimensions of the elements, etc. Each of these quantities are assigned a set of values by the designer. The engineer in construction or the builder is expected to maintain the specified values within a set of tolerances. Suppose the designer specifies a concrete of grade M 40 in his design computations. It is then expected of the builder to produce that grade of concrete. However, one should realise that no one can produce a concrete having a cube strength exactly equal to 40 MPa. A certain amount of variations from the main value is expected even under ideal circumstances. The aim of the builder is therefore to produce a concrete within certain variations. Similarly the sizes and dimensions of the elements of a structure will vary depending on the controls imposed in the production. It has been found that the frequency distributions of quantities such as the strengths of materials, sizes of elements, etc., follow a normal distribution. The satisfactory and safe production of a material can be ensured rationally by assigning a probability of attaining the specified value.

Let f_k = a specified value called the characteristic value of the strength of a material

$\quad\ f_m$ = mean value of the data of the actual strength of the material

$\quad\ s$ = standard deviation of the strength

Then the probabilitythat the strength is likely to be less than the specified value can be expressed as

$$p_f = \phi \left(\frac{f_k - f_m}{s} \right) = P(f < f_k) \qquad (7.1.1)$$

where

$$P_f = \text{probability of the strength less than } f_k$$

$$\phi = \text{cumulative distribution function.}$$

Equation 7.1 can be inversed to determine the relation between f_k and f_m. That is

$$\frac{f_k - f_m}{s} \leq \phi^{-1} (pf)$$

The value $(f_k - f_m)$ is certainly negative for low probability of failure, therefore the value $\phi^{-1}(p_f)$ can be assigned a value which is certainly negative.

$$\frac{f_k - f_m}{s} \leq - k$$

in which k = positive definite and defines the accepted probability of failure of the character.

$$= \phi^{-1} (P_f)$$

The above equation can be rearranged as

$$f_m \geq f_k + k_s \qquad (7.1.2)$$

The efforts of the builder in quality control is therefore to produce a material which follows Eq. 7.1.2. An accepted probability of failure can be assigned depending on the nature of importance of the structure. Most codes of practice in the design of concrete structures assign a five per cent accepted probability that the strength of the structural material falls below the specified value, then the value of k for such an acceptance is 1.65. This value can be obtained from the normal distribution function. The accepted probability of failure of the material in turn influences the reliability of the structure.

It is possible to design different type of structures by permitting different values of k. One can assign a higher probability of failure to temporary structures, or godowns such as cement godowns, while assigning a lesser probability of failure in the case of more important structures such as nuclear containment vessels, etc. Till such time when more realistic values could be set to k, one may want to settle for a single value of k. Hence Eq. 7.1.2 can be expressed as

$$f_m = f_k + 1.65 \, s$$

in which a 5 per cent accepted probability of failure is assigned.

This equation provides for certain flexibility in the quality of production of the material. Consider the case of the quality control in concrete. If the specified grade of concrete is M40, then the above equation can be expressed as

$$f_m = 40 + 1.65 \, s$$

The quality control is usually reflected through the standard deviation s. Lower the value of s, higher is the quality control, therefore a builder who maintains a low value of s has the advantage of settling for a lower mean value f_m. One can also interpret the other way to the advantage of a small builder or a builder who is building in difficult environment. He can relax on the quality control by paying more in the quantity of cement so as to attain a higher mean value of the strength.

Sometimes, Eq. 7.1.2 is interpreted in the reverse way. Anticipating the mean value and the standard deviation of the concrete, the designer can assign a characteristic value. In the case of concrete, one can always lay down the characteristic value and expect the builder to meet the accepted probability of failure. However, in the case of strength of the steels, the builder has no control in the production of the material, even though he may have the choice of selecting s brand of steel. In such materials, a reverse process can be applied by selecting the characteristic value from the mean value of the data on the strength of the material, even though such a thing is possible but highly improbable in normal practice. Hence, invariably, the characteristic values are specified by the designer and the builder meets the demand either by control in the field or by appropriate choice of the material from the market. In the case of concrete, the normally accepted characteristic strengths are:

$$f_{ck} = 30 \text{ to } 60 \text{ MPa at intervals of 5 MPa}$$

in which

f_{ck} = characteristic strength of concrete, defined by the compressive strength of 150 mm cube tested in moist condition after 28 days of water curing or equivalent.

The characteristic strength of high tensile steel used for prestressing is defined as the ultimate strength or proof stress of the wire and it is normally denoted by

$$f_y = 1400 \text{ MPa for bars and}$$

$$= 1500, 1600, 1650 \text{ and } 1700 \text{ MPa for wires}$$

In the case of untensioned reinforcement bars, the characteristic strength is defined by the value of the proof stress at 0.2 per cent permanent strain; and it is usually denoted by

$$f_p = 250 \text{ MPa for mild steel and}$$

$$= 415 \text{ and } 500 \text{ MPa for HYSD bars and wires}$$

Typical statistical data of the strengths some materials is given in Table 7.1.1.

One can extend the philosophy to any type of property such as dimensions, diameter of bars, etc. and express in common terms as

$$X_m = X_k + ks \qquad (7.1.3)$$

where X = any property such as strength of material, dimension of the elements, distance of the reinforcement in which the mean value has to be higher than the specified value.

The active forces on a structure should be treated differently from the passive resistance. The active forces are the loads including dead, live, dynamic, environmental, prestressing, etc. The active forces may be called loads for the sake of simplicity. In load study and treatment one should consider the maximum loads measured at the natural frequency and establish a characteristic load. One should measure a maximum live load in a year and establish the mean value of the maximum loads for a number of years. Similarly loads such as wind, earthquake, etc. could be established. The characteristic load is then defined as the load which shall not be exceeded by a certain accepted probability. Therefore the probability of failure of the loads can be defined by

TABLE 7.1.1 Typical statistical field data

Property	$\dfrac{X_m}{X_k}$	$\dfrac{100s}{X_m}$	P_f	% sets satisfying IS-code
1. *Concrete strengths*				
M 15 (nominal mix)	1.60	24	0.0332	97
M 20 (nominal mix)	1.45	19	0.0476	95
M 25 (design mix)	1.21	12	0.0808	80
M 35 (design mix)	1.21	13	0.0983	91
2. *Steel strength*				
HYSD 415	1.04	5.5	0.2451	–
HYSD 485	1.09	4.3	–	–
HTS 1500	1.05	3.1	0.0823	–
HTS 1600	1.13	1.0	10^{-24}	–
3. *Concrete dimensions*				
Thickness				
25 to 40 mm	1.03 to 1.06	5.0		
140 to 200 mm	0.99 to 1.03	1.0		
Depth of steel				
130 to 200 mm	1.03 to 1.18	6.0		
4. *Diameter of HTS*				
ϕ_k = mm	1.01	0.5		

Explanation of table 8.1.1.

X_m = mean value of the property

X_k = characteristic value (specified value)

s = standard deviation

P_f = the probability that the property will not satisfy the specified value. It is computed using the actual frequency distribution of the property. All properties except M IS nominal mix strength, follow the normal distribution at one per cent significance level.

Sets: A set normally consists of three specimens and in each group many sets are included. The last column gives the percentage of the sets satisfying IS specification based on the acceptability criterion for 3 cubes. Each of the groups contains more than 30 sets from a given project. Some of the groups contain more than one project data. The data was collected between 1965 and 1972 in northern India.

$$p_f = \phi\left(\frac{F_m - F_k}{s}\right)$$

or
$$F_k = F_m + s_k \tag{7.1.4}$$

in which k = the accepted probability that the load is likely to exceed the specified value. For a five per cent accepted failure, the value of k is 1.65.

One can see that the characteristic value is to be more than the measured mean value and it is influenced by the variation.

Having established a characteristic value, the designer adopts a design value of the property giving either reduction or enhancement to the property depending on its influence on the overall safety of the structure. Such factors are called partial safety factors. In the case of strength characteristics, the design strength is specified by:

$$f_d = \frac{k_a f_k}{\gamma_m} \tag{7.1.5}$$

in which γ_m = the partial safety factor applied to the materials

  k_a = the ratio of the actual strength of the material in the element to that of the characteristic value

and f_d = design strength

The characteristic strength of the concrete is associated with 150 mm cubes tested at 28 days of water curing. The actual strength of the concrete in an element is not the same as that of the cube and bears a certain relation depending on the size of the element, curing period, type of the force acting on the element. The coefficient k_a can be expressed as

$$k_a = k_p k_g k_f k_s \tag{7.1.6}$$

where k_p = ratio of the prism strength to the characteristic strength and it is = 0.67 (or 2/3) for cube characteristic and slenderness of prism more than 4; and = 0.85 for cylinder characteristic

  k_g = age factor and it is the ratio of the strength of the concrete at the age of loading to the 28th days strength.

  k_f = force effect factor which depends on whether the force is axial or bending compression and it is = 1 for bending compression and = 0.8 for axial compression

  k_s = size effect factor, it is usually taken as 1 unless the size of the element is very small or very large.

In the case of steel, the characteristic strength is usually obtained from the uniaxial tension test on the same or similar diameter bar. In such a case, the value of k_a is equal to one. In case one wants to extrapolate the strength of a higher diameter bar from the tests of lower diameter bars, then the actual strength factor k_a comes into force.

The partial safety factor depends on the type of material and limit state condition. Normally recommended partial safety factors for materials are given in Table 7.1.2.

The design load is obtained by applying some safety factor to the characteristic loads and it is given by

$$F_d = \gamma_f F_k$$

or when more than one type of load acts, the above equation can be generalised as

$$F_d = \Sigma \gamma_{fi} F_{ki} \tag{7.1.7}$$

in which

γ_{fi} = partial safety factor applied to the appropriate load.

TABLE 7.1.2 Partial safety factors (γ_m) for materials

Character	Limit state	γ_m
1. Reinforcement steels (mild steel, HYSD)	Collapse	1.15
2. High tensile steel wires	Collapse	1.15
(pretensioned wires)	Serviceability	1.67
	Transfer	1.25
3. Concrete	Collapse	1.5
	Serviceability	2.0

The partial safety, factor should normally depend on the nature of the load, the load combinations and the type of the limit state Table 7.1.3 gives the partial safety factors applied to the loads.

TABLE 7.1.3 Partial safety factors applied to loads (γ) in buildings

Load combination	Limit state of					
	Collapse			Serviceability		
	DL	LL	WL/EL	DL	LL	WL/EL
DL + LL	1.5	1.5	0	1.0	1.0	0
DL + + LL + WL/EL	1.2	1.2	1.2	1.0	0.8	0.8
DL + WL/EL	1.5 or 0.9*	0	1.5	1.0		1.0

DL = dead load. LL = live load, WL/EL = wind or earthquake load.
*0.9 is applied when the dead load has stabilising or compensating effect when compared with WL.

The design criterion is to proportion the member size and provide reinforcement such that the resistance of the section or the structure is always more than or at the most equal to the force from the external sources. That can be expressed as

$$R \geq F \tag{7.1.8}$$

in which

R = resistance or capacity

F = force

The resistance or capacity of a section is obtained with the analysis of the section whereas the force is obtained from the overall structural analysis of the structure. An elastic analysis is applied in the limit state of serviceability – and a collapse or limit analysis (plastic analysis) is applied to the limit of collapse. One can see that trie safety of the structure is built up on a four-tier system:

1. A preassigned probability that the strength of the material or similar element property does not fall below the characteristic value.
2. An accepted probability that the actual loads acting on the structure do not exceed the characteristic values.
3. Partial safety factors applied to the characteristic strengths.
4. Partial safety factor applied to the characteristic loads.

Therefore the overall reliability of the structure is doubly ensured. If one analyses the probability of failure of the structure either in the collapse limit state or in the serviceability limit state, he will find that the probability of failure of the structure would be less than one in a million. And in fact it was found that in some cases the failure probability was even less than 10^{-24}.

Let the resistance of a structure be probablistic and be expressed as a continuous frequency distribution function by a curve as shown in Figure 7.1.1. The frequency distribution curve for resistance follows a normal distribution. The probability of failure of the structure for a deterministic load can be obtained by integrating the area under the curve up to the load point. This is illustrated in Figure 7.1.1. If the load is also treated as probablistic, then Figure 7.1.2 illustrates the probability of failure of the structure. A rigorous analysis of probability of failure of prestressed

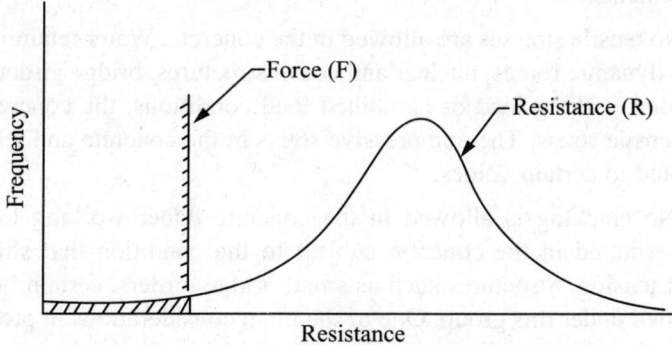

Figure 7.1.1 Probability of failure for deterministic load.

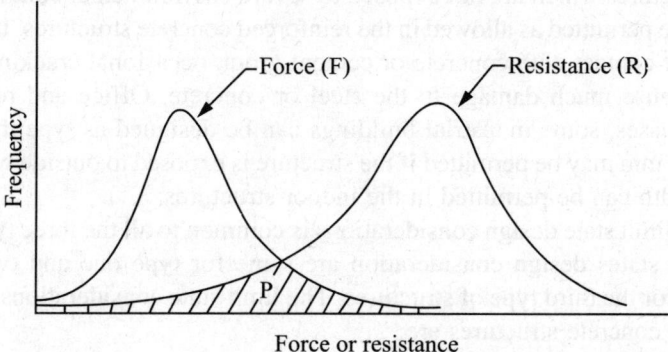

Figure 7.1.2 Probability of failure for probabilistic loads.

concrete structures designed and produced as per the Indian code specification have shown a wide range of probability of failures. Table 7.1.4 gives the probability of failures under different limit state conditions of different types of beams.

TABLE 7.1.4 Probability of failure of PSC structures*

Limit state	Statically	
	Determinate	Indeterminate
1. Collapse at transfer of prestress	10^{-9} to 10^{-6}	10^{-11} to 10^{-8}
2. Collapse under working load	10^{-8} to 10^{-6}	10^{-22} to 10^{-16}
3. Deflection	10^{-11} to 10^{-9}	10^{-12} to 10^{-11}
4. Cracking	10^{-8} to 10^{-6}	10^{-9} to 10^{-6}

*Study made under the supervision of the author at IIT, Kanpur.

7.2. Types of Prestressed Concrete Structures

Prestressed concrete structures are normally classified into three groups based on the functional and design requirements.

Type one: No tensile stresses are allowed in the concrete. Water retaining structures, and those subjected to dynamic forces, nuclear and ocean structures, bridge girders, etc., fall under this category. Under working load or combined load conditions, the concrete should not be subjected to any tensile stress. The compressive stress in the concrete and tensile stress in the steel are also limited to certain values.

Type two: No cracking is allowed in the concrete under working loads, and limited tensile stress is permitted in the concrete subject to the condition that stress is within the allowable flexural tension. Structures such as small bridge girders, certain building elements, etc., can be classified under this group. One of the main considerations in prestressed concrete structures is to see that the stress corrosion is not initiated in the steel.

Type three: Cracked section concrete structures. In the case of certain type of indoor or even outdoor structures which are not exposed to severe environmental conditions, cracking in the concrete can be permitted as allowed in the reinforced concrete structures. If the pretensioned wires are in direct contact with concrete or cement grout, occasional cracking of the concrete is not likely to cause much damage to the steel or concrete. Office and residential type of buildings, roof trusses, some industrial buildings can be designed as type three structures. A crack width of 0.1 mm may be permitted if the structure is exposed to outside weather conditions 0.2 mm crack width can be permitted in the indoor structures.

The collapse limit state design considerations is common to all the three types of structures. The service limit states design consideration are same for type one and two structures and slightly different for the third type of structures. The limit state considerations normally applied to the prestressed concrete structures are:
   1. Collapse limit state : (Also called strength limit state)
         Bending

Shear

Torsion

Bursting and bond

2. Serviceability limit state:

Stresses

Deflections

Crack width in the case of type three structures

3. Durability limit state:

Applied to all types through specifications.

7.3. Strength Limit State in Flexure

The collapse limit state in flexure is based on the following assumptions:

1. Plane sections before bending remain plane even at the time of collapse. This controls the compatibility of strains in the concrete and steel.

2. The maximum compressive strain in concrete is limited to a preassigned value. This value varies from 0.003 to 0.006 depending upon the percentage of compression reinforcement. Normally a value of 0.0035 is accepted in prestressed concrete members.

3. The maximum tensile stress in the steel is limited to slightly more than the initial yield stress. Since there is no fixed yield strain in high tensile steels, a strain corresponding to 0.2 per cent residual strain is taken as the limit of strength of the steel. At this strain, the steel has a tendency to flow and consequently the secondary compression in concrete will get generated rather too fast. The failure of the section at this strain is by the crushing of the concrete. The strain in steel at 0.2 per cent permanent set can be expressed as

$$\epsilon_{yp} = 0.002 + \frac{f_p}{E_s}$$

in which

f_p = proof stress in pretensioned steel

= same as characteristic strength

E_s = modulus of elasticity of steel

A design yield strain in steel can be obtained by applying a partial safety factor of 1.15 to the strength of the steel. Therefore the strain in steel at the collapse limit state is

$$\epsilon_y = 0.002 + \frac{f_p}{1.15 E_s} \tag{7.3.1}$$

Figure 7.3.1 illustrates the typical stress strain variation in steel and the corresponding yield strain.

The modulus of elasticity of high tensile steels vary from 195 000 to 210 000 MPa. However a value of 200 000 MPa can invariably be taken.

4. The stress distribution in the compression zone of the concrete can be idealised as rectangular or trapezoidal or parabolic-cum-rectangular as shown in Figure 7.3.2. The

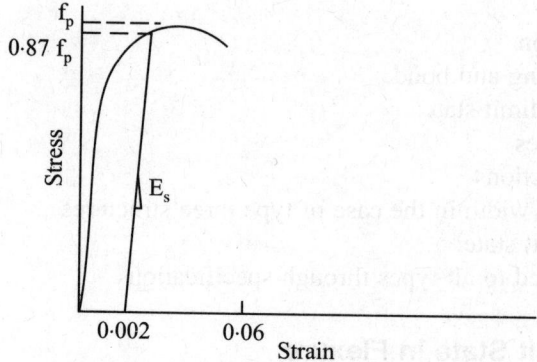

Figure 7.3.1 Stress-strain curve HTS.

most common idealisations are rectangular and parabolic cum rectangular. The two idealised stress distributions along with the centroid of action when acting on a rectangular section are shown in Figure 7.3.3. The idealisation of the rectangular stress distribution is very practicable to apply to any type of cross section. However it may introduce more pronounced errors in sections having larger widths at the neutral axis. Therefore a parabolic-cum-rectangular stress distribution appears to be slightly better when compared with the rectangular pattern.

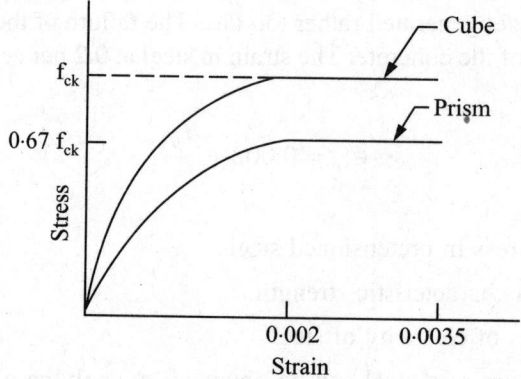

Figure 7.3.2 Idealised Stress-strain relation of concrete.

5. The prism compressive strength of the concrete in flexure is taken equal to 0.67 times that of the 150 mm cube strength. The strength of 150 mm cube is normally taken as the characteristic value.
6. A material partial safety of 1.5 is applied to concrete and 1.15 is applied to steel.
7. The tensile strength of concrete is negligible when compared with the other force contributions.
8. No slip in tensioned steel is assumed in the case of pretensioned or grouted post-tensioned constructions. However the free sliding of the wires can be assumed in the case of the unbonded construction.

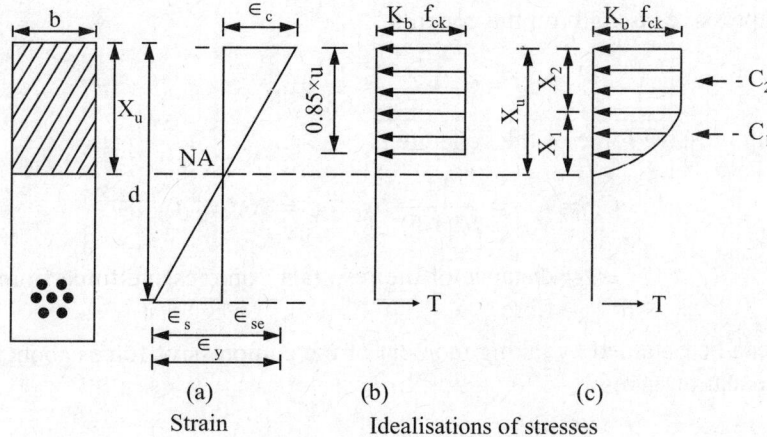

(a)   (b)   (c)

Strain   Idealisations of stresses

Figure 7.3.3 (a) Idealised stress blocks in compression, (b) Rectangular,
(c) Parabolic-cum-Rectangular.

The design moment capacity of a section at collapse limit state is computed using the principles of statics and compatibility of strains. Consider the case of a rectangular section subjected to flexure. The depths of the parabolic and rectangular portions of the stress block are:

$$x_1 = \frac{0.0020}{0.0035} x_u = \frac{4x_u}{7} \tag{7.3.2}$$

$$x_2 = \frac{0.0015}{0.0035} x_u = \frac{3x_u}{7} \tag{7.3.3}$$

The total force contributions from the parabolic and rectangular portions of the stress block are:

$$C_1 = \frac{2}{3}(x_1 k_p f_{ck}) b$$

$$= \frac{2}{3}\left(\frac{4}{7}\right)\left(\frac{2}{3}\right) bx_u f_{ck}$$

$$= \frac{16}{63}(bx_u f_{ck})$$

$$C_2 = (x_2 k_p f_{ck}) b = \left(\frac{3}{7}\right)\left(\frac{2}{3}\right)(bx_u f_{ck})$$

$$= \left(\frac{2}{7}\right)(bx_u f_{ck}) \tag{7.3.4}$$

in which k_p = ratio of the compressive strength of the concrete in flexure to that of the cube and it is taken equal to 2/3 or 0.67.

The total compressive strength on the concrete

$$C = C_1 + C_2 = \frac{34}{63} bx_u f_{ck} \tag{7.3.5}$$

The design compressive force of the concrete is

$$C_d = \frac{C}{\gamma_m} = \frac{34}{63(1.5)} bx_u \, f_{ck} = 36 \, bx_u \, f_{ck} \tag{7.3.6}$$

Let $\qquad x_c$ = distance of the resultant compressive force from the extreme fibre

The x_c value can be obtained by taking moment of the compressive forces about the top fibre. The moment equation gives

$$Cx_c = C_2 \left(\frac{x_2}{2} \right) + C_1 \left(x_2 + \frac{3x_1}{8} \right)$$

or $\qquad x_c \left(\frac{34}{63} \right)(bx_u f_{ck}) = \left(\left(\frac{3}{14} \right)\left(\frac{2}{7} \right) + \left(\frac{3}{7} + \frac{12}{56} \right)\left(\frac{16}{63} \right) \right) bx_u^2 f_{ck}$

This results in

$$x_c = 0.416 \, x_u \cong 0.42 \, x_u \tag{7.3.7}$$

The design moment caused by the compressive force about the centre of steel is

$$M = C_d \, (d - x_c) \tag{7.3.8}$$

The value of x_u can be obtained from the compatibility of strains consideration. From Figure 7.3.3(a),

$$\frac{\epsilon_c}{\epsilon_c + \epsilon_s} = \frac{x_u}{d}$$

or $\qquad x_u = \frac{\epsilon_c d}{\epsilon_c + \epsilon_s} \tag{7.3.9}$

in which $\qquad \epsilon_c$ = compressive strain in extreme fibre of the concrete

$\qquad \epsilon_s$ = the compatible strain in steel

The total strain in the pretensioned steel is

$$\epsilon_y = \epsilon_s + \epsilon_{se} \tag{7.3.10}$$

in which $\qquad \epsilon_{se}$ = effective prestrain in steel

Assuming the simultaneous occurrence of the crushing strain in the extreme fibre concrete and yield strain in steel, the compatibility of strains (Eq. 7.3.9) gives

$$x_u = \frac{\epsilon_c d}{\epsilon_c + \epsilon_y - \epsilon_{se}}$$

The normally accepted strains at failure of a balanced section are:

$$\epsilon_c = 0.0035$$

$$\epsilon_y = 0.0020 + \frac{0.87\, f_p}{E_s} \tag{7.3.11}$$

$$\epsilon_{se} = 0.004 \text{ and } E_s = 200000 \text{ MPa}$$

The substitution of the above values in Eq. 7.3.11 gives us

$$x_u = \frac{35d}{15 + 0.0435\, f_p} \tag{7.3.12}$$

The balanced moment capacity of a rectangular section can be expressed by using Eqs. 7.3.6 to 7.3.12 as

$$M_r = bd^2 f_{ck} \tag{7.3.13}$$

where

$$K = \frac{12.6}{14 + 0.0435\, f_p}\left(1 - \frac{14.7}{15 + 0.0435\, f_p}\right) \tag{7.3.14}$$

The area of pretensioned steel can be obtained from the equilibrium of forces on the section, that is,

$$C_d = T_d \tag{7.3.15}$$

in which $\quad T_d =$ design tension force

$$= \frac{A_p f_p}{\gamma_m} \tag{7.3.16}$$

where $\qquad A_p =$ area of pretensioned steel

$\qquad\qquad f_p =$ characteristic strength of pretensioned steel

and $\qquad\quad \gamma_m =$ partial safety factor for steel and usually taken as 1.15

Equations 7.3.16, 7.3.15 and 7.3.6 give

$$A_p = 1.15(0.36)\frac{bx_u f_{ck}}{f_p} \tag{7.3.17}$$

Substitution of Eq. 7.3.12 in Eq. 7.3.17 given

$$A_p = \frac{14.49bd}{15 + 0.0435 f_p}\left(\frac{f_{ck}}{f_p}\right) \tag{7.3.18}$$

The ratio of pretensioned steel to the effective depth are of the concrete can be obtained from the above equation. Thus,

$$\frac{A_p}{bd} = \frac{15.49}{15 + 0.0435\, f_p}\left(\frac{f_{ck}}{f_p}\right) = p_o\, \frac{f_{ck}}{f_p} \tag{7.3.19}$$

in which

$$p_o = \frac{14.49}{15 + 0.0435 f_p}$$

Various important coefficients associated with the balanced section for different characteristic strengths of steel are given in Table 7.3.1.

One can observe from table 7.3.1, the neutral axis distance is in the range of $0.4d$, the lever arm is in the range of 0.82 and the ratio of reinforcement is about $0.17 f_{ck}/f_p$. The percentage of reinforcement is in the range of 0.15 to 0.35.

Type 3 beams are often under-reinforced and type one and two beams are likely to be over-reinforced. A section can be said to be under-reinforced if the steel reaches the yield stress before the crushing in the concrete is initiated. It is given by:

$$\frac{A_p}{bd} < p_o \frac{f_{ck}}{f_p} \tag{7.3.20}$$

TABLE 7.3.1 Design coefficients

f_p (MPa)	$\dfrac{x_u}{d} = k_3$		K	p_o
1400	0.461	0.80	0.133	0.19
1500	0.436	0.82	0.128	0.18
1600	0.414	0.83	0.123	0.17
1700	0-393	0.83	0.118	0.16
1750	0.384	0.84	0.116	0.16

The over-reinforced section in the one in which the crushing of the concrete takes place before yielding of the steel and it is given by:

$$\frac{A_p}{bd} \geq p_o \frac{f_{ck}}{f_p} \tag{7.3.21}$$

The code of practice in prestressed concrete recommends that the amount of prestressing steel in a section be limited to the following constraint

$$\frac{A_p}{bd} \leq 0.24 \frac{f_{ck}}{f_p} \tag{7.3.22}$$

In the case of under-reinforced section the moment capacity is given by

$$M_r = f_{pu} A_p (d - 0.42 x_u) \tag{7.3.23}$$

where $\qquad f_{pu} = \dfrac{\alpha f_p}{\gamma_m} = \dfrac{\alpha f_p}{1.15}$

α = coefficient in the range of 0.9 to 1 depending on the % steel whereas in the case of over-reinforced sections, the design moment capacity be taken equal to that of the balanced section. In other words, the area of steel in excess of the balanced proportion of the steel should be neglected.

7.4. Limit State of Strength-Shear Capacity

Uncracked section: The shear capacity of a section is directly related with the maximum tensile strength or the diagonal tension in the concrete. The axial force due to prestressing force, off-sets the principal tensile stress, therefore the prestressed concrete beams have higher shear-resisting capacity when compared with the corresponding reinforced concrete beams. The principal tensile stress in an uncracked section is given by

$$f_1 = \frac{-f_{cp} + \sqrt{f^2_{cp} + 4\tau^2}}{2} \tag{7.4.1}$$

in which f_{cp} = axial prestress compression on the section

$$= \frac{P_e}{A}$$

f_1 = principal tensile stress

τ = shear stress on the section

The above equation can be rearranged as

$$\tau = \sqrt{f_1^2 + f_1 f_{cp}} \tag{7.4.2}$$

The shear failure takes place when the principal tensile stress on the section reaches the tensile capacity of the section:

$$f_1 = f_t$$

in which f_t = the tensile strength of the concrete

The tensile strength of the concrete is usually taken from 0.24 to 0.35 times $\sqrt{f_{ck}}$. For the present purpose it is taken as

$$f_t = 0.24\sqrt{f_{ck}} \tag{7.4.3}$$

The shear strength of an uncracked section is equal to

$$V_r = \tau b_w D \tag{7.4.4}$$

in which D = overall depth and b_w = web thickness

The design shear strength is obtained by dividing the ultimate capacity by the partial safety factor of 1.5. In addition when a beam is subjected to combined bending, axial force and shear, an allowance is made to the effective axial stress. Hence an effective axial stress of 0.8 f_{cp} is taken in the shear strength can therefore be expressed from Eqs. 7.4.2 and 7.4.5 as

$$V_t = \frac{\tau b_w D}{\gamma_m} = 0.67\, b_w D \sqrt{f_1^2 + 0.8 f_{cp} f_1} \tag{7.4.5}$$

In case of prestressed concrete beams having an inclined cable or tendon in which the tendon profile is to balance the shear, the component of the tendon force can be accounted while computing the effective shear force. That is

$$V_e = V_u - P_e \sin \theta \tag{7.4.6}$$

where V_u = static shear force at the section

P_e = effective prestress

θ = slope of the tendon

In beams, most of the tendons have shallow curvature, therefore the effective shear force on the section can be computed by approximating $\sin \theta$ equal to θ in radians.

$$V_e = V_u - P_e \theta \tag{7.4.7}$$

Shear capacity of cracked section: When a section has developed inclined cracking, the shear capacity is influenced by the shear stress and the moment acting on the section. The design shear capacity of a cracked section which is given by

$$V_r = \left(1 - \frac{0.55 f_{pe}}{f_p}\right) b_w d_{\tau_e} + \frac{M_o V_u}{M_u} \quad \text{and} \tag{7.4.8}$$

$$V_r \geq 0.b_w d \sqrt{f_{ck}} \tag{7.4.9}$$

in which

τ_c = design shear strength of the concrete and its values are given in table 7.4.1.

f_{pe} = effective prestress

M_o = moment needed to produce zero tensile stress in the concrete and it is given by

$$M_o = 0.8 f_{pt} \frac{I}{y_d} \tag{7.4.10}$$

f_{pt} = compressive stress caused by the prestress at depth d

y_d = distance of the effective steel from CG. This is equal to $d - x_u$.

TABLE 7.4.1 Design shear strength τ_c (in N/mm²)

100 Ap	Concrete grade		
$\dfrac{}{b_w d}$	30	35	40 and above
0.25	0.37	0.37	0.38
0.50	0.50	0.50	0.51
0.75	0.59	0.59	0.60
1.00	0.66	0.67	0.68
1.25	0.71	0.73	0.74
1.50	0.76	0.78	0.79
2.0	0.84	0.86	0.88
2.5	0.91	0.93	0.95

M_u = bending moment on the section

f = moment of inertia of the section

The effective shear force on the cracked section is same as the static force without any reduction due to the inclination of the cables.

$$V_e = V_u \qquad (7.4.11)$$

7.5. Design for Limit State of Strength in Shear

A prestressed concrete section must be designed to withstand the ultimate shear force on the section with or without shear reinforcement. The sections are classified into four groups based on the design requirement for shear reinforcement.

Group 1: No shear reinforcement needed   If the ultimate shear force on the section is less than fifty per cent of the shear capacity or if the members are of minor importance, no shear reinforcement is needed. That is if

$$V_e \leq 0.5\ V_r \qquad (7.5.1)$$

Group 2:   Nominal shear reinforcement

If $$0.5\ V_r \leq V_r \qquad (7.5.2)$$

then only a nominal shear reinforcement is required. The spacing of the reinforcement is limited to

$$s_v \leq \frac{0.87\ A_{sv} f_y}{0.4\ b_w} \qquad (7.5.3)$$

$$\leq 0.7\ d_t \qquad (7.5.4)$$

$$\leq 4b_w \qquad (7.5.5)$$

in which

A_{sy} = total sectional area of the stirrup legs

f_y = characteristic strength of the stirrup steel limited to 415 N/mm^2

d_t = effective depth of beam in shear

= D for uncracked section and

= d for cracked section

Group 3:   Shear stirrups needed

If $V_r < V_e \leq V_{cmax}$ $\qquad (7.5.6)$

then stirrups are provided. The value of V_{cmax} is given by

$$V_{cmax} = \tau_{cmax}\ b_w d \qquad (7.5.7)$$

in which

τ_{cmax} = ultimate shear strength of the concrete with transverse reinforcement and its values are given in table 7.5.1.

TABLE 7.5.1 Maximum shear strength of concrete with stirrups (τ_{cmax}) in N/mm^2)

Concrete grade	M30	M35	M40	M45	M50	M55
τ_{cmax}	3.5	3.7	4.0	4.3	4.6	4.8

The shear reinforcement be provided as

$$s_p \leq \frac{0.87\, A_{gy} f_y d_t}{V_e - V_r} \text{ or, subject to} \tag{7.5.8}$$

$$\leq 0.75\, d_t \text{ (for } V_e \leq 1.8\, V_r) \tag{7.5.9}$$

$$\leq 0.75\, d_t \text{ (for } V_e > 1.8\, V_r) \tag{7.5.10}$$

$$\leq 4b_w$$

Group 4: Revision of section for shear

If $V_{cmas} \leq V_e$ $\hspace{4cm}$ (7.7.11)

then the sectional dimensions must be modified so as to increase, the ultimate shear capacity of the section.

7.6. Design for Limit State of Strength Torsion

A torsional force on a member has the tendency to cause shear stress and it ultimately produces a skew plane of failure. The torsional moment is invariably accompanied by shear and/or bending moment. Therefore one should design a section for combined action of all the forces. The transverse shear and bending moment are uncoupled therefore the design against the above two forces is carried out independently. However, the stresses by torsion are coupled with those caused by the transverse shear or bending moment. Consequently the design for torque is combined with the other forces. The torsional moment on a section is divided into equivalent moment and shear forces for the purpose of design. The equivalent moment of the torque is given by

$$M_t = T_u \sqrt{(1 + 2D / b_w)} \tag{7.6.1}$$

in which

D = overall depth and

T_u = ultimate torsion on the section

The member must be designed to withstand effective bending moments

$$M_{e1} = M_u + M_t \tag{7.6.2}$$

and $\hspace{4cm}$ $M_{e2} = M_u - M_t$ $\hspace{4cm}$ (7.6.3)

The latter case becomes effective if M_u is less than M_t. In such a case the beam should also be designed to resist an effective beading moment acting on the section but on an axis perpendicular to the moment M. That effective transverse moment is given by

$$M_{e_3} = M_t \left(1 + \frac{x_1}{2e}\right)^2 \left(\frac{D + 2b}{b - 2D}\right) \frac{b}{D} \qquad (7.6.4)$$

in which

$$e = \frac{T_u}{V_u} \qquad (7.6.5)$$

x_1 = smaller dimension of the loop of the stirrup

The longitudinal reinforcement must be provided to resist M_{e1}, M_{e2} and M_{e3}, acting separately. In case M_u is positive bending moment, then the reinforcement has to be provided at the bottom fibres to resist M_{e1}, at top fibres to resist M_{e2} and both side faces to resist M_{e3}. The size of the depth and width are normally governed by M_{e1} and M_{e3}. In case M_u is larger than M_t (which is the case in most beams), the effective moment that governs the design is M_{e1}.

Equivalent shear force and transverse reinforcement: The torsional capacity of a prestressed concrete is enhanced by the existence of the axial force on the section. The torsion capacity of the concrete in the section is given by

$$T_{cr} = T_r \left(\frac{e}{e + e_c}\right) \qquad (7.6.6)$$

in which
T_{cr} = torsional capacity of a plain concrete section

$$e = \frac{T_u}{V_u}$$

$$e_c = \frac{T_r}{V_r}$$

Let

$$f_{cp} = \frac{P_e}{A}$$

The torsional capacity of an open flange section in an uncracked state can be obtained by the tension failure criterion. The shear stress which initiates a tension failure when the section is subjected shear and axial stress was shown to be

$$\tau = \sqrt{f_1^2 + f_1 f_{cp}} \qquad (7.4.2)$$

where f_1 = principal tensile stress and it is equated to the tensile strength of the concrete which is

$$f_t = 0.24 \sqrt{f_{ck}} = f_1 \qquad (7.4.3)$$

The torsional capacity of the open web section is

$$T_\gamma = \frac{J_\tau}{\gamma_m b} = \frac{\tau}{3b\gamma_m} \Sigma b_i^3 D_i \qquad (7.6.7)$$

where b_i, D_i = sizes of the elements of the section,

$$b = \text{smallest of } b'_1 s$$

$$J = \text{torsional constant}$$

Eqs. 7.4.2, 7.4.3 and 7.6.7 give

$$T_r = \frac{0.67}{3b}\sqrt{\left(0.057b\,f_{ck} + 0.24f_{cp}\,\sqrt{f_{ck}}\right)\Sigma b_i^3 D_i} \tag{7.6.8}$$

The Indian code of practice on prestressed concrete structures recommends the value of T_c as

$$T_r = 1.5\sqrt{(f_{ck} + 12f_{cp})}\,\Sigma b_i^2 D_i\left(1 - \frac{b_i}{3D_i}\right) \tag{7.6.9}$$

For consistency's sake, the equation recommended by the code will be used in the design.

The equivalent shear force on the section is given by

$$V_{cp} = V_r\left(\frac{e}{e + e_c}\right) \tag{7.6.10}$$

The torsional capacity of a box section can be obtained as

$$T_r = \frac{A_0 \tau}{\gamma_m} \tag{7.6.11}$$

in which

A_0 = area enclosed by the centre line of the walls of the box section

The substitution of the value of τ for tension failure in the above expression gives

$$T_{cr} = 0.67\,A_0 b\sqrt{f_1^2 + 0.8f_t f_{cp}} \tag{7.6.12}$$

To be on the safer side, the value of f_{ca} is modified as $0.8\,f_{ca}$ (T_r = torsional resistance = T_{cp}). Only nominal stirrups are to be provided in the section if

$$T_{cp} \geq T_u \tag{7.6.13}$$

and

$$V_{cp} \geq V_u \tag{7.6.14}$$

Otherwise the reinforcement is governed by the force design considerations. The stirrups or transverse reinforcement shall be designed to resist the torsion and the spacing of the the stirrups shall be limited to the smallest of the values given below

$$s_y \leq \frac{1.5\,A_{sy}b_1 d_1 f_{yv}}{M_t} \tag{7.6.15}$$

or

$$A_{sy} \geq \frac{1.15}{f_y d_1}\left((V_u - V_{cp}) + \frac{2(T_u - T_r)}{b_1}\right)s_v \tag{7.6.16}$$

or

$$s_v \leq \frac{b_1 + d_1}{4} + \phi \tag{7.6.17}$$

or

$$s_v \leq 200\text{ mm} \tag{7.6.18}$$

or
$$s_v \leq \frac{0.87 \, A_{gy} f_y}{0.4 b_w} \qquad (7.6.19)$$

in which

b_1 and d_1 = centre to centre distances between the corner bars in horizontal and vertical directions.

ϕ_v = diameter of the stirrup

Equations 7.6.17 to 7.6.19 are associated with the minimum reinforcement requirements.

The torsional reinforcement must be continued up to $(D - b_w)$ plus the bond length distance beyond which it is no longer required so as to avoid the skew failure mode.

7.7. Limit State Strength at Transfer Condition

During transfer condition, the concrete might not have gained its full strength and the prestressing force is at its maximum with very little external load acting on the structure. Therefore the prestressing force at transfer must be treated as an external load along with the dead load.

The ultimate tendon force
$$P_u = \gamma_{fp} A_p f_b \qquad (7.7.1)$$

in which

γ_{fp} = partial safety factor taken equal to 1.0 to 1.2.

The tension produced in the tendons by the dead load moment can be approximately taken as

$$F_g = \frac{\gamma_{fd} M_g}{(d - 0.42 \, x_u)} \qquad (7.7.2)$$

where γ_{fd} is taken in the range of 0.9 as the dead load moment is opposite to that caused by the tendons. The net compressive force to be resisted by the concrete section is

$$F_u = P_u - F_g \qquad (7.7.3)$$

The area of the concrete needed to withstand the above force with the centre of compression assumed at the centroid of the steel is

$$A_{ct} \geq \frac{\gamma_m F_u}{k_g f_{ci}} \qquad (7.7.4)$$

where f_{ci} = concrete cube strength at the time of transfer of prestress.

In a rectangular section, the effective concrete area on either side of the prestressing force having its centroid at the steel level is

$$A_{ct} = 2(D - d)b \qquad (7.7.5)$$

Therefore the design requirement is

$$2(D - d) \, b \geq \frac{\gamma_m F_u}{k_p f_{ci}} \qquad (7.7.6)$$

$$\geq \frac{1.5 \, (3)}{2} \frac{F_u}{f_{ci}}$$

or
$$(D - d)b \geq \frac{1.125\, F_u}{f_{ci}} \tag{7.7.7}$$

in which $\gamma_m = 1.5$ and $k_p = 2/3$

The design for shear at transfer follows the same procedure as discussed earlier. Usually shear does not govern the design in the transfer condition, therefore one need not necessarily be concerned about shear force at transfer.

The transfer limit state of collapse need not be imposed if the allowable stresses at transfer are satisfied.

7.8. Flanged Sections and Their Moment Capacity

The bending moment capacity of flanged sections can be obtained by considering in two parts: one corresponding to the contribution of the web-portion and the other corresponding to the portion of the flanges projecting beyond the width of the web. In case the neutral axis falls in the flange, the moment capacity is exactly same as that of a rectangular section with a width equal to the flange width. Therefore the moment capacity of a flanged beam is given by:

$$M_r = K\, b_f d^2\, f_{ck} \text{ if } x_u \leq t \tag{7.8.1}$$

in which

b_f = width of the compression flange

t = thickness of the compression flange

If $x_u \geq t$ then the moment capacity is given by

$$M_r = K b_w\, d^2 f_{ck} + 0.45\,(bf - bw)\, t(d - 0.5t)f_{ck} \tag{7.8.2}$$

In this expression, the design strength of the concrete in the flange portion is taken equal to $0.45\, f_{ck}$ as the stress on the flange is close to the rectangular stress distribution. The area of the tensioned steel required to balance the compressive force is obtained by the equilibrium of the forces on the section and it is

$$\frac{A_p f_{pu}}{\gamma_m} = 0.36\, b_w\, x_u\, f_{ck} + 0.45\,(b_f - b_w)\, t\, f_{ck}$$

The rearrangement of the above equation with the partial safety factor for steel as 1.15 gives

$$A_p = 1.15(0.36\, b_w x_u + 0.45\,(b_f - b_w)\, t)\, \frac{f_{ck}}{f_p} \tag{7.8.3}$$

where
$$x_u = \frac{35\, d}{15 + 0.0435\, f_p} \tag{7.8.12}$$

In flanged sections, the thickness of the flange is obtained by other considerations associated with the design of the slab or other practical considerations. The width of the flange (b_f) is governed by the shear lag considerations and it is limited to:

$b_f \leq$ spacing of the beams

$\leq \dfrac{L_0}{6} + b_w + 6t$

$\leq b_w + 12t$

In case of the isolated beams, the effective width of the flange is restricted to

$$b_f \leq \frac{L_0}{4 + L_0 / b_0} + b_w \qquad (7.8.4)$$

where L_0 = distance between the two points of contraflexure

= effective span in. simply supported 'beams

= 2 (effective span) in cantilever beams

= 0.7 (effective span) in continuous beams and frames

b_0 = actual width of the flange

The values of t and b_f can be taken as known values in the above expressions. Therefore the design variables are: d, b_w and A_p. For an assumed value of b_w, one can obtain the design values d and A_p from the design consideration

$$M_r \geq M_u \qquad (7.8.5)$$

and Eq. 7.8.3.

The width of the web is usually selected or designed based on transfer load condition or by shear or practical considerations.

Design for transfer condition: In case of T-beams, the design for transfer condition is exactly the same as that of a rectangular section which was discussed in section 7.8. If the beam is an I-section, then the size of the bottom flange is governed by the transfer condition. The centroid of the steel can be assumed to be at a higher level than the middle line of the bottom flange in most beams, that is

$$(D - d) > 0.5\, t_b \qquad (7.8.6)$$

in which t_b = thickness of the flange under compression in transfer condition

The area of the concrete in compression at transfer load condition is

$$A_{ct} = 2\,(D - d)\, b_w + (b_b - b_w)\, t_b \qquad (7.8.7)$$

where b_b = width of the flange under consideration at transfer.

The subscript b is used for convenience as it indicates the bottom flange in simply supported beams. The design criterion can be expressed as

$$A_{ct} = 2(D - d)\, b_w + (b_b - b_w)\, t_b \geq \frac{2.25\, F_u}{f_{ci}} \qquad (7.8.8)$$

The value of b_b can be obtained from the above equation for selected values of t_b and b_w.

The design of flanged sections to resist shear force and torsion is exactly the same as that was discussed in the earlier portion of this chapter.

Approximate design formulae

In the case of flanged beams, the contribution of the compressive force in the web below the flange depth is usually small and therefore it can be neglected. Then Eq. 7.8.2 reduces to

$$M_r = 0.45\, b_f\, t(d - 0.5t)f_{ck}$$

248

The effective depth for balanced section can be obtained by equating M_r to M_e and this condition gives

$$d = 0.5t + \frac{M_c}{0.45\, b_f t\, f_{ck}} \qquad (7.8.9)$$

By the same reason, the amount of steel required from Eq. 7.8.3 reduces to (for balanced section only)

$$A_p = 0.518\, b_f\, t\, \frac{f_{ck}}{f_{pu}} \qquad (7.8.10)$$

In case the section is not balanced the area of pretensioned steel is given by

$$A_p = \frac{1.15\, M_u}{(d - 0.5t)\, f_{pu}} \qquad (7.8.11)$$

7.9. Limit State of Serviceability—Allowable Stresses

Prestressed concrete structures of types one and two have to be designed on a no-crack basis. This automatically enforces an allowable tensile stress consideration. The maximum compressive stress on the concrete in working load or transfer load condition should also be limited to the allowable values. Table 7.9.1 gives the allowable compressive stresses in concrete. The allowable tensile stresses in the second type of structures are limited to 3 MPa in service load condition

TABLE 7.9.1 Allowable bending compressive stresses in concrete

A. *Transfer condition*

(1) **Post-tensioned** construction (σ_{aci})

Grade	M30	35	40	45	50	55	60
$\dfrac{\sigma_{aci}}{f_{ci}}$	0.540	0.512	0.483	0.455	0.427	0.398	0.37

(2) Pretensioned construction (σ_{aci})

Grade	–	–	M40	45	50	55	60
$\dfrac{\sigma_{aci}}{f_{ci}}$	–	–	0.510	0.492	0.475	0.458	0.44

B. *Working load condition* (σ_{ace})

Grade	M30	35	40	45	50	55	60
$\dfrac{\sigma_{ace}}{f_{ck}}$ zone*1	10.41	0.398	0.387	0.375	0.363	0.352	0.340
zone*2	0.350	0.337	0.323	0.310	0.297	0.283	0.27

*zone 1–where compressive stresses are not likely to increase
*zone 2–where compressive stresses are likely to increase
Note: Allowable direct compressive stress is 0.8 times of that in bending.

and 4.5 MPa in temporary short duration load condition. However such an increase in stress is permitted when the stress caused by the permanent load is compressive. This implies that no increase in the tensile stress is permitted in handling and transfer of prestress load conditions.

In the case of type 3 structures, cracking is allowed in the concrete. A hypothetical tensile is computed in such type of structures assuming that the section is uncracked. The hypothetical allowable tensile stresses are defined for a given crack width. Table 7.9.2 gives the allowable tensile stresses in concrete.

The hypothetical allowable tensile stress in type 3 structures can be increased further in case an additional non-tensioned reinforcement is placed in the tension zone of the concrete section. The stress can be increased by 4 MPa in pretensioned and grouted construction for 1 per cent of the additional reinforcement. Further increase in the hypothetical stresses is allowed for higher percentages of reinforcement subject to the condition that the maximum hypothetical stress is limited to 25 per cent of the characteristic stress.

TABLE 7.9.2 Allowable bending tensile stresses in plain concrete (stress in MPa or N/mm^2) (σ_{ati} or σ_{ate})

Type and construction	Grades of concrete				
	M30	M35	M40	M45	M50
(1) *Type* I	.... zero				
(2) *Type* 2					
(a) Transfer and working load conditions	....		3.0	....	
(b) Temporary load condition	....		4.5	....	
(3) *Type* 3					
Post-tensioned and grouted					
Crack width = 0.1 mm	3.2	3.4	4.1	4.4	4.8
= 0.2 mm	3.8	4.4	5.0	5.4	5.8
Pretensioned					
Crack width = 0.1 mm	–	–	4.1	4.4	4.8
= 0.2 mm	–	–	5.0	5.4	5.8

Assuming a positive bending moment (that is compression is caused at the top fibres of the concrete by external loads), the actual stresses on the section and the governing condition can be expressed as follows:

At transfer condition

$$\frac{P_i}{A} + \frac{P_i e y_b}{I} - \frac{M_g y_b}{I} = \sigma_{ci} \leq \sigma_{aci} \tag{7.9.1}$$

$$-\frac{P_i}{A} + \frac{P_i e y_t}{I} - \frac{M_g y_t}{I} = \sigma_{ti} \leq \sigma_{ati} \tag{7.9.2}$$

At working load condition

$$\frac{P_e}{A} - \frac{P_e e y_t}{I} + \frac{M_w y_t}{I} = \sigma_{ce} \leq \sigma_{ace} \qquad (7.9.3)$$

$$-\frac{P_e}{A} - \frac{P_e e y_b}{I} + \frac{M_w y_b}{I} = \sigma_{ce} \leq \sigma_{ate} \qquad (7.9.4)$$

in which

P_i = initial prestress

P_e = effective prestress after losses

e = eccentricity of the steel

y_b and y_i = distances of the extreme bottom and top fibres from CG

$I = I_g$ = second moment of area of the gross section

M_g = bending moment at the section caused by the loads at the time of transfer (excluding the prestressing force). This moment is mostly due to the self weight.

M_w = bending moment at the section caused at the effective working loads.

σ_{aci} = compressive stress at transfer (+ ve)

σ_{ti} = tensile stress at transfer (+ ve)

σ_{ce} = compressive stress at working load (+ ve)

σ_{te} = tensile stress at working load (+ ve)

σ_{aci}, σ_{ate} are the allowable stresses

7.10. Limit State of Cracking

Cracking of concrete is not permitted in types 1 and 2 of prestressed concrete structures. Only type 3 structures are allowed to crack in concrete and the crack width is indirectly controlled by the hypothetical tensile stress in concrete. Therefore crack width does not enter into the design computations of the prestressed concrete structure.

7.11. Limit State of Deflection

The deflections of all types of prestressed concrete beams are limited to maximum allowable values given in Table 7.11.1. The deflections in types one and two prestressed concrete structures can be calculated using gross second moment of the area in the case of bonded construction. The deflections in the type-3 structures are computed based on the cracked section. The effective sectional moment of inertia of a cracked section can be taken as

$$I_e = \frac{I_{cr}}{1.2 - \dfrac{jM_{cr}}{M}\left(1 - \dfrac{x_u}{d}\right)\dfrac{b_w}{b_f}} \qquad (7.11.1)$$

such that

$$I_{cr} \leq I_e \leq I_g \qquad (7.11.2)$$

where I_g = sectional moment of inertia of the gross section

I_{cr} = sectional moment of inertia of the cracked section about the neutral axis

M_{ct} = bending moment which produces cracking in the plain concrete section

$$= \frac{I_g f_{cr}}{y_t}$$

M = bending moment due to external load

j = lever arm ratio

The deflection caused by shrinkage can be expressed as

$$v_{cs} = \frac{k_3 k_4 \in_{cs} L^2}{D} \tag{7.11.3}$$

in which

k_3 = bending moment coefficient for the beam under UDL and it is

k_3 = 0.5 for cantilever beams

= 0.125 for simply supported beams

= 0.086 for partly continuous beams

= 0.063 for fully continuous beams

$$k_4 = \frac{0.72\,(p_t - p_c)}{\sqrt{p_t}} \le 1 \tag{7.11.4}$$

for $0.25 \le p_t - p_c \le 1$ and

$$= \frac{0.65\,(p_t - p_c)}{\sqrt{p_t}} \le 1 \text{ for } p_t - p_c \ge 1 \tag{7.11.5}$$

$$p_t = \frac{100\,A_{st}}{bd} \tag{7.11.6}$$

$$p_c = \frac{100\,A_{sc}}{bd} \tag{7.11.7}$$

The long term deflection including the effect of creep can be computed by multiplying the short term deflection by a factor $(1 + c_c)$.

Deflection due to creep alone is

$$v_{cre} = c_c v_d \tag{7.11.8}$$

where

c_c = net creep coefficient

v_d = short term deflection due to fixed loads

TABLE 7.11.1 Allowable deflections

Load condition	Allowable deflection
1. Final deflection including of all effects (creep, temperature, shrinkage, etc.) measured from the cast level.	$L/250$
2. Deflection including all effects after positioning of permanent loads.	$L/350$ or 20 mm
3. Total upward deflection or camber	$L/300$

7.12. Durability Limits State

Durability is the character of a structure to withstand environmental and repeated type of forces with the least maintenance. Prestressed concrete structures are usually built to last at least one hundred years with the least maintenance. The durability of a structure is reflected through the following parameters:

(1) Increases with increase in cement content; however, increase of the cement content increases shrinkage, consequently shrinkage surface cracks by themselves are not bad for structural strength but they decrease the durability of the structure: therefore, care must be taken to minimise the shrinkage through water cement ratio, curing and contraction joints; and minimum cement must be provided for a given exposure conditions. Table 7.12.1 gives the minimum cement contents.

(2) Decrease in the water cement ratio increases the durability of the concrete. While water content helps in workability and absorbing heat of hydration in concrete, it decreases the strength and increases shrinkage and void ratio in the concrete. Therefore a maximum water cement ratio is specified for a given cement content which is given in Table 7.12.1.

(3) The quality, the quantity, the size and grading of the aggregate have a marked influence on the durability. The crushing strength of the aggregate must be good. The impact value of the aggregate should be less than 30 and the aggregate should be free from not only sulphate compounds but must not have more than 2 per cent in foreign

TABLE 7.12.1 Minimum cement content (in kg/m^3 of concrete) and maximum water-cement ratio (w/c)

Exposure condition	Cement	w/c
1. *Mild*– completely protected against weather–(non corrossive indoors)	300	0.60
2. *Moderate*–sheltered from heavy rain and wind–buried or submerged concrete	300	0.55
3. *Severe*–exposed to saline weather and corrossive fumes	360	0.45
4. *Very corrosive*–under sea water or chemical fumes	400	0.42

Note: Minimum size of aggregate is 20 mm.

material for mild exposure condition and it should be within one per cent in severe exposed condition. The grading should be even. The maximum size of the aggregate has better lasting and smaller air entrainment property. Since the size of the aggregate also controls the cover and spacing of the bars, a compromise must be accepted. The minimum of the maximum size of the aggregate for prestressed concrete should be 20 mm. Wherever possible, a higher size such as 40 mm can be used.

(4) *Workability of the concrete:* It helps in placing the concrete evenly with minimum air content. A workability of 10 to 35 mm slump or 0.8 to 0.9 compaction factor is recommended for prestressed concrete.

(5) *Cover to the reinforcement:* Damage to the concrete exposed to severe conditions is often caused by corrosion of the reinforcement either main or nominal. A minimum cover to the reinforcement must be observed so as to eliminate corrosion of any reinforcement or metal inserts in the concrete. Corrosion increases the volume of the metal and consequently generates splitting in the concrete. A minimum of 25 mm cover should be provided in all the prestressed concrete constructions. A minimum cover of 50 mm in case of buried or under water structuress, and 75 mm in ocean structures should be provided.

7.13. Design Procedure

The recommended procedure in limit state design of types 1 and 2 prestressed concrete structures is:

1. The governing *moment at collapse* is computed from the given data and an assumed self-weight.

   Let M_c = collapse moment caused by the external forces

2. The effective depth of steel and other details are computed from the *collapse limit state design* criterion.

For rectangular sections:

$$Kbd^2 f_{ck} \geq M_c \qquad (7.13.1)$$

For flanged sections:

$$(Kb_w d^2 + 0.45 (b_f - b_w) t (d - 0.5t)) f_{ck} \geq M_c \qquad (7.13.2)$$

Usually b or b_w are assumed and b_f and t are either given or assumed.

3. Also estimate the pretensioned force.

4. Apply the *limit state of collapse at transfer load condition* and obtain other dimensions.

For rectangular sections:

$$(D - d)b \geq \frac{1.125 F_u}{f_{ci}} \qquad (7.13.3)$$

For flanged sections

$$(D - d) b_w + 0.5 (b_b - b_w) t_b \geq \frac{1.125 \, F_u}{f_{ci}} \qquad (7.13.4)$$

The overall depth D and the bottom flange dimensions b_b and t_b are to be computed. Sometimes, b_w is controlled by this condition

5. Check for limit state of collapse in shear

Compute the effective shear force and the nominal shear stress at the critical section. In case the shear stress is less than the ultimate shear strength (Table 7.5.1), design the shear reinforcement. There are three possible cases of shear stress limitations, so select the appropriate case and the diameter of the stirrups, and then design the spacing of the stirrups. Usually the effective depth designed by the limit state of collapse due to the moment is adequate to withstand the ultimate shear stress. Estimate the percentage of reinforcement for determination of the shear strength. Check whether the assumed self-weight is equal to or less than the actual self-weight. Compute the actual collapse moment and revise the section in case of unsafe variation in the assumed self-weight.

6. Design for prestressing steel

The area of the pretensioned steel is computed as under, *For rectangular section*

$$A_p = \frac{M_u}{(d - 0.42 x_u) f_{pu}} = \frac{M_u}{jd \, f_{pu}} \qquad (7.13.5)$$

In case of balanced flanged sections

$$A_p = 1.15 \left(0.36 b_w x_u + 0.45 (b_f - b_w) t\right) \frac{f_{ck}}{f_{pu}} \qquad (7.13.6)$$

Select the appropriate number of wires or cables and ensure that adequate space and cover are available to the steel.

7. Design of serviceability limit state of stresses

Compute the critical working, moments and effective prestress at working load,

$$P_e = \frac{A_p f_p}{1.67} = 0.6 \, A_p f_p \qquad (7.13.7)$$

Compute P_i after estimating or assuming losses in prestress.

Determine the maximum compressive and tensile stresses in the concrete and check whether they are within the allowable limits. Otherwise redesign the section.

8. Design of serviceability limits state of deflection

Compute the gross moment of inertia of the section (must have been computed in the design of the allowable stress limit state) and deflections under transfer and working load conditions. Ensure the deflections are with the limits.

9. *Detailing*

Tile sectional dimensions are rounded off to the nearest 5 mm in case of small sections and 10 mm in the case of large sections (if it is not already done earlier) and the placement details of the reinforcement are to be shown. Also provide nominal reinforcement as required. The profile of the cable must also be fixed.

10. *End zone design*

It is discussed in later chapters.

7.14. Design of Short Span Beams

There could be many ways of interpreting short spans, and what is a short span for one type of construction, may appear to be a medium or a long span for some other type. Irrespective of the types of construction, one can define a span as a short one if the self-weight of the beam is much less when compared with the superimposed load on it. In the case of prestressed concrete structures, spans up to 20 metres can be considered as short spans, whereas in the *RCC* construction even 20 m has to be treated as a long span. The self weight of a short span beams could be less than fifty per cent of the superimposed load. In such cases, the tensile stresses developed in the beam at the working load condition, dominate the design criterion. A high percentape of prestressing force needed to compensate such high tensile stresses sometimes produce adverse effects at the initial condition.

Example 7.14.1 *Small span simply supported post-tensioned beam of type 2*

A simply supported beam of effective span 10 m is subjected to a superimposed dead and live loads of 8 and 12 *k*N/m. Design a bonded post-tensioned prestressed concrete beam of type 2.

Design data

Effective span	$= L$	$= 10$ m
Superimposed dead load	$= w_{sd}$	$= 8$ *k*N/m
Live load	$= w_1$	$= 12$ kN/m
	Grade of concrete = M35	

A rectangular section is to be designed. The cube strength at the time of transfer prestress is 30 MPa. The allowable stresses in MPa are:

$$\sigma_{aci} = 15.36, \sigma_{ati} = 3$$

$$\sigma_{ace} = 15.36, \sigma_{ate} = 3$$

Design bending moment on the beam

The maximum bending moment occurs at the mid span, so the mid span section is designed for moment. For the purpose of assigning a self-weight to the beam, the approximate dimensions are assumed and the self-weight is estimated. Let $b = L/30$ and $D = L/16$, then the self-weight for short span beams can be estimated as:

$$w_g = \frac{\gamma L^2}{480} = \frac{25(100)}{480} = 5.2 \ kN/m$$

where γ = unit weight of the concrete taken as 25 kN/m^3

The characteristic load is

$$w_k = (w_g + w_{sd} + w_1) = (5.2 + 8 + 12) = 25.2 \ kN/m$$

The design bending moment at collapse limit state is

$$M_c = \gamma_f w_k \frac{L^2}{8} = 1.5(25.2) \frac{(100)}{8} = 472.5 \ kNm$$

Design for collapse limit, state of moment

The moment capacity of a balanced rectangular section and the corresponding neutral axis and lever arms are taken from Table 7.3.1 for $f_p = 1500$ MPa, and they are:

$$K = 0.128 \ N/mm^2, \qquad x_u = 1.436 \ d,$$

$$j = 0.82 \qquad \text{and} \qquad p_0 = 0.18$$

Assuming a balanced section, the design criterion for moment capacity is

$$Kbd^2 f_{ck} \geq M_c$$

Let

$$b = \left(\text{about} \ \frac{L}{30} \right) = 0.35 \ m$$

Then we have the effective depth as

$$d = \sqrt{\left(\frac{M_c}{Kbf_{ck}} \right)} = \sqrt{\left(\frac{472500}{0.128(0.35)(35)(10^6)} \right)} = 0.547 \ m$$

Estimate the area of pretensioned steel as

$$A_p = p_0 \ bd \ \frac{f_{ck}}{f_p}$$

$$= 0.18(0.35) \ (0.547) \ \frac{(35)}{1500} = 0.000807 \ m^2$$

The limit of ultimate prestressing force is

$$P_u = \gamma_f A_p f_p = 1.2(0.000807)(1500) = 1.4526 \ MN$$

in which $\gamma_f = 1.2$. In large span beams γf can be taken between 1.0 and 1.2.

Design for limit state at transfer condition

The bending moment due to self-weight is

$$M_g = \frac{w_g L^2}{8} = \frac{500}{8} \ kNm$$

The axial tension in the cables caused by the self weight moment is

$$F_g = \gamma_f \frac{M_g}{jd} = \frac{\gamma_f \, 500000 \, (10^{-6})}{(8)(0.82)(1.549)} = 0.125 \ MN$$

in which $\gamma_f = 0.9$

and $F_u = P_u - F_g = 0.449 - 0.125 = 1.324 \ MN$

The design condition is

$$(D - d) \, b \geq \frac{1.125 \, F_u}{f_{ci}}$$

where $f_{ci} = 30 \ MPa$

Therefore the value of D which is the overall depth of the beam is given by

$$D = \frac{1.125 \, F_u}{b f_{ci}} + d = \frac{1.125(1.324)10^6}{0.35(30)(10^6)} + 0.549$$

$$= 0.142 + 0.549 = 0.691 \ m$$

The actual self weight of the beam is

$$w_g = \gamma b D = 25(0.35)(0.691) = 6.3 \ kN/m$$

Since the increase in the total load as a result of increase in the self-weight is about 3 per cent, the effective depth can be increased marginally.

Use $D = 0.7 \ m$, $d = 0.56 \ m$ and $b = 0.35 \ m$.

Let $wg = 25(0.7)(0.35)$ $= 6.125$ $= 6 \ kN/m$

$w_t = w_g + w_{sd} + w_l$ $= 6 + 8 + 12$ $= 26 \ kN/m$

$$M_c = \frac{\gamma_f w_t L^2}{8} \qquad = 1.5(26)\frac{(100)}{8} \qquad = 487.5 \ kNm$$

The resisting moment capacity of the section is

$$M_r = 0.128 \, bd^2 f_{ck} = 0.128(0.35)(0.56)2(35)/1000$$

$$= 491.7 \ kNm$$

Since M_r is larger than M_c, the section is adequate against limit state of collapse in moment.

Limit state design for collapse in shear

The critical shear failure section is at a distance d from the effective support. The working shear force at that section is

$$V_i = \frac{w_t L}{2} - w_i \, d = 26(5 - 0.56) = 115.44 \; kN$$

Let the prestressing cables be parabolic in profile with its centroid coinciding with that of the section at the end section. Then the maximum sag of the cable is

$$g = d - 0.5D = 0.56 - 0.35 = 0.21 \text{ m}$$

The slope of the cable at distance d from the support is

$$\theta = \frac{4g}{L^2} (L - 2d) = \frac{4}{100} (10 - 1.12)(0.21) = 0.075 \;\; \text{radians}$$

The component of the cable force in the transverse direction

$$P_e \; \theta = \frac{A_p f_{pu}}{\gamma f_p} (\theta) = \frac{0.000807 \, (1500)}{1.67} (0.075) = 0.0544 \text{ MN}$$

in which γ_{fp} for prestressing force is taken as 1.67. The effective transverse shear force on the section is

$$V_e = V_i - P_e \, \theta = 1.5V - 54.4 \; kN$$

$$= 1.5(115.44) - 54.4 = 118.76 \; kN = 0.11876 \text{ MN}$$

The shearr capacity of the uncracked section (from Eq. 7.4.5) is given by

$$V_r = 0.67 b_w D \sqrt{(f_i^2 + 0.8 \, f_{cp} f_t)}$$

where

$$f_t = 0.24 \sqrt{f_{ck}} = 1.42 \text{ N} / \text{mm}^2$$

and

$$f_{cp} = \frac{P_e}{A} = \frac{0.723}{(0.35)(0.7)} = 2.95 \text{ MPa}$$

The substitution of various quantities in the shear capacity expression given

$$V_r = 0.67(0.35)(0.7) \sqrt{(1.42^2 + 0.8(2.95) \, (1.42))} = 0.38 \;\; \text{MN}$$

The shear capacity of the section (0.38 MN) is more than twice the effective collapse shear force on the section. Therefore no shear reinforcement need to be provided in the beam.

Design of prestressing steel

The area of prestressing steel needed is given by

$$A_p = \frac{r_m M_u}{j d f_p} = \frac{(1.15)4.87500}{0.82(0.56)(1500)} = 814 \text{ mm}^2$$

Provide 19 numbers of 7 mm diameter wires; the actual area of pretensioned steel provided is

$$A_p = 19(38.48) = 731 \text{ mm}^2$$

The prestressing force at transfer condition (also called the initial prestressing force) is

$$P_i = A_p f_{pi} = A_p (0.7 f_p)$$
$$= 731 (0.7)(1500) = 767600 \text{ N}$$

(The prestressing force is rounded off to 100 N)

Let the loss of prestress be $0.15 f_p = 225$ MPa

The effective prestressing force is then

$$P_e = P_i - 0.15 f_p A_p$$
$$= 767600 - 0.15 (1500)(731) = 603\ 100 \text{ N}$$

Note: The loss of prestress in post-tensioned construction is around 200 to 250 MPa. Therefore an average value of 225 MPa is acceptable. One can calculate the actual losses in prestressing force and check with the assumption. The limit state design at transfer condition was checked with prestressing steel area as 805 mm^2 which is about ten per cent more than that actually provided. However if the actual area provided is less then that used in the limit state of collapse at transfer, then the overall depth is likely to be overestimated. In such a case the overall depth can be recomputed from the limit state of collapse at transfer.

Design for serviceability limit state–allowable stresses

The properties of the section neglecting the effect of the steel are:

$$A = 0.35 (0.7) = 0.245 \text{ m}^2$$
$$Z = \frac{bD^2}{6} = \frac{0.35(0.7)^2}{6} = 0.02858 \text{ m}^3$$

The forces on the section are:

$$M_g = \frac{w_g L^2}{8} = \frac{26(100)}{8} = 75 \text{ } kNm$$

$$M_c = \frac{w_t L^2}{8} = \frac{26(100)}{8} = 325 \text{ } kNm$$

$$P_i = 767600 \text{ N}, P_e = 603100 \text{ N and } e = 0.21 \text{ m}$$

The allowable stresses on the concrete at service load are:

$$\sigma_{aci} = 0.512 (30) = 15.36 \text{ MPa}$$
$$\sigma_{ace} = 0.337 (35) = 11.80 \text{ MPa}$$

The actual stresses on the section are:

The compressive stress at transfer (units are MN and m)

$$\sigma_{ci} = \frac{P_i}{A} + \frac{P_i e}{Z} - \frac{M_g}{Z}$$

$$= \frac{0.7676}{0.245} + \frac{0.7676(0.21)}{0.02858} - \frac{0.075}{0.02858}$$

$$= 0.13 + 5.64 - 2.62 = 6.15 < 15.35 \text{ MPa}$$

260

The tensile stress at transfer is:

$$\sigma_{ti} = \frac{P_i}{A} + \frac{P_i e}{Z} - \frac{M_g}{Z}$$

$$= -3.13 + 5.64 - 2.62 = -0. < 3 \text{ MPa}$$

The compressive stress at the effective working load condition is:

$$\sigma_{ce} = \frac{P_e}{A} - \frac{P_e e}{Z} + \frac{M_e}{Z}$$

$$= \frac{0.6031}{0.245} - \frac{0.6031\,(0.21)}{0.02858} + \frac{0.325}{0.02858}$$

$$= 2.46 - 4.43 + 11.37 = 9.40 < 11.80 \text{ MPa}$$

The tensile stress at the effective working load condition is:

$$\sigma_{te} = -\frac{P_e}{A} - \frac{P_e e}{Z} + \frac{M_e}{Z}$$

$$= -2.46 - 4.43 + 11.37 = 4.48 > 3 \text{ MPa}$$

The tensile stress in concrete at effective working load is higher than the allowable value, whereas the other stresses have been found to be far less than the allowable values. One way of satisfying the allowable tensile stress limitation at effective working load is to increase the effective prestressing force. The last expression used to compute the tensile stress will give an indication of how much increase in the prestressing force is to be affected. It can be observed that the net stress produced by prestressing force is –6.89 MPa whereas the tensile stress caused by the working load was 11.37 MPa. Therefore the prestressing force must be increased by (11.37 – 3.0)/6.89 = 1.2 times the original value. Therefore the member of wires to be provided are;

$$N = 1.2(1.9) \approx 23$$

Such an increase in the prestress will violate the limit state of collapse at transfer condition. This violation is admissible provided the allowable stresses are limited. The area of prestressing steel is

$$A_p = 23\,(38.48) = 885 \text{ mm}^2$$

Allowing an initial prestress of 0.7 f_p and 225 MPa as loss of prestress, we have

$$P_i = A_p(0.7 f_p) = 885\,(0.7)(1500) = 929250 \text{ N}$$

$$P_e = P_i - 225\,A_p = 730100 \text{ N}$$

The actual stresses under different load conditions are (the force in MN and sizes in m)

$$\sigma_{ci} = \frac{P_i}{A} + \frac{P_i e}{Z} - \frac{M_g}{Z}$$

$$= \frac{0.9292}{0.245} + \frac{0.9292(0.21)}{0.02858} - \frac{0.075}{0.02858}$$

$$= 3.79 + 6.83 - 2.62 = 8.00 < 15.36 \text{ MPa}$$

$$\sigma_{ti} = -3.79 + 6.83 - 2.62 = 0.42 < 3 \text{ MPa}$$

$$\sigma_{ce} = \frac{P_e}{A} - \frac{P_e e}{Z} + \frac{M_e}{Z}$$

$$= \frac{0.7301}{0.245} - \frac{0.7301(0.21)}{0.02858} + \frac{0.325}{0.02858}$$

$$= 2.98 - 5.36 + 11.37 = 8.99 < 11.8 \text{ MPa}$$

$$\sigma_{te} = -2.98 - 5.36 + 11.37 = 3.03 \approx 3.0 \text{ MPa}$$

The violation in the allowable stress criterion in the second decimal place can be ignored as the stresses are rounded up to the second decimal place. All the stresses are within the permissible limits, therefore the design is acceptable.

Serviceability limit state for deflection

Net creep coefficient = 1.6

Second moment of the area of the section neglecting the effect of steel is

$$I = \frac{bD^3}{12} = \frac{0.35\,(0.7)^3}{12} = 0.01004 \text{ m}^4$$

$$E_e = 5700\sqrt{35} = 33720 \text{ MPa}$$

The equivalent distribution loads of the cables at initial stage and at transfer stage are computed for parabolic profile and these values are:

$$w_{pi} = \frac{8 p_i g}{L^2} = \frac{8(929.2)(0.21)}{100} = 15.61 \text{ } kN/m$$

$$w_{pe} = \frac{8 P_e g}{L^2} = \frac{8(730.1)(0.21)}{100} = 12.26 \text{ } kN/m$$

The effective distributed loads acting at transfer and working load condition, and also the effective permanent loads are:

$$w_i = w_g - w_{pi} = 6 - 15.61 = -9.61 \text{ } kN/m$$

$$w_e = w_t - w_{pe} = 26 - 12.26 = 13.74 \text{ } kN/m$$

$$w_{ed} = w_d - w_{pe} = 6 + 8 - 12.26 = 1.74 \text{ } kN/m$$

All the loads are uniformly distributed. Therefore direct substraction is admissible. The maximum deflection either upward or downward is caused at the mid span. The deflection at the initial stage is

$$v_i = \frac{5 w_i L^4}{384 \, E_c I} = \frac{5(-9610)\,(10^4)}{384(33720)\,(10^6)\,(0.010004)} = -0.0037 \text{ m}$$

The negative sign indicates an upward deflection. The allowable upward deflection in the beam is

$$v_{ai} = \frac{L}{300} = \frac{10}{300} = 0.033 \text{ m}$$

The deflection at working load condition is

$$v_e = \frac{5w_e L^4}{384 E_c I} = \frac{5(13740)(10^4)}{384(33720)(10^6)(0.010004)} = 0.0053 \text{ m}$$

Deflection due to effective permanent load is

$$v_{ed} = \frac{5w_{ed} L^4}{384 E_c I} = 0.0006 \text{ m}$$

The total deflection including the effect of creep is

$$v_t = v_e + c_c v_{ed} = 0.0053 + 1.6(0.0006) = 0.0063 \text{ m}$$

The allowable total deflection is

$$v_{at} = \frac{L}{250} = 0.04 \text{ m}$$

The deflections are within the limits.

Arrangement of wires

The 23 wires can be arranged in the following way: Three sets of four wires in the first row and two sets of four wires and one set of three wires in the second row. The minimum width needed to accommodate the 2nd set of wires can be computed by assuming clear spacing between the groups as 50 mm and side cover as 40 mm. Therefore the minimum width for the above spacing is

$$b_m = 2(50) + 2(40) + 3(14) = 222 \text{ m}$$

The width provided is more than adequate. The depth of the concrete available below the *CG* of the steel is

$$d_s = 700 - 560 = 140 \text{ mm}$$

Even after arranging two rows of the groups of wires, there will be at least 90 mm cover to the steel. A minimum cover of 40 mm is

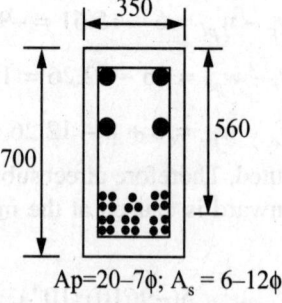

Ap=20–7ϕ; A_s = 6–12ϕ

Figure 7.14.1 Illustrates the section of Example 7.14.1

Example 7.14.2 *Small span simply supported post-tensioned beam of type 1*

Design data

Effective span	= L = 10 m
Superimposed dead load	= w_{sd} = 8 kN/m
Live load	= w_1 = 12 kN/m

$$f_{ck} = 35 \text{ MPa}, f_p = 1500 \text{ MPa}$$

Strength of concrete at transfer = f_{ci} = 30 MPa
Allowable stresses arc:

$$\sigma_{act} = 15.36 \text{ MPa}, \sigma_{ace} = 11.8 \text{ MPa}; \sigma_{ati} = \sigma_{ace} = 0$$

Note : The statement of this problem is same as that of example 7.14.1 except that this beam is of type 1. The design for collapse limit state is exactly the same for all the types. Therefore the reader is referred to example 7.14.1 in which the following computations were given.

(a) Design moment on the beam

(b) Design for collapse limit state of moment

(c) Design for collapse limit state at transfer

(d) Limit state design for collapse in shear

(e) Design of prestressing steel

(f) Part of the design for the serviceability limit state in allowable stresses.

The design details upto serviceability limit state are:

$$b = 0.35, \qquad D = 0.7 \text{ m}, \quad g = sag = 0.21 \text{ m}$$

$$d = 0.56 \text{ m}, \quad e = 0.21 \text{ m}$$

$$P_t = 0.7676 \text{ MN}, P_e = 0.6031 \text{ MN}$$

The actual stresses computed for the above design parameters are (from the example 7.14.1):

$$\sigma_{ci} = 6.15 < \sigma_{aci} = 15.36 \text{ MPa}$$

$$\sigma_{ti} = -0.11 < \sigma_{ati} = 0$$

$$\sigma_{ce} = 9.40 < \sigma_{ace} = 11.80 \text{ MPa}$$

$$\sigma_{te} = -2.46 - 4.43 + 11.37$$

$$= 4.48 > \sigma_{ace} = 0$$

The tensile stress in concrete at the working load exceeds the allowable value. It can also be observed that the tensile stress caused by the working load is far higher than the compressive stress caused by the prestressing force. Therefore one should increase the prestressing force proportional to the ratio of the tensile stress by the working load to the compressive stress caused by the prestressing force. This ratio is 11.37/(2.46 + 4.43) = 1.65. Such an increase in the prestressing force is likely to increase the compressive stresses, and the tensile stress at. initial condition. The collapse limit state at the initial condition will be violated, however, this can be overlooked if the stresses arc within the allowable values. Looking at all these conditions, one should modify the section along with an increase in the prestress. The following modification are affected:

$$b = 0.35 \text{ m}, d = 0.6 \text{ m}, D = 0.76 \text{ m}$$

$$g = e = 0.6 - 0.76/2 = 0.22 \text{ m}$$

$$P_i = 1.4(0.7676) = 1.07464 \text{ MN}$$

$$P_e = 1.4(0.6031) = 0.4434 \text{ MN}$$

The number of 7 mm wires

Then
$$Z = \frac{bD^2}{6} = 0.03369 \text{ m}^3$$

$$A = bD = 0.266 \text{ m}^2$$

There is a marginal change in the bending moments due to the change in the dead load and so they are taken from the previous example.

$$M_g = 0.075 \text{ MNm and } M_e = 0.325 \text{ MNm}$$

The stresses in the section in different load conditions are:

$$\sigma_{ci} = \frac{P_i}{A} + \frac{P_i e}{Z} - \frac{M_g}{Z}$$

$$= \frac{1.67464}{0.266} + \frac{1.67464(0.22)}{0.03369} - \frac{0.075}{0.03369}$$

$$= 4.04 + 7.02 - 2.23 = 8.83 < \sigma_{aci}$$

$$= \sigma_{ti} = -4.04 + 7.02 - 2.23 = 0.75 > \sigma_{ati} = 0$$

$$\sigma_{ti} = \frac{P_e}{A} - \frac{P_e e}{Z} + \frac{M_e}{Z}$$

$$= -\frac{0.84484}{0.266} - \frac{0.84434(0.22)}{0.03369} + \frac{0.325}{0.03369}$$

$$= -3.17 - 5.51 + 9.65 = 0.97 > \sigma_{ate}$$

The tensile stresses exceed the allowable values, therefore, a modification has to be done. The increase in P_i will increase the tensik stress at transfer. Therefore increase the size of the beam.

Let
$$d = 0.65 \text{ m}, b = 0.35 \text{ m and } D = 0.86 \text{ m}$$

The increase in depth of the beam increases the self-weight from 6 kN/m to 7.3 kN/m. The corresponding bending moments are

$$M_g = 0.090 \text{ MNm and } M_e = 0.342 \text{ MNm}$$

Also increase the magnitude of prestressing force so as to compensate the tensile stress at working load.

Let the number of 7 mm HTS wires be = N = 30

The area of pretensioned steel is

$$A_p = N(38.48) = 30(38.48) = 1154.4 \text{ mm}^2$$

The initial and effective prestressing forces are (with 225 MPa as loss of prestress).

$$P_i = A_p(0.7f_p) = 1154.4(0.7)(1500) = 1212120 \text{ N}$$

$$P_e = P_i - 225(A_p) = 952300 \text{ N}$$

The properties of the section are

$$A = 0.35(0.86) = 0.301 \text{ m}^2$$

$$Z = \frac{0.35(0.86)^2}{6} = 0.0431 \text{ m}^2$$

$$g = e = 0.65 - 0.86/2 = 0.22 \text{ m}$$

The stresses in MPa in the section are computed and they are

$$\sigma_{ci} = \frac{P_i}{A} + \frac{P_i e}{Z} - \frac{M_g}{Z}$$

$$= \frac{1.21212}{0.301} + \frac{1.21212(0.22)}{0.04314} - \frac{0.09}{0.04314}$$

$$= 4.03 + 6.18 - 2.09 = 8.12 \text{ MPa} < \sigma_{aci}$$

$$\sigma_{ti} = -4.03 + 6.18 - 2.09 = 0.06$$

$$\sigma_{te} = -\frac{P_e}{A} - \frac{P_e e}{Z} + \frac{M_e}{Z}$$

$$= -\frac{0.9523}{0.301} - \frac{0.9523(0.22)}{0.04314} + \frac{0.342}{0.04314}$$

$$= -3.16 - 4.85 + 7.93 = 0.08 > \sigma_{ate}$$

$$\sigma_{ce} = 3.26 - 5.00 + 7.93 = 6.19 < \sigma_{ace}$$

The compressive stresses satisfy the allowable stresses. The actual tensile stress is 0.08 MPa against an allowable value of zero. The excess stress is in tbe second decimal position therefore it can be ignored as the stresses are rounded upto second decimal position.

It was observed in the previous example that the actual deflections are far too less than the allowable values. It can easily be extrapolated that the deflections in this case of the beam with increased size in the section and prestressing force will also be far less than the allowable value. Hence the section satisfies

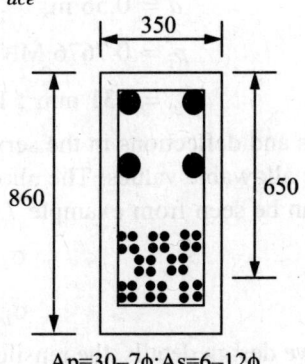

Figure 7.14.2 Illustrates the section of beam. Example 7.14.2

the design requirement. The final design parameters are

$$b = 0.35 \text{ m}, D = 0.86 \text{ m}, \qquad d = 0.65 \text{ m}$$

$$A_p = 1154.4 \text{ mm}^2, \qquad P_i = 1.221212 \text{ MN}$$

Example 7.14.3 *Small span simply supported post-tensioned beam of type 3*

A simply supported beam of effective span 10 m is subjected to superimposed dead and live loads of 8 and 12 kN/m respectively. Design a bonded post-tensioned prestressed concrete beam of type 3–namely a cracked sectioned beam.

Design data

Effective span $\qquad = L = 10$ m

Superimposed dead load $= w_{sd}$ 8 kN/m

Live load $\qquad = w_1 = 12$ kN/m

Allowable crack width $= 0.2$ m

The characteristic and allowable stress in MPa are

$$f_{ck} = 35 \; f_{ci} = 30$$

$$\sigma_{aci} = 15.36 \; \sigma_{ati} = 4.4$$

$$\sigma_{ace} = 11.8 \; \sigma_{ate} = 4.4$$

Note: Beams of types f and 2 for the same data have been designed in examples 7.14.2 and 7.14.1 respectively. Since the limit state design at collapse of all the types of the beam is the same, we can reproduce the design details from example 7.14.1. The serviceability of limit state design varies for each case, and therefore this example checks for the serviceability limit state of the type 3 beam. The design details based on collapse limit state in bending, shear and transfer are:

$$b = 35 \text{ m}; \qquad\qquad D = 0.7 \text{ m}$$

$$d = 0.56 \text{ m}; \qquad\qquad g = e = 0.21 \text{ m}$$

$$p_i = 0.7676 \text{ MN}; \qquad P_e = 0.6031 \text{ MN}$$

$$A_p = 731 \text{ mm}^2; \; 19 \text{ numbers of } 7 \text{ mm wires}$$

The stresses and deflections in the serviceability state have to be checked and compared with those of the allowable values. The allowable tensile stress for a crack Width of 0.2 mm is 4.4 MPa. As can be seen from example 7.14.1, the actual tensile stresses in the concrete are:

$$\sigma_{ti} = -0.11 \le 4.4 \text{ MPa}$$

$$\sigma_{te} = 4.48 \ge 4.4 \text{ MPa}$$

For the above design details, the tensile stress in service exceeds the allowable value, therefore the design needs modification. Since the stress exceeds only marginally, increase the pre-tensioning. Instead of 19 numbers of 7 mm wires, provide 20 numbers of the wires. Therefore the area tensioned stress is

$$A_p = 20(38.48) = 769.6 \text{ mm}^2$$

The pretressing forces at transfer and effective load conditions, allowing 225 MPa as loss of prestress are;

$$P_i = A_p(0.7 f_p) = 769.6(0.7)(1500) = 808080 \text{ N}$$

$$P_e = A_p(0.7f_p - 225)$$

$$= 769.6(0.7(1500) - 225) = 634920 \text{ N}$$

The tensile stresses at the initial and effective load conditions are:

$$\sigma_{ti} = \frac{-P_i}{A} + \frac{P_i e}{Z} - \frac{M_g}{Z}$$

$$= \frac{-0.80808}{0.245} + \frac{0.80808(0.21)}{0.02858} - \frac{0.075}{3.02858}$$

$$= -3.30 + 5.94 - 2.62 = -0.02 \le 4.4 \text{ MPa}$$

$$\sigma_{te} = \frac{-P_e}{A} - \frac{P_e e}{Z} + \frac{M_e}{Z}$$

$$= -\frac{0.63492}{0.245} - \frac{0.63492(0.21)}{0.02858} + \frac{0.325}{0.02858}$$

$$= -2.59 - 4.67 + 11.37 = 4.11 < \sigma_{ate} = 4.4 \text{ MPa}$$

The stresses at the serviceability limit state are satisfied. The deflection of the beam at the initial and working load needs to be checked. (The collapse limit state at transfer load condition can be ignored if the section is designed with the tensile stress as a criterion.)

There is no tensile stress in the beam at transfer of prestress with the result that the section should be treated as uncracked section. The camber of the beam for uncracked section was found in example 7.14.1 and shown to be far too small when compared with the allowable value. Therefore the deflection at working load using a cracked section is computed and checked in this example. One should compute the actual neutral axis in the cracked state and determine the second moment of area of the cracked section. This implies that the flexural tensile strength of the concrete must be assumed. For the sake of simplicity, the moment of inertia of the section can be taken equal to the average of the gross and cracked section, where the cracked section at the limit state of strength is considered. The neutral axis distance is given by (Table 7.3.1).

$$x_u = 0.436d = 0.436(0.56) = 0.244 \text{ m}$$

The second moment of area of the cracked section is

$$I_{cr} = \frac{bx_u}{3} = \frac{0.35(0.244)^3}{3} = 0.001695 \text{ m}^4$$

The flexural strength of the concrete is

$$f_{ck} = 0.7 \sqrt{f_{ck}} = 0.7\sqrt{35} = 4.14 \text{ MPa}$$

The flexural moment capacity of the section is

$$M_{cr} = f_{cr} \, Z = 4.14 \, \frac{(0.35)(0.7)^2}{6} = 0.118 \text{ MNm}$$

$$\frac{jM_{cr}}{M} = (0.817) \, \frac{(0.118)}{0.325} = 0.297$$

The gross moment of inertia is

$$I = \frac{0.36 \, (0.7)^3}{12} = 0.01 \text{ m}^4$$

The effective second moment of the area of the section is

$$I = \frac{I_{cr}}{1.2 - \dfrac{jM_{cr}}{M}\left(1 - \dfrac{x_u}{d}\right)\dfrac{b_w}{b_f}}$$

$$= \frac{0.001695}{1.2 - 0.297 \, (1 - 0.436)(1)} = 0.001642 \text{ m}^4$$

The effective moment of inertia is also subject to

$$I_{cr} \le I_e \le I$$

Therefore select

$$I_e = \frac{I_{cr} + I}{2} = \frac{0.001695 + 0.01}{2} = 0.0058 \text{ m}^4$$

The effective upward load caused by the cables is

$$w_{pe} = \frac{8P_e g}{L^2} = \frac{8(0.63492)(0.21)}{100} = 0.0107 \text{ MN/m}$$

The working load acting downwards is (vide example 7.14.1)

$$w_i = 26 \text{ kN/m} = 0.026 \text{ MN/m}$$

The effective load acting downwards is

$$w_e = w_t - w_{pe} = 0.026 - 0.0107 = 0.0153 \text{ MN/m}$$

Effective load excluding live load is

$$= 0.0153 - 0.026 - 0.0107 = 0.0033 \text{ MN/m}$$

The downward deflection due to *LL* is

$$v_l = \frac{5w_l L^4}{384 F_c \, I_e}$$

$$= \frac{5(0.012)(1000)}{384(5700)\left(\sqrt{35}\right)(0.0058)} = 0.008 \text{ m}$$

The allowable deflection $= v_a = \dfrac{L}{350} = 0.0286$ m

The deflection under dead load is

$$v_{ed} = \frac{5(0.0033)(10000)}{384(5700)\left(\sqrt{35}\right)(0.0058)} = 0.002 \text{ m}$$

The total deflection due to creep and load can be computed for a given creep coefficient. Let the net creep coefficient be $= 1.2$. Then the total deflection including the effect of the live load in creep gives

$$v_t = v_{ed}(1.2) + v_l = 0.011 \text{ m}$$

The deflection due to shrinkage is negligible. The allowable total deflection is

$$v_{at} = \frac{L}{250} = 0.04 \text{ m}$$

The deflections of the beam under various load conditions are less than the allowable values, therefore the beam is acceptable under serviceability limit state of deflection. Figure 7.14.3 illustrates the section.

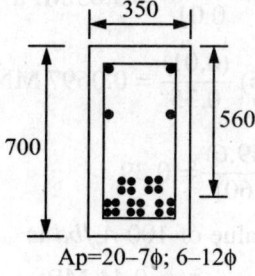

350

560

700

Ap=20–7ϕ; 6–12ϕ

Figure 7.14.3 Section of beam. Example 7.14.3 (Type 3).

Design for shear at limit state of strength

The critical shear force occurs at a distance d from the support and it is

$$V = w_t \left(\frac{L}{2} - d \right)$$

$$= 26(5 - 0.56) = 115.44 \text{ } kN$$

The effective shear force with a partial safety factor of 1.5 is

$$V_u = \gamma_f V = 1.5\,(115.44) = 173.16 \text{ } kN$$

The shear capacity of the section is given by

$$V_{cr} = \left(1 - 0.55 \frac{f_{pc}}{f_p} \right) \tau_c \, bd + \frac{M_0 V}{M}$$

where

$$\frac{f_{pe}}{f_p} = 0.6$$

$$M = \frac{w_t d}{2}(L - d) = \frac{26(0.56)}{2}(10 - 0.56) = 68.72 \; k\text{Nm}$$

$$M_0 = f_{pt} \frac{I}{y}$$

$$f_{pt} = \frac{P_e e^2}{I}$$

$$I = \frac{bD^3}{12} = 0.35 \frac{(0.7)^3}{12} = 0.01 \; \text{m}^4$$

$$P_e = A_p \, (0.6 f_p) = 769.6(0.6) \frac{1500}{10^6} = 0.69264 \; \text{MN}$$

$$f_{pt} = 0.69264 \frac{(0.21)^2}{0.01} = 3.05 \; \text{MPa}$$

$$M_0 = 0.8(3.05) \frac{(0.01)}{0.35} = 0.0697 \; \text{MNm}$$

$$\frac{100 A_p}{bd} = \frac{100 \, (769.6)}{350 \, (560)} = 0.39$$

The shear strength for the above value of 100 A_p/bd is

$$\tau_c = 0.44 \; \text{MPa}$$

The shear capacity of the cracked section is

$$V_{cr} = (1 - 0.55(0.6)) \, (0.44) \, (0.35) \, (0.56)$$

$$+ \, 0.0697 \frac{(0.11544)}{0.06872} = 0.1749 \; \text{MN}$$

The limiting shear capacity is

$$V_{cr} > 0.1 \, bd \, \sqrt{f_{ck}} = 0.1(0.35) \, (0.56) \, \sqrt{35} = 0.116 \; \text{MN}$$

$$V_r = V_{cr} = 0.1749 \; \text{MN}$$

This value is almost equal the effective shear force, therefore only nomial shear reinforcement is to be designed. Select two-legged 8 mm stirrups, then

$$A_{sy} = 2(50) = 100 \; \text{mm}^2$$

The minimum transverse reinforcement demands

$$s_v = \frac{0.87 \, A_{sy} f_{yv}}{(0.4) \, b} = \frac{0.87(100) \, (415)}{0.4 \, (350)} = 258 \; \text{mm}$$

Provide two-legged stirrups at 250 mm c/c. of 8 mm dia

Note: It is quite possible that the section under investigation is uncracked as it is close to the support. In such a case, the shear stress design could be based on an uncracked section. The present investigation is on the safer side.

Example 7.14.4 *Cost comparison of small span post-tensioned beams*

A simply supported beam of effective span 10 m is designed ior superimposed dead and live loads of 8 and 12 kN/m. The beam is designed as types 1, 2 and 3, and then the costs are compared.

The design data is:

$$L = 10 \text{ m}$$

$$w_{sd} = 8 \ k\text{N/m}; \ w_l = 12 \ k\text{N/m};$$

$$f_{ck} = 35 \text{ MPa}; \ f_p = 1500 \text{ MPa}$$

The unit costs of different items including supplying, placing, casting and curing are:

Cost of M35 concrete = Rs. 1500/m^2

Cost of form work = Rs. 150/m^2

Cost of pretensioned steel = Rs. 3000/kN

The beam has been designed in examples 7.14.1 to 7.14.3 and the design details are given in Table 7.14.1 along with cost comparisons.

TABLE 7.14.1 Cost comparisons of post-tensioned small span beams (units in m)

type	Design details			Quantities		Costs (in Rs.)				
	b	D	N 7ϕ	Concrete (m^3)	Formwork (m^2)	Concrete	Formwork	Steel	Total	Rel.
1	0.35	0.86	30	3.01	20.7	4515	3105	2720	10340	1.0
2	0.35	0.70	23	2.45	17.5	3575	2625	2085	8385	0.81
3	0.35	0.70	20	2.45	17 5	3675	2625	1800	8100	0.78
						(excluding cost of stirrups)				

Rel. = relative cost in comparison with that of the concrete.

7.15. Short Span Axially Prestressed Beams

Many short span structures such as slabs, canopy cantilever beams, poles, sleepers, etc., are often axially prestressed. Some beams or poles subjected to reversal of stresses have to be axially prestressed. In such cases the prestressing wires are placed on two or four faces such that the net effect is axial. It is desirable to place the wires close to the surface rather than to the centroid even in the axially prestressed members. The strength capacity of the section is proportional to the lever arm with respect to the tension reinforcement therefore the placing of the reinforcement near the surface helps. However, in the serviceability limit state condition, the position of the pretensioned steel in axially prestressed members has very little effect.

Example 7.15.1   Axially pretensioned poles–types 1

An electric line pole is 7.5 m in length and is subjected to 1800 N wind force at a height 6 m above GL. The pole projects 0.3 m above the level of the wires and embedded 1.2 m into the ground. Design a pretensioned pole such that it can withstand 1800 N wind force on the wires and (1800/4) N force along the direction of the wires.

Design Data

Cantilever span $= L = 7.5 - 1.2 - 0.3 = 6$ m

Bending force $= W_k = 1800$ N

$f_{ck} = 40$ MPa, $f_{ci} = 30$ MPa   and $f_p = 1600$ MPa

Design the pole for no tension condition with no deflection limitation.

Let the width of the pole be 100 mm and constant throughout. The depth of the pole is the least at the top and increases towards the ground. Since the weight of the pole is acting along the axis of the pole, it causes only axial stress but no bending stress. The axial stress due to the self-weight of the pole is small when compared with the bending stress and so it is neglected in the calculations. The amount of steel provided at each face would be equal to that required to withstand the wind load in one direction.

The design moment on the beam

The maximum bending moment occurs at the ground level of the pole and it is caused by the wind load. The design bending moment with a partial safety factor of 1.5 is

$$M_u = \gamma_f W_k L = 1.5(1800)(6) = 16200 \text{ Nm}$$

Design for collapse limit state

The design coefficients for a balanced section with $f_p = 1600$ MPa are taken from Table 7.3.1 and they are:

$$K = 0.123, \qquad\qquad x_u = 0.414\, d,$$

$$j = 0.83 \qquad \text{and} \qquad p_0 = 0.17$$

Let $\qquad\qquad b = 0.1$ m

then equating the moment capacity to the maximum moment, we have

$$M_r = Kbd^2 f_{ck} \geq M_u$$

or $\qquad\qquad d = \sqrt{\dfrac{M_u}{Kb\, f_{ck}}} = \sqrt{\dfrac{16200}{0.123(0.1)(40)10^6}} = 0.181$ m

The area of pretensioned steel at one face is

$$A_p = p_0 bd \frac{f_{ck}}{f_p} = 0.17(100)(180)\left(\frac{40}{1600}\right) = 77.6 \text{ mm}^2$$

The number of 5 mm wire on each face = 4

$$A_p \text{ (provided)} = 4(19.4) = 77.6 \text{ mm}^2$$

Design for limit state at transfer

The strength of the concrete at the initial condition is

$$f_{ci} = 30 \text{ MPa}$$

Minimum area of concrete needed to withstand the prestressing force with a partial safety factor of 1.2 applied to prestressing force is given by

$$bD_{min} = \frac{P_u}{h_p f_{ci}} = \frac{\gamma_f(2A_p)}{k_p f_{ci}}$$

$$= \frac{1.2(2)(77.6)(1600)}{0.67(30)(10^6)} = 0.014825 \text{ m}^2$$

The minimum depth required for this force is

$$D_{min} = \frac{0.014825}{0.1} = 0.148 \text{ m}$$

Use the size of the section at top as

$$b = 0.1 \text{ m and } D = 0.15 \text{ m}$$

Moment capacity

The moment capacity of the section in the lateral direction should not be less than one-fourth of that in the transverse direction. The area of pre-tensioned steel on the side faces will be equal to that on the main faces. Assuming the wires are placed at the corners, the effective depth in this direction is

$$d_y = 100 - 30 = 70 \text{ mm}$$

The moment capacity is

$$M_{ry} = 0.87\, A_p f_p j d_y$$

$$= 0.87\,(77.6)(1600)\,(0.83)\,(0.07) = 6276 \text{ Nm}$$

$$\frac{M_{ry}}{M_u} = \frac{6276}{16200} = 0.387$$

Since M_{ry} is more than $0.25\, M_u$, the section is adequate.

Limit state design–shear collapse

The design shear force on the pole at any height is the same and it is

$$V = 1800 \text{ N}$$

The profile of the wires is linear and no balancing effect is given by the wires. The effective design shear is

$$V_e = V_u - \gamma_f V = 1.5 \ (1800) = 2700 \text{ N}$$

The shear capacity of the uncracked section is given by

$$V_r = 0.67 \ bD \ \sqrt{(f_t^2 + 0.8 \ f_{cp} f_t)}$$

in which

$$f_t = 0.24 \ \sqrt{f_{ck}} = 0.24 \ \sqrt{40} = 1.51 \text{ MPa}$$

$$f_{cp} = \frac{P_e}{A} = \frac{2A_p(0.6 f_p)}{bD}$$

$$D = D_{min} = 0.15 \text{ m}$$

and

$$f_{cp} = \frac{2(77.6)(0.6)(1600)}{100(150)} = 9.93 \text{ MPa}$$

Then

$$V_r = 0.67(0.1)(0.15) \ \sqrt{(1.51^2 + 0.8(0.151)(9.93))} = 0.0379 \text{ MN}$$

since shear capacity of the section is far higher than the design shear force, there is no need to provide transverse reinforcement.

Prestressing force

Let the allowable stress in steel at the initial stress condition be

$$\sigma_{asi} = 0.7 \ f_p = 1120 \text{ MPa}$$

The loss of prestress in the pretensioned condition is estimated at 240 MPa; thus the effective prestress is

$$\sigma_{se} = 1120 - 240 = 880 \text{ MPa}$$

The prestressing forces at the initial and effective prestress stages are

$$P_i = 2 \ A_p \ \sigma_{asi} = 2(77.6)(1120) = 173824 \text{ N}$$

$$P_e = 2(77.6)(880) = 136576 \text{ N}$$

Serviceability limit state–Allowable stresses

The allowable axial compressive stresses in the pretensioned concrete taken from the Table 7.9.1 are:

$$\sigma_{aci} = 0.51 \ f_{ci} = 0.51 \ (30) = 15.3 \text{ MPa}$$

$$\sigma_{ace} = 0.323 \ f_{ck} = 0.323 \ (40) = 12.92 \text{ MPa}$$

$$\sigma_{ati} = \sigma_{ate} = 0$$

At the initial condition, there is only the prestressing force and that too is an axial force, therefore only compressive stress is produced. And it is critical at top section.

$$\sigma_{ci} = \frac{P_i}{A} = \frac{0.173824}{1.1(0.15)} = 11.59 < \sigma_{aci} = 15.3 \text{ MPa}$$

At working load condition, the compressive stress near the load point is

$$\sigma_{ce} = \frac{P_e}{A} = \frac{0.136576}{0.1(0.15)} = 9.11 < \sigma_{ace} = 12.92 \text{ MPa}$$

The bending moment at ground level caused by the wind load is

$$M = W_k L = 1800(6) = 10800 \text{ Nm}$$

The size of the section at ground level is

$$b = 0.1 \text{ m and } D = d + \text{cover} = 0.181 + 0.029 = 0.21 \text{ m}$$

The compressive and tensile stresses in concrete at the ground level of the pole under working load are:

$$\sigma_{ce} = \frac{P_e}{A} + \frac{M}{Z} = \frac{0.136}{0.1(0.21)} + \frac{0.0108(6)}{(0.1)0.21)(0.21)}$$

$$= 6.5 + 14.69 = 21.19 > \sigma_{ace} = 12.92$$

$$\sigma_{te} = -6.5 + 14.69 = 8.19 > \sigma_{ate}$$

The actual compressive and tensile stresses at the working load condition are far higher than the allowable values. Therefore the section has to be redesigned for serviceability condition. One can set up the design criteria and derive the sectional properties.

The design criteria are:

$$\sigma_{ce} = \frac{P_e}{A} + \frac{M}{Z} \le \sigma_{ace} \tag{1}$$

$$\sigma_{te} = -\frac{P_e}{A} + \frac{M}{Z} \le 0 \tag{2}$$

Adding the above two inequalities, we have

$$\frac{2M}{Z} \le \sigma_{ace}$$

or

$$Z \ge \frac{2M}{\sigma_{ace}} \tag{3}$$

From Inequality (2) we have

$$P_e \ge \frac{MA}{Z} = \frac{6M}{D} \tag{4}$$

Inequality (3) can be rearranged as

$$D^2 > \frac{12M}{b\,\sigma_{ace}} \tag{5}$$

Substitution of different values in the above expression gives

$$D \geq \sqrt{\frac{12(0.0108)}{0.1(12.92)}} = 0.317 \text{ m}$$

Let $\qquad D = 0.32 \text{ m}$

then $\qquad P_e = \dfrac{6M}{D} = \dfrac{6(0.0108)}{0.32} = 0.2025 \text{ MN}$

The area of pretensioned steel required is

$$2A_p = \frac{P_e}{\sigma_{ase}} = \frac{0.2025\,(10^6)}{880} = 230 \text{ mm}^2$$

Provide, 12 *number of 5 mm wires*, 6 on each face, then,

$$2A_p = 19.6(12) = 235.2 \text{ mm}^2$$

Since the prestressing force is increased, the minimum area of the concrete required at the transfer condition will also be increased. The prestressing force at the transfer condition is

$$P_i = 2A_p\,\sigma_{asi} = (235.2)(1120) = 263424 \text{ N}$$

The minimum depth needed is

$$D_{\min} = \frac{P_i}{b\,\sigma_{aci}} = \frac{0.263424}{0.1(15.3)} = 0.172 \text{ m}$$

Provide depth of the section at top as 0.175 m.

Check for stress at the ground level at working load condition needs to be done. The effective prestressing force is

$$P_e = 2A_p(\sigma_{ase}) = 235.2(880) = 206976 \text{ N}$$

The compressive and tensile stresses at the working load are:

$$\sigma_{ce} = \frac{P_e}{A} + \frac{M}{Z} = \frac{0.206976}{0.1(0.32)} + \frac{6(0.0108)}{0.1(0.32)(0.32)}$$

$$= 6.47 + 6.33 = 12.80 < \sigma_{ace} = 12.92$$

$$\sigma_{te} = -6.47 + 6.33 = -0.14 < \sigma_{ate} = 0$$

The size of the section at the ground level is acceptable as the stresses are less than those allowable.

Final design details

The dimensions at the top of the pole are:

$$b = 100 \text{ mm and } D = 175 \text{ mm}$$

The dimensions of the section of the pole at *GL* are:

$$b = 100 \text{ mm and } D = 320 \text{ mm}$$

Provide 6 numbers of 5 mm wires on each face with a cover of 25 mm. Figure 7.15.1 gives the details.

Example 7.15.2   Axially pretensioned pole-type 2

An electric line pole 7.5 m in length is subjected to 1800 N transverse force at 0.3 m from the top. A length of 1.2 m of the pole is embedded in the ground. Design a pretensioned concrete pole to withstand the wind force. The strength of the pole in the lateral direction should be at least one-fourth of that in the transverse direction.

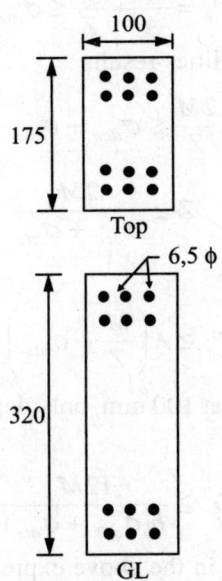

Figure 7.15.1 Sectional details of Example 7.15.1

Design data

$$L = 7.5 - 0.3 - 1.2 = 6 \text{ m}$$

$$= W_k = 1800 \text{ N}, f_{ck} = 40 \text{ MPa}, f_{ci} = 30 \text{ MPa and } f_p = 1600 \text{ MPa}$$

The design of the section for limit state of collapse is same for all the three types. The pole was already designed as type 1 structure in example 7.15.1. The design up to the serviceability limit state is exactly the same in both cases, and so the design details of cpllapse limit state are taken from example 7.15.1. They are:

$$b = 100 \text{ mm}, D = 210 \text{ mm and } P_i = 173824 \text{ N}$$

The actual stresses under working load are:

$$\sigma_{ce} = 21.19 > \sigma_{ace} = 12.92 \text{ MPa}$$

$$\sigma_{te} = 8.19 \geq \sigma_{ate} = 3 \text{ MPa}$$

Since the stresses at the working load condition are higher than those allowable, the section needs to be modified.

Serviceability limit state–allowable stresses

The design criterion of the stress limitation is

$$\sigma_{ce} = \frac{P_e}{A} + \frac{M}{Z} \le \sigma_{ace} \tag{1}$$

$$\sigma_{te} = \frac{P_e}{A} + \frac{M}{Z} \le \sigma_{ate} \tag{2}$$

The addition of the above two inequalities results

$$\frac{2M}{Z} \le \sigma_{ace} + \sigma_{ate}$$

or

$$Z \ge \frac{2M}{\sigma_{ace} + \sigma_{ate}} \tag{3}$$

The value of P_e from Inequality (2) is

$$P_e \ge A\left(\frac{M}{Z} - \sigma_{ate}\right) \tag{4}$$

Since the width of the section is fixed at 100 mm, only depth of the section is to be determined. From Inequality (3), we have

$$D^2 \ge \frac{12M}{b(\sigma_{ace} + \sigma_{ate})} \tag{5}$$

The substitution of various quantities in the above expression gives

$$D^2 \ge \sqrt{\frac{12(0.0108)}{(0.1)(12.92 + 3)}} = 0.285 \text{ m}$$

Use

$$D = 0.290 \text{ m}$$

Then

$$P_e \ge A\left(\frac{M}{Z} - \sigma_{ate}\right)$$

$$\ge 0.1 (0.29)\left(\frac{0.0108(6)}{0.1 (0.29)(0.29)} - 3\right) = 0.13645 \text{ MN}$$

The total area of tensioned steel needed is

$$2A_p = \frac{P_e}{\sigma_{ase}} = \frac{0.13645 (10^6)}{880} = 155 \text{ mm}^2$$

Provide 8 numbers of 5 mm wires at 4 on each face. Then

$$2A_p \text{ (provided)} = 8(19.6) = 156.8 \text{ mm}^2$$

$$P_i = 156.8 (1120) = 175616 \text{ N and}$$

$$P_e = 156.8(880) = 137984 \text{ N}$$

The actual stresses in the section at GL under the working load condition are:

$$\sigma_{ce} = \frac{P_e}{A} + \frac{M}{Z} = \frac{0.137984}{0.029} + \frac{0.0108(6)}{(0.1)(0.29)(0.29)}$$

$$= 4.76 + 7.71 = 12.47 < \sigma_{ace} = 12.94$$

$$\sigma_{te} = -4.76 + 7.71 = 2.95 < \sigma_{ate} = 3$$

The stresses in the section at GL are within the limits.

The section at the top of the pole needs to be designed for allowable stress condition. The allowable stress in the concrete at the initial stage is:

$$\sigma_{aci} = 0.51 f_{ci} = 0.51(30) = 15.3 \text{ MPa}$$

The minimum depth of the section needed is

$$D_{min} = \frac{P_i}{b\sigma_{aci}} = \frac{175616}{100(15.3)} = 115 \text{ mm}$$

The section at top of the pole is : $b = 100$ mm and $D = 115$ mm
The section at the GL of the pol is: $b = 100$ mm and $D = 290$ mm
Provide 4 numbers of 5 mm wires on each face with 25 mm cover. Figure 7.15.2 gives the details.

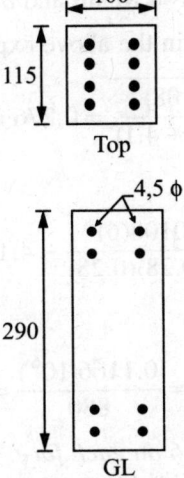

Figure 7.15.2   Example 7.15.2

Example 7.15.3   Axially pretensionedpole–type 3

Redesign the electric pole of example 7.15.1 as type 3 prestressed concrete structure. Allowable crack width = 0.1 mm.

Dseign data from example 7.15.1 is;

$$L = 6 \text{ m}, W_k = 1800 \text{ N}$$

$$f_{ck} = 40 \text{ MPa}, f_{ci} = 30 \text{ MPa and } f_p = 1600 \text{ MPa}$$

The collapse limit state design of types 1 and 3 structures is the same. The design details based on the limit state of collapse are

$$b = 100 \text{ mm}, \ D = 210 \text{ mm and } P_l = 173\ 824 \text{ N}$$

It can also be seen from example 7.15.1 that the stresses at working load have exceeded the permissible values allowed under type 3 structures. The allowable stresses are:

$$\sigma_{aci} = 0.51 f_{ci} = 15.3 \text{ MPa}$$

$$\sigma_{ace} = 0.323\ (40) = 12.92 \text{ MPa}$$

$$\sigma_{ate} = 4.1 \text{ MPa for } 0.1 \text{ mm crack width}$$

The design parameters based on the stress limitations were derived in example 7.15.2 and they are:

$$D^2 \geq \frac{12M}{b(\sigma_{ace} + \sigma_{ate})}$$

$$P_e \geq A\left(\frac{M}{Z} - \sigma_{ate}\right)$$

The values of M and b taken from the example 7.15.1 are:

$$M = 0.108 \text{ MNm and } b = 0.1 \text{ m}$$

The substitution of various quantities in the above expressions gives

$$D = \sqrt{\frac{12(0.0108)}{0.1(12.92 + 4.1)}} = 0.276 \text{ m}$$

Let

$$D = 0.28$$

$$P_e \geq 0.028 \left(\frac{0.0108(6)}{0.1(0.28)(0.28)} - 4.1\right) = 0.1166 \text{ MN}$$

Area of the tensioned steel needed is

$$2A_p = \frac{P_e}{\sigma_{ase}} = \frac{0.1166(10^6)}{880} = 132.5 \text{ mm}^2$$

provide 12 numbers of 4 mm wires at 6 on each face.
The total area of the tension steel provided is

$$2A_p = 12(12.56) = 150.72 \text{ mm}^2$$

and

$$P_e = 150.72(880) = 132633 \text{ N}$$

$$P_i = 150.72(1120) = 168806 \text{ N}$$

The stresses in the concrete at the working load condition are:

$$\sigma_{ce} = \frac{P_e}{A} + \frac{M}{Z} = \frac{0.132633}{0.028} + \frac{0.108(6)}{0.1(0.28)(0.28)}$$

$$= 4.74 + 8.27 = 13.01 < \sigma_{aei}$$

$$\sigma_{te} = -4.74 + 8.27 = 3.53 < \sigma_{ate}$$

The section at the *GL* is adequate. The minimum depth required at the top of the section is

$$D_{min} = \frac{P_i}{b\sigma_{aci}} = \frac{168806}{100(15.3)} = 110 \text{ mm}$$

The section at the top of tne pole is:

$$b = 100 \text{ mm and } D = 110 \text{ mm}$$

The section at the *GL* of the pole is

$$b = 100 \text{ mm and } D = 280 \text{ mm}$$

12 numbers of 4 mm bars, 6 at each face.

Figure 7.15.3 gives the design details.

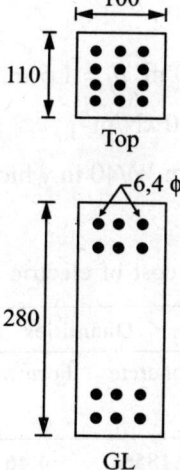

Figure 7.15.3   Example 7.15.3.

Example 7.15.4   Comparison of costs of type 1, 2 and 3 poles

Examples 7.15.1, 7.15.2 and 7.15.3 illustrate the design of pretensioned concrete pole under types 1, 2 and 3 classifications. A comparison of cost of the poles is made in this example. The unit costs of concrete, form work and steel including supplying, placing in position and finishing are:

Cost of concrete (M40)   = Rs 1625/m^2

Cost of formwork   = Rs 150/m^2

Cost of HT steel   = Rs 3000/kN

Length of the pole = L_0 = 7.5 m.

The cost comparison is given in Table 7.15.1

7.16. Medium Span Beams

The medium spans in prestressed concrete beams fall in the range of 20 m to 40 m in the case of simply supported beams and 15 m to.30 m in the case of cantilever beams. Many structures built in prestressed concrete may fall under this category. Bridges, roof beams are commonly built in such medium and even in long spans. The self-weight of such beams is still less than the superimposed load in most cases. The tensile stresses are likely to dominate the design criterion and the limit state of allowable stresses still plays an important role in the final design details. Flanged sections are normally recommended for medium and long span beams.

Example 7.16.1 30 m span simply supported beam–type 1

A beam of effective span of 30 m is subjected to a live load of 20 kN/m. The width and thickness of the top flange are given as 1.5 m and 0.2 m respectively. Design a post-tension beam with M 45 concrete.

Design data

Simply supported span $= L - 30$ m, $b_f = 1.5$ m, $t = 0.2$ m

Live load $= w_l = 20$ kN/m

Assume the self-weight of the beam $W_l/40$ in which W_l is the total load in kN on the span and the self-weight is in kN/m.

TABLE 7.15.1 Comparison of cost of electric pole designed as type 1, 2 and 3

Type	Dimensions				Quantities		Costs			
	Areas (m^2)		Bars		Concrete	Formwork	Concrete	Formwork	Steel	Total
	Top	Base	Dia	No.						
1	0.0175	0.032	5	12	0.1856	4.463	278	669	415	1362
2	0.0115	0.029	5	8	0.1519	3.788	228	568	277	1073
3	0.0110	0.028	4	12	0.1463	3.675	219	551	264	1034
WSD	0.0152	0.0275	5	14	0.150	4.703	225	705	485	1415

*The pole was designed by working stress design using the allowable stress of the earlier code.

Self-weight $w_g = \dfrac{W_l}{40} = \dfrac{20(30)}{40}$ $= 15$ kN/m

Total working load $= 35$ kN/m

Quality of the concrete $= f_{ck}$ $= 45$ MPa

Quality of steel $= f_p$ $= 1500$ MPa

Strength of concrete at transfer $= f_{ci}$ $= 40$ MPa

The allowable stresses for type 1 structure with $f_{ck} = 45$ MPa are taken from Table 7.9.1 and they are:

$$\sigma_{aci} = 0.455 \, f_{ci} = 18.2 \text{ MPa}$$

$$\sigma_{ace} = 0.375 \, f_{ck} = 16.88 \text{ MPa}$$

$$\sigma_{ati} = \sigma_{ate} = 0$$

The design coefficients for balanced section are taken from Table 7.3.1 and they are:

$$K = 0.128, \, p_o = 0.18, \, k_3 = 0.436$$

Limit state of strength

The collapse moment with a partial safety factor of 1.5 applied to the loads is

$$M_c = \gamma_f \, \frac{wL^2}{8}$$

$$= 1.5(35) \, \frac{(30)\,(30)}{8} = 5906.25 \text{ } k\text{Nm}$$

The moment capacity of a flanged beam is taken from Eq. 7.8.2 and it is

$$M_r = Kb_w d^2 f_{ck} + 0.45(b_f - b_w) \, t(d - 0.5t) f_{ck}$$

Let the thickness of the web in the range of

$$b_w = 0.10 + \frac{L}{30} = 0.1 + 0.06 = 0.16 \text{ m}$$

Then the moment capacity reduces to

$$M_r = 0.128(0.16)(45) \, d^2 + 0.45(1.5 - 0.16) \, (0.2)(d - 0.1)(45)$$

$$= 0.9216 \, d^2 + 5.43 \, d - 0.543 \text{ MNm}$$

Equating the moment capacity to the collapse moment, we have

$$0.9216 \, d^2 + 5.43 \, d - 0.543 = 5.9063$$

or
$$d^2 + 5.892 \, d - 7.0 = 0$$

The solution of the above quadratic equation gives

$$d = \frac{-4.892 + 7.92}{2} = 1.02 \text{ m}$$

Since the beam is of type 1 and the working load stresses normally govern the design, use $d = 1.24$ and the cover beyond the centre of steel as 0.21 m. Hence

$$d = 1.24 \text{ m and } h = 1.24 + 0.21 = 1.45 \text{ m}$$

The area of tensioned steel is given by

$$A_p = \frac{1.15 \, M_c}{(d - 0.5t) \, f_p} = 3972 \text{ mm}^2$$

Since the allowable tensile stress normally governs the design, provide a higher pretensioned steel than actually needed.

Provide 137 numbers of 7 mm wires, then the area of pretensioned steel is

$$A_p = 137 \ (38.48) = 5272 \ \text{mm}^2$$

Design for transfer condition

The ultimate force in the tension with partial safety factor of 1.0 is

$$P_u = A_p f_p = 0.005272 \ (1500) = 7.908 \ \text{MN}$$

The approximate bending moment caused by the dead load is

$$M_g = \frac{w_g L^2}{8} = \frac{(0.015)(900)}{8} = 1.6875 \ \text{MNn}$$

The tensile force caused by the self-weight is

$$F_g = \frac{\gamma_f \ M_g}{(d - 0.42x_u)} = \frac{0.9(1.6875)}{1.24 - 0.5406} = 2.1715 \ \text{MN}$$

$$F_u = P_u - F_g = 7.908 - 2.1715 = 5.7365 \ \text{MN}$$

The area of the concrete needed to withstand the pre-tensioning is

$$A_{ct} = \frac{\gamma_m F_u}{k_p f_{cl}} = \frac{1.5(5.7365)}{0.67(40)} = 0.3211 \ \text{m}^2$$

Depth of the web below the *CG* of the steel is 0.21 m, therefore the area of the web which contributes to the concrete action is

$$A_{ctw} = b_w(2)(0.21) = 0.0672 \ \text{m}^2$$

The area of the bottom flange overhangs should be, equal to

$$A_{cb} = 0.3211 - 0.0672 = 0.2539 \ \text{m}^2$$

Let the thickness of the bottom flange be equal to

$$t_b = 0.4 \ \text{m}$$

then the overhang of the bottom flange is

$$b_b - b_w = \frac{A_{ci}}{t_b} = \frac{0.2539}{0.4} = 0.635$$

or

$$b_b = 0.635 + 0.16 = 0.795$$

Use

$$b_b = 0.8 \ \text{m}$$

Since the shear stress is not so governing in the prestressed concrete beams, it is discussed after checking for the serviceability of the allowable stresses.

Serviceability in allowable stresses: In prestressed concrete structures, the allowable stresses appear to control the design. The properties of the uncracked section are computed in the

Table 7.16.1. Figure 7.16.1(a) gives the notations and the number sizes obtained from the limit state of collapse are:

$$h = 1.45$$

$$b_t = b_f = 1.5 \text{ m}$$

$$t_t = t = 0.2 \text{ m}$$

$$b_b = 0.8$$

$$t_b = 0.4 \text{ m}$$

$$b_w = 0.16 \text{ m}$$

$$h_w = h - t_b - t_t = 0.85 \text{ m}$$

The distance of y_i is measured from the extreme top fibre while computing moment of areas, etc. in Table 7.16.1. Columns 2, 3, 4 and 5 are computed first using the above data. A_i refers to the area of the segment under consideration and it is obtained by multiplying the width by depth and placed in column 2.

$$l_{ci} = \text{second moment of area of ith element about its own axis}$$

The distance of the ccmroid (y_t) of the section is computed after taking the first moment of areas and it is measured from the top:

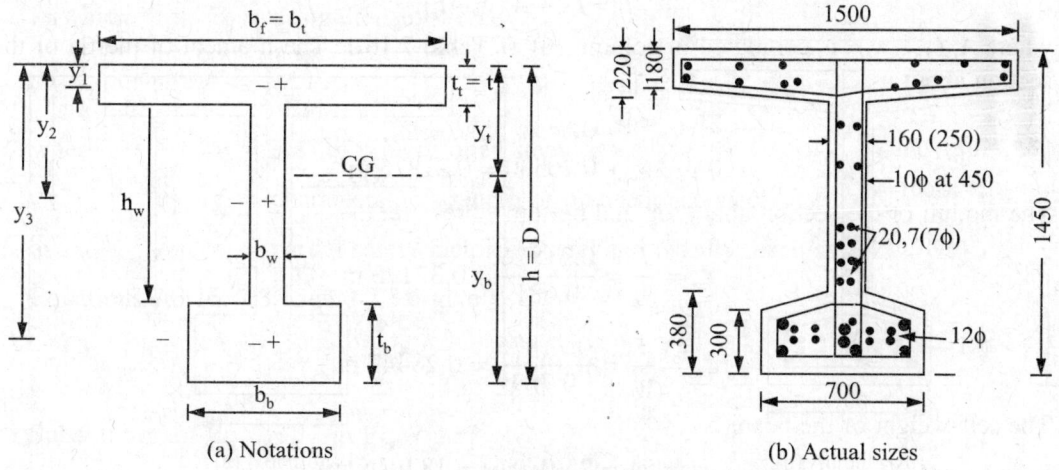

(a) Notations   (b) Actual sizes

Figure 7.16.1 (a) & (b)   Section of Example 7.16.1

Using the values from the table, we have

$$y_i = \frac{\Sigma A_i y_t}{\Sigma A_i} = \frac{0.515}{0.756} = 0.681 \text{ m}$$

TABLE 7.16.1 Sectional properties of the preliminary section (units in metres)

Segment (1)	A_i (2)	y_i (3)	$A_i y_t$ (4)	I_{ci} (5)	$A_i(y_i - y_t)^2$ (6)
Top-flange	0. 300	0.100	0.030	0.00100	0.10127
Web	0.136	0.625	0.085	0.00819	0.00043
Bottom flange	0.320	1.250	0.400	0.00427	0.10360
Total	0.756		0.515	0.01346	0.20530

TABLE 7.16.2 Sectional properties of the modified section (units in metres)

Segment	A_i	y_i	$A_i y_i$	I_{ci}	$A_i(y_i - y_i)^2$
Top flange	0.300	0.100	0.0300	0.00100	0.08522
Web	0.144	0.650	0.0936	0.00972	0.00004
Bottom flange	0.245	1.275	0.3124	0.00250	0.10098
Total	0.689		0.4360	0.01322	0.18624

$$y_b = h - y_t = 1.45 - 0.681 = 0.769 \text{ m}$$

The transformed moment of inertia of each segment is computed as

$$I_i = I_{ci} + A_i(y_t - t_i)^2$$

where $A_i (y_i - y_i)^2$ is computed in column (6) of Table 7.16.1. The moment of inertia of the section about its CG is:

$$I = \Sigma[(I_{ci} + A_i (y_t - y_i)^2]$$
$$= 0.01346 + 0.20530 = 0.21876 \text{ m}^4$$

The moduli of the section about top and bottom fibres are:

$$Z_t = \frac{I}{y_t} = \frac{0.21876}{0.681} = 0.32123 \text{ m}^3$$

$$Z_b = \frac{I}{y_b} = \frac{0.21876}{0.769} = 0.28447 \text{ m}^3$$

The self-weight of the beam is

$$w_g = \gamma A = 25(0.756) = 18.9 \text{ } kN/m$$

The values assumed was 15 kN/m and so the moment calculations are revised here. The bending moments due to self-weight and service load are:

$$M_g = \frac{w_g L^2}{8} = 18.9 \frac{(900)}{8} = 2126.25 \text{ } kNm$$

$$M_e = \frac{(w_g + w_l)L^2}{8} = 38.9 \frac{(900)}{8} = 4376.25 \text{ } kNm$$

The' initial prestressing force is

$$P_i = A_p\,(0.7f_p) = 0.005272(0.7)\,(1500) = 5.5356 \text{ MN}$$

Allowing a loss of prestress of 240 MPa in the post-tensioned construction, the effective prestressing force is

$$P_e = P_i - A_p\,(240)$$

$$= 5.5356 - 0.002572(240) = 4.2703 \text{ MN}$$

The eccentricity of the steel at mid-span is

$$e = d - y_t = 1.24 - 0.681 = 0.559 \text{ m}$$

The actual compressive and tensile stresses in the section at the initial load conditions are:

$$\sigma_{ci} = \frac{P_i}{A} + \frac{P_i e}{Z_b} - \frac{M_g}{Z_b}$$

$$= \frac{5.5356}{0.756} + \frac{5.5356(0.559)}{0.28447} - \frac{2.12625}{0.28447}$$

$$= 7.32 + 10.88 - 7.47 = 10.73 \text{ MPa} < \sigma_{aci}$$

$$\sigma_{ti} = \frac{P_i}{A} + \frac{P_i e}{Z_t} - \frac{M_g}{Z_t}$$

$$= -7.32 + 9.63 - 6.62 = 4.31 < \sigma_{ati} = 0$$

The compressive and tensile stresses at the working load condition are:

$$\sigma_{ce} = \frac{P_e}{A} - \frac{P_e e}{Z_t} + \frac{M_g}{Z_t}$$

$$= \frac{4.2703}{0.756} - \frac{2.3871}{0.32123} + \frac{4.37625}{0.32123}$$

$$= 5.65 - 7.43 + 13.62 = 11.84 < \sigma_{ace} = 16.88 \text{ MPa}$$

$$\sigma_{te} = -\frac{P_e}{A} - \frac{P_e e}{Z_b} + \frac{M}{Z_b}$$

$$= -5.65 - \frac{2.3871}{0.28447} + \frac{4.37625}{0.28447}$$

$$= -5.65 - 8.39 + 15.38 = 1.34 \text{ MPa} > \sigma_{ate} = 0$$

The tensile stress at working load exceeds the allowable value, whereas the rest of the stresses are less than the allowable stresses. One can decrease the size of the bottom flange marginally and increase the area of the pretensioned steel.

Let
$$b_b = 0.70 \text{ m}$$

$$t_b = 0.35 \text{ m}$$

and
$$A_p = 140(38.48) = 5387.2 \text{ mm}^2$$

then
$$h_w = h - t_b - t_l = 1.45 - 0.35 - 0.2 = 0.90 \text{ m}$$

This modification has a tendency to increase the tensile and compressive stresses at the initial load condition and decrease the tensile stress in the effective load condition. This is desirable without over doing. The properties of the revised section are computed with the revised dimension and are given in Table 7.16.2.

We can compute the distance of the centroid from Table 7.16.2 and it is

$$y_t = \frac{\Sigma A_i y_i}{\Sigma A_i} = \frac{0.436}{0.689} = 0.633 \text{ m}$$

then
$$y_b = h - y_t = 1.45 - 0.633 = 0.817 \text{ m}$$

The total moment of inertia ot the section is computed using Table 7.16.2:

$$I = \Sigma I_{ci} + \Sigma A_i (y_i - y_i)^2$$

$$= 0.01322 + 0.18624 = 0.19946 \text{ m}^4$$

The self-weight of the beam is

$$w_g = \gamma A = 25(0.689) = 17.225 \text{ } kN/m$$

The bending moment due to the self-weight is

$$M_g = 17.225 \frac{(900)}{8} = 1937.8 \text{ } k\text{Nm} \text{ and}$$

$$M_e = (w_g + w_l) \frac{L^2}{8} = 37.225 \frac{(900)}{8} = 4187.8 \text{ } k\text{Nm}$$

The initial and effective prestressing forces are:

$$P_i = A_p(0.7 f_p) = 0.0053872(0.7)(1500) = 5.6503 \text{ MN}$$

$$P_e = P_i - A_p(240) = 4.3573 \text{ MN}$$

and
$$e = d - y_t = 1.24 - 0.633 = 0.607 \text{ m}$$

$$P_i e = 3.4297 \text{ MNm}$$

$$P_e e = 2.6449 \text{ MNm}$$

$$Z_b = \frac{I}{y_b} = \frac{0.19946}{0.817} = 0.2441 \text{ m}^3$$

$$Z_t = \frac{I}{y_t} = \frac{0.19946}{0.633} = 0.3151 \text{ m}^3$$

The compressive and tensile stresses in the concrete of the revised section are computed using the above set of results.

$$\sigma_{ci} = \frac{P_i}{A} + \frac{P_i e}{Z_b} - \frac{M_g}{Z_b}$$

$$= \frac{5.6503}{0.689} + \frac{3.4297}{0.2441} - \frac{1.9378}{0.2441}$$

$$= 8.20 + 14.05 - 7.94 = 14.31 < \sigma_{aci} = 18.2 \text{ MPa}$$

$$\sigma_{ti} = \frac{P_i}{A} + \frac{P_i e}{Z_t} - \frac{M_g}{Z_t}$$

$$= -8.20 + 10.88 - 615 = -3.47 < \sigma_{ati} = 0$$

$$\sigma_{ce} = \frac{P_e}{A} - \frac{P_e e}{Z_t} + \frac{M_e}{Z_t}$$

$$= \frac{4.3573}{0.689} - \frac{2.6449}{0.3151} + \frac{4.1878}{0.3151}$$

$$= 6.32 - 8.39 + 13.29 = 11.22 < \sigma_{ace} = 16.88 \text{ MPa}$$

$$\sigma_{te} = \frac{P_e}{A} - \frac{P_e e}{Z_b} + \frac{M_e}{Z_b}$$

$$= -6.32 - \frac{2.6449}{0.2441} + \frac{4.1878}{0.2441}$$

$$= -6.32 - 10.84 + 17.16 = 0.1 > \sigma_{ate} = 0$$

Only the tensile stress in concrete at working load exceeds the allowable value by 0.1 MPa which is very small. And this can be compensated by further decreasing the size of the bottom flange marginally. Such a thing can be achieved while modifying the ideal rectangular flanges into tapered ones as shown in the Figure 7.16.2(b).

Serviceability in deflections: The prestressing cables are laid in oparabolic profile with zero eccentricity at the end section. Therefore the sag of the cable is

$$g = e = 0.607 \text{ m}$$

The upward balancing forces caused by the cable during intial and effective load conditions are:

$$w_{bi} = \frac{8 P_i g}{L^2} = \frac{8 P_i e}{L^2} = \frac{8(3.4297)}{900} = 0.03048 \text{ MN/m}$$

$$w_{be} = \frac{8 P_e g}{L^2} = \frac{8 P_e e}{L^2} = \frac{8(2.6449)}{900} = 0.02351 \text{ MN/m}$$

The effective loads at the initial and effective working load conditions are:

$$w_i = w_g - w_{bi} = 0.017225 - 0.03048 = -0.013255 \text{ MN/m}$$

$$w_e = w - w_{be} = 0.037225 - 0.02351 = 0.013715 \text{ MN/m}$$

The negative sign indicates that the load is acting upwards. The Young's modulus of the concrete is

$$E_c = 5700\sqrt{45} = 38237 \text{ MPa}$$

The short range deflections at the initial and effective load conditions are:

$$v_i = \frac{5w_iL^4}{384E_cI} = \frac{5(-0.013255)(900)(900)}{384(38237)(0.19446)} = -0.0188 \text{ m}$$

$$v_e = \frac{5w_eL^4}{384E_cI} = \frac{5(0.013715)(810000)}{384(38237)(0.19446)} = 0.01945 \text{ m}$$

Assuming that the structure is loaded after six months of construction, the creep coefficient can be taken as

$$c_c = 1 + 1.1 = 2.1$$

The total deflection of the beam including creep effect is

$$v_i = c_e v_e = 2.1(19.45) = 40.85 \text{ mm}$$

The allowable upward deflection at the initial stage and the total downward deflection during working load are:

$$v_{ai} = \frac{L}{300} = \frac{30000}{300} = 100 \text{ mm}$$

$$v_{ae} = \frac{L}{250} = \frac{30000}{250} = 120 \text{ mm}$$

The actual deflections are far below the permitted values.

Check for shear strength

The critical shear plane is at a distance d from the support and the shear force at that section is

$$V = w_i \left(\frac{L}{2} - d \right) = 37.225 (15 - 1.24) = 512.216 \text{ } kN$$

The slope of the cable near the support is

$$0 = \frac{4g}{L} = \frac{4(0.607)}{15}$$

The component of the cable force in the transverse direction is

$$V_p = P_e\theta = \frac{4P_eg}{L} = \frac{4P_ee}{L} = \frac{4(2.6449)}{15} = 0.7053 \text{ MN}$$

The effective transverse shear force for design is
$$V_e = \gamma_f V - V_p = 1.5(0.512216) - 0.7053 = 0.063 \text{ MN}$$
The shear capacity of the section is
$$V_c = 0.67\, b_w D \sqrt{(f_1^2 + 0.8 f_{cp} f_t)}$$
in which
$$f_t = 0.24\sqrt{f_{ck}} = 1.61 \text{ MPa}$$
$$f_{cp} = \frac{P_e}{A} = \frac{4.3573}{0.689} = 6.32 \text{ MPa}$$
$$D = h = 1.45 \text{ m}$$
The substitution cf the various quantities gives
$$V_{ec} = 0.67(0.16)(1.45)\sqrt{(1.61)^2 + 0.8\,(6.32)\,(1.61)} = 0.509 \text{ MN}$$
The shear capacity is far higher than the shear force, so provide only nominal shear reinforcement. Let two legged 10 mm mild steel bars be used as the stirrups. Then the area of the stirrup steel is
$$A_{sv} = 2(78.5) = 157 \text{ mm}^2$$
The maximum spacing of the mild steel stirrups is
$$S_{sv} = \frac{0.87 A_{sy} f_{yv}}{0.4 b_w} = \frac{157(250)(0.87)}{0.4(160)} = 533 \text{ mm}$$
Provide two-legged 10 ϕ stirrups at 450 mm *c/c*.

Figure 7.16 1(b) gives the details of the reinforcement and the sizes of the elements.

Example 7.16.1 30 m span simply supported beams–Type 3

A beam of effective span 30 m is subjected to a live load of 20 *k*N/m. The width and the thickness of the top flange are given as 1.5 m and 0.2 m respectively. Design a post-tensioned concrete beam with M 45 grade concrete.

Design data is:

$L = 30$ m, $b_f = 1.5$ m, $t = 0.2$ m

$w_l = 20$ *k*N/m, $f_{ck} = 45$ MPa, $f_{ci} = 40$ MPa and

$f_p = 1500$ MPa

The allowable stresses in the concrete for type 3 structures are:
$$\sigma_{aci} = 0.455\, f_{ci} = 18.2 \text{ MPa}$$
$$\sigma_{ace} = 0.375\, f_{ck} = 16.88 \text{ MPa}$$
$$\sigma_{ati} = 4.1;\ \sigma_{ate} = 4.4$$

Note: Example 7.16.1 gives the design details of a type 1 beam of the same span and loads. The limit state design for strength is exactly the same in this case also. Therefore the reader can see the example up to calculation of stresses in the concrete of the preliminary section. The stress computed in the previous example are repeated here and compared with the allowable values.

$$\sigma_{ci} = 10.73 < \sigma_{aci} = 18.2 \text{ MPa}$$

$$\sigma_{ti} = -4.31 < \sigma_{ati} = 4.1 \text{ MPa}$$

$$\sigma_{ce} = 11.84 < \sigma_{ace} = 16.88 \text{ MPa}$$

$$\sigma_{te} = 1.34 < \sigma_{ate} = 4.4 \text{ MPa}$$

The design details provided in example 7.16.1 are more than those required for the limit state of strength.

It can be seen that all the stresses are far less than the allowable values, therefore some reduction in the size of the bottom fibre and web can be affected.

The amount of prestressing force can be decreased as the strength limit state demands. Therefore assume the following dimensions :

104 numbers of 7 ϕ HTS wires. $A_p = 4002 \text{ mm}^2$

$$h = 1.24 + 0.16 = 1.40 \text{ m}$$

$$b_b = 0.65 \text{ m}, t_b = 0.3 \text{ m}$$

Therefore

$$h_w = h - t_b - t_t = 1.40 - 0.5 = 0.9 \text{ m}$$

Service Limit State Design for Stresses : The properties of the modified section are computed in Table 7.16.3 in the same manner as in Example 7.16.1.

$$y_t = \frac{\Sigma A_i y_i}{\Sigma A_i} = \frac{0.3672}{0.639} = 0.57 \text{ m}$$

$$y_b = 1.4 - 0.575 = 0.825 \text{ m}$$

TABLE 7.16.3 Sectional properties

Segment	A_i	y_i	$A_i y_i$	I_{ci}	$A_i(y_i - y_i)^2$
Top flange	0.300	0.10	0.0300	0.00100	0.06769
Web	0.144	0.65	0.0936	0.00972	0.00081
Bottom flange	0.195	1.25	0.2436	0.00146	0.08885
Total	0.639		0.3672	0.01218	0.15735

$$I = \Sigma I_{ci} + \Sigma A_i (y_i - y_i)^2$$

$$= 0.01218 + 0.157.35 = 0.16953 \text{ m}^4$$

$$Z_t = \frac{I}{y_t} = \frac{0.16953}{0.575} = 0.29483 \text{ m}^3$$

$$Z_b = \frac{I}{y_b} = \frac{0.16853}{0.825} = 0.20549 \text{ m}^3$$

The self-weight of the beam is

$$w_g = \gamma A = 25(0.639) = 15.975 \text{ } k\text{N/m}$$

The bending moments caused by the self-weight and working loads are:

$$M_g = \frac{w_g L^2}{8} = \frac{15.975(900)}{8} = 1797.2 \text{ } k\text{Nm}$$

$$M_e = \frac{w_e L^2}{8} = \frac{35.975(900)}{8} = 4047.2 \text{ } k\text{Nm}$$

The initial and effective prestressing forces are computed for 104 numbers of 7 wires and they are:

$$P_i = A_p(0.7f_p) = 0.004002(0.7)(1500) = 4.202 \text{ MN}$$

$$P_e = P_i - 240 \text{ } A_p = 3.242 \text{ MN}$$

$$e = d - y_t = 1.24 - 0.575 = 0.665$$

$$P_i e = 2.794 \text{ MNm and } P_e e = 2.155 \text{ MNm}$$

The stresses caused by .the loads in the concrete are:

$$\sigma_{ti} = -\frac{P_i}{A} + \frac{P_i e}{Z_t} - \frac{M_g}{Z_t}$$

$$= \frac{-4.202}{0.639} + \frac{2.794}{0.29483} - \frac{1.7972}{0.29483}$$

$$= -6.58 + 9.48 - 6.10$$

$$= -3.20 < \sigma_{ati} = 4.1 \text{ MPa}$$

$$\sigma_{ci} = \frac{P_i}{A} + \frac{P_i e}{Z_b} - \frac{M_g}{Z_b}$$

$$= 6.58 + \frac{2.794}{0.20549} - \frac{1.7972}{0.20549}$$

$$= 6.58 + 13.60 - 8.75 = 11.43 < \sigma_{aci} = 18.2 \text{ MPa}$$

$$\sigma_{ce} = \frac{P_e}{A} - \frac{P_e e}{Z_t} + \frac{M_e}{Z_t}$$

$$= \frac{3.242}{0.639} - \frac{2.156}{0.29483} + \frac{4.0472}{0.29483}$$

$$= 5.07 - 7.31 + 13.73 = 11.49 < \sigma_{ace} = 16.88 \text{ MPa}$$

$$\sigma_{te} = -\frac{P_e}{A} - \frac{P_e e}{Z_b} + \frac{M_e}{Z_b}$$

$$= -5.07 - \frac{2.156}{0.20549} + \frac{4.0472}{0.20549}$$

$$= -5.07 - 10.49 + 19.70 = 4.14 < \sigma_{ate} = 4.4 \text{ MPa}$$

The stresses are less than the allowable values. It can be seen from the previous example that the deflections and shear stresses are far below the allowable limits, therefore the beam can be considered safe in such limit state. The beam details are given in Figure 7.16.2.

Note: The design of the beam for the limit state of strength is same for all the types of beams. The main variation is in the allowable tensile stresses in the limit state of serviceability. The minimum size based on the limit state of strength results into a tensile stress of 4.14 MPa. This value is within the limits for type 3 structures but above the allowable value of the type 2 structure. The type 2 structures require marginal increase in the prestressing force. Table 7.16.4 gives the comparison of costs of type 1 and 3 structures.

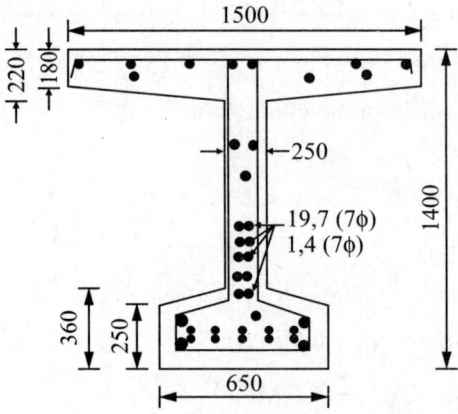

Figure 7.16.2 Section of beam in Example 7.16.2

TABLE 7.16.4 Comparison of costs

Type of structure	Cost in Rupees			
	Concrete	Form work	Steel	Total
Type 1	33588	33058	44466	111112
Type 3	31153	32143	28080	91376

Unit costs including supplying and placing of materials in position are:

$$\text{Cost of concrete} = \text{Rs. } 1625/\text{m}^3$$

$$\text{Cost of form work} = \text{Rs. } 150/\text{m}^2$$

$$\text{Cost of steel} = \text{Rs. } 3000/k\text{N}$$

7.17. Long Span Beams

Long span structures tend to influence the design by self-weight as much as the live load or even more. The self-weight of the beams is likely to be at least equal to or even more than the superimposed loads. Therefore the design is governed by the compressive stresses rather than the tensile stresses.

Example 7.17.1 Long span beam Type 1

Design a 50 m span, simply supported post-tensioned prestressed concrete beam with M45 concrete. The superimposed load on the beam is 20 kN/m. The depth of the beam is not to exceed 2.4 m.

Design data

$$\text{Span} = L = 50 \text{ m}, \ w_g = 20 \ k\text{N/m}$$

$$f_{ci} = f_{ck} = 45 \text{ MPa}, f_p = 1500 \text{ MPa}$$

The allowable stress Taken from table 7.9.1 are:

$$\sigma_{aci} = 0.455 \, f_{ci} = 20.475 \text{ MPa}$$

$$\sigma_{ace} = 0.375 \, f_{ck} = 16.88 \text{ MPa}$$

$$\sigma_{ati} = \sigma_{ate} = 0$$

Design coefficients for balanced section taken from Table 7.3.1 are:

$$K = 0.128 \text{ MPa}, \ p_0 = 0.18, \ k_3 = 0.436$$

Limit state of strength

Let the self-weight of the beam be assumed as

$$W_g = \frac{W_g L^2}{2200} = 20(50) \left(\frac{2500}{2200} \right) = 1136 \ k\text{N}$$

in which

$$W_g = \text{total self-weight}$$

$$W_g = \text{total superimposed load}$$

The collapse moment using a partial load factor of 1.5 is

$$M_c = \gamma f \, \frac{WL}{8} = 1.5 \, (1136 + 20(50)) \, \frac{50}{8} = 20025 \ k\text{Nm}$$

The effective depth of the steel and the area of pre-tensioned steel are computed by using approximate and accurate formulae.

Using approximate formulae Eqs. 7.8.9 and 7.8.10, we have

$$d = 0.5t + \frac{M_c}{0.45 b_f \, t f_{ck}}$$

Let the dimensions of the top flange be selected as

$$t = 0.3 \text{ m and } b_f = 1.6 \text{ m};$$

then from the above equation we have

$$d = 0.15 + \frac{20.025}{0.45(1.6)(0.3)45} = 2.21 \text{ m}$$

The approximate area of pre-tension steel from Eq. 7.8.10 is

$$A_p = 0.518 \, b_{fb} \, \frac{f_{ck}}{f_{pu}} = 0.518(1.6)(0.3) \frac{(45)}{1500} = 0.007459 \text{ m}^2$$

Consider the more accurate formulae given by Eqs. 7.8.2 and 7.8.3 for determination of d and A_p.

$$K b_w d^2 f_{ck} + 0.45 \, (b_f - b_w) \, t \, (d - 0.5t) f_{ck} = M_c$$

Substituting the various quantities in the equation, we have

$$0.128(0.18)(45) \, d^2 + 0.45(\, 1.6 - 0.18)(0.3)(d - 0.15)45 = 20.025 \text{ MNm}$$

in which b_w is assumed as 180 mm.

The above equation reduces to

$$1.0368 d^2 + 8.6265 d - 21.319 = 0$$

or

$$d = \frac{-8.6265 + \sqrt{(8.6265^2 + 4(21.319)(1.0368)}}{2(1.0368)} = 1.994 \text{ m}$$

$$A_p = 1.15(0.36 \, b_w x_u + 0.45 \, (b_f - b_w) \, t) \frac{f_{ck}}{f_{pu}}$$

$$= 1.15(0.36(0.18)(0.436)(1.99)$$

$$+0.45 \, (1.6 - 0.18) \, (0.3)) \frac{45}{1500}$$

$$= 0.008446 \text{ m}^2 = 8446 \text{ mm}^2$$

The effective depth obtained by the more accurate formula is less than the approximate one. This will happen invariably and the extent of the difference depends on the order of the thickness of the web. It is desirable to select the depth based on the approximate formula as it reduces the area of the pre-tensioned steel.

Use $d = 2.21$ m; then the area of the pre-tensioned steel can be computed from the formula of the under reinforced beams which is given in Eq. 7.8.11.

$$A_p = \frac{1.15 \, M_c}{(d - 0.5t) \, f_{pu}} = \frac{1.15(20.025)}{(2.21 - 0.15)(1500)} = 0.007453 \text{ m}^2$$

This area checks closely with that already calculated as approximate area.

Use 194 numbers of 7 mm wires, then the actual area of the pre-tensioned steel provided is

$$A_p = 194(38.48) = 7465 \text{ mm}^2$$

Design for transfer condition

The ultimate strength of tendons with partial safety factor of 1.0 is

$$P_u = A_p f_p = 0.007465(1500) = 11.20 \text{ MN}$$

Tensile force due to self-weight is

$$F_g = \frac{\gamma_f M_g}{(d - 0.5t)} = \frac{0.9(1.136)(50)}{2.06(8)} = 3.102 \text{ MN}$$

The ultimate compressive force on the bottom zone is

$$F_u = P_u - F_g = 11.20 - 3.102 = 8.098 \text{ MN}$$

The compression area needed to withstand the pretensioning force (for $\gamma_m = 1$) is

$$A_{ct} = \frac{\gamma_m F_u}{k_p f_{ci}} = \frac{(8.098)}{(0.67)(5)} = 0.2686 \text{ m}^2$$

Let the overall depth be assumed as $h = 2.4$ m.
Then the depth of the web below the effectivo depth of steel is

$$= h - d = 2.4 - 2.21 = 0.19 \text{ m}$$

The area of the web effective in resisting the axial force is

$$A_{ctw} = b_w (2) (h - d) = 0.18(2)(0.19) = 0.0684 \text{ m}^2$$

Area of the overhangs of the bottom flange needed is

$$A_{cb} = A_{ct} - A_{ctw} = 0.2686 - 0.0684 = 0.2002 \text{ m}^2$$

Use

$$b_b = 1.0 \text{ m and } t_b = 0.26 \text{ m}$$

which will give the area of the overhang of the bottom flange as

$$(b_b - b_w) t_b = 0.82 (0.26) = 0.2132 \text{ m}^2$$

Limit state of strength in shear is discussed later as it does not normally govern the design of the section.

Serviceability in allowable stresses

The sizes of different segments of the I-beam at this stage are:

$$b_t = b_f = 1.60 \text{ m}; \ t_i = t = 0.3 \text{ m}$$

$$b_b = 1.0 \text{ m}; \ t_b = 0.26 \text{ m}, \ b_w = 0.18 \text{ m}$$

$$h_w = h - t_t - t_b = 2.4 - 0.56 = 1.84 \text{ m}$$

$$A_p = 7465 \text{ mm}^2$$

Figure 7.16.1(a) gives the general notations. The properties of the section such as the distance of CG from the top fibre, second moment of area, etc., are computed through Table 7.17.1. The notations are:

A_i = area of ith segment (such as top flange, web, bottom flange)

y_t = distance of the CG of the ith segment from the top fibre

I_{ci} = moment of inertia of the ith segment about its own axis

TABLE 7.17.1 Properties of the preliminary section (in units of metres)

Segment	A_i	y_i	$A_i y_i$	I_{ci}	$A(y_i - y_i)^2$
Top flange	0.4800	0.15	0.0720	0.00360	0.34273
Web	0.3312	1.12	0.4041	0.09344	0.01677
Bottom flange	0.2600	2.27	0.5902	0.00146	0.42266
Total	1.0712		1.0663	0.09850	0.78216

The distance of the centroid from the top fibre is

$$y_t = \frac{\Sigma A_i y_t}{\Sigma A_i} = \frac{1.0663}{1.0712} = 0.995 \text{ m}$$

Then

$$y_b = 2.4 - 0.995 = 1.405$$

The moment of inertia about the transformed axis is computed for each segment in Table 7.17.1.

The total second moment of area is

$$I = \Sigma I_{ci} + \Sigma A_i(y_i - y_i)^2 = 0.09850 + 0.78216 = 0.88066 \text{ m}^4$$

The moduli of section about the top and bottom fibres are

$$Z_t = \frac{I}{y_t} = \frac{0.88060}{0.995} = 0.8851 \text{ m}^3$$

$$Z_b = \frac{I}{y_b} = \frac{0.88060}{1.405} = 0.6268 \text{ m}^3$$

The self-weight of the beam is

$$w_g = \gamma A = 25 \,(1.0712) = 25.709 \text{ } kN/m$$

The bending moments caused by the self-weight and working loads are:

$$M_g = \frac{w_g L^2}{8} = \frac{0.025709 \,(2500)}{8} = 8.034 \text{ MNm}$$

$$M_e = \frac{w_e L^2}{8} = \frac{0.045709 \,(2500)}{8} = 14.284 \text{ MNm}$$

The initial and effective prestressing forces using 220 MPa as the loss of prestress are:

$$P_i = A_p(0.7f_p) = 0.007465 \ (0.7)(1500) = 7.8382 \text{ MN}$$

$$P_e = P_i - 220A_p = 7.8382 - 220(0.007465) = 6.1959 \text{ MN}$$

The eccentricity of the cables at mid span is

$$e = d - y_t = 2.21 - 0.995 = 1.215 \text{ m}$$

$$P_ie = 7.8382 \ (1.215) = 9.5234 \text{ MNm}$$

$$P_ee = 6.1959(1.215) = 7.5280 \text{ MNm}$$

The stresses caused in the concrete at the initial and effective load stage are:

$$\sigma_{ci} = \frac{P_i}{A} + \frac{P_ie}{Z_b} - \frac{M_g}{Z_b}$$

$$= \frac{7.8382}{1.0712} + \frac{9.5234}{0.6268} = \frac{8.034}{0.6268}$$

$$= 7.32 + 15.19 - 12.82 = 9.69 < \sigma_{aci} = 20.475 \text{ MPa}$$

$$\sigma_{ti} = -\frac{P_i}{A} + \frac{P_ie}{Z_t} - \frac{M_g}{Z_t}$$

$$= -7.32 + \frac{9.5234}{0.8851} - \frac{8.034}{0.8851}$$

$$= 7.32 + 10.76 - 9.08 = -5.64 < \sigma_{ati} = 0$$

$$\sigma_{ce} = -\frac{P_i}{A} + \frac{P_ee}{Z_t} - \frac{M_e}{Z_t}$$

$$= \frac{6.1959}{1.0712} - \frac{7.528}{0.8851} + \frac{14.284}{0.8851}$$

$$= 5.78 - 8.51 + 16.14 = 13.41 < \sigma_{ace} = 16.88$$

$$\sigma_{te} = -\frac{P_e}{A} - \frac{P_ee}{Z_b} + \frac{M_e}{Z_b}$$

$$= -5.78 - \frac{7.528}{0.6268} + \frac{14.284}{0.6268}$$

$$= -5.78 - 12.01 + 22.79 = 5.00 > \sigma_{ate} = 0$$

The tensile stress at working load condition exceeds the permissible value; therefore the design needs modification. An extra pretensioning will improve the situation. The total compressive force caused by prestressing is 17.72 MPa whereas the total tension caused by BM is 22.79. Therefore 17.72 be raised to 22.79 MPa which means the number of wires must be proportionally increased.

The number of wires required are:

$$N = 194 \ (22.79)/17.72 = 247$$

The area of the tensioned steel is

$$A_p = 247 \, (38.48) = 9504 \text{ mm}^2$$

The pretensioned forces are

$$P_i = 0.7 \, f_p \, A_p = 0.7(1500)(0.009507) = 9.9792 \text{ MN}$$

$$P_e = P_i - 220 \, A_p = 9.9792 - 220 \, (0.009504) = 7.8883 \text{ MN}$$

$$\frac{P_i}{A} = 9.32; \quad \frac{P_e}{A} = 7.36$$

$$P_i e = 9.9792 \, (1.215) = 12.1247 \text{ MNm}$$

$$P_e e = 7.8883 \, (1.215) = 9.5843 \text{ MNm}$$

$$\frac{P_i e}{Z_b} = 19.34; \quad \frac{P_i e}{Z_t} = 13.70$$

$$\frac{P_e e}{Z_b} = 15.29; \quad \frac{P_e e}{Z_t} = 10.83$$

The stresses in the concrete are obtained by suitably substituting the above values in the expressions for stresses computed earlier.

$$\sigma_{ci} = 9.32 + 19.34 - 12.82 = 15.84 < \sigma_{aci}$$

$$\sigma_{ti} = 9.32 + 15.29 - 9.08 = -3.11 < \sigma_{ati}$$

$$\sigma_{ce} = 7.36 - 10.83 + 16.14 = 12.67 < \sigma_{ace}$$

$$\sigma_{te} = -7.36 - 15.29 + 22.79 = 0.14 \simeq 0$$

The stresses are within the limits, except for a marginal tension in the working load condition.

Check for shear capacity

$$f_{cp} = \frac{P_e}{A} = 7.36 \text{ MPa}$$

The characteristic and design loads are:

$$w_{ek} = 0.045709 \text{ MN/m}$$

$$w_t = \gamma_f w_{ek} = 1.5(0.45709) = 0.06856 \text{ MN/m}$$

The shear force at distance d from the support is

$$V_e = w_t \left(\frac{L}{2} - d \right) = 0.06856(25 - 2.21) = 1.56248 \text{ MN}$$

Diagonal tensile strength of the concrete is:

$$f_t = 0.24 \sqrt{45} = 1.61 \text{ MPa}$$

The shear capacity of the section is

$$V_r = 0.67b_w D\sqrt{(f_t^2 = 0.8f_{cp}f_t)}$$

$$= 0.67(0.18)(2.4)\sqrt{(1.61^2 + 0.8(7.36)(1.61)}$$

$$= 1.0 \text{ MN}$$

Since the shear capacity is less than the shear force, shear reinforcement is to be provided. Select two-legged 12 mm dia HYSD bars. The area of the stirrup steel is

$$A_{sv} = 2(113) = 226 \text{ mm}^2 \text{ and the spacing of the bars is}$$

$$s_v = \frac{0.87\,A_{sy}f_y d}{V - V_r} = \frac{0.87(226)(415)(2210)}{562480} = 320 \text{ mm}$$

Provide two-legged 12 mm dia HYSD bars stirrups at 320 mm spacing.

Example 7.17.2   Long span beam – Type 3

Design a 50 m span, simply supported post-tensioned prestressed concrete beam with M 45 concrete for the following design data

Design data

$$\text{Span} = L = 50 \text{ m}, \qquad w_g = 20 \text{ kN/m}$$

$$f_{ci} = f_{ck} = 45 \text{ MPa and} \qquad f_p = 1500 \text{ MPa}$$

The allowable stresses (in MPa) for 0.1 mm crack width are:

$$\sigma_{aci} = 0.455\,f_{ci} = 20.475, \quad \sigma_{ace} = 0.375\,f_{ck} = 16.880$$

$$\sigma_{ati} = \sigma_{ate} = 4.4$$

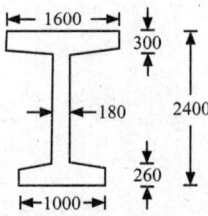

Let $\qquad b_f = b_t = 1.6 \text{ m}, t = 0.3 \text{ m, and } b_w = 0.18 \text{ m.}$

Limit State of Strength in Flexure

The design for limit state of strength is same for all the three types of beams. So the design computations made for type 1 beam in example 7.17.1 will hold good in this case also. Therefore the design details and the allowable stresses obtained from example 7.17.1 are:

$$b_f = b_t = 1.6 \text{ m}, t = 0.3 \text{ m}, y_t = 0.995, h = 2.4 \text{ m}$$

$$b_b = 1.0 \text{ m}, t_b = 0.26 \text{ m}, y_b = 1.405 \text{ m}$$

$$b_w = 0.18 \text{ m}, \ h_w = h - t_t - t_b = 2.4 - 0.56 = 1.84 \text{ m}$$

$$A_p = 7465 \text{ mm}^2, \ e = 1.215 \text{ m}$$

$$P_i = 7.8383 \text{ MN, and } P_e = 6.1959 \text{ MN}$$

The actual stresses in the concrete for the above details are computed in example 7.17.1 and they are (in MPa):

$$\sigma_{ci} = 9.69, \qquad \sigma_{ti} = -5.64$$

$$\sigma_{ce} = 13.41, \qquad \sigma_{te} = 5.0$$

It can be seen from the above stresses that all the stresses except tensile stress at working load are within the allowable values. The allowable tension is 4.4 MPa against an actual stress of 5 MPa. It is therefore recommended that the pretensioning be increased marginally. Increase the number of HTS wires from 194 to 204. Then

$$A_p = 204(38.48) = 7850 \text{ mm}^2$$

$$P_i = 0.7 f_p A_p = 0.7(1500)(0.00785) = 8.2425 \text{ MN}$$

$$P_e = P_i - 220 \, A_p = 6.5155 \text{ MN}$$

$$P_i e = 10.0146 \text{ MNm}, \ P_e e = 7.9163 \text{ MNm}$$

The bending moments on the beam and the properties of the section are taken from the previous example and they are:

$$M_g = 8.034 \text{ MNm}, \ M_e = 14.284 \text{ MNm}$$

$$Z_b = 0.6268 \text{ m}^3, \ Z_t = 0.8851 \text{ m}^3 \text{ and } A = 1.0712 \text{ m}^2$$

The actual stress (in MPa) in the concrete section are:

$$\sigma_{ci} = \frac{P_i}{A} + \frac{P_i e}{Z_b} - \frac{M_g}{Z_b} = 7.69 + 15.98 - 12.82 = 10.85$$

$$\sigma_{ti} = -\frac{P_i}{A} + \frac{P_i e}{Z_t} - \frac{M_g}{Z_t} = -7.69 + 11.32 - 9.08 = -5.45$$

$$\sigma_{ce} = \frac{P_e}{A} - \frac{P_e e}{Z_t} + \frac{M_e}{Z_t} = 6.08 - 8.94 + 16.14 = 13.28$$

$$\sigma_{te} = -\frac{P_e}{A} - \frac{P_e e}{Z_b} + \frac{M_e}{Z_b} = -6.08 - 12.63 + 22.79 = 4.08$$

All the stresses are within the limits.

Design for shear

The maximum shear force occurs at a distance d from the support and it is:

$$V_e = w_t \left(\frac{L}{2} - d \right) = 0.06856(25 - 2.21) = 1.5625 \text{ MN}$$

The section at the critical shear force is subjected to very small bending moment therefore it will be an uncracked section. The shear capacity of the uncracked section is:

$$V_r = 0.67\, b_w D \sqrt{(f_t^2 + 0.8\, f_{cp} f_t)}$$

$$= 0.67(0.18)\,(2.4)\,\sqrt{(1.61^2 + 0.8(7.36)\,(1.61)} = 1\ \text{MN}$$

As the shear resistance of the concrete is less than the shear force, transverse reinforcement be provided. Select two-legged 12 mm HYSD bars, then the spacing of the stirrups is given by

$$s_v = \frac{0.87\, A_{sv} f_y d}{V_e - V_r} = \frac{0.87(226)(415)(2210)}{562480} = 320\ \text{mm}$$

Example 7.17.3 Design of cantilever beam

A 700 *kN* tracked vehicle of wheel base 4 m is moving on a cantilever bridge girder of effective span 30 m. Design a single box section of the girder with top slab width as 7.5 m and 250 mm thick as a type three construction.

Design data:

Cantilever span = $L = 30$ m, $W_l = 700\ kN$

$$f_{ci} = f_{ck} = 45\ \text{MPa}\ f_p = 1500\ \text{MPa}$$
$$b_t = 7.5\ \text{m and}\ t_t = 0.25\ \text{m}$$

Load base = $a = 4$m

Crack width allowed = 0.1 mm

The allowable stresses are:

$$\sigma_{acl} = 0.455\, f_{ci} = 20.475\ \text{MPa},\ \sigma_{ace} = 0.375\, f_{ck} = 16.88\ \text{MPa}$$
$$\sigma_{ati} = \sigma_{ate} = 4.4\ \text{MPa}$$

Design for flexural strength

The maximum bending moment due to the live load occurs when the vehicle is at the free end and it is

$$M_l = W_l \left(L - \frac{a}{2} \right)$$

$$= 0.700\,(30 - 2) = 19.6\ \text{MNm}$$

For the purpose of computing the self-weight of the beam, the following dimensions be assumed.

$$b_b = b_f = 5\ \text{m},\ t_b = t = 0.4\ \text{m},\ b_w = 0.15\ \text{m}$$

Depth of web at free end = $h_{wi} = 1$m

Depth of web at fixed end = $\dfrac{L}{15} = 2$m

The bending moment caused by the self-weight is

$$M_g = \frac{\gamma L^2}{2}\left[b_t t_t + b_b t + \frac{b_w}{2}(h_{w1} + 2h_{w2})\right]$$

$$= \frac{0.025(900)}{2}\left[7.5(0.25) + 5(0.4) + \frac{0.15}{3}(4)\right] = 45.84 \text{ MNm}$$

The collapse BM with a partial safety factor of 1.5 is

$$M_c = 1.5(45.84 + 19.6) = 98.16 \text{ MNm}$$

The effective depth of the steel is

$$d = 0.5t + \frac{M_c}{0.45\, b_f t f_{ck}} = 0.2 + \frac{98.16}{0.45(5)(0.4)(45)} = 2.49 \text{ m}$$

Let $d = 2.55$ m then the area of the tensioned steel is given by

$$A_p = \frac{1.15\, M_c}{(d - 0.5t)\, f_p} = \frac{1.15(98.16)(10^6)}{2.25(1500)} = 30190 \text{ mm}^2$$

Provide 66 numbers of 12 wire stands with 7 mm wires, then

$$A_p = 66(12)(38.48) = 30476 \text{ mm}^2$$

Only some cables originate from the free end of the girder.

The self-weight of the girder is now computed with the following details.

$$b_t = 7.5,\ t_t = 0.25 \text{ m},\ b_b = 5 \text{ m},$$

$$t_b \text{ (at free end)} = 0.3 \text{ m},\ t_b \text{ (support)} = 0.4 \text{ m}$$

$$h = \text{overall depth} = d + 0.15 = 2.7 \text{ m}$$

$$p_0 = \text{overall depth at free end} = 1.5 \text{ m}$$

$$b_w = \text{thickness of each web} = 0.2 \text{ m}.$$

The depths of the web at the free and fixed ends are :

$$h_{w1} = 1.5 - 0.3 - 0.25 = 0.95 \text{ m}$$

$$hw_2 = 2.7 - 0.4 - 0.25 = 2.05 \text{ m}$$

The total weight of the girder is

$$W_g = 0.025\left[7.5(0.25) + \frac{5(0.4 + 0.3)}{2} + 2(0.2)\frac{(2.05 + 0.95)}{2}\right]30$$

$$= 1.40 + 1.313 + 0.45 = 3.16 \text{ MN}$$

where 0.025 MN/m^3 is the density of the prestressed concrete.

The corresponding bending moment due to the self-weight is

$$M_g = 1.4(15) + 0.025(5)(0.3)(30)(15)$$

$$+ 0.025\frac{(5)(0.1)(30)}{2}\frac{(10)}{3} + 0.025\frac{(2.05)(0.2)(30)}{2}\frac{(30)}{3}$$

$$+ 0.025 \frac{(0.95)(0.2)(30)}{2} \frac{60}{3} = 41.0 \text{ MNm}$$

$$M_e = M_g + M_i = 41.0 + 19.6 = 60.6 \text{ MNm}$$

Design for transfer condition

The ultimate tensile force in the cables with partial safety factor as 1.2 is

$$P_u = 1.2 \, A_p f_p = 1.2 \frac{(30\,190)(1500)}{10^6} = 54.342 \text{ MN}$$

The compensating tensile force produced in the cables by the self-weight moment is

$$F_g = \frac{0.9 \, M_g}{0.95 \, d} = \frac{0.9(41.0)}{0.95(2.55)} = 15.23 \text{ MN}$$

The total compression force to compensate the tensile force is

$$F_u = C_{dt} = P_u - F_g = 54.342 - 15.23 = 39.112 \text{ MN}$$

Area of the concrete required to resist the compression at transfer condition is

$$A_{ct} = \frac{C_{dt}}{0.67 \, f_{ck}} = \frac{39.11}{0.67(45)} = 1.30 \text{ m}^2$$

Area of the top flange = 7.5(0.25) = 1.875 m²

The area of the web which comes into action to resist the compression is

$$A_{cw} = 2(h - d)2b_w = 2(2.7 - 2.25)(2)(0.2) = 0.28 \text{ m}^2$$

This value is more than the required value

Design for serviceability limit state

The sectional properties at the support are computed in table 7.17.2.

Depth of web = 2.7 − 0.25 − 0.4 = 2.15 m

Area of two webs = 2(2.15)(0.2) = 0.9 m²

The centroidal depth computed using Table 7.17.2 is

$$y_t = \frac{\Sigma A_i y_i}{\Sigma A_i} = \frac{6.382}{4.775} = 1.34 \text{ m}$$

and

$$y_b = h - y_t = 1.36 \text{ m}$$

TABLE 7.17.2 Sectional pioperties (in m units)

Segment	A_i	y_i	$A_i y_i$	I_{ci}	$A_i(y_i - y_i)^2$
Top flange	1.875	0.125	0.2343	0.01	2.767
Webs	0.900	1.275	1.1475	0.38	0.003
Bot flange	2.000	2.500	5.0000	0.03	2.69
Total	4.775		6.382	0.42	5.45

The moment of inertia of the section is

$$I = \Sigma(I_{ci} + A_i (y_i - y_i)^2) = 0.42 + 5.45 = 5.87 \ m^4$$

$$Z_b = \frac{I}{y_b} = 4.32 \ m^3, \ Z_t = \frac{I}{y_t} = 4.38 \ m^3$$

The initial and the effective prestressing forces are:

$$P_i = 0.7f_p \ A_p = 0.7(0.030476)(1500) = 32 \ MN$$
$$P_e = P_i - 220 \ A_p = 25.295 \ MN$$

Further,
$$e = d - y_b = 2.55 - 1.36 = 1.190 \ m$$
$$P_ie = 38.008 \ MNm, \ P_ee = 30.10 \ MNm$$

The stresses in the concrete are:

$$\sigma_{ci} = \frac{P_i}{A} + \frac{P_ie}{Z_t} - \frac{M_g}{Z_t}$$

$$= \frac{32}{4.775} + \frac{38.08 - 41.0}{4.38}$$

$$= 6.7 - 0.67 = 6.03 \le \sigma_{aci}$$

$$\sigma_{ti} = -\frac{P_i}{A} + \frac{P_ie}{Z_b} - \frac{M_g}{Z_b}$$

$$= -6.7 + \frac{38.0.08 - 41.0}{4.32} = -6.03 \ MPa < \sigma_{ati}$$

$$\sigma_{ce} = \frac{P_e}{A} + \frac{M_e - P_ee}{Z_b}$$

$$= \frac{25.295}{4.775} + \frac{60.6 - 30.10}{4.32}$$

$$= 5.3 + 7.08 = 12.38 < \sigma_{ace}$$

$$\sigma_{te} = \frac{P_e}{A} + \frac{M_e - P_ee}{Z_t}$$

$$= -5.3 + 6.96 = 1.66 \ MPa < \sigma_{ate}$$

All the stresses are within the allowable limits.

Design for shear

The critical shear force occurs at a distance d from the support. This shear force with a partial safety factor of 1.5 is

$$V_e = 1.5W_l + W_g \left(\frac{L - d}{L}\right)$$

$$= 1.5 \left(0.7 + 3.16 \left(\frac{27.65}{30} \right) \right) = 5.42 \text{ MN}$$

The shear capacity of the cracked section is given by

$$V_r = \left(1 - 0.55 \frac{f_{pe}}{f_{pu}} \right) \tau_c b_w d + \frac{M_0 V}{M}$$

where

M = BM at distance d from the support and that due to the self-weight is prorated for simplicity

$$= W_l \left(L - d - \frac{a}{2} \right) + M_g \frac{(L - d)}{L}$$

$$= 0.7(30 - 2.35 - 2) + 41.0 \frac{(27.65)}{30}$$

$$= 17.955 + 37.788 = 55.743 \text{ MNm}$$

f_{pi} = stress at the level of steel caused by the prestressing force only

$$= \frac{P_e}{A} + \frac{P_e e^2}{I}$$

$$= 5.3 + \frac{30.10 \, (1.190)}{5.87} = 11.4 \text{ MPa}$$

$$M_0 = 0.8 \, f_{pt} \frac{I}{y_t} = 40.0 \text{ MNm}$$

$$\frac{100 A_p}{b_w d} = \frac{3047600}{2(200)\,(2550)} = 3.0 \text{ per cent}$$

The corresponding shear capacity of the concrete is

$$\tau_c = 1.01 \text{ MPa}$$

The shear capacity of the section is

$$V_r = (1 - 0.55(0.6))(1.01)(2)(2.55) + \frac{(40.0)(5.42)}{(55.743) \, 1.5} = 6.0 \text{ MN}$$

Since the shear force is less than the shear capacity, the transverse reinforcement must be nominal. Select two-legged 12ϕ HYSD bars as stirrups; then, the area of the stirrup in each web is

$$A_{sv} = 2 \, (2)(113) = 452 \text{ mm}^2$$

Provide two-legged 12ϕ *stirrups in each web at 175 mm spacing.* Figure 7.17.2 gives the sectional details.

Note: The bridge girder should also be designed for torsion, transverse bending moment and for deflections. This portion of the design is not included here.

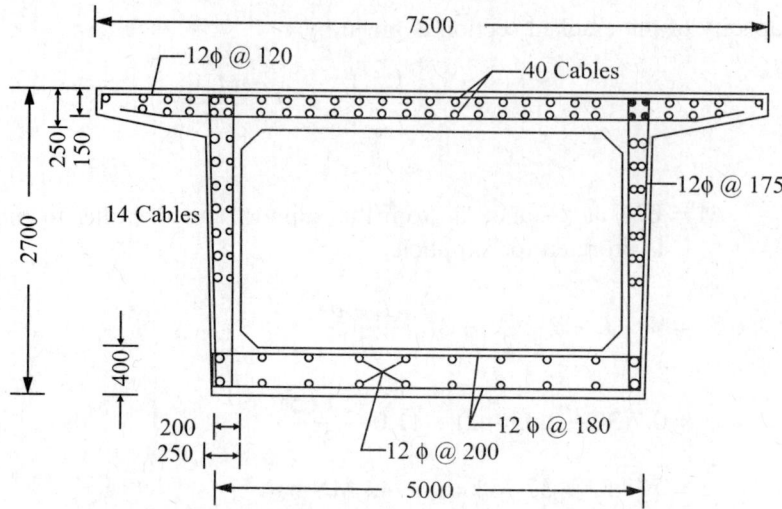

Figure 7.17.2 Sectional details (dimensions in mm).

Note : Comparison of details and cost of different types of beams are given in chapter 9.

PROBLEMS

7.1 The stress strain in compression in concrete is idealised by a triangular-cum-rectangular distribution with the following limits:

(for a rectangular cross section)

Crushing stress = $k_p f_{ck} = 0.67 f_{ck}$

Crushing strain at the extreme concrete fibre = 0.0035

Strain at the interface of the triangular and rectangular stress
$= 0.002$

Strain in steel $= \epsilon_y = 0.002 + \dfrac{0.87 f_p}{E_s}$

$\epsilon_{se} = 0.004$

Determine the following for balanced section:

  (a) Neutral axis
  (b) Total compression on concrete
  (c) Distance of the centroid of the compression on concrete
  (d) Resisting moment capacity of the section
  (e) Percentage of reinforcement

(f) Compare this moment capacity with that when the stress strain diagram was assumed as parabolic cum rectangular (as given in the text)

7.2 Assume the stress strain diagram of compression in concrete is linear till failure strain of 0.0035 and stress of $k_p f_{ck}$. Compute the following for a rectangular section:

(a) Neutral axis
(b) Total compression on concrete
(c) Resisting moment capacity
(d) Percentage of reinforcement

The strain in steel is

$$\epsilon_y = 0.002 + \frac{0.87 \, f_p}{E_s}$$

$$\epsilon_{se} = 0.004$$

Compare this moment capacity of the section with that calculated based on parabolic cum rectangular stress.

Compare the same moment capacity of the section with that computed in problem 7.1.

7.3 Vide the section in the text on the computation of the moment capacity of a rectangular section. In that it was assumed the $\epsilon_{se} = 0.004$. Compute the neutral axis and the moment capacity in case the strain ϵ_{se} is taken as 0.005 instead of 0.004. Compare the results with those obtained with $\epsilon_{se} = 0.004$.

7.4 Idealise the stress-strain diagram, of concrete in compression as parabolic with the following limits:

$$\epsilon_{se} = 0.004$$

$$\epsilon_c = 0.0035$$

$$\epsilon_y = 0.002 + \frac{0.87 \, f_p}{E_s}$$

$$\sigma_{si} = 0.7 f_p$$

Compute the following and compare with the corresponding values for the stress-strain diagram given in the code (that is parabolic cum rectangular stress block).

(a) Neutral axis
(b) Moment capacity of rectangular section.
(c) Percentage of steel

7.5 A simply supported beam of span 8 m is subjected to a live load of 20 kN/m. Design a rectangular sectioned, post-teusioned type 1 beam with the following data by the limit states design.

$$f_{ck} = 35 \text{ MPa}, f_p = 1500 \text{ MPa}$$

$$\sigma_{aci} = 14 \text{ MPa}$$

$$\sigma_{ace} = 12 \text{ MPa}$$

$$v_a = \frac{L}{350}$$

$$\tau_0 = 0.4 \text{ MPa}$$

7.6 Repeat solving the problem 7.5 for a post-tensioned typs 2 beam in which the allowable tensile stress in the concrete be taken as 3 MPa instead of zero.

7.7 Repeat solving the probleni 7.5 for a post-tensioned type 3 beam in which the allowable tensile stress in the concrete be taken as 4.4 MPa instead of zero.

7.8 A cantilever beam of span 6 m is subjected to a live load of 20 kN/m. Design a pretensioned type 2 concrete beam with the following data by limit states design:

$$f_{ck} = 40 \text{ MPa}, \qquad f_p = 1600 \text{ MPa}$$

$$\sigma_{aci} = 16.5 \text{ MPa}, \qquad \sigma_{ace} = 14 \text{ MPa}$$

$$\sigma_{ate} = \sigma_{ati} = 3 \text{ MPa}; \qquad \tau_c = 0.6 \text{ MPa}$$

Design a rectangular sectioned beam with a width equal to 400 mm.

7.9 Repeat the problem 7.8 with an inverted T-beam section instead of rectangular section. The width and thickness of the flange in compression be taken as 700 mm and 100 mm.

7.10 A simply supported beam of span 40 m is subjected to a live load of 20 kN/m. Design a post-tensioned type 1 beam of I-section with the allowable stresses given in the book:

Use $f_{ck} = 40$ MPa

$f_p = 1500$ MPa and loss of prestress as 220 MPa.

7.11 The stress-strain diagram of concrete in compression is idealised as parabolic with the following limits :

Crushing strain in concrete = 0.0035

Effective prestrain in steel = ϵ_{se} = 0.004

Yield strain in steel $= \epsilon_y = 0.002 + \dfrac{0.87\, f_p}{E_s}$

The section is triangular with its apex in compression and width at steel level is b.

$$k_p f_{ck} = 0.67 f_{ck}$$

Compute the moment capacity of the triangular section.

7.12 Repeat solving the problem 7.11 with an inverted triangular section where the apex is in the tension zone instead of the compression zone.

7.13 Repeat solving the problem 7.11 with a trapezoidal section instead of a triangular section.

7.14 Design the beam mentioned in the problem 7.8 with a triangular section with the width of the triangle limited to 600 mm.

7.15 Design the beam mentioned in the problem 7.8 with an inverted triangle section having a width equal to 600 mm.

7.16 Design the beam mentioned in the problem 7.5 with a trapezoidal section having widths at top and bottom as 300 mm and 400 mm respectively.

7.17 Design the beam mentioned in, the problem 7.5 with a trapezoidal section having the width of the section at top and bottom as 400 mm and 300 mm respectively.

7.18 Design the beam mentioned in the problem 7.5 with axial prestressing without any eccentricity.

7.19 Design the beam mentioned in the problem 7.8 with axial prestressing without any eccentricity.

CHAPTER 8

Ultimate Load Design of Prestressed Concrete Beams

8.1. Introduction

Structural Safety in working stress design is assured by not exceeding a permissible stress which is equal to 0.3 to 0.7 times the ultimate stress. The factor of safety applied to stresses will not assure the same factor of safety against ultimate loads since the stress and external load relations do not vary linearly till ultimate load. Propagation of cracks in concrete also plays an important role in connecting the sectional properties with the external load and thus affecting the factor of safety provided in the permissible stresses. It is therefore desirable to predict the actual failure of the member in order to establish the structural safety of the construction.

Ultimate resistance of a section must be more than the force coming on the section. The safety of the structure is assured by selecting a design load as a small multiple of actual load such that the design strength is more than the actual force on the structure. The philosophy of the ultimate strength design can be stated as

$$M_r \geq M_u = F_g M_g + F_1 M_1 \qquad (8.1.1)$$

where M_r = resisting moment capacity of the section, F_g and F_1 are load factors, M_g and M_1 are moments caused by the dead and live loads respectively on the section. Now one is posed with the problem of calculating resisting capacity of a section and selection of load factors to ensure safety. Equation 8.1.1 ensures safety of the section against flexure. Similar types of equations can be generated for safety of a section against shear, bond, torsion and axial forces. This chapter discusses design of sections for ultimate moment only.

8.2. Resisting Moment Capacity of a Section

Resisting moment capacity of a section is also called ultimate moment capacity. A section is said to have failed in strength at a moment when all the material in the section has exhausted its strength or strain limitations. Plane section before bending will remain plane even at the time of failure of the section if the failure is caused by flexure. There are two types of failures in flexure. One initiated by yielding of steel which is called tension failure and the other is initiated by crushing of concrete which is called primary compression failure. Tension failure is also called secondary compression failure because the section collapses by crushing of concrete even though the steel has yielded as the strain limitation on concrete is about ten per cent of the strain limitation on steel. High tensile steel fractures around 4 per cent elongation whereas concrete gets crushed at 0.4 per cent strain.

As the bending moment on a section increases, the strain in steel and concrete increase and then reach their limits at the time of failure. Figure 8.2.1 (b) illustrates the strains on cross section, 8.2.1 (c) illustrates actual stress distribution on concrete at the time of failure and figure 8.2.1 (d) gives the idealised stress distribution on concrete for the purpose of moment calculations. It is assumed that the reader is familiar with calculation of ultimate moment capacity of reinforced concrete sections so not much explanation is given in this section. The ultimate moment calculation of bonded prestressed concrete section is very similar to that of reinforced concrete. Prism strength of concrete in flexure is about 0.67 times the strength of 15 cm cube so when computing the ultimate moment capacity of a section, the ultimate strength of concrete is taken as $0.67\,f_{ck}$.

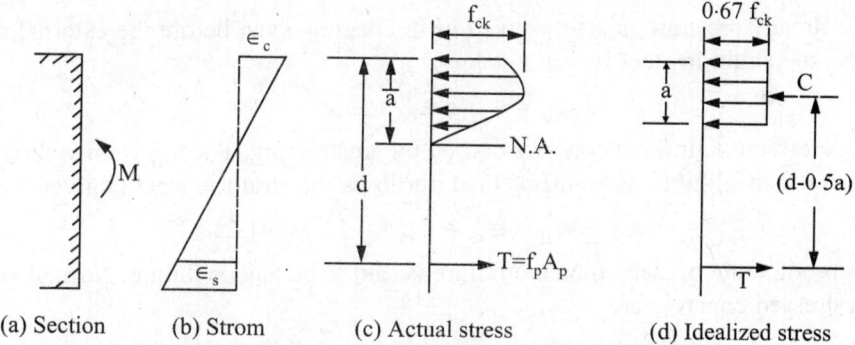

(a) Section   (b) Strom   (c) Actual stress   (d) Idealized stress

Figure 8.2.1 Stress distribution at failure.

8.3. Moment Capacity of a Section Initiated by Yielding of Tension Steel (Tension Failure)

(a) **Moment calculation for rectangular sections** : Assuming the steel is bonded and has reached the yield strain before the crushing of the concrete, the equilibrium of forces at failure as seen from figure 8.2.1 (d) for a rectangular section are

$$A_p f_p = (0.67\,f_{ck})ba \tag{8.3.1}$$

$$M_r = A_p f_p \left(d - \frac{a}{2} \right) \tag{8.3.2}$$

where b = width of the section, d = distance of steel from the extreme compression fibre, and a = depth of the idealised compressive stress. From eq. 8.3.1

$$a = \frac{A_p f_p}{0.67 b f_{ck}} \tag{8.3.3}$$

Substituting the value of 'a' from eq. 8.3.3 in the eq. 8.3.2 one gets

$$M_r = A_p f_p \left(d - \frac{0.74 A_p f_p}{b f_{ck}} \right) \tag{8.3.4}$$

Indian code of practice rationalises the ultimate moment resistance as

$$M_r = A_p f_p \left(d - \frac{0.75 A_p f_p}{b f_{ck}} \right) \tag{8.3.5}$$

The neutral axis distance is assumed as

$$a = 1.25a \tag{8.3.6}$$

Let ϵ_c = strain in extreme fibre of concrete which is equal to crushing strain of concrete, then from geometry of similar triangles of strains in figure 8.2.1 (b), the corresponding strain at steel level is

$$\epsilon_{sl} = \frac{\epsilon_c}{\bar{a}} \left(d - \bar{a} \right) = \frac{\epsilon_c (d - 1.25a)}{1.25a}$$

There is already prestrain in steel as well as in concrete even before the external moment is applied. So the strain in steel is

$$\epsilon_s = \epsilon_{se} + \epsilon_{ce} + \epsilon_{sl}$$

in which ϵ_{ce} = effective strain in concrete caused by prestressing. The ϵ_{ce} is only about 10 per cent of ϵ_{se} so it is negligible. For all practical purposes the strain in steel is given by

$$\epsilon_s = \epsilon_{se} + \epsilon_{sl} = \epsilon_{se} + \epsilon_c(d - 1.25a)/1.25a \tag{8.3.7}$$

If $\epsilon_s \geq \epsilon_y$ = proof strain of steel, then the failure is said to be tensile failure. Normal values of strain in prestressed concrete are

$$\epsilon_{se} = 0.004 \text{ to } 0.005, \ \epsilon_c = 0.0035, \ \epsilon_y = 0.006 \text{ to } 0.009 \tag{8.3.8}$$

Example 8.3.1 A rectangular cross section 20 cm by 43 cm is prestressed with 4 cm^2 high tensile steel at a depth of 35 cm from top. Find the ultimate moment capacity of the section if f_{ck} = 35, f_p = 1500 MPa.

Solution

The solution of the problem is illustrated in three steps:
(i) calculate the neutral axis,
(ii) check for type of failure, and
(iii) calculate the ultimate moment.

From eq. 8.3.3

$$a = \frac{A_p f_p}{0.67 b f_{ck}} = \frac{4(150000)}{0.67(20)(3500)} = 12.6 \text{ cm}$$

$$\bar{a} = 1.25a = 36.8 \text{ cm}$$

From eq. 8.3.7

$$\epsilon_{se} = 0.004 + 0.003 \ (35 - 15.8) \ /15.8$$

$$= 0.0078 > \epsilon_y$$

Since $\epsilon_s \geq \epsilon_y$, the failure is by yielding of steel and the resisting moment is obtained from eq. 8.3.2

$$M_r = A_p f_p \left(d - \frac{a}{2} \right) = 4(150000)(35 - 6.3) = 17,220,000 \text{ Ncm}$$

(b) Moment calculation if neutral axis is in the top flange of I- or T- sections: The ultimate resistance moment of T- or I-sections can be calculated on the same basis if the neutral axis is within the flange. In such a case the formulae used in the earlier section will hold good with '*b*' equal to the width of the flange. If the neutral axis lies in the web then a portion of the web contributes towards the moment resistance. The moment calculations are illustrated through examples.

Example 8.3.2   Calculate the ultimate moment capacity of the section shown in figure 8.3.1.

$$f_{ck} = 45 \text{ MPa} = 4500 \text{ N/cm}^2, \qquad f_p = 1500 \text{ MPa} = 150000 \text{ N/cm}^2,$$

$$t = 36 \text{ cm}, \qquad b_f = 137 \text{ cm}, d = 218 \text{ cm}$$

The section is same as that of example 3.10.1 except $t = 36$ cm instead of 28.8 cm.

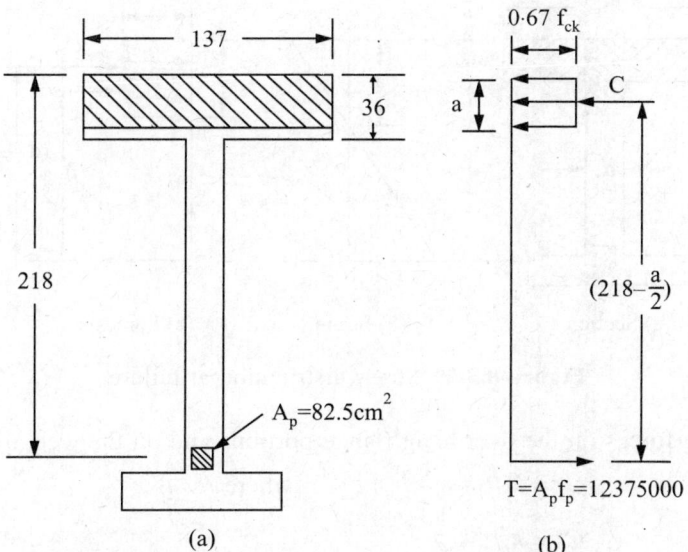

(a)   (b)

Figure 8.3.1   Example 8.3.1 (Dimensions in cm).

Solution

Assume that the neutral axis is in the top flange, then from eq. 8.3.3

$$a = \frac{f_p A_p}{0.67 f_{ck} b_t} = \frac{82.5(150000)}{0.67(4500)(137)} = 30 \text{ cm}$$

$$\bar{a} = 1.25a = 36.8 \text{ cm}$$

The neutral axis distance is slightly more than t_1 but 'a' is less than t_i so the assumption is acceptable

$$\epsilon_s = \epsilon_{se} + \epsilon_c (d - 1.25a) / 1.25a$$

$$= 0.004 + 0.0035 (218 - 36.8) /36.8 = 0.0188 > \epsilon_y$$

The failure is by yielding of the steel so the resisting moment is obtained from eq. 8.3.2

$$M_r = A_p f_p (d - a/2) = 82.5(150000)(218 - 14.75)$$

$$= 2580,000,000 \text{ Ncm}$$

(c) Moment calculation if neutral axis is in web of I- or T-sections: Let the flange thickness be small so that the neutral axis falls in the web as shown in figure 8.3.2, then the equilibrium of forces from figure 8.3.2 (c) are

$$T = C_1 + C_2$$

where C_1 = compressive force resisted by the overhang portion of the flanges

   C_2 = compressive force resisted by the web area

   T = total tension force

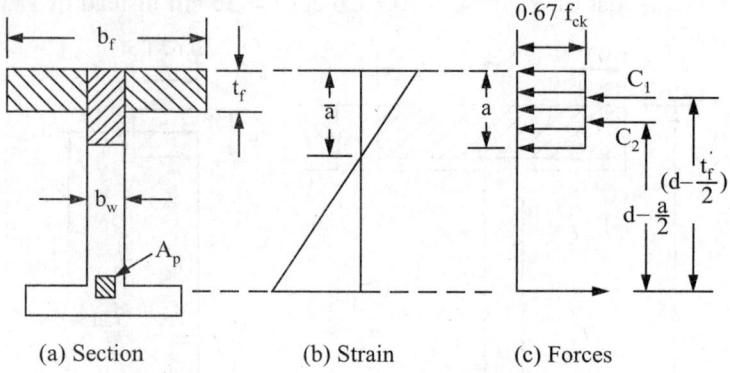

(a) Section   (b) Strain   (c) Forces

Figure 8.3.2 Stress distribution at failure.

The compressive forces on the over hang flange portions and on the web are :

$$C_1 = k_p f_{ck} (b_f - b_w) t \qquad \text{where } t = t_f$$

$$C_2 = k_p f_{ck} b_w a$$

The total tension in the cable at ultimate force is

$$T = A_p f_p$$

Equating the compression to tensile force, we have

$$a = \frac{A_p f_p}{0.67 b_w f_{ck}} = \frac{\left(b_f - b_w\right)t}{b_w}$$

$$= \frac{A_p f_p}{0.67 b_w f_{ck}} + t - \frac{b_f t}{b_w}$$

in which $k_p = 0.67$ where b_f and t are width and thickness of flange respectively. They could be either top or bottom flange dimensions depending on the location of the steel.

Moment of the forces C_1, C_2 and T about a point passing through the action of C_2 gives

$$M_r = T\left(d - \frac{a}{2}\right) + C_1\left(\frac{a}{2} - \frac{t}{2}\right) \tag{8.3.10}$$

$$M_r = A_p f_p\left(d - \frac{a}{2}\right) + 0.67 f_{ck} t\left(b_f - b_w\right)(a - t)/2 \tag{8.3.11}$$

Example 8.3.3 Calculate the ultimate moment capacity of the section of example 3.10.1

Solution

$$b_f = b_t = 137 \text{ cm}, \ t = 28.8 \text{ cm}, \ d = 218 \text{ cm}, \ b_w = 17 \text{ cm}.$$
$$f_p = 150\ 000 \text{ N/cm}^2$$
$$A_p = 82.5 \text{ cm}^2$$

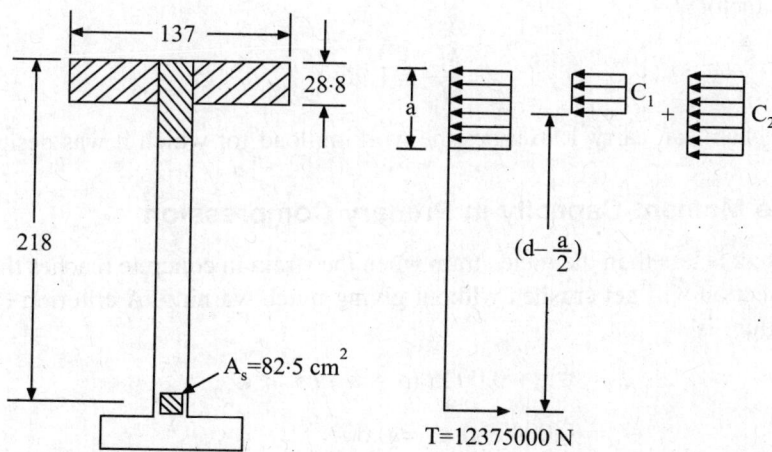

Figure 8.3.3 Example 8.3.3 (Dimensions in cm).

$$f_{ck} = 4500 \text{ N/cm}^2$$

$$T = A_p f_p = 82.5(150\ 000) = 12{,}375{,}000 \text{ N}$$

Let the neutral axis be in the top flange, then from eq. 8.3.3

$$a = \frac{A_p f_p}{0.67 b_f f_{ck}} = 30 \text{ cm} > t = 28.8$$

Since $a > t$ the neutral axis is in the web, so using eq. 8.3.9

$$a = \frac{A_p f_p}{0.67 f_{ck} b_w} + t - \frac{b_f t}{b_w}$$

$$= 241 + 28.8 - 232 = 37.8 \text{ cm}$$

$$\epsilon_s = 0.004 + 0.0035[218 - 1.25(37.8)\ /1.25(37.8)] > 0.006$$

So the failure is by yielding of the steel.

The resisting moment of the section from eq. 8.3.11 is

$$M_r = A_p f_p \left(d - \frac{a}{2} \right) + \left(0.67 f_{ck}\ t \left(b_f - b_w \right) \right)(a - t)\ /2$$

$$= (12\ 375\ 000\ 218 - 18.9) + 0.67(4500)(28.8)(120)(37.8 - 28.8)/2$$

$$= 2510\ 751\ 800 \text{ Ncm}$$

The M_r in example 8.3.2 in which $t_f = 36$ cm is 2,580,000,000 N cm which is slightly higher than the value of this example.

The working moment on the cross section from example 3.10.1 is

$$M_w = (w_g + w_1)\ L^2/8 = (21000 + 20000)(50)(5000)/8$$

$$= 1{,}280{,}000{,}000 \text{ Ncm}$$

Combined load factor

$$= \frac{M_r}{M_w} = 1.96$$

This means the beam can carry 1.96 times the working load for which it was designed.

8.4. Ultimate Moment Capacity in Primary Compression

If the strain in steel is less than the yield strain when the strain in concrete reaches the crushing strain then the section will get crushed without giving much warning. A criterion for primary compression failure is

$$\epsilon_s = \epsilon_{se} + 0.0035\ (d - \bar{a})\ /\bar{a} < \epsilon_{yp} \qquad (8.4.1)$$

Let

$$\epsilon_{se} = 0.004 \text{ and } \epsilon_y = 0.007$$

then

$$0.0035(d - \bar{a}) \simeq 0.003\,\bar{a}$$

or
$$\bar{a} \simeq \frac{0.003d}{0.06} = 0.5d$$

$$\bar{a} \simeq \frac{0.5d}{1.25} = 0.40d \qquad (8.4.2)$$

If the neutral axis falls at a distance more than $0.5d$ from compression fibre then the beam is likely to fail by primary compression.

The value of 'a' for rectangular or heavy flanged section can be obtained from the eq. 8.3.3. The substitution of 'a' from the eq. 8.3.3 in the eq. 8.4.2 gives

$$a = \frac{A_p f_p}{0.67 f_{ck} b} \simeq 0.40d$$

or
$$\frac{A_s}{bd} \frac{f_p}{f_{ck}} \simeq (0.40)(0.67) = 0.268 \qquad (8.4.3)$$

Assuming $d \simeq 0.9h$ the above equation works out

$$\frac{A_s}{bd}\left(\frac{f_p}{f_{ck}}\right) \simeq 0.24 \qquad (8.4.4)$$

The above equation gives approximate relation of steel for which a compression failure takes place. Approximate percentage of prestressing steel in rectangular section for which a compression failure may take place varies from 0.6 to 1.0. In the case of I or T beams the area of concrete is expressed as

$$A = c_a b_f h$$

Equation 8.4.4 on substitution of value of $b_f h$ gives

$$\frac{A_s}{A} \simeq \frac{0.24}{c_a}\left(\frac{f_{ck}}{f_p}\right) \qquad (8.4.5)$$

It can be seen from table 2 of appendix A, value of c_a varies from 0.2 to 0.4. Approximate percentage of steel in flange section varies 1 to 4 per cent for tension failure. Very rarely the beams are provided with such high percentage of steel so the compression failure is not usually a criterion in design even though one must check for such possibility. The ultimate moment capacity should be calculated by using the actual value of σ_s given by strain instead of f_p. The following equations which are obtained by replacing f_p by σ_s in the previous derivations are to be used for ultimate strength of a section failing compression.

$$\sigma_s = \phi_s\,(\in_s)$$

$$\sigma_s = \phi_s\,[\in_{se} + 0.003\,(d - 1.25a)/1.25a] \qquad (8.4.6)$$

(a) For rectangular or heavy flanged sections say $t_f \geq d/4$

$$a = \frac{A_p \sigma_s}{0.67 f_{ck}} \qquad (8.4.7)$$

$$M_r = A_p \sigma_s\left(d - \frac{a}{2}\right) \qquad (8.4.8)$$

(b) For the neutral axis falls in the web of flanged section

$$a = \frac{A_p \sigma_s}{0.67 b_w f_{ck}} + t - \frac{b_f t}{b_w} \qquad (8.4.9)$$

$$M_r = A_p \sigma_s \left(d - \frac{a}{2} \right) + 0.67 f_{ck} t \left(b_f - b_w \right)(a - t)/2 \qquad (8.4.10)$$

This set of equations requires stress-strain relation of steel up to ultimate stress which is not a linear relation. Most codes specify a limiting moment capacity in case of compressive failure criterion. For a balanced section in which the limiting strains in concrete and steel reach at the same time, the neutral axis distance as per eq. 8.4.2 is

$$a = 0.40 \, d$$

The limiting moment for a rectangular section is

$$M_r = 0.67 b a f_{ck} \left(d - \frac{a}{2} \right) = 0.22 b d^2 f_{ck} \qquad (8.4.11)$$

Considering the possibility of the steel not attaining its full capacity and uncertainties in the materials, one can suggest a limiting moment for rectangular section as

$$M_r = 0.1875 \, b d^2 f_{ck} \qquad (8.4.12)$$

and for flanged sections

$$M_r = 0.1875 \, b_w d^2 f_{ck} + 0.55 t \, (b_f - b_w)(d - t/2) f_{ck} \qquad (8.4.13)$$

The resisting moment capacities suggested by the Indian Road Congress for post-tensioned bonded construction are given below
(1) Tension criterion,

$$M_r = 0.9 \, A_p f_p \qquad (8.4.14)$$

(2) Compression criterion

$$M_r = 0.176 \, b d^2 f_{ck} \quad \text{for rectangular sections} \qquad (8.4.15)$$

$$M_r = 0.176 \, b d^2 f_{ck} + 0.533 \, t (b_f - b_w)(d - t/2) f_{ck} \qquad (8.4.16)$$

for flange sections.

Most codes specify material reduction factors less than one which multiply the resisting moment of eq. 8.4.11 to account for uncertainties in materials and manufacturing techniques. The yield strain in steel and the crushing strain of concrete affects the balanced moment capacity of a section.

Example 8.4.1 The beam section of the example 8.3.3 is provided with 87.5 cm^2 of steel instead of 82.5 cm^2. Calculate the ultimate moment capacity of the section :

$$t = t_f = 28.8 \text{ cm}, \; b_f = b_t = 137 \text{ cm}, \; b_w = 17 \text{ cm}$$

$$d = 218 \text{ cm}, \; A_p = 87.5 \text{ cm}^2, \; A = 8750 \text{ cm}^2,$$

$$f_{ck} = 4500 \text{ N/cm}^2, \; f_p = 150\,000 \text{ N/cm}^2$$

Solution

$$A_p/A = 87.5 / 8750 = 0.01$$

The percentage reinforcement is 1 so the compression failure may occur. Assuming $a > t$

$$a = \frac{A_p f_p}{0.67 b_w f_{ck}} + t - \frac{b_f t}{b_w} = 38.8 \text{ cm}$$

As indicated in example 8.3.3 the steel will not develop its full capacity. The moment capacity of the section from the eq. 8.4.13

$$M_r = 0.1875(17)(218)^2(4500) + 0.55(28.8)(120)(203.6)(4500)$$

$$= 680\,000\,000 + 1585,000,000 = 2265\,000\,000 \text{ Ncm}$$

8.5. Moment Capacity of Unbonded Construction

Most of the beams in practice are bonded either originally or through grouting. There are very rare cases in which the steel cables in post-tensioned construction are not grouted thus leaving the cables unbonded with the concrete section. Such unbonded wires have to be greased or oiled at regular intervals so to avoid corrosion. Corrosion of high tensile steel wires which are under tension can cause a serious damage in splitting the wires. Such a phenomenon of splitting of the wires under tension due to corrosion is called stress corrosion. Therefore, high tensile steel are either grouted or continuously greased.

In an unbonded construction, the wires are free to slide and the strain in the wires will get evenly distributed even if one section is under critical bending moment. Since the stress throughout the wire will reach the maximum stress, the total elongation in the cable in an unbonded construction will be much more than that of a bonded construction under identical circumstances. Excessive elongation in steel causes large deflection and higher strains in concrete. Very few wide cracks will be generated in an unbonded construction whereas many fine cracks will be generated in a bonded construction. The excessive elongation of the wires and wide cracks which increase the strains in concrete will reduce the ultimate moment capacity of a cross section. The ultimate moment capacity of a section of an unbonded construction is about 70 to 95 per cent of that of a bonded construction. In the case of structures with free sliding of the wires, the bond reduction factor of 0.7 will hold good whereas in a structure without free sliding of the wires a reduction factor of 0.95 can be applied. If k_b is the bond reduction factor, the ultimate strength of an unbonded prestressed concrete beam is given by

$$M_{ru} = k_b M_r \qquad (8.5.1)$$

where k_b varies from 0.7 to 0.95 as discussed above and M_{ru} is the moment resisting of an unbonded beam.

8.6. Load Factors

Many kinds of loads act on structures. These loads can broadly be classified into four groups such as : (1) fixed loads often called dead loads which are caused by weight of the structure, (2) live loads which move on and out of the structure without any accelerating forces,

(3) dynamic forces which are caused by the acceleration of external bodies or structures themselves, and (4) environmental forces. All these loads may not be acting on the structure all the time and the chances of all the critical loads acting on the structure at a time is very very small. A structure must be designed such that the resisting capacity is more than the external forces acting on the structure. To ensure a safety to the structure the design leads must be higher than the actual loads. The ultimate load on a structure is defined as

$$UL = F_g DL + F_1 LL \tag{8.6.1}$$

or

$$UL = F_g DL + F_{l2} LL + F_w WL \tag{8.6.2}$$

or

$$UL = F_g DL + F_{l2} LL + F_q QL \tag{8.6.3}$$

where

UL = design ultimate load which is only an imaginary load

DL = dead load

LL = live load including impact

WL = wind load

QL = earthquake load

F_g = load factor for dead loads

F_l = load factor for live load when no wind or earthquake loads act

F_{l2} = load factor for live load when wind or earthquake loads act

F_w, F_q = load factors for critical wind and earthquake loads respectively

The dead loads are known reasonably well. So load factor for dead load can be slightly more than one. The live loads cannot be defined as precisely as the dead loads so the load factor for live load should be more than that of the dead loads. The load factors for wind or earthquake can only be nominal since the wind or earthquake forces are taken from the most severe cases. And the survival of the structure is the only criterion under such severe conditions whereas the structure must be in serviceable condition under working loads. The following are the ranges of the load factors.

$$F_g = 1.1 \text{ to } 1.5, \; F_1 = 1.4 \text{ to } 2.5, \; F_{l2} = 1.0 \text{ to } 2$$

$$F_q = F_w = 0.5 \text{ to } 1.5$$

The Indian Road Congress suggests the following ultimate load for bridges

$$UL = 1.5 \, DL + 2.5 LL \tag{8.6.4}$$

if DL and LL are acting in the same direction

or

$$UL = 1.0 DL + 2.5 LL \tag{8.6.5}$$

if DL and LL are acting in opposite directions.

Example 8.6.1 Calculate the load factor available for live load of the section discussed in example 8.3.3, if $F_g = 1.5$.

Solution

The working moments on the section and ultimate moment capacity of the section from example 8.3.3 are

$$M_g = w_g L^2/8 = 21000(50)(5000)/8 = 656,000,000 \text{ Ncm}$$

$$M_1 = W_1 L^2/8 = 20000(50)(5000)/8 = 625,000,000 \text{ Ncm}$$

$$M_r = 2448,000,000 \text{ Ncm}$$

The design criterion is

$$M_r \geq 1.5 M_g + F_1 M_1$$

So $\qquad\qquad F_1 \leq (M_r - 1.5 M_g) \; M_l = (2448 - 984)/625 = 2.02$

The beam designed by the working strength design has 1.5 as load factor against *DL* and 2.02 as load factor against *LL*.

8.7. Ultimate Strength Design of Rectangular Cross Sections

In ultimate strength design method, the loads are multiplied by the corresponding load factors to calculate the ultimate loads. The structure is then analysed by elastic structural analysis and ultimate moments are computed at the critical sections. The sections are then designed so that the resisting moment capacity is more than the ultimate moment acting at the section. The ultimate moment equal to the resisting moment implies that the section has reached its limit but do not necessarily mean the structure has collapsed. If the ultimate limit load on a structure is defined as the load at which the structure collapses then the ultimate load on the critical section is same as the ultimate limit load of the structure in statically determinate structures. A redistribution of moments takes place when a critical section in a statically indeterminate structure yields and the structure continues to carry more load till enough number of sections fail so as to form a collapse mechanism. Therefore, the ultimate limit load on a statically indeterminate structure is more than the ultimate load of a critical section. Ultimate strength design procedure is carried in three steps :

(1) calculate the ultimate loads from working loads and load factors,
(2) calculate the critical moments at critical sections by elastic structural analysis using the ultimate loads, and
(3) design the sections such that the resisting moments are more than the ultimate moments acting on the sections.

The design criterion is

$$M_r \geq M_u \qquad\qquad\qquad (8.7.1)$$

$$M_r = 0.67 bdf_{ck} \, (d - a/2) \simeq 0.1875 bd^2 f_{ck}$$

for balanced sections. So eq. 8.7.1 gives

$$d^2 \geq M_u /0.1875 b \, f_{ck} \qquad\qquad\qquad (8.7.2)$$

The value of d can be obtained by assuming b. The area of steel is given by

$$A_p \geq \frac{M_u}{f_p\left(d - \dfrac{a}{2}\right)} = \frac{4M_u}{3df_p} \tag{8.7.3}$$

Example 8.7.1 Design a 9 m long prestressed concrete pole for power lines. Use the following data. A transverse ultimate load of 3000 N is acting at 0.6 m from top of the pole. 1.5 m length of the pole is embedded into the ground. 5 mm wires and M 42 concrete are used in the pole.

Solution

The dead weight of the pole will not cause bending moment as the pole is upright. Let the width of the pole is 10 cm and $f_p = 1600$ MPa.

$$M_u = w_u L = 3000(9 - 0.6 - 1.5) = 20700 \text{ Nm}$$

The effective depth of the section from the eq. 8.7.2 is

$$d \geq \sqrt{\frac{2070000}{0.1875(10)42000}} = 16.5 \text{ cm}$$

Equation 8.7.3 gives

$$A_s \geq \frac{4M_u}{3df_p} = \frac{4(2070000)}{3(16.5)(160000)} = 1.04 \text{ cm}^2$$

Use 5 numbers of 5 mm wire. Area of steel provided is

$$A_p = 5(0.196) = 0.980 \text{ cm}^2$$

Since area of steel provided is less than the minimum required, increase the depth of the section proportional to the decrease in the area of steel

$$d = \frac{16.5(1.04)}{0.98} = 17.6 \text{ cm}$$

Let the clear cover be 2 cm so the overall depth of the pole is

$$D = 17.6 + (0.5/2) + 2 \simeq 20 \text{ cm}$$

Select the overall size of the pole as 10 cm by 20 cm at ground level and 10 cm by 15 cm at top with 5 numbers of 5 mm wires on each face of the pole. The transverse load can act in either direction so prestressing is to be provided on both the faces. The steel at the compressive face will not contribute any towards the capacity as it is pre-tensioned to a strain more than the crushing strain of concrete.

Example 8.7.2 Design the electrical pole of example 3.4.1 by ultimate strength design. $w_w = 1800$ N, $L = 6$ m, $M42$ concrete.

Solution

As per the Indian code of practice for poles, IS 1678-1960 the load factor is 2.5

$$w_u = 2.5\,(1800) = 4500 \text{ N}$$

$$M_u = 4500\,(600) = 270\,0000 \text{ Ncm}$$

Let $\qquad b = 10$ cm

$$d \geq \sqrt{\frac{2700000}{0.1875(10)(4200)}} = 18.5 \text{ cm}$$

$$A_s \geq \frac{4M_u}{3df_p} = \frac{4(2700000)}{3(18.5)(160000)} = 1.22 \text{ cm}^2$$

Use 6 numbers of 5 mm wires on either face, then

$$A_p = 0.196(6) = 1.176 \text{ cm}^2$$

Since A_s selected is less than the minimum, increase 'd' proportionately

$$d = \frac{18.5(1.22)}{1.176} = 19.2 \text{ cm}$$

Adopt $\qquad D = 19.2 + 0.25 + 2 \simeq 22$ cm

Provide 10 cm by 22 cm section at the ground level and 10 cm by 16 cm at top with 6 numbers of 5 mm wires on each face with 2 cm cover. It can be seen that the cross section based on working strength design (example 3.4.1) is bigger than that designed by the ultimate strength design.

Example 8.7.3 One way roof slab of effective span 5 m is supporting a live load of 2000 N/m^2. Design a pre-tensioned concrete slab with 5 mm wires $f_p = 16\,0000$ N/cm^2. (This example is same as example 3.5 1.)

Solution

Let $\qquad h = 12$ cm, then for 1 m width of the slab

$$w_g = 1(0.12)\,(24000) = 2880 \text{ N/m and}$$

$$w_1 = 2000 \text{ N/m}$$

Let $\qquad F_g = 1.5$ and $F_1 = 2.2$

$$w_u = 1.5w_g + 2.2w_1 = 8720 \text{ N/m}$$

$$M_u = \frac{w_u L^2}{8} = \frac{8720(25)(100)}{8} = 2725000 \text{ Ncm}$$

$$d \geq \sqrt{\left[\frac{2725000}{0.1875(100)(4500)}\right]} = 5.7 \text{ cm}$$

$$A_p \geq \frac{4(2725000)}{3(5.7)(160000)} = 3.98 \text{ cm}^2$$

Using 5 mm wires, the number of wires required are

$$n = \frac{3.98}{0.196} = 20$$

$$A_p = 20\,(0.196) = 3.92 \text{ cm}^2$$

$$h = 5.7 + 0.25 + 2 = 8 \text{ cm}$$

Use 8 cm thick slab with 20 numbers of 5 mm wires per metre width. (The ultimate strength design gives $h = 8$ cm and 20 numbers of 5 mm wires whereas working strength design with axial prestressing gives $h = 10.7$ cm with 40 numbers of 5 mm wires.) The slab at transfer with 20 wires at 5.7 cm eccentricity will develop excessive tension on top fibre so provide mild steel reinforcement to take the entire tensile force on the top. Provide 6 mm MS bars at 25 cm spacing on top with 2 cm cover. Check for safety against crushing at transfer

$$a = \frac{A_p f_p}{0.67 f_{ck} b} = \frac{(3.92)(16,0000)}{(0.67)(100)(4500)} = 2.08 \text{ cm}$$

The thickness below the CG of steel is more than the depth of compression. At the time of transfer concrete around the steel will be under compression. As there is enough concrete section available even below the CG of steel, the section will not fail by crushing of concrete at the time of transfer of prestress.

8.8. Design of Flanged Sections

Cross sections of long span girders are economically viable if they are flanged type rather than solid rectangular type. Box type of flanged sections are preferable. Either open or closed type of flanged sections can be designed by ultimate strength design by assuming the width of the flange. The design consists of determination of one flange thickness to resist ultimate load and the other flange thickness to resist possible failure at transfer condition. As a starting point assume the neutral axis is in the flange and use the rectangular section formula to calculate the flange thickness. If $a \geq d/5$ then fix the flange at minimum practicable limits and then find the neutral axis distance and area steel.

Example 8.8.1 Example 3.10.1 is reworked on ultimate strength design basis. Assume $F_g = 1.5$ and $F_1 = 2.5$.

Solution

The data of the problem from example 3.10.1 is

$$L = 50 \text{ m}, \quad w_1 = 20000 \text{ N/m}, \quad f_p = 150000 \text{ N/cm}^2$$

$$f_{ck} = 4500 \text{ Ncm}^2$$

Assume the top flange width

$$b_t = 140 \text{ cm and } w_g = 20000 \text{ N/m}$$

$$w_u = 1.5(20000) + 2.5(20000) = 80000 \text{ N/m}$$

$$M_u = \frac{w_u L^2}{8} = \frac{80000(2500)(100)}{8} = 2500\ 000\ 000 \text{ Ncm}$$

Assuming the neutral axis falls in the top flange, the effective depth of the beam can be obtained from the eq. 8.7.2 is

$$d \geq \sqrt{\frac{M_u}{0.1875 b_t f_{ck}}}$$

$$\geq \sqrt{\frac{250\ 000\ 0000}{0.1875(140)(4500)}} = 146 \text{ cm}$$

Approximate $a = 0.48\ d = 0.48(146) = 68$ cm. The flange thickness cannot be 68 cm, so increase 'd' by about 50 per cent,

Let d ≃ 1.5(146) = 219 cm

or   d = 215 cm

Assume $a \geq t$

$$M_r = 0.67\ ba f_{ck} \left(d - \frac{a}{2} \right) \geq M_u$$

$$0.67(140)(4500)(a) \left(215 - \frac{a}{2} \right) \geq 2500000000$$

$$a \left(215 - \frac{a}{2} \right) \geq \frac{2500000000}{0.67(140)(4500)} = 5850$$

$$a^2 - 430a + 11700 \geq 0$$

$$a \geq \frac{430 - \sqrt{430^2 - 46800}}{2} = 28 \text{ cm}$$

Adopt $t_f = 28 \text{ cm} = a$

Distance between the tension and compressive forces

$$jd = d - \frac{a}{2} = 215 - 14 = 201 \text{ cm}$$

So $j = \frac{201}{215} = 0.93$

In the case of flange sections the value of 'j' is in the range of 0.9 to 0.95

$$A_s = \frac{M_u}{jdf_p} = \frac{2500000000}{201(150000)} = 83 \text{ cm}^2$$

Let
$$h \simeq 1.1d \simeq 240 \text{ cm}$$

Minimum web thickness $= 10 + h/400 = 10.6$ cm

Assume
$$b_w = 16 \text{ cm}$$

Minimum compressive area required to resist the pre-tension force is $= \dfrac{83(150000)}{0.68(4500)} = 4060 \text{ cm}^2$

Design of bottom flange is

Bending moment acting on the beam at transfer of prestress stage is M_g and it is

$$M_g = \frac{20000(2500)(100)}{2} = 625000000 \text{ Ncm}$$

Assuming the crushing will take place in the lower face, the lever arm at transfer condition (jd_t) is

$$jd_t = \frac{M_g}{T} = \frac{625000000}{83(150000)} = 50 \text{ cm}$$

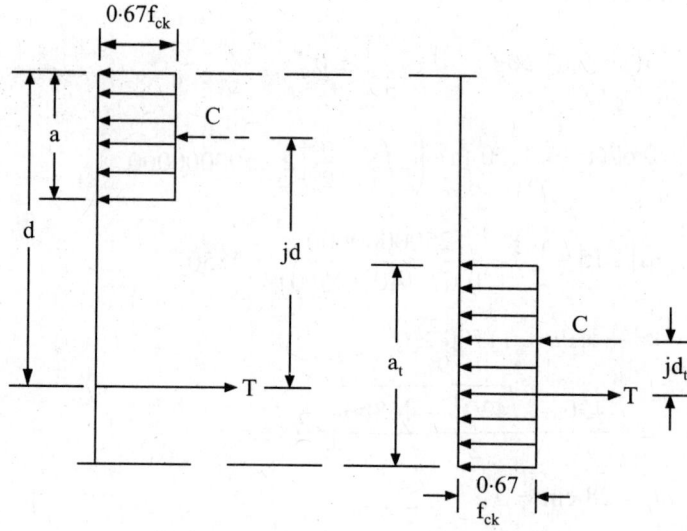

Figure 8.8.1 Stress distribution.

As seen from figure 8.8.1 (b), the value of a_t, that is, the depth of compression of concrete at transfer condition is more than jd_t.

Assume the total web from bottom fibre to the bottom of top flange is in compression. Then area of web under compression is

$$A_w = 16 (240 - 27) = 3408 \text{ cm}^2$$

Area required to take the total ultimate tension of the steel is 4060 cm^2 so area of bottom flange needed is

$$A_{cb} = 4060 - 3408 = 652 \text{ cm}^2$$

So use 60 cm by 15 cm bottom flange. Check for self weight of the beam

$$w_g = 25000 \ [140 \times 28 + 60 \times 15 + 16 \ (240 - 27 - 15)]$$
$$= 19750 \text{ N/m}$$

Since actual value of w_g is approximately equal to the assumed value the design need no further revision. The self-weight of concrete obtained by USD is 19750 N/m whereas that by WSD it is 21000 N/m. There is about 13.6 per cent of reduction in weight of the beam when the design is done by USD. The sectional details of the beam are given in figure 8.8.2.

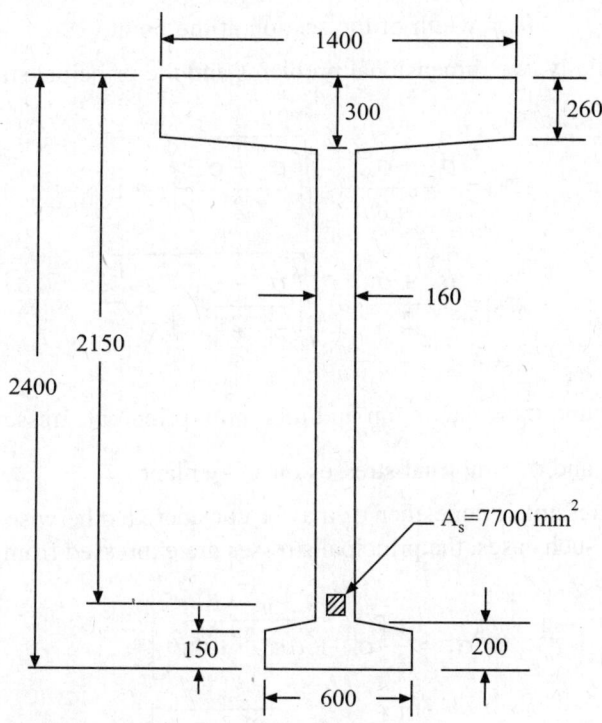

Figure 8.8.2   Example 8.8.1 (Dimensions in mm).

8.9.   Principal Tensile Stress

The previous chapters discuss analysis and design of prestressed concrete beams subjected to bending moments. Bending moments in beams are usually accompanied by shear forces, so such members are designed for combined bending and shear. Because of simplicity and limited theoretical and experimental data, design of members subjected to combined bending and shear is done independently for bending and for shear. This chapter presents some methods of analysis and design for combined shear and moment forces. Ultimate load capacity based on shear moment failure is also discussed in this chapter.

Uncracked sectional properties are used while calculating the shear stress at working load.

Let τ = shear stress

V_e = effective shear force

Q_y = first moment of area about the point under consideration

Then the shear stress τ for a prismatic member is given by

$$\tau = \frac{V_e Q_y}{I b_y} \tag{8.9.1}$$

where b_y = width of the section at the point.

Beams are essentially two dimensional problems and the principal stresses in plane stress problem are given by

$$\sigma_1 = \frac{\sigma_x + \sigma_y}{2} + \sqrt{\left(\frac{\sigma_x - \sigma_y}{2}\right)^2 + \tau^2} \tag{8.9.2}$$

$$\sigma_2 = \frac{\sigma_x + \sigma_y}{2} - \sqrt{\left(\frac{\sigma_x - \sigma_y}{2}\right)^2 + \tau^2} \tag{8.9.3}$$

where

σ_1 and σ_2 = maximum and minimum principal stresses

σ_x and σ_y = normal stresses on $x - y$ plane

If transverse prestressing is done, then σ_y may be considered, otherwise σ_y may be neglected as compared to σ_x. In such cases, the principal stresses are expressed from eqs. 8.9.1 and 8.9.2 as

$$\sigma_1 = \frac{1}{2}\left[\sigma_x + \sqrt{\sigma_x^2 + 4\tau^2}\right] \tag{8.9.4}$$

$$\sigma_2 = \frac{1}{2}\left[\sigma_x - \sqrt{\sigma_x^2 + 4\tau^2}\right] \tag{8.9.5}$$

The principal stress plane is given by

$$\tan 2\alpha = \frac{(2\tau)}{\sigma_x} \tag{8.9.6}$$

where α = angle of principal stress plane.

One of the principal stresses is likely to be tensile and will act on a plane making angle 'α' with the horizontal axis. For very efficient use of shear steel, reinforcement should be laid at an angle 'α' with the horizontal. However, the practical and economical considerations often suggest vertical stirrups.

In case the tendon is inclined, the vertical component of the tendon force will resist a part of the shear force and it is given by

$$V_p = P_e \frac{dy}{dx} \qquad (8.9.7)$$

where $\qquad\qquad\qquad\qquad V_p$ = shear force resisted by the tendon.

The effective shear force at any section is given by

$$V_e = (V - V_p) \qquad (8.9.8)$$

where $\qquad\qquad\qquad\qquad V$ = total shear force.

Let $\qquad\qquad\qquad\qquad A_{sv}$ = area of vertical stirrup

$\qquad\qquad\qquad\qquad\qquad \sigma_{sv}$ = allowable stress in steel stirrups

$\qquad\qquad\qquad\qquad\qquad s_v$ = spacing of the stirrups

Area of shear steel is given by

$$A_{sv} = \frac{\sigma_1 (s_v b_w)}{\sigma_{sv}} \qquad (8.9.9)$$

If A_{sv} is already selected, then the spacing of the stirrups is given by

$$s_v = \left(\frac{A_{sv} \sigma_{sv}}{\sigma_1 b_w} \right) \qquad (8.9.10)$$

8.10. Failure Due to Shear

Failure of a prestressed concrete beam caused by shear can be divided into two groups :

    (i) failure caused by pure shear, and
    (ii) failure caused by combined shear and flexure.

Failure caused by pure shear practically never occurs in prestressed concrete structures. Failure caused by the combined action of shear and bending moment is common. Therefore, it is desirable to design or check beams for "shear-moment failure".

It has been observed that a vertical crack starts at the lower fibre during overload conditions at maximum bending moment location. In simply supported beam the vertical cracking propagates towards supports as the load increases. Then shear-moment cracks began to develop. Shear-moment cracks are normally developed at a continuation of vertical cracks and those cracks in general are from 'h to $4h$' distance from a simple support or from '$0.5h$ to h' distance from intermediate supports. Shear-moment cracks develop large deformation, distorting the horizontal reinforcement. The location and development of cracks affect the deformation characteristics of the beam. Specifically, diagonal cracks develop some kind of rotational hinges about which the cracked elements rotate. A typical crack development and beam rotation is shown in figures. 8.10.1 and 8.10.2.

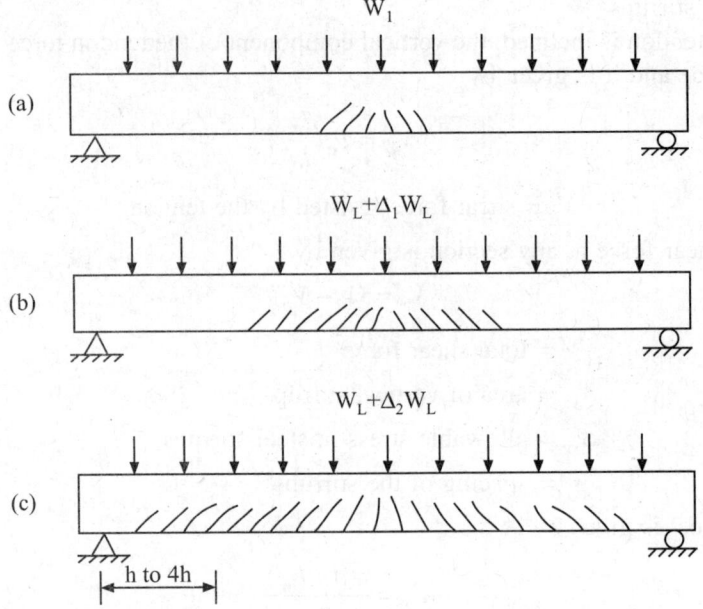

Figure 8.10.1 Typical crack development. (a) Crack initiation; (b) Vertical crack propagation; and (c) Inclined shear-moment cracks.

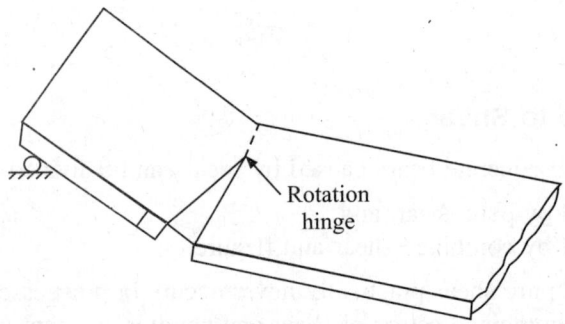

Widening of shear-moment crack (rotation hinge)
Figure 8.10.2 Hinge development due to inclined cracks just before failure.

8.11. Compressive Strength of Concrete Subjected to Combined Bending and Shear

Ultimate compressive stress available in beams is less than f_{ck} because of shape, length and other segregation effects. Ultimate concrete stress is generally given by $k_p f_{ck}$ where k_p is a factor approximately in the range of 0.67. If a beam is subjected to both bending as well as shear forces, then the allowable ultimate compressive stress is further reduced. An approximate interaction formula has been suggested based on experimental data and it is

$$f_{cs} = \left(k_p f_{ck}\right) \bigg/ \left[1 + 3.2\left(\frac{Vh}{M}\right)\right] = k_p k_7 f_{ck} \tag{8.11.1}$$

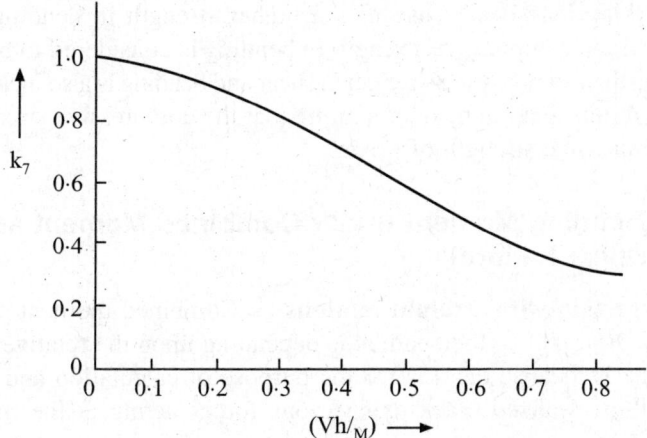

Figure 8.11.1 Interaction curve for combined shear and bending moment.

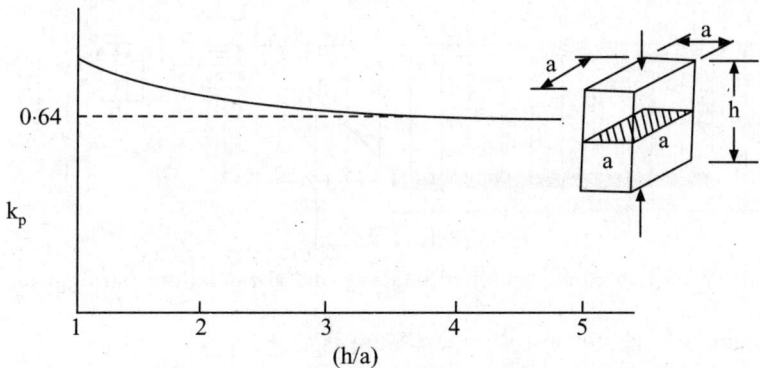

Figure 8.11.2 Concrete prism strength.

$$k_p = 1 \Big/ \left[1 + 3.2 \left(\frac{Vh}{M} \right) \right]$$ (8.11.2)

where f_{cs} = ultimate concrete stress in a beam subjected to combined bending and shear

$k_p f_{ck}$ = ultimate concrete stress in prism

Interaction curve for k_7 is shown in figure 8.11.1. The values of Vh/M are in the order of 0.1 to 0.4 in most of the practical problems.

Value of k_p decreases as the length to side ratio increases. In case of concrete subjected to pure bending, ultimate bending strength in compression is assumed to be higher than that of a prism strength. The relation is given by

$$f_{cb} = k_p k_6 f_{ck}$$ (8.11.3)

where f_{cs} = compressive strength of concrete in bending

k_6 = non-dimensionalised coefficient and its value is suggested to be about 1.2 to 1.25

The practice in working load design assumes a higher strength in bending as compared to direct compression. If the compressive strength in bending is considered to be higher than that in direct compression, then compressive strength in shear and bending is also increased accordingly. All the calculations in this text are made assuming that the compressive strength in bending is the same as the compressive strength of prism.

8.12. Ultimate Bending Moment under Combined Moment and Shear (Moment Shear Failure)

Rectangular cross section with straight tendons : Combined moment and shear crack is generally inclined at 20 to 40° to the beam axis depending upon the relative magnitude of the quantities. This crack is idealised at 45° for the purpose of calculation and to be on the safe side of the design. An idealised crack aud various forces acting at the time of failure are shown in figure 8.12.1.

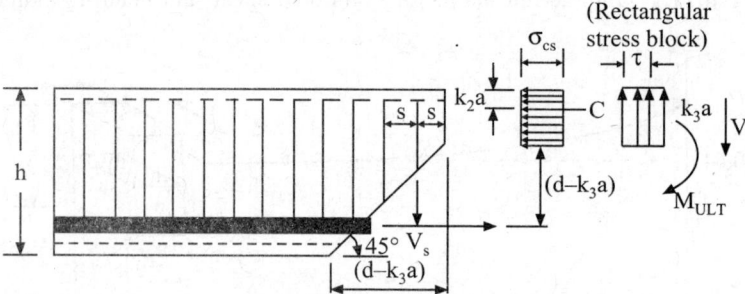

Figure 8.12.1 Forces at combined bending and shear failure (straight tendon).

Vertical component of the stirrup forces at failure is

$$V_s = \left[\frac{(d - k_3 a)}{s} A_{sv} f_y \right]$$

$$= d \left[\frac{1 - k_3 k_u}{s} \right] A_{sv} f_y \qquad (8.12.1)$$

where V_s = shear force resisted by the stirrup steel.

Equations of equilibrium at ultimate load capacity of the section are

$$f_p A_p = k_1 k_7 k_3 k_u b d f_{ck} \qquad (8.12.2)$$

$$Mu = \left[\frac{d}{s} (1 - k_3 k_u) A_{sv} f_y \right] \left(\frac{d - k_3 a}{2} \right) + (k_1 k_7 f_{ck})(k_3 k_u d)(b)(d - k_2 a)$$

$$= k_1 k_7 k_3 k_u (1 - k_2 k_u) b d^2 f_{ck} + \tfrac{1}{2}(1 - k_3 k_u)^2 \frac{d^2}{s_v} A_{sv} f_y$$

$$\in = \in_{se} + \in_{ce} + \in_u \left(\frac{1}{k_u} - 1 \right)$$

Above set of equations gives the ultimate moment for a beam of rectangular cross section subjected to combined bending and shear.

Example 8.12.1 A 8 m long prestressed concrete beam of rectangular cross section (30 cm by 80 cm) is provided with a straight cable of 10 cm² having 20 cm eccentricity. Mild steel stirrups of 1 cm wires at 15 cm are provided to resist shear. Determine the ultimate capacity of the beam.

$$f_{ck} = 45 \text{ MPa}$$
$$f_c = 1500 \text{ MPa}$$
$$k_1 = 0.85, k_3 = 0.85, k_2 = 0.5 \text{ and } k_7 = 1$$

Solution

From eq. 8.12.2

$$k_u = (Af_p)/(k_1 k_7 k_3 b d f_{ck}) = 0.295$$

$$k_3 a = k_3 k_u d = 0.295 \times 0.85 \times 80 = 20 \text{ cm}$$

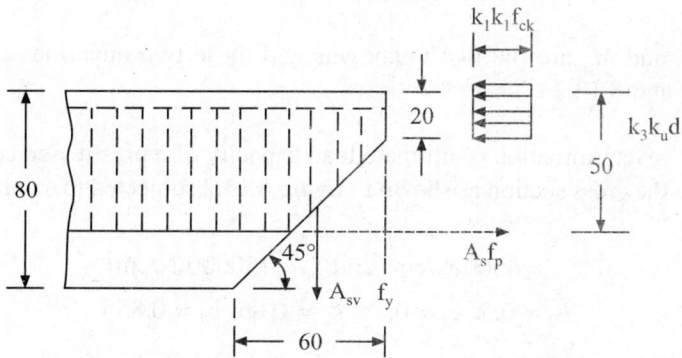

Figure 8.12.2 Example 8.12.1 (Dimensions in cm).

The following computations are made with reference to figure 8.12.2

$$A_{sv} = \left(\frac{60}{15}\right)\frac{\pi}{4} = 3.14 \text{ cm}^2$$

Let the ultimate stress of mild steel be 300 MPa the ultimate moment capacity is

$$M_u = 150000(10) \times 50 + 3.14 \times 30000 \times 30 = 77.83 \times 10^6 \text{ Ncm}$$

8.13. Ultimate Moment under Combined Bending and Shear of I-girder

There are two possible cases of failure in I-girders: (i) neutral axis within the top flange, and (ii) neutral axis in the web of the girder. If the neutral axis is in the top flange, then ultimate moment is given by the previous section by substituting b_i for b.

Figure 8.13.1 gives complete details at the instant of failure of an I-section where the neutral axis is in the web portion.

Equilibrium equations are

$$A_p f_p = k_1 k_7 k_3 k_u f_{ck} db_w + k_4 k_7 t_t (b_t - b_w) f_{ck} \tag{8.13.1}$$

$$M_u = k_1 k_7 k_3 k_u (1 - k_2 k_u) b_w d^2 f_{ck} + k_4 k_7 (d - k_5 t_t) (b_t - b_w) t f_{ck}$$

$$+ [A_{sv} f_y (1 - k_3 k_u)^2 / 2s] d^5 + A_s f_p \left(\frac{dy}{dx}\right) d \tag{8.13.2}$$

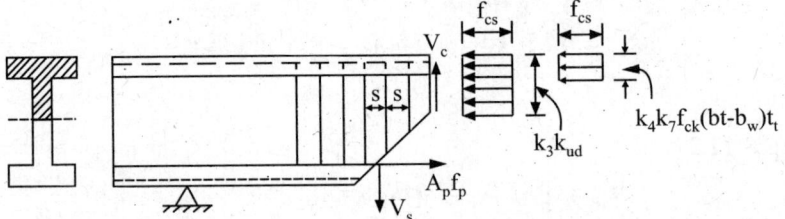

Figure 8.13.1 Forces at combined bending and shear failure
(I-section with neutral axis in the web).

The values of k_u and M_u are the two unknowns and these two quantities could be obtained from eqs. 8.13.1 and 8.13.2

Example 8.13.1 Determination of ultimate load capacity of a prestressed concrete cantilever beam of 8 m with the cross section as shown in figure 8.13.2 subjected to a uniformly distributed load. Given data :

$$f_{ck} = 4500 \text{ N/cm}^2 \text{ and } f_p = 150000 \text{ N/cm}^2$$

$$k_2 = 0.5, \ k_3 = 0.85, \ k_1 = 0.85, \ k_4 = 0.85$$

Figure 8.13.2 Example 8.13.1 (Dimensions in cm).

Solution

The section at support is subjected to maximum shear and bending moment, so the allowable compressive stress depends upon the relative ratio

$$\frac{Vh}{M} = \frac{wLh}{wL^2/2} = \frac{2h}{L} = \frac{2(0.8)}{8} = 0.2 \text{ and}$$

$$k_7 = \frac{1}{1 + 3.2 \times 0.2} = 0.61$$

Assuming the neutral axis in the web portion, from eq. 8.13.1

$$k_u = [A_p f_p - k_4 k_7 t_b (b_b - b_w) f_{ck}] / (k_1 k_7 k_3 f_{ck} d b_w) = 0.18$$

$$k_3 k_u d = 12.2 (> t_b)$$

$$\in_s = 0.0044 + 0.0035 \left(\frac{72 - 12.2}{12.2} \right) = 0.0193 > \in_{yp}$$

$$M_u = k_1 k_7 f_{ck} [(t_b b_b) \times (72 - 5) + b_w (12.2 - t_b)(62 - 1.1)]$$

$$= 9637,300 \text{ Ncm}$$

$$w_u = (M_u \times 2)/(L^2) = \frac{9637300 \times 2}{800 \times 800} = 300 \text{ N/cm}$$

$$= 30000 \text{ N/m}$$

8.14. Location of Failure Section

In the last section the failure was assumed to occur at a section where the value of k_7 is calculated from the shear and moment relation at the section. In any problem, it is necessary to locate a section which is likely to fail and establish the value of k_7 at that section. In statically determinate structures, location of shear moment failure can be estimated or derived. An exampie of a well-defined situation is shown in Figure 8.14.1, where shear force and bending moment are maximum at a particular section. Hence shear-moment failure will also occur at that section.

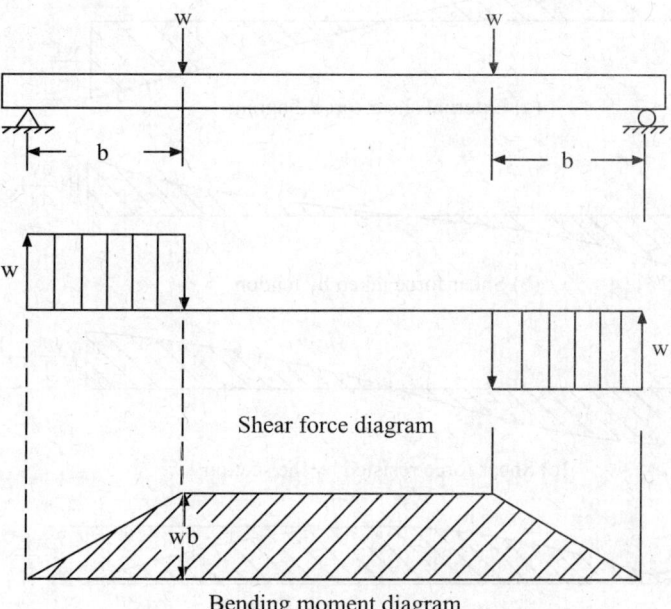

Figure 8.14.1   Defined problem of shear moment failure (failure occurs at section cc where bending moment and shear force are maximum).

In some problems, the zone of failure is not well-defined. In a simple problem like a simply supported beam subjected to uniform loading, maximum pending moment occurs at mid-span although the maximum shear force is near the support. Investigations have to be made to locate the maximum combined effect. Since the section will develop cracks due to bending, it is difficult to establish a generalised principal stress equation of the beam. Independent moment and shear force diagrams of a simply supported beam are shown in figure 8.14.2. The figure also gives the effective shear resisted by the cross section and tendon. Location of maximum damage due to the combined effect of moment and shear is rather undefined. In such a case as this, failure could be either : (i) moment failure, or (ii) shear-moment failure. Shear-moment failure is likely to occur near a heavy concentrated load if it is present. Trial analysis generally gives a satisfactory result even though it may not give an exact section where the failure, occurs.

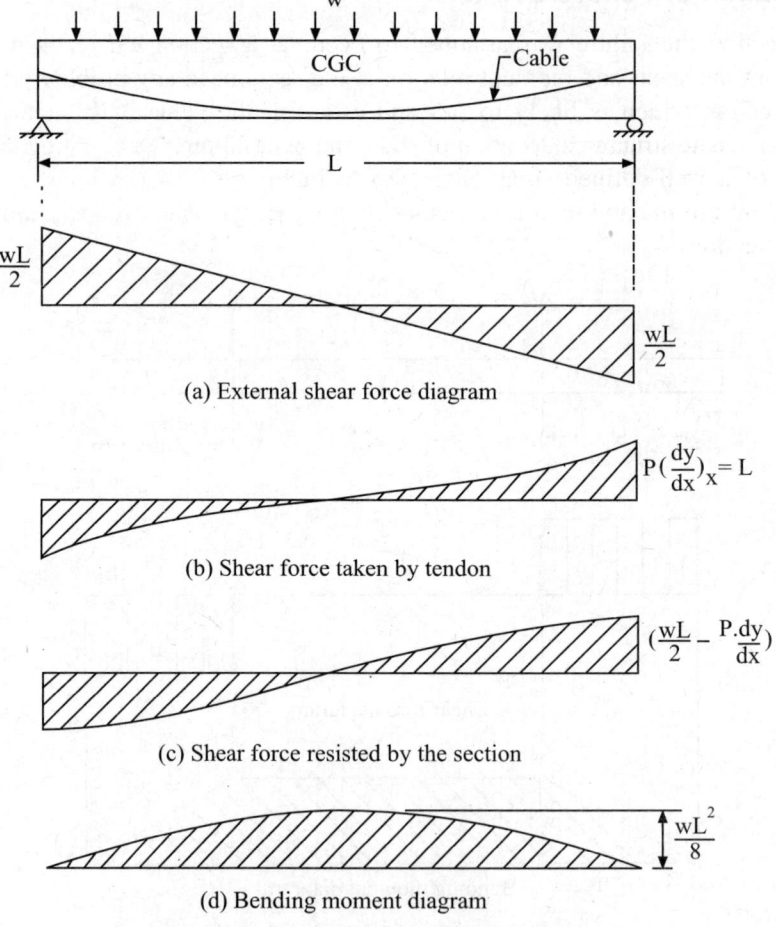

(a) External shear force diagram

(b) Shear force taken by tendon

(c) Shear force resisted by the section

(d) Bending moment diagram

Figure 8.14.2   Shear and moment diagrams.

Location of shear moment failure in a statically indeterminate structure is not easy to predict because of the nature of the shear and bending moment distribution. The effects of secondary bending force and the curvature in the cable make the problem more complicated. Even though a particularly critical section exists at which the failure starts, the real failure of a structure is not due to failure of a single section but due to a localised effect. An investigation near the critical section gives fairly close approximation of failure load. Shear and moment are maximum near the intermediate supports so the shear-moment failure is likely to occur at about $h/2$ to $2h$ from an intermediate support. A statically indeterminate structure collapses when sufficient number of sections yield into a collapsable mechanism.

8.15. Suggestions from Experimental Investigations

Extensive experimental investigations have been done on shear failures and a large amount of data is being accumulated on the subject. In spite of the large experimental data, it is still difficult to suggest any exact procedure for design against shear-moment failure. Design is done independently for bending moment and shear force while shear-moment failure load is computed based on the designed section. Effect of different parameters on the shear-moment failure, as suggested by McGregor, Sozen and Siess are given here to help in the design of shear reinforcement.

(i)   Effect of prestress : Increase in prestressing force increases the inclined cracking load. Even though certain amount of increase in cracking load is obtained by increasing the prestressing force, the warning against ultimate failure is reduced. A typical behaviour of cracking load relative to effective prestressing force is shown in figure 8.15.1.

(ii)   Effect of web thickness : The inclined cracking load appears to be independent of the thickness of the web. In most cases, the web thickness is in a narrow range so that three dimensional effects do not come into play. Web could be treated as a plane stress problem without any loss of accuracy for design purposes.

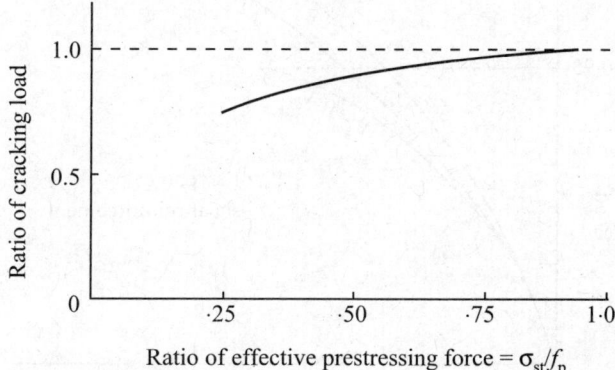

Figure 8.15.1   Effect of prestress on cracking load.

(iii) Effect of web reinforcement : Rate of propagation of the shear crack depends on web reinforcement. The initiation of a crack starts near the bottom fibre and is generally independent of the vertical (web) reinforcement. However, the propagation of the crack depends upon the shear reioforcement. Ultimate shear-moment capacity could be increased by increasing the shear reinforcement.

In beams without web reinforcement, the beam distorts and causes kink in the main prestressing force. The vertical component of the prestressing force at the kink contributes towards shear capacity. The kink action reduces the ultimate bending moment capacity even though it increases the ultimate shear capacity. The overall ultimate capacity of the beam is affected by the kink. The deformation of a beam is also affected by vertical reinforcement. A typical load deflection curve with different percentages of shear reinforcement is shown in figure 8.15.2. It can be seen that the increase of shear reinforcement decreases the deflection to some extent and increases the ultimate load capacity of the beam. Even in the case of problems where the reinforcement is not needed based upon the working load design, it is still desirable to provide some nominal shear reinforcement. It will increase the ultimate capacity of the beam and also avoid a sudden failure of the beam due to fast propagation of inclined crack.

(iv) Effect of vertical prestressing: Vertical prestress reduce the tensile principal stress as can be seen from the following equation :

$$\sigma_1 = \frac{\sigma_x + \sigma_y}{2} - \left[\frac{\left(\sigma_x + \sigma_y\right)^2}{4} + \tau^2 \right]^{1/2} \qquad (8.15.1)$$

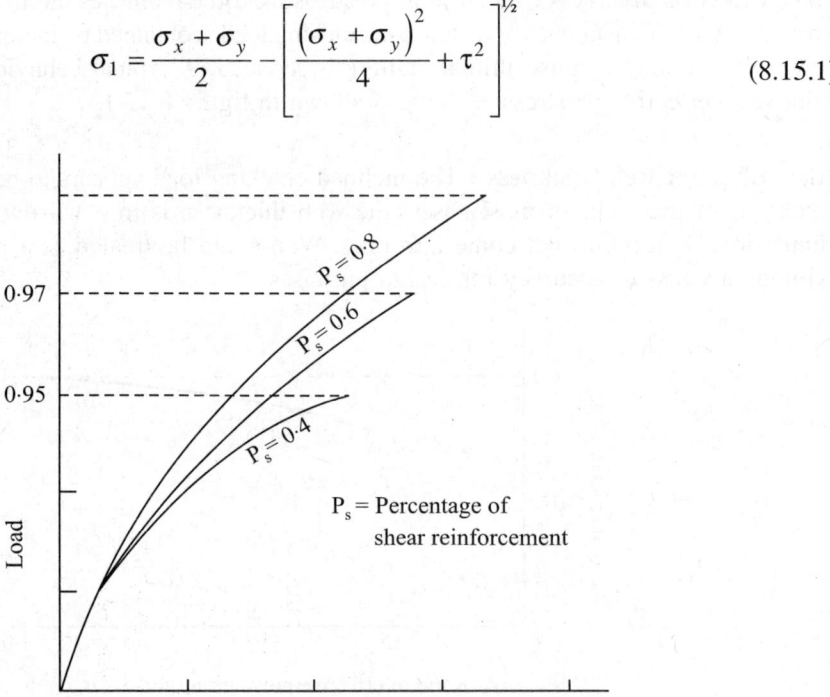

Figure 8.15.2 Qualitative behaviour of a prestressed concrete beam based on shear reinforcement.

Effect of σ_y is same as of σ_x as far as the principal stresses are concerned. The inclined crack load decreases with increased vertical prestressing. The shear-moment failure load could be increased by about 5 to 10% by introduction of vertical prestressing, whereas the moment failure load is not affected by the vertical prestressing force.

(v)   Effect of draping of longitudinal steel: Upward inclination of longitudinal reinforcement resists vertical shear. If the failure is due to shear domination, then the ultimate capacity could be increased by draping of the wires. If the failure is a moment failure, draping of wires decreases the ultimate load capacity.

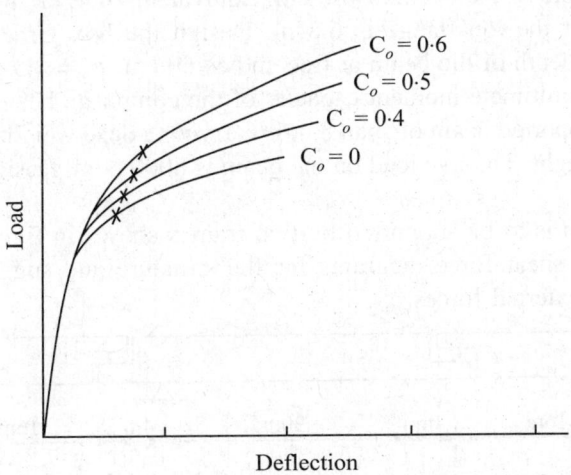

Figure 8.15.3   Typical behaviour of prestressed concrete beam with vertical prestressing.

$$x\text{-Cracking load } \frac{\sigma_{sv}}{f_{sv}} = C_0$$

PROBLEMS

8.1 A rectangular section of 20 cm by 40 cm is provided with 6 cm² of HT steel. Calculate the ultimate resisting capacity of the section if $f_{ck} = 40$ MPa and $f_p = 1500$ MPa.

8.2 Calculate the moment capacity of the section of problem 8.1 if area of steel is 12 cm² instead of 6 cm².

8.3 A hollow box section with 60 cm by 40 cm outside dimensions and 40 cm by 20 cm inside dimensions is provided with 40 numbers of 5 mm high tensile steel wires at top and bottom flanges. Calculate the ultimate moment capacity of the cross section. $f_{ck} = 45$ MPa.

8.4 Calculate the ultimate moment capacity of the box section of example 3.4.1.

8.5 The sizes of an idealised I-section are $b_t = 150$ cm, $t = 20$ cm, $b_b = 80$ cm, $t_b = 15$ cm, $b_w = 16$ cm, $h = 220$ cm, $d = 190$ cm, and $A_s = 80$ cm². Calculate the moment capacity of the section if $f_{ck} = 45$ MPa and $f_p = 1500$ MPa.

8.6 An electric pole of total 8 m length is embedded 1.5 m in ground. A working transverse load of 1800 N is applied at 0.3 m from top of the pole. Design the cross section assuming $F_1 = 2.5$, $f_{ck} = 40$ MPa and $f_p = 1600$ MPa.

8.7 A simply supported one way slab of 4 m span is subjected to UDL of 4000 N/m². Design the slab with M45 concrete.

8.8 A bridge girder of simply supported span 20 m is subjected to a UDL of 15000 N/m live load. The width of the top flange is 120 cm. Design the beam with M35 concrete. Assume an I-section; and $F_g = 1.5$ and $F_1 = 1.5$.

8.9 A cantilever bridge girder is to be made of a single hollow box section. The span is 20 m and depth of the beam is 180 cm. Equivalent UDL on the beam is 60000 N/m. The width of the top flange is 6.8 m. Design the box girder with M35 concrete. Assume the depth of the beam at free end as 60 cm. $F_g = 1.5$ and $F_1 = 2.5$.

8.10 Calculate the ultimate moment capacity of the composite beam of example 4.4.1.

8.11 A simply supported beam of span 8 m is carrying a dead weight 4000 N/m in addition to its self weight. The live load on the beam is 8000 N/m. Design the beam with M35 concrete.

8.12 A bridge slab is to be supported by two frames shown in figure 8.1. Draw bending moment and shear force diagrams for the structure and suggest a cable profile to balance the external forces.

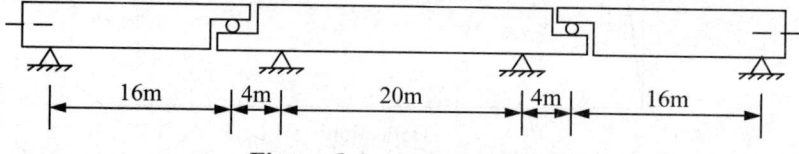

Figure 8.1 Problem 8.12.

8.13 The beam of problem 8.12 is subjected to a distributed load of 30000 N/m. Design prestressing and shear steel for the beam. Determine the ultimate strength of the cantilever and the overhang beam sections.

CHAPTER 9

Limit State Design of Partially Prestressed Concrete Beams

9.1. Introduction

Prestressed concrete beams are normally designed such that the total bending moment is resisted by pretensioned steel. Such beams are called *fully prestressed* beams which were discussed in this book at length. If a beam is designed such that the tension force required to generate a moment to resist the external bending moment is achieved partly by tensioned and partly by non-tensioned steel, such a construction is called *partially prestressed concrete* construction. The tensioned steel is invariably the high tensile steel whereas the non-tensioned is usually high yield deformed bars. In beams where the dead weight is small when compared with the live load, the pre-tensioned steel is likely to cause excessive tensile stresses at the no live load condition. It may even cause high compressive forces in the initial load condition. Even if multistage prestressing is adopted, the problem still exists if the live load is not present on the beam. If sufficient dead load is not present all the time on the beam, the prestressing will cause excessive creep deflections which might lead to collapse. Such a phenomenon is likely to occur in roof beams and short bridge girders.

Therefore one can provide non-tensioned steel as a reinforcement in addition to the prestressing force in beams and such a reinforcement which is passive, will be activated during the live load condition. In addition, in the under-reinforced beams the concrete is not utilised to the fullest extent possible. By providing additional non-tensioned reinforcement, one can gain full advantage of the concrete. It is also possible that the partially prestressed concrete construction could be economical in some situations. Only type 3 prestressed concrete can be considered for possible partially prestressed concrete construction.

9.2. Balanced Moment Capacity of Rectangular Sections

The maximum strain permitted in the pre-tensioned high tensile steels at the limit state of strength is:

$$\epsilon_p = 0.002 + \frac{0.87\ f_p}{E_s} \tag{9.2.1}$$

In most high tensile steels it is in the range of 0.007 to 0.009. The effective prestrain in steel is in the range of 0.0035 to 0.005 whereas the precompressive strain in concrete or nontensioned steel at the level of tensioncd steel is usually of the order of 0.0004. The difference between the strains at proof stress and effective prestressed is in the range of 0.0035 to 0.0055. If the nontensioned reinforcement is placed near or beyond the pretensioned steel, then the strain to which such steel is subjected to is in the range of 0.003 to 0.005. The strains at yield of mild steel and high yield strength deformed bars is given by

$$\epsilon_y = 0.002 + \frac{0.87 f_y}{E_s} \tag{9.2.2}$$

And this is approximately equal to 0.003 for mild steel and 0.004 for high yield deformed bars. Hence one can see that the nontensioned steel will also yield at the limit state of strength of the pretensioned steel if it is placed near or beyond the level of the tensioned steel.

Figure 9.2.1(a) illustrates a typical cross section of a partially prestressed concrete beam.

Let A_p = area of the tensioned steel

A_{st} = area of the nontensioned steel

d = distance of the tensioned steel from the extreme compression fibre

d_s = distance of nontensioned steel from the extreme compression fibre.

The strain variation along the depth of the section is shown in Figure 9.2.1(b).

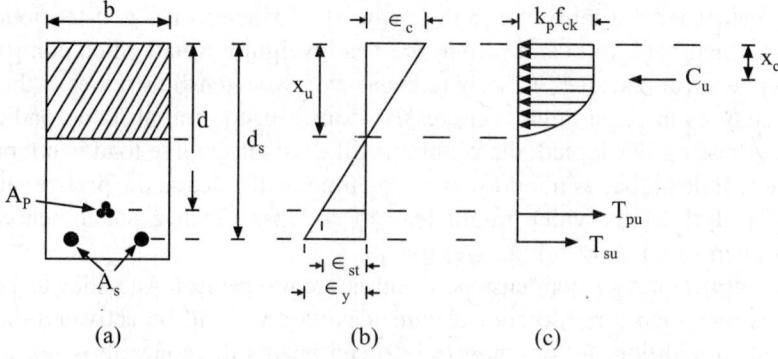

Figure 9.2.1   Strain and stress distributions at failure.

Let a parabolic-cum-rectangular stress distribution be assumed on the compression face of the concrete as indicated in Figure 9.2.1(c).

From Chapter 7, it is seen that the total ultimate compressive force on the concrete is given by

$$C_u = 0.54 \, b_{xu} f_{ck} \tag{9.2.3}$$

where

$$x_u = \frac{0.0035}{0.0015 + \dfrac{0.87 \, f_p}{E_s}} \tag{9.2.4}$$

The value of x_u for different steels was already given in Table 7.3.1. The design compressive force is:

$$C_d = \frac{C_u}{\gamma_m} = 0.36 \, b \, x_u f_{ck} \tag{9.2.5}$$

where γ_m = partial safety factor applied to materials and it is equal to 1.5 for concrete.

The moment equilibrium about the centre of tensioned steel gives the resisting moment capacity of the section. Thus

$$M_r = C_d (d - x_c) + T_{sd} (d_s - d) \tag{9.2.6}$$

where x_c = distance of the CG of the compression from extreme compression fibre and = $0.42 \, x_u$

T_{sd} = design tension force in the nontensioned steel and it is given by

$$T_{sd} = \frac{T_{su}}{\gamma_m} = \frac{A_{st} f_y}{\gamma_m} \tag{9.2.7}$$

in which $\gamma_m = 115$ and d_s is assumed to be equal to or greate than d. The yielding of steel is assumed at the limit state of strength.

Equation 9.2.6 can be expressed as

$$Mr = Kbd^2 f_{ck} + 0.87 f_y A_{st} (d_s - d) \tag{9.2.8}$$

where $$K = C_d \left(1 - \frac{x_c}{d} \right) \text{ and it is given in Table 7.5.}$$

Let $$p_s = \frac{A_{st}}{bd} = \text{nontensioned steel ratio.} \tag{9.2.9}$$

Then Eq. 9.2.8 can be modified by

$$M_r = bd^2 \, f_{ck} \left[K + 0.87 \, p_s \left(\frac{d_s}{d} - 1 \right) \frac{f_y}{f_{ck}} \right] \tag{9.2.10}$$

The above equation is helpful in the design.

The equilibrium of limit forces on the cross section gives

$$C_u = T_{pu} + T_{su} \tag{9.2.11}$$

where $$T_{pu} = A_p f_p \text{ and } T_{su} = A_{st} f_y$$

But unfortunately, the limit state design suggests the equilibrium, of design forces rather than the limit forces. Therefore the balancing of the design forces on the section gives

$$T_{pd} + T_{sd} = C_d$$

or
$$0.87 \, (A_p f_p + A_{st} f_y) = 0.36 \, b \, x_u f_{ck} \qquad (9.2.12)$$

The above equation can be rearranged as

$$A_p = \frac{(0.414 \, b \, x_u \, f_{ck} - A_{st} f_y)}{f_p}$$

$$= 0.414 \, bd \, \frac{x_u}{d} \, \frac{f_{ck}}{f_p} - \frac{A_{st}}{bd} \, bd \, \frac{f_y}{f_p}$$

$$= \frac{bd \, (p_o f_{ck} - p_s f_y)}{f_p} \qquad (9.2.13)$$

where
$$p_o = 0.414 \, \frac{x_u}{d} \qquad (9.2.14)$$

The design consideration is that

$$M_t \geq M_c$$

in which M_c = collapse moment
This results in

$$d \geq \sqrt{\frac{M_c}{bf_{ck} \left(K + 0.87 \, p_s \, (c_s - 1) \, \dfrac{f_y}{f_{ck}} \right)}} \qquad (9.2.15)$$

where $c_s = \dfrac{d_s}{d}$ and it can be assumed in the range of 1 to 1.15, usually 1.1. Equations 9.2.15 and 9.2.14 give the design parameter for preassigned nontensioned steel.

9.3. Moment Capacity of Flanged Sections

Flanged sections are very common in prestressed concrete construction. A typical idealised section is shown in Figure 9.3.1 and it is similar to that given in Chapter 7. The moment capacity of a flanged section is computed neglecting the contribution of the web below the flange level in case the NA is below the flange. If the neutral axis falls in the compression flange, then the moment capacity is exactly same as that given for rectangular sections with b replaced by b_f.

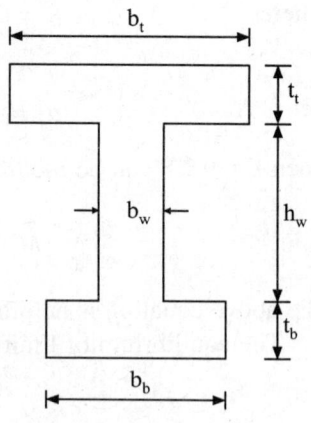

Figure 9.3.1 Notations.

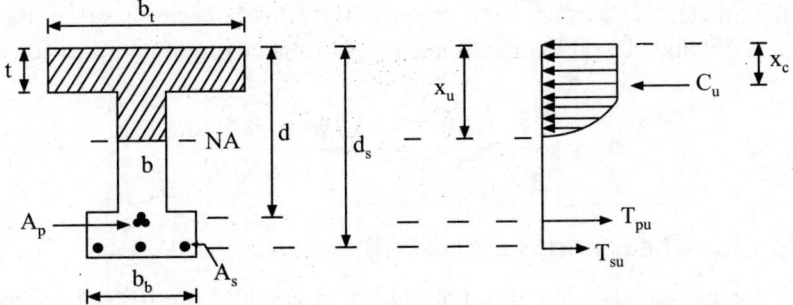

Figure 9.3.2 Stress distribution at failure.

Let $x_u \geq t$; then the design compressive force in the flange alone is

$$C_d = 0.45\, f_{ck} b_f t \tag{9.3.1}$$

The design tension forces are

$$T_{pd} = 0.87\, f_p\, A_p$$

$$T_{sd} = 0.87\, f_y\, A_{st}$$

The equilibrium of moments about the pretensioned steel gives the moment capacity of the section and it is

$$M_r = 0.45\, f_{ck}\, b_f t\, (d - 0.5t) + 0.87\, f_y\, A_{st}\, (d_s - d) \tag{9.3.2}$$

The above equation can be rearranged as

$$M_r = 0.45\, f_{ck}\, b_f t \left(\left(1 + 2p_s\, (c_s - 1)\, \frac{f_y}{f_{ck}} \right) d - 0.5t \right) \tag{9.3.3}$$

The design criterion that $M_r \geq M_c$ gives

$$d \geq \frac{0.5t}{\alpha} + \frac{M_c}{0.45 f_{ck} b_f t \left(1 + 2p_s(c_s - 1)\, \dfrac{f_y}{f_{ck}} \right)} \tag{9.3.4}$$

where

$$p_s = \frac{A_{st}}{b_f t} \quad \text{and} \quad \alpha = 1 + 2p_s\, (c_s - 1) f_y / f_{ck} \tag{9.3.5}$$

Usually b_f, t and A_{st} are assumed or given, so the depth of the section can be obtained from Eq. 9.3.4.

The area of tensioned steel can be obtained from the equilibrium of the design forces.

$$0.87\, (A_p f_p + A_{st} f_y) = 0.45\, f_{ck}\, b_f t$$

$$A_p = 0.518\, b_f t\, f_{ck} - A_{st} f_y$$

$$= 0.518\, b_f t\, f_{ck} \left(1 - 1.93\, p_s\, \frac{f_y}{f_{ck}} \right) \tag{9.3.6}$$

This is for a balanced section only, whereas for an unbalanced section, the area of the reinforcement is obtained by taking the moment about 'the centre of compression and it results into

$$A_p = \frac{1.15\, M_c}{f_p\,(d - 0.5t)} - \frac{A_{st}\,(d_s - 0.5t)\, f_y}{(d - 0.5t)\, f_p} \qquad (9.3.7)$$

9.4. Design for Shear and Serviceability

The design for shear and serviceability limit states in partially prestressed concrete beams is same as that discussed in Chapter 7 for type 3 structures. The design for shear can be treated as independent of the reinforcement so that no additional requirement or constraint exists over and above what was discussed in type 3 structures. The allowable bending compressive stresses in concrete are same as those discussed in Chapter 7. The allowable bending tension in concrete of the partially prestressed concrete structures is more than that in type 3 structures and these stresses are given in Table 9.4.1. The allowable stresses are proportional to the nontensioned reinforcement and are limited to a maximum value of $0.25\, f_{ck}$.

TABLE 9.4.1 Hypothetical allowable bending tension in concrete of partially prestressed concrete (stresses in MPa)

Construction	Crack width (mm)	Grades of concrete				
		M 30	M 35	M 40	M 45	M 50
A Minimum allowable for zero nontensioned steel						
(1) Pretensioned	0.1	–	–	4.1	4.4	4.8
	0.2	–	–	5.0	5.4	5.8
(2) Post-tensioned grouted	0.1	3.2	3.6	4.1	4.4	4.8
	0.2	3.8	4.4	5.0	5.4	5.8
(3) Pretensioned and the tendons are	0.1	–	–	5.3	5.8	6.3
positioned close to tension zone	0.2	–	–	6.3	6.8	7.3

The deflection computations and the limits are also same as those discussed in Chapter 7 for type 3 structures. The most critical point for discussion in deflection calculations is the effective moment of inertia. It is not only cumbersome to compute the neutral axis distance, but there are many uncertainties associated with the cracked section. The moment of inertia including the effect of the reinforcement in the cracked section is much higher than the corresponding value without the effect of the reinforcement. The percentage of total reinforcement in partially prestressed concrete structures will be around 0.5 to 1.5 and so its effect on the sectional inertia would be at least 20 to 50 per cent of the cracked concrete section. Sometimes the moment of inertia of the cracked section including the effect of the reinforcement could be more than the moment of inertia of the gross section without the reinforcements. Looking at the complexity of the problem along with the assumptions, it is recommended that the effective moment of inertia of the section could be selected as

$$I_c = \frac{I_{cr} + I_g}{2} \qquad (9.4.1)$$

where

I_g = gross moment of inertia

I_{cr} = moment of inertia of the cracked section at the limit state of collapse.

Increase the allowable stress at the rate of 4 MPa for 1 per cent nontensioned reinforcement subject to a maximum total of $0.25\,f_{ck}$.

TABLE 9 4.2 Design shear strengths in MPa

$100\,A_p/bd$	M 30	M 35	M 40 and above
0.25	0.37	0.37	0.38
0.50	0.50	0.50	0.51
0.75	0.59	0.59	0.60
1.00	0.66	0.67	0.68
1.25	0.71	0.73	0.74
1.50	0.76	0.78	0.79
2.00	0.84	0.86	0.88

9.5. Design Examples

Example 9.5.1 A simply supported beam of effective span 10 m is subjected to superimposed dead and live loads of 8 and 12kN/m respectively. Design a bonded post-tensioned prestressed concrete beam with 0.5 per cent nontensioned reinforcement.

Design data

Effective span $= L = 10$ m

Superimposed dead load $= w_{sd} = 8$ kN/m

Live load $= W_l = 12$ kN/m

Characteristic strengths and allowable stress in MPa are:

$$f_{ci} = 30, f_{ck} = 35, f_y = 415, f_p = 1500$$

$$\sigma_{aci} = 15.36, \sigma_{ace} = 11.8 \text{ and crack width} = 0.1 \text{ mm}$$

Allowable hypothetical tensile stresses for 0.5 per cent nontensioned steel are:

$$\sigma_{ati} = 4.4 + 2 = 6.4 = \sigma_{ate}$$

Design for bending moment and collapse limit

For the purpose of preliminary design, assume the self-weight of the beam in the range of

$$w_g = \frac{\gamma L^2}{480} = \frac{25(100)}{480} = 5.2 \text{ } kN/m$$

Let $w_g = 6$ kN/m, then the total load is:

$$w_t = w_g + w_{sd} + w_l = 6 + 8 + 12 = 26 \text{ } kN/m$$

The limiting critical bending moment with a partial safety factor of 1.5 applied to loads is

$$M_c = \gamma_f \frac{w_t L^2}{8} = \frac{1.5(26)(100)}{8} = 487.5 \ k\text{Nm}.$$

The design coefficients for balanced section are taken from Table 7.3.1 and they are:

$$K = 0.128 \ \text{N/mm}^2$$

$$x_u = 0.436 \ d; \ j = 0.82 \ \text{and} \ p_0 = 0.18$$

The design criterion for moment is that the resisting moment capacity should be more than the collapse moment and it can be expressed as

$$M_r = bd^2 \ f_{ck} \ (K + 0.87 p_s) \left(\frac{d_s}{d} - 1 \right) \frac{f_y}{f_{ck}} \geq M_c = 487500 \ \text{Nm}$$

Let

$$b = 0.35 \ \text{m}$$

The values of p_s and d_s/d are selected as

$$p_s = \frac{0.5}{100} = 0.005$$

$$\frac{d_s}{d} = 1.05 \ \text{and} \ \frac{f_y}{f_{ck}} = \frac{415}{35} = 11.857$$

Therefore the substitution of various quantities in the moment expression gives

$$M_r = 0.128(0.35)(35)(10^6)(1 + 6.8(0.005)(0.05)(11.875)) \ d^2 = 1.60(10^6) \ d^2$$

This value should be more than the collapse moment

$$1.6 \ (10^6) \ d^2 \geq M_c = 487 \ 500 \ \text{Nm} \ \text{or}$$

$$d \geq 0.552 \ \text{m}$$

The area of the pretensioned steel is given by Eq. 9.2.13

$$A_p = \left(p_o \frac{f_{ck}}{f_p} - \frac{p_s \ f_y}{f_p} \right) bd$$

$$= (0.18) \frac{(35)}{1500} - \frac{(0.005)(415)}{1500} (350)(552) = 549 \ \text{mm}^2$$

Use 15 numbers of 7 mm diameter wires. The actual area of pretensioned steel provided is

$$A_p = 15 \ (38.48) = 577.2 \ \text{mm}^2$$

Design for limit state of transfer condition

Since the effective depth is 0.552 m, the self-weight of the beam will be in the range of

$$w_g = 0.35(0.7)(25) = 6.125 \ k\text{N/m}$$

(in which $b = 0.35$ m, and $D = 0.7$ m assumed)
The bending moment due to the self-weight is

$$M_g = \frac{w_g L^2}{8} = 6.125 \frac{(100)}{8} = 76.56 \ k\text{Nm}$$

The axial tension force caused in the cable by the self-weight

$$F_g = \gamma_f \frac{M_g}{jd} = \frac{(0.9)76.56}{(8)0.82(0.557)} = 150.8 \ kN$$

where $\gamma_f = 0.9$ for dead load

The ultimate force in the cables is

$$P_u = \gamma_f A_p f_p = 1.2 \ (577.2) \frac{(1500)}{1000} = 1039 \ kN$$

where $\gamma_f = 1.2$

The ultimate compressive force on the bottom face at transfer limit state is:

$$F_u = P_u - F_g = 1039 - 150.8 = 888.2 \ kN$$

To ensure no crushing at the bottom at the transfer condition, we have

$$k_p f_{ci} \ 2(D - d) \ b = F_u$$

where $2 \ (D - d) \ b$ is the effective area under compression.

Then
$$D - d = \frac{F_u}{2k_p f_{ci} b} = \frac{0.8882}{2(0.67)(30)(0.35)} = 0.043$$

Use
$$d = 0.56 \ \text{m}, \ d_s = 1.05d = 0.59 \ \text{m}$$

$$D = d_s + \text{minimum cover} = d + 0.063$$

Use
$$D = 0.64 \ \text{m}$$

Design for serviceability limit state–allowable stresses

The properties of the section, etc. (all units are in MN and MPa) are:

$$A = 0.35(0.64) = 0.224 \ \text{m}^2$$

$$Z = 0.35(0.64)^2/6 = 0.02389 \ \text{m}^3$$

$$P_l = 0.7 \ f_p A_p = 0.7 \ \frac{(1500)(577.2)}{10^6} = 0.60606 \ \text{MN}$$

$$P_e = P_i - 225(A_p)(10^{-6}) = 0.47619 \ \text{MN}$$

$$e = 0.22 \ \text{m}$$

$$\frac{P_l}{A} = 2.70, \ \frac{P_e}{A} = 2.13$$

$$\frac{P_i e}{Z} = 5.58, \ \frac{P_e e}{Z} = 4.39$$

$$M_g = \frac{w_g L^2}{8} = \frac{6(100)}{8} = 75 \ kNm = 0.075 \ \text{MNm}$$

$$M_e = \frac{w_t L^2}{8} = \frac{26(100)}{8} = 325 \ kNm = 0.325 \ \text{MNm}$$

$$\frac{M_g}{Z} = \frac{0.075}{0.02389} = 3.14 \ \text{and} \ \frac{M_e}{Z} = \frac{0.325}{0.02389} = 13.6$$

Tne hypothetical stresses on the concrete section are: (all in MPa)

$$\sigma_{ci} = \frac{P_i}{A} + \frac{P_i e}{Z} - \frac{M_g}{Z}$$
$$= 2.70 + 5.58 - 3.14 = 5.14 < \sigma_{aci} = 15.36$$

$$\sigma_{ti} = -\frac{P_i}{A} + \frac{P_i e}{Z} = \frac{M_g}{Z}$$
$$= 2.70 + 5.58 - 3.14 = -0.26 < \sigma_{ati} = 6.4$$

$$\sigma_{ce} = \frac{P_e}{A} - \frac{P_e e}{Z} + \frac{M_e}{Z}$$
$$= 2.13 - 4.39 + 13.6 = 11.34 < \sigma_{ace} = 11.8$$
$$\sigma_{te} = -2.13 - 4.39 + 13.6 = 7.08 > \sigma_{ate} = 6.4$$

All the stresses except the tensile stress under the effective working load are less than the allowable values. To overcome the above problem increase the number of pretensioned wires from 15 to 18. Then

$$A_p = 18 \, (38.48) = 692.6 \text{ mm}^2$$

$$P_i = 0.7 \, A_p f_p = 8.7272 \text{ MN}$$

$$P_e = P_t - 225 \, A_p/10^6 = 5.713 \text{ MN}$$

As only two of the stresses are sensitive to this change, only these two can be verified and they are:

$$\sigma_{ce} = \frac{P_e}{A} - \frac{P_e e}{Z} + \frac{M_e}{Z}$$
$$= 2.55 - 5.26 + 13.6 = 10.89 < \sigma_{ace}$$
$$\sigma_{te} = -2.55 - 5.26 + 13.6 = 5.79 < \sigma_{ate} = 6.4$$

The serviceability limit state for deflection

The effective sectional moment of the cracked section is taken on the average of the cracked and gross sectional moment of inertia.

$$I = \frac{bD^3}{12} = 0.35 \frac{(0.64)^3}{12} = 0.007646 \text{ m}^4$$

$$I_{cr} = \frac{b \, x_u^3}{12} = 0.35 \frac{(0.436)(0.56)^2}{12} = 0.000425 \text{ m}^4$$

$$I_e = \frac{I + I_{cr}}{2} = 0.00404 \text{ m}^4$$

The effective upward load caused by the cables is

$$w_{pe} = \frac{8 \, P_e g}{L^2} = \frac{8(0.5713)(0.22)}{100} = 0.0100 \text{ MN/m}$$

The effective permanent load acting on the beam is

$$w_{ed} = w_d - w_{pe}$$

$$= (0.006 + 0.008) - 0.0100 = 0.004 \text{ MN/m}$$

The total deflection with a creep coefficient of 1.2 is

$$v_t = 1.2 \, v_{ed} + v_l$$

$$= (1.2 \, w_{ed} + w_l) \frac{5L^4}{384 \, E_c \, I_e}$$

$$= \frac{5(0.0168)(10000)}{384(5700\sqrt{35})(0.00404)} = 0.016 \text{ m}$$

The total allowable deflection is

$$V_{at} = \frac{L}{250} = 0.04 \text{ m}$$

The actual deflection is less than the allowable value. Similarly the camber and the live load deflections are within the limits.

Nontensioned steel

The area of the nontensioned steel was assumed as 0.5 per cent, therefore it is equal to

$$A_{st} = \frac{0.5}{100} \, bd = \frac{0.5(350)(560)}{100} = 980 \text{ mm}^2$$

Provide 5 numbers of 16 φ bars. Then

$$A_{st} \text{ (provided)} = 5 \, (201) = 1005 \text{ mm}^2$$

Limit state design for shear

The critical shear failure section is at distance *d* from the effective support and the statical shear force at this section is

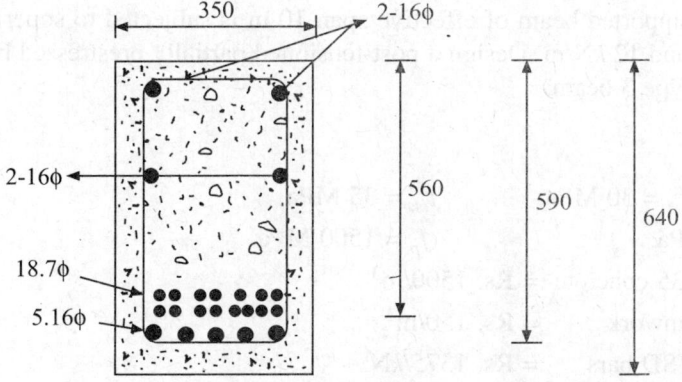

Figure 9.5.1 Example 9.5.1

$$V = \frac{w_t\, L}{2} - w_i d = 26(5 - 0.56) = 115.44\ kN$$

The cables are assumed to be laid in a parabolic cable profile with its centroid at the support coinciding with that of the section. Then the sag of the cable is

$$g = d - 0.5D = 0.56 - 0.32 - 0.22\ \text{m}$$

Ignore the contribution of the vertical component of the cable force in resisting the shear force, then the effective collapse shear force is

$$V_e = \gamma_f\, V = 1.5(115.44) = 173.16\ kN$$

The section near the support is subjected to only a nominal bending moment therefore it can be treated as uncracked. The shear resistance of the uncracked section is

$$V_r = 0.67b\, D\sqrt{\left(f_t^2 + 0.8\, f_t f_{cp}\right)}$$

where
$$f_{cp} = \frac{P_e}{A} = 2.55\ \text{MPa}$$

$$f_t = 0.24\,\sqrt{f_{ck}} = 1.42\ \text{MPa}$$

Then
$$V_r = 0.67(0.35)(0.46)\left(\sqrt{4.913}\right) = 0.33\ \text{MN}$$

As the shear capacity is more than the shear force, provide only nominal shear reinforcement. The minimum requirements demand

$$s_{vm} = \frac{0.87\, A_{sv}\, f_{yv}}{(0.4)\, b} = \frac{0.87(100)(415)}{0.4(350)} = 258\ \text{mm}$$

or
$$s_{vm} = 0.75\, d_s = 442\ \text{mm}$$

Use two-legged 8ϕ stirrups at 250 mm spacing.

Example 9.5.2 Comparison of type 3 and partially prestressed concrete beams.

A simply supported beam of effective span 10 m is subjected to superimposed dead and live loads of 8 and 12 kN/m. Design a post-tensioned partially prestressed beam and compare its cost with a type 3 beam.

Design data

$L = 10$ m, $f_{si} = 30$ MPa, $f_{ck} = 35$ MPa

$f_y = 415$ MPa, $f_p = 1500$ MPa

Cost of M 35 concrete = Rs. $1500/\text{m}^3$

Cost of formwork   = Rs. $150/\text{m}^2$

Cost of HYSD bars   = Rs. $1375/kN$

Cost of HTS wires   = Rs. $3000/kN$

(Costs include supply, placing, etc.)

The beams have been designed in examples 7.14.3 and 9.5.1 and the corresponding quantities are computed and given in Table 9.5.1.

TABLE 9.5.1   Quantity and cost comparisons

	b	D	Concrete	Steel(HTS) No Dia		HYSD No Dia		Length
Type 3	0.35	0.70	2.45	20,	7φ	40,	8φ	1.1m
Partial	0.35	0.64	2.24	18,	7φ	40,	8φ,	1.1m
						5,	16φ,	10 m

	Concrete	Formwork	HTS	HYSD	Total
Type 3	3675	2625	1800	213	8313
Partial	3360	2445	1620	1470	8895

Example 9.5.3   Axially pretentioned poles–Type partially prestressed

An electric line pole is 7.5 m in length and is subjected to 1800 N wind force at a height 6 m above GL. The pole projects 0.3 m above the level of the wires and embedded 1.2 m into the ground. Design a pretensioned pole such that it can withstand 1800 N wind force on the wires and (1800/4) N force along the direction of the wires. Non-tensioned steel on each face = 0.25 per cent.

Design data

Cantilever span = L = 7.5 – 1.2 – 0.3 = 6 m

Bending force = W_k = 1800 N

f_{ck} = 40 MPa, f_{ci} = 30 MPa and f_p = 1600 MPa

Design the pole for no tension condition rectangular section.

Let the width of the pole be 100 mm and constant throughout. The depth of the pole is the least at the top and increases towards the ground. The weight of the pole is acting along the axis of the pole, therefore it causes only axial stress and no bending stress. The axial stress due to the self-weight of the pole is small when compared with the bending stress so it is neglected in the calculations. The amount of steel provided on each face would be equal to that required to withstand the wind load in one direction.

The design moment on the beam

The maximum bending moment occurs at the GL of the pole and it is due to the wind load. The design bending moment with a partial safety factor of 1.5 is

$$M_u = \gamma f \, W_k L = 1.5 \, (1800) \, (6) = 16200 \text{ Nm}$$

Design for collapse limit state

The design coefficients for a balanced section with f_p = 1600 MPa are taken from Table 7.3.1 and they are:

$$K = 0.123, \ x_u = 0.414 \ d, \ j = 0.83 \text{ and } p_o = 0.17$$

Let
$$b = 0.1 \text{ m}, p_s = 0.25 \text{ per cent on each face and}$$

$$f_y = 415 \text{ MPa}$$

Crack width = 0.1 mm and d_s = 1.05 d

The allowable tensile stresses on the concrete for 0.5 per cent nontensioned reinforcement are:

$$\sigma_{ati} = \sigma_{ate} = 4.1 + 2 = 6.1 \text{ MPa}$$

The resisting moment capacity of the section is

$$M_r = K \, bd^2 \, f_{ck} \left(1 + \frac{0.87}{K} \, p_s \left(\frac{d_s}{d} - 1 \right) \frac{f_y}{f_{ck}} \right)$$

Substituting the appropriate quantities, we get

$$M_r = 0.123(0.1)(40) \left(1 + \frac{0.87}{0.123} \, (0.0025) \, (0.05) \, (415) / 40 \right) d^2$$

$$= 0.4965 \ d^2 \text{ MNm}$$

Equating the above to the collapse moment, we have

$$0.4965 \ d^2 = 0.0162 \text{ MNm}$$

or
$$d = 0.181 \text{ m}$$

The area of pretensioned steel is

$$A_p = \left(p_o \frac{f_{ck}}{f_p} - p_s \frac{f_y}{f_p} \right) bd$$

$$= (0.17(40) - 0.0025 \, (415)) \, (100)(181)/1600$$

$$= 61 \text{ mm}^2$$

Provide 3 numbers of 5 mm wirer on each face then

$$A_p \text{ (provided)} = 3 \, (19.6) = 58.8 \text{ mm}^6$$

Design for limit state at transfer

$$f_{ci} = 30 \text{ MPa}$$

$$P_u = \gamma_f \, (2 \, A_p) f_p = 2.4 \, \frac{(58.8) \, (1600)}{10^6} = 0.2258 \text{ MN}$$

Minimum area of the concrete needed to withstand the force is

$$b \, D_{\min} = \frac{P_u}{k_p f_{ci}} = \frac{0.2258}{0.67(30)} = 0.0112 \text{ m}^2$$

or $\qquad\qquad D_{min} = 0.112$ m

Use size of the section at top of the pole as 100 by 125 mm.

Limit state design–shear failure

The effective shear force which is constant over the length with 1.5 as load factor is

$$V_e = 1.5 \ (1800) = 2700 \ \text{N} = 0.0027 \ \text{MN}$$

The shear capacity is matched near the top of the pole where the section is the least and shear force is the maximum

$$f_t = 0.24 \ \sqrt{f_{ck}} = 1.51 \ \text{MPa}$$

$$f_{cp} = \frac{P_e}{A} = \frac{2A_p \ (0.6f_p)}{b \ D_{min}} = \frac{2(58.8) \ (0.6) \ (1600)}{1600 \ (125)} = 9.03 \ \text{MPa}$$

The shear capacity is

$$V_e = k_p \ bD \ \sqrt{(f_1^2 + 0.8 \ f_{cp} \ f_t)}$$

$$= 0.67 \ (0.1) \ (0.125) \sqrt{1.51^2 + 0.8(9.03)(1.51)}$$

$$= 0.0304 \ \text{MN}$$

The shear capacity of the pole at the top is about 10 times the shear force. Therefore it is not necessary to check for shear at the base where the section is likely to be a cracked one with an effective depth of 0.181m. There is no need for transverse reinforcement either.

Serviceability limit state–allowable stresses

The allowable stresses are:

$$\sigma_{aci} = 0.51 \ f_{ci} = 15.3 \ \text{MPa}$$

$$\sigma_{ace} = 0.323 \ f_{ck} = 12.92 \ \text{MPa}$$

$$\sigma_{ati} = \sigma_{ate} = 6.1 \ \text{MPa}$$

$$P_i = 0.7 \ f_p \ (2A_p) = 0.7(1600) \ \frac{(2)(58.8)}{10^6} = 0.1317 \ \text{MN}$$

$$P_e = P_i - 225(2A_p) = 0.1052 \ \text{MN}$$

The overall depth of the section is

$$D = d_s + 35 \ \text{mm} = 1.05d + 30 = 225 \ \text{mm}$$

$$A = bD = (0.1)(0.225) = 0.0225 \ \text{m}^2$$

$$Z = \frac{bD^2}{6} = 0.000844 \ \text{m}^3$$

$$\frac{P_i}{A} = \frac{0.1317}{0.0225} = 5.85 \text{ MPa}$$

$$\frac{P_e}{A} = \frac{0.1052}{0.0225} = 4.68 \text{ MPa}$$

$$\frac{M}{Z} = \frac{WL}{Z} = \frac{0.0018\,(6)}{0.000844} = 12.8 \text{ MPa}$$

$$\sigma_{ct} = 5.85 \text{ MPa}$$

The axial compressive stress at the top section is

$$\sigma_{cl} = \frac{P_i}{A} = \frac{0.1317}{0.0112} = 15.3 \text{ MPa}$$

$$\sigma_{ce} = 4.68 + 12.8 = 17.48 \text{ MPa} > \sigma_{ace} = 12.92$$

$$\sigma_{te} = 4.68 + 12.8 = 7.12 > \sigma_{ate} = 6.1$$

Redesign the section. The design criterion is

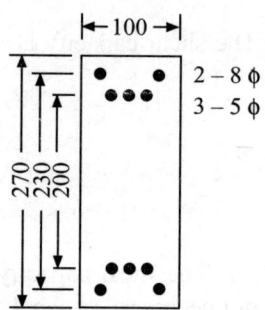

$$\sigma_{ce} = \frac{P_e}{A} + \frac{M}{Z} \le \sigma_{ace}$$

$$\sigma_{te} = \frac{P_e}{A} + \frac{M}{Z} \le \sigma_{ate}$$

From the above two inequalities, we have

$$Z \ge \frac{2M}{\sigma_{ace} + \sigma_{ate}} = \frac{2(0.0018)(6)}{19.02} = 0.001136 \text{ m}^3$$

$$P_e \ge \frac{A}{2}\,(\sigma_{ace} - \sigma_{ate}) = bD(3.41) \text{ MN}$$

Simplification gives

$$\frac{bD^2}{6} = Z = 0.001136 \text{ m}^3$$

or $D = 0.26 \text{ m}$

and $P_e = 3.41\,(b)\,(D) = 3.41\,(0.1)(0.26) = 0.089 \text{ MN}$

Select $D = 270 \text{ mm and } P_e = 0.1052 \text{ MN}$

$$\frac{P_e}{A} = \frac{0.1052}{0.027} = 4.9 \text{ MPa}$$

$$\frac{M}{Z} = \frac{0.0018\,(6)\,(6)}{(0.1)\,(0.27)\,(0.27)} = 8.9 \text{ MPa}$$

$$\sigma_{ce} = 3.9 + 8.9 = 12.8 < \sigma_{ace} = 12.92$$

$$\sigma_{te} = -3.9 + 8.9 = 5 < \sigma_{ate} = 6.1$$

HYSD reinforcement is

$$A_{st} = p_s\, bd = 0.005\,(100)(270 - 40) = 115 \text{ mm}^2$$

Provide 2 numbers of 8 mm bars on each face.

Example 9.5.4   Cost comparisons of pole

Compare the dimensions and cost of an electric pole of 7.5 m in total length carrying a lateral load of 1800 N at 6 m above ground level designed by :

(a) Working stress design;

(b) Ultimate strength design;

(c) Limit states design as type 1, type 2 and type 3; and

(d) Partially prestressed LSD method.

The unit cost of the supplying, placing etc., of the materials are:

Cost of M 35 concrete   = Rs. $1500/\text{m}^3$

Cost of M 40 concrete   = Rs. $1625/\text{m}^3$

Cost of shuttering, formwork   = Rs. $150/\text{m}^3$

Cost of HTS work   = Rs. $3000/k$N

The example has been worked out by different methods. The comparison of the dimensions arrived by different methods is given in Table 9.5.2.

TABLE 9.5.2   Quantity comparisons of pole ($b = 0.1$ m)

Method of design	Depth = D		Concrete		HTS	HYSD
	Top	Bottom	Vol.	Form	No. Dia	No. Dia
1. WSD	0.152	0.275	0.15	4.703	14, 5φ	
2. LSD–type 1	0.175	0.32	0.1856	4.463	12, 5φ	
3. TlSD–type 2	0.115	0.29	0.1519	3.788	8, 5φ	
4. LSD–type 3	0.110	0.28	0.1463	3.675	12, 4φ	
5. LSD–PPSC	0.125	0.27	0.1383	3.03	6, 5φ	4, 6φ
6. USD	0.160	0.22	0.1425	2.95	12, 5φ	

Vol = Volume in m^3, Form = formwork in m^2.

TABLE 9.5.3   Cost comparisons

Method of Design	Cost in Rupees			
	Concrete	Formwork	Steel	Total
1. WSD	240	705	380	1425
2. LSD–type I	303	670	414	1387
3. LSD–type 2	248	568	276	1092
4. LSD–type 3	238	553	264	1055
5. LSD–PPSC	225	455	270	950
6. USD	213	440	414	1067

Example 9.5.5 Thirty metres span simply supported beam partially prestressed

A beam of an effective span of 30 m is subjected to a live load of 20 kN/m. The width and thickness of the top flange are given as 1.5 m and 0.2 m respectively. Design a post-tension beam with M45 concrete with 0.5 per cent nontensioned HYSD bars.

Design data

$$\text{Simply supported span} \quad = L = 30 \text{ m}$$
$$\text{Live load} \quad = w_1 = 20 \text{ } kN/m$$

Assume the self-weight of the beam in the range of $W_1/40$ in which $W_1 =$ is the total load in kN on the span and the self-weight is in kN/m.

$$\text{Self-weight} = w_g = \frac{W_1}{40} = \frac{20(30)}{40} = 15 \text{ } kN/m$$

$$\text{Total working load} \quad\quad = 35 \text{ } kN/m$$
$$\text{Quality of the concrete} \quad f_{ck} \quad = 45 \text{ MPa}$$
$$\text{Quality of steel (HTS)} \quad = f_p \quad = 1500 \text{ PMa}$$
$$\text{Quality of HYSD bars} \quad = f_y \quad = 415 \text{ MPa}$$
$$\text{Strength of concrete at transfer} \quad = f_{ct} = 40 \text{ MPa}$$

The allowable stresses in concrete with 0.5 per cent nontensioned steel for 0.1 mm crack width are:

$$\sigma_{aci} = 0.455 \, f_{ci} = 18.2 \text{ MPa}$$

$$\sigma_{ace} = 0.455 \, f_{ck} = 16.88 \text{ MPa}$$

$$\sigma_{ati} = \sigma_{ate} = 6.4 \text{ MPa}$$

The design coefficients for the balanced section are taken from Table 7.3.1 and they are

$$K = 0.128, \, p_0 = 0.18, \, k_3 = 0.436$$

Limit state of strength

The collapse moment with a partial safety factor of 1 5 applied to the loads is

$$M_c = \gamma_f \frac{wL^2}{8}$$

$$= 1.5(35) \frac{(30)(30)}{8} = 5906.25 \text{ } kNm = 5.9063 \text{ MNm}$$

The moment capacity of the flanged beam is equated to the design moment, the depth of the section from Eq. 9.3.4 is

$$d > \frac{0.5t}{\alpha} + \frac{M_c}{0.45 f_{ck} b_f t \, (1 + 2p_s \, (c_s - 1) \, f_y / f_{ck})}$$

in which
$$p_s = 0.5 \text{ per cent} = 0.005$$
$$b_f = 1.5 \text{ m}, \qquad t = 0.2 \text{ m}$$
Let
$$c_s = 1.05$$
$$\alpha = (1 + 2p_s(c_s - 1) f_y / f_{ck})$$
$$= (1 + 0.01\,(0.05)\,415/45) = 1.005$$

The value of d is

$$d \geq \frac{0.1}{1.005} + \frac{5.9063}{0.45(45)(1.5)(0.2)(1 + 0.01(0.05)9.222)} = 1.07 \text{ m}$$

(The value of x_u will be larger than t).

In most cases the working load stresses normally govern the design, therefore use $d - 1.24$ and the cover beyond the centre of steel as 0.21 m. Hence

$$d = 1.24 \text{ m and } h = 1.24 + 0.21 = 1.45 \text{ m}$$

The area of tensioned steel is given by Eq. 9.3.7 as

$$A_p = \frac{1.15 M_c}{f_p(d - 0.5t)} - \frac{(d_s - 0.5t) f_y A_{st}}{(d - 0.5t) f_p}$$

in which
$$d_s = 1.05\, d = 1.302 \text{ m}$$
$$A_{si} = p_s\, b_f t = 0.005(1.5)(0.2)(10^6) = 1500 \text{ mm}^2$$

Then we have
$$A_p = 3480 \text{ mm}^2$$

Provide 91 numbers of 7 mm wires, then the area of pretensioned steel is
$$A_p = 91(38.48) = 3502 \text{ mm}^2$$

Design for transfer condition

The ultimate force in the tendon with partial safety factor of 1.0 is
$$P_u = A_p f_p = 0.003502(1500) = 5.253 \text{ MN}$$

Approximate bending moment caused by the dead load is

$$M_g = \frac{w_g L^2}{8} = \frac{(0.015)(900)}{8} = 1.6875 \text{ MNm}$$

The tensile force caused by the self-weight is

$$F_g = \frac{\gamma_f M_g}{(d - 0.42 x_u)} = \frac{\gamma_f M_g}{0.82 d}$$

$$= \frac{(0.9)(1.6875)}{1.0168} = 1.4937 \text{ MN}$$

$$F_u = P_u - F_g = 5.2530 - 1.4937 = 3.7593 \text{ MN}$$

The area of the concrete needed to withstand the pretensioning is

$$A_{ct} = \frac{\gamma_m F_u}{k_p f_{ct}} = \frac{1.5(3.7593)}{0.67(40)} = 0.2104 \text{ m}^2$$

Depth of the web assumed below the GC of the steel is 0.21 m, therefore the area of the web which contributes to the resistance action is

$$A_{ctw} = b_w (2) (0.21) = 0.0672 \text{ m}^3$$

The area of the bottom flange overhangs (A_{cb}) should be equal to

$$0.2104 - 0.0672 = 0.1432 \text{ cm}^2$$

Let the thickness of the flange $t_b = 0.3$ m
   Then the overhang of the bottom flange is

$$b_b - b_w = \frac{A_{cb}}{t_b} = \frac{0.1437}{0.3} = 0.479 \text{ m}$$

or
$$b_b = 0.479 + 0.16 = 0.639$$

Use
$$b_b = 0.6 \text{ m}$$

Serviceability in allowable stresses

In prestressed concrete structures the allowable stress appear to control the design. The properties of the uncracked section are computed in Table 9.5.6. Figure 9.3.1 gives the notations. The member size3 obtained from the limit state of collapse are:

$$h = 1.45 \text{ m}$$
$$b_t = b_f = 1.5 \text{ m}; \, t_t - t = 0.2 \text{ m}$$

TABLE 9.5.4 Sectional properties (units in m)

Segment	A_i	y_i	$A_i y_i$	I_{ci}	$A_i(y_i - \dot{y}_t)^2$
(1)	(2)	(3)	(4)	(5)	(6)
Top flange	0.300	0.100	0.0300	0.00100	0.06769
Web	0.152	0.675	0.1026	0.01143	0.00152
Bottom flange	0.180	1.300	0.2310	0.00135	0.09461
Totals =	0.632		0.3636	0.01378	0.16382

$$b_b = 0.6 \text{ m}; \qquad t_b = 0.3 \text{ m}$$
$$b_w = 0.16 \text{ m}; \qquad h_w = h - t_b - t_t = 0.95 \text{ m}$$

The distance y_t is measured from the extreme top fibre while computing moment of areas, etc. in Table 9.5.4.

Columns 2, 3, 4 and 5 are computed first using the above data. A_i refers to the area of the segment of the section under consideration and it is obtained by multiplying the width by depth and placed in column 2.

I_{ci} = second moment of area of ith element about its own axis.

The distance of the cetitroid (y_t) of the section is computed first after taking the first moment of areas and it is measured from the top of the section: Using the values from the table, we have

$$y_t = \frac{\Sigma A_i y_i}{\Sigma A_i} = \frac{0.3636}{0.632} = 0.575 \text{ m}$$

$$y_b = h - y_t = 1.45 - 0.575 = 0.875 \text{ m}$$

The transformed moment of intertia of each segment is computed as

$$I_t = I_{ci} + A_i (y_t - y_i)^2$$

where $A_i (y_t - y_t)^2$ is computed in column (6) of Table 9.5.4

The moment of inertia of the section about its own *CG* is

$$I = \Sigma(I_{ci} + A_i (y_t - y_t)^2)$$
$$= 0.01378 + 0.16382 = 0.1776 \text{ m}^4$$

The moduli of the section about the top and bottom fibres are:

$$Z_t = \frac{I}{y_t} = \frac{0.1776}{0.575} = 0.3089 \text{ m}^3$$

$$Z_b = \frac{I}{y_b} = \frac{0.1776}{0.875} = 0.203 \text{ m}^3$$

The self-weight of the beam is

$$W_g = \gamma A = 25 \ (0.632) = 15.8 \ kN/m$$

The value assumed was 15 *k*N/m as against 15.8 *k*N/m. So the moment calculations are revised. The bending moments due to self-weight and service load are:

$$M_g = \frac{w_g L^2}{8} = \frac{15.8(900)}{8} = 1777.5 \ kNm$$

$$M_e = \frac{(w_g + w_l)L^2}{8} = 35.8 \ \frac{(100)}{8} = 4027.5 \ kNm$$

The initial prestressing force is

$$P_i = A_p \ (0.7f_p) = 0.003502 \ (0.7)(1500) = 3.6771 \text{ MN}$$

Allowing a loss of prestress of 240 MPa in the post-tensioned construction, the effective prestressing force is

$$P_e = P_i - A_p \ (240) = 3.6771 - 0.8405 = 2.8366 \text{ MN}$$

The eccentricity of the steel at mid span is

$$e = d - y_t = 1.24 - 0.575 = 0.665 \text{ m}$$

The actual compressive and tensile stresses in the concrete section at the initial load conditions are:

$$\sigma_{ci} = \frac{P_i}{A} + \frac{P_i e}{Z_b} - \frac{M_g}{Z_b}$$

$$= \frac{3.6771}{0.632} + \frac{3.6771\,(0.665)}{0.203} - \frac{1.7775}{0.203}$$

$$= 5.82 + 12.05 - 8.76 = 9.11 < \sigma_{aci} = 18.2 \text{ MPa}$$

$$\sigma_{ti} = -\frac{P_i}{A} + \frac{P_i e}{Z_t} - \frac{M_g}{Z_t}$$

$$= -\frac{3.632}{0.632} + \frac{3.6771\,(0.665)}{0.3089} - \frac{1.7775}{0.3089}$$

$$= -5.82 + 7.92 - 5.75 = -3.65 < \sigma_{ati} = 6.4 \text{ MPa}$$

The compressive and tensile stress at the working load condition on the concrete are:

$$\sigma_{ce} = \frac{P_e}{A} - \frac{P_e e}{Z_t} + \frac{M_g}{Z_t}$$

$$= \frac{2.8366}{0.632} - \frac{2.8366\,(0.665)}{0.3089} + \frac{4.0275}{0.3089}$$

$$= 4.49 - 6.11 + 13.04 = 11.42 < \sigma_{ace} = 16.88 \text{ MPa}$$

$$\sigma_{te} = -\frac{P_e}{A} - \frac{P_e e}{Z_t} + \frac{M_g}{Z_b}$$

$$= -4.49 - \frac{2.8366\,(0.665)}{0.203} + \frac{4.0275}{0.203}$$

$$= -4.49 - 9.29 + 19.84 = 6.06 < \sigma_{ate} = 6.4 \text{ MPa}$$

Serviceability in deflections

The prestressing cables are laid in parabolic profile with zero eccentricity at the end section. Therefore the sag of the cable is

$$g = e = 0.665 \text{ m}$$

The upward balancing forces caused by the cable during initial and effective load conditions are:

$$w_{bi} = \frac{8P_i g}{L^2} = \frac{8P_i e}{L^2} = 0.0217 \text{ MN/m}$$

$$w_{be} = \frac{8P_e g}{L^2} = \frac{8P_e e}{L^2} = 0.0168 \text{ MN/m}$$

The effective loads at the initial and effective permanent load conditions are:

$$w_i = w_g - w_{bi} = 0.0158 - 0.0217 = -0.0059 \text{ MN/m}$$

$$w_{ed} = w_g - w_{be} = 0.0158 - 0.0168 = -0.001 \text{ MN/m}$$

$$A_{st} = 24 - 12\phi, \; A_p = 91 - 7\phi$$

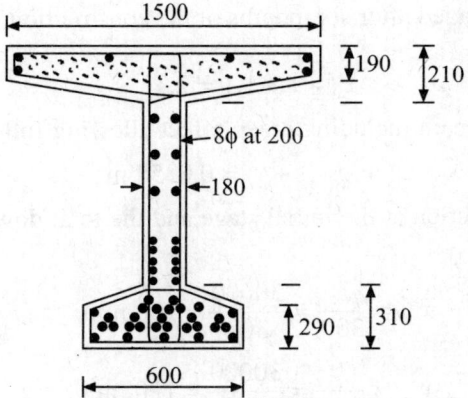

Figure 9.5.2   Example 9.5.5

The negative sign indicates the load is acting upwards. The Young's modulus of the concrete is

$$E_c = 5700 \sqrt{45} = 38237 \text{ MPa}$$

As discussed earlier, the moment of inertia of the cracked section at limit state of collapse about the NA is calculated. The neutral axis distance from the top fibre is

$$x_u = 0.436 \, d = 0.541 \text{ m}$$

The moment of inertia of the cracked section about NA is

$$I_{cr} = \frac{1.5(0.2)^3}{2} + 1.5 \, (0.2) \, (0.541 - 0.1)^2$$

$$+ \frac{0.16 \, (0.541 - 0.2)^3}{3} = 0.0655 \text{ m}^4$$

The effective moment of inertia of the section is

$$I_e = \frac{I + I_{cr}}{2} = \frac{0.1776 + 0.0655}{2} = 0.1216 \text{ m}^6$$

The short range deflections at the initial and effective permanent load conditions are:

$$v_t = \frac{5 w_t L^4}{384 \, E_c I_e}$$

$$= \frac{5(-0.0059) \, (900) \, (900)}{384 \, (38237) \, (0.1216)} = -0.013 \text{ m}$$

$$v_{ed} = \frac{5 w_{ed} L^4}{384 \, E_c I_e} = \frac{-5(0.001) \, (810000)}{384(38237) \, (0.1216)} = -0.002 \text{ m}$$

Deflection due to the live load is

$$v_l = \frac{5 w_l L^4}{384 \, E_c I_e} = \frac{5(0.02)(810000)}{384(38237)(0.1216)} = 0.04 \text{ m}$$

Assuming the structure is loaded after six months of the construction, the creep coefficient can be taken as

$$c_c = 1 + 1.1 = 2.1$$

The total deflection of the beam including creep effect allowing full live load is

$$v_i = c_c v_{ed} + v_l = 0.0358 \text{ m}$$

The allowable upward deflection at the initial stage and the total downward deflection during working load are:

$$v_{ai} = \frac{L}{300} = \frac{30000}{300} = 100 \text{ mm}$$

$$v_{ae} = \frac{L}{250} = \frac{30000}{250} = 120 \text{ mm}$$

The actual deflections are less than the allowable values.

Design for shear strength

The critical shear plane is at a distance d from the support and the shear force at that section is

$$V = w_t \, (L/2 - d)$$

$$= 35.8 \, (15 - 1.24) = 492.61 \text{ } kN = 0.49261 \text{ MN}$$

The design shear force with partial load factor of 1.5 is

$$V_d = 1.5 \, V = 0.739 \text{ MN}$$

Shear force balanced by the cable

$$= V_{pe} = w_{be} \, (L/2 - d) \, \gamma f$$

$$= 0.0162(15 - 1.24)(0.9) = 0.2 \text{ MN}$$

Net design shear force is

$$V_e = V_d - V_{pe} = 0.539 \text{ MN}$$

The section where the critical shear force is acting is near the support at which point the bending moment is nominal. The section can be treated as uncracked and the shear resistance of the section is

$$V_r = 0.67 \, b_w D \sqrt{\left(f_i + 0.8 \, f_{cp} \, f_t \right)}$$

$$= 0.67 \, (0.16) \, (1.45) \, \sqrt{(1.61 + (0.8) \, (4.49) \, (1.61))}$$

$$= 0.4498 \text{ MN}$$

Select two-legged 8 mm bars as stirrups.

Then

$$A_{sv} = 2(50) = 100 \text{ mm}^2$$

The maximum spacing of the stirrups is

$$S_{vm} = \frac{A_{sv} f_y}{0.4 b_w} = \frac{100(415)}{0.4(160)} = 648 \text{ mm}$$

The required spacing

$$s_v = \frac{0.87 \, A_{sv} f_y d}{V_e - V_r} = 501 \text{ mm}$$

Provide two-legged 8 stirrups at 200 mm spacing

The area of nontensioned steel is given by

$$A_{st} = 0.005, \, b_t \, t = 1500 \text{ mm}^2$$

Use 6 numbers of 20 mm HYSD bars, the actual area of HYSD bars is

$$A_{st} \text{ (provided)} = 6(314) = 1884 \text{ mm}^2$$

The depth of the nontensioned steel is

$$d_s = 1.05, \, d = 1.302 \text{ m}$$

$$\text{extra nominal } 18\text{--}12 \, \phi$$

Example 9.5.6 Cost comparisons of 30 m span of type 3 and partially prestressed concrete beams

A beam of effective span 30 m was designed as a type 3, partially prestressed beam. The design details of the two solutions are given in Table 9.5.5.

TABLE 9.5.5 Comparison of details

Detail (units)	Type 3	Partially prestressed
b_f (m)	1.5	1.5
t (m)	0.2	0.2
b_b (m)	0.65	0.6
t_b (m)	0.3	0.3
h (m)	1.40	1.45
A_c (m^2)	0.639	0.632
A_s (mm^2)		3818
A_p (mm^2)	4002	3502
σ_{ci} (MPa)	11.43	9.11
σ_{ti} (MPa)	-3.2	-3.65
σ_{ce} (MPa)	11.49	11.42
σ_{te} (MPa)	4.14	6.06

TABLE 9.5.6 Cost comparisons

Type	Cost of Rupees			
	Concrete	Formwork	Steel	Total
Type 1	33588	33058	44466	111112
Type 2	31153	32143	28080	91376
Partial	30810	30510	24742	98426
			+12364	

Unit cost of materials including supplying, placing, etc., are:

Cost of concrete (M 45)   = Rs. 1625/m^3

Cost of formwork   = Rs. 150/m^2

Cost of HTS   = Rs. 3000/kN

Cost of HYSD   = Rs. 1375/kN

Table 9.5.6 gives the comparison of costs based on examples 7.16.2 and 9.5.5.

Example 9.5.7   Design of long span beam

Design a 50 m, simply supported, partially post-tensioned prestressed concrete beam with M 45 concrete with the help of the following data.

Design data:

Span = L = 50 m

w_g = 20 kN/m = 0.02 MN/m; $W_s = w_s L = 0.02(50)$

$f_{ci} = f_{ck}$ = 45 MPa and f_p = 1500 MPa

Percentage of non-tensioned HYSD steel = 0.5. The allowable stresses (in MPa) for 0.1 mm crack width are:

$$\sigma_{aci} = 20.475; \ \sigma_{ace} = 16.88$$

$$\sigma_{ati} = \sigma_{ate} = 4.4 + 2 = 6.4$$

Also assume $b_f = b_t = 1.6$ m; $t_t = t = 0.3$ m

Let $b_w = 0.18$ m

Limit state of strength in flexure

Let the total self-weight of the beam be assumed as

$$W_g = \frac{W_s L^2}{2200} = \frac{0.020(50)(2500)}{2200} = 1.136 \text{ MN}$$

Total load = $W = W_g + W_s = 1.136 + 0.02(50) = 2.136$ MN

The collapse moment on the beam with a partial safety factor of 1.5 is

$$M_c = \gamma_f \frac{WL}{8} = 1.5(1.136 + 1.0)\frac{(50)}{L} = 20.025 \text{ MNm}$$

Let the depth of the nontensioned steel is at $1.05d$. The effective depth of prestressed steel is (for $c_s = 1.05$)

$$d > \frac{0.5t}{\alpha} + \frac{M_c}{0.45 \, f_{ck} b_f t \left(1 + 2p_s (c_s - 1)\dfrac{f_y}{f_{ck}}\right)} = 2.20 \text{ m}$$

Use $\qquad d = 2.21$ m

Then $\qquad d_s = 1.05(2.21) = 2.32$ in

Let $\qquad h = 2.32 + 0.08 = 2.4$ m

$$A_{st} = p_s b_f t = 0.005 \ (1.6) \ (0.3) = 0.0024 \ \text{m}^2 = 2400 \ \text{mm}^2$$

The area of tbe pretensioned steel is given by

$$A_p = \frac{1.15 \ M_c}{f_p(d - 0.5t)} - \frac{(d_s - 0.5t)f_y A_{st}}{f_p(d - 0.5t)}$$

$$= \frac{1.15(20.025)(10^6)}{1500(2.06)} - \frac{(2.17)(415)(2400)}{1500(2.06)}$$

$$= 6670 \ \text{mm}^2$$

Provide 174 numbers of 7 mm wires, the actual area of the tensioned steel is

$$A_p = 174(38.48) = 6695 \ \text{mm}^2$$

Also provide 8 numbers of 20 nontensioned HYSD bars; then

$$A_{st} = 8(314) = 2512 \ \text{mm}^2$$

Design for transfer condition

The ultimate tensile force in the pretensioned steel is

$$P_u = A_p f_p = 0.006695(1500) = 10.0425 \ \text{MN}$$

The tensile force caused by the dead load moment with partial safety factor of 0.9 is

$$F_g = \frac{0.9 M_g}{(d - 0.5t)} = \frac{0.9 W_g L}{8(d - 0.5t)}$$

$$= \frac{0.9(1.136) \ (50)}{2.06 \ (8)} = 3.102 \ \text{MN}$$

$$F_u = P_u - F_g = 10.0425 - 3.102 = 6.9405 \ \text{MN}$$

The compression area needed to withstand the pretensioned force at transfer condition is

$$A_{ct} = \frac{F_u}{k_p f_{ci}} = \frac{6.9405}{0.67 \ (45)} = 0.2302 \ \text{m}^2$$

The overall depth of the section is

$$h = 2.4 \ \text{m}$$

The depth of the web below the level of steel is

$$h - d = 2.4 - 2.21 = 0.19 \ \text{m}$$

Effective area of the web in resisting the axial force is

$$A_{ctw} = b_w \ 2 \ (h - d) = 0.18 \ (2) \ (0.19) = 0.0684 \ \text{m}^2$$

Area of the overhang flanges needed is

$$A_{cb} = A_{ct} - A_{ctw} = 0.2302 - 0.0684 = 0.1618 \text{ m}^2$$

Provide $\qquad b_b = 1.0 \text{ m}; t_b = 0.26 \text{ m}$

then $\qquad (b_b - b_w) t_b = 0.82 (0.26) = 0.2131 \text{ m}^2$

Serviceability limit state design-stresses

The different segment sizes of the I-section are (all in m units)

$$b_f = 1.6, t = 0.3, A_{tf} = 0.48 \text{ m}^2$$
$$b_b = 1.0, t_b = 0.26, A_{bf} = 0.26 \text{ m}^2$$
$$h_w = h - t - t_b = 1.84, A_w = 1.84(0.18) = 0.3312 \text{ m}^2$$

Table 9.5.7 gives the properties of the section. The centroidal distance from the top is

$$y_t = \frac{\Sigma A_i y_i}{\Sigma A_i} = \frac{1.0663}{1.0712} = 0.995 \text{ m}$$

Then $\qquad y_b = h - y_t = 1.405 \text{ m}$

The gross moment of inertia of the section is

$$I = \Sigma I_{ci} + \Sigma A_i (y_i - y_t)^2 = 0.88066 \text{ m}^4$$

$$Z_t = \frac{I}{y_t} = 0.8851 \text{ m}^3$$

TABLE 9.5.7   Sectional properties (units are in m)

Segment	A_i	y_i	$A_i y_i$	I_{ci}	$A_i(y_i - y_t)^2$
Top flange	0.4800	0.15	0.0720	0.00360	0.34273
Web	0.3312	1.12	0.4041	0.09344	0.01677
Bot. flange	0.2600	2.27	0.5902	0.00146	0.42266
Total	1.0712		1.0663	0.0985	0.78216

$$Z_b = \frac{I}{y_b} = 0.6268 \text{ m}^3$$

Self-weight of the beam is

$$W_g = A = 25 \ (1.0712) = 26.78 \ kN/m$$

The bending moments due to self-weight and working load are:

$$M_g = \frac{w_g L^2}{8} = \frac{0.02678(2500)}{8} = 8.3688 \text{ MNm}$$

$$M_e = \frac{w L^2}{8} = \frac{0.04678(2500)}{8} = 14.6188 \text{ MNm}$$

The initial and effective prestressing forces using 220 MPa as the loss of prestress are:

$$P_i = A_p \, (0.7 f_p) = 0.00669(0.7)(1500) = 7.0298$$

$$P_e = P_i - 220 \, A_p = 7.0298 - 220 \, (0.006695) = 5.5569 \text{ MN}$$

The eccentricity of the cables at mid span is

$$e = d - y_t = 2.21 - 0.995 = 1.215 \text{ m}$$

$$P_i e = 7.0298(1.215) = 8.5412 \text{ MNm}$$

$$P_e e = 5.5569 \, (1.215) = 6.7516 \text{ MNm}$$

The stresses in MPa caused in the concrete at the initial and effective load stages are:

$$\sigma_{ci} = \frac{P_i}{A} + \frac{P_i e}{Z_b} - \frac{M_g}{Z_b}$$

$$= \frac{7.0298}{1.0712} + \frac{8.5412}{0.6268} - \frac{8.3688}{0.6268}$$

$$= 6.56 + 13.63 - 13.35 = 6.84 < \sigma_{aci} = 20.4$$

$$\sigma_{ti} = \frac{P_i}{A} + \frac{P_i e}{Z_t} - \frac{M_g}{Z_t}$$

$$= -6.82 + 9.65 - 9.45 = -6.62 < \sigma_{ati}$$

$$\sigma_{ce} = \frac{P_e}{A} - \frac{P_e e}{Z_t} + \frac{M_e}{Z_t}$$

$$= 5.18 - 7.63 + 16.52 = 14.07 < \sigma_{ace} = 16.88$$

$$\sigma_{te} = \frac{P_e}{A} - \frac{P_e e}{Z_b} + \frac{M_e}{Z_b}$$

$$= -5.18 - 10.77 + 23.32 = 7.37 > \sigma_{ate} = 6.4$$

The tensile stress at the working load condition exceeds the permissible value; therefore the design needs modification. An extra pretensioning will improve the situation. The total compressive force caused by prestressing is 15.95 MPa whereas the total tension is 23.32 MPa. This means the number of wires must be proportionally increased to have a net stress of 6.4 MPa.

The number of wires required are:

$$N = 174 \, (23.32 - 6.4)/15.95 = 185$$

The area of the tensioned steel is

$$A_p = 185 \, (38.48) = 7119 \text{ mm}^2$$

The pretensioned forces are

$$P_i = 0.7 \, f_p A_p = 0.7 \, (1500) \, (0.007119) = 7.475 \text{ MN}$$

$$P_e = P_i - 220 \, A_p = 7.475 - 220 \, (0.007119) = 5.909 \text{ MN}$$

$$\frac{P_i}{A} = 6.98; \frac{P_e}{A} = 5.52$$

$$\frac{P_i e}{Z_b} = 14.49; \frac{P_i e}{Z_b} = 10.26$$

$$\frac{P_e e}{Z_t} = 11.45; \frac{P_e e}{Z_t} = 8.11$$

The stresses in the concrete are obtained by suitably substituting the above values in the expressions for stresses computed earlier.

$$\sigma_{ci} = 6.98 + 14.49 - 13.35 = 8.12 < \sigma_{aci}$$

$$\sigma_{ti} = 6.98 + 11.45 - 9.45 = -4.98 < \sigma_{ati}$$

$$\sigma_{ce} = 5.52 - 8.11 + 16.52 = 13.93 < \sigma_{ace}$$

$$\sigma_{te} = -5.52 - 11.45 + 23.32 = 6.35 < \sigma_{ate}$$

The stresses are within the limits.

Check for deflection

The cables are laid in a parabolic profile with zero eccentricity at the support ends of the beam. Then the sag of the cable is

$$g = e = 1.215 \text{ m}$$

The upward balancing force at the initial and effective load stages are:

$$w_{bd} = \frac{8 P_i g}{L^2} = \frac{8(7.475)(1.215)}{2500} = 0.029 \text{ MN/m}$$

$$w_{be} = \frac{8 P_e g}{L^2} = \frac{8(5.909)(1.215)}{2500} = 0.023 \text{ MN/m}$$

The net permanent loads at the initial and long term stages are (in MN/m)

$$w_l = w_g - w_{bi} = 0.02678 - 0.029 = -0.00222$$

$$w_{ed} = w_g - w_{be} = 0.02678 - 0.023 = 0.00378$$

$$E_c = 5700\sqrt{45} = 38237 \text{ MPa}$$

$$x_u = 0.436d = 0.964 \text{ m}$$

The moment of inertia of the cracked section about NA

$$I_{cr} = \frac{1.6(0.3)^3}{12} + 1.6(0.3)(0.964 - 1.15)^2 + \frac{0.18(0.064 - 0.3)^2}{3} = 0.3392 \text{ m}^4$$

The effective moment of inertia of the section in the cracked stage is

$$I_e = \frac{0.88066 + 0.3392}{2} = 0.61 \text{ m}^4$$

The short range camber in the beam at the initial stage is

$$v_i = \frac{5 w_i L}{384\, E_c\, I_e} = \frac{5(0.00222)\,(2500)\,(2500)}{384\,(38237)\,(0.61)} = 0.007\ \text{m}$$

whereas the allowable carrtber is $L/300 = 0.167$ m which is much higher than the actual value and hence safe. The permanent deflection with a creep coefficient of 2.1 is

$$v_{cd} = \frac{c_c\,(5 w_{ed}\,L^4)}{384\, E_c\, I_e} = \frac{2.1(5)\,(0.0038)\,(6250000)}{384(38237)(0.61)} = 0.028\ \text{m}$$

The deflection due to the live load is

$$v_l = \frac{5 W_i L^4}{384\, E_c\, I_e} = \frac{5(0.02)(625000)}{384(3823)(0.61)} = 0.070\ \text{m}$$

The total downward deflection is

$$v_t = v_{ed} + v_l = 0.098\ \text{m} = 98\ \text{mm}$$

The allowable deflection is

$$v_a = \frac{L}{250} = \frac{50}{250} = 0.2\ \text{m} = 200\ \text{mm}$$

The total deflection is less than the allowable value; therefore the deflection limitations are within the limits.

Design for shear

The critical shear plan is at a distance d from the support and the static shear force at this section with 1.5 m partial safety factor is

$$V = 1.5 w_t \left(\frac{L}{2} - d \right)$$

$$= 1.5[0.04678(25 - 2.21)] = 1.5693\ \text{MN}$$

The effective shear force is

$$V_e = V - 0.9\, V_{pe} = 1.5693 - 0.9\, w_{be}\, (L/2 - d) = 1.0975\ \text{MN}$$

The shear capacity of the uncracked section is ,

$$V_r = 0.67\, b_w\, D\sqrt{(f_1^2 + 0.8\, f_{cp} f_i)}$$

$$= 0.67(0.18)(2.4)\sqrt{(1.61^2 + 0.8(5.18)(1.61)} = 0.88\ \text{MN}$$

The shear capacity is less than the shear force. That is why transverse reinforcement be provided. Let two-legged 8 mm bars be provided; then the area and the spacing of the stirrups are:

$$A_{sv} = 2(50) = 100\ \text{mm}^2$$

$$s_v = \frac{0.87\, A_{sv}\, f_y d}{V_e - V_r} = \frac{0.87(100)(415)(2210)}{1097500 - 880000} = 367\ \text{mm}$$

374

The maximum spacing of the stirrups is 450 mm or 576 mm as computed below:

$$s_{vm} = \frac{A_{sv}\, f_y}{0.4 b_w} = \frac{100\,(415)}{0.4(180)} = 576 \text{ mm}$$

Provide two legged 8 mm bars at 450 mm c/c

The area of the non-tensioned reinforcement is

$$A_{st} = 0005,\, b_f\, t = 0.005\,(1600((300) = 2400 \text{ mm}^2$$

Provide 8 numbers of 20 ϕ HYSD bars.

$$A_{st} \text{ (provided)} = 8(314) = 2512 \text{ mm}^2$$

$$d_s = 1.05\; d = 2.32 \text{ m}$$

9.6. Comparison of Design Detail by Different Methods

Example 9.6.1 Comparison of design details of 50 m span simply supported beam subjected 20 kN/m; f_{ck} = 45 MPa, f_p = 1500 MPa.

The design has been carried by the following methods:

(1) Working stress method
(2) Composite cross-section
(3) Ultimate strength design
(4) Limit states design–type 1
(5) Limit states design–type 3
(6) Limit states design–(partially)

TABLE 9.6.1 Comparison of design details of 50 m span beam (h = 2.4 m)
(dimensions in mm) Example 9.6.1

Method	b_f	t	b_w	b_b	t_b	A_c	A_{cs}	A_p	A_{st}
1. WSD	1370	288	165	685	192	875 000	0	8250	0
2. WSD (composite)	1180	192	187	702	192	780 000	160 000	8380	0
3. USD	1400	280	160	600	150	758 300	0	8300	0
4. LSD (type 1)	1600	300	180	1000	260	1071 200	0	9504	0
5. LSD (type 3)	1600	300	180	1000	260	1071 200	0	7465	0
6. LSD (partial)	1600	300	180	1000	260	1071 200	0	7119	2512

The design details of the beam by various methods are given in Table 9.6.1. The comparison of cost are made in Table 9.6.2 with the following rates of materials including supplying and placing in position (the design examples have already been discussed earlier in Chapters 3, 4, 7, 8 and 9).

Cost of M 20 concrete = Rs. 1125/m³
Cost of M 45 concrete = Rs. 1625/m³
Cost of HYSD bars = Rs. 1375/kN

Cost of HTS wires   = Rs. 3000/kN

Cost of formwork   = Rs. I50/m^2

Example 9.6.2   Comparison of cost with respect to variation in span

Three different, simply supported beams of span 10 m, 30 m and 50 m are designed to resist alive load of 20 kN/m in each case. The examples have been discussed in Chapter 7 and this chapter and the details are taken from the same. The variation of the cost of the beams with respect to the span are given in

TABLE 9.6.2   Comparison of costs (in rupees)

Method	Concrete	Formwork	Steel	Total
1. WSD	71095	51575	96540	219210
2. WSD (composite)	63375 + 9000	52575	98055	223005
3. USD	61625	52750	97125	211500
4. LSD–type 1	87025	60300	111195	258520
5. LSD–type 3	87025	60300	87345	234670
6. LSD–Partial	87025	60300	83295 +13557	244095

Table 9.6.3. Figure 9.6.1 illustrates the cost variation per metre span.

TABLE 9.6.3   Cost comparisons (in rupees)

Item	Span 10 m	30 m	50 m
	Type 1		
Concrete	4515	33 588	87 025
Formwork	3105	33 058	60 300
Steel	2720	44 466	111 195
Total	10340	111 112	258 520
Per metre span	1034	3704	5170
	Type 3		
Concrete	3675	31 153	87 025
Formwork	2625	32 143	60 300
Steel	2013	28 080	87 345
Total	8313	91 376	234 670
Per metre span	831	3046	4693

	Partially prestressed		
Concrete	3340	30 810	87 025
Formwork	2445	30 510	60 300
Steel	3090	37 106	96 852
Total	8895	98 426	244 117
Per metre span	890	3281	4884

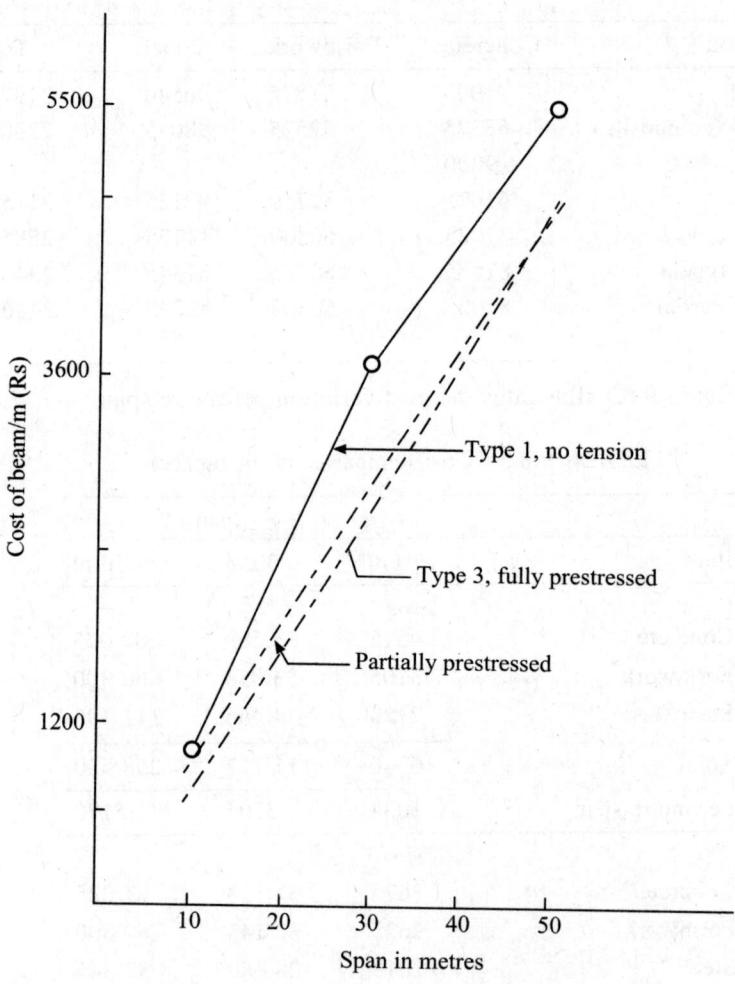

Figure 9.6.1 Variation of cost of beam with span.

PROBLEMS

9.1. A simply supported beam of span 8 m is subjected to a load of 20 kN/m. Design a partially prestressed post-tensioned concrete beam with the following data:

$$f_{ck} = 35 \text{ MPa}, f_p = 1500 \text{ MPa},$$

$$\sigma_{aci} = 14 \text{ MPa}, \sigma_{ace} = 12 \text{ MPa}, \tau_c = 0.4 \text{ MPa}$$

$$\sigma_{ate} = \sigma_{ati} = 6.4 \text{ MPa}$$

The percentage of non-tensioned steel = 0.5.

9.2 Design the beam mentioned in the problem 9.1 with the percentage of the non-tensioned steel as one instead of 0.5.

9.3 A cantilever beam of span 6 m is subjected to a live load of 20 kN/m. Design a pre-tensioned partially prestressed concrete beam of rectangular section with the following limit states design.

$$f_{ck} = 40 \text{ MPa}, f_p = 1600 \text{ MPa}$$

$$\sigma_{aci} = 16.5 \text{ MPa}, \sigma_{ace} = 14 \text{ MPa}$$

$$\sigma_{ate} = \sigma_{ati} = 6.4 \text{ MPa}$$

$$\tau_c = 0.6 \text{ MPa}$$

The width of the section be = 400 mm.

Percentage of nontensioned steel = 0.5.

9.4 Design the beam mentioned in the problem 9.3 with percentage of the nontensioned steel as 1.0 instead of 0.5, and then compare the relative merits.

9.5 A simply supported beam of span 40 m is subjected to a load of 20 kN/m. Design a post-tensioned partially prestressed concrete I-sectioned beam with allowable stresses given in the book for 0.1 mm crack width. Loss of prestress can be taken as 220 MPa.

CHAPTER 10

Design of Prestressed Concrete Slabs

10.1. Introduction

There are several types of slab constructions, the designs of which present interesting and challenging problems. The three major groups of slab construction are :

   (i)   beam and slab construction;
  (ii) flat slab-panel construction; and
 (iii) flat plate construction.

The different types of slab constructions are shown in figure 10.1.1. Any slab may be classified as one-way or two-way slab based upon the structural action of the slab. If the span ratio (also called as aspect ratio) is large, the slab acts in one direction as a very wide beam. If the width of the slab is small as compared to the length, most of the load is carried by the short span. Such a slab is called a one-way slab and designed as a very wide beam. If the span ratio is in the neighbourhood of 1, the load is carried by both the spans and the slab is called two-way slab. The inherent continuity of the structure which has a built-in load distributing and sharing capacity gives a safety much higher than the safety factor adopted in working stress design. This chapter discusses only an approximate analysis of slabs so that the readers who are not familiar with classical theory of plates can still design the slabs.

10.2. One-way Slab

The bending moment at any point of a slab is represented by bending moments along the two perpendicular directions and twisting moment in the plane. Neglecting the Poisson's effect, the resisting moments along the two spans can be written as

$$M_x = -\frac{Et^2}{12}\frac{\partial^2 w}{\partial x^2} = \frac{Et^3}{12R_x} \qquad (10.2.1)$$

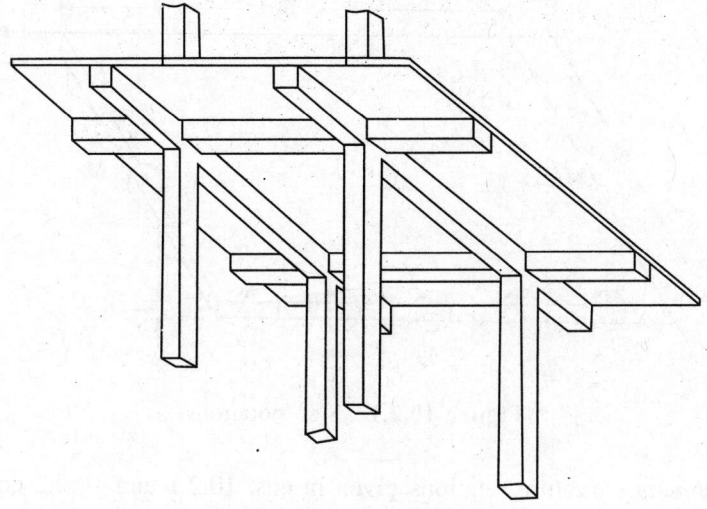

(a) Beam and slab construction.

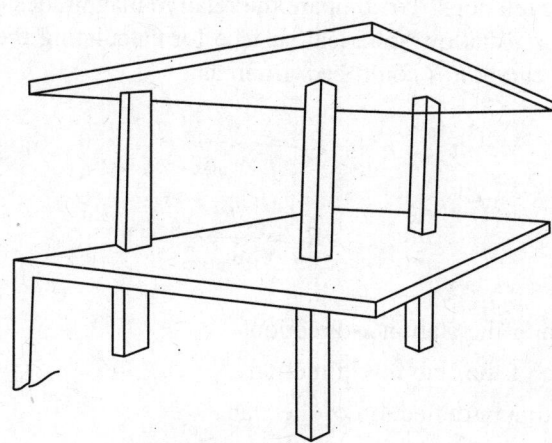

(b) Flat slab construction.

Figure 10.1.1 Different types of slab construction.

$$M_y = -\frac{Et^3}{12}\frac{\partial^2 w}{\partial x^2} = \frac{Et^3}{12R_y} \qquad (10.2.2)$$

where w = deflection of the slab

R_x, R_y = radii of curvature of the bent slab in x and y planes

M_x = bending moment on x-plane

M_y = bending moment on y-plane

t = thickness of the slab

380

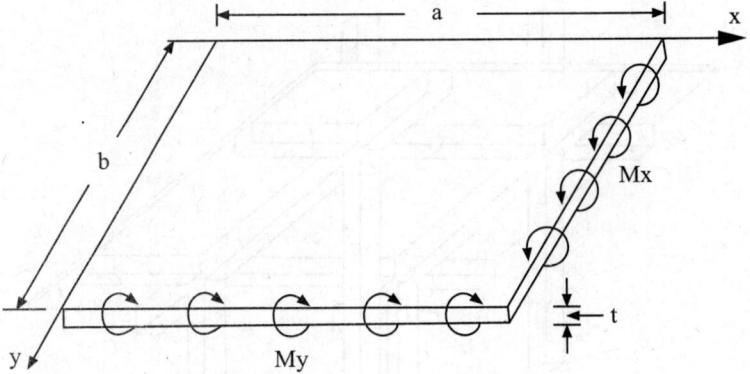

Figure 10.2.1 Slab notations.

The bending moment curvature relations given in eqs. 10.2.1 and 10.2.2 could be obtained using a simple beam theory. Figure 10.2.1 gives the sign conventions assumed in arriving at the moment curvature relations. To compare the relative magnitudes of the bending moments, let the slab be bent in a shallow spherical surface for calculating the curvature. From figure 10.2.2 the maximum curvatures could be written as

$$R_x \simeq \frac{a^2}{8w_{max}}$$

$$R_y \simeq \frac{b^2}{8w_{max}}$$

where

a = span of the slab in x-direction

b = span of the slab in y-direction

w_{max} = maximum deflection of the slab

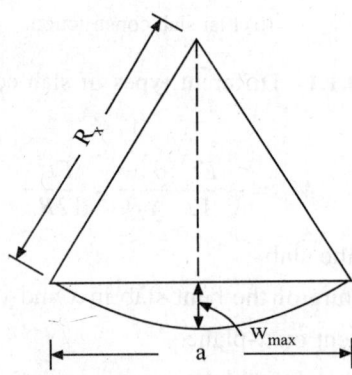

Figure 10.2.2 Curvature deflection relation.

On substitution of the curvature relations in eqs. 10.2.1 and 10.2.2, the ratio of the moments works out to be

$$\frac{M_y}{M_x} = \frac{a^2}{b^2} \tag{10.2.3}$$

Equation 10.2.3 illustrates an approximate relative magnitude of bending moments in two directions. It can be seen that the moment in short span direction is much greater than that in the long span direction. For a slab in which the span ratio is more than two, most of the load is carried by the short span, and the slab should be designed as one-way slab. An illustrative example of one-way slab is given below.

Example 10.2.1   Design a prestressed concrete slab 5 m by 20 m to carry a live load of 9000 N/m2. The slab is simply supported on four sides.

Given data is

$$f_{ck} = 45 \text{ MPa}, f_p = 1500 \text{ MPa and } \eta = 0.85$$

Solution

The aspect ratio of the slab is 4 so the slab should be designed as one-way slab. Load balancing method is used by considering one unit width of the slab.

Let
$$t = \frac{\text{short span}}{40} = 12.5 \text{ cm, use } t = 12 \text{ cm}$$

All calculations are made considering one cm width of the slab with a parabolic cable profile.

Let
q_l = intensity of live load

q_g = intensity of self load

q_t = intensity of total load, then

q_l = 90 N/m

$q_g = 0.01 \times 0.12 \times 250 = 28.8$ N/m

$q_t = (q_l + q_g) = 118.8$ N/m

Let the load balanced at working load condition be

$$Q_{bw} = 75 \text{ N/m and } e = 3.0 \text{ cm, then}$$

$$P_e = \frac{q_{bw}b^2}{8e} = 7810 \text{ N/cm width}$$

$$P_l = \frac{7810}{\eta} = 9200 \text{ N/cm width}$$

$$q_{bt} = \frac{q_{bw}}{\eta} = 88 \text{ N/m}$$

Check for stresses at transfer condition :

Load not balanced = 28.8 − 88 = −59.2 N/m

$$\sigma_{ci} = \frac{P_t}{A} + \frac{M}{Z} = \frac{9200}{12} + \frac{59.2 \times 2500 \times 6}{8 \times 12 \times 12}$$

$$= 1374 \text{ N/cm}^2$$

$$\sigma_{ti} = -\frac{9200}{12} = \frac{59.2 \times 2500 \times 6}{8 \times 12 \times 12} = -158 \text{ N/cm}^2 \text{(Comp)}$$

Check for stresses at working load :

Load not balanced = 118.8 − 75 = 43.8 N/m

$$\sigma_{ce} = \frac{P_e}{A} + \frac{M}{Z} = \frac{7810}{12} + \frac{43.8 \times 2500 \times 6}{8 \times 12 \times 12}$$

$$= 110.6 \text{ N/cm}^2$$

$$\sigma_{te} = \frac{P_e}{A} + \frac{M}{Z} = -194 \text{ N/cm}^2 \text{(Comp)}$$

All the stresses are compressive and less than the permissible stresses; hence the section is suitable. The eccentricity at the mid-span is taken as 3 cm which gives a cover of 3 cm to centre of the steel. The clear cover provision is satisfied for 7 mm wires.

$$A_p = \frac{P_i}{0.7 f_p} = \frac{9200}{105000} = 0.0875 \text{ cm}^2$$

Use 7 mm wires at 4.5 cm spacing.

Note : (a) Minimum clear cover of 2 cm is desired.

(b) Any additional reinforcement will lead to excessive stresses at transfer condition, so the steel requirement should be kept close to the design.

(c) A nominal longitudinal reinforcement of mild steel of 6 mm at 30 cm spacing along the long span should be used for temperature and secondary stresses.

There are two practical limitations in adopting load balancing design for slab constructions: (i) setting up a curved profile for a thin slab is difficult and more so if the eccentricity of the cables have to be changed along the span; (ii) in a slab which is supported on four edges, the maximum bending moment occurs at the middle of the short span and decreases slowly towards the edge. To adjust the cables and cable profiles to suit the varying bending moment along the long span is not easy. Straight cables in the slab may prove to be more economical in many small constructions.

Example 10.2.2 Design a simply supported slab to cover an area of 20 m by 5 m and to carry a load of 9000 N/cm^2, using a straight cable profile. Compare this design with the design made by load balancing method of example 10.2.1.

Given data

$$f_{ck} = 45 \text{ MPa}, f_p = 1500 \text{ MPa}, \eta = 0.85$$

Solution

Let the thickness $= t = \dfrac{\text{span}}{40} = 12.5$ cm,

then the self-weight is

$$q_g = 0.125 \left(\frac{25000}{100} \right) = 300 \text{ N/m/cm}$$

$$q_t = 120 \text{ N/m/cm width}$$

Let the cable be placed at 2.0 cm eccentricity and with no tensile stress at working load. This leads to (by considering 1 cm width)

$$\sigma_{te} = -\frac{P_e}{A} - \frac{P_e}{Z} + \frac{M}{Z} = 0$$

$$= -P_e \left[\frac{1}{12.5} + \frac{2 \times 6}{12.5 \times 12.5} \right] + \frac{120 \times 2500 \times 6}{8 \times 12.5 \times 12.5} = 0$$

or $\quad P_e = 9200$ N/cm and

$$P_i = 9200/\eta = 10800 \text{ N/cm}$$

$$\sigma_{ci} = \frac{P_t}{A} + \frac{P_t e}{Z} - \frac{M_g}{Z} = 1370 \text{ N/cm}^2$$

$$\sigma_{ti} = -\frac{P_i}{A} + \frac{P_t e}{Z} - \frac{M_g}{Z} = -410 \text{ N/cm}^2 \text{ (Comp)}$$

$$\sigma_{ce} = \frac{P_e}{A} - \frac{P_e e}{Z} + \frac{M_t}{Z} = 1430 \text{ N/cm}^2$$

$$\sigma_{te} = 0$$

All the stresses at the middle of the span are within the limits. The eccentricity at the edge of the span is 2 cm so it is desirable to check for stresses at the end section. Eccentricity is within the kern distance so no tensile stresses are developed. The compressive stress at transfer will dominate and is given by

$$\sigma_{ci} = \frac{P_i}{A} + \frac{P_i e}{Z} = 1693 \text{ N/cm}^2$$

All the stresses at any cross section are within the permissible limits.

$$A_p = \frac{P_i}{0.7 f_p} = \frac{10800}{105000} = 0.103 \text{ cm}^2$$

Use 7 mm wires at 3.75 cm apart. A nominal mild steel reinforcement of 6 mm wires at 30 cm apart should be provided in the longitudinal direction.

A straight cable profile increases the area of steel and concrete requirements by 17.5% and 4.17% respectively as compared to balanced load design (curved cable profile). If a controlled curved cable profile can be set at less expense, then the curved cable profile is more economical.

10.3. Two-way Slab

The external load on a slab is distributed on both the spans if the pan ratio of the slab is in the range of 1 to 2. In such slab construction; prestressing is to be done in both the directions. If z_x and z_y are cable profiles in x and y directions, the effective vertical component of the cable force is then given by

$$q_b = P_{ia} \frac{d^2 z_x}{dx^2} + P_{ib} \frac{d^2 z_y}{dy^2} \qquad (10.3.1)$$

where P_{ib} = prestressing force in short span

P_{ia} = prestressing force in long span

q_b = balancing load

Let g_b = sag of the cable in short span direction

g_a = sag of the cable in long span direction

Then for parabolic cable profiles the balancing load is given by

$$q_b = P_{ia} \frac{d^2 z_x}{dx^2} + P_{ib} \frac{d^2 z_y}{dy^2} = 8 \left[\frac{P_{ia} g_a}{a^2} + \frac{P_{ib} g_b}{b^2} \right] \qquad (10.3.2)$$

The load that is not balanced should be considered for calculation of bending moment stresses and deflections.

There are several approximate formulae developed for calculation of bending moments in slabs. The moment coefficient of slabs based on orthotropic plate theory are used in the present calculations. In the absence of these coefficients, the designer can use the regular plate theory coefficients. A modified orthotropic plate theory (10.1) divides the slab into middle and end strips and suggests bending moment coefficients. The strip partitions are shown in figure 10.3.1 and the bending moment coefficients are given in figure 10.3.2. The design bending moments are given by

$$M_{b1} = \beta_1 q b^2$$
$$M_{b2} = \beta_2 q b^2$$
$$M_{a1} = \beta_3 q b^2 \qquad (10.3.3)$$
$$M_{a2} = \beta_4 q b^2$$
$$M_{ab} = \beta_5 q b^2$$

where β_1, β_2, β_3, β_4 and β_5 are the bending moment coefficients (figure 10.3.2).

M_{b1} = maximum bending moment for middle half of the short span

M_{b2} = maximum bending moment for extreme quarter of the short span

M_{a1} = maximum bending moment for middle half of the long span

M_{a2} = maximum bending moment for the extreme quarter of the long span

M_{ab} = maximum twisting moment which occurs at the corners

b = short span length

The design of two-way slab is illustrated through an example.

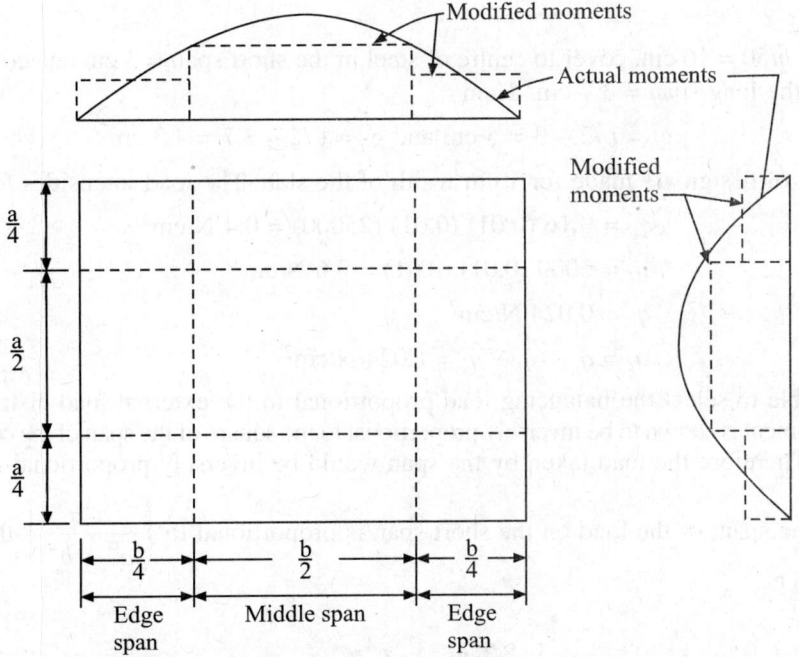

Figure 10.3.1 Modified and actual moments in a two-way slab.

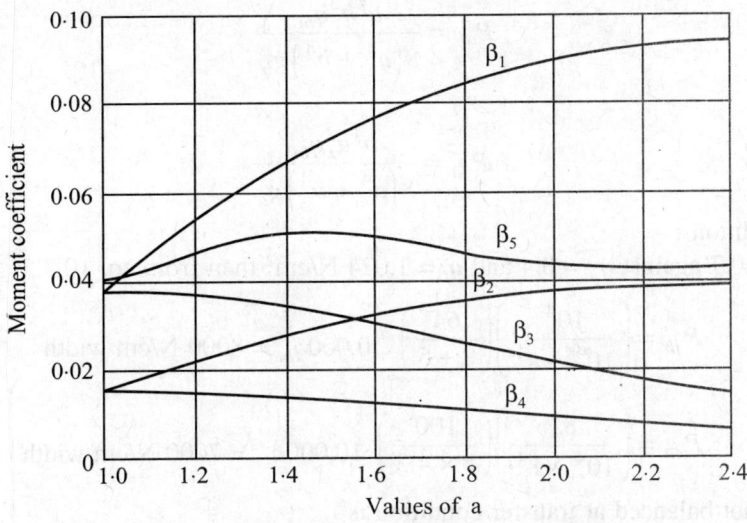

Figure 10.3.2 Moments coefficient graphs.

Example 10.3.1 Design of a 8m by 10 m simply supported slab to carry a load of 6000N/m².

Given data:

$$f_{ck} = 45 \text{ MPa}, f_p = 1500 \text{ MPa}$$

Solution

Let $t = b/50 = 16$ cm, cover to centre of steel in the short span = 3 cm and cover to centre of steel in the long span = 3.7 cm. Then

$$g_b = t/2 - 3 = 5 \text{ cm and } g_a = t/2 - 3.7 = 4.3 \text{ cm}$$

The details of design are made for 1 cm width of the slab. The load intensities for 1 cm² are

$$q_g = 0.16 \, (0.01) \, (0.01) \, (25000) = 0.4 \text{ N/cm}^2$$

$$q_l = 6000 \, (0.01) \, (0.01) = 0.6 \text{ N/cm}^2$$

Let

$$q_s = 0.024 \text{ N/cm}^2$$

Then

$$q_t = q_g + q_l + q_s = 1.024 \text{ N/cm}^2$$

It is desirable to select the balancing load proportional to the external load distribution. The bending moment is shown to be inversely proportional to the square of the span of the corresponding direction. Therefore the load taken by the span would be inversely proportional to the fourth power of the span, or the load on the short span is proportional to $\left(\dfrac{a^4 q}{a^4 + b^4}\right)$ the balancing load is $\dfrac{8 P_{ib} g_b}{b^2}$

$$\frac{8 P_{ib} g_b}{b^2} = \frac{a^4 q_{bi}}{a^4 + b^4} \tag{10.3.4}$$

or

$$P_{ib} = \frac{a^4 b^2 q_{bi}}{8\left(a^4 + b^4\right) g_b} \tag{10.3.5}$$

Similarly

$$P_{ia} = \frac{b^4 a^2 q_{bi}}{8\left(a^4 + b^4\right) g_a}$$

Transfer condition :

Let $q_{bi} = 0.7$ against $q_g = 0.4$ and $q_t = 1.024$ N/cm² then from eq. 10.3.5

$$P_{ib} = \left(\frac{10^4}{10^4 + 8^4}\right)\left(\frac{64}{8 + 5}\right) 10,000 q_{bi} = 8000 \text{ N/cm width}$$

$$P_{ia} = \left(\frac{8^4}{10^4 + 8^4}\right)\left(\frac{100}{8 \times 4.3}\right) 10,000 q_{bi} = 7000 \text{ N/cm width}$$

Load that is not balanced at transfer condition is

$$= 0.4 - 0.7 = -0.3 \text{ N/cm}^2$$

The bending moment coefficients for two-way slab from figure 10.3.2 can be obtained as (for $a/b = 1.25$)

$$\beta_1 = 0.055, \ \beta_2 = 0.025, \ \beta_3 = 0.038$$

$$\beta_4 = 0.018, \ \beta_5 = 0.046$$

Therefore corresponding effective bending moments from eq. 10.3.3 are

$$M_{b1} = 0.055 \ (0.3) \ (640 \ 000) = 10560 \ \text{Ncm/cm}$$

Similarly the other BM in Ncm/cm are:

$$M_{b2} = 5050$$
$$M_{a1} = 7650$$
$$M_{a2} = 3650$$
$$M_{ab} = 9600$$

Check for Stresses at Transfer Condition

$$\sigma_{ci} = \frac{P_{ib}}{A} + \frac{M_{b1}}{Z} = \frac{8000}{16} + \frac{10500(6)}{256}$$
$$= 500 + 248 = 748 \ \text{N/cm}^2 = 7.48 \ \text{MPa}$$
$$\sigma_{ci} = -500 + 248 = -252 \ \text{N/cm}^2 = -2.52 \ \text{MPa}$$

Check for Stresses at Working Load Condition

$$P_{eb} = \eta P_{ib} = 6800 \ \text{N/cm}$$

unbalanced load $= q_t - 0.85 \ q_{bi} = 1.024 - 0.85 \ (0.7) = 0.429$

$$\sigma_{ce} = \frac{P_{eb}}{A} + \frac{M_{b1}}{Z} = 425 + 352 = 777 \ \text{N/cm}^2$$
$$\sigma_{te} = -425 + 352 = -73 \ \text{N/cm}^2$$

As the stresses at the critical section are much below the limiting stresses, it is not necessary to check for the stresses at the other sections. Since the stresses are much below the allowable limits, the thickness of the slab could be reduced to 15 cm. With 15 cm thick slab, the stresses are found to be within the permissible limits. The area of steel may be obtained as

$$A_{sb} = \frac{P_{ib}}{\sigma_{st}} = \frac{8000}{105000} = 0.071 \ \text{cm}^2/\text{cm width}$$

$$A_{sa} = \frac{P_{ia}}{\sigma_{st}} = \frac{7000}{105000} = 0.067 \ \text{cm}^2/\text{cm width}$$

in which A_{sa} and A_{sb} are the reinforcement in the long and short span directions.

Mild steel reinforcement resist the twisting moment at the corners has to be provided. *Check for deflection* : Elastic theory of plates can be applied to prestressed concrete slabs for deflection measurements. The deflection coefficients can be taken from the theory of plates. In the present section, the deflection coefficients are taken from "Theory of plates and shells" by Timoshenko (10.2).

Let

δ_t = maximum deflection at transfer condition, and

δ_e = maximum deflection at working load condition

The deflection for a plate of aspect ratio 1.25 is given as

$$\delta = \frac{0.006\, qb^4}{E_c I} \quad \text{then at transfer condition}$$

$$\delta_t = \frac{0.006 \times (-0.3) \times 800^4}{(1/12)15^3 \times 3.82 \times 10^6} = -0.76 \ \text{(upward) cm}$$

$$\delta_e = \frac{0.006 \times (0.429)\,800^4}{(1/12)15^3 \times 3.82 \times 10^6} = 1.06 \ \text{cm}$$

(The unbalanced loads are taken same as the original because there is not much deviation in the final section.)

A reinforced cement concrete slab will deflect by about 2.3 cm for the same live load conditions assuming as uncracked section. So the deflection of prestressed concrete slab is small as compared to a reinforced concrete slab. This behaviour is very useful for large spans.

Ultimate load capacity of a slab is much higher than the ultimate capacity of the section. A slab will not fail by yielding at any one particular cross section but fails by yielding at several sections in a pattern so that the total slab forms into a collapsible mechanism. Calculation of the ultimate load of a slab by yield line theory is not discussed in this section.

10.4. Prestressed Concrete Beam and Slab Construction

Prestressed concrete beam and slab construction is not very popular because of the constructional difficulties and economical considerations. Prestressing operations in beam and slab construction are too involved to be very economical. Use of precast prestressed beams appear to be a more logical method of construction and has gained some popularity in the field. In such a case of construction, the design could be done independently for beams and for slab, treating the beam as a composite construction.

10.5. Prestressed Concrete Flat Slab

Prestressed concrete flat slab construction gained great importance because of several reasons that yielded in economy. Flat slab constructions has the adaptability for any type of building requirements, ease in form work, casting of concrete and prestressing operations (10.3 to 10.7). Lift slab construction which forms a part of flat slab construction has gained importance because of the ease in prestressing and form work operations. Lift slab construction is the one in which all the floors are cast in sequence at ground level, prestressed and lifted to the corresponding floor level by proper hoisting arrangements. There are three positive advantages in lift slab construction: (i) the flat surface of the slab gives an advantage and economy in casting the slabs at one level, one over the other with practically no money and time involved in shuttering arrangements; (ii) prestressing operations do not involve in any temporary hoisting and supporting arrangements; and (iii) the self weight of the slab is small as compared to any other construction. The methods for lift slab construction are changing fast to suit the availability of equipment and personnel.

Lifting of the slab is done with a mechanical or hydraulic jacking system using the building columns as supports. Fastening the slab to the columns and other details need special care.

Design of prestressed concrete flat slab : Flat slabs are invariably continuous so it is desirable to discuss the design of continuous flat slabs. Analysis of continuous flat plate supported on column presents a complex statically indeterminate structural problem. An exact analysis is difficult, so an approximate analysis based on wide beam theory is presented here. Approximate moment coefficients are available in "Theory of plates and shells" by Timoshenko (10.8) and also in the code of practice of plain an reinforced concrete (10.9) which can be used in design. The flat slab could be treated as two sets of continuous beams, each set spanning in one direction. Figure 10.5.1 illustrates the approximation made. The approximation of beam theory assumes that the fiat slab is treated as a wide beam rigidly supported along the column line.

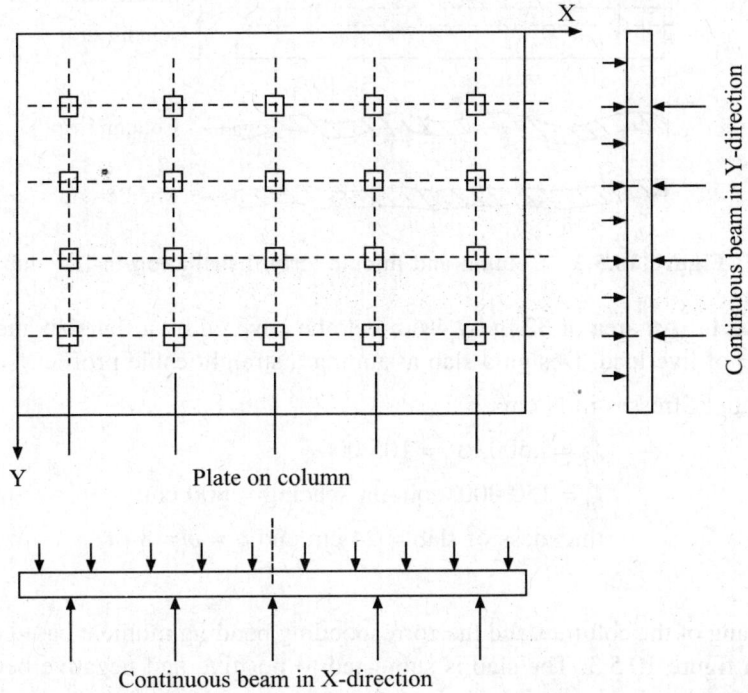

Figure 10.5.1 Flat slab on columns approximated as beams in two directions.

The design bending moments either at support or at middle of any span are computed based on continuous beam theory treating the total width of the slab as the width of the beam. The total bending moment is then redistributed along the width giving a higher weightage to the column strip as compared to the middle strip. The weightage of column and middle strip bending moments depend upon the number of columns along the width of the assumed continuous beam. The division of strips, its qualitative behaviour and approximate bending moments along the width of the slab are shown in figure 10.5.2.

Two methods of design are discussed in this section: (i) if the cable profile could be made in a curved form, then load balancing method of design is used; (ii) if a straight cable profile is adopted, then simple beam column theory is adopted. The bending moment profile is not very well-known in a flat slab and even if an approximate moment profile is determined, it changes very rapidly so the adjustment cable profile to suit the bending moment profile in balancing the load is not an easy practicable task and the loss of prestress due to curvature is relatively high. Hence a straight cable profile appears to be a more realistic and practicable solution in flat slab.

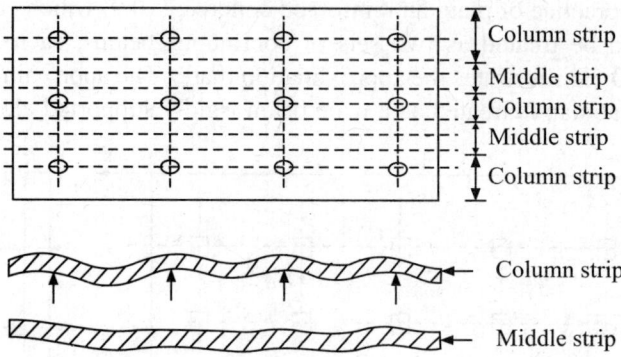

Figure 10.5.2   Column and middle strip distribution in flat slab.

Example 10.5.1   An area of 32 m by 40 m is to be covered by a flat slab and it is subjected to 5000 N/m^2 of live load. Design a slab assuming a straight cable profile.

Given data : Stresses in N/cm^2

$$f_{ck} = 4500, \sigma_{st} = 105\,000,$$
$$f_p = 150\,000, \text{column spacing} = 800 \text{ cm},$$
$$\text{thickness of slab} = 24 \text{ cm and } a = b = 8 \text{ m}$$

Solution

The spacing of the columns and the corresponding bending moment based on beam theory are shown in figure 10.5.3. The slab is subjected to positive and negative bending moments and the negative bending moment dominates the design. So adopt a negative eccentricity of 1 cm straight cable.

$$q_g = 25000(0.24) = 6000 \text{ N/m}^2$$
$$q_1 = 5000 \text{ N/m}^2$$
$$\therefore \qquad q_t = 11000 \text{ N/m}^2$$

Design of column strip : It is assumed that 75 per cent of the moment is taken by the column strip and 25 per cent by the middle strip. The critical negative moments at support B are

$$M^c{}_{Be} = \frac{41}{384} \times 11000 \times 8 \times 800 \times 0.75 = 550,000 \text{ kg cm/50 cm}$$

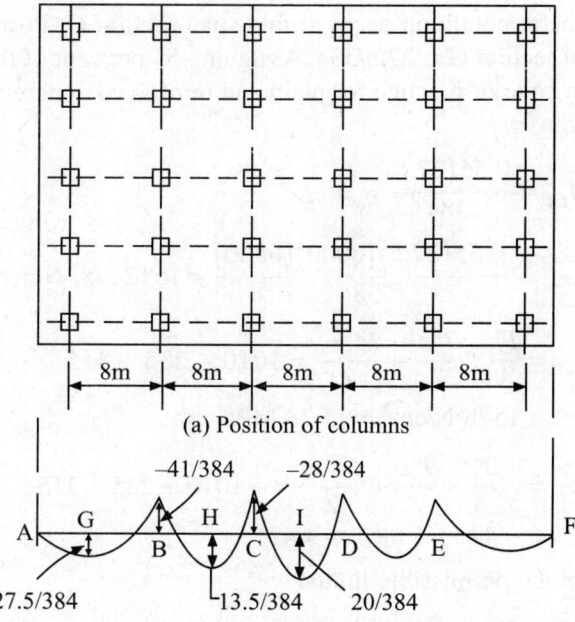

8m   8m   8m   8m   8m

(a) Position of columns

−41/384   −28/384

27.5/384   13.5/384   20/384

(b) Bending moments coefficients (qL^2)

Figure 10.5.3 Flat slab examples.

$$M^m{}_{Be} = \frac{41}{384} \times 11000 \times 8 \times 800 \times 0.25 = 184{,}000 \text{ kg cm/50 cm}$$

Let
$$\frac{P_e}{A} = 0.2 \, f_{ck} \text{ then}$$

$$P_e = (0.2)\,(4500)\,(24)\,(50) = 1026{,}000 \text{ N/50 cm and}$$

$$P_i = \frac{102\,6000}{0.85} = 1210\,000 \text{ N/50 cm}$$

Check for stress of the column strip at working load:

Critical stresses at working load occur at support B so a check on stresses on section at support B is made

$$\sigma_{ce} = \frac{P_e}{A} - \frac{P_e e}{Z} + \frac{M^c{}_{Bw}}{Z} = \frac{102\,6000}{(50)(20)} - \frac{102\,6000\,(6)}{(50)(24)(24)}$$

$$+ \frac{550\,0000\,(6)}{(50)(24)(24)} = 858 - 215 + 1150 = 1783 \text{ N/cm}^2$$

$$\sigma_{te} = -\frac{P_e}{A} - \frac{P_e e}{Z} + \frac{M^c{}_{Bw}}{Z} = -858 - 215 + 1150 = 77 \text{ N/cm}^2$$

Critical stresses at transfer condition occur at mid-span section G. Positive bending moment coefficient at mid-span section G is 27.5/384. Assigning 55 per cent of the *BM* to column strip as suggested by Indian code of practice for plain and reinforced concrete design, the *BM* at G during transfer condition is

$$M_{Gi} = \frac{0.55\,(27.5)}{384}\,q_g b^2$$

$$= \frac{0.55\,(27.5)\,(6000)\,(6400)}{384} = 1512\,500 \text{ Ncm/50 cm}$$

$$\sigma_{ci} = \frac{P_i}{A} + \frac{P_i e}{Z} + \frac{M^c{}_{Gi}}{Z} = 1010 + 235 + 315$$

$$= 1578 \text{ N/cm}^2 = 15.78 \text{ MPa}$$

$$\sigma_{ti} = -\frac{P_i}{A} + \frac{P_i e}{Z} + \frac{M^c{}_{Gi}}{Z} = -1010 + 235 + 315$$

$$= -442 \text{ N/cm}^2 = -4.42$$

The stresses are within the permissible limits.

$$A_p = \frac{P_i}{\sigma_{sl}} = \frac{121\,0000}{105\,000} = 11.5 \text{ cm}^2/50\text{cm}$$

Use 60 wires of 7 mm diameter per metre width in two or three rows.

Design of middle strip : The bending moments at the middle strip are 3/7 of the column strip. Therefore the prestressing force could be about half of the column strip and the stresses be checked.

Let
$$P_e = \frac{102\,6000}{2} = 513\,000 \text{ N /50 cm}$$

Check for stresses at support of middle strip :

$$\sigma_{te} = -\frac{P_e}{A} - \frac{P_e e}{Z} + \frac{M^m{}_{Bw}}{Z} = -426 - 108 + 460 = -74 \text{ N/cm}^2$$

$$\sigma_{ce} = 426 - 108 + 460 = 788 \text{ N/cm}^2$$

$$\sigma_{ti} = -\frac{P_i}{A} + \frac{P_i e}{Z} - \frac{M^m{}_{Bt}}{Z} < \text{(permissible)}$$

$$\sigma_{ci} = \frac{P_i}{A} - \frac{P_i e}{Z} - \frac{M^m{}_{Bt}}{Z} < \text{(permissible)}$$

Check for stresses at middle of the middle strip :

$$\sigma_{te} = -\frac{P_e}{A} + \frac{P_e e}{Z} + \frac{M^m{}_{Ge}}{Z} = -426 + 108 + 310 = 8 \text{ N/cm}^2 \text{ (permissible)}$$

$$\sigma_{ce} = \frac{P_e}{A} + \frac{P_e e}{Z} + \frac{M^m{}_{Ge}}{Z} < \text{(permissible)}$$

Use 30 wires of 7 mm diameter for one metre width. Figure 10.5.4 gives the steel distribution in one half of the slab.

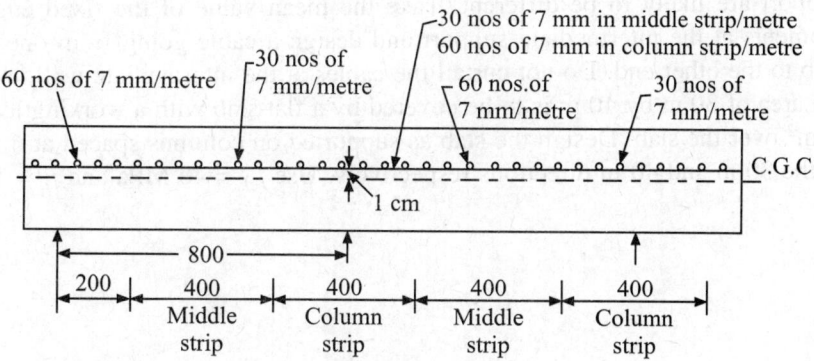

Figure 10.5.4   Reinforcement in slab Example 10.5.1

PROBLEMS

10.1 There are three parallel walls of 20 m long and spaced at 3 m apart with two end walls. Design a prestressed concrete slab to cover the area. Use f_{ck} = 40 MPa, f_p = 1500 MPa. The working load on the slab be taken as 5000 N/m².

10.2 Design a two-way slab to cover an area of 4 m by 6 m and to carry a working load of 6000 N/m². The slab be designed as prestressed in both the directions with straight cables, Use f_{ck} = 40 MPa and f_p = 1500 MPa.

10.3 A small building consists of three rooms and the plan of which is shown in figure 10.1. The live load to be carried by the roof is 2400 N/m². Design a single flat slab to cover the entire area. Use f_{ck} = 40 MPa and f_p = 1500 MPa. *Hint* : The slab is to be assumed as simply supported on the outer walls. The continuity of the slab over the intermediate walls be idealised as fixed edge support. The appropriate bending moment

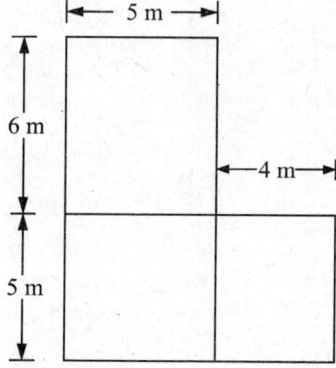

Figure 10.1   Problem 10.3.

coefficients for slabs with different edge conditions may be taken from the code specifications. The fixed end bending moments for two adjacent slabs at the common support are likely to be different. Take the mean value of the fixed edge bending moments at the intermediate support and design a cable going from one end of the slab to the other end. Do not curtail the cables at the intermediate wall level.

10.4 An area of 30 m by 40 m is to be covered by a flat slab with a working load of 4000 N/m^2 over the slab. Design the slab as supported on columns spaced at 6 m and 8 m in the 30 m and 40 m directions respectively. Use f_{ck} = 40 MPa and f_p = 1500 MPa.

CHAPTER 11

Design of Folded Plates and Shell Structures

11.1. Introduction

The efficiency of a structure depends upon the effective use of materials in the structure. Shell structures have an inherent advantage in their geometry to distribute the load effectively to all the material instead of concentrating the load resistances to any particular portion. The load carrying capacity of the shells depends basically on the shape of the structure rather than on the mass of materials used. Even though the shells are very efficient structures, their use is limited to specific purposes due to other utility criteria. They are generally used as roof structures and sometimes as foundation structural elements. Conventional shells such as cylindrical and spherical shells are not very difficult to analyse for given standard boundary conditions. Shells of double curvature and shells of arbitrary curves accommodate the imagination of architects but are rather difficult to analyse. Approximate methods or experimental techniques are used as a basis of design in such complicated shell structures. Analysis of prestressed concrete shell structures presents much more sophistication as compared to the reinforced concrete shells. The use of prestressing is commonly adopted for conventional shells such as cylindrical, spherical and folded plates without much difficulty. This chapter also presents some simplified and approximate analyses which are good for most of the practical problems. The book is designed essentially for those who are not familiar with the theory of shells. Readers who would like to have more detailed information are referred to some of the standard books on shell theory (11.1 to 11.6).

11.2. Design of Folded Plates

Folded plates may be broadly classified into three groups such as long, intermediate and short folded plates. When the length to raise the ratio of the folded plate is greater than 10, it is

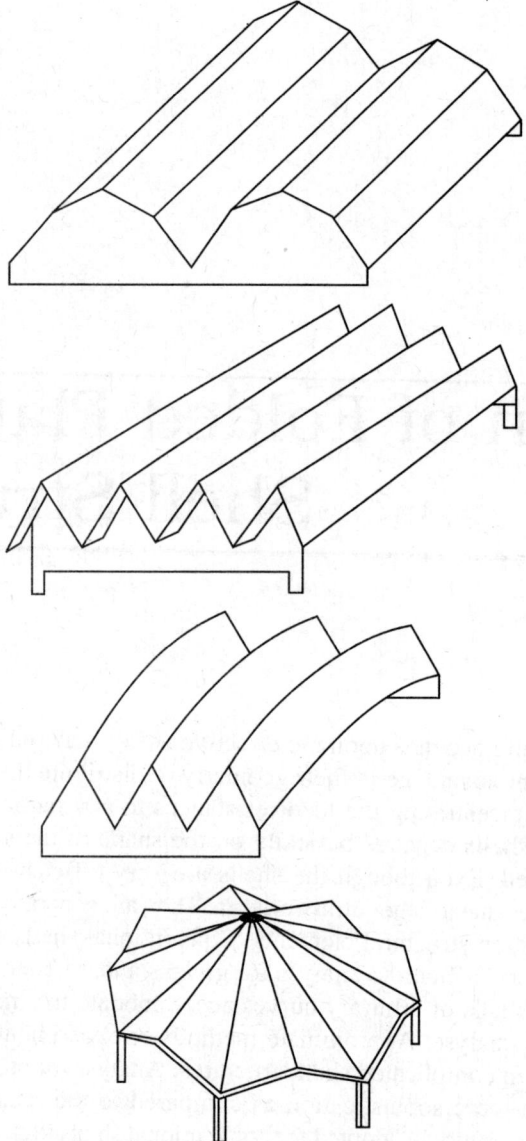

Figure 11.2.1 Typical folded plates.

called long plate; otherwise it falls into one of the two remaining categories. Methods of analyses of folded plates are generally classified as: (i) beam method; (ii) folded plate theory neglecting the relative joint displacements; (Hi) folded plate theory considering relative joint displacements; and (iv) elasticity approach. Beam method which comes out of simple statics, uses the beam formula treating the folded plate as a beam. This method, which is based on linear stress variation in cross section of the fold, neglects the transverse slab bending moments and therefore the method is dependable for long folded plates. Folded plate theory neglecting the relative joint displacements assumes that the plates are relatively deep in their own plane so as to provide transverse supporting reactions at the folds. Thus the folded plate in the transverse direction is treated as continuous one-way slab. This method, even though it makes a better approximation than beam theory, is still an approximate method and likely to give transverse bending moments of the slab higher than the actual ones. The third folded plate theory which considers the relative joint displacements, is much improved and more accurate. It can be applied to most of the folded plates except very short plates. The fourth method which is the most accurate of all, treats the structure as a continuum and uses an elasticity approach. The presentation of this method is beyond the scope of this book. The first two methods are fairly straightforward and do not need much explanation.

The derivation apparently looks pretty involved. However the solution could be obtained from the series of simultaneous equations, or through an iterated scheme.

The load balancing method is a good approach to the design of folded plates. Folded plates are to be designed for transfer load and working load conditions. The external load acting on the folded plate is essentially due to self weight of the plate with relatively small live load. These two loads could be taken as acting vertically downwards and resolved into perpendicular coordinates, one normal to the plate and another in the plane of the plate. Span of the slab is usually small and it is also not convenient to impose prestressing force along the slab profile. Therefore, the slab is designed as a simple reinforced concrete slab and the prestressing is done in the longitudinal direction of the plate to compensate the beam action (11.8–11.12).

Example 11.2.1   A folded plate is to be designed to cover an area of 32 m by 35 m. The service and superimposed loads are 750 and 750 N/m² respectively. The length of the folded plate is to be taken as 32 m.

Solution

An inverted U type of fold with 7 m fold width as shown in Figure 11.2.2 is selected. Five folds of this type will cover 35 m, width and also give a wider top flange, it is desirable to have a wider top flange to meet the working load compressive force at the top fibre. The span to fold width of the folded plate is 32/7 and span to height ratio is 32/2. Therefore the folded plate can be treated as a long folded plate .and be designed as a beam.

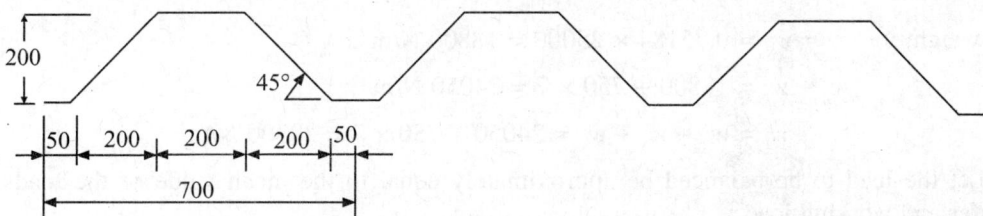

Figure 11.2.2   Typical folded plate selected for the roof (Dimensions in cm).

The thickness of the flanges and web are usually governed by practical limitations. So the sizes of the various elements are assumed subject to practical limitations and shown in Figure 11.2 3. Various sectional properties of the fold, such as area, moment of inertia, are computed in Table 11.2.1.

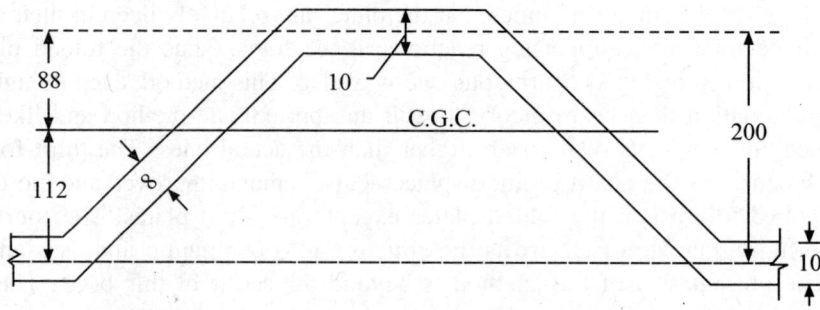

Figure 11.2.3 Folded slab dimensions (Dimensions in cm).

TABLE 11.2.1 Sectional properties of a fold of 7 m width

Section	Area	Distance from bottom flange	Moment of the area	Moment of inertia about CGC
	cm^2	cm	cm^3	cm^4
1. Top flange	$200 \times 10 = 2000$	200	400,000	$2000 \times 88^2 = 154.88 \times 10^5$
2. Inclined web	$(282.4 \times 8) \times 2$ $= 2 \times 2259.2$ $= 4518.4$	100	451,840	$\left(\dfrac{1}{12} \times 2259.2 \times 200^2\right.$ $\left. + 2259.2 \times 12^2\right) \times 2$ $= 157.15 \times 10^5$
3. Bottom flange	$(50 \times 10) \times 2 = 1000$	0	0	1000×112^2 $= 125.44 \times 10^5$
Total	7518.4		851,840	437.47×10^5

$$\bar{y} = \frac{851.840}{7518.4} = 112 \text{ cm}$$

The loads are taken for one fold width instead of the intensity of load.

$$w_t = q \times 7$$

Self weight
$$= w_g = 0.75184 \times 25000 = 18800 \text{ N/m}$$

$$w_g + w_t = 18800 + 750 \times 7 = 24050 \text{ N/m}$$

$$w_t = w_g + w_s + w_1 = 24050 + 750 \times 7 = 29300 \text{ N/m}$$

Let the load to be balanced be approximately equal to the mean value of the loads at transfer and working load

$$w_{be} = \frac{24050 + 29300}{2} \approx 26000 \text{ N/m}$$

Some of the prestressing cables can be straight in the bottom flange and the remaining have parabolic profile in the plane of the inclined web. The vertical component of the balancing cable force in the web should be treated as the balancing load. The relative sizes of the bottom flange and the web suggest that about two-thirds of the cables could be accommodated in the web portion and the remaining in the bottom flange.

Let the effective sag in the inclined plane of the web = 1 m. Equating the load balancing forces in the vertical direction, gives

$$\frac{1}{\sqrt{2}}\left(\frac{2}{3}\right)\left(\frac{8P_e g}{L^2}\right) = w_{be}$$

or

$$P_e = \frac{26600 \times 3 \times \sqrt{2} \times 31 \times 32}{2 \times 8 \times 1} = 7222\ 670 \text{ N/fold width}$$

$$P_i = \frac{P_e}{\eta} = \frac{7222\ 670}{0.85} = 8497\ 268 \text{ N/fold width}$$

One-third of P_i is placed in the bottom flange and the remaining two-third is placed in the two web elements.

Check for stresses at transfer

Load not balanced at transfer condition is

$$w_i = 24050 - \frac{26600}{0.85} = -7244 \text{ N/fold}$$

$$\sigma_{ci} = \frac{P_i}{A} + \frac{My_b}{I} = \frac{8497\ 260}{7518 - 4} + \frac{7244\ (32)(32)(100)(112)}{8(437.47)10^5}$$

$$= 1130 + 237 = 1367 \text{ N/cm}^2 = 13.67 \text{ MPa}$$

in which $M = \dfrac{w_i L^2}{8}$

$$\sigma_{ti} = -1130 + \frac{My_b}{I} = -1130 + \frac{237(88)}{112} = -944 \text{ N/cm}^2$$

Check for stresses at working load

Effective load not balanced at working is

$$w_e = 29300 - 26600 = 1700 \text{ N/fold}$$

Net bending moment is

$$M = \frac{w_e L^2}{8} = \frac{1700(32)(32)(100)}{8} = 21760\ 000 \text{ Ncm /fold}$$

$$\sigma_{ce} = \frac{P_i}{A} + \frac{My_t}{I} = 960 + 44 = 1004 \text{ N/cm}^2$$

$$\sigma_{te} = -960 + \frac{44(112)}{88} = -904 \text{ N/cm}^2$$

use M 35 concrete with 7 mm HTS wires,

The area of steel required is

$$A_p = \frac{P_i}{0.7f_p} = \frac{8497\,260}{0.7(150000)} = 46.64 \text{ cm}^2 \text{ /fold}$$

Provide 40 numbers of 7 ϕ wires in the bottom segment and 41 in each of the inclined webs.

This example gives the design of the longitudinal prestressing force only. The folded plate is to be treated as a continuous beam supported at each of the fold, and be designed as a reinforced concrete slab. The end diaphragms should be designed to resist the shear force and the affect of the prestressing force.

11.3. Introduction to Prestressed Concrete Shells

The aim of this book is only to present a basic concept of design of simple shell structures such as spherical and cylindrical shells. Prestressed concrete cylindrical water tanks have become very popular so an attempt is made on the design of cylindrical water tanks. Because spherical domes are used for buildings and tank roofs, a typical design of a spherical shell is also presented. Cylindrical shell roofs are prestressed in the longitudinal direction as done iu the folded plate construction. The edge beams of cylindrical barrel which are generally subjected to high tensile and torsional stresses are prestressed to increase the working strength and stiffness. A detailed design of prestressed concrete cylindrical shell is discussed here.

11.4. Design of Cylindrical Tanks

Four types of cylindrical tanks are in current use. Their classification is essentially based on structural behaviour caused by the boundary conditions at the base of the tank. Theoretically, the base of a tank could be fixed even though it is practically impossible to enforce a rigid restraint against rotation at the base. The footing or the bottom slab and the soil on which the footing rests are elastic materials, so only an elastic restraint is offered against rotation at the base of the cylindrical wall. Design of such tanks should be done on partially restrained base condition. An illustration of this example is shown in Figure 11.4.1. Hinged base (Figure 11.4.2), which is free to rotate about the joint, is reasonable to obtain in practice, and if designed

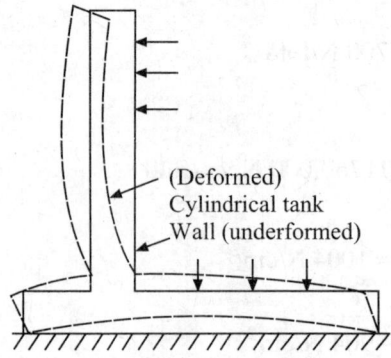

Figure 11.4.1 Fixed base which essentially acts as partially fixed base.

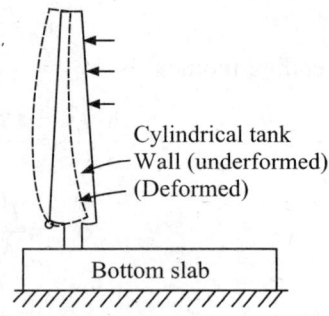

Figure 11.4.2 Hinged base.

as an idealised hinge, it is safe even if some partial restrains are enforced at the joint. The third type of base, the sliding base as shown in Figure 11.4.3, is very simple in its structural design but not so easy to attain in practice. There will always be some frictional restrain between the base and the wall which will introduce secondary bending moments. The fourth type of base, as shown in Figure 11.4.4, is an extension of sliding base with more flexibility for rotation and sliding.

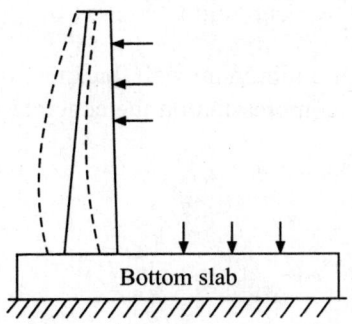

Figure 11.4.3 Sliding base.

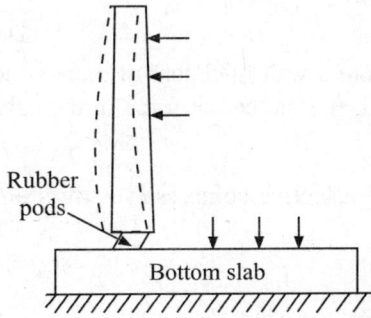

Figure 11.4.4 Base on rubber pads.

Design example 11.4.1   Design of a prestressed water tank wall for storage of 2×10^9 litres of water. The tank may be assumed to be free at top and fixed or hinged at the base.

Given data:

$$f_{ck} = 45 \text{ MPa}, f_p = 1500 \text{ MPa}, \eta = 0.8$$
$$\sigma_{aci} = 0.5 \, f_{ck}, \, \sigma_{ace} = 0.1 \, f_{ck} = 450 \text{ N/cm}^2$$
$$\sigma_{si} = 0.7 \, f_p = 105\,000 \text{ N/cm}^2; \, m = 5.5$$

Solution

Let the height (h) to diameter (D) ratio of the tank be 0.25 i.e. ($h/D = 0.25$).

$$\text{Volume} = \left(\frac{\pi D^2 h}{4} \right) = 2 \times 10^6 \text{ m}^3 \text{ by substituting } D = 4h$$

$$D^3 = \frac{4 \times 4 \times 2,000,000}{\pi}$$

or
$$D = 21.6 \text{ m}$$
$$R = 10.8 \text{ m}$$

and
$$h = 5.4 \text{ m}$$

Determination of prestressing force: The formulae developed in Chapter 6 based on load balance technique are used here

$$p = \frac{\sigma_{aci}}{\sigma_{si}} = 0.021$$

$$A_c = \frac{qR(1 - pm)}{\eta p(\sigma_{si} - \sigma_{ace})}$$

where the hydrostatic pressure at base is

$$q = \gamma h = 0.01(540) = 5.4 \text{ N/cm}$$

then

$$A_c = \frac{5.4\,(1080)(1 - 0.116)}{0.8(0.021)(105000 - 450)} = 3 \text{ cm}^2/\text{cm}$$

Therefore a wall thickness of 3 cm is adequate, however a minimum wall thickness of 10 cm be used. If P is the net force in the cables, then the net compression on the concrete is

$$C_c = P - qR$$

and the effective compressive stress on concrete is

$$\sigma_{ce} = \frac{C_c}{A_c} = \frac{P - qR}{A_c} \leq \sigma_{ace} \tag{11.4.1}$$

From Eq. 6.3.10 we have

$$P = E_s A_p \left(\epsilon_{se} + \frac{qR}{A_c E_c} \right) \tag{11.4.2}$$

From Eq. 11.4.1 we have

$$P = \sigma_{ace} A_c + qR = 450(10) + 5.4\,(1080) = 10330 \text{ N}$$

From Eq. 11.4.2, we have

$$A_p = \frac{10330}{84000 + 3208} = 0.118 \text{ cm}^2/\text{cm}$$

It may be observed that by increasing the area of concrete, the area of steel requirement is also increased, to keep a minimum of 450 N/cm² of compression at working load condition.

$$P_i = A_p \sigma_{si} = 0.118(105000) = 12400 \text{ N/cm}$$

$$\sigma_{ci} = \frac{P_i}{A_c} = \frac{12400}{10} = 1240 \text{ N/cm}^2$$

Bending reinforcement: There is always a radial thrust inward and it is maximum at transfer condition. The fixed bending moment and shear force at the base of the wall are given by Timoshenko (11.13) as

$$M_0 = \frac{q_r R t}{\sqrt{12(1 - v^2)}} \left(1 - \frac{1}{h\beta} \right) \tag{11.4.3}$$

$$Q_0 = \frac{q_r R t}{\sqrt{12(1 - v^2)}} \left(2\beta - \frac{1}{h} \right) \tag{11.4.4}$$

where

$$M_0 = \text{fixed end bending moment}$$

$$Q_0 = \text{shear force at fixed base}$$

$$\beta = 4\sqrt{\left[\frac{3(1-v^2)}{R^2 t^2}\right]}$$

$$q_r = \text{radial thrust}$$

The hydrostatic thrust 'q' should be replaced by radial thrust caused due to prestressing force and such a balancing condition gives

$$q_r = \frac{P_t}{R} = \frac{12400}{1080} = 11.5 \text{ N/cm}^2$$

Substitution of various quantities in eqs. 11.43 and 11.4.4 gives $\gamma M_0 = -69850$ N cm/cm, and $Q_0 = -21.3$ N/cm.

The fixed end bending moment is 69850 N cm/cm and is too high for 10 cm thick slab. Thus the tank should be designed with hinged base rather than fixed base.

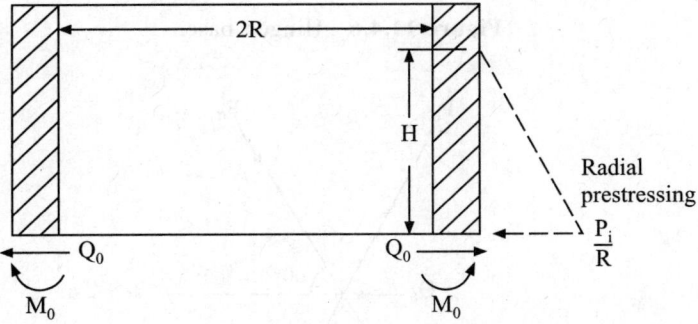

Figure 11.4.5 Fixed end forces at transfer.

Design of reinforcement at hinged base: Let Q_0 be shear force at hinged base, then it can be obtained from "Theory of plates and shells" by Timoshenko (11.13) that

$$Q_0 = \frac{q_r \beta^3 t^2 R^2}{6t\sqrt{(1-v^2)}} = 460 \text{ N/cm}$$

The reinforcement should be designed so as to take all the shear forces. A typical reinforcement for hinged base is shown in Figure 11.4.6. Assuming that the weight of the wall is transferred directly to base but not through the reinforcement, the force in the inclined wire due to shear force could be obtained from Figure 11.4.7.

Let there be two inclined wires at the base and depending upon the direction of shear force, one wire will be in tension and the other in compression. Let F_0 be the force (compression or tension) in the wire then

$$4F_0 \cos 60 = Q_0 = 460$$

$$F_0 = 115 \text{ N}$$

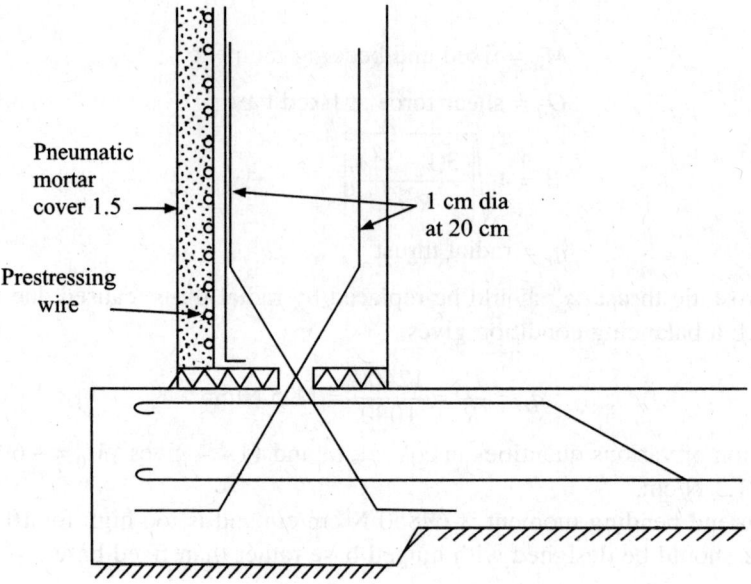

Figure 11.4.6 Hinged base.

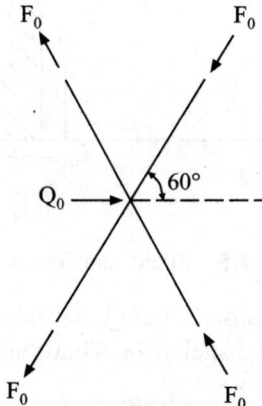

Figure 11.4.7 Forces in reinforcement at hinge.

$$\text{Area of steel} = \frac{115}{14000} = 0.009 \text{ cm}^2/\text{cm width}$$

Provide a minimum shear key mild steel reinforcement of 1 cm wire at 20 cm.

Note: Minimum thickness of the pneumatic mortar in between any two successive wire layers should be at least 8 mm. The thickness of the pneumatic mortar or the wire diameter should not be counted in the thickness of the wall which is generally indicated by core of the wall. The splicing of the wires should be designed properly. A minimum of about 0.15 per cent reinforcement has to be provided to take care of the thermal stresses.

11.5. Dome-ring Construction

A typical spherical dome with edge ring beam is shown in Figure 11.5.1. Forces due to membrane analysis are superimposed on the forces due to bending and horizontal line loads on the dome and then the design is done for the resulting forces.

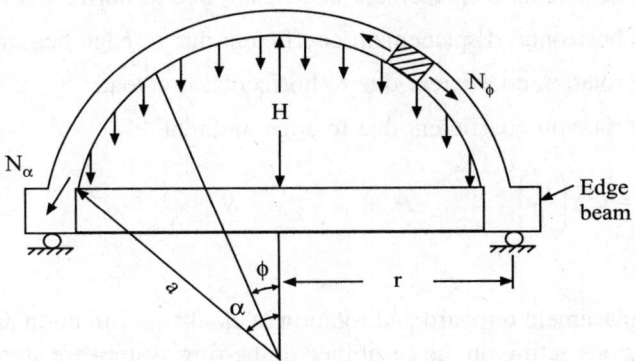

Figure 11.5.1 Dome with ring beam.

The resultant forces and deformations due to membrane theory are given by (11.14) as

$$N_{\phi m} = -\frac{aq}{1 + \cos\phi} \tag{11.5.1}$$

$$N_{\theta m} = aq\left(\frac{1}{1 + \cos\phi} - \cos\phi\right) \tag{11.5.2}$$

$$v_{0d} = \frac{a^2 q}{Eh}\left(\frac{1 + v}{1 + \cos\alpha} - \cos\alpha\right)\sin\theta \tag{11.5.3}$$

$$\theta_{0d} = \frac{-aq}{Et}(2 + v)\sin\alpha \tag{11.5.4}$$

where

a = radius of the sphere

$N_{\phi m}$ = meridional stress resultant at ϕ on the surface of the shell

$N_{\phi m}$ = hoop stress resultant at ϕ on the surface of the shell

q = load intensity proportional to the self weight of the shell

t = thickness of the shell

v_{0d} = horizontal displacement of the dome due to external load

θ_{0d} = rotation of the dome due to external load

v = Poisson's ratio

Subscript '0' indicates external load and 'd' indicates the dome.

The stress resultants and displacements due to constant line loads acting along the boundary of the shell are given in table 11.5.1.
where

M_ϕ = bending moment on the meridian

V_{hd} = edge horizontal displacement coefficient due to horizontal line load

V_{md} = edge horizontal displacement coefficient due to edge bending moment

θ_{hd} = edge rotation coefficient due to horizontal line load

θ_{md} = edge rotation coefficient due to edge moment

$$\lambda^4 = 3(1 - v^2)\left(\frac{a}{t}\right)^2$$

$$\psi = \alpha - \phi$$

Horizontal displacement outward and rotation in clockwise direction are taken as positive.

The force notations acting on the combined dome-ring system are shown in Figure 11.5.2 and the corresponding displacements due to these forces are given in Table 11.5.1.

TABLE 11.5.1   Stress resultants and displacement coefficients due to line loads

$N_\phi = -\sqrt{2e^{-\lambda\psi}}\,\cot\phi\sin\left(\lambda\psi - \dfrac{\pi}{4}\right)\sin\alpha$	$-\dfrac{2}{a}\lambda e^{-\lambda\psi}\cot\phi\sin\lambda\psi$
$N_0 = -2\lambda\,e^{-\lambda\psi}\sin\left(\lambda\psi - \dfrac{\pi}{2}\right)\sin\alpha$	$-\dfrac{2\sqrt{2}}{a}e^{-\lambda\psi}\sin\left(\lambda\psi - \dfrac{\pi}{4}\right)\lambda^2$
$M_\phi = \dfrac{ae^{-\lambda\psi}}{\lambda}\sin\lambda\psi\sin\alpha$	$\sqrt{2e^{-\lambda\psi}}\sin\left(\lambda\psi + \dfrac{\pi}{4}\right)$
v $v_{hd} = \dfrac{2a\lambda}{Et}\sin^2\alpha$	$v_{md} = \dfrac{2\lambda^2}{Et}\sin\alpha$
θ $\theta_{hd} = \dfrac{2\lambda^2}{Et}\sin\alpha$	$\theta_{md} = \dfrac{4\lambda^3}{Eat}$

The ring beam is subjected to membrane thrust of the shell and the line load due to bending forces. The membrane thrust which acts eccentrically on the ring cross section may be subdivided into centroidal thrust and bending moment as shown in Figure 11.5.3. These forces cause deformations of the ring, their relation is given in eq. 11.5.5. Compatible conditions at the junction of the ring beam and the shell give sufficient conditions to evaluate the edge corrections, and thus the resultant forces acting on the shell could be developed.

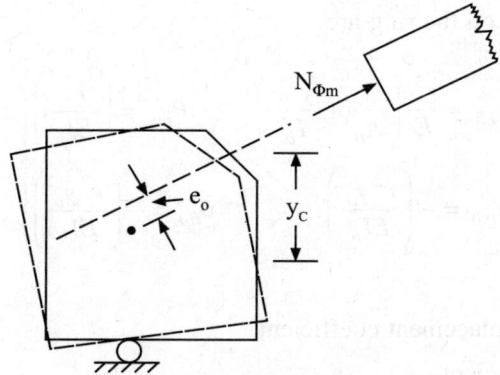

Figure 11.5.2 Dome-ring force system.

The deformations caused by edge thrust on the ring beam are

$$v_{0b} = \frac{r^2}{E}\left(\frac{\cos\alpha}{A_b} + \frac{y_0 e_0}{I_b}\right) N_\alpha$$

(The inward deflections and anticlockwise rotations of the ring beam are treated as positive).

$$\theta_{0b} = -\frac{r^2 e_0}{EI_b} N_\alpha$$

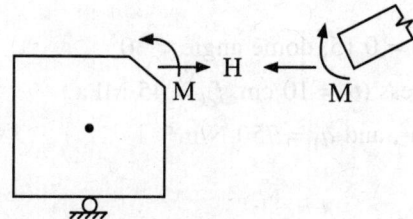

(a) Membrance force on the beam

(b) Correction forces at the interface

(a) Effective membrane equivalent forces on the ring beam.
(b) Bending correction forces at the junction of the ring and shell.

Figure 11.5.3 Ring beam interface.

Deformation coefficients on the ring are

$$v_{hb} = \frac{r^2}{E}\left(\frac{1}{A_b} + \frac{y^2_o}{I_b}\right), \qquad \theta_{mb} = \left(\frac{r^2}{EI_b}\right)$$

$$v_{mb} = -\left(\frac{r^2 y_o}{EI_b}\right), \qquad \theta_{hb} = -\left(\frac{r^2 y_o}{EI_b}\right)$$

(11.5.5)

where

v = horizontal displacement coefficient

θ = rotation coefficient

y_o = distance of the junction of ring beam and dome from the centroid of the ring beam

e_o = perpendicular distance of the membrane force from centroid

A_b = area of ring beam

I_b = moment of inertia of ring beam about vertical plane

The subscript 'h' or 'm' indicates horizontal forces or moments respectively, 'b' indicates the beam.

The deformation compatibility relation between the ring beam and the dome are given by Eq. 11.5.6. This set of equations when solved with the appropriate coefficients will result in the moment M_1 and thrust H_1 for which the beam to be designed.

$$(v_{od} + v_{ob}) + (v_{hd} + v_{hb})H_1 + (v_{md} + v_{mb})M_1 = 0$$
$$(\theta_{od} + \theta_{ob}) + (\theta_{hd} + \theta_{hb})H_1 + (\theta_{md} + \theta_{mb})M_1 = 0$$

(11.5.6)

An illustrative example is given with complete details.

Example 11.5.1 An area of 30 m diameter is to be covered by a dome. Design a prestressed concrete spherical dome with ring beam for a live load intensity of 2000 N/m^2 of the dome surface acting vertically downwards.

Given data:

Poisson's ratio = 0.15, dome angle = 30°,

average thickness (t) = 10 cm, f_{ck} = 45 MPa

q_s = 1150 N/m^2, and q_1 = 750 N/m^2

Solution

r = radius of the ring beam = 15 m

a = radius of sphere = 15/sin 30 = 30 m

q_g = 2500 N/m^2, $q_t = q_g + q_1 + q_g$ = 4400 N/m^2

$$N_\alpha = -\frac{aq}{1 + \cos\alpha} = -\frac{30q}{1 + 0.867} = -16q$$

$$v_{od} = \frac{a^2}{t}\left(\frac{1+v}{1+\cos\alpha} - \cos\alpha\right)(\sin\alpha)\left(\frac{q}{E}\right) = -1130.4\ (q/E)$$

$$\theta_{od} = \frac{a}{t}(2+v)(\sin\alpha)(q/E) = -322.5\left(\frac{q}{E}\right)$$

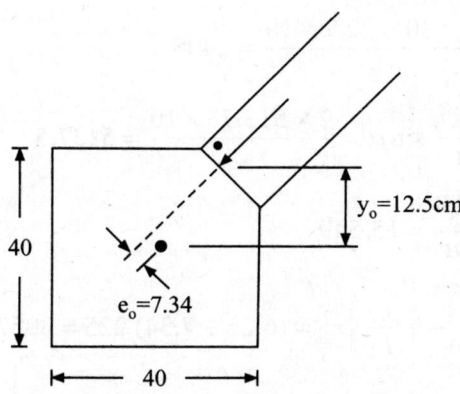

Figure 11.5.4 Ring beam.

Let the ring beam be square in cross section of size 0.4 m.

$$A_b = 0.4 \times 0.4 = 0.16\ \text{m}^2$$
$$I_b = (1/12)(0.4 \times 0.4)^3 = 0.00213\ \text{m}^4$$
$$v_{ob} = -\frac{30 \times 225}{1.867}\left(\frac{0.867}{0.16} + \frac{0.125 \times 0.0734}{0.00213}\right)\frac{q}{E}$$
$$= -35{,}178\ (q/E)$$
$$\theta_{0b} = \frac{225 \times 0.0734}{0.00213} \times 16\left(\frac{q}{E}\right) = 124{,}000\ \frac{q}{E}$$
$$v_o = (v_{0d} + v_{0b}) = -(1130.4 + 35{,}178)\frac{q}{E}$$
$$= -36{,}308.4\ \frac{q}{E}$$
$$\theta_o = (\theta_{od} + \theta_{ob}) = (-322.5 + 124{,}000)\frac{q}{E}$$
$$= 123{,}677.5\ \frac{q}{E}$$
$$\lambda^4 = 3(1 - v^2)\left(\frac{a}{t}\right)^2$$

$$= 3\,(1 - 0.0225)\left(\frac{3000}{10}\right)^2 = 263{,}925$$

$$\lambda^2 = 513.73 \text{ or } \lambda = 22.7$$

$$v_{hd} = \frac{2a\lambda}{Et}\sin^2\alpha \text{ or}$$

$$Ev_{hd} = \frac{2 \times 30 \times 22.7 \times 10}{4} = 3405$$

$$Ev_{md} = \frac{2\lambda^2}{t}\sin\alpha = \frac{2 \times 513.73 \times 10}{2} = 5137.5$$

$$E\theta_{md} = \frac{4\lambda^2}{at} = 15{,}549$$

$$Ev_{hb} = \left(\frac{1}{A_b} + \frac{y_o^2}{I_b}\right)r^2 = (6.25 + 7.34)\,225 = 3057.8$$

$$Ev_{hb} = \frac{-r^2 y_o}{I_b} = -\frac{225 \times 0.125}{0.00213} = -13{,}204$$

$$E\theta_{mb} = \frac{r^2}{I_b} = \frac{225}{0.00213} = 105{,}634$$

$$Ev_h = E\,(v_{hd} + v_{hb}) = 6463$$

$$Ev_m = E\,(v_{md} + v_{mb}) = -8067$$

$$E\theta_m = E\,(\theta_{md} + \theta_{mb}) = 121{,}183$$

Compatibility eq. 11.5.6 yields

$$6463H_1 - 8067M_1 - 36{,}308q = 0$$

$$-8067H_1 + 121{,}183M_1 - 123{,}678q = 0$$

$$H_1 - 1.248\,M_1 = 5.617q$$

$$\frac{-H_1 + 15.02\,M_1 = -15.33q}{13.772\,M_1 = -9.713q}$$

$$M_1 = -\frac{9.713}{13.772}\,q = -0.705q = 3100 \text{ N cm/cm}$$

$$H_1 = (5.617 - 1.248 \times 0.705)q = 4.74q$$

TABLE 11.5.2 Coefficients of resultant forces due to edge

											Coefficient of H			Coefficient of M		
	ψ		λψ								N_ϕ -0.707	N_e -22.7	M_ϕ $.655$	N_ϕ 1.52	N_e -49	M_ϕ 1.414
φ	Deg	Rad	Deg	Rad	$e^{-\lambda\psi}$	cot φ	sin $\lambda\psi$	sin $(\lambda\psi-\pi/4)$	sin $(\lambda\psi+\pi/4)$	sin $(\lambda\psi-\pi/2)$	$(6)(7)(9)$	$(6)(11)$	$(8)(6)$	$(6)(7)(8)$	$(6)(9)$	$(6)(10)$
(1)	(2)	(3)	(4)	(5)	(6)	(7)	(8)	(9)	(10)	(11)	(12)	(13)	(14)	(15)	(16)	(17)
30	0	0	0	0	1.0	1.732	0	-.707	-.707	-1.0	.865	22.7	0	0	34.2	1.0
28	2	.0349	45.4	0.804	0.449	1.881	.7169	.014	.999	-.697	-.0083	7.15	.210	-.92	-.307	0.63
24	6	.1047	136.2	2.4	0.091	2.246	.6769	.999	-.042	.736	-.143	-1.53	.0405	-.21	-4.45	-.005
20	10	.1745	227	4.01	.0018	2.748	-.7547	-.669	-.997	.665	.0002	-.027	-.0009	.0056	.006	0
10	20	.3491	434	8.0	.0002	5.675	.9903	.799	.602	.139	0	0	0	0	0	0
0	30	.5286	681	12.0	0	∞	544	-.978	.208	-.839	0	0	0	0	0	0

Note: Columns (12), (13) etc. are obtained by the multiplication of the coefficient and the appropriate columns listed therein.

Design of prestressing and reinforced steel   The horizontal thrust on the ring beam is

$$= H_1 = 4.74q = 20850 \simeq 21000 \text{N/m (for } q = q_t)$$

Use 18 Nos. of 7 mm wires, then prestressing force $P_e = 18 \times 34500 = 620000$ taking 34500 N as the effective prestressing capacity of 7 mm wire. It is now necessary to calculate the stress resultants due to thrust caused by the actual prestressing force and also due to the moment caused by the eccentricity of the prestressing force. It is desirable to provide the cables so that there is a counteracting eccentricity. Let the prestressing be done in one layer of cables with an eccentricity of 6 cm as shown in Figure 11.5.5.

The horizontal thrust H_2 and the moment M_2, have to be calculated by considering prestressing force as an external force. The corresponding deformation coefficients are computed in order.

$$(Ev_{hb}) = r^2 \left(\frac{1}{A_b} + \frac{y_0 e}{I_b} \right) H_p$$

$$= 225 \left(6.25 + \frac{0.14 \times 0.06}{0.00213} H_p \right) = 2239 \, H_p$$

$$(E\theta_{hb}) = \left(\frac{-r^2 y_0 e}{I} \right) H_p = -888 \, H_p$$

The influence coefficients are taken from the previous section are:

$$E_{vh} = 6463, \quad E\theta_h = -8067, \quad E\theta_m = 121183$$

The compatibility equation yields

$$6463 \, H_2 - 8067 \, M_2 = -2293 \, H_p$$

$$-8067 \, H_2 + 121183 \, M_2 = -888 \, H_p$$

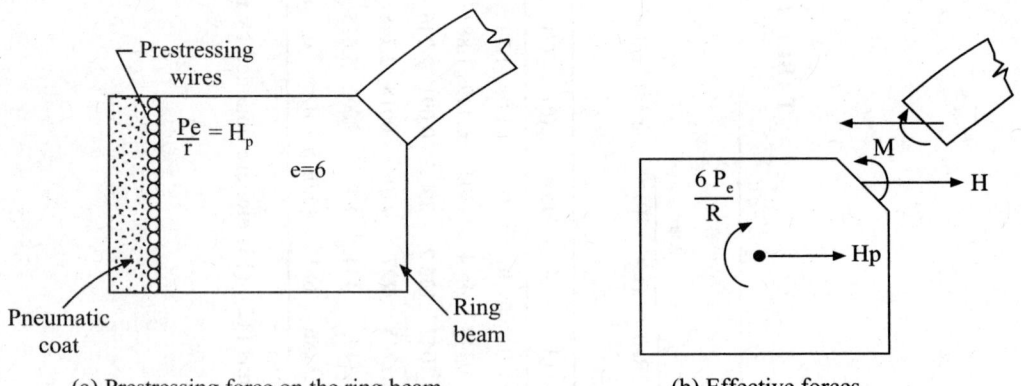

(a) Prestressing force on the ring beam.   (b) Effective forces

Figure 11.5.5   Edge beam.

or
$$H_2 - 1.248\, M_2 = -0.355\, H_p$$
$$\underline{-H_2 + 15.22\, M_2 = 0.110\, H_p}$$
$$13.972\, M_1 = -0.245\, H_p$$

$$M_2 = \frac{-0.245 H_p}{13.972} = -0.018\, H_p$$

where H_2 and M_2 are the interacting forces caused by the prestressing of effective prestress.

TABLE 11.5.3 Stress resultants in dome-ring without prestressing force (kN/m or kNm/m)

ϕ	N_ϕ Membrane	H_0	M_0	Total	N_θ Membrane	H_0	M_0	Total	M_ϕ H_0	M_0	Total
30	−70.8	18.2	0	−52.6	−43.6	476.7	−116.9	316.5	0	−3.1	−3.1
28	−70.1	−0.1	−2.9	−73.1	−46.5	150.2	−1.0	102.7	4.4	−2.0	2.4
24	−69.0	−3.0	−0.7	−72.7	−51.6	−32.1	1.0	−82.7	0.9	0	0.9
20	−68.1	0	0	−68.1	−56.0	−0.5	4.5	−52.0	0	0	0
10	−66.5	0	0	−66.5	−63.5	0	0	−63.5	0	0	0
0	−66.0	0	0	−66.0	−66.0	0	0	−66.0	0	0	0

This table is prepared using table coefficients in table 11.5.2
$$H_2 = (-0.355 - 1.2248 \times 0.018)\, H_p$$
$$= -0.377\, H_p$$

where
$$H_p = \frac{P_e}{r} = \frac{620000}{15} = 41330 \text{ N/m}$$

then
$$M_2 = 740 \text{ N m/m}$$
$$H_2 = -0.377 \times 41330 = -15580 \text{ N/m}$$

The stress resultants are now computed in Table 11.5.4 using Table 11.5.2

Design of ring beam: The forces acting on the ring beam at working load condition are shown in Figure 11.5.5. The ring beam is subjected to axial thrust and twisting moment for which the beam is designed.

Design of dome reinforcement: The dome has to be designed to resist the axial thrust, hoop tension and the bending moment. The critical section is essentially at the junction of the shell and ring beam, so the thickness of the shell near the edge is increased to give higher flexural stiffness. As the disturbance damps from the edge beam, the thickness of the shell is gradually decreased from 13 cm at the edge beam to 10 cm at 5 m from edge

$$\text{Maximum axial force} = N_\phi = 73.8 \text{ } kN/m$$

$$\text{Axial stress} = \left(\frac{73800}{100 \times 13}\right) = 56 \text{ N/cm}^2$$

No meridinal reinforcement is required except for shrinkage and-temperature effects. Provide 0.18% reinforcement for shrinkage

$$A_s = 0.0018 \times 100 \times 13 = 2.34 \ cm^2/m$$

Use 10 numbers of 6 mm mild steel wire per metre width.

Hoop reinforcement: There is hoop compression in the dome, therefore provide only the nominal circumferential reinforcement on the meridinal one.

Bending reinforcement: Maximum bending moment

$$= 2360 \ Nm/m$$

Let the effective depth = 11 cm

TABLE 11.5.4 Stress resultants at working load (kN/m, kNm/m)

ϕ	N_ϕ				N_θ				M_ϕ			
	Dome* Ring	H_2	M_2	Total	Dome	H_2	M_2	Total	Dome	H_2	M_2	Total
30	−52.6	−13.5	0	−66.1	316.5	−357.0	28.0	−12.5	−3.1	0	0.7	−2.4
28	−73.1	0	−0.7	−73.8	102.7	−117.0	−0.2	−14.5	2.4	−3.3	0.5	−0.4
24	−72.7	2.2	−0.2	−70.7	−82.7	23.8	−3.3	−62.2	0.9	−0.6	0	0.3
20	−68.1	0	0	−68.1	−52.0	4.2	0	−47.8	0	0	0	0
10	−66.5	0	0	−66.5	−63.5	0	0	−63.5	0	0	0	0
0	−66.0	0	0	−66.0	−66.0	0	0	−66.0	0	0	0	0

*The stress resultants taken from Dome-ring analysis of table 11.5.3 stress resultants due to H_2 and M_2 are obtained using table 11.5.2

TABLE 11.5.5 Stress resultants at initial condition (kN/m, kNm/m)

ϕ	N_ϕ				N_ϕ				M_ϕ			
	Dome*	H_i^+	M_i^+	Total	Dome	H_i	M_i	Total	Dome	H_i	M_i	Total
30	−28.7	−15,9	0	−34.6	172.7	−420.2	33.0	−214.5	−1.7	0	0.9	−0.8
28	−40.0	0.2	−0.8	−40.6	57.1	−137.7	−0.3	−80.9	1.3	−3.9	0.6	−2:0
24	−39.7	2.6	−0.2	−37.3	− 8.2	28.0	−3.9	15.9	0.5	−0.7	0	−0.2
20	−37.2	0	0	−37.2	−30.9	4.9	0	−26.0	0	0	0	0
10	−36.3	0	0	−36.3	−34.6	0	0	−34.6	0	0	0	0
0	−36.0	0	0	−36 0	−36.0	0	0	−36.0	0	0	0	0

*The external load at initial condition is 2.4 kN/m^2. So the stress resultants are computed by prorating those in table 11.5.3 by (2.4/4.4) as those computed in table 11.5.3 are for 4.4kN/m^2.
+The initial prestress is estimated at $P_e/0.85$, so the stress resultants caused by the H_i and M_i are computed by dividing those in table 11.5.4 by 0.85.

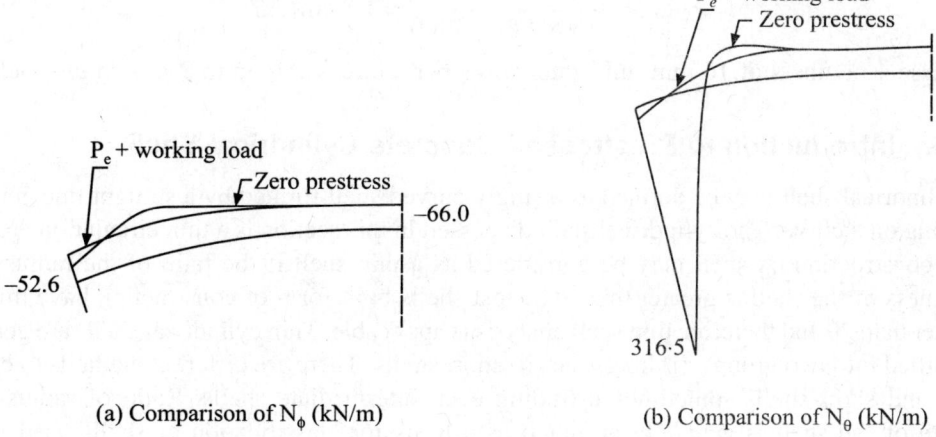

(a) Comparison of N_ϕ (kN/m)

(b) Comparison of N_θ (kN/m)

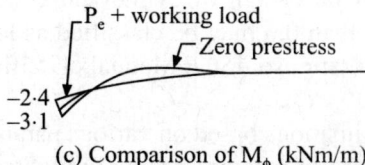

(c) Comparison of M_ϕ (kNm/m)

(a) Comparison of values of N_ϕ for different conditions.
(b) Comparison of N_ϕ, prestressed and non-prestressed.
(c) Comparison of M_ϕ, prestressed and non-prestressed.

Figure 11.5.6 Comparison of stress resultants.

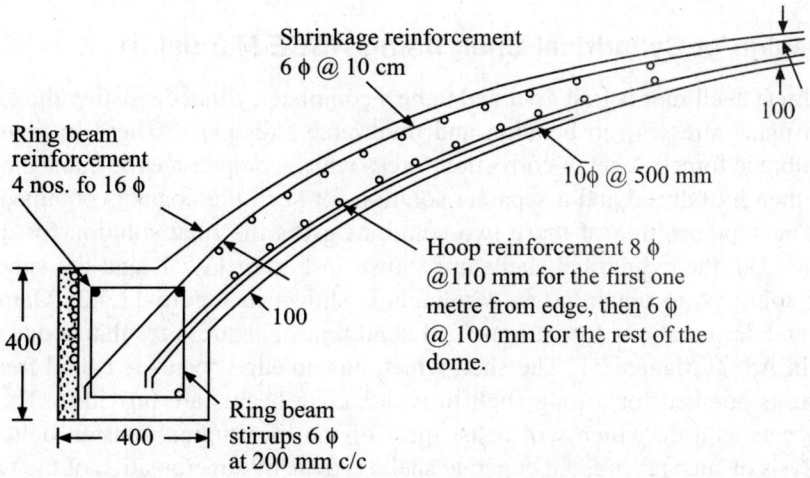

Figure 11.5.7 Reinforcement details (all dimensions in mm).

$$A_p = \frac{236000}{0.9 \times 11 \times 14000} = 1.8 \text{ cm}^2/\text{m}$$

Use 4 numbers of 10 mm mild steel wires per metre width up to 2 m into the shell.

11.6. Introduction to Prestressed Concrete Cylindrical Shell

A cylindrical shell may be defined as a singly curved shell formed by a straight line generator moving on a curve. The cylindrical shell discussed in this section is a thin circular one used for roof construction. A shell may be considered as a thin shell if the ratio of the radius to the thickness of the shell is greater than 20. Most shells built for roof construction have this ratio greater than 20 and therefore thin shell analysis is applicable. Thin cylindrical shells are generally classified into two groups: (i) long, and (ii) short shells. There are different methods to classify long and short shells sometimes including even intermediate shells. Ratio of radius to the length of the shell is generally accepted as a basis for classification of shells used in roof construction. *ASCE* Manual on design of cylindrical concrete shell roofs suggests that the shells having R/L ratio less than 0.6 may be classified as long shells, otherwise short barrels. There is no exactness to this ratio, so *ASCE* Manual classification may be considered good for practical purposes.

Several types of approximations based on various parameters were made in the analysis of cylindrical shells. Each theory or solution has its limitations and advantages. Any attempt to discuss the cylindrical shell analysis is beyond the scope of the book since the main aim of this book is to present the design of cylindrical shell in a simplified manner so that readers, who have not studied any analysis of shell structures, can follow and do the design computations.

ASCE Manual 31 presents certain coefficients of forces and deformations of the shell for different types of load conditions. These coefficients are used for the design of the shell. Rational use of these coefficients needs certain understanding in the basic analysis and notations.

11.7. Design of Cylindrical Shell Using *ASCE* Manual 31

The cylindrical shell roof is first assumed to be a complete cylinder resisting the external loads through in plane stresses (no bending and transverse stresses). These in plane forces are called membrane forces. A set of correction forces which compensate the unbalanced boundary forces are then introduced and a separate solution for such line loads is obtained by bending analysis. The superposition of these two solutions gives the final solution for design. Basic notations used in the cylindrical shells are shown in Figure 11.7.1, and the superposition of two linear solutions to obtain stress resultants is shown in Figure 11.7.2. Membrane stress resultants and deformations for various load conditions and stress resultants due to line loads are given in ASCE Manual 31. The shell which has no edge beams is called free edge shell. Prestressing is adopted for a long shell in which edge beams are provided. The edge beam essentially acts as a tie which will resist most of the longitudinal tension developed in the shell. Analysis of such prestressed concrete shells is done by superposition of the two solutions. Table 4 Appendix A give some relevant *ASCE* Manual 31 coefficients.

(i) The cylindrical shell and the prestressed concrete edge beams are analysed as two independent structural systems.

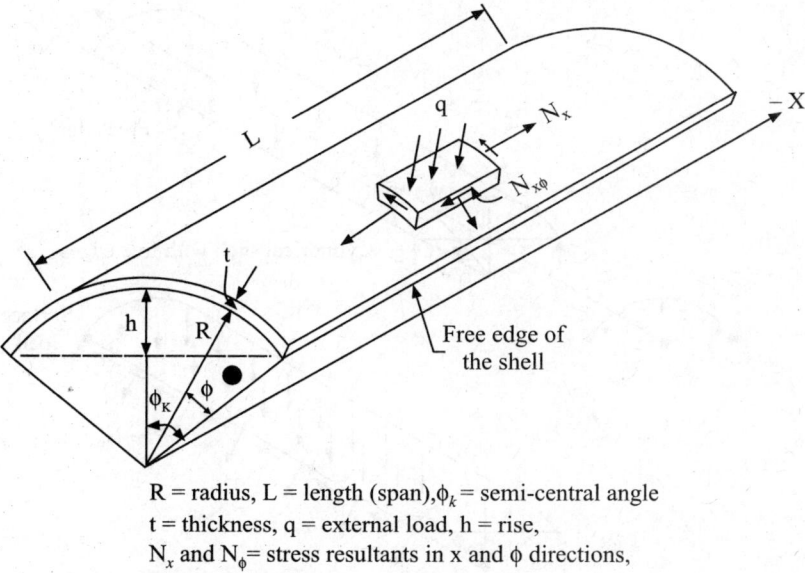

R = radius, L = length (span), ϕ_k = semi-central angle
t = thickness, q = external load, h = rise,
N_x and N_ϕ = stress resultants in x and ϕ directions,
N_x = shear force

Figure 11.7.1   Shell dimensions and notations.

(ii) Four possible forces–(a) vertical force (V_1), (b) horizontal force (Hi) (c) transverse moment on the curved axis of the shell or torque on the beam (M_1), and (d) the force along the longitudinal axis (S_1), i.e., shear force on the shell and longitudinal force on the beam–are likely to act at the intersection of the shell and the edge beam. This set of forces are introduced in the two independent shell and beam systems. The shell and the beams are analysed independently for this set of line loads. The free body diagrams of the two independent systems of forces on the shell are shown in figures 11.7.2 and 11.7.3. The line forces at the intersection of the shell and beam are obtained through the four deformation compatibility conditions. Each of the algebraic sums of: (i) vertical, (ii) horizontal, (iii) rotational, and (iv) extensional longitudinal strain deformations must be equal to zero. Poisson's ratio for concrete is small so the fourth condition of compatibility of longitudinal strain may be assumed same as longitudinal stress equivalence.

Let w, v and θ be the displacements taken positive in the directions of forces V_1, H_1 and M_1 respectively. The first subscript '0' indicates deformations of the individual shell and beam systems subjected to external loads. The first subscripts v, h, m and s (for some deformation quantities) indicate deformations due to vertical, horizontal, moment and shear line forces respectively. The second subscript or 'b' indicates shell or beam system.

The compatibility equations using the notations as described above may now be written as:

$$(w_0 + w_r + w_h + w_m + w_s)_s + (w_0 + w_v + w_h + w_m + w_s)_b = 0 \qquad (11.7.1)$$

$$(v_0 + v_v + v_h + v_s)_s + (v_0 + v_v + v_h + v_m + v_s)_b = 0 \qquad (11.7.2)$$

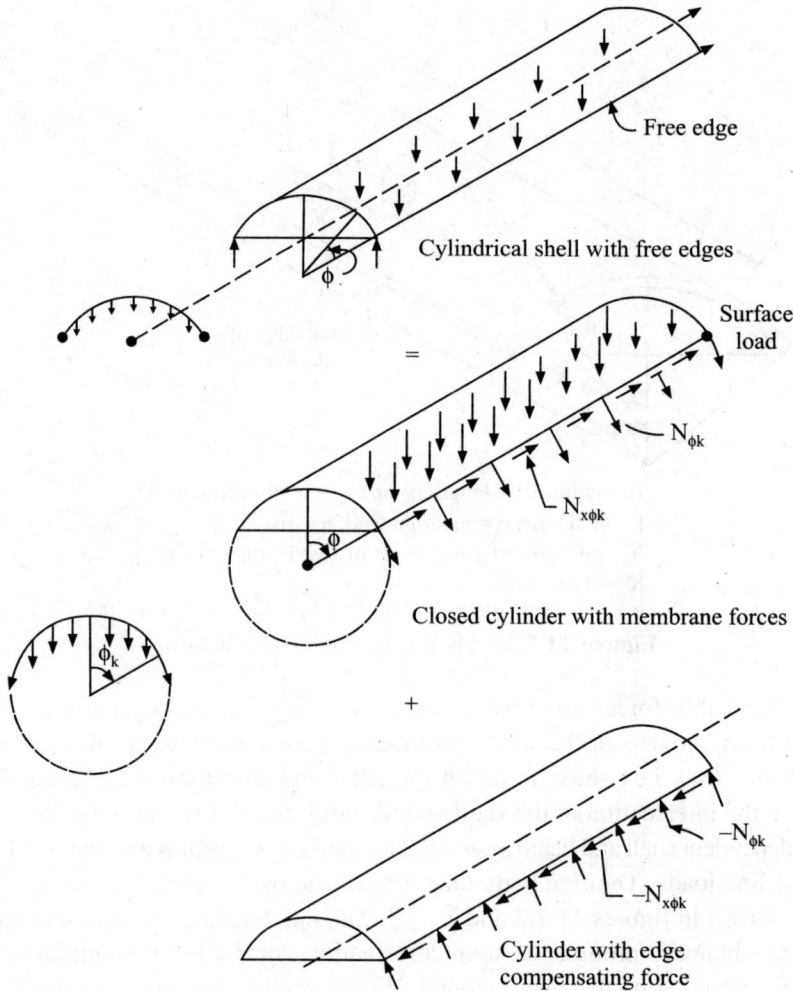

Figure 11.7.2 Supposition of two shells to get free edge conditions of a shell.

$$(\theta_0 + \theta_v + \theta_h + \theta_m + \theta_s)_s + \theta_0 + \theta_v + \theta_h + \theta_m + \theta_s)_b = 0 \qquad (11.7.3)$$

$$(s_0 + \sigma_v + \sigma_h + \sigma_m + \sigma_s)_s - (\sigma_0 + \sigma_v + \sigma_h + \sigma_m + \sigma_s)_b = 0 \qquad (11.7.4)$$

The deformations of the shell for various load conditions can be obtained directly from *ASCE Manual 31*. So only the deformations of prestressed concrete beam system are derived in this section.

The deformation compatibility at the intersection of beam and shell should be satisfied for all values of *x*. To meet this restraint, the external loads are expressed in Fourier expansion, with the result that the deformations will also be of the same form.

Let $\qquad\qquad\qquad\qquad q(x) =$ intensity of load

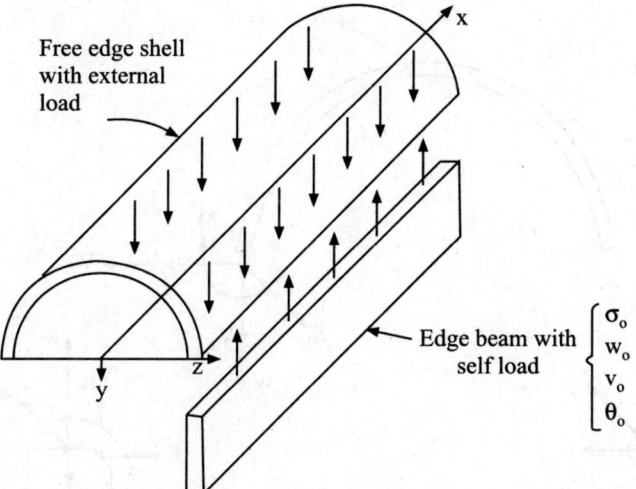

Free edge shell and edge beam with external force

+

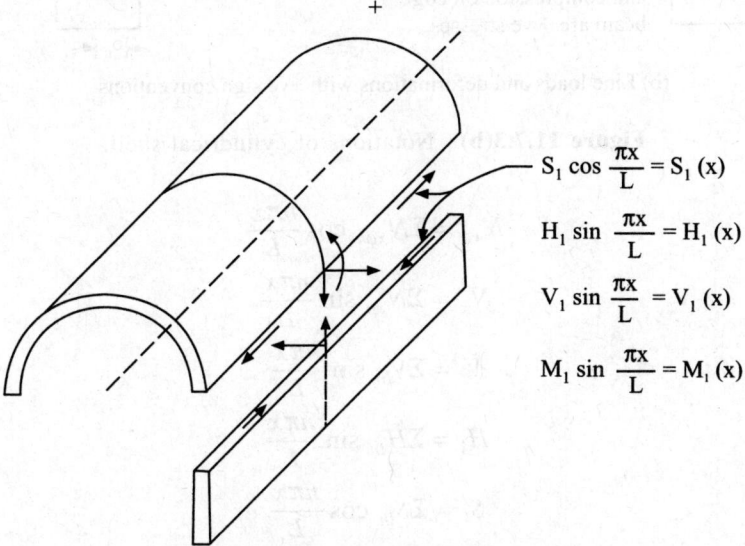

$$S_1 \cos \frac{\pi x}{L} = S_1(x)$$

$$H_1 \sin \frac{\pi x}{L} = H_1(x)$$

$$V_1 \sin \frac{\pi x}{L} = V_1(x)$$

$$M_1 \sin \frac{\pi x}{L} = M_1(x)$$

Shell and edge beam under line loads
(a) Cylindrical shell with edge beam compatible forces

Figure 11.7.3(a) Notations of cylindrical shell.

Then q may be expressed in Fourier expansion as

$$q(x) = \Sigma q_n \sin \frac{n\pi x}{L} \tag{11.7.5}$$

The corresponding edge forces and deformations are of the same form

$$N_\phi = \Sigma N_{\phi n} \sin \frac{n\pi x}{L}$$

420

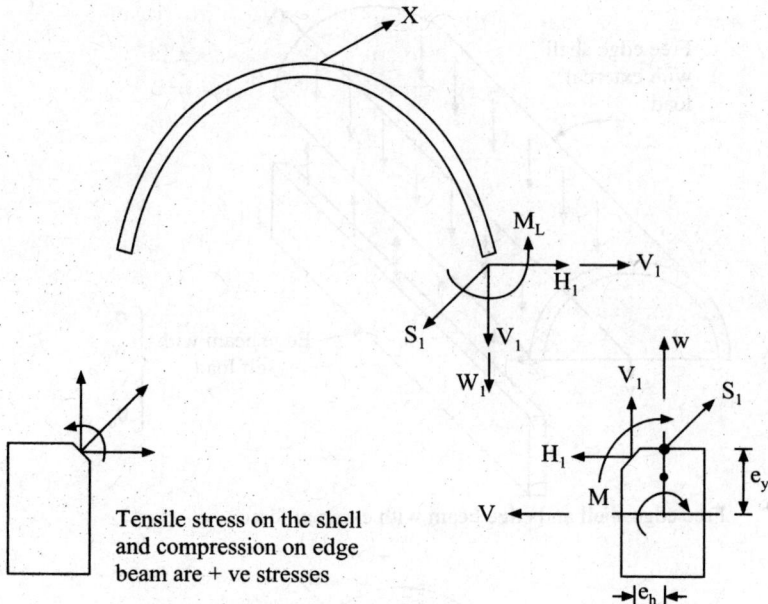

(b) Line loads and deformations with +ve sign conventions.

Figure 11.7.3(b)   Notations of cylindrical shell.

$$N_{x\phi} = \Sigma N_{x\phi n} \cos \frac{n\pi x}{L}$$

$$N_x = \Sigma N_{xn} \sin \frac{n\pi x}{L}$$

$$V_t = \Sigma V_{ln} \sin \frac{n\pi x}{L}$$

$$H_1 = \Sigma H_{ln} \sin \frac{n\pi x}{L}$$

$$S_1 = \Sigma S_{ln} \cos \frac{n\pi x}{L}$$

$$M_1 = \Sigma M_{ln} \sin \frac{n\pi x}{L}$$

$$w = \Sigma M_n \sin \frac{n\pi x}{L}$$

$$v = \Sigma v_n \sin \frac{n\pi x}{L}$$

$$\theta = \Sigma \theta_n \sin \frac{n\pi x}{L}$$

(11.7.6)

If $q(x) = q$ (constant), then expansion of Fourier series gives

$$q(x) = \frac{4}{\pi} q \Sigma \frac{1}{n} \sin \frac{n\pi x}{L}$$

(11.7.7)

Tensile stress on the shell
and compression on edge
beam are + ve stresses

It is desirable to work with as many terms of the series as possible. However, it was observed that approximation of the load with first term of the series gave fairly good results, so the design is presented using this approximation. Hence

$$q(x) = \frac{4}{\pi} q \sin \frac{\pi x}{L} \tag{11.7.8}$$

Similarly other quantities are expressed through first term of the series.

Force and deformations computed due to a uniform load intensity should always be multiplied by $(4/\pi \sin \pi x/L)$ if first term approximation is made for the uniform load. The line loads at the intersection of the shell and edge beam are always taken in the form as given in eq. (11.7.6) in which case the value $(4/\pi)$ does not appear.

Deformation of edge beam

Let q_{gb} = self weight intensity of beam acting downward

q_{pb} = prestressing force balancing component which is given by $= \dfrac{8P_e g}{L^2}$

P_e, g and L are already defined in Chapter III for load balancing design. The effective balancing force acting upward is given by

$$q_b(x) = q_{pb} - q_{gb} = \left(\frac{8P_e g}{L^2} - q_{gb} \right) \tag{11.7.9}$$

q_b is assumed to be uniform and then approximated by the first term of the series as

$$q_b(x) = \frac{4}{\pi} q_b \sin \frac{\pi x}{L} \tag{11.7.10}$$

External forces acting on the edge beam are shown in Figure 11.7.4 and the corresponding deformations can easily be obtained using

$$q(x) = \frac{4}{\pi} q \sin \frac{\pi x}{L}$$

$$M_x = \frac{4}{\pi} \left(\frac{qL^2}{\pi^2} - P_e e_0 \right) \sin \frac{\pi x}{L} \tag{11.7.11}$$

$$w_{0b} = \frac{4}{EI_h \pi} \left(\frac{q_b L^4}{\pi^4} - \frac{P_e e_0 L^2}{\pi^2} \right) \sin \frac{\pi x}{L} \tag{11.7.12}$$

$$\theta_{0b} = v_{0b} = 0$$

where I_h = moment of inertia of the beam about horizontal axis.

The deformations caused by various line loads are derived independently.

(i) *Deformations due to* $V \left(V_1 \sin \dfrac{\pi x}{L} \right)$: Considering Figure 11.7.5a

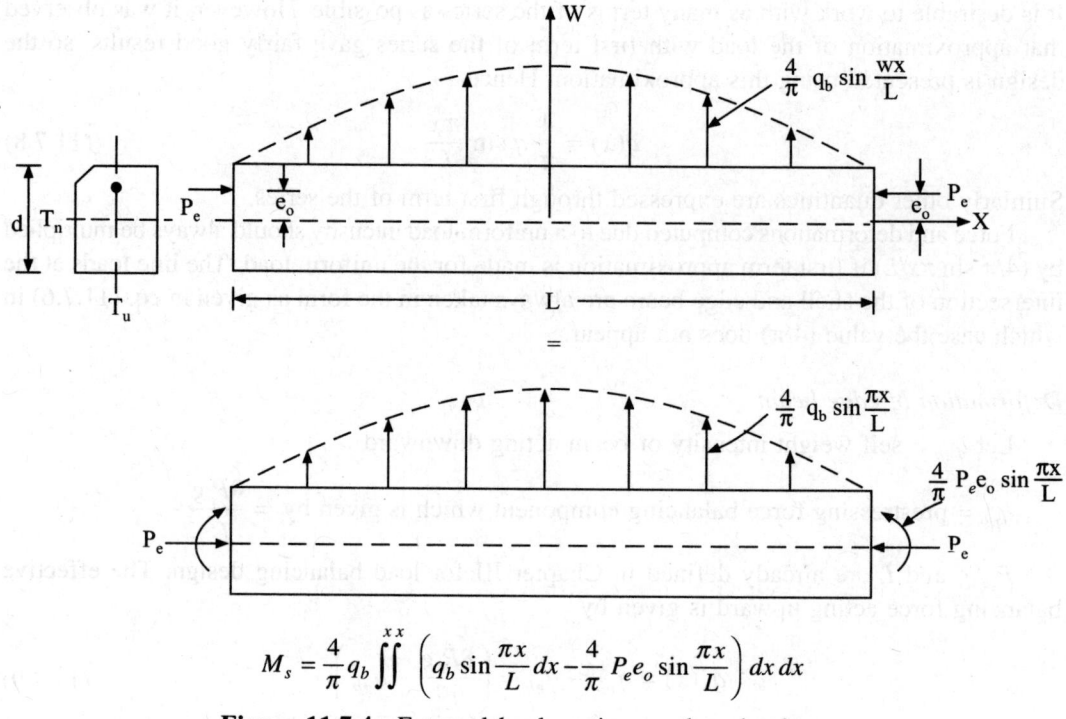

$$M_s = \frac{4}{\pi} q_b \int\limits^{xx}_{\ } \left(q_b \sin \frac{\pi x}{L} dx - \frac{4}{\pi} P_e e_o \sin \frac{\pi x}{L} \right) dx\, dx$$

Figure 11.7.4 External loads acting on the edge beam.

Torque on the beam at $x = \int\limits_0^x V_i e_h \sin \frac{\pi x}{L} dx = \frac{L}{\pi} V_i e_h \left(1 - \cos \frac{\pi x}{L} \right)$

where e_h = horizontal distance of the shell beam intersection point from centroid of the beam.

Rotation due to torque $= \frac{1}{GJ} \int\limits_0^x \left(\frac{V_i e_h L}{\pi} \right) \left(1 - \cos \frac{\pi x}{L} \right) dx = \frac{L V_i e_h}{\pi GJ} \left(x - \frac{L}{\pi} \sin \frac{\pi x}{L} \right)$

The twist at the middle section of the beam is given by substituting $x = L/2$ in the above expression.

$$\theta_{vb} = \frac{L^2 V_i e_h}{\pi GJ} \left(\frac{1}{2} - \frac{1}{\pi} \right) = (\pi - 2) \frac{L^2 e_h V_t}{2\pi^2 GJ} \tag{11.7.13}$$

where GJ = torsional rigidity of the beam

$$w_{vb} = \frac{V_1 L^4}{\pi^4 EI_h} + \theta_{rb}\, e_h = \frac{L^2}{\pi^2} \left(\frac{L^2}{\pi^2 EI_h} + \frac{(\pi - 2) e_h^2}{2GJ} \right) V_1 \tag{11.7.14}$$

$$v_{vb} = -\theta_{vb}\, e_v = \frac{-(\pi - 2) L^2}{2\pi\, GJ} e_v e_h\, V_1 \tag{11.7.15}$$

$$\sigma_{vb} = -\frac{Me_v}{I_h} = -\frac{L^2 e_v V_1}{\pi^2 I_h} \tag{11.7.16}$$

where e_v = vertical distance of the shell beam intersection point from the centroid of the beam. Compressive stress is treated as positive.

(ii) *Deformations due to horizontal load* $H_1 \left(H_1 \sin \dfrac{\pi x}{L} \right)$: Considering Figure 11.7.5b,

$$\text{Torque on edge beam } = -\int_0^x \left(H_1 e_v \sin \frac{\pi x}{L} \right) dx = \frac{-L}{\pi} H_1 e_v \left(1 - \cos \frac{\pi x}{L} \right)$$

Rotation of the beam at middle section (θ_{hb}) is

$$\theta_{hb} = -\frac{1}{GJ} \int_0^{L/2} \frac{L}{\pi} H_1 e_v \left(1 - \cos \frac{\pi x}{L} \right) dx = -\frac{(\pi-2)}{2\pi^2 GJ} L^2 e_v H_1 \tag{11.7.17}$$

$$w_{hb} = \theta_{hb} e_h = \frac{-(\pi-2) L^2}{2\pi^2 GJ} e_v e_h H_1 \tag{11.7.18}$$

$$v_{hb} = \frac{H_1 L^4}{\pi^4 EI_v} - \theta_{hb} e_v = \frac{L^2}{\pi^2} \left[\frac{L^2}{\pi^2 EI_v} + \frac{(\pi-2)}{2GJ} e_v^2 \right] H_1 \tag{11.7.19}$$

(iii) *Deformations due to Twist* $M_1 \left(M_1 \sin \dfrac{\pi x}{L} \right)$: Considering Figure 11.7. 5c,

$$\text{Torque on the edge beam } = \int_0^x M_1 \sin \frac{\pi x}{L} dx = \frac{L}{\pi} \left(1 - \cos \frac{\pi x}{L} \right) M_1$$

$$\theta_{mb} = \frac{L}{\pi} \left(\frac{M_1}{GJ} \right) \int_0^{L/2} \left(1 - \cos \frac{\pi x}{L} \right) dx = \frac{(\pi-2) L^2}{2\pi^2 GJ} M_1 \tag{11.7.20}$$

$$w_{mb} = \theta_{mb} e_h = \frac{(\pi-2) L^2}{2\pi GJ} e_h M_1 \tag{11.7.21}$$

$$v_{mb} = -\theta_{mb} e_v = \frac{-(\pi-2) L^2 e_v M_1}{2\pi^2 GJ}$$

$$\sigma_{mb} = 0$$

(iv) *Deformation due to shear forces* $S_1 \left(S_1 \sin \dfrac{\pi x}{L} \right)$: Considering Figure 11.7.5d, axial force at distance x is

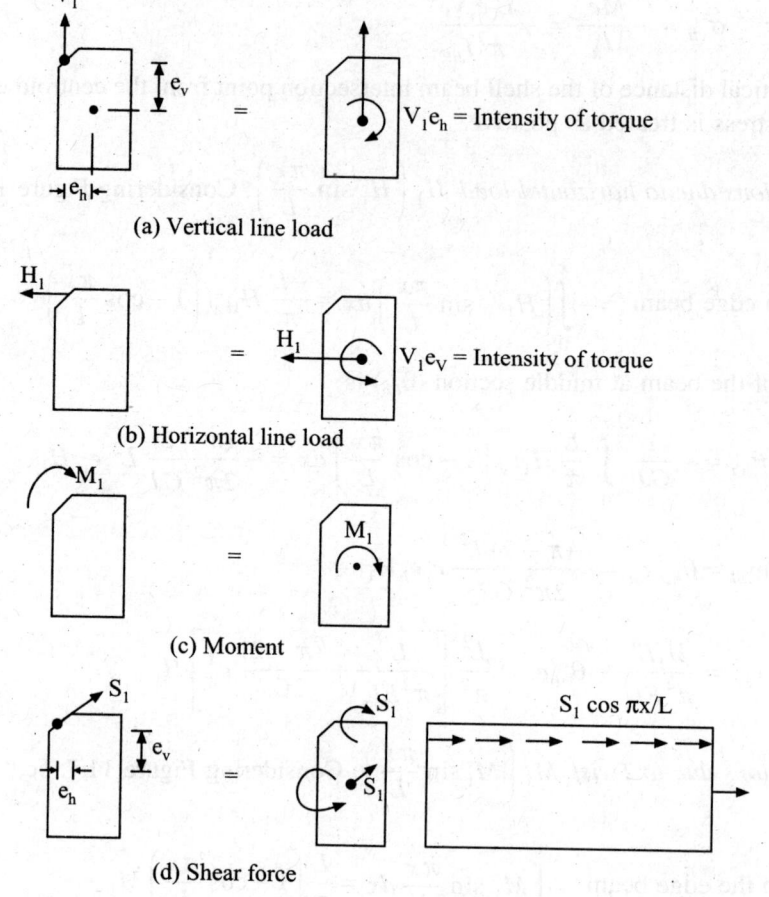

Figure 11.7.5 Various line loads acting on the edge beam at the intersection of the shell and beam.

$$F_x = \int_0^x S_1 \cos \frac{\pi x}{L} \, dx = \frac{LS_1}{\pi} \sin \frac{\pi x}{L}$$

$$M_{xh} = F_x e_v = \frac{L}{\pi} e_v S_1 \sin \frac{\pi x}{L}$$

$$M_{xv} = F_x e_h = \frac{L}{\pi} e_h S_1 \sin \frac{\pi x}{L}$$

$$0_{sb} = 0$$

$$w_{sb} = -\frac{L}{\pi} e_v S_1 \left(\frac{L^2}{\pi^2 EI_h} \right) = \frac{L^3 e_v S_1}{\pi^3 EI_h} \tag{11.7.22}$$

$$v_{sb} = -\frac{L^3 e_h S_1}{\pi^2 E I_v} \tag{11.7.23}$$

$$\sigma_{sb} = \frac{F_x}{A} = \frac{M_{xh}e_v}{I_h} + \frac{M_{xv}e_h}{I_v} = \frac{L}{\pi}\left(\frac{1}{A} + \frac{e_v^2}{I_h} + \frac{e_h^2}{I_v}\right)S_1 \tag{11.7.24}$$

All the deformations due to live loads are shown in matrix form in eq. 11.7.25. This matrix is a stiffness matrix and so symmetric. Symmetry in the stiffness coefficients can be obtained from Maxwells reciprocal theorem.

Example 11.7.1 Design of a prestressed concrete cylindrical shell to cover an area of about 40 m by 16 m.

Basic dimensions

Let span of the shell, (L) be 40 m; then the chord of the shell will be 16 m. Some basic dimensions such as rise, thickness of the shell and edge beam dimensions have to be assumed based on architectural and practical considerations.

Let rise of the shell, (h) = 3 m; then the radius (R) of the shell will be given by ($2R - h$) $h = 8 \times 8$ from which

$$R = 12.17 \text{ m, } \sin \phi_k = (8/12.17) = 0.658 \text{ or } \phi_k = 41°$$

Value of ϕ_k is chosen as 45° so that the coefficients of *ASCE* Manual 31 can be used directly.

$$R = 8 \text{ cosec } \phi_k = 11.31 \text{ m}$$

Let the thickness of shell = 15 cm,
width of the beam = 15 cm, and
depth of the beam = 200 cm

∴   (R/L) = 0.283 or (L/R) = 3.54, and (L/R)^2 = 12.5

Membrane analysis

Weight of the shell = 0.15 × 25000 = 3750 N/m^2

Surface load on the shell including
live load of 750 N/m^2 distributed $\Bigg\}$ = 1000 N/m^2
like the self weight

Total load on the shell = q = 3600 N/m^2

$$qR = 3600 (11.31) = 40\ 700 \text{ N/m}$$

The membrane stress resultants at the free edge of the shell are to be computed and they must be nullified by line forces at the free edges in case of free edge shell. Tables 4 and 5 of appendix A give the membrane forces caused in the shell for different values of ϕ. Since the line loads to be superimposed are of sinusoidal in nature, one must select a compatable set of

$$
\begin{bmatrix} w \\[2pt] v \\[2pt] \theta \\[2pt] \sigma \end{bmatrix}
=
\begin{bmatrix}
\dfrac{L^2}{\pi^2}\left(\dfrac{L^2}{\pi^2 EI_h}+\dfrac{(\pi-2)e_{h^2}}{2GJ}\right) & \dfrac{-L^2(\pi-2)e_v e_h}{2\pi^2 GJ} & \dfrac{L^2(\pi-2)e_h}{2\pi^2 GJ} & -\dfrac{L}{E\pi}\left(\dfrac{L^2 e_v}{\pi^2/h}\right) \\[14pt]
\dfrac{-L^2(\pi-2)e_v e_h}{2\pi GJ} & \dfrac{L^2}{\pi^2}\left(\dfrac{L^2}{\pi^2 EI_v}+\dfrac{(\pi-2)e_v^2}{2GJ}\right) & -\dfrac{L^2(\pi-2)e_v}{2\pi^2 GJ} & -\dfrac{L}{E\pi}\left(\dfrac{L^2 e_h}{\pi^2 I_v}\right) \\[14pt]
\dfrac{L^2(\pi-2)e_h}{2\pi^2 GJ} & -\dfrac{L^2(\pi-2)e_v}{2\pi^2 GJ} & \dfrac{L^2(\pi-2)}{2\pi GJ} & 0 \\[14pt]
\dfrac{L^2 e_h}{\pi^2 I_h} & -\dfrac{L^2 e_h}{\pi^2 I_v} & 0 & \dfrac{L}{\pi}\left(\dfrac{1}{A}+\dfrac{e_v^2}{I_h}+\dfrac{e_h^2}{I_v}\right)
\end{bmatrix}
\begin{bmatrix} V \\[2pt] H \\[2pt] M \\[2pt] S \end{bmatrix}
$$

(11.7.25)

stresses from the membrane. Even though Table 4 gives the membrane stress resultants for uniformly distributed load along the length but the coefficients from Table 5 are chosen so as to match with the compensating line loads which are of sinusoidal in nature. The actual load is uniform where as the coefficients of Table 5 are for sinusoidal load. Therefore the coefficients of Table 5 are multiplied by $(4/\pi)$ by approximating the UDL by a sinusoidal load.

The stress resultants at the free edge are indicated by a subscript k and the coefficients are indicated by the appropriate column or by C_{tm} where the subscript i-reads x or s or ϕ depending on N_x or $N_{x\phi}$ or N_ϕ respectively. (Vide Table 5 for the appropriate coefficients).

$$N_{xk} = \frac{4qR}{\pi}\left(\frac{L}{R}\right)^2 \text{ (col. 7)}$$

$$= 53730\ (12.51)\ (-0.1433) = -96320 \text{ N/m}$$

Please note that Col. 7 refers to the coefficient given in the column of the appropriate table, in this case it is in Table 5 for $\phi_k - \phi = 45°$. Similarly other columns.

$$N_{x\phi k} = \frac{4qR}{\pi}\left(\frac{L}{R}\right)^2 \text{ (col. 8)}$$

$$= 53730\ (3.54)\ (-0.4502) = -85630 \text{ N/m}$$

$$N_{\phi k} = \frac{4qR}{\pi} \text{ (col. 9)}$$

$$= 53730\ (-0.707) = -38000 \text{ N/m}$$

Similarly the deformation (w_{osm}) at the middle of the free edge due to membrane forces is

$$Ew_{osm} = \frac{4qR}{\pi}\left(\frac{L^4}{R^3 t}\right)\left[\left(\frac{2R}{L\pi}\right)^2 + \frac{2}{\pi^4} + \left(\frac{R}{L}\right)^4\right] \text{(Col.10)}$$

$$= (53730)\ 11,800\ (0.0324 + 0.0205 + 0.0064)(0.5)$$

$$= 1.88(10^7)$$

N_x and N_ϕ at $x = L/2$ and $N_{x\phi}$ at $x = 0$ due to membrane forces are calculated using the coefficients of the Table 5 and are shown in Table 11.7.1. These membrane stress resultants will be used for final stress resultants.

TABLE 11.7.1 Membrane stress resultants

ϕ	N_x at $(x = L/2)$		N_ϕ at $(x = L/2)$		$N_{x\phi}$ at $(x = 0)$	
	Coeff. (C_{xm})	Force	Coeff. $(C_{\phi m})$	Force	Coeff. (C_{sm})	Force
45	−0.2026	−136100	− 1.0	−53730	−0.0555	−0
30	−0.1957	−131540	−0.9659	−51900	−0.1648	−31350
20	−0.1837	−123480	−0.9063	−48700	−0.2690	−51170
10	−0.1660	−111580	−0.8191	−44010	−0.3652	−69460
0	−0.1443	−96320	−0.7071	−38000	−0.4502	−85630

Correction to membrane edge forces: Slender edge beam will have negligible stiffness against horizontal and torsional forces. The problem with slender edge beams then reduces to two free edge boundary conditions:

  (i) free horizontal displacement, and
  (ii) free torsional rotation.

Only the remaining two compatibility conditions have to be satisfied in the present shell as the edge beam is very slender.

Correction forces applied to force a free edge condition on the shell are $-N_{\phi k}$ and $-N_{x\phi k}$ and these forces are resolved in the vertical and horizontal directions.

$$V_m = -N_{\phi k} \sin \phi_k = (38000)\, 0.707 = 26870$$

$$H_m = -N_{\phi k} \cos \phi_k = 26870$$

$$S_m = -N_{x\phi k} = 85630$$

The particular shell parameters are $(R/t) = 75.4$ and $(R/L) = 0.283$ whereas the closest corresponding values in the Table 6 are 100 and 0.3 respectively.

Tables 6 and 7 of appendix A give the coefficients to obtain the stress resultants and deformations respectively caused by the edge forces. The edge forces are the vertical force (V_L), horizontal force (H_L) shear force (S_L) and bending moment (M_L). The vertical, horizontal and moment vary as sinusoidal whereas the shear force varies as cosine curve. This set is compatible where V_L, H_L, S_L and M_L correspond to the maximum ordinates. The subscript 'L' is only a dummy variable indicating a line load. It takes the notations as $L = m$ corresponds to membrane edge forces, and $L = 1$ or 2 corresponds interaction force between the edge of the shell and the edge beam. A stress resultant or a deformation caused by the edge line loads can be expressed as

$$X = C_{xo}\, (C_{xV}\, V_L + C_{xH}\, H_L + C_{xS}\, S_L + C_{xM}\, M_L)$$

where X = dummy variable and it is equal to N_x, $N_{x\phi}$, N_ϕ, w, v, etc.

  C_{xo} = common coefficient such as (L/R), $(L/R)^2$ $(L^4/R^3 t)$ etc. as seen in the Tables 6 and 7. These common coefficients are indicated in the notations of the Tables 6 and 7.

  C_{xV} = coefficients due to the vertical load and they are given in the first four columns of the tables.

  C_{xH} = coefficients due to the horizontal load at the edge and they are given in columns 5 to 8.

  C_{xS} = coefficients due to the shear load at the edge and they are given in columns 9 to 12.

  C_{xM} = coefficients due to the moment at the edge and they are given in columns 13 to 16.

For convenience of interpretation, the coefficient is replaced by the column number. For example

$$C_{xV} = (\text{col. 1}) \text{ or (col. 2) or (col. 3) or (col. 4).}$$

If one is interested in calculating the vertical displacement (w) which is considered downward as positive, it can be obtained by using Table 7. The vertical, horizontal and rotational deformations form a set and they can be caused by any of the four lines. Table 7 of the appendix A gives the information about the deformation. The vertical deformation caused by the edge loads can be expressed as

$$w = \frac{L^4}{R^3 tE} = \left[(\text{col. } 1)\, V_L + (\text{col. } 4)\, H_L + (\text{col. } 7)\, S_L + (\text{col. } 10)\, \frac{M_L}{R} \right]$$

One must select the appropriate value of (R/t) in the table 7.

In the present case, $R/L = 0.3$ and (R/t) = 100 are chosen, then enter table 7 for the above set and select the semicentral angle $\phi_k = 45°$, then the coefficients in the columns (1), (4), (7) and (10) are:

$$(\text{Col. } 1) = 46.06$$
$$(\text{Col. } 4) = -30.12$$
$$(\text{Col. } 7) = 1.073$$
$$(\text{Col. } 10) = -127.9$$

The vertical deflection caused by the edge loads is given by

$$Ew = \frac{L^4}{R^3 t} (46.06\, V_L - 30.12\, H_L + 1.073\, S_L - 127.9\, M_L)$$

where E = Youngs modulus of the material. If we are interested in calculating the deflections due to the membrane forces at the free edge, then we have

$$V_L = V_m = 26870\ N$$
$$H_L = H_m = 26870\ N$$
$$S_L = S_m = 85630\ N$$
$$M_L = M_m = 0$$

The substitution of various values in the expression for vertical deflection gives

$$Ew_{OSL} = 543360\, V_m + 12660\, S_m - 355\,320\, H_m = 614(10^7)$$

Similarly the tensile force at the middle point of the free edge caused by the line loads introduced at the free edge is obtained from Table 6.

$$N_{xL} = \left(-\frac{L}{R} \right)^2 \left[(\text{col. } 1)\, V_L + (\text{col. } 5)\, H_L + (\text{col. } 9)\, S_L + (\text{col. } 13)\, \frac{M_L}{R} \right]$$

$$= \left(-\frac{L}{R} \right)^2 (11.2L\, V_m - 5.257\, H_m + 0.874\, S_m) = 2950000\ N/m$$

In case of a free edge shell, that is when no edge beam is provided, the net effects such as deformations or stress resultants are obtained by superposition of the quantities in the membrane analysis and those caused by the compensating line loads equivalent and opposite to the membrane forces at the free edge. The net vertical deflection at the middle point of the free edge of the shell is given by

$$Ew_{os} = E(w_{oSm} + w_{osL}) = (1.88 + 614)\ 10^7 = 616(10^7)$$

Similarly the axial tension at the middle point of the free edge is

$$N_{ox} = N_{xl} + N_{xk} = 2950\ 000 - 96\ 320 = 2854\ 000 \text{ N/m}$$

The corresponding Tensile stress is

$$\sigma_{os} = \frac{N_{ox}}{t} = \frac{2854000}{0.15} = 1900\ 000 \text{ N/m}^2; A_{st} = 12203 \text{ mm/m}$$

The subscript s indicates the shell and o indicates the free edge.

Corrections to the shell deformations due to the provision of edge beam

Let there be edge beams provided at the so called free edges of the shell. That is at $\phi = \phi_k = 45°$, two longitudinal edge beams are provided to support the shell. The edge beam offers resistance thus causing interacting forces between the shell and the beam. Let the edge beam be thin and deep, and having negligible torsional and horizontal resistance, which means that the interacting moment and horizontal force on the edge beam must be zero. Therefore, out of the four interacting force at the interface of the shell and the beam, the above two forces must be zero. In other words the boundary condition of the shell with edge beam results into

$$M_1 = M_L = 0$$
$$H_1 = H_L = 0$$
$$V_1 = V_L = \text{exists}$$
$$S_1 = S_L = \text{exists}$$

The stress resultants caused by the line loads can be obtained from Table 6 of the appendix A for $R/t = 100$, $R/L = 0.3$ and $\phi_k = 45°$.

Let N_{xV} = axial force due to V_1

N_{xS} = axial force due to S_1

The total axial force in the shell caused by the V_1 and S_1 are:

$$(N_{xV} + N_{xs})_s = \left(\frac{L}{R}\right)^2 [(\text{col. 1}) V_1 + (\text{col. 9}) S_1] = 12.5\ (11.24\ V_1 + 0.8743\ S_1)$$

(vide Table 6)

The corresponding axial stress at the intersecting edge middle point is

$$\sigma_{vs} + \sigma_{ss} = \frac{(N_{xv} + N_{xs})_s}{t} = 83.3\ (11.24V_1 + 0.8743\ S_1)$$

Similarly the vertical deflection at the middle point of the edge of the shell is

$$E(w_{vs} + w_{ss}) = \frac{L^4}{Rt} [(\text{col. 1}) V_1 + (\text{col. 7}) S_1]$$
$$= 11,800\ (46.06\ V_1 + 1.073\ S_1)$$

(vide Table 7 for the column coefficients)

The interacting line load force V_1 and S_1 can be obtained from the compatibility condition of the vertical deflection and the axial strain (or equivalent axial stress) at the interface. The deformation and the axial stress in the edge beam caused by various effects are computed here.

Deformations of the edge beam due to external loads

Self weight of beam = 0.15 × 2 (25000) = 7500 N/m

Membrane correction force in the vertical direction is 26870 N/m acting vertically downward. To obtain internal forces, a vertical balancing force of about 26000 N/m should be applied through the prestressed concrete beam.

Let the effective balancing load = 26000 = q_b

Actual balancing prestressing force = 26000 + 7500 = 33500 N/m

Let the sag of the cable, (g) = 120 cm with eccentricity at the middle as 80 cm.

The-eccentricity of the cable at the end of the beam is = e_0 = 120 – 80 = 40 cm above the centroid of the beam section. This eccentricity introduces a bending moment of $P_e e_0$ causing compression on the top fibre of the beam. The vertical distance of the point of intersection of the shell and the beam from the centroid of the beam is = e_v = 94.7 cm. The effective uniformly distributed load of q_b acting vertically upwards and a constant bending moment of $-P_e e_0$ is converted into the first term of the Fourier series. The corresponding bending moment causing tension on the top fibre of the beam is

$$M_{ob} = \frac{4}{\pi}\left[\frac{L^2 q_b}{\pi^2} - P_e e_0\right]\sin\frac{\pi x}{L}$$

The stress at the mid-span of the beam at the interface of the shell and the beam caused by the prestressing and self weight of the beam is (σ_{ob})

$$\sigma_{ob} = \left[\frac{M_{ob}}{I_b}e_v - \frac{P_e}{A_b}\right]$$

where A_b = area of the beam section = 0.15 (2) = 0.3 m^2 and I_b = moment of inertia of the beam = 0.15 (2)^3/12 = 0.1 m^4. The vertical deflection (w_{ob}) at mid span of the beam caused by the effective vertical load q_b and the moment $-P_e e_0$ is obtained from

$$Ew_{ob} = -\frac{1}{I_b}\iint M_{ob}\,dx\,dx = \frac{4}{\pi I_b}\left[\frac{L^4 q_b}{\pi^4} - \frac{P_e e_0 L^2}{\pi^2}\right]$$

Similarly the mid-span bending moments caused by the $V_1\sin\dfrac{\pi x}{L}$ and $S_1\cos\dfrac{\pi x}{L}$ in the beam are

$$M_{vb} = \frac{L^2 V_1}{\pi^2}\ \text{and}$$

$$M_{sb} = \frac{LS_1 e_v}{\pi}$$

432

The corresponding deflections are given by

$$Ew_{vb} = \frac{L^4 V_1}{\pi^4 I_b}$$

and

$$Ew_{sb} = -\frac{L^2 e_v S_1}{\pi^3 I_b}$$

The bending stresses at the interface of the beam and the shell caused by V_1 and S_1 are:

$$\sigma_{vb} = \frac{L^2 e_v V_1}{\pi^2 I_b} \quad \text{and}$$

$$\sigma_{sb} = \frac{L}{\pi}\left(\frac{1}{A_b} + \frac{e_v^2}{I_b}\right) S_1$$

The prestressing force corresponding to the balancing force is given by

$$(8gP_e/L^2) = 33500$$

$$\therefore \qquad P_e = \frac{(32200)\,40 \times 40}{8 \times 1.2} = 5583000 \text{ N}$$

$$Ew_{ob} = \frac{4}{\pi I_b}\left(\frac{L^4 q_b}{\pi^4} - \frac{P_e e_0\, L^2}{\pi^2}\right)$$

$$= \frac{4}{\pi}(262,800\, q_b - 362 \times 10^7) = 660 \times 10^7$$

$$\sigma_{ob} = \frac{4e_v}{\pi I_b}\left(\frac{L^2}{\pi^2}\, q_b - P_e e_0\right) - \frac{P_e}{A_b}$$

$$= \frac{4}{\pi}(1540\, q_b - 21,215,000) - 18,610,000$$

$$= 20(10^6)$$

$$Ew_{vb} = \frac{L^4 V_1}{\pi^4 I_b} = 263,000\, V_1$$

$$Ew_{sb} = \frac{L^3 e_v}{\pi^3 I_b}\, S_1 = -19610\, S_1$$

$$\sigma_{vb} = \frac{L^2 e_v}{\pi^2}\, V_1 = 1540\, V_1$$

$$\sigma_{sb} = -\frac{L}{I\pi}\left(\frac{1}{A} + \frac{e_v^2}{I_b}\right) S_1 = -157\, S_1$$

Compatibility conditions: The vertical deflection compatibility yields

$$w_{os} + w_{vs} + w_{ss} + w_{ob} + w_{vb} + w_{sb} = 0$$

$$616 \times 10^7 + 11{,}800(46.01\ V_1 + 1.073\ S_1) + 660 \times 10^7 + 263{,}000\ V_1 - 19{,}610\ S_1 = 0$$

or $\qquad\qquad\qquad\qquad 0.8065\ V_1 - 0.007\ S_1 = -12760$

The longitudinal strain compatibility yields

$$\sigma_{os} + \sigma_{vs} + \sigma_{ss} - (\sigma_{ob} + \sigma_{sb} + \sigma_{vb}) = 0$$

$$28070000 + 125\ (11.24\ V_1 + 0.8743\ S_1) - 6340000 - 1620\ V_1 + 170\ S_1 = 0$$

or $\qquad\qquad\qquad\qquad -0.215\ V_1 + 0.280\ S_1 = -21730$

Solution of the above equations gives

$$V_1 = -12000 \text{ N/m}$$

$$S_1 = -86300 \text{ N/m}$$

Final stress resultants: There are two sets of correction line loads to be applied to the membrane solutions:

  (i) Correction line load forces to develop free edge in the membrane solution.
  (ii) Correction line load forces due to compatibility of the edge beam.

The final correction forces are

$$V = V_m + V_1 = 26000 - 12000 = 14000 \text{ N/m}$$

$$H = H_m + H_1 = 26000 \text{ N/m}$$

$$S = S_m + S_1 = 83000 - 86800 = -3800 \text{ N/m}$$

Stress resultants in the shell due to these correction forces are given in Table 11.7.2. Table 11.7.3 give final resultant forces in the shell which are obtained from Tables 11.7.1 and 11.7.2.

The stress resultants caused by the membrane action within the shell are given by:

$$N_{xm} = \frac{4}{\pi} qR \left(\frac{L}{R}\right)^2 C_{xm} \sin\frac{\pi x}{L}$$

$$N_{x\phi} = \frac{4}{\pi} qR \left(\frac{L}{R}\right) C_{sm} \cos\frac{\pi x}{L}$$

$$N_{\phi} = \frac{4}{\pi} qR\, C_{\phi m} \sin\frac{\pi x}{L}$$

where C_{xm}, C_{sm} and $C_{\phi n}$ are the coefficients taken from the Table 4 for appropriate angle ϕ.

The stress resultants caused by the line loads, V, H and S are given by:

$$N_{xl} = \left(\frac{L}{R}\right)^2 [C_{XV}\, V + C_{X_H}\, H + C_{XS}\, S] \sin\frac{\pi x}{L}$$

$$N_{x\phi l} = \left(\frac{L}{R}\right) [C_{SV}\, V + C_{SH}\, H + C_{SS}\, S] \cos\frac{\pi x}{L}$$

TABLE 11.7.2 Stress resultants doe to correction forces (N/m)

	N_x at $(x = L/2)$		$N_{x\phi}$ at $(x = 0)$		N_ϕ at $(x = L/2)$		M_ϕ at $(x = L/2)$	
ϕ	C_{XF}	$C_{XF}F$	C_{SF}	$C_{SF}F$	$C_{F\phi}$	$C_{\phi F}F$	C_{mF}	$C_{mF}F$
	Due to vertical edge load $(F = V = 14000)$							
45	1.400	19700	0	0	−1.178	−16600	−0.257	−3620
30	−1.044	−14600	0.432	6100	−1.454	−20500	−0.230	−3250
20	−3.110	−43800	−0.788	−11100	−1.326	−18800	−0.174	−2450
10	−0.946	−13300	−2.214	−31200	− .3300	−4600	−0.090	−12.6
0	11.240	157300	0	0	0.700	9950	0	0
	Due to horizontal edge load $(F = H = 26000)$							
45	− 1.970	−51100	0	0	0.306	7930	0.159	4140
30	−0.047	−1220	−1.061	−27600	0.839	21800	0.155	4030
20	1.941	50500	0.505	−13160	1.279	33300	0.134	3490
10	1.513	39300	0.633	16500	1.158	30100	0 082	2130
0	−5.257	−137000	0	0	0.707	18400	0	0
	Due to shear edge load $(F = S = -3800)$							
45	−0.119	440	0	0	0.072	−280	−0.006	20
30	−0.077	290	−0.085	330	−0.033	120	−0.004	10
20	0.038	−140	−0.101	380	0.024	− 90	−0.002	0
10	0.321	−1220	−0.013	50	0.024	− 90	0	0
0	0.870	−3270	0.300	−1140	0	0	0	0

*where F indicates either V, or H or S depending on the table location.

$$N_\phi = [C_{\phi V}\ V + C_{\phi H}\ H + C_{\phi S}\ S]\sin\frac{\pi x}{L}$$

$$M_\phi = R[C_{MV}\ V + C_{MH}\ H + C_{MS}\ S]\sin\frac{\pi x}{L}$$

where C_{XV} = coefficient corresponding longitudinal stress and associated with vertical load.

Similarly the other coefficients are defined. The first subscript refer to the nature of stress resultant and the second subscript refers to the line load force causing the stress. These coefficients are to be taken from Table 6 for appropriate values of R/L, R/t, ϕ_K and ϕ.

Reinforcement details

The reinforcement is to be provided as per the stress resultants calculated in Table 11.7.3. The longitudinal stress near the crown is compressive and it decreases towards the edge of the shell. The nature of the stress changes from compression to tension at 20° from the edge of the shell.

TABLE 11.7.3 Final stress resultants (N/m)

0		Due to force			Final
	Membrane	V_1	H_1	S_1	

N_x at $(x = L/2)$

[Common multiplier = $(L/R)^2$ = 12.5]

45	−132000	19700	−51100	440	−519000
30	−126500	−14600	− 1220	290	−295600
20	−119100	−43800	50500	−140	− 37100
10	−108000	−13300	39300	−1220	201700
0	− 93000	157300	−137000	−3270	−119000

$N_{x\phi}$ at $(x = 0)$

[Common multiplier – L/R = 3.54]

45	−1020	0	0	0	−1020
30	−30200	6100	−27600	330	−105100
20	−49500	−11100	−13160	380	−134000
10	−67200	−31200	16500	50	−119000
0	−83000	0	0	−1140	− 87000

N_ϕ at $(x = L/2)$

[Common multiplier = 1]

45	−52000	−16600	7930	−280	− 60950
30	−50100	−20500	21800	120	− 48680
20	−47100	−18800	33300	− 90	− 32690
10	−42500	−4600	30100	− 90	− 17070
0	−36700	9950	18400	0	− 86050

M_ϕ at $(x = L/2)$

[Common multiplier = R = 11.31]

45	0	−3620	4140	2	6100
30	0	−3250	4030	1	8930
20	0	−2450	3490	0	11650
10	0	−1260	2100	0	9840
0	0	0	0	0	0

The maximum compressive stress which occurs at the crown is

$$\sigma_{ob} = \frac{N_x}{t} = \frac{519000}{10(100)} = 519 \text{ N/cm}^2$$

The concrete used in the shell portion is of M 20 and that used in the edge beam is of M 35.

The permissible bending compressive stress in the shell concrete is 700 N/cm^2, therefore the actual bending stress is within the permissible limits. It is also necessary to check whether the actual compressive stress is less than the permissible buckling stress. The buckling stress σ_{cr} in long cylindrical shells is given by

$$\sigma_{cr} = \frac{0.2Et}{R}$$

$$E = 5700\sqrt{f_{ck}} = 25490 \text{ MPa}$$

$$\sigma_{cr} = \frac{0.2(2.54)(10)^6(10)}{1131} = 4490 \text{ N/cm}^2$$

The shell is safe against elastic buckling.

A nominal reinforcement of 0.15 per cent is to be provided in the compression zone of the shell. The nominal longitudinal reinforcement from crown to 5000 mm from the edge is:

$$A_{st1} = \frac{0.15}{100}(10)(100) = 1.5 \text{ cm}^2/\text{m}$$

*Use 8 mm diameter bars at 150 mm spacing in the longitudinal direction from crown 5000 mm from the edge.

Maximum tensile stress occurs around 10° from the edge of the shell and it is 201 700 N/m. The tensile force at the edge is 141 100 N/m.

The average tensile force = (201 700 + 119 100)/2 = 160 350 N/m

Assuming mild steel reinforcement is used in the shell, the permissible tensile stress in steel is taken as 14000 N/cm^2.

Area of longitudinal tension steel is

$$A_{st2} = \frac{160\,350}{14000} = 11.45 \text{ cm}^2/\text{m}$$

*Provide 14 mm diameter longitudinal bars at 250 mm spacing at top and bottom surface of the shell starting from edge to 25° of the shell (which is 5000 mm of arc length). 25% of the longitudinal reinforcement is curtailed at $L/4 = 10$ m from support and another 25% at $L/6 = 6.67$ m from support.

(b) *Shear Reinforcement:*

The maximum shear force which occurs about 20° from the edge and at the support is = 134000 N/m.

$$\text{Maximum shear stress} = \frac{134\,000}{10(1000)} = 134 \text{ N/cm}^2$$

Shear reinforcement must be provided to resist the diagonal tension.

The maximum diagonal tension force = 134000 N/cm^2.

$$\text{Area of steel required} = A_{st3} = \frac{134000}{14000} = 9.6 \text{ cm}^2/\text{m}.$$

*Provide 12 mm diameter bars at 120 mm spacing in the middle surface of the shell at 45° to the longitudinal axis. This spacing is at 20° from the edge. The spacing of the bars is gradually increased to 200 mm towards the crown. The shear reinforcement is provided from end support to L/5 = 8 m towards the mid span.

(c) *Transverse reinforcement*

The maximum bending moment on the shell occurs at about 20 from the edge of the shell at mid span. The bending moment decreases towards the crown and it is all the way positive.

The maximum bending moment = M = 11650 Nm/m

$$\text{Effective depth needed} = d = \sqrt{\frac{M}{Kb}} = \sqrt{\left[\frac{1165000}{125(100)}\right]} = 11.5 \text{ cm}$$

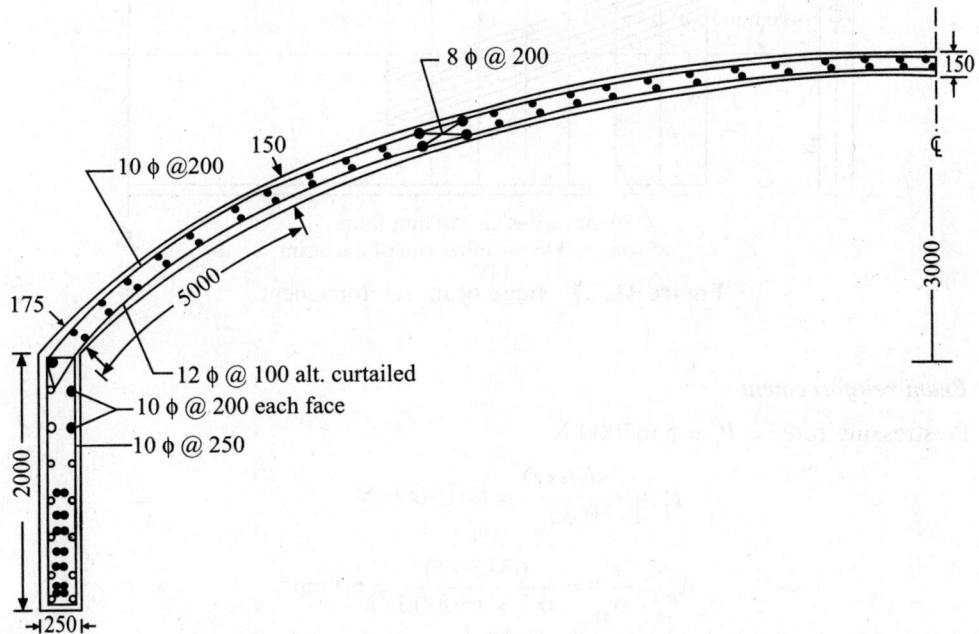

Figure 11.7.6 Reinforcement details of the cross-section at mid-span of the shell (the dimensions are in mm).

It is necessary to thicken the shell near the edge up to 20° (4000 mm) from edge. Let the thickness at the edge is 175 mm and it is gradually reduced to 150 mm at 4000 mm from the edge.

Since the thickness available at the maximum bending moment zone is less than that is required for a balanced section, the section is designed as doubly reinforced section.

$$\text{Tension steel required} = \frac{1165000}{14000(0.865)(9)} = 10.8 \text{ cm}^2/\text{m}$$

*Provide 12 mm dia bars at 100 mm spacing at the bottom face of the shell. These bars extend from the edge beam till 5000 mm into the arch. Then only the alternate bars are taken up to the crown. Provide 10 mm diameter bars at 200 mm spacing at the top surface of the shell and they are curtailed at 5000 mm from the edge beam. This spacing of the reinforcement is adapted at the middle span and it is gradually increased to 2.5 times at the end supports.

No reinforcement is needed for circumferential membrane stress as it is throughout compressive. The doubly reinforced section is designed such that combined compressive stress due to bending and axial thrust is with in the permissible limits.

Figure 11.7.6 illustrates the reinforcement of the shell.

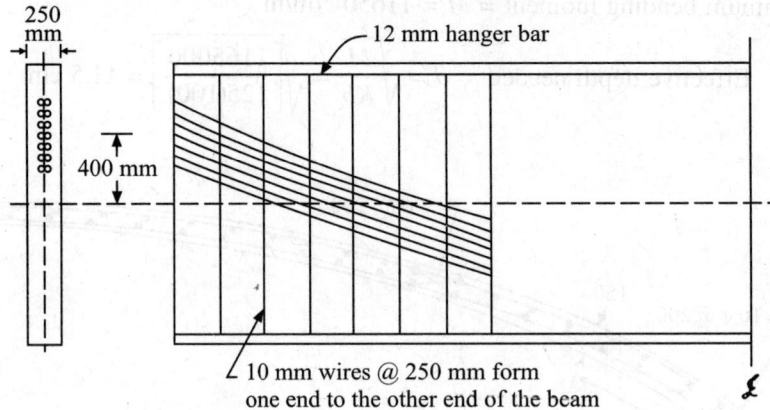

Figure 11.7.7 Edge beam reinforcement.

(d) *Beam reinforcement*

Prestressing force = P_e = 5367000 N

$$P_t = \frac{5367000}{0.85} = 6313000 \text{ N}$$

$$A_p = \frac{P_t}{\sigma_{st}} = \frac{6313000}{0.7 \times 150000} = 60 \text{ cm}^2$$

Use 14 numbers of 12 number 7 mm cables. Provide also nominal shear and torsional reinforcement. Use 10 mm shear stirrups at 25 cm apart with 4 numbers of 12 mm corner bars. The reinforcement details in the beam are shown in Figure 11.7.7.

11.8. Design of Cylindrical Shells by Using Design Coefficients

It has been found (11.17) that the design variables such as the radius of the shell, the prestressing force and the depth of the edge beam converge to a set of values when programmed for minimum cost of the shell. Typical variation in the cost of materials of a set of shells is given in Table 11.8.1 from which it can be observed that the cost of concrete varies from 50 to 60 per cent of the total cost of the materials. Cost of high tensile steel varies 15 to 40 per cent of the total cost of the materials. The mild steel reinforcement required for critical shear force and transverse bending moment is mostly nominal in prestressed concrete shells. The critical shear force and transverse bending moments are limited to a small zone. Tables 11.8.2 and 11.8.3

TABLE 11.8.1 Relative cost of materials in cylindrical shells

Chord (B) (m)	$\frac{L}{B}$	Cost of concrete			Cost of steel				Cost of prestress steel	Total cost*
		Shell	Edge beam	Total	Longi-tudinal	Trans-verse	Shear	Total		
10	2	0.42	0.08	0.50	0.04	0.05	0.01	0.10	0.10	0.70
(t = 7.5 cm)	3	0.62	0.19	0.81	0.06	0.07	0.03	0.16	0.28	1.25
(q = 2500 N/m²)	4	0.83	0.39	1.22	0.08	0.09	0:03	0.20	0.62	2 04
	5	1.04	1.04	2.08	0.09	0.23	0.03	0.35	1.43	3.86
15	2	0.94	0.18	1.12	0.08	0.11	0.05	0.24	0.37	1.73
(t = 8.5 cm)	3	1.41	0.63	2.04	0.13	0.18	0.10	0.41	0.86	3.31
(q = 2750 N/m*)	4	1.86	1.47	3.33	0.17	0.84	0.10	1.11	2.75	7.19
20										
(t = 10 cm)	2	1.68	0.40	2.08	0.15	0.28	0.13	0.56	0.76	3.40
(q = 3000 N/m²)	3	2.52	0.97	3.49	0.23	0-49	0.23	0.95	2.07	6.51

*Cost of material is given in equivalent volume of mild steel in m².
q = Total load per unit surface area.

TABLE 11.8.2 Design Variables for outer shells

Chord width (m)	$\frac{L}{B}$	R (cm)	d (cm)	P_t (kN)	$\frac{N_x}{qL}$	$\frac{N_\phi}{qB}$	$\frac{N_{x\phi}}{qB}$	1000 M_ϕ/qB^2		
								Edge	Middle	Crown
10	2	1093	40	790	-3.21	-1.68	-4.50	0.76	-3.74	-6.67
(t = 7.5 cm)	3	1052	60	1430	-3.87	-1.61	-5.65	0.03	4.29	-6.25
(q = 2500 N/m²)	4	1047	97	2340	-5.03	-1.52	-6.81	-2.82	5.82	0.65
	5	1033	207	4290	-4.63	-1.18	-6.64	-15.08	15.41	20.68
15	2	1533	60	1870	-3.88	-1.51	-4.46	-0.36	3.75	-4.03
(t = 8.5 cm)	3	1544	140	2870	-3.85	-1.51	-5.07	-3.11	4.62	-1.50
(q = 2750 N/m²)	4	1565	246	7710	-3.48	-1.51	-5.23	24.00	20.84	31.74
20										
(t = 10 cm)	2	1933	102	2870	-2.90	-1.41	-4.00	-1.04	3.80	-3.42
(q = 3000 N/m²)	3	1946	162	5180	-3.59	-1.46	-5.12	-1.71	4.56	-3.25

TABLE 11.8.3 Design variables for inner shells

Chord (m)	$\dfrac{L}{B}$	R (cm)	d (cm)	P_t (kN)	$\dfrac{N_x}{qL}$	$\dfrac{N_\phi}{qB}$	$\dfrac{N_{x\phi}}{qB}$	1000 M_ϕ/qB^2		
								Edge	Middle	Crown
10	2	1161	20	760	−3.37	−1.74	−4.75	−8.41	2.49	−2.52
(t = 7.5 cm)	3	1147	40	1520	−4.77	−1.70	−6.35	−7.51	2.78	−1.98
(q = 2500 N/m²)	4	1030	60	2400	−5.56	−1.52	−7.36	−6.25	3.59	−0.99
	5	1071	179	4210	−4.62	−1.48	−7.12	−4.20	2.02	−0.60
15	2	1670	58	2010	−4.21	−1.59	−4.68	−6.12	1.58	−1.93
(t = 7.5 cm)	3	1649	112	3160	−4.33	−1.50	−5.15	−7.02	3.17	−2.30
(q = 2750 N/m²)	4	1632	215	6530	−3.85	−1.52	−6.18	−4.70	1.43	−1.46
20	2	2067	100	2870	−2.58	−1.49	−4.02	−6.37	1.76	−2.17
(t = 10 cm)	3	2050	156	5370	−3.77	−1.49	−5.31	−6.49	2.01	−2.27
(q = 3000 N/m²)										

give the design details of outer and inner barrel shells with 10 m, 15 m and 20 m chord widths. These details can be used directly in the design of prestressed concrete shells. In the case of shells having chord widths other than those specified, the details of design can be obtained by interpolation. The design details are obtained for specific loads which are commonly used in India. The loads acting on the shell and the thickness used in the analysis are also shown in the tables. The width of edge beam is kept to a minimum of 15 to 20 cm and the sag of the cable as 0.4 depth of the beam.

Even though the Tables 11.8.2 and 11.8.3 give details for specific shells the accuracy of the results obtained by an interpolation for other shells has been tested and found to give acceptable results. An illustrative example of design of a shell explains the method of design of a cylindrical shell using the fables.

Example 11.8.1 Design a prestressed concrete single roof shell to cover an area of 15 m by 37.5 m.

Solution

Assumptions and data:

Let the concrete used in shell and edge beam correspond to M 20 and M 35 respectively.

$$t = 8.5 \text{ cm, width of edge beam} = 20 \text{ cm}$$

$$\sigma_{st} = 90000 \text{ N/cm}^2$$

$$q_g = 0.085 \times 25000 = 2125 \text{ N/m}^2$$

$$q_l = 625 \text{ N/m}^2$$

$$q_t = q_g + q_l = 2750 \text{ N/m}^2$$

$$L/B = 37.5/15 = 2.5$$

Interpolation of details of the design.

The details are taken from Table 11.8.2 since the shell is a single shell. An interpolation for $L/B = 2.5$ is made from $L/B = 2$ and $L/B = 3$ in which $B = 15$ m. The interpolated values are given in Table 11.8.4.

TABLE 11.8.4 Interpolation of design details
B = 15 m and $L/B = 2.5$

Details		$L/B = 2$	$L/B = 3$	$L/B = 2.5$
Radius R (in metres)		15.33	15.54	15.44
Depth of edge beam d (m)		0.60	1.40	1.00
Prestress force P_t (kN)		1870	2870	2370
N_x/qL		−3.88	−3.85	−3.86
N_ϕ/qB		−1.51	−1.51	−1.51
$N_{x\phi}/qB$		−4.46	−5.07	−4.77
$1000\, M_\phi/qB^2$	Edge	−0.36	−3.11	−1.24
	Middle	3.75	4.62	4.19
	Crown	−4.03	−1.50	−2.77

Design details

The forces and area of prestressing steel are computed using the Table 11.8.4.

$$N_x = -3.86(2750)(37.5) = -400\ 000 \text{ N/m}$$

$$\sigma_x = -\frac{400000}{8.5(100)} = -470 \text{ N/cm}^2 \text{ (permissible)}$$

$$N_\phi = -1.51(2750)(15) = -62200 \text{ N/m}$$

$$\sigma_\phi = -\frac{62200}{8.5(100)} = -73 \text{ N/cm}^2 \text{ (permissible)}$$

$$N_{x\phi} = -4.77(2750)(15) = -197000 \text{ N/m}$$

$$\sigma_{x\phi} = -\frac{197000}{8.5(100)} = -232 \text{ N/cm}^2$$

Provide nominal longitudinal reinforcement and then design shear reinforcement at the end quarter of the shell to resist the shear force given above.

$$M_\phi \text{ at edge} = -1.24(2750)(225)/1000 = -760 \text{ N m/m}$$

$$M_\phi \text{ at middle} = 3.75\ (2750)(225)/1000 = 2590 \text{ N m/m}$$

$$M_\phi \text{ at crown} = -2.77(2750)(225)/1000 = -1710 \text{ N m/m}$$

Bending moment at crown and at middle of the shell requires reinforcement so provide 12 mm bars at 16 cm at bottom fibre and 10 mm bars at 16 cm at the top fibre at the mid-span of the shell and gradually increase the spacing to 25 cm at the ends.

$$P_t = 2370 \text{ kN in each beam}$$

$$A_p = \frac{2370000}{90000} = 26.3 \text{ cm}^2$$

Provide 70 numbers of 7 mm wires with a sag of $0.4d = 40$ cm in each edge beam.

Size of the edge beam = 20 cm by 100 cm.

11.9. Some Recommended Specifications for Shell and Folded Plate Structures

The following guidelines are recommended for designers and builders. There is much more flexibility in special cases.

11.9.1 *Materials*

M 15 and M 20 concretes are usually recommended in the RCC shell construction and M 35 to M 45 concrete is recommended in prestressed concrete construction. The maximum size of the aggregate is 12 mm for 60 mm thick shell and 20 mm for thicker ones. Mild steel plain or deformed bars of small diameters can be used. High tensile steel wires of 5 mm are desirable; however, 7 mm wires can be used in shells having thickness more than 100 mm thick in prestressed concrete elements.

11.9.2 *Dimensions*

Minimum thickness of cast-in-situ singly curved shells to be 50 mm and doubly curved ones to be 40 mm. Thickness of precast shells can be as low as 25 mm. Thickness of folded plates is to be more than 70 mm. Span to total rise ratio of the shells can be around 4 to 15 unless otherwise needed. The semi-central angle in the case of cylindrical shells can be between 30 and 45°. The level of working should not coincide with the centre of curvature of the shell. The width of the edge beams of the diaphragms should be limited to three times the thickness of the shell. The thickness of the shell near the edges may be of the order of 30 per cent of the shell thickness. The distance through which thickening is recommended be between $0.4 \sqrt{Rt}$ and $0.8 \sqrt{Rt}$.

11.9.3 *Reinforcement Details*

Maximum sizes of the bars in the unthickened shells are: 10 mm for 50 mm thick shell, 12 mm for 65 mm thick one and 16.mm for shells having thickness more than 65 mm. Larger diameter bars are permissible in the thickened portion. The total depth occupied by the reinforcement in the thickened portion shall be not more than three times the maximum size of the recommended bar. The minimum diameter of the plain bar is 5 mm. The spacing of the bars should not be more than five times the thickness of the shell. Minimum clear cover must be 12 mm or the nominal size of the bar. Minimum of 0.15 per cent of reinforcement in each direction or 8 mm bars at 150 mm spacing is recommended.

11.9.4 *Construction*

Expansion joints be provided in shells having lengths more than 40 m. Double columns and diaphragms are to be provided at the expansion joint. The construction joints be made along the curved lengths where there is minimum shear force. Shells can be cured by spraying of water over gunny bags layed over the shell for a minimum period of two weeks. The decentering may commence when the concrete has gained a strength equal to twice the stress caused by the dead loads alone. This may be done after 14 days for shell portion and 21 days for beam and diaphragm portions.

PROBLEMS

11.1 An area of 20 m by 30 m is to be provided by a folded plate roof. The length of the folded plate be taken as 30 m and the folds as inverted U type of 5 m fold width. The rise of the fold should not exceed 2 m. Design the folded plate using $f_{ck} = 40$ MPa and $f_p = 1500$ MPa for a service load of 750 N/m^2.

11.2 A circular tank is to be designed for storage of 650,000 litres of water. Design the complete details of the water tank listed below using $f_{ck} = 40$ MPa, and $f_p = 1500$ MPa.
  (a) Vertical walls of the tank as prestressed with base hinged or fixed.
  (b) Top ring beam as prestressed concrete beam.
  (c) Top dome as ordinary reinforced concrete in the radial direction. Hoop prestressing steel if desirable.

11.3 Design a 30 m long and 10 m wide circular cylindrical shell for a service load of 750 N/m^2. Provide an edge beam of depth not exceeding 100 cm. The shell is to be supported by end diaphragms resting on columns.

CHAPTER 12

Bond in Prestressed Concrete

12.1. Introduction

Bond between steel and concrete in prestressed concrete construction exists on two different basis. In pre-tensioning system, bond is used as a means of transferring the prestressing force of steel to the concrete section while in post-tensioning system, bond is found to be necessary for two different purposes:

  (i) protection against corrosion; and
  (ii) increase in ultimate strength.

    Importance of bond in pre-tensioned construction was recognised from the very beginning, whereas the need for bond in post-tensioned construction has come into light with the progress in the methods of constructions.

12.2. Bond in Pre-tensioned Construction

Bond in pre-tensioned construction is furnished by two factors. When a prestressing cable is tensioned, it is accompanied by a reduction in area of cross section of steel. As the tendon is released from jacks, a part of the lateral strain is recovered at the ends of the cable. Due to this increased cross sectional area, frictional bond is developed between steel and concrete. A second factor contributing to the bond is the adhesive property between the two materials as, the concrete hardens. The bond of wires at the edges is of great importance since there is no special provision to transfer the prestressing force.

    When a wire is released from the prestressing bed, it tends to regain its original length. The force at the very end of the cable has to be zero since it is free; therefore the original area of the cross section is restored at the face. Due to the increase in the cross section of the tendon, there will be a high restraining force between the tendon and concrete. This phenomenon of recovery of lateral contraction develops a wedge action at the end of the cable through which the prestressing force is transferred. This type of self anchoring property was discussed by Hoyer and is called "Hoyer effect". The Hoyer effect is illustrated in Figure 12.2.1. The bond length needed to transmit the complete prestressing force is called "transmission length" (l_t). The bond stress along the transmission length is maximum immediately after the free surface and slowly evens out to uniform at the end of transmission length. The bond and other stress distributions along the transmission length are shown in Figure 12.2.2.

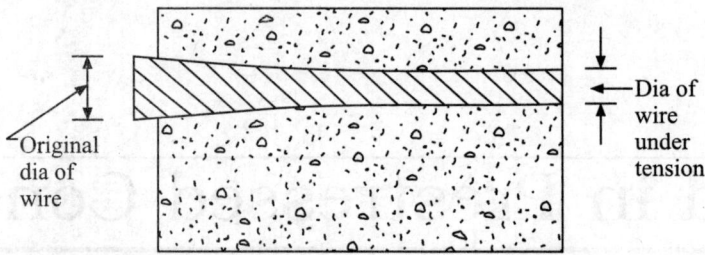

Figure 12.2.1 Illustration of Hoyer effect in pre-tensioned construction.

It is seen from Figure 12.2.2 that compressive as well as tensile stresses exist in the transmission zone, so transverse reinforcement has to be provided to resist the tensile f ·ͻ

Bond length is generally equal to transfer or transmission length and was derived by Hoyer as

$$l_t = \frac{D}{2\mu}(1 + m_c)\left(\frac{n}{m_s} - \frac{\sigma_{si}}{E_c}\right)\frac{\sigma_{se}}{(2\sigma_{si} - \sigma_{se})} \quad (12.2.1)$$

where D = diameter of the wire

μ = coefficient of friction

m_c = Poisson's ratio for concrete

m_s = Poisson's ratio for steel

n = modular ratio

Let

$m_c = 0.1$, $m_s = 0.3$

$E_c = 3(10^6)$ N/cm², $E_s = 20(10^6)$ N/cm²

$\sigma_{se} = 0.6 f_p$, $\sigma_{st} = 0.7 f_p$

Then $l_t = \dfrac{D}{2\mu}(1.1)\left(\dfrac{20}{0.9} - \dfrac{\sigma_{st}}{E_c}\right)\left(\dfrac{6}{8}\right)$

$\dfrac{\sigma_{st}}{E_c}$ is small as compared to $\dfrac{20}{0.9}$, so neglecting $\dfrac{\sigma_{st}}{E_c}$

$$l_t = \frac{55}{6} \times \frac{D}{\mu} \simeq \frac{9D}{\mu} \quad (12.2.2)$$

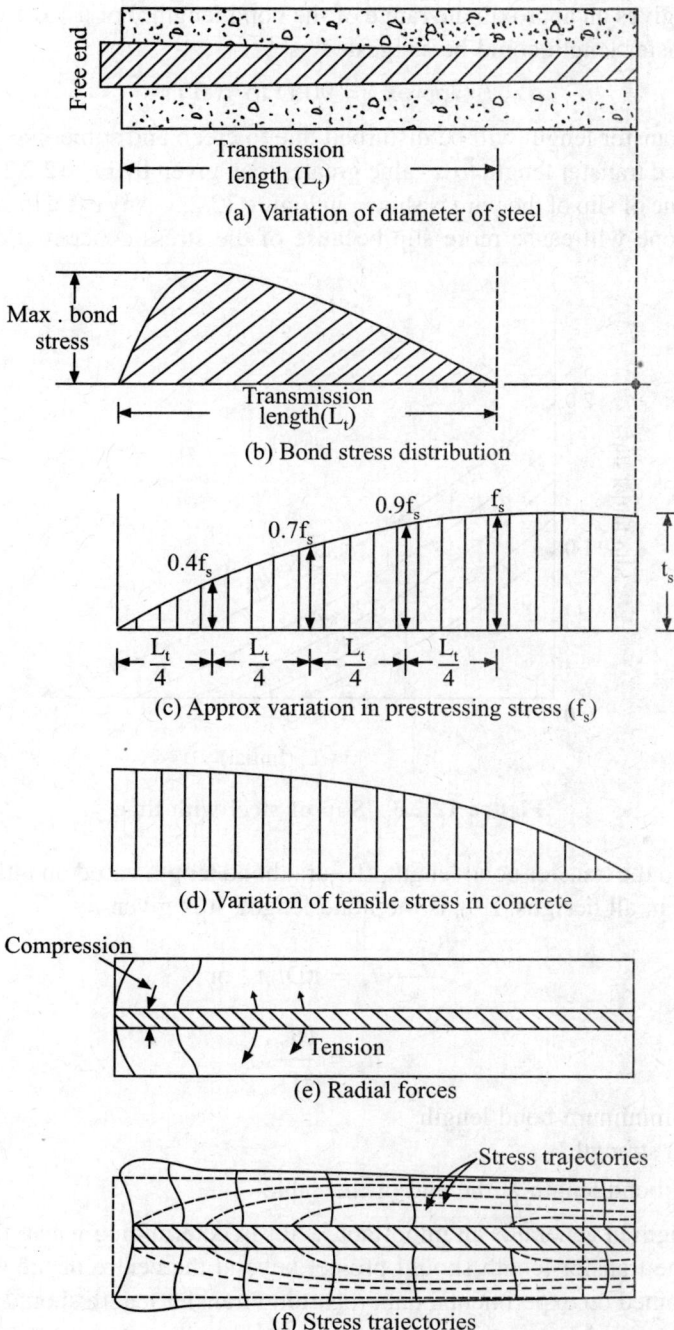

(a) Variation of diameter of steel

(b) Bond stress distribution

(c) Approx variation in prestressing stress (f_s)

(d) Variation of tensile stress in concrete

(e) Radial forces

(f) Stress trajectories

Figure 12.2.2 Stress variations along the transmission length.

Equation 12.2.2 gives an approximate range of the bond length. For $\mu = 0.1$ or in that order, an approximate transfer length could be suggested as

$$l_t \text{ (approx.)} = 90 \text{ D to } 100 \text{ D}$$

The initial transfer length will be disturbed due to creep and sometimes due to slip which forces the required transfer length to a value greater than given by Eq. 12.2.1. The approximate experimental value of slip of the wire is shown in Figure 12.2.3. Any extra loading or disturbance in the transfer zone will cause more slip because of the stress concentrations existing at the end.

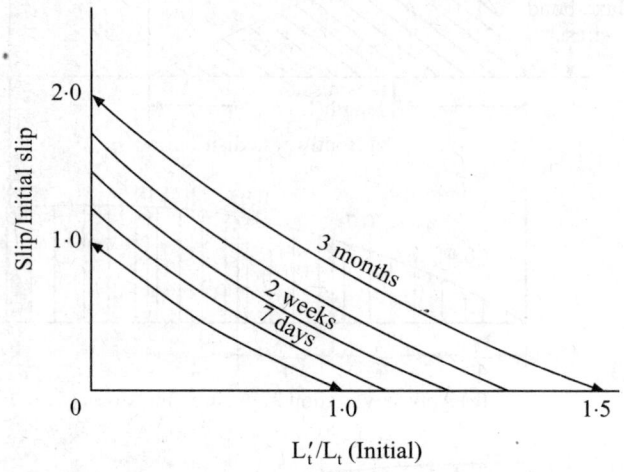

Figure 12.2.3 Slip of steel with time.

In addition to the transfer bond length, flexural bond length based on ultimate stress should also be checked in all designs. If l_b is the bond length, it is given by

$$\frac{\pi D^2}{4} f_p = \pi D l_b \tau_b \text{ or}$$

$$l_b = \frac{D f_p}{4 \tau_b} \tag{12.2.3}$$

$$l_\mu = l_b \text{ or } l_b \text{ minimum bond length} \tag{12.2.4}$$

where τ_b = bond strength

l_μ = total bond length at the end of the cable

Transfer length in general is an important factor in determining where the end support has to be placed. The transfer length should project beyond the centre of the support. Following conclusions obtained on experimental data regarding transfer length should be observed while designing pre-tensioned beams (12.1 to 12.3):

(i) if the transfer length is defined as $l_t = \alpha D$, then α is found to be a function of D which increases with D. In general α varies from 50 to 200 and for small wires it could be taken as 100;

   (ii)   deformed or indented bars require less transfer length;

  (iii)   every care should be taken to get good compaction at the ends;

  (iv)   it is desirable to use high strength concrete at end zones;

   (v)   transfer of prestress should be done gradually instead of cutting all the wires suddenly;

  (vi)   heavy concentration of wires should be avoided;

 (vii)   sufficient transverse reinforcement should be provided to take care of the radial stresses;

(viii)   transfer length should be beyond the support point; and

  (ix)   repeated loading increases the transfer length so additional allowance in transfer length should be given for creep as well as for repeated loads.

Example 12.2.1 Check the bond length of beam in example 3.10.1. Consider the beam as pre-tensioned concrete beam.

Given data:

    Span = 50 m, A_p = 268 numbers of 7 mm wire,
    coefficient of friction = 0.1, f_{ck} = 4500 N/cm^2

$$f_p = 150000 \text{ N/cm}^2,$$

$$E_c = 3.82 \times 10^6 \text{ N/cm}^2, E_s = 21 \times 10^6 \text{ N/cm}^2,$$

Poisson's ratio of concrete = m_c = 0.1, Poisson's ratio of steel = 0.3 and allowable Bond stress = 100 N/cm^2.

Solution

    On substitution of the respective quantities as given in the data, the eq. 12.2.1 gives

$$l_t = 53 \text{ cm}$$

Bond length at ultimate load is given by eq. 12.2.3 and works out to be l_b = 65 cm.

    Since the beam is simply supported of 50 m span, the stress in steel at the end of the beam is not equal to the ultimate stress of the steel. Hence the ultimate bond length be taken as 53 cm. There should be at least 55 cm projection of the beam beyond the centre of the support.

12.3. Bond in Post-tensioned Construction

Effect of bond in post-tensioned construction has two distinct purposes:

    (i) protection against stress corrosion, and (ii) increase in ultimate strength. In an unbonded construction, if any moisture enters into the duct either through a crack or through some leak, the entrapped moisture will cause corrosion to the high tensile steel wires. As the diameter of the wires is small, any small amount of rusting of the wires reduces the effective area considerably. Corrosion once started will increase rapidly and develop cracks. Sometimes a small corrosion of high tensile wire under tension will cause splitting of the wires called *stress corrosion*. Grouting of the ducts after anchoring of the prestressed wires reduces the corrosion considerably. The alkaline property of cement grout tends to prevent the development of any corrosion. The grouting has to be done as early as possible so that rusting of wires does not start at any time.

    Effect of bond on ultimate strength is an interesting phenomenon. As soon as a crack is developed at maximum moment location, strain in steel is increased considerably. In an unbonded

450

construction, this increased strain is more or less evenly distributed along the length of the cable thus causing excessive deformation of the structure. Due to this large deformation, cracks open much wider reducing the compressive zone. Once the compressive area of concrete is reduced, ultimate strength is also reduced; whereas in bonded construction, the initial crack at the critical section does not affect the strain in steel at any other point and more cracks of small width are formed in the maximum bending moment zone. Because of the large number of cracks, the deformation is evenly distributed thus keeping the propagation of the cracks as low as possible; and at any section, the compressive area is not reduced considerably. General analytical calculation gives only one ultimate strength for either bonded or unbonded construction since it does not take into consideration the spacing of the cracks. Ultimate strength of unbonded construction in general is less than the computed value. It all depends upon the spacing and number of cracks; fewer cracks, the less is the ultimate strength. Ultimate strength of members without bond is about 5 to 30% less than those with bond. A typical crack formation in bonded and unbonded construction is shown in Figure 12.3.1.

Once a duct is grouted to establish a bond between steel and concrete, the differential force in the cable has to be controlled by bond. There are two layers of bonding media in post-tensioned construction:

 (i) bond between the steel and the sheath or duct, and
 (ii) bond between the sheath and the concrete.

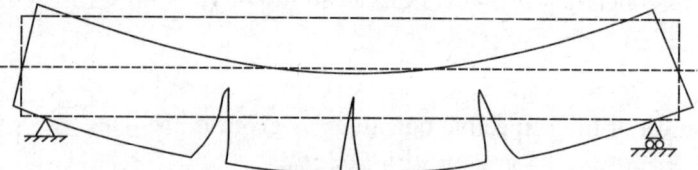

(a) Small number of large cracks in construction without bond.

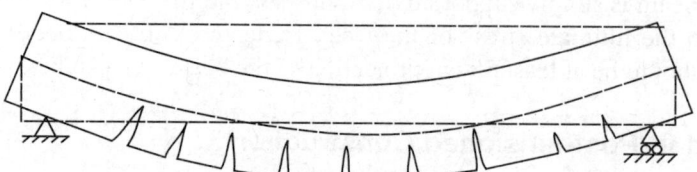

(b) Large number of small cracks in bonded construction.

Figure 12.3.1 Crack formation in bonded and unbonded construction.

Absolute bond is difficult to attain through grouting even though the method of grouting sounds fairly simple. Therefore allowable bond stress in such a bonded construction should be low. A certain amount of testing is desirable to test the effectiveness of the grouting material on any particular major job before assuming any allowable stresses. Bond stress is obtained in a similar way as is done in reinforced concrete construction (Figure 12.3.2).

$$(P_2 x_c - P_1 x_c) = V dx$$

$$(P_2 - P_1) = \frac{V dx}{x_c}$$

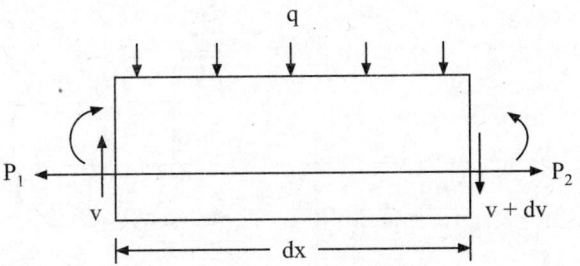

Figure 12.3.2 Element forces.

where x_c = distance of steel from centre of compression

Assuming a uniform bond stress

$$(P_2 - P_1) = \tau(\Sigma 0)dx$$

where $(\Sigma 0)$ = total perimeter of the steel

$$\tau = \left(\frac{V}{x_c \Sigma 0}\right) \tag{12.3.1}$$

Allowable bond stress in prestressed concrete will be less than that of the reinforced concrete construction as the bonding of tendons is less. Similarly, bond between the sheathing and the concrete is also different.

An allowable bond stress of about 100 N/cm² is generally permitted in static loading conditions. In bridges, where a repetition of load comes more often, the permissible bond stress is about 50 N/cm² for M 35 concrete.

PROBLEMS

12.1 Determine the bond length of a pre-tensioned concrete beam of 100 cm by 60 cm in cross section, provided by four prestressing cables of 400000 N each. Each cable consists of 12 numbers of 7 mm wires.

12.2 A cantilever beam of span 6 m is under 10000 N/m uniformly distributed load. Design a pre-tensioned concrete beam with 4500 N/cm² concrete. Also design the bond length based on prestress and ultimate strength bond criteria.

CHAPTER 13

Anchorage of Prestressing Cables

13.1. Post-tensioned Construction

Transfer of prestressing force from the cable to the concrete section in post-tensioned construction is done at some selected areas such as end of the beam. As a consequence of the transfer of force, stress concentrations are developed at the transfer zone resulting in secondary stress which dominates the design criteria of the zone. Anchor plates are used as a medium of transfer of the prestressing force. The size of the plate and type of anchoring arrangements govern the stresses in concrete at the transfer zone. The prestressing force applied through the anchor plate propagates into the beam along the axis and after some distance, a linear stress distribution is developed. The zone in which the stress variation reaches a linear distribution over the cross section is called "transmission zone"; the length of this zone is called "transmission length". A typical representation of transmission of force is shown in Figure 13.1.1. Since the application of any concentrated load over a cross section generates a transmission zone before a linear distribution of the stress is achieved over the cross section, the determination of stress distribution within the transmission zone is difficult even though it could be analysed by elastic and photoelastic theory. The stress distribution within the transmission zone is analysed with reasonable accuracy based upon elastic plate theory; the results are checked with photoelastic experiments. The material along the axis of the concentrated force is subjected to transverse tensile force which is called "bursting force" and the surface at end section just adjacent to the anchor plate is subjected to tensile force which is called "spalling force". Only the design aspects of the transmission zone with a qualitative study of the stresses are discussed in this chapter. A general discussion on the type of stress distribution for various cross sections and application of loads is not possible, so some practical cases are presented. For further discussion the reader is referred to Prestress Concrete by Guyon (13.1).

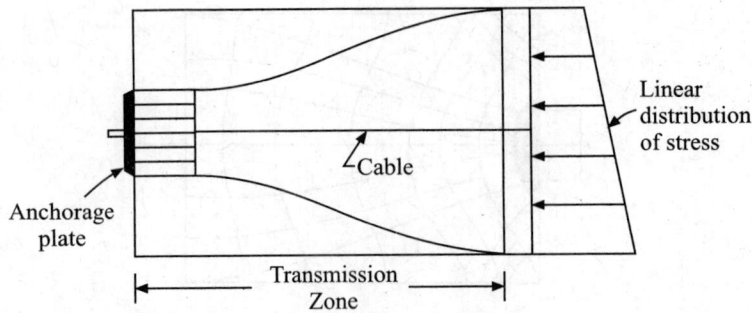

Figure 13.1.1 Transmission zone.

13.2. Prestressed Cable at the Centroidal Axis

The existence of transverse tensile force is illustrated through Figure 13.2.1. Let the section *AA* be the end section at which the prestressing force is applied through an anchor plate and let *BB* be the end of transmission zone at which the stress distribution is linear over the entire section. Considering a free body diagram of an element of the transmission zone cut by a horizontal plane '*CC*' it could be seen that there should be some tensile as well as shear stress in the horizontal plane to keep the body in equilibrium. Typical stress trajectories for concentrated force are shown in Figure 13.2.2.

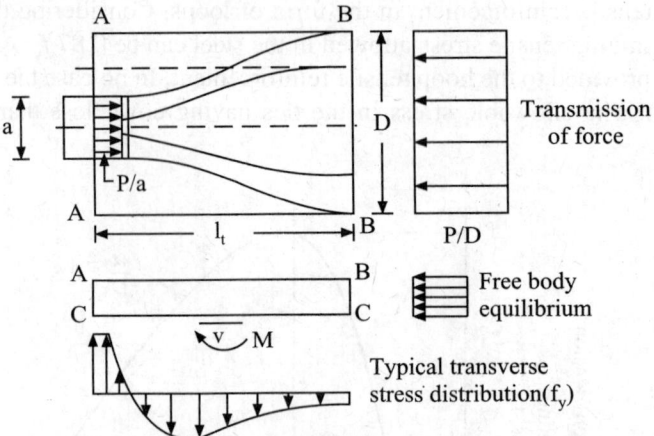

Figure 13.2.1 Transmission zone-stress distribution.

The transmission length and the stress distribution depends upon the depth of the anchor plate relative to that of the beam. The transmission length and the transverse tensile force (*bursting force*) are found to be inversely proportional to the ratio of depth of the anchor plate to depth of the beam. A typical representation of bursting force is approximated by a straight line (13.2) and expressed as

$$F_{bst} = 0.32\left(1 - \frac{0.93a}{D}\right)P \qquad (13.2.1)$$

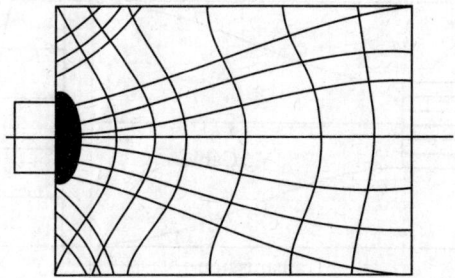

Figure 13.2.2 Principal stress trajectories.

where P = axial force, a = depth of anchor plate,

D = depth of beam, and F_{bst} = bursting force.

It is also important to find the position of zero and the maximum transverse tension to locate the reinforcement properly. Figure 13.2.3 gives the location of zero and maximum tensile stresses along the axis of the tendon. Even though the transmission length varies with depth of the anchor plate, the maximum transmission length is found to be approximately equal to the depth of the beam.

Even though the bursting tensile force varies parabolically in a region of $0.8D$, it is often more convenient and practicable to treat it as evenly distributed over the region of $0.1\,D$ to D and provide the tensile reinforcement in the form of loops. Considering the limit state of the cable force, the limiting tensile stress allowed in the steel can be $0.87\,f_y$. An adequate cover of about 50 mm be provided to the hoop tensile reinforcement. In no case the cover should not be less than 25 mm. The allowable stress in the ties having cover less than 50 mm should be

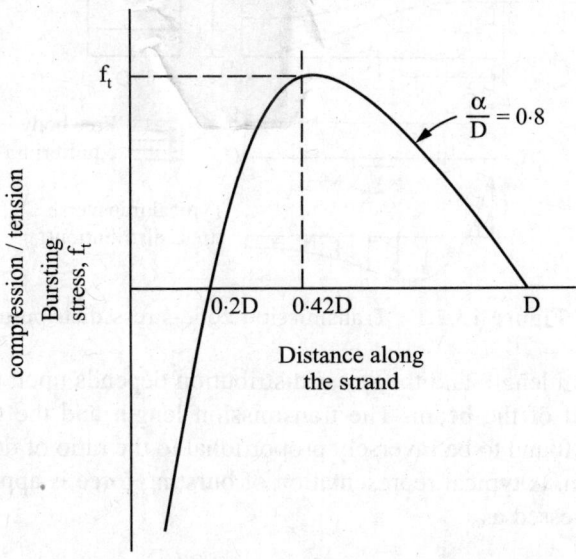

Figure 13.2.3 Approximate variation of bursting force.

reduced to 200 MPa for a cover of 25 mm. The ties loops should be evenly spaced over a distance 0.1 D to D from the bearing plane.

The concrete next to the bearing plate is subjected to a high bearing pressure normal to the plane of the bearing surface. Further very high transverse compressive force is generated immediately in the vicinity of the bearing plate. This type of high transverse compressive force generates spalling stresses just adjacent to the bearing plate. A provision should be made to prevent spalling of the concrete next to the bearing plates. The size of the bearing plate should be such that the bearing stress on the concrete is limited to the following:

$$\sigma_{ab} = 0.48 f_{ci} \sqrt{\frac{A_{br}}{A_{pum}}} \text{ or} \qquad (13.2.1)$$

$$\leq 0.8 f_{ci}$$

where

σ_{ab} = allowable bearing pressure after accounting for losses, etc.

A_{br} = bearing area

A_{pun} = punching area

However the bearing pressure at the time of anchorage can be increased by 25 per cent subject to the limit of f_{ci}.

The effective punching area is the area of the bearing of the anchorage devise whereas the bearing area is the maximum area of that portion of the member which is geometrically similar and concentric to the effective punching area. The bearing area is usually larger or equal to the punching area. The bearing plates or cones are made smaller than the size of bearing surface. In case the anchorages are placed close to each other, the bearing area of one anchorage should not overlap the other, and in such cases, the centre to centre distance of the anchorage devises should be taken as the limiting depth The design is illustrated with examples.

Example 13.2.1 A rectangular cross sectioned beam of size 300 by 800 mm is subjected to an initial anchorage force of 1600 kN. Assume the force is at the centroid of the section.

Design data

$$f_{ck} = 30 \text{ MPa}, f_{ci} = 30 \text{ MPa}$$

$$f_y = 415 \text{ MPa (for ties)}$$

The cables are placed symmetrically over a mild steel plate in an area of 200 by 350 mm.

Let $$P_{oe} = 1600 \ k\text{N}$$

Design of anchor plate

To start with we do not have the size of the anchor plate to determine the punching area which in turn establishes the allowable bearing pressure. Therefore select the allowable bearing pressure as:

$$\sigma_{ab} = 0.65 f_{ck} = 0.65(30) = 19.5 \text{ MPa}$$

The contact area needed is

$$A_b = \frac{P_{oe}}{\sigma_{ab}} = \frac{1600000}{19.5} = 82051 \text{ mm}^2$$

Select the width of the plate less than the width of the beam. Let the width of the plate be

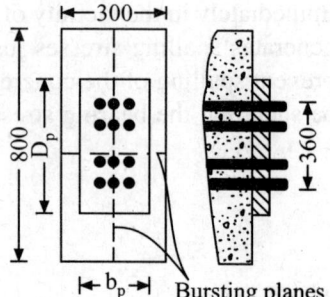

$$b_p = 240 \text{ mm}$$

then the depth of the plate is

$$a = \frac{A_b}{b_p} = \frac{82051}{240} = 342 \text{ mm}$$

Select

$$b_p = 240 \text{ mm}, \ a = 400 \text{ mm}$$

so that the size of the plate is more than the area occupied by the cables (which was given as 200 mm by 350 mm).

The punching area is equal to

$$A_{pun} = b_p a = 96000 \text{ mm}^2$$

The effective bearing width is same as the width of the beam

$$b_b = b = 300 \text{ mm}$$

the projection of the bearing width beyond the width of the plate is 300 – 240 = 60 mm. The effective bearing depth is then equal to

$$D_b = a + 60 = 460 \text{ mm}$$

The bearing area is then equal to

$$A_{bt} = b_b D_b = 300(460) = 138000 \text{ mm}^2$$

The allowable bearing pressure is

$$a_{ab} = 0.48 \, f_{ci} \sqrt{\frac{A_{br}}{A_{pun}}}$$

$$= 0.48(30) \sqrt{\frac{138000}{96000}} = 17.25 \text{ MPa}$$

The actual bearing stress is

$$a_b = \frac{P_{oe}}{b_p a} = \frac{1600000}{96000} = 16.67 \text{ MPa}$$

The bearing pressure is less than the allowable bearing pressure therefore the size of the plate is adequate. The thickness of the plate is decided by the bending moment acting on the plate. The projection of the plate beyond the cable cones acts as a cantilever subjected by the bearing pressure. The cantilever span of the plate is

$$L = \frac{(b_p - 200)}{2} = 20 \text{ mm},$$

or
$$= \frac{a - 350}{2} = 25 \text{ mm}$$

Select
$$L = 25 \text{ mm}$$

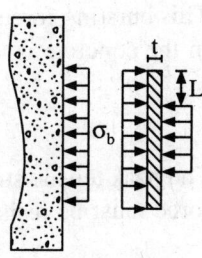

Bearing stress

The beading moment on the plate is

$$M = \frac{\sigma_b L^2}{2} = \frac{16.67(25)^2}{2} = 5209 \text{ Nmm/mm}$$

The bending moment capacity of a plate is given by

$$M_b = \sigma_s Z_s$$

where σ_s = allowable bending stress in steel

Z_s = modulus of the section per unit width

$$= \frac{t^2}{6}$$

t = thickness of the plate

The value of allowable bending stress in mild steel plate is 165 MPa = σ_s.

Equating the moment capacity to the external bending moment, we have

$$\sigma_s \frac{t^2}{6} = M = 5209 \text{ Nmm/mm}$$

or
$$t = \sqrt{\frac{6(5209)}{165}} = 13.76 \text{ mm}$$

*Use 14 mm thick mild steel plate 250 by 400 mm.

Design of hoop reinforcement

The bursting force on the beam section is given by

$$F_{bat} = 0.32 \left(1 - \frac{0.93a}{D} \right) P$$

or
$$= 0.32 \left(1 - \frac{0.93 b_p}{b} \right) P$$

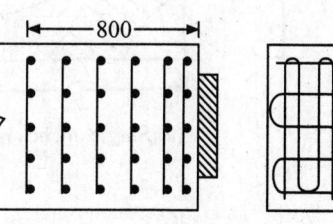

End zone loops

where P = axial force

$\quad\quad$ = anchorage force = P_0

a = 400 mm, $\quad D$ = 800 mm

b_p = 250 mm, $\quad b$ = 300 mm

The first expression governs the bursting force. So

$$F_{bat} = 0.32 \left(1 - 0.93 \frac{(400)}{800} \right) \quad (1600)$$

$$= 280 \text{ kN}$$

This bursting force is to be resisted by an area equal to 0.9Db. Hence the average tensile stress in the concrete is

$$\sigma_t = \frac{F_{bst}}{0.9Db} = \frac{280000}{0.9(800)(300)} = 1.3 \text{ MPa}.$$

The peak tensile stress in concrete will be about 1.8 times the average value. The total bursting force must be resisted by steel only. The total area of steel for ties is

$$A_{sv} = \frac{F_{bst}}{0.87f_y} = \frac{280000}{0.87(415)} = 780 \text{ mm}^2$$

*Provide six 8 mm stirrup loops between 0.1D or 80 mm and 800 mm from the end of the beam. Also provide one 8 mm stirrup loop at 35 mm from the edge. The size of the loop be 306 − 100 = 200 mm by 800 − 100 = 700 mm.

13.3. Symmetric Multiple Cables Causing Axial Thrust

When separate anchor plates distributed over the depth of the beam are used for different cables, the transmission zone and distribution of stress are altered considerably. The even distribution of the anchor plates reduces the transmission length and also the bursting forces. A typical behaviour of the multiple cable force is shown in Figure 13.3.1.

The bursting force could be calculated by dividing the total depth into a number of equivalent prism depths. A simple example illustrating the procedure is given below.

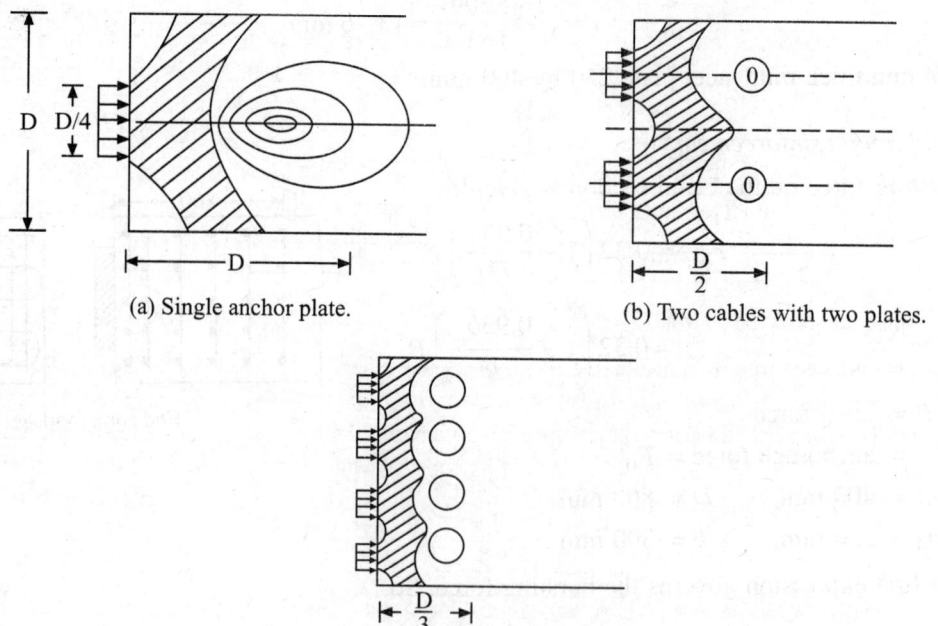

(a) Single anchor plate.

(b) Two cables with two plates.

(c) Four cables with four plates.

Figure 13.3.1 Approximate isobars for transverse tensile stress.

Example 13.3.1 A prestressed beam of 80 cm by 30 cm cross section is subjected to two symmetric cables each with 1200 kN transfer prestressing force. An anchor plate of 20 cm by 30 cm is provided for each cable. Determine the reinforcement in the transmission zone.

Solution

The size of the anchor plate is already given; so only the end zone stirrups have to be designed.

The bearing pressure on the concrete is

$$\sigma_b = \frac{P_e}{200(300)} = \frac{1200000}{60000} = 20 \text{ MPa}$$

The bursting force is given by

$$F_{bst} = 0.32 \left(1 - 0.93 \frac{a}{D} \right) P_0,$$

or

$$= 0.32 \left(1 - 0.93 \frac{b_p}{b} \right) P_0$$

where D = effective bearing depth = $\dfrac{h}{2}$ = 400 mm

$$\frac{a}{D} = \frac{300}{400} = \frac{3}{4}$$

$$\frac{b_p}{b} = \frac{200}{300} = \frac{2}{3}.$$

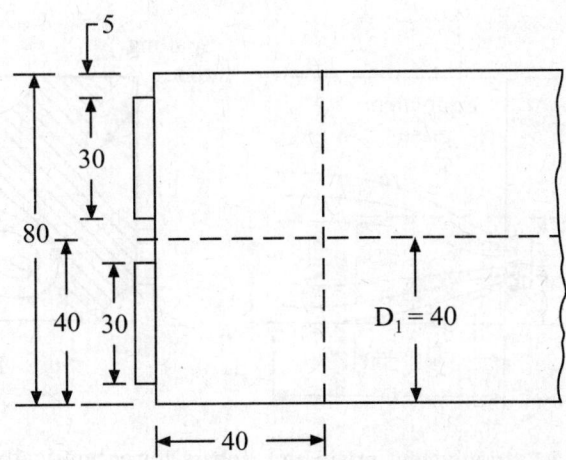

Figure 13.3.2 Transmission zone of Example 13.3.1

Therefore the second expression controls the design

$$F_{bst} = 0.32 \left(1 - \frac{0.93 \, (2)}{3} \right) (1200000)(1.2)$$

$$= 175 \; 104 \; N$$

By providing HYSD bars as stirrups, the area of the reinforcement is

$$A_{st} = \frac{F_{bst}}{0.87 \, f_y} = \frac{175 \, 104}{0.87 \, (415)} = 485 \; mm^2$$

*Provide five 8 mm diameter HYSD bar loops in the end 400 mm zone

13.4. Cable with Eccentricity

Eccentrically acting prestressing force would cause reorientation in the principal stress trajectories and also in the magnitude of bursting force. A typical representation of isobars and lines of compression are shown in Figure 13.4.1 from which it may be observed that the bursting force is shifted towards the eccentricity. Guyon has suggested an "equivalent prism" method in which a prism is selected where the axis of the load is the axis of the bursting line and the depth of the prism is twice the smallest distance from the load axis to the edge of the concrete. The stress analysis is to be made based on the equivalent prism. This method is found to be satisfactory even though it has some disadvantages in limiting cases. The design of transmission zone is best shown through an illustrative example.

Example 13.4.1 A beam of cross section 80 cm by 30 cm is subjected to an eccentric prestressing force of 1600 *k*N with 10 cm eccentricity. Assuming an anchor plate of size 40 cm by 20 cm, determine the bursting stress and the necessary reinforcement.

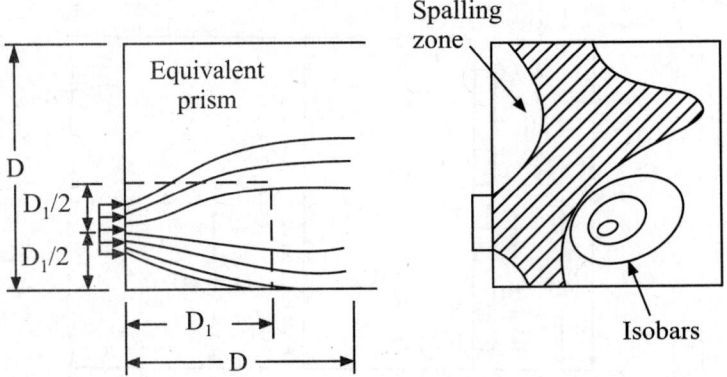

Figure 13.4.1 Equivalent prism and isobars for eccentrically acting force.

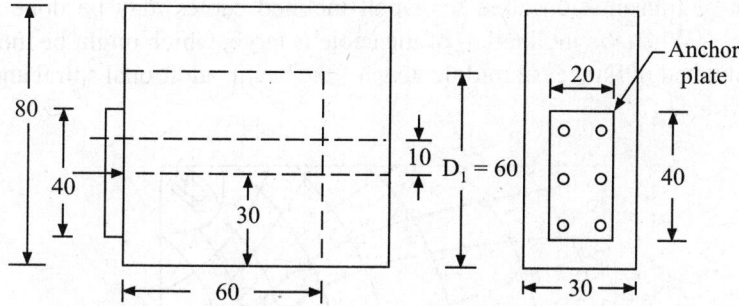

Figure 13.4.2 Section of example 13.4.1 (Dimensions in cm).

Solution

    Vertical plane reinforcement:

    Depth of equivalent prism = D_1 = 60 cm

$$\frac{a}{D_1} = \frac{40}{60} = 0.667$$

Procedure as adopted in the previous section is followed from here onwards taking only the width of the anchor plate as effective width.

    The bursting force is

$$F_{bst} = 0.32\left(1 - 0.93\,\frac{a}{D_1}\right)P_0$$

$$= 0.31\,(1 - 0.93\,(0.667))\,1.2\,(1600)$$

$$= 233\text{ kN}$$

The area of the stirrups is

$$A_{st} = \frac{F_{bst}}{0.87\,f_y} = \frac{233000}{0.87\,(415)} = 645\text{ mm}^2$$

*Provide seven 8 mm diameter. HYSD bar loops in the end 600 mm zone

13.5. Inclined Prestressing Cable

Most of the practical problems are associated with inclined cable at the end section. The effect of the inclination of the cable is to reorient the stress trajectories and the isobars so that the symmetric axis of the stress trajectories is along the inclined cable axis. The inclination of the cable is generally small, so the reorientation of the axis is not very appreciable. Typical stress trajectories for an inclined cable are shown in figure 13.5.1. If the inclination is small, as is the case in many practical problems, the bursting force is not much altered; however the spalling forces are increased appreciably. The effect of the vertical component of the cable force will generate secondary splitting stresses which may be neglected for small inclinations of the

462

cables. Design of transmission zone for small inclined cables may be done as adapted for horizontal forces. When the inclination of the cable is large, which might be the case when the cables are terminated in flanges at middle zone of the beam, additional spiral and hoop stirrups should be provided.

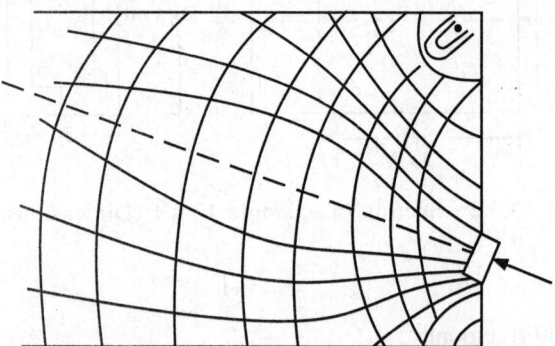

Figure 13.5.1   Typical stress trajectories for an inclined cable.

13.6.   Spalling Stresses

Spalling stresses are developed on the edge surfaces of the beam adjacent to the anchor plate. High bearing stresses cause higher lateral tensile strains which are probably the main reasons for spalling of the concrete. The spalling stresses are directly proportional to the bearing stresses imposed and inversely proportional to the interval of the bearing forces. In most cases, the spalling stresses are in the region of 0.3 P/A to 0.4 P/A and, if the intensity of bearing stress is very high, then the spalling stress could be very high. A typical variation of spalling stresses are shown in figure 13.6.1. Since the spalling stress is only a localised character at the surfaces adjacent to the anchor plates, the total amount of reinforcement involved is not considerable. A high percentage of small diameter loop stirrup reinforcement should be provided against spalling.

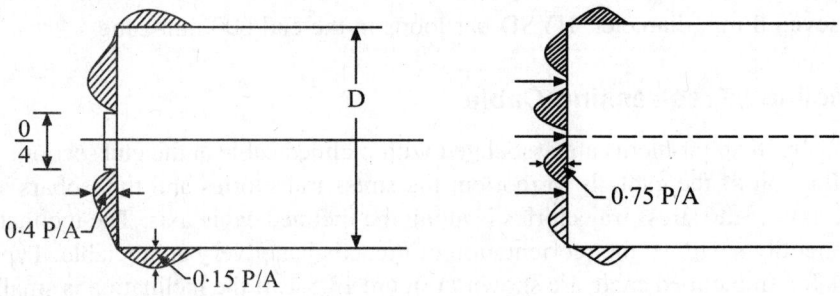

Figure 13.6.1   Typical spalling stresses.

PROBLEMS

13.1 A post-tensioned concrete beam having end section 100 cm by 160 cm is provided with 4 cables of 400 *k*N each. The cables are spaced in a vertical line so that the centroid of the cables coincide with the centre of gravity of the section. Design anchor plates, vertical and horizontal bursting reinforcement and spalling reinforcement.

13.2 Post A-tensioned concrete beam having end section of 60 cm by 100 cm is provided with two cables of 300 *k*N each. The two cables are spaced at 20 cm on either side of the centroid of the beam with 4° inclination. Design the anchor plates, the bursting and spalling reinforcement.

CHAPTER 14

Some Points of Interest in Prestressed Concrete Construction

14.1. High Tensile Steels

High tensile steels do not have a clear yield point like the one which is noticeable in mild steels. The specifications on the permissible stresses are referred with respect to ultimate stress or proof stress. The proof stress is usually defined as the stress at which 0.2 per cent permanent strain is noticed. Because of the brittle nature of the high tensile steel and because of the high ratio of they permissible stresses in construction, it is necessary that sample tests on ultimate strength and percentage elongation are conducted on each coil of the wire. It is desirable to have at least 3 to 4 percentage of elongation at the time of breaking of the wire. The mean value of the ultimate strength of the wire should not be less than the characteristic strength specified. The wires are subjected to maximum stress at the time of anchorage or jacking. The probability of failure of a high tensile steel at the time of jacking is estimated 5 to 9 in 100,000 wires, whereas the probability of failure of the wire by strength considerations at the working load conditions 3 to 5 in 10^9 wires. The estimate is based on the theory of probability using a random data generated subject to code specifications of the materials. It can be stated that every wire is proto-tested at the time of transfer of prestress from jack to the structure, so, if a wire can stand the jacking force, it will not fail at working load conditions by strength consideration. The failure of a wire, if at all it occurs during working load condition, it must be due to environmental conditions such as corrosion or creep. The engineer-in-charge of a project must see that every wire is prestressed as per specifications through which indirect proto-testing of each wire or cable is achieved.

Wires are usually supplied in coils by the manufacturers. The diameter of the roll must be more than 100 D in which D is the diameter of the wire. A roll having diameter less than 100 D will develop strains more than 0.005 cm/cm. The usual diameter of the rolls is about 150 D. The rolls of the wires must be wrapped in damp proof paper before supplying to the users. The coil of wire must be stored in damp proof room and should be taken out in the field only at the time of actual use. Exposure of the wire to mosit air will cause corrosion of the wire. The moisture along with some chemicals and oxygen in the air act on the surface of the wire resulting in corrosion. Presence of cracks or honeycomb in the concrete will also permit corrosion of wires. The rust on the surface, if any, should be cleaned before the wires are used in the construction. Oil should not be applied to the wires as the oil reduces the bond capacity of the wires. Deep rusting of wires is dangerous in two ways. The effective area of cross section is reduced considerably since the wires themselves are thin and secondly the wires are liable to special type of corrosion called *stress corrosion*. The presence of rusting in a groove even if the groove is very small, will cause brittleness to a high tensile wire under tension. The crack on the surface of the wire penetrates into the wire and splits the wire which is under tension. Such a fracture occurs suddenly. Stress corrosion may occur in a coiled wire to a limited extent but it will be accelerated by high tensile stresses. It is also reported that certain heat treated wires are more prone to stress corrosion as compared to as-drawn wires. The wires should not be exposed to even light acids as the presence of acids will result in deep cut rusting.

The wires should not be left in the field or in the ducts for long time without grouting. In the case of post-tensioned construction, the grouting should be done within few hours or at the most few days after tensioning. A minimum number of two grout holes one on either end of the duct are necessary. One for grouting and another for allowing air to escape as the grouting is done. The second grout hole also helps in ensuring the filling of the duct with grouting mixture. Grouting mortar should come out through the other hole uninterrupted if the duct is completely filled. In some cases, specially in long metallic ducts, the cement mortar is likely to leak into the ducts while casting the concrete thus causing obstruction to the easy flow of grout.

Therefore it is desirable to have more than two grout holes in long spans. If it is observed that there is some clogging in the duct which will be known if the grout does not come out of the open hole but squeezes out of the grouting hole, the actual location of the clogging must be identified. Grout holes must be drilled on either side of the clogged portion and the entire duct be grouted. The alkaline cement along with water and oxygen form a thin protective coating on the surface of the high tensile wires. Chloride admixtures or additives in cement must be avoided in the prestressed concrete to minimise the corrosion of steel.

14.2. Concrete

High strength concrete with low shrinkage and creep characteristics should be used. The properties of the concrete are governed by the cement and water content in the concrete in prestressed concrete construction. High water-cement ratio not only gives low strength but increases shrinkage and creep. The water-cement ratio should be kept below 0.45 and preferably in the range of 0.38 to 0.42. Such a low water-cement ratio decreases the workability of concrete therefore the aggregate must be graded well and fine clay and dust particles should be avoided. The concrete must be compacted by use of vibrators. Immersion vibrators should be used in the case of large sections whereas shutter vibrators should be used in the case of small sections. Higher water-cement ratio also causes segregation of particles while vibrating the concrete, therefore low water-cement ratio is preferred.

The strength of the concrete increases with increase in the cement content but the increased cement content also produces higher shrinkage and creep. Therefore it is desirable to select an appropriate cement content which will give high strength at the same time will not produce high shrinkage. As the water-cement ratio decreases, the cement content should be increased to maintain good workability of the concrete. Approximate quantity of cement content is in the order of 350 to 600 kg/m^3 of concrete. Quick setting cements are not very suitable as high heat of hydration is caused in the early stages of concrete. Because of low water-cement ratio such heat of hydration is not absorbed easily. However quick setting cements can be used with controlled cooling of the concrete. Air entraining agents have very little use in prestressed concrete construction, however the additive which decreases the surface tension of the water can be used provided they are not harmful to steel.

Concrete must be kept moist and a minimum temperature of 10°C be maintained at least for ten days. Any type of curing such as sprinkling or immersion are recommended for 28 days but not less than 10 days. In case the concrete is cured by sprinkling of water it must be covered with water absorbing material such as gunny bags. The concrete must be covered immediately after the setting so as to avoid loss of water by evaporation. Steam curing is also acceptable, sometimes it is more economical as the time required for steam curing is much less than the period of water cubing. 24 hours of steam curing at about 100°C under atmospheric pressure is approximately equal to 28 days of water curing. In case of mass manufacturing of small prestressed concrete components such as electric poles, railway sleepers, small beam and slab elements, these can be cured economically by steam. Concrete gains about seventy per cent of the 28 days strength within the first seven days of water curing or 10 hours of steam curing. To save space and time it is sometimes desirable to transfer prestress at the end of seven days of water curing or 10 hours steam curing. But in such a case, the maximum compressive stress on the concrete should not exceed fifty per cent of the actual strength of the concrete at the time of transfer of prestress. The rate of gain of strength decreases with the decrease in atmospheric temperature when water curing is used. The rate of gain is about ten per cent less at 6 to 10°C temperature as compared to 25 to 30°C atmospheric temperature. So seven days of curing may not be enough in winter season.

Depending on the economics of construction, it is sometimes desirable to transfer prestress after three to four days of water curing. Concrete gains about 50 per cent of the 28 days strength in three to four days curing by water or five to six hours of steam curing. The specified strength of concrete can be called as characteristic strength. The actual strength of the concrete can be selected more than the characteristic strength so as to release the prestressing bed and equipment at an early age. One can use higher cement content than specified and gain seventy per cent of the 28 days specified strength within first three days. It is not desirable that the prestress force is transferred to the member within the first four days of curing of the concrete unless early setting cements are used. A minimum of six hours of steam curing excluding the presteam period is desirable.

It has been found from the field data and random data generated by computer satisfying the quality concrete specifications that the mean value of the concrete is about 10 to 15 per cent more than the characteristic strength of the concrete. It was also observed that the strength of the field cured concrete having some supervision is around 10 to 15 per cent less than the laboratory cured concrete. While designing a concrete mix in a laboratory for prestressed concrete construction, it is desirable to design a concrete which will give 15 to 20 per cent higher strength in the laboratory as compared to the characteristic strength. However, in the case of builders having good experience, a concrete which gives about ten to fifteen per cent higher strength in the laboratory as compared to that of the characteristic should be enough. The quality of cement plays a very important role in high strength concrete. There is a limit on strength of concrete beyond which it is very difficult to attain higher strength of concrete. The normal portland cement currently available could give as high as 50 N/mm^2 concrete as per the tests conducted at the structural engineering laboratory of Indian Institute of Technology, Kanpur.

Concrete construction and the curing of concrete have become very common in India, so it is not necessary to discuss the methods of curing. However, few points of interest and observation are given here. Preservation of moisture in concrete is the most important factor in absorbing the heat of hydration generated during hardening of the concrete. Green concrete should be covered with one to two layers of gunny bags after one to four hours of casting of the concrete. One hour in the case of summer season and two hours in other seasons are desirable. The initial setting time of the concrete should be completed and the surface should get little hardened so that the gunny bags covering the surface should not leave marks on the concrete. The concrete should then be left overnight or say about 5 to 10 hours before sprinkling of the water on the gunny bags. In the case of beams and columns, the form work could be left as it is so as to prevent escape of moisture from the concrete. Immersion of hardened concrete in water is not essential, just sprinkling of water at 2 to 4 hours intervals on the gunny bags so as to keep the bags wet is enough to provide good curing. Immersion of the hardened concrete in water is about the best but even constant sprinkling of water so as to keep the surface wet is enough. Most important thing for excellent curing is to see that no moisture escapes from the concrete and the exposed surfaces are supplied with constant moisture during first seven days. Unless otherwise the concrete mix is designed, water curing should be continued for 28 days. If the concrete gains the characteristic strength in less than 28 days of curing, the curing can be discontinued. Dry concrete shows higher strengths as compared to the wet concrete of the same mix. Therefore concrete specimens (cubes/cylinders) cured by sprinkling of water are likely to record higher compressive strengths as compared to those immersed in water as later are water saturated. To predict a clear representative strength, it is desirable that the concrete cubes even if cured by sprinkling of water be immersed in water format least an hour before testing.

Steam curing is not commonly used in India. Generation of steam needs initial investment towards equipment and recurring expenses for fuel. In the case of mass manufacturing of prestressed concrete elements steam curing may prove economical. The products if cured by sprinkling of the water require large working capital locked up in the products which are under curing. The steam curing even though is expensive but turns out the products at a highly accelerated phase. Concrete will gain the strength of 28 days water curing within 24 hours of steam curing. Proper timings should be maintained so as to gain the full strength without damaging the concrete. Two to three hours of surface drying must be given as a presteam curing period so as to avoid boiling of water in the concrete. The steam is let in into the curing chambers or under the tarpaulins in which the specimens are kept. The temperature of steam is kept around 100°C and it should be let in slowly so that the fresh concrete specimens are not suddenly exposed to excessive heat in short period. The temperature around the specimen can be raised to 70-90 °C in about one hour time. Rise in temperature along with moisture in the steam accelerates the chemical reaction in the concrete and hardening of concrete takes place very fast. Five to six hours of steam curing will generate about fifty per cent of 28 days water curing strength. Ten hours of steam curing gives about seventy per cent of the strength of the concrete and 24 hours of steam curing gives total 28 days water cured concrete strength. About 30 to 60 minutes of post-steam-curing period should be allowed before the steam cured concrete is exposed to the atmosphere in the tropical climate zones. If the steam is let in without proper presteam curing period, the top surface of the concrete literally gets cooked and about one to two centimetres thick concrete will get disintegrated. The concrete can be left in the formwork with one face exposed to steam. Rapid temperature changes in the concrete develop fine hair cracks on the surface.

14.3. Bond Failure

The probability of failure of steel even if the beam is designed as an under-reinforced beam is about one in hundred of the probability of failure of concrete by secondary compression. According to the probability of failure study on prestressed concrete beams designed by working stress method, the probability of failure of concrete is about 100 to 1000 times the failure of steel by ultimate stress. Several prestressed concrete beams of different cross sections and portal frames designed as under-reinforced and tested in the laboratory showed failure by secondary crushing of concrete or bond failure. Bond failure was very common in pretensioned concrete beams. The bond strength of concrete with high tensile wire is small and if proper transfer prestress procedure is not used, the bond will get further decreased. Wire should usually be cut during the transfer of prestress at the noncritical end of the beam section. For example, the wires should be cut at the top end of a prestressed concrete electric pole rather than the base end. It is even desirable to use a relatively stronger concrete at the ends of the elements so as to ensure better bonding.

14.4. Fire Resistance

Structures must be capable of resisting fire to a limited time in which the limit of the time vary with the functional requirement of the building. A building must be able to withstand fire for a minimum period of 30 minutes. Some public buildings must be able to withstand the fire for about 90 to 120 minutes. Fire in a building produces excessive heat and causes large deformations in a short period and a total collapse of the structure if the fire is not controlled in the limited time. As the concrete gets heated by the fire, the moisture in the concrete is converted into steam and consequently the steam burst out of the concrete causing spalling of the concrete. The rate of spalling of the concrete depends on the moisture content, size of the element, temperature and the amount of reinforcement in the outer cover of the concrete. Spalling of the concrete may start around 100°C if there is no reinforcement. An outer mesh of reinforcement prevents the concrete against cracking due to temperature rise. If a fine mesh is provided within 15 mm cover of the concrete, the spalling of the concrete within the mesh may take place at much higher temperature. Steel cables which are pretensioned will expand extensively at high temperature not only because of the thermal expansion but also due to the decrease in the modulus of elasticity. Steel is likely to reach a semi-plastic stage around 400°C so the structure becomes unserviceable at 400°C. The cover to steel must be designed such that the steel will not be subjected to a temperature of 300 to 400°C before the minimum fire risk period. The following are some of the recommendations of fire protection in prestressed concrete construction.

1. A minimum of 20 mm cover should be provided so as to provide about 30 minutes of safety period after break of fire.
2. The minimum thickness of the cover must be increased from 20 to 50 mm to increase the safety period from 30 to 90 minutes.
3. A fine wire mesh is to be provided within the cover so as to reinforce the concrete against spalling if the safety period of fire protection is more than 30 minutes.
4. The fire safety period can be increased by providing 15 to 25 mm thick gypsum concrete or vermiculite concrete cover. This low thermal conductivity cover has to be rigidly attached to the concrete cover of the structure.
5. In the case of fire damaged structures in which only spalling of the concrete has taken place with minor damage to the cables, they can be made reserviceable by providing a cover of cement plaster of about 20 to 30 mm thick. A set of U-shaped steel studs must be imbedded into the concrete so as to make the cover an integral part of the concrete.

14.5. Handing and Transportation

The most critical conditions in the life of a prestressed concrete structure are the handling conditions. The period of handling is usually short and it is at a time when the concrete is still gaining strength. Transfer of prestress on the concrete structure or handling of the prestressed or partly prestressed elements in the casting and curing yards, moving of the elements to the site and erection of the elements into the position are the main handling operations. The permissible stresses in handling are usually 25 per cent more than those at working load. In addition, the elements are not interconnected into a structural system during the handling condition. Therefore all things add up the risk level of damage during handling. Location of pickup points at the time of transportation must be specified by the designer and they must be adhered to. The prestressed concrete elements are under high initial stresses. Any wrong handling will lead to total collapse. Take an example of a prestressed concrete pole. It is usually cast flat on the ground in which the depth of the pole is flat on the ground. The width of the pole is usually small as compared to the depth and it may not carry its own weight if lifted widthwise at base. The self weight of the pole does not come into the moment calculations at working load as the pole is placed in the upright position. Therefore in such situations the designer's specifications of handling have to be followed. Very thin slab strip if carried with arbitrary pickup points, may deflect considerably and fall under its own weight. Long prestressed concrete beam when placed in position alongwith other beams has high lateral stability in the integrated system. However a single beam in transit has much less lateral stability and may buckle under its own weight. The transverse buckling similar to the column buckling is almost impossible in prestressed concrete beams. However I-beams having thin open web sections are liable to lateral or torsional buckling. The level of the lateral buckling load decreases if the beam is carried by supporting at the upper flange. The failure by lateral instability is always sudden and will not give any notice; therefore, the field engineer transporting or erecting a heavy I-girder must take special precaution against instability failure. The handling and erection of small precast elements in a developing country like India is either manual or semi-mechanical. The chances of failure of an element in such situation is more as compared to those in service condition.

CHAPTER 15

Design of Prestressed Concrete Bridge

15.1. Introduction

Prestressed concrete is used extensively in bridge construction, it is therefore desirable to present a detailed design of bridge. Bridge engineering forms a subject by itself and difficult to cover in this book. However a simple example illustrating various aspects of the design is discussed here. Further this example is made complete covering entire design of a real life structure. The specifications on loads and materials for bridge design are not dealt here. Design of a bridge normally consists three parts namely:

(a) Superstructure,
(b) Sub structure, and (c) Foundation.

The superstructure can further be divided into sub groups as:

(1) Deck slab,
(2) Cross beams (if any),
(3) Main beam, and
(4) Bearings.

The substructure and foundaton consist of:

(1) Well or pile foundation, and
(2) Pier, and pier cap.

The design of a well has several elements such as:

(a) Bottom plug,

(b) Cutting edge (or curb),

(c) Steining,

(d) Top plug, and

(e) Well cap.

and similarly the pier design has:

(a) Pier cap, and

(b) Pier stem.

Figure 15.1 illustrates different structural components of a typical bridge. A very brief discussion on materials and other specifications with special reference to the Indian Road Congress Code on Prestressed Concrete bridge, IRC 18—1985 is made here.

15.2. Materials for Prestressed Concrete Bridge

This section is specially written for bridges and partly overlaps with some of the text presented earlier. Any of the following steels can be used as tensioned reinforcement:

(a) Plain wires confirming to IS: 1785

(b) High tensile steel bars (IS: 2090)

(c) Cold drawn indented wires (IS: 6003)

(d) Stress relieved strands.(IS: 6006).

Any of the standard reinforcement used in RCC construction can also be used as untensioned or nominal reinforcement. Ordinary Portland Cement (OPC) (IS: 269) or Rapid Hardening Portland Cement (RH PC) (IS: 8041) or High Strength Portland Cement (HSPC) (IS: 8112) can be used in the construction. A nominal size 20 mm crushed stone or gravel confirming to IS: 383 can be used as a coarse aggregate and fine aggregate with FM in the range of 2 or more is recommended.

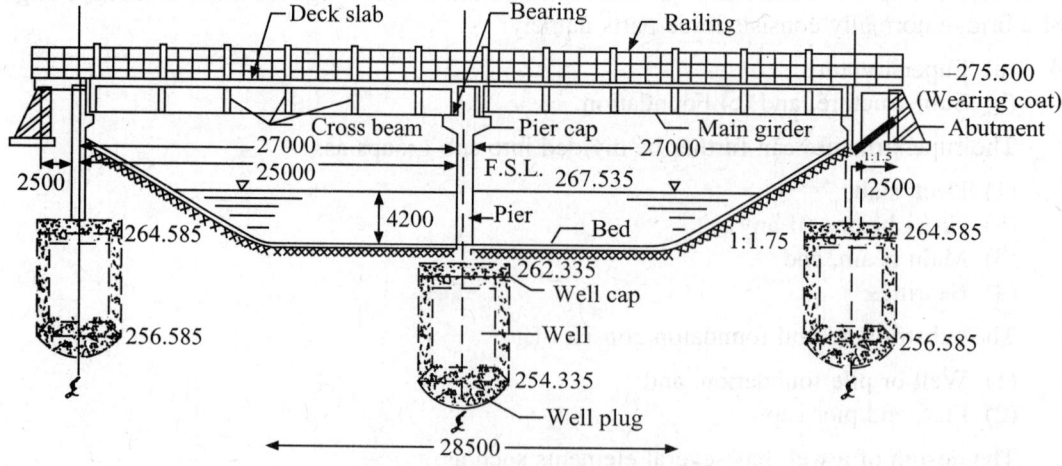

Figure 15.1 Elevation of bridge.

Concrete mix must be a designed with minimum characteristic strength of 35 N/mm^2 using a minimum of 360 kg cement per m^3 of concrete and a maximum of 550 kg/m^3. Normally recommended water cement ratio is 0.35 to 0.45.

15.3. Permissible Stresses (N/mm^2)

(a) Compressive stresses is concrete:

   (1) Temporary loading condition: $\sigma_{ach} = 0.5\, f_{ei} \leq 20$ (transport, handling, erection)

   (2) Full transfer of prestress = $\sigma_{aci} = 0.45\, f_{ci} \leq 20$

   (3) Under service loads = $\sigma_{ace} = 0.33\, f_{ck}$

(b) Tensile stresses in concrete:

   temporary tensile stress = $\sigma_{ati} = \sigma_{aci}/10$ and at service load = $\sigma_{ate} = 0$

(c) Maximum bearing stress $f_b = 0.48\, f_{ci}\sqrt{\dfrac{A_2}{A_1}}$ or $\leq 0.8\, f_{ci}$

where A_1 = square equivalent bearing area

   A_2 = maximum square area contained by the member concentric with A_1. However the value of f_b can be suitably enhanced by providing hoop reinforcement at anchorage.

(d) Tensile stresses in HTS wires.

   (1) Temporary stress (transfer condition) = $\sigma_{asi} = 0.7\, f_p$

   (2) Ahchorage stress = $\sigma_{asa} \leq 0.8\, f'_p$ or $\leq 0.95\, f_p$

15.4. Limiting Requirements

(a) Cement

   Minimum cement   = 360 kg/m^3

   Maximum cement   = 550 kg/m^3

   Maximum W/C   = 0.45

(b) *Minimum Untensioned* reinforcements are:

   Vertical direction in webs   = 0.3 per cent (MS bars) or

   = 0.18 per cent (HYSD bars)

   Longitudinal steel   = 0.25 per cent (MS bars) or

   = 0.15 per cent (HYSD bars)

   (for $f_{ck} \leq 45$ MPa)

   Minimum diameter of bars   = 10 mm (MS)

   = 8 mm (HYSD)

   Maximum spacing of reinforcement = 200 mm

(c) Cover and spacing of tensioned steel:

The nominal clear cover beyond sheathing is:

50 mm in moderate exposure

60 mm in severe exposure

(d) Clear cover to untensioned steel:

30 mm in moderate exposure for $f_{ck} < 40$ MPa

25 mm in moderate exposure for $f_{ck} \geq 40$ MPa

40 mm in severe exposure for $f_{ck} < 40$ MPa

30 mm in severe exposure for $f_{ck} \geq 40$ MPa

subject to a minimum of twice the diameter of the bars

(e) Minimum clear distance between cables is: ≥ 50 mm

$\geq$ diameter of duct

≥ 10 mm

+ largest size of aggregate

(f) Splay of cables:

Splay of the cable near the anchorages should be less or equal to 1 in 6 (that is 1 in 6 or 1 in 7 etc.). The radius of the cable in elevation shall not be less that 1.5 m + 700 d_s where d_s = diameter of the cable or wire.

(g) Thickness of deck slab:

Minimum thickness of the deck slab should be 150 mm in normal exposure and 180 mm in aggressive environment. Minimum thickness at the tip of the cantilever in any environment can be 150 mm.

(h) Thickness of web:

Minimum thickness of web should be 150 mm + diameter of duct. In case of box girders, the minimum thickness is also controlled by $h/36 + 2$ (clear cover) + diameter of the duct where h = over all depth of beam.

15.5. Design Example

An example of design of a complete bridge is given in this section. Some of the obvious points such as computation of weight moment of inertia, bending moments etc. are not given to minimize the volume of calculations. This is a design undertaken for UP Irrigation department for construction of a bridge across parallel Ganga Canal:

A bridge whose general elevation is shown in Figure 15.1 consists of the following:

1. *Statement of the Problem*

The main spans of 25 m waterway (each) and two end spans of RCC slab of 2.5 m effective span. The main girders are of prestressed while the cross beams, deck, end spans, piers and well foundations are of reinforced concrete. The bridge is to be designed to carry a national highway loading.

The entire design is divided into number of sections which are listed as 1, 2.... etc. for convenience, these sections are:
1. Statement of problem,
2. Design of continuous RCC Deck slab etc.
2. Design of Continuous Rcc Deck Slab

2. *Design of Continuous RCC Deck Slab*

2.1 *Assumed data* (vide Figures 15.1 and 15.2)

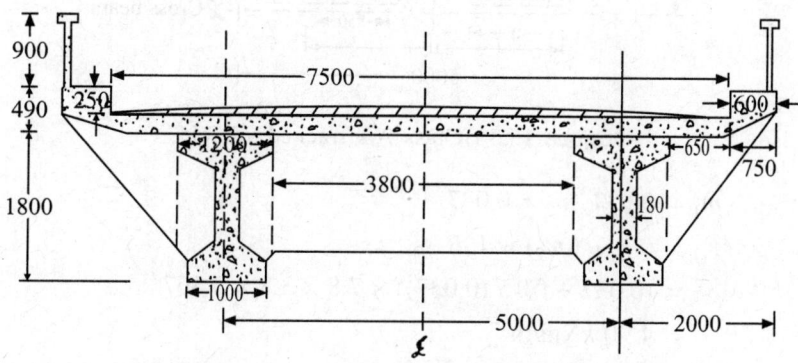

Figure 15.2 Deck slab (assumed).

(a) Spacing of middle cross beam = 25/6 = 4.167 m = 4167 mm

(b) Spacing of main beam (two main beams) = 5000 mm

(c) Average thickness of wearing coat = 75 mm

(d) Slab thickness = 240 mm

(e) Width of web of cross beam (T-beam) = 300 mm

(f) Clear long span = $(4.167 - 0.300)\,10^3\,B$ = 3867 mm

(g) Clear short span = $(5.0 - 1.2)\,10^3 = L$ = 3800 mm

(h) Weight of wearing coat = 0.075×24 = 1.8 kN/m^2

(i) Weight of slab = 0.24×25 = 6.0 kN/m^2

Dead load = w = 7.8 kN/m^2

2.2 *Bending moment due to dead load* (see Figure 15.3)

Aspect ratio of slab = $k = L/B = 3.8/3.86 = 0.98$. $1/k = 01.02$

Bending moment coefficients from Pigeaud's curve for UDL are:

Notations k, u, v, m_1, m_2 etc. are same as those given in the Pigeaud's curves.

$(v/L = u/B = 1)$

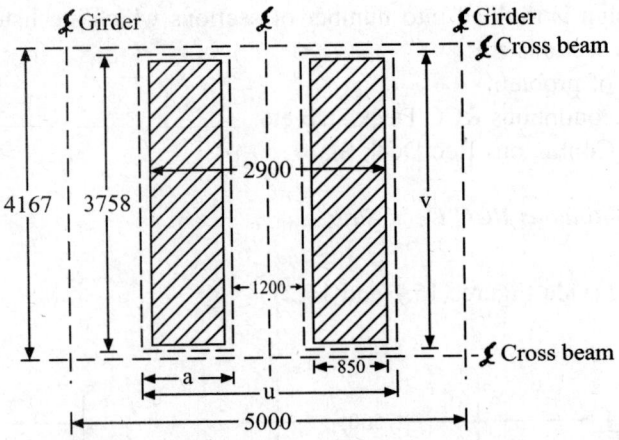

Figure 15.3 Class AA tracked vehicle.

$$m_1 = 0.037; \; m_2 = 0.037$$

$$
\begin{aligned}
M_{11} &= (m_1 + \mu m_2) \, w \, L \, B \\
&= (0.037 + 0.15 \, (0.037)) \times 7.8 \times 3.8 \times 3.867 \\
&= 4.90 \; k\text{Nm/m}
\end{aligned}
$$

$$
\begin{aligned}
M_{21} &= (m_2 + \mu m_1) \, w \, L \, B \\
&= (0.037 + 0.15 \, (0.037)) \times 7.8 \times 3.8 \times 3.867 \\
&= 4.90 \; k\text{Nm/m}
\end{aligned}
$$

2.3 *Bending moment due to moving loads*

(a) *Class AA load (tracked vehicle vide Figure 15.3 for arrangement)*
Panel of slab = 4.167 × 5.0 m
Impact factor = 25 per cent
Contact width = 850 mm

$$a = \sqrt{[(x + 2D)^2 + H^2]}$$

$$a = \sqrt{[(850 + 2(75))^2 + 240^2]} = 1028 \text{ mm}$$

$$b = \sqrt{[(3600) + 2 \times 75)^2 + 240^2} = 3758 \text{ mm}$$

$$u = 2900 - 850 + a = 3078 \text{ mm}$$

(or one can also use $u = 2900 + 2(75 + 240) = 3530$ mm
Lower value of the two is used.

$$v = \sqrt{[(3.6 + 0.15)^2 + 0.24^2]} = 3.758$$

$$k = 0.98; \; u/B = 3078/5000 = 0.616$$

$$v/L = 3758/4167 = 0.902$$

From Pigeaud's curves we get

for $\qquad k = 1, m_1 = m_2 = 0.054$

and $\qquad k = 0.9; m_1 = 0.061; m_2 = 0.048$

Use the moment coefficients as

for $\qquad k = 0.98; m_1 = 0.054; m_2 = 0.052$

Applying continuity correction of 0.8 and with load as 700 kN (Class–AA)

$$M_{22} = 0.8 \,(0.053 + 0.15(0.054)(\,1.25)(700)$$

$$= 42.78 \text{ kNm/m}$$

$$M_{12} = 0.8(0.054 + 0.15(0.053))(1.25)(700)$$

$$= 43.37 \text{ kNm/m}$$

(b) *Class AA wheeled vehicle (vide Figures 15.4 and 15.5 for load locations)*
Bending moment due to frontal wheel marked 1 in Figure 15.4

$$W = 2(62.5) \times (1.25) = 156.25 \text{ kN}$$

$$u = \sqrt{[(1.3 + 0.15)^2 + (0.24)^2]} = 1.47 \text{ m}$$

$$v = \sqrt{[(0.3)^2 + (0.24)^2]}$$

$$= 0.384 \text{ m}$$

(Note: $0.3 = b + 0.15 = 0.15 + 0.15$)

$$u/B = 1470/5000 = 0.29$$

$$u/L = 0.384/4.167 = 0.10$$

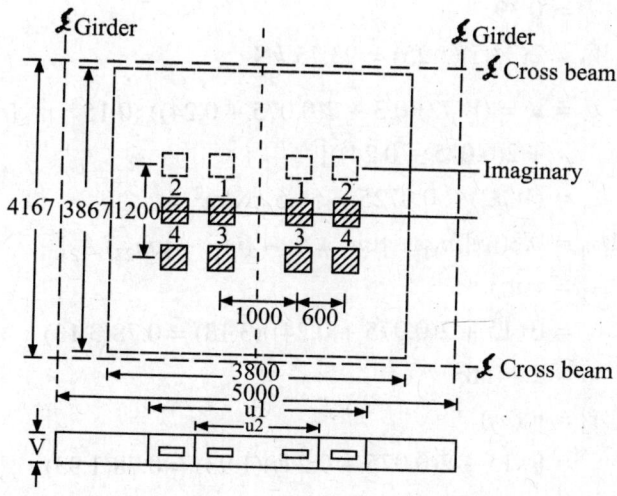

Figure 15.4 Class AA wheeled vehicle.

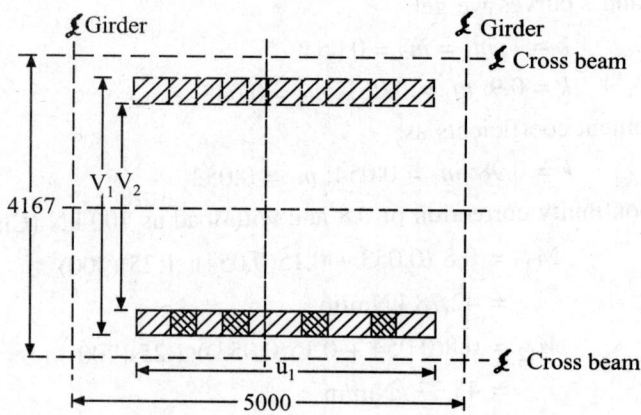

Figure I5.5 Class AA wheeled vehicle.

From the Pigeaud's curves we get

$$m_1 = 0.19 \text{ and } m_2 = 0.19$$
$$M_{13} = M_{23} = 0.8(0.19(1 + 0.15)\ 156.25)$$
$$= 22.31\ k\text{Nm/m}$$

Moments due to wheels marked 2 are:

$$u_1 = 2.2 + 2(0.15 + 0.075 + 0.24) = 3.13 \text{ m}$$
$$u_2 = 1.0 + 2(0.15 + 0.075 + 0.24) = 1.93 \text{ m}$$
$$B = 5.0 \text{ m}$$
$$u_1/B = 0.63$$
$$u_2/B = 0.39$$
$$W_1 = 2(37.5)(1.25) = 93.75\ k\text{N}$$

Intensity

$$= w = (W_1/2)(0.3 + 2(0.075 + 0.24))\ (0.15$$
$$+ 2(0.075 + 0.24))]$$
$$= (W_1/2) \times 0.0725 = 64.6\ k\text{N/m}^2$$
$$M_{14} = 0.8(w)[m_{11} + \mu m_{12})\ A_1 - (m_{21} + \mu m_{22})\ A_2]$$
$$A_1 = v(u_1)$$
$$= (0.15 + 2(0.075 + 0.24))(3.13) = 0.78(3.13)$$
$$= 2.44 \text{ m}^2$$
$$A_2 = v_2(u_2)$$
$$= (0.15 + 2(0.075 + 0.24))(1.93) = 0.78(1.93)$$
$$= 1.505 \text{ m}^2$$
$$v_1 = v_2 = 0.78;\ v_1/L = 0.19$$

$$m_{11} = m_{12} = 0.12$$

$$m_{21} = m_{22} = 0.145$$

$$M_{14} = M_{24} = 0.8(64.6)(1.15(0.12)\,2.44$$

$$- 1.15(0.145)(1.505))$$

$$= 4.43 \; k\text{Nm/m}$$

Consider loads at wheels marked 3 and 4 (vide Figures 15.4 and 15.5)

$$W = 2(62.5 + 37.5(1.25)) = 250 \; kN$$

$$u_1 = 3.13 \; \text{m}$$

$$v_1 = 1.2 + (0.15 + 0.075 + 0.24) \times 2 = 2.13 \; \text{m}$$

Intensity
$$= w = \frac{W}{\text{Width} \times \text{Breadth}} = \frac{250}{2.5 \times 2 \times 0.15} = 333.3 \; k\text{N/m}^2$$

$$u_1/B = 0.63; \; v_1/L = 2.13/4.167 = 0.511$$

$$u_2/B = 0.63; \; v_2 / L = \frac{2.13 - 0.15}{4.167} = 0.48$$

$$A_1 = 3.13 \times 2.13 = 6.667 \; \text{m}^2$$

$$A_2 = 3.13 \times 1.197 = 6.197 \; \text{m}^2$$

$$m_{11} = 0.078 = m_{12}$$

$$m_{21} = m_{22} = 0.080$$

$$M_{15} = M_{25} = 0.8(333.3)\,[1.15(0.078)6.667$$

$$- 1.15(0.08)(6.197)]$$

$$= 7.44 \; k\text{Nm}$$

Hence total moment due to all the wheeled loads combined is

$$M_{16} = M_{13} + M_{14} + M_{15} = 22.31 + 4.43 + 7.44$$

$$= 34.18 \; k\text{Nm/m}$$

$$M_{26} = M_{23} + M_{24} + M_{25} = 34.18 \; k\text{Nm/m}$$

2.4 *Final bending moments and design*

From comparison we can see that the moments from the tracked vehicle are greater than those caused by the wheeled vehicle, therefore

$$M_1 = M_{11} + M_{12} = 4.9 + 43.37 = 48.27 \; k\text{Nm/m}$$

$$M_2 = M_{21} + M_{22} = 4.9 + 42.78 = 47.68 \; k\text{Nm/m}$$

Using M 20 concrete: (Revised code suggest M 35)

$$d = \sqrt{\frac{M}{kb\sigma_{cb}}} = \sqrt{\frac{48270}{0.13(7)}} = 230 \text{ mm (required)}$$

Use $t = 250$ mm; $d = 250 - 25 - 10 = 215$ mm

as d provided is less than the required, compression sleel is to be provided

$$A_{st} = M/j \, d\sigma_{st} = \frac{48270}{0.86(215)(200)} = 1305 \text{ mm}^2/\text{m}$$

Hence provide 20 mm ϕ bars at 200 mm spacing both ways with alternate bars cranked. Provide 12 ϕ at 400 mm at top (HYSD bars).

2.5 Cantilever slab (see Figure 15.6)

Since the cantilever distance is 650 mm beyond this edge, the AA load cannot be accommodated as the minimum edge distance is 1200 mm. Therefore design the cantilever for other class A loading

Live load of class A wheel = 57 kN

Width of dispersion = $e = 1.2x + w$

$$x = 500/2 = 250 \text{ mm}$$

Type spread = $w = 250 + 2(75) = 400$ mm

$$e = 1.2 \times 250 + 400 \text{ mm} = 700 \text{ mm}$$

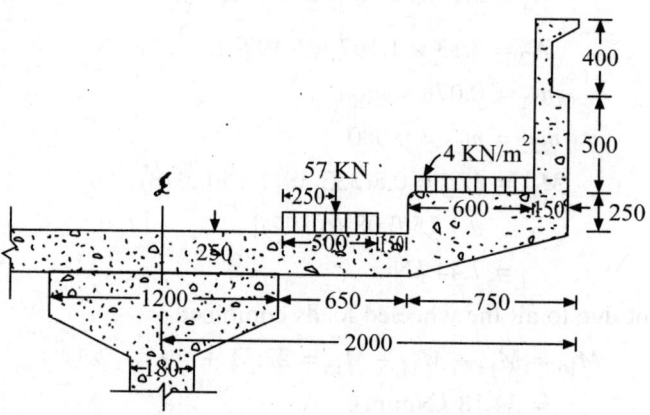

Figure 15.6 Contilever slab.

At the edge near end of the beam:

$$e = 700/2 + 250/2 = 350 + 125 = 475 \text{ mm}$$

Impact factor $I = 1 + 4.5/(6 + 2) = 1.6$

(a) Moment due to live load (concentrated and UDL on footpath) is:

$$M_1 = \frac{1.6 \times (57)(0.25)}{0.475} + 0.6(4)(0.65 + 0.3)$$

$$= 50.20 \text{ } k\text{Nm/m}$$

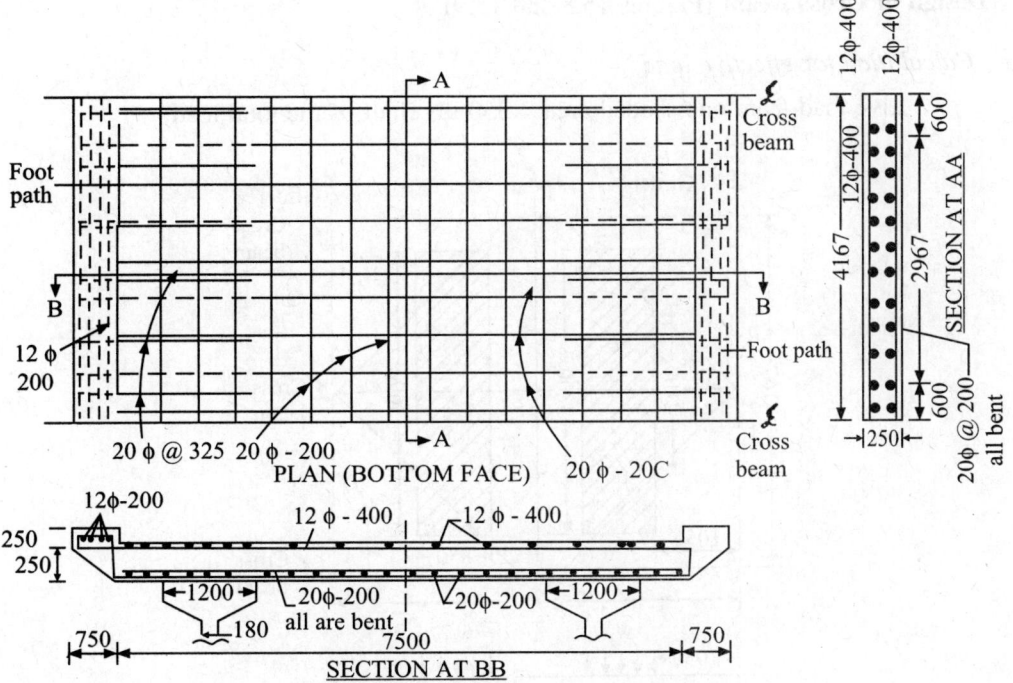

Figure 15.7 Deck slab (main).

(b) Moment due to dead load: (Slab and footpath)

$$M_2 = [0.25 \, (0.65/2)(0.65) + 0.25(0.75)(0.65$$

$$+ \, 0.25/2 + 0.25(0.75)(0.65 + 0.375]25 = 8.23 \; k\text{Nm/m}$$

(c) Moment due to railing:

$$M_3 = 0.9 \, (0.15)(1.25 + 0.075)(25) = 3.57 \; k\text{Nm/m}$$

(d) Total moments

$$M = M_1 + M_2 + M_3 = 62.00 \; k\text{Nm/m}$$

(using age factor = 1.15)

(e) Doubly reinforced slab design is adopted for cantilever portion.

$$A_{st} = \frac{M}{jd\sigma_{st}} = \frac{62.00 \times 10^6}{0.866(215)(200)} = 1697 \; \text{mm}^2/\text{m}$$

cranked bars 20 ϕ at 400 mm and 12 ϕ at 400 are already available.

Add 20 ϕ at 400 mm extra at the top

$A_{min} = 0.2(2500) = 500 \; \text{mm}^2/\text{m}$; (Add extra 20 ϕ at the rate of 325 at bottom and hooked up)

Use 12 ϕ at 200 mm longitudinal + partly at bottom.

3. Design of Cross Beam (Figures 15.8 and 15.9)

3.1 *Calculation for effective load*

Live load from one shaded area = 350 kN (half of the total load)

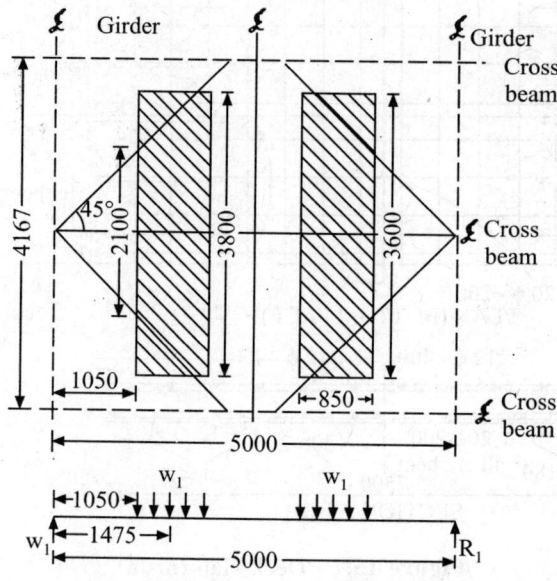

Figure 15.8 Class AA tracked vehicle.

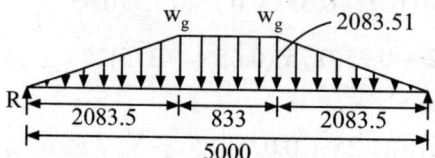

Figure 15.9 Dead load.

$$\text{One load on cross beam} = \frac{350}{(3600 \times 850)} \times \left(\frac{2100 + 3800}{2}\right)$$

= 286.00 *k*N (also equal to reaction)
Impact factor = 0.25

Therefore W_1 = 1.25 × 287 kN (each of the two loads) = R_1

(a) Bending moment $M_1 = R_1 a$ = 1.25 × 287 × 1.050 = 377 *k*Nm (at 1.05 m)

Bending moment at mid span = M_2 = 1.25(287) (1.475) = 529 *k*Nm.

The greater of the two controls the design. (M_2 always controls).

(b) Dead load moment: $W_g = 2 \times (0.3 \times 25 \times 2.0835) = 31.25$ kN/m

Reaction $R = 2 \times \left[\dfrac{0.3 \times 25}{1000 \times 2} \left(\dfrac{2083.5}{2} + 833 + \dfrac{2083.5}{2} \right) \right] = 45.57$ kN

Dead load moment at the centre

$$M_D = 45.57 \times 2.5 - W_g \left(\dfrac{0.833}{2} \right) \dfrac{0.833}{4}$$

$$- W_g \times \dfrac{2.083}{2} \left(\dfrac{2.0835}{3} + \dfrac{0.833}{2} \right)$$

$$= 113.93 - 38.88 = 75.00 \; kNm$$

Total moment = 529 + 75 = 604 kNm

There is another possible critical location of tracked vehicle. One of the track is right at mid span of the cross beam and the other at the appropriate distance as shown in Figure 15.10. The effective load from the second track on to the cross beam is

$$W_L = \dfrac{350}{(3600 \times 850)} \dfrac{2(850 \times 900)}{2}$$

$$= 87.5 \; kN; \text{ The reaction } R' \text{ is}$$

$$R' = \dfrac{1.25(87.5 \times 0.45 + 350 \times 2.5)}{5} = 1.25 \times 182.875$$

Max moment = $1.25 \times (182.875 \times 2.5) = 1.25 \left(\dfrac{350}{2} \right) \left(\dfrac{0.85}{4} \right) = 525$ KNm

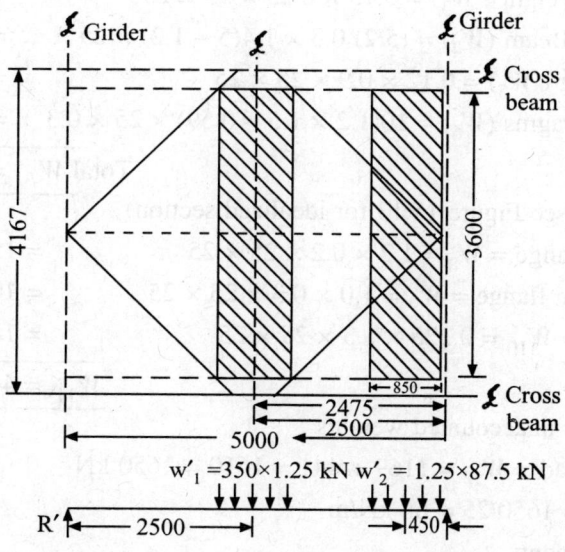

Figure 15.10 Class AA wheeled vehicle.

Dead Load moment = 75 kNm

Total moment 525 + 75 = 600.00 kNm

The critical moment is from case one and it is 604 kNm.

$$M_r = kbd^2\sigma_{cb} \times \text{age factor}$$

$$= 0.13(0.3)(1.3)2(7)(1.15)(1000) = 530 \text{ } k\text{Nm}$$

$$A_{st} = \frac{604 \times 10^6}{0.866(1300)(200)} = 2682 \text{ m}^2$$

Use 6 nos 25 ϕ at bottom ($p_t = 0.7$ per cent)

2 nos 25 ϕ at top

4 nos 12 ϕ at mid height

Shear: $V = 350 \times 1.25$

$$\tau_v = \frac{350(1.25) \times 10^3}{300(10)^3 \times 1.3} = 1.12 \text{ N/mm}^2$$

Use 10 ϕ at 175 mm as stirrups.

4. Design of Main Beam

4.1 *Dead load moments (vide Figure 15.12–15.15)*

(a) Weights (for 25 m length i.e. effective span on one beam)

(i)   Footpath $(W_1) = 0.75 \left(\dfrac{0.25 + 0.50}{2} \right) \times 25 \times 25$ = 176 kN

(ii) Wear Coat $(W_2) = 3.75 \times 0.075 \times 25 \times 24$ = 169 kN

(iii) Slab Weight (W_3) = $3.75 \times 0.25 \times 25 \times 25$ = 568 kN

(iv) Cross Beam $(W_4) = (5/2)\ 0.3 \times 1.4(5 - 1.2) \times 25$ = 100 kN

(v)   Railing (W_5) = $0.12 \times 0.9 \times 25 \times 25$ = 68 kN

(vi) Diaphragms $(W_6) = 10(1.2 \times 1.5 - 0.59) \times 25 \times 0.3$ = 70 kN

Total W_7 = 1169 kN

Self Weight (see Figure 15.13 for idealized section)

(i)   Top flange = $W_8 = 1.2 \times 0.2 \times 25 \times 25$ = 150 kN

(ii) Bottom flange = $W_9 = 1.0 \times 0.3 \times 25 \times 25$ = 108 kN

(iii) Web = $W_{10} = 0.225 \times 1.3 \times 25 \times 25$ = 183 kN

W_{11} = 441 kN

Add extra for unaccounted weights

Total dead load = W_{12} = 1169 + 441 = 1610 $\simeq$ 1650 kN

Load/meter = 1650/25 = 66 kN/m

Bending moment:

$M_s = W_7 L/8 = 1166 \times 25/8 = 3653$ kNm.

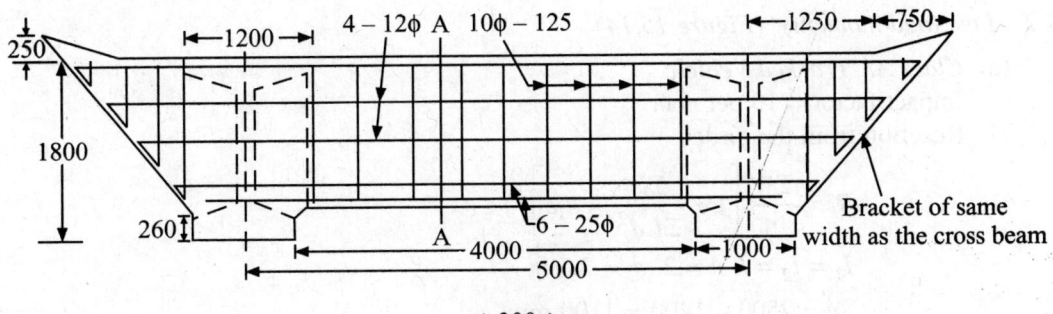

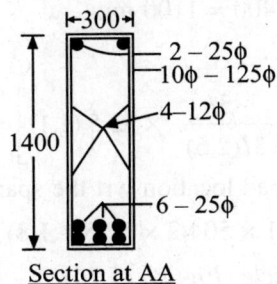

Figure 15.11 Cross beam.

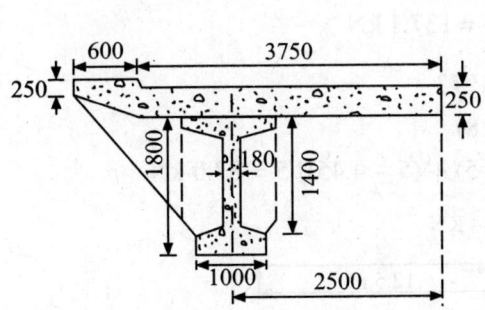

Figure 15.12 Dead load.

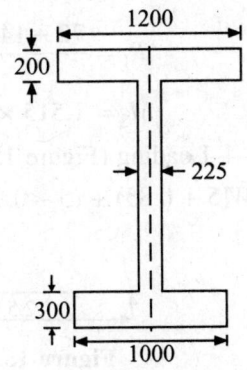

Figure 15.13 Self weight.

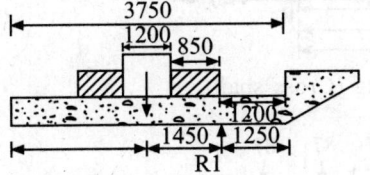

Figure 15.14 Girder reaction.

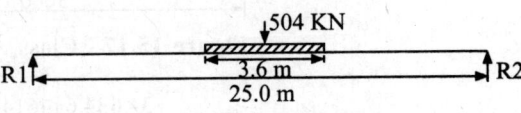

Figure 15.15 Load on beam.

4.2 *Live load moments: (Figure 15.14)*

(a) *Class AA Tracked Vehicle*
   Impact factor = 10 per cent
   Reaction from the girder

$$R_1 = \frac{\Sigma W}{n}\left[1 + \frac{\Sigma I_i}{\Sigma I\, d_i^2} \times d_i e\right]$$

$I_1 = I_2 = I$, $n = 2$, $d_i = 2.5$ m

$e = 2500 - 1400 = 1100$ mm

$\Sigma W = 700$ kN

$$R_1 = \frac{700}{2}\left[1 + \frac{3I}{3I(2.5)^2} \times 12.5\,(1.1)\right] = 504\ kN$$

Vide Figure 15.15 for load location wrt the span, the moment is

$$M_1 = (1.1 \times 504/2 \times 12.5 - 1.8) = 2966 \text{ kNm}$$

(b) *Class–A A wheeled vehicle (Figure 15.16)*
   Impact factor = $4.5/(6 + L) = 0.1515$

$$R_a = 400(3.6)/5 = 288 \text{ kN}$$

$$R_1 = \frac{72 + 144(12.5 - 1.2)}{25} = 137.1 \text{ kN}$$

$$M_2 = 1.515 \times R_1 \times L/2 = 1973$$

(c) Class–4–Loading (Figure 15.17 and 15.18)

$$R_a = W[5 + 0.85) + (5 - 0.95) + (5 - 2.65) + (5 - 4.45)]/5 = 87.0 \text{ kN}$$

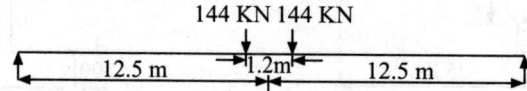

Figure 15.16   Class AA wheeled vehicle.

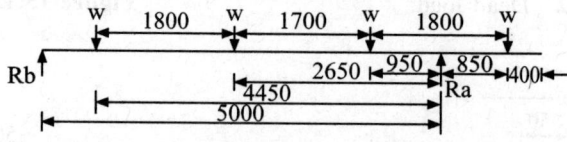

Figure 15.17   Class A loading (Cross span).

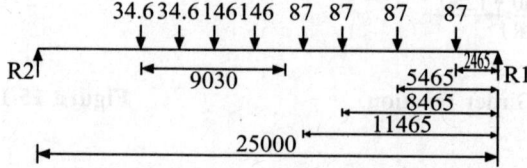

Figure 15.18   Class A loading (Main span).

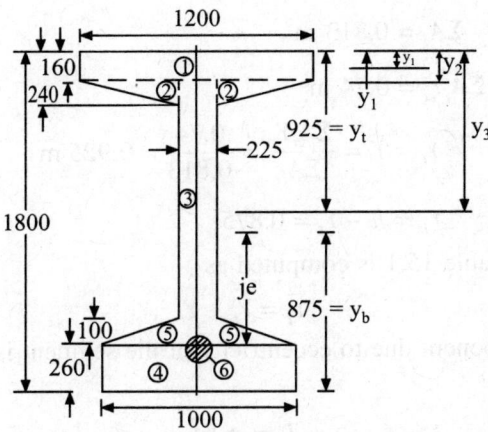

Figure 15.19 Beam CS (Initial).

$$R_1 - 362.12 \text{ kN}$$

$$M_3 = 1.1515\ 362.12,(12.115 - 87)[(12.115 - 2.465)$$

$$+ (12.115 - 5.465) + (12\ 115 - 8.465)] = 3031 \text{ kNm}$$

The maximum *BM* due to *LL* = M_1 – 3031 kNm

4.3 *Sectional properties of the beam and composite section*

The sectional properties are computed through Table 15.1.

TABLE 15.1 Girder Properties (Vide Figure 15.19)

Section	A_i	Y_i	A_iY_i	$\overline{Y}_i$	I_{ci}	$A_i\overline{Y}_i^2$
(1) Top flange 1.2(0.16)	0.192	0.08	0.0154	0.845	0.0004	0.1371
(2) Triangles of top flange $\dfrac{(0.975)(0.08)}{2}$	0.037	0.187	0.007	0.738	0.0000	0.0201
(3) Web 1.38 × 0.225	0.311	0.85	0.264	0.075	0.0493	0.0017
(4) Bottom flange 1 × 0.26	0.26	1.67	0.434	– 0.745	0.0020	0.1443
(5) Bottom flange triangle 0.775 (0.1)/2	0.038	1.51	0.058	0.575	0.0000	0.0125
(6) Duct (3%)	–0.025	1.54	–0.038	–0.30	–· ·	–0.017
Sum =	0.813	0.740			0.0520	0.299

$$\Sigma A_i = 0.813 \text{ m}^2$$

$$\Sigma A_i Y_i = 0.74 \text{ m}^3$$

$$Y_t = \bar{Y} = \frac{\Sigma A_i Y_i}{\Sigma A_i} = \frac{0.74}{0.813} = 0.925 \text{ m}$$

$$Y_b = h - Y_t = 0.875$$

The column (5) of Table 15.1 is computed as

$$\bar{Y}_i = Y_t - Y_i$$

and then the inertia component due to eccentricity of the segment is computed in column (6) as:

$$I_i = A_i \bar{Y}_i^2$$

and final moment of inertia of the beam section about *CG* is

$$I_b = \Sigma(I_{ci} + A_i \bar{Y}_i^2) = 0.052 + 0.299 = 0.351 \text{ m}^4$$

The moment of inertia of the composite section is computed after the transformed area of the cast-in-situ concrete slab. The gross width of the flange chosen was controlled by the overhang of 650 mm beyond the top flange on the outer side.

Width of top flange = 1.2 + 2(0.65) = 2.7 m

Thickness of slab = 0.25 m

Effective thickness $= 0.25 \sqrt{(E_{20}/E_{35})} = 0.25\sqrt{(20/35)} = 0.189$ m

Use $t_c = 0.190$ m

The properties of the composite section are computed in Table 15.2 in the same lines as Table 15.1.

$$Y_{tc} = \frac{\Sigma A_i Y_i}{\Sigma A_i} = 0.52 \text{ m}$$

$$Y_{bc} = 1.8 - 0.52 = 1.28 \text{ m}$$

$$\bar{Y}_i = \bar{Y}_{ic} - Y_i$$

$$I_c = \Sigma(I_{ci} + A_i \bar{Y}_i^2) = 0.054 + 0.652 = 0.706 \text{ m}^4$$

The allowable stresses (N/mm²) used are:

$$\sigma_{aci} = 0.5(0.7)(35) = 12.25$$

(Transfer of prestress at 21 days of curing so a factor of 0.7 is applied)

$$\sigma_{ace} = 0.33(35) = 11.5$$

$$\sigma_{ati} = \sigma_{ate} = 0$$

The prestressing of the cables is to be carried out in two stages. In the first stage, also called initial stage, fifty-three per cent of prestress is applied with an assumed initial eccentricity of sixty-four per cent of the final ecentricity. The loss of prestress is assumed as 25 per cent.

TABLE 15.2 Properties of Composite Section

Section	A_i	Y_i	A_iY_i	$\bar{Y}_i$	I_{ci}	$A_i\bar{Y}_i^2$
Top flange	0.192	0.080	0.0154	0.44	0.0004	0.038
Top flange	0.037	0.187	0.0070	0.28	0.0000	0.003
Web	0.311	0.850	0.2640	−0.33	0.0493	0.036
Bottom flange	0.260	1.670	0.4340	−1.15	0.0020	0.345
Bottom flange	0.038	1.510	0.0580	−0.987	0.0000	0.037
Duct	0.025	1.54	−0.0380	0.83	–	0.017
Slab (2.7(0.19))	0.513	−0.095	−0.0487	0.615	0.0015	00.200
	1.326		0.6917		0.054	0.652

Its stage prestress force = kp_i
Ist stage eccentricity = je

where P_i = total of initial prestress transferred to the beam

e = eccentricity of the total prestress

= Y_{bc}–distance of CGS from bottom fibre = $(1.28 - 0.45) = 0.83$ m

k = ratio of the first stage prestress force to the total prestress force

j = ratio of the eccentricities of the first and total prestress forces.

The following initial values arc assumed

$$k = 0.53 \text{ and } j = 0.64$$

(Actually these values are iterated ones)
The bounds of prestress force are first estimated.
The self weight bending moment is

$$M_g - (0.813)(25)(625)/8 = 1588 \text{ kNm}$$

The bending moments due to superimposed and live loads and that at working load condition are:

$$M_s = 3653 \text{ kNm}; M_1 = 3031 \text{ kNm}$$

$$M_w = M_g + M_s + M_1 = 1588 + 3653 + 3031 = 8272 \text{ kNm}$$

4.4 Bounds for Prestress

$$\frac{kP_i}{A} + \frac{kP_iY_{bje}}{I_b} - \frac{M_gY_b}{I_b} \leq 12.25 \tag{A}$$

$$\frac{-kP_i}{A} + \frac{kP_iY_i je}{I_b} - \frac{M_gY_i}{I_b} \leq 0 \tag{B}$$

Addition of (A) and'(B) gives $jP_i \le (12.25\, I_b/h + M_g)/kj$

$$P_i e \le 12230 \le \left[\frac{12.25(1000)(0.354)\,/\,1.8 + 1588}{(0.53)(0.64)} \right]$$

$$P_i < 12230/0.83 = 14735 \text{ kN} \le 12230$$

Lower bound is given by:

$$\frac{\eta P_i}{A} - \frac{\eta P_i e\, Y_{tc}}{I_c} + \frac{M_w Y_{tc}}{I_c} \le 11.5 \qquad (C)$$

$$\frac{\eta P_i}{A} - \frac{\eta P_i e\, Y_{bc}}{I_c} + \frac{M_w Y_{bc}}{I_c} \le 0 \qquad (D)$$

Addition of (C) and (D) gives

$$P_i \ge \left(M_w - 11.5\frac{I_c}{h} \right) / \eta e$$

$$\ge \left(8272 - \frac{11.5 \times 1000 \times 0.706}{1.8} \right) / 0.75 \times 0.83$$

$$P_i \ge 6042 \text{ kN}$$

It is desirable to use a value close to the lower bound one.

4.5   *Frozen stress + Additional stresses due to M_s, M_1, and $(1 - k)\, P_i$ are:*

$$\frac{\eta k P_i}{A} - \frac{\eta P_i jek Y_t}{I_b} + \frac{M_g Y_t}{I_b} + \frac{\eta (1 - k)\, P_i}{A_c}$$

$$\frac{-\eta (1 - k) P_i \overline{X} Y_{tc}}{I_c} + \frac{(M_1 + M_s)\, Y_{tc}}{I_c} \le \sigma_{ace} \qquad (E)$$

$$\frac{\eta k P_i}{A} - \frac{\eta k P_i je Y_b}{I_b} + \frac{M_g Y_b}{I_b} - \frac{\eta (1 - k)\, P_i}{A_c}$$

$$\frac{\eta (1 - k) p \overline{X} Y_{bc}}{I_c} + \frac{(M_a + M_s)\, Y_{bc}}{I_c} \le 0 \qquad (F)$$

At this stage the grouting of the ducts has been done hence the effect of the duct holes on the cross sectional area and the moment of inertia is neglected. Therefore

$A_c = 1.326 + 0.025 = 1.351$ m^2

$I_c = 0.6520 + 0.017 = 0.6690$ m^4

$\overline{x} = $ distance of initial eccentricity from the centroid of the tensioned steel to be computed from

$$e \text{ and } ej$$

$\bar{e}$ = difference between the *CG* of the beam and composite section

$= 0.925 - 0.5 = 0.425$ m

$$P_i e = kP_i(ej + \bar{e}) + (1 + k)p_i\bar{x} \quad \text{or}$$

$$\bar{x} = \frac{e - k(ej + \bar{e})}{1 - k}$$

$$= \frac{1.086 - 0.53(1.086(0.64) + 0.425)}{0.47} = 1.061 \text{ m}$$

Adding Eqs *E* and *F* and reassembling them gives

$$P_i \geq \frac{M_g + (M_1 + M_s)I_b / I_c - \sigma_{ace}I_b / h}{\eta[(1 - k)\,\bar{x}I_b / I_c + kje]}$$

$$\geq \frac{1588 + 6684(0.351)/0.669 - 11500(0.351)/1.8}{0.75[0.47(1.061)(0.351)/0.669 + 0.53(0.64)/1.086]} = 6626 \text{ kN}$$

This value is slightly more than the lower bound computed. Therefore use P_i in the range of 6107 kN,

$$A_p \text{ (rev)} > \frac{6626}{0.7 f_p} = \frac{6627}{0.7(1.47)} = 6439 \text{ mm}^2$$

Use 192 wires of 7 mm diameter in 16 cables of each 12 wires

$$A_p \text{ (Provided)} = 16(12)(38.48) = 7388.16 \text{ mm}^2$$

$$P_i = A_p(0.7f_p) = 7602 \text{ MkN} = 7.602 \text{ MN}$$

4.6 *Check for actual stresses*

It is now necessary to check whether the actual stresses are within the allowable values. The two stages are investigated,

(a) Initial stresses

$$\sigma_{ti} = \frac{-kP_i}{A} + \frac{kP_i y_i e}{I_b} = \frac{M_g y_t}{I_b}$$

$$= \frac{-0.53(7.602)}{0.813} + \frac{0.53(7.602)(0.925)(1.086((0.64)}{0.351} - \frac{1.588(0.925)}{0.351}$$

$$= -4.96 + 7.38 - 4.19 = -1.77 \text{ N/mm}^2 < \sigma_{ati}$$

(b) Final stage

$$\sigma_{ce} = \frac{\eta kP_i}{A} - \frac{\eta kP_i jey_t}{I_b} + \frac{M_g y_t}{I_b} + \frac{(1 - k)P_i}{A_c}$$

$$- \frac{\eta(1 - k)P_i \bar{x}y_{tc}}{I_c} + \frac{(M_a + M_s)y_{tc}}{I_c}$$

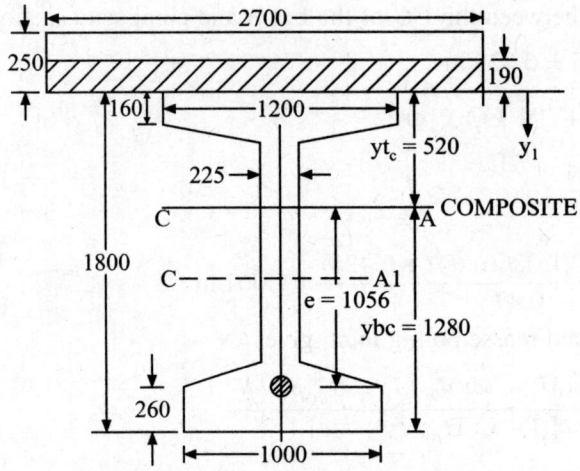

Figure 15.20 Composite section.

$$= \frac{0.75(0.53)(7.602)}{0.813}$$

$$- \frac{0.75(0.53)(7.602)(0.64)(1.086)(0.925)}{0.351}$$

$$+ \frac{1.588(0.925)}{0.351} + \frac{0.75(0.47)(7.602)}{1.326}$$

$$- \frac{0.75(0.47)(7.602)(1.061)(0.52)}{0.706} + \frac{6.684(0.52)}{0.706}$$

$$= 3.71 - 5.53 + 4.19 + 2.02 - 2.09 + 4.93$$

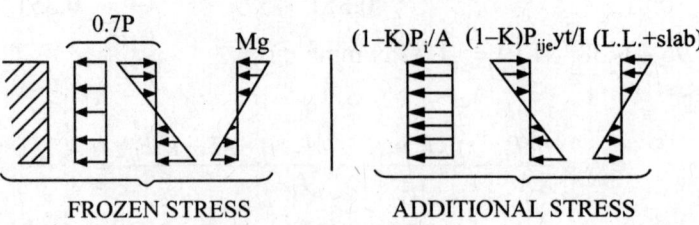

Figure 15.21 Stresses.

$$= 7.23 < \sigma_{ace} = 115$$

$$\sigma_{te} = -3.71 - 5.530(0.875)/0.925 + 4.19\,(0.875)/0.925$$

$$-2.02 - 2.09\,(1.28)/0.52 + 4.93\,(1.28)/0.52$$

$$= 0.0\ \text{N/mm}^2$$

All stresses except tension at working are less than the allowable values. Provide 17 cables of 12 wires each then the value of P_i is

$$P_i = 17(12)(38.48)(1.47(0.7)) = 8077\ \text{kN}$$

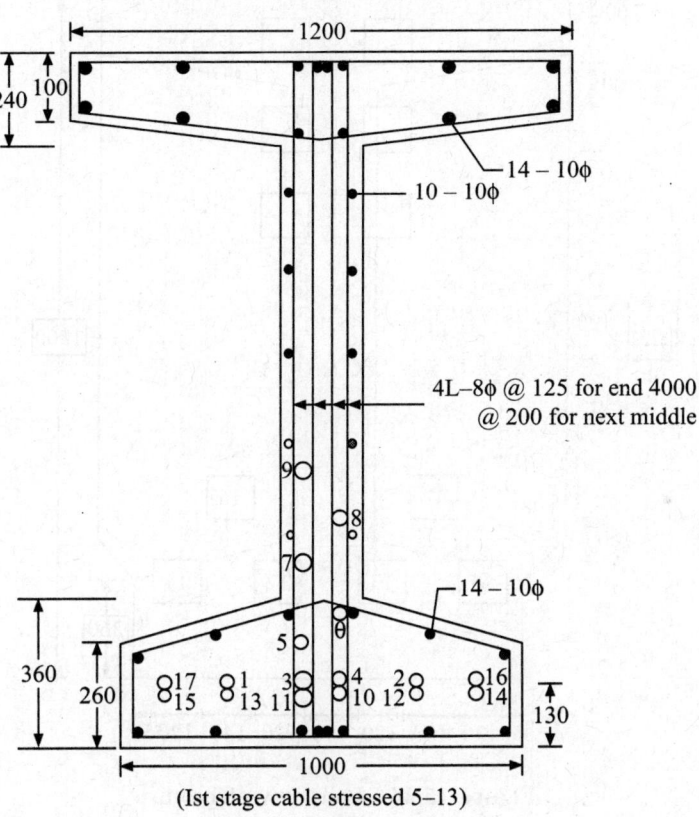

Figure 15.22 Main beam section at mid span.

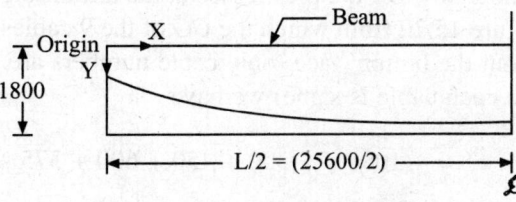

Figure 15.23 Cable profile.

494

The 17 cables are placed at raid section as shown in Figure 15.22 and end section as shown in Figure 15.24 and general profiles shown in Figure 15.23. Normally a stand-by cable is needed in case, some wires give-up during stressing. By providing 17 cables we have enough margin of extra prestress..

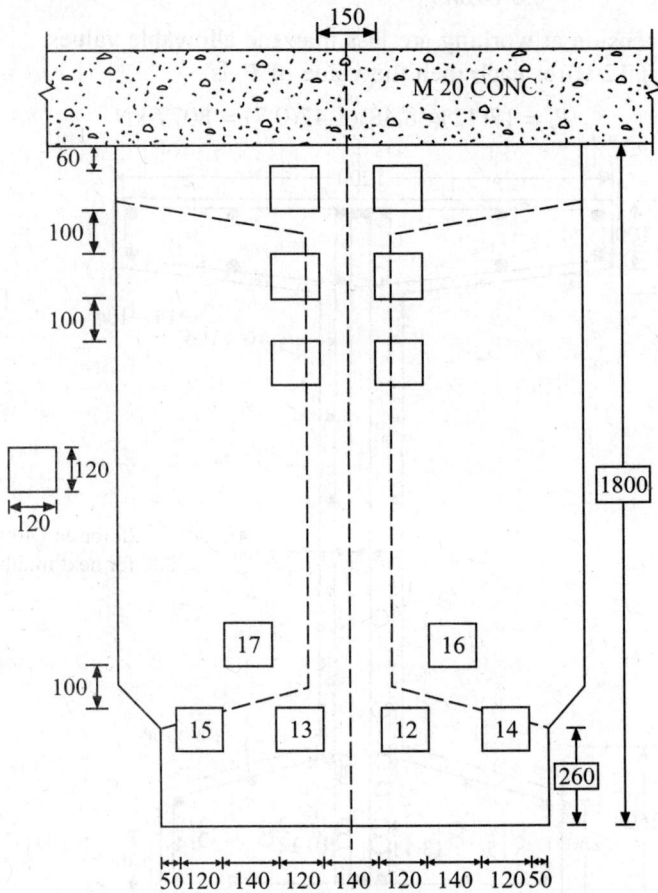

Figure 15.24 End view of beam.

4.7 *Eccentricity computation*

Initial stage: 9 cables numbered 1 to 9 are prestressed at the initial stage. The actual spacing of the cables is shown in Figure 15.22 from which the *CG* of the 9 cables is computed. By taking moment of the cables about the bottom face (only cable numbers are used instead of force in each cable as the force in each cable is same) we have

$$9(y_b - e_j) = 4(1000 + 1(300 + 450 + 600 + 375 + 525)$$

$$e_j = 875 - (4 \times 100 + 2250)/9 = 580$$

The value of ej provided is 580 mm as against an assumed value of $0.64(1068) = 683$ mm. Since the stresses at the initial stage are far below the allowable ones, there should not be any problem. The stress should be recomputed using the actual e_j.

TABLE 15.3 Coordinates of Cables

X Cables	0	1000	2000	3000	4000	5000	6000	7000	8000	9000	10000	11000	12000	12800
1	560	717	860	991	1108	1214	1307	1386	1454	1508	1551	1579	1596	1600
2	560	731	888	1031	1161	1277	1378	1466	1540	1600	1645	1677	1696	1700
3	120	340	541	725	891	1039	1170	1282	1376	1453	1512	1553	1576	1582
4	120	354	770	766	944	1102	1241	1361	1462	1544	1607	1651	1676	1682
5	–	0	243	465	666	844	1002	1138	1252	1244	1416	1465	1493	1500
6	–	–	0	256	487	694	875	1032	1164	1270	1353	1410	1442	1450
7	–	–	–	0	271	513	726	910	1064	1190	1286	1353	1391	1400
8	–	–	–	–	0	289	544	764	948	1099	1213	1294	1339	1350
9	–	–	–	–	–	0	312	581	808	992	1132	1231	1286	1300
10	340	547	737	911	1467	1207	1330	1436	1526	1598	1653	1692	1714	1719
11	340	532	709	869	1014	1144	1258	1356	1439	1506	1558	1594	1614	1619
12,13 14,15	1540	1583	1627	1670	1670	1670	1670	1670	1670	1670	1670	1670	1670	1670
16,17	1320	1413	1507	1600	1603	1606	1609	1612	1615	8618	1621	1624	1627	1630

TABLIE 15.4 Extensions of Cables

S. No.	Cable	Length L (mm)	Extension Δc (mm)	Extension +(mm)	Remarks
1	1	25684	132	139	
2	2	25701	132	139	
3	3	25767	133	139	
4	4	25790	133	139	
5	5	23790	122	129	
6	6	21794	112	118	
7	7	19799	102	107	
8	8	17806	92	96	
9	9	15815	82	85	
10	11	25727	132	139	
11	10	25748	132	139	
12	12, 13 14, 15	25601	132	138	
13	16, 17	25608	132	138	

$$\Delta c = (0.7 \times 1470 \times 1/200000)L$$
$$= \Delta c(1 + 0.05)$$

The final eccentricity is calculated similarly and it is

$$17(y_{bc} - e) = 12(130) + 1(300 + 450 + 600 + 375 + 525)$$

$$e = 1280 - \frac{12 \times 130 + 2250}{17} = 1056$$

The actual value provided is 1056 mm as against an assumed value of 1068 mm. Figures 15.22 to 15.29 illustrate various details of the reinforcement.

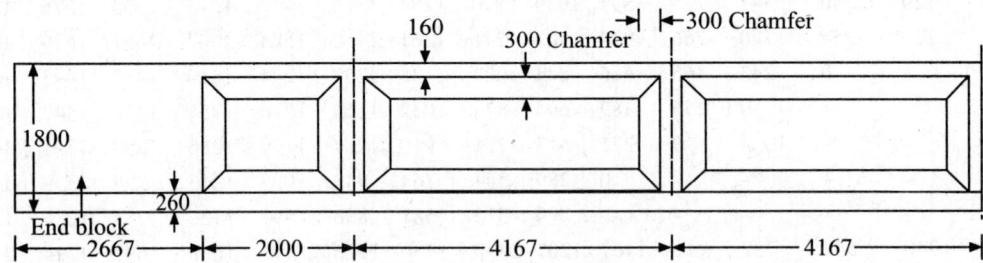

Figure 15.25 Part longitudinal view.

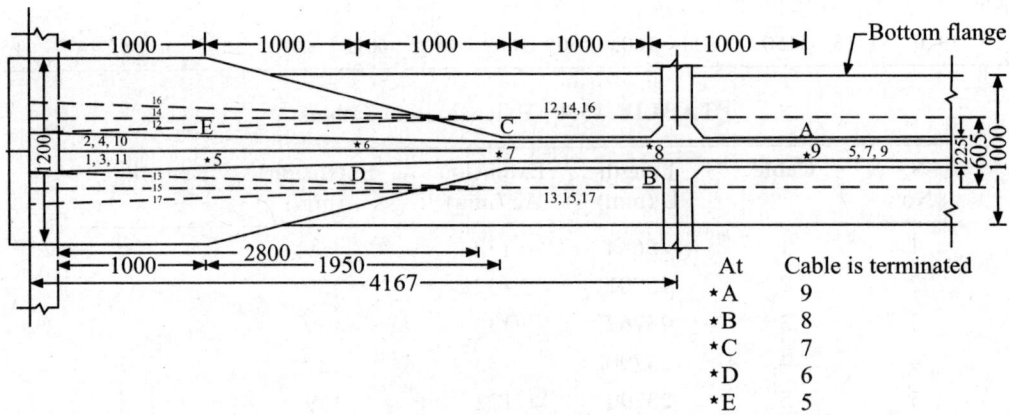

At	Cable is terminated
⋆A	9
⋆B	8
⋆C	7
⋆D	6
⋆E	5

Figure 15.26 Plan (cables).

5. Design of Middle Pier Cap *(Figure 15.30 and 15.31)*

Figure 15.30 illustrates the general layout and profile of the cap at the middle pier. All calculations are made with reference to this figure.

Load calculations

There are seven cross beams for each girder and two rest on pier top, hence there are in all 14 cross beams. Hence total load due to cross beams and slab is:

Total load from slab and beam = 1850 kNm

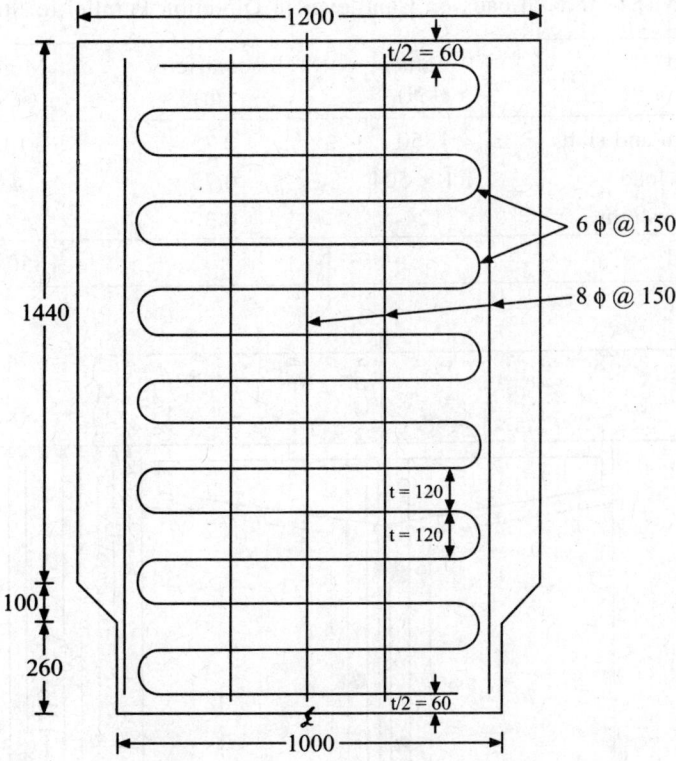

Figure 15.27 Spiral reinforcement.

TABLE 15.5 Load on Cantilever in the Traffic Direction

Load	Value (kN)	Eccentricity (m)	Moment (kNm)
Beam and slab	(1850)2/2	1−0.6=0.4	740
Live load	722	1−0.6=0.4	285
Self weight	$\dfrac{0.8(0.75 + 1.25)\,6.2 \times 25}{2}$ $= 124$	$\dfrac{(2 \times 0.75 + 1.25)0.8/3}{(0.75 + 1.25)}$	46
Sum =			1071 kNm

Live Load Calculations (see Figure 15.32)

$$R_1 = \frac{700 \times 1.1(25 - (1.8 - 0.3))}{25} = 722 \text{ kN}$$

TABLE 15.6 Loads on Cantilever in Direction Parallel to Stream

Load	Value (kN)	Eccentricity (m)	Moment (kNm)
Beam and slab	1850	0.75	1388
Live load	1.1×504	0.75	416
Self weight	124	0.37	46
Total			1850 kNm

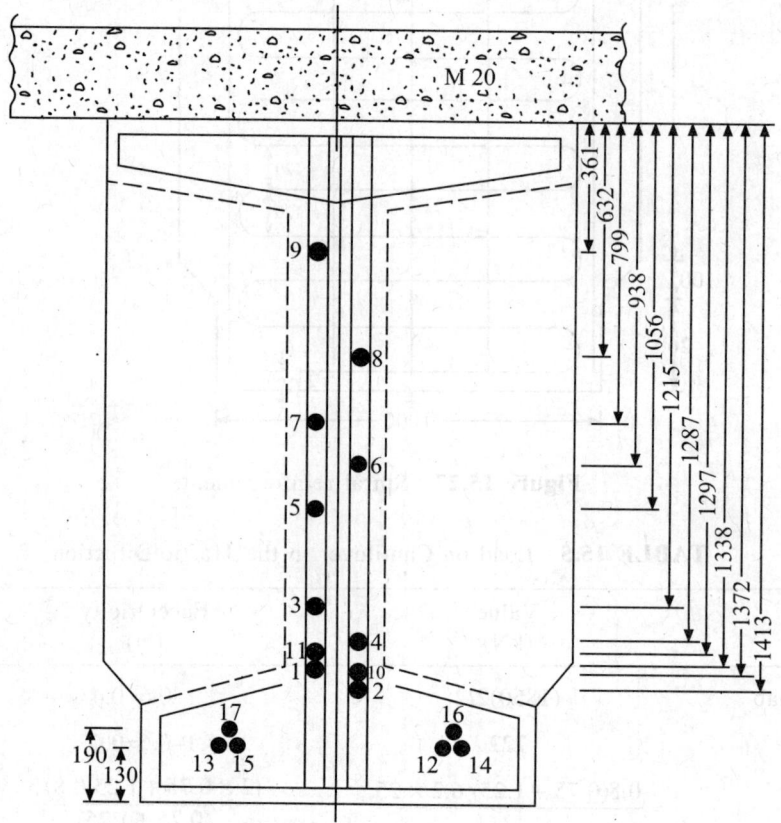

Figure 15.28 Section at 1/4 span (6400 mm from end).

Design

Case I Earthquake along the stream

$$M = 1071 \times 10^6 \, \text{Nmm}$$

$$b = 6200 \, \text{mm} \qquad \sigma_{si} = 190 \, \text{N/mm}^2$$

$$d = 1250 \, \text{mm}$$

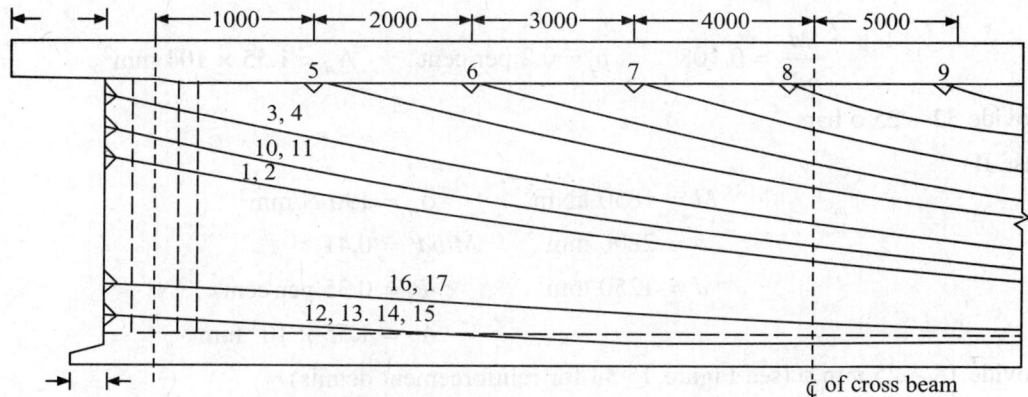

Figure 15.29   Cable profiles at end.

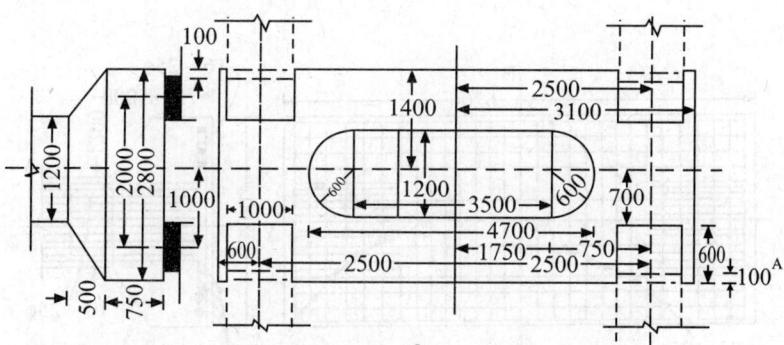

Figure 15.30   Pier and pier cap plan layout.

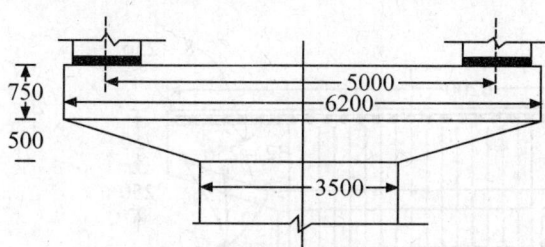

Figure 15.31   Pier cap.

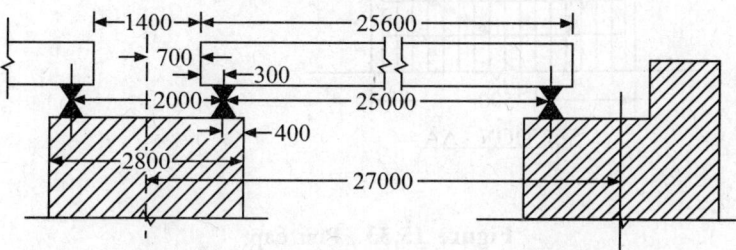

Figure 15.32   Beam with abutment & pier.

$$\frac{M}{bd^2} = 0.108 \qquad p_i = 0.2 \text{ per cent} \qquad A_{si} = 1.55 \times 104 \text{ mm}^2$$

Provide 32 – 25 φ bars

Case II

$$M = 1850 \text{ kNm} \qquad \sigma_{st} = 190 \text{ N/mm}^2$$
$$b = 2800 \text{ mm} \qquad M/bd^2 = 0.41$$
$$d = 1250 \text{ mm} \qquad P_i = 0.35 \text{ per cent}$$
$$A_{si} = 1.55 \times 10^4 \text{ mm}^2$$

Provide 18 – 25 mm φ (see Figure 15.33 for reinforcement details)

6. Design of Central Pier (Refer Figure 15.34)

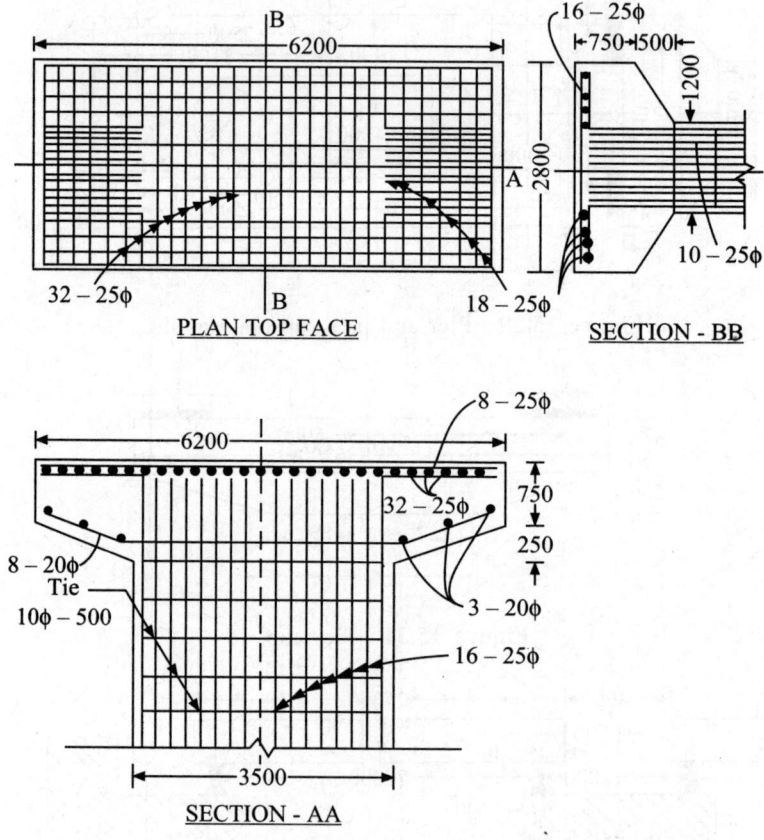

Figure 15.33 Pier cap.

6.1 Weight Calculations

Main beam $= W_5 = 0.813(25.6)(25)2 = 1040$ kN

$$\text{Pier Cap } = W_6 = 25\left[\frac{((3.5(1.2) + (6.2)(2.8))(0.5)}{2} + 6.2(2.8)0.75\right]$$

$$= 25(5.39 + 13.02) = 460 \text{ kN}$$

Pier self weight $= W_7 = 3.5(1.2)(10.965 - 1.25)(25) = 1021$ kN

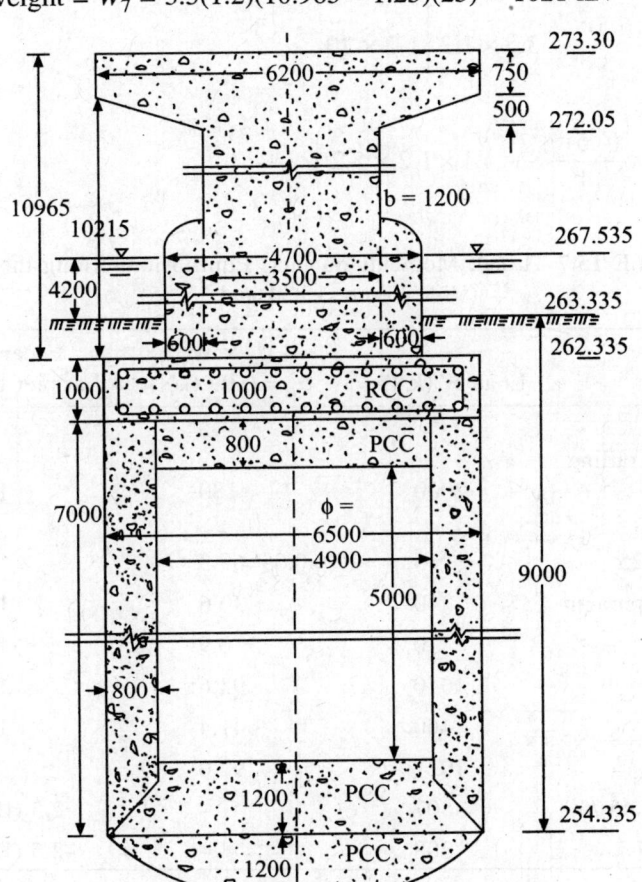

Figure 15.34 Cross section of a well.

Well cap $= W_9 = (\pi)(6.5)^2 \times 1 \times 25/4$ $= 830$ kN

Top plug $= W_{10} = (\pi)(4.9)^2 \times 0.8 \times 25/4$ $= 377$ kN

Steining $= W_{11} = (6.5^2 - 4.9^2) \times 6.4 \times 25/4$ $= 2292$ kN

Bottom plug $= W_{12} = \pi[(5.7)^2(1.2) \times 24 + (2/3)(6.5)^2(1.2) \times 24]/4$ $= 1402$ kN

Water in well = $W_{13} = \pi(4.9)^2(5)10/4$ = 942 kN

Weight of water above well

$$W_{14} = \left[\pi \frac{(6.5)^2}{4} - 3.5 \times 1.2\right] 5.2 \times 10 \qquad = 1507 \text{ kN}$$

Weight of soil abpve well

$$W_{15} = \left[\pi \frac{(6.5)^2}{3} - 3.5 \times 1.2\right] 1.2 \times 20 \qquad = 580 \text{ kN}$$

$$\text{Buoyancy} = 10\left(\pi \frac{(6.5)^2}{4} 8 + 4.1 \times 1.2 \times 5.2\right) \qquad = 2911 \text{ kN}$$

TABLE 15.7   Load, Moments on Pier: Equilibrium Along the Stream
(Vide Figure 15.34 and 15.35)

Load	Load W (kN)	$\alpha_h \times W$ $\alpha_h = 0.09$ (kN)	Eccentricity at pier base (m)	Moment (kNm)
Slab + Wearing coat + Footpath + railing $W_1 = 2(1169-170)$	2000	180	13.5	2430
Live load + Impact $W_2 = 770(25-1.8)/25$	715	770×0.09=69.3	13.7	950
Cross beam + Diaphragm	340	30.6	12.5	382
Bearing	10	0.9	11.3	10
Main beam	1040	93.6	12.5	1170
Pier cap	460	41.4	10.4	430
Pier	1021	92.0	5.2	460
Eccentric 715(5−1.45)/5	507	–	2.5 (Hor)	1270
Live load 715−507	208	–	−2.5 (Hor)	−520
Totals	5995	507.8		6122

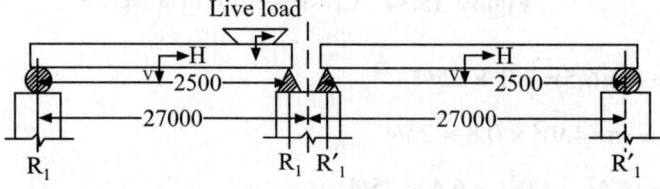

Figure 15.35   Earthquake across the stream.

Case 2. The equilibrium perpendicular to the stream. An increase in reaction due to live load and that due to dead weights on one span is counter balanced by another span, adjacent to it, therefore, the horizontal force on the central pier is = Total horizontal force of two spans due to dead load − Horizontal force absorbed by the rollers + Additional load due to class AA load on one span.

Total eccentricity of loads above bearing is

$$e = \left[\frac{2000(0.3 + 1.8 + 0.2) + 1040(0.3 + 0.9) + 340(0.3 + 0.9)}{2000 + 1040 + 340} \right]$$

$$e = \frac{6256}{3380} = 1.85 \text{ m above bearing}$$

$$R_1 = 3380/2 \pm \frac{0.09 \times 3380 \times 1.85}{25} = 1690 \pm 22.5 \text{ kN}$$

$R_{1max} = 1712.5$ kN and $R_{1min} = 1667.5$ kN

Horizontal reaction absorbed by the roller = μR_{1min}

$$= 0.3(1667.5) = 500 \text{ kN}$$

Therefore horizontal force at pier

$$= \frac{0.09 \times 3380}{2} = 152 \text{ kN}$$

Horizontal reaction due to live is computed as below:
Height of the force from roller level is

$$e_1 = (0.3 + 1.8 + 0.3) = 2.4 \text{ m}$$

The reaction due to seismic action on the live load is

$$R_1' = (770 - 722) \pm 770 \times 2.4 \times 0.09/25 = 48 \pm 6.6$$

$$R_{1max} = 54.6 \text{ and } R_{1min} = 41.4 \text{ kN}$$

Horizontal reaction absorbed by the roller:

$$\mu R_{1min} = 0.3 \times 41.4 = 12.4 \text{ kN}$$

Since

$$\mu R_{1min} < \frac{0.09 \times 770}{2} = 35 \text{ kN}$$

Horizontal reaction due to live load on the hinged end is

$$(0.09 \times 770 - \mu R_{1min}) = (70 - 12.4) = 57.6 \text{ kN}$$

Hence net horizontal force on the central pier is

$$= 304 + 57.6 = 360 \text{ kN}$$

Where the seismic force from superstructure is

$$= 109(2000 + 340 + 1040) = 0.09(3380) = 304 \text{ kN}$$

TABLE 15.8 Load and Moment on Pier Case 2

Load	Value W (kN)	Horizontal load (kN)	Fxcentricity at pier (m) Base	Moment (kNm)
Superstructure	2000			
(slab + wear)	340		10.97+ 0.3 +	
coat + Railing	1040		1.86 = 12.46	
main and cross beam	3380	304.2		3790
Live load	722	57.6	10.97 + 0.3 + 2.4	788
	+ 6.6		=13.67	
	729			
Bearing	10	0.9	11.3	10
Pier cap	460	41.4	10.4	430
Pier	1021	92.0	5.2	460
Eccentricity live load	722	–	1.5	722
W_{16}	= 6322 kN			M = 6200 kNm

6.3 Design

Since the pier section is critical in case II, it need be designed for, M_w = 6200 kNm and W = 6322 kN

The factored forces are

$$M_c = 1.5 \times M_w = 9300 \text{ kNm}$$

and

$$P_u = 1.5 \times W = 9483 \text{ kN}$$

The assumed properties are

$$f_{ck} = 20.0 \text{ N/mm}^2, f_y = 415 \text{ N/mm}^2$$

$$b = 3500 \text{ mm}, D = 1200 \text{ mm}$$

$$d' = 40/2 + 16 + 80 = 116 \text{ mm (cover)}$$

$$d'/D = 0.1$$

$$\frac{P_u}{f_{ck}bD} = \frac{9483 \times 10^3}{20(3500)(1200)} = 0.11$$

$$\frac{M_u}{f_{ck}bD^2} = \frac{9300 \times 10^6}{20(1200)^2(3500)} = 0.092$$

from IS: SP–16 handbook we have

$$P/f_{ck} = 0.04 \text{ or } p = 0.04(20) = 0.8$$

$$A_{st} = 3500 \times 1200 \times 0.8/100 = 3.36 \times 10^4 \text{ mm}^2$$

Use 68 numbers of 25 mm ϕ bars (vide Figure 15.36 for details)

7. Stability of Well at Central Pier

7.1 *Moments due to soil resistance (vide Figure 15.37)*

$$W_{17} = \text{Load on Pier} = W_{16} + W_9 + W_{10} + W_{11} + W_{12} + W_{13} + W_{14} + W_{15}$$

$$= 6322 + 830 + 377 + 2292 + 1402 + 1507 + 580 = 14{,}252 \text{ kN}$$

(vide IRC–45/1972 for stability of well)
various quantities required for stability analysis are

$$W_{17} = 14{,}252 \text{ kN}, \ B = 6.5 \text{ m}, \ Q = 0.5$$

(from Tables 2 of IRC: 45–1972)

$$\Phi = 31°, \ \delta = 6°, \ D = 8.0 \text{ m}$$

The moment resistance of the base about the point of rotation is

$$M_b = QWB\,\phi = 0.5(1.1)(14252)6.5\tan 31°$$

$$= 30164 \text{ kNm}, \ L = 6.5 \text{ m}, \ \gamma = 10 \text{ kN/m}^3$$

$$K_p = 3.12, \ K_a = 0.32$$

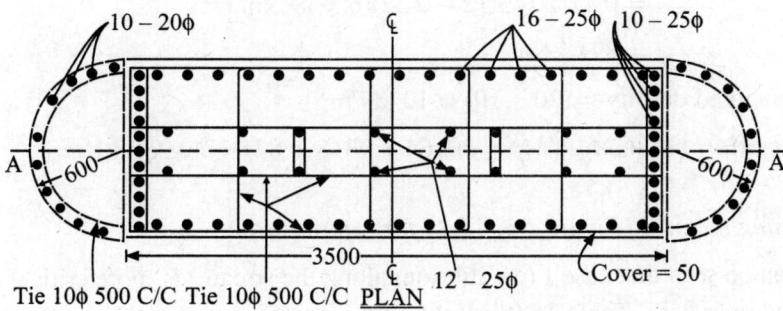

Tie 10ϕ 500 C/C   Tie 10ϕ 500 C/C   PLAN

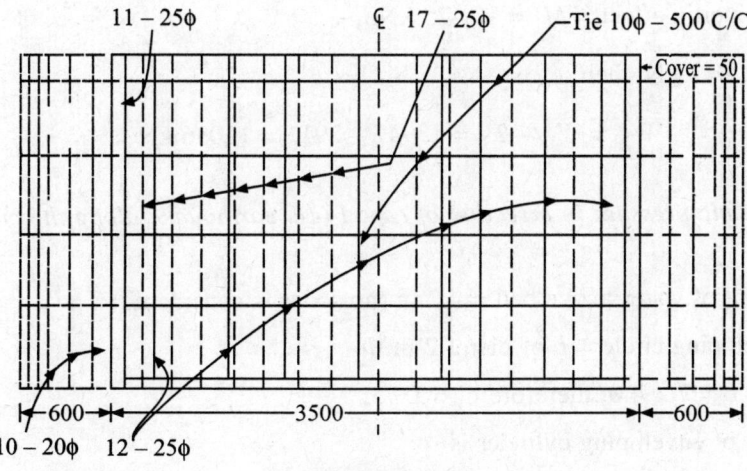

SECTION-AA

Figure 15.36 Pier details.

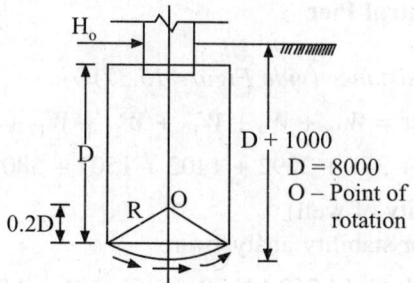

Figure 15.37 Calculations for soil resistance.

The soil resisting moment is

$$M_s = 0.1\gamma D^3(K_p - K_a)L = 0.1(10)(8)^3(3.12 - 0.32)6.5 = 9318 \text{ kNm}$$

The frictional resisting moment is

$$M_f = 0.11\gamma(K_p - K_a)B^2D^2 \sin\delta$$
$$= 0.11(10)(3.12 - 0.32)(6.5^2)8^2 \sin 6°$$
$$= 871 \text{ kNm}$$

where γ = submerged density = (20 – 10) = 10 kN/m³

$$M_r = (M_s + M_b + M_f) = 30164 + 9318 + 871 = 40,353 \text{ kNm}$$

7.2 *Overturning moment (Figure No. 15.34 for reference)*

Since it can be seen that case I (equilibrium along the stream) is more critical only that is being discussed over here. Table 15.9 list out all the quantities.

Moment due to inertia effect = M_1 = 10,425 kNm

Buoyancy F_B = 2911 kN, then

$$W_{19} = W_{18} - F_B = 13947 - 2911 = 11036 \text{ kN}$$

7.3 *Hydrodynamic pressure in direction of case I (i.e. earthquake along the current) (vide Figure 15.38)*

h = Height of water above bed = 4.2 m (no scour is considered)

r = Enveloping circle = b of pier 1.2 m Φ

h/r = 4.2/0.6 = 7 > 4.0; therefore C_e = 0.73

Weight of water of enveloping cylinder is

$$W_e = (\pi/4)((1.2)^2 \times 4.2 \times 10) = 47.5 \text{ kN}$$

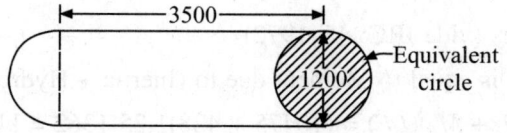

Figure 15.38 Hydrodynamic pressure.

TABI.F 15.9 Moment Computations on Central Well

Load type	Value (W) (kN)	$\alpha_h \times W$ or Horizontal loads	Eccentricity $0.2 \times D$ above the base (m)	Moment (kNm)
Slab + Wcarcoat + footpath (W_1)	2000	180	19.9	3582
Live load + Impact (W_2)	722	770 × 0.09		
		69.3	21.1	1462
Crossbeam + diaphragms (W_3)	340	30.6	18.9	578
Bearing (W_4)	10	0.9	17.7	16
Main beam (W_5)	1040	93.6	18.9	1741
Pier cap (W_6)	460	41.4	16.8	696
Pier (W_7)	1024	92.2	11.4	1051
Eccentric Live Loads (W_8)	507	–	(2.5) (Horizontal)	1268
	208		(–2.5)	–511
Well cap (W_9)	830	74.7	5.9	441
Top plug (W_{10})	377	33.9	5.0	170
Steining (W_{11})	2292	206.3	2.7	557
Bottom plug (W_{12})	1402	126.2	–2.2	–278
Water in well (W_{13})	942	85.0	1.7	144
Water above Well (W_{14})	1507	–	–	–
Soil above well (W_{15})	290	46.7	–	–
Total W_{18}	13947			M = 10,425 kNM

Horizontal force = $C_e \alpha_h W_c$ = .73(0.09)47.5 = 3.12 kN

It acts at ($C_4 H$) above soil level (vide IS 1893) for C_4)

$$= 0.4286(4.2) = 1.8 \text{ m}$$

Moment due to hydrodynamic force is then

$$M_n = 47.5(1.8 + 1 + 8 - 1.8) = 47.5(9) = 428 \text{ kNm}$$

7.4 *Stability calculations* (vide IRC: 45–1972)

Overturning moment is equal to Moment due to (Inertia + Hydrodynamic) load factor

$$= (M_I + M_h)(LF) = (10475 + 428)1.25 \ 1362 = \text{kNm}$$

Restoring moment with reduction factor is

$$M_R = (M_s + M_b + M_f)(0.7) = 40353(0.7) = 28247 \text{ kNm}$$

The overturning moment is less than the stabilizing moment.

7.5 *Stresses in steining*

Axial and bending stresses in the steining are critical at about 2/3 depth of the well. The forces and moments at about 2/3 depth are computed in Table 15.10 below.

TABLE 15.10 Case I Equilibrium Along the Stream up to Pier Base $2D/3 = 2(8)/3 \simeq 5$ m

Load	Vertical load (kN)	Horizontal (kN)	Eccentricity (m)	Moment (kNm)
Moment at base of pier (see Table 15.7)				
All loads above pier				6122
base	5995	507.8	5.0	2539
Well cap (W_9)	830	75.0	4.5	336
Top plug (W_{10})	377	33.9	3.6	122
Steining (W_{11})5/8	1432	129.0	2.0	260
Weight water above				
well(W_{14})	1507	–	–	–
Weight of Soil above				
Well (W_{15})	290	–	–	–
W_0 =	10431	$H_0 = 746$		$M_0 = 9379$

Moment due to soil pressure

At a depth of $2D/3 = 2(8)/3 = 5.33$ m below the well cap

$$M_s = \gamma(2/3)D(K_p - K_a)(2/3)D^2/6$$

$$= (2/3) \ 10(8) \ (2.35 - 0.31)(8)^2/6 = 517 \text{ kNm}$$

$$M_E = \text{Moment due to inertia + Hydrodynamic − Soil pressure}$$

$$= 9379 + 428 - 517 = 9290 \text{ kNm}$$

Total load $\quad = W = 10431$ kN

area of cross section $= A = (\pi/4)(6.5^2 - 4.9^2) = 14.32$ m^2

$$I = (\pi/64)(6.5^4 - 4.94) = 39.32 \text{ m}^4$$

$$\frac{W}{A} \pm \frac{M_E\, y}{I} = \frac{10431}{14.32} \pm \frac{9290(3.25)}{59.32} = 728 \pm 509$$

$$\sigma_1 = 1237 \text{ kN/m}^2 = 1.24 \text{ N/mm}^2$$

$$\sigma_2 = +\,279 \text{ kN/m}^2 \text{ (safe)}$$

Use M 15 concrete and
20 Φ at 600 mm spacing at each face in vertical direction and 16 Φ at 400 mm on each face circumferentially.

7.6 *Bearing pressures*

Net allowable bearing capacity at RL 274.000 is = 120 kN/m²

Factor of safety used = 3

RL of base of well = 254.335 m

Weight of soil over burden at base = $(274 - 254.335)20 = 393.3$ kN/m³

Where 20 kN/m² is the unit weight of the compacted soil.

Allowable net bearing capacity at base under seismic condition

$$= 120(1.2) + 393.3 = 537.3 \text{ kN/m}^2$$

Ultimate bearing capacity = $\sigma_u = 120(3) + 393.3 = 753.3$ kN/m²

Axial load from well = $W_{18} = 13947$ kN

Bending moment due to inertia = 10475 kNm

Restoring moment from soil = $M_s = 6790$ kNm

Net moment = $M = M_I = M_s = 10475 - 6790 = 3685$ kNm

The area of cross section and moment of inertia of the cross section at base are:

$$A_f = (\pi/4)(6.5)^2(1.1) = 36.5 \text{ m}^2$$

$$I_f = (\pi/64)(6.5)^4(1.1) = 96.38 \text{ m}^4$$

The bearing pressure caused in seismic load condition are:

$$\sigma = \frac{W}{A_f} \pm \frac{My}{I_f} = \frac{13947}{36.5} \pm \frac{3685(3.25)}{96.38} = 382 \pm 124$$

$$\sigma_1 = 506 \text{ kN/m}^2 < 537.3$$

$$\sigma_2 = 258 \text{ kN/m}^2 > 0 \text{ no tension}$$

The stresses are with in the limits.

DESIGN OF WELL CAP

Basic width of pier = 3.5 m
Width of two end curbs = 1.2 m

Effective width of the pier over the well cap is

$$= 3.5 + 1.2 = 4.7 \text{ m}$$

Assuming the thickness of the well cap as 1.0 m, the width over which the load is dispersed at the bottom face of the well cap (or it is same as top of steining of the well, see Figure 15.34)

$$B = 4.7 + 2(1) = 6.7 \text{ m}$$

The diameter of the centre of the thickness of the well is

$$D_e = \frac{6.5 + 4.9}{2} = 5.7 \text{ m}$$

This value is less than the dispersion width, therefore the load from the pier is directly transmitted to the steining without causing bending in the well cap. The well cap can be designed wfth nominal reinforcement.

9. Design of Well Steining and Curb

9.1 *Steining*

(a) Local Bending moment on steining is

$$M = \frac{prt}{\sqrt{12}}\left(1 - \frac{1}{H\beta}\right); H = 8 \text{ m}$$

Where p = Radial thrust

$= \gamma$ (HFL + total depth of well − Bottom plug thickness − centre of curb) + $C_a(\gamma_s - \gamma)$total depth of well − Bottom plug thickness − Centre of curb)

$= 10(4.2 + 9 - 1.2 - 0.6) + 0.32(10) \times (09 - 1.2 - 0.6)$

$= 114 + 23 = 137 \text{ kN/m}^2$

$$\beta = 4\sqrt{\frac{3}{R^2 t^2}} = 4\sqrt{\frac{3}{3.25^2 \times 0.8^2}} = 2.89$$

$$M_{11} = \frac{137 \times 6.5 \times 0.8}{\sqrt{12}}\left(1 - \frac{1}{8 \times 2.89}\right) = 197 \text{ kNm/m}$$

(b) Radial thrust from the bottom plug

$$T_1 = \frac{pDC^2}{16\gamma_0}$$

γ_0 = vertical height of inverted sag of curve

$= 1.2 + 1.2 = 2.4 \text{ m}$

p = Actual Bearing pressure:

$$= \frac{\text{Total vertical load} - 50\text{per cent of Buoyancy}}{\text{Area}}$$

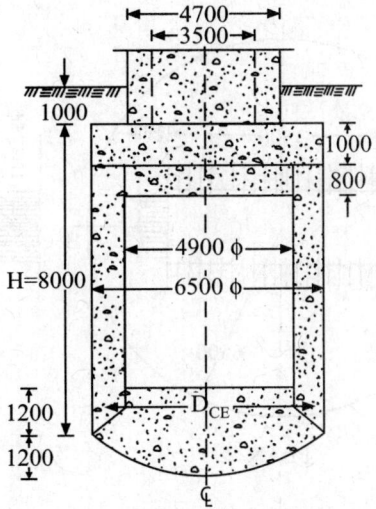

Figure 15.39 Outline of the well.

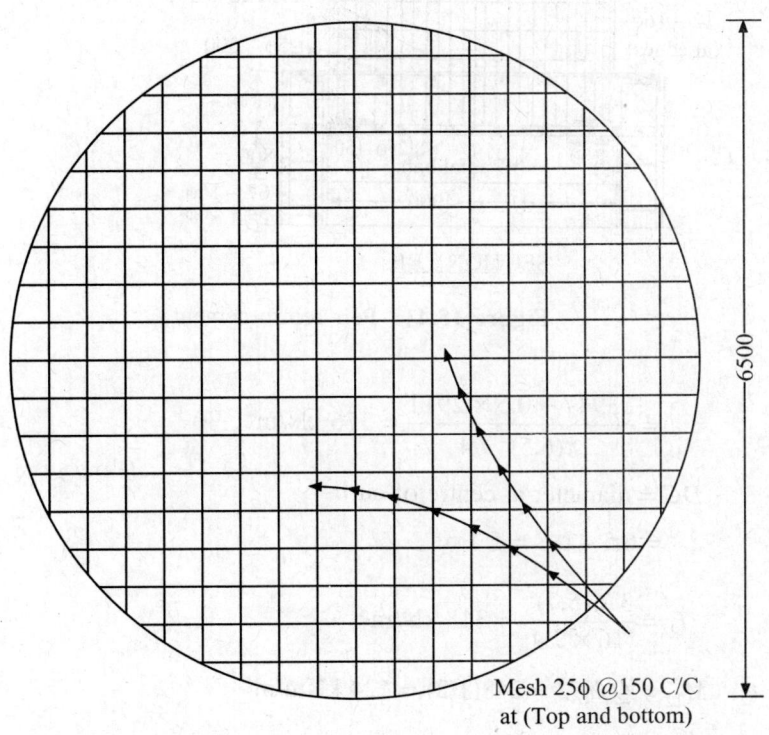

Mesh 25φ @150 C/C
at (Top and bottom)

Figure 15.40 Well cap reinforcement.

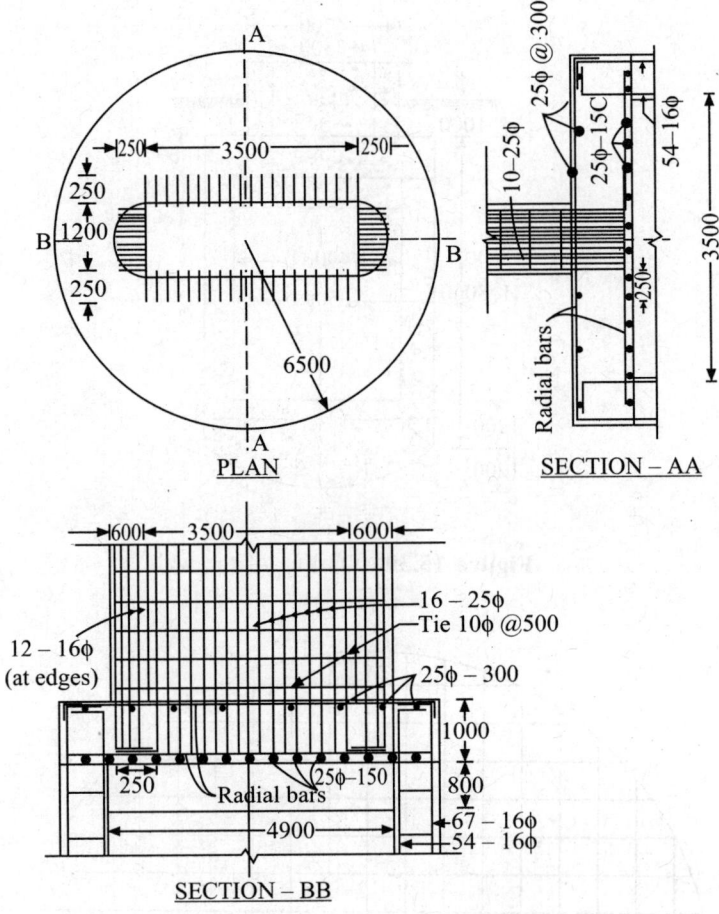

Figure 15.41 Peir reinforcement.

$$= \frac{13947 - 0.5 \times 2911}{\pi(6.5)^2 / 4} = 376 \text{ kN/m}^2$$

DC = diameter at centre of curb

$= 6.5 - 0.8 = 5.7$ m

$$T_1 = \frac{376(5.7)^2}{16 \times 2.4} = 318 \text{ kN/m}$$

$M_{12} = Q_1 h_0/2 = 248(1/2) = 124$ kNm/m

Since $M_{11} > M_{12}$ hence M_{11} controls the design

$$M_{\text{design}} = 197 \text{ kNm/m}$$

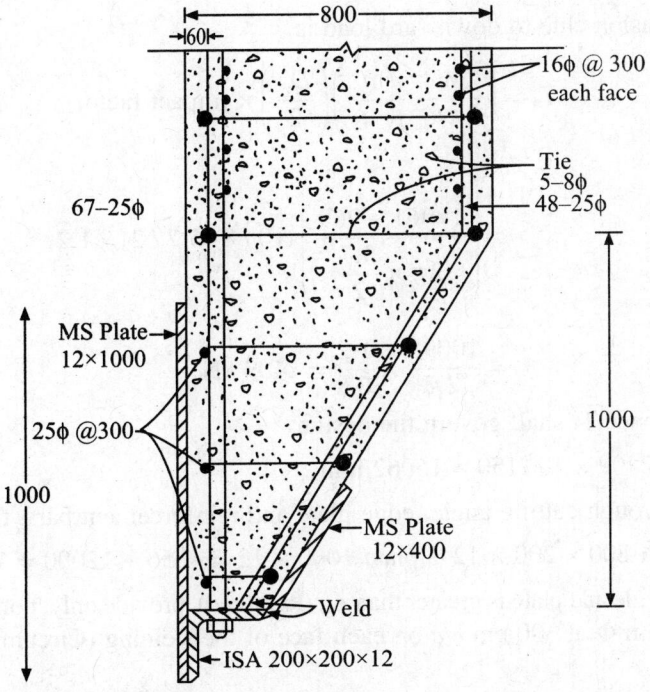

Figure 15.42 Details of well steining & cutting edge.

$$A_{st} = \frac{197 \times 10^6}{0.866 \times (700)(230)} = 1413 \text{ mm}^2/\text{m}$$

Provide 25 mm Φ at 330 mm *c/c*
Details are given in Figure 15.42.

9.2 *Curb design*

(a) Hoop tension is

$$T_1 = \frac{D_c}{4}\left(\frac{pD_c^2}{4r_0} - 0.5c_a(\gamma_s - \gamma)(D^2 + (D - h_c)^2\, h_c)\right)$$

$D_c = 5.7$ m; $c_a = 0.32$

$p = 376$ kN/m^2; $(\gamma_s - \gamma) = 10$ kN/m^3

$r_0 = 2.4$ m; $D = 6.5$ m

$h_c = 1.0$ m

$$T_2 = \frac{5.7}{4}\left[\frac{376 \times 5.7^2}{4 \times 2.4} - 0.5 \times 0.32(10)(6.5^2 + (6.5-1)^2 1)\right]$$

$$= \frac{5.5}{4}(1175) = 1615 \text{ kN}$$

In the figure: 800, 60, 16φ @ 300 each face, Tie 5–8φ, 48–25φ, 67–25φ, MS Plate 12×1000, 25φ @300, 1000, MS Plate 12×400, Weld, ISA 200×200×12, 1000

(b) Hoop tension clue to downward load is

$$T_2 = \left(\frac{W}{2\pi R_c}\right)\left(\frac{h_c}{t}\right)\left(\frac{D_c}{2}\right) \times \text{Impact factor}$$

$$= \left(\frac{11061 - 969}{2\pi \times \dfrac{5.7}{2}}\right)(10/8)(5.7/2) \times 1.2$$

$$= \frac{10098}{2\pi} \times \frac{1.2}{0.8} = 2409 \text{ kN}$$

Since $T_2 > T_1$. Hence Ti shall govern the design

A_{st} (Required) $= 2409 \times 10^3/150 = 16062$ mm^2.

A_{st} provided is through cutting angle, edge plate and reinforcement bars, they are:

Angle *ISA* 200 × 200 × 12 + plate 1000 × 12 = 4656 + 12000 = 16656 mm^2

Since steel from angle and plate is greater than A_{st} (Required), provide only, nominal reinforcement. Provide 16 mm Φ at 300 cm *c/c* on each face of the steining (Circumferential bars).

9.3   *Bottom plug*

The bending moment on the bottom plug is

$$M = \frac{3p_o D_c^2}{16}$$

where

$\qquad p_0$ = net up lift pressure

$\qquad$ = [Total vertical load –50 percent Buoyancy
$\qquad$ –plug weight – weight of water in the well]/area

$$= \frac{16947 - (2911/2) - 942 - 1402}{(\pi/4)(6.5)^2}$$

$$= \frac{10147}{33.18} = 305 \text{ kN/m}^2$$

$$M = \frac{3 \times 305 \times (4.9)^2}{16} = 1376 \text{ kNm/m}$$

for P.C.C. (*M* 15), the allowable flexure is

$$\sigma_{all} = 0.35\sqrt{f_{ck}} = 1.36 \text{ N/mm}^2 \text{ and}$$

thickness required $\qquad = \sqrt{\dfrac{6 \times M}{b \times \sigma_{all}}} = \sqrt{\dfrac{6 \times 1364 \times 10^6}{1000 \times 1.36}} = 2460 \text{ mm}$

The bottom plug has an arch action and it is rather too thick, most of the load is transmitted by thrust or by shear but not so much by bending. The bending moment computed is going to be far higher than the real moment. Hence a thickness of 2400 mm is adequate.

10. End Well Foundation (figure 15.43)

The outer wells support the main span and the short end span. The well along with pier is illustrated in Figure 15.43. Stability of the well is computed through Table 15.11.

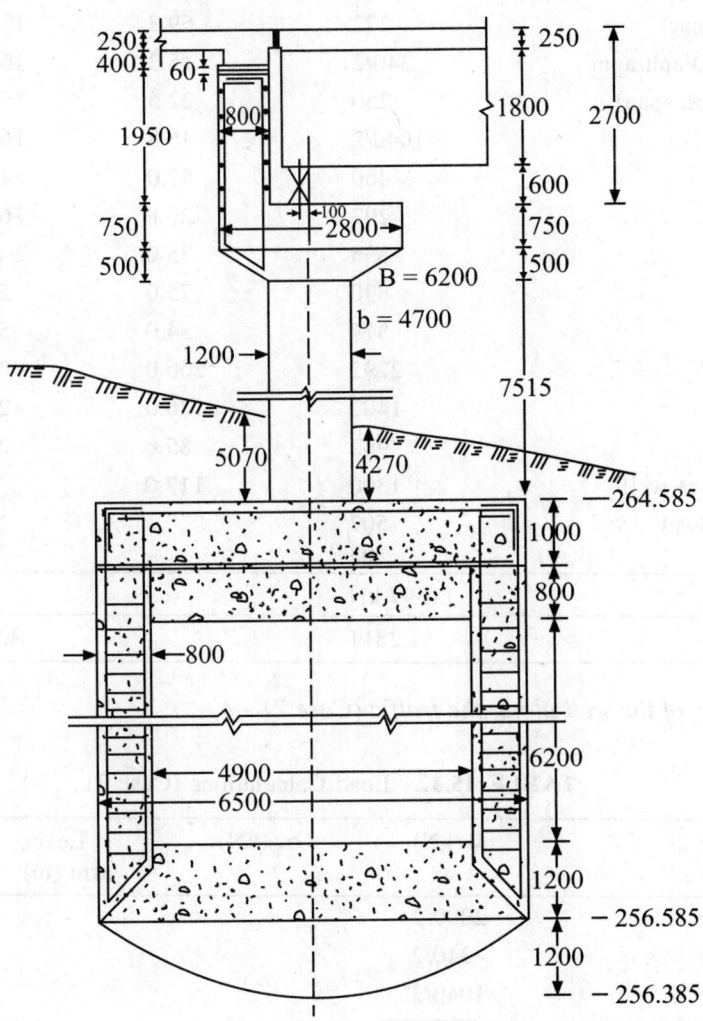

Figure 15.43 End Well.

10.1 *Stability analysis of the end well across the bridge*

TABLIE 15.11 Earthquake Along the Stream Direction (Vide Table 11.9)

Load type	W (kN)	α_W(kN)	0.2D above base e(m)	Moment (kNm)
Slab + wearing coat + railing (25 m span)	2000/2	90	17.0	1530
Live load + Impact	272	69.3	17.2	1192
Cross beam + Diaphragm	340/2	15.3	16.0	245
Deck slab (2.5 m span)	250	22.5	16.8	378
Main gifder	1040/2	46.8	16.0	740
Pier cap	460	42.0	14.9	626
Pier cap leg	297	26.8	16.2	434
Pier	835	75.0	10.2	765
Well cap	830	75.0	5.9	442
Top plug	377	34.0	5.0	170
Steining	2292	206.0	2.7	557
Bottom plug	1402	126.0	−2.2	−278
Water in well	942	85.0	2.7	144
Wt. of soil above well	1300	117.0	–	–
Eccentric live load	507	–	2.5	1268
	208	–	−2.5	−520
Sum =	11397 kN			7702 kNm
Buoyancy	2811	–	3.25	9461

10.2 *Stability of the well along the traffic (Case 2)*

TABLE 15.12 Load Calculations (Case 2)

Load type	W(kN)	α_W(kN)	Lever arm (m)	Moment (kNm)
Super structure	2000/2			
upto bearing	340/2			
	1040/2			
	1690	152	17.00	2584
Live load	722.0			
	+6.6			
	729	57.6	17.2	991

Eccentric live load	722	–	0.1	72.2
Deck slab (2.5 m span)	250	22.5	16.80	378.0
Pier cap leg	297	26.8	16.2	434
Eccentric load wrt centre				
(i) main span	1690	–	0.1	169
(ii) small span	250	–	1.0	250
(iii) extra pier	297	–	1.0	297
Pier cap	460	42.0	14.9	626
Pier	835	75.0	10.2	765
Well cap	830	75.0	5.9	443
Top plug	377	34.00	5.0	170
Steining	2292	206.3	2.7	557
Bottom plug	1042	94.0	–2.2	–207
Water in the well	942	84.8	1.7	144
Wt. of soil above well	1300	117.0	–	–
Total	11044 kN			7673 kNm
Buoyancy	2911	–	6.5/2	9460 kNm

10.3 Soil resistance

$$W = 11397 \text{ kN}, \ B = 6.5 \text{ m}, \ Q = 0.5, \ \Phi = 18°,$$

$$\delta = 6°, \ D = 8.0 \text{ m}$$

$$M_b = QWB \tan \Phi = 11397(0.5)(6.5) \tan 18° = 12035 \text{ kNm}$$

$$L = 6.5 \text{ m}$$

$$\gamma = 10 \text{ kN/m}^3$$

$$K_p = 1.48$$

$$K_a = 0.51$$

$$M_3 = 0.1\gamma D^3(K_p - K_a)\, L = 0.1(10)(8)^3(1.48 - 0.5)6.5 = 3229 \text{ kNm}$$

$$M_f = 0.11\gamma(K_p - K_a)B^2D^2 \sin \delta$$

$$= 0.11(10)(1.48 - 0.51)(6.5)^2(8)^2 \sin 6° = 301 \text{ kNm}$$

$$M_r = (M_s + M_b + M_f) = 15565 \text{ kNm}$$

Case 1

Overturning Moment on the well is

$$M_0 = 7702 \times 1.25 = 9628 \text{ kNm}$$

Restoring Moment of the well is

$$M_r = 0.1 \, M_r = 0.7 \times 15565 = 10897 \text{ kNm}$$

Since $M_r > M_0$, the well is safe against overturning.

The stability of the well in case 2 is safe as the overturning moment is still smaller. The bearing pressure is also within the limits.

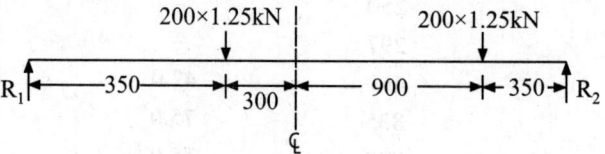

Figure 15.44 Class AA wheeled vehicle.

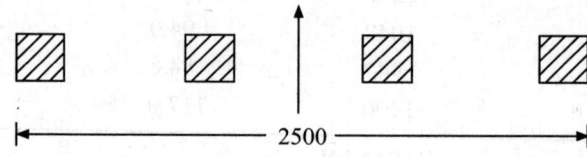

Figure 15.45 Wheel loads

11. Design of Deck Slab of End Span

11.1 *Class–A A wheeled vehicle (vide Figure 15.44 and 15.45)*

$$\text{Reaction} = R_1 = \frac{250(350 + 1250 + 300)}{2500} = 190 \text{ kN}$$

$$M_{\text{max}} = R_1(950) = 180.5 \text{ kN}$$

Width of slab resisting bending moment due to above loading is

$$(2500 + 300 + (6)300) = 4900 \text{ mm}$$

$$M_1 = 180.5/4.9 = 35.74 \text{ kNm/m}$$

11.2 *Dead load effect*

$W =$ Weight of slab and footpath is

$$= 25 \times 0.3 \times 9 + 25 \times 0.25 \times 0.8 \times 2$$

$$= 67.5 + 10 = 77.5 \text{ kN/m and bending moment is}$$

$$M_2 = 77.5 \times (2.5)^2/8 = 60.54 \text{ kNm/9 m width or}$$

$$M_3 = 60.5/9.0 = 6.73 \text{ kNm/m}$$

Note: Since the span is less than the length of class *AA* tracked vehicle, this load will not be critical.

11.3 *Check for class A; wheeled vehicle (Figure 15.4b and 15.47)*

Effective width of slab = (150 + 250 + 1800 + 1700 + 1800
+ 500/2 + 300) + 6(300) 8050 mm

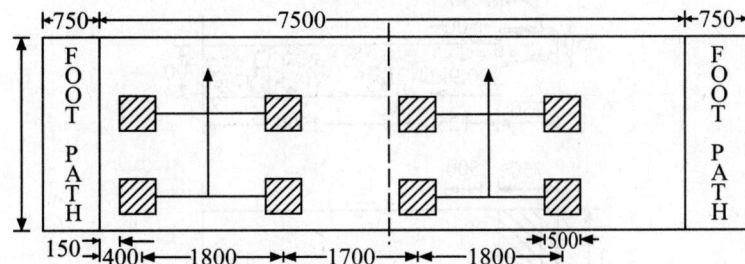

Figure 15.46 Class A vehicle loading.

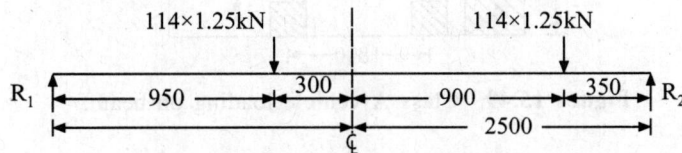

Figure 15.47 Loads on end span.

$$R_1 = \frac{142.5\ 350 + 1250 + 300}{2500} = 108.3 \text{ kNm}$$

$$M_4 = 108.3/8.05 = 13.5 \text{ kNm/m}$$

Hence class–*AA* wheeled vehicle will dominate

$$BM = M = M_1 + M_3 = 35.74 + 6.73 = 42.5 \text{ kNm/m}$$

thickness of deck = 250 mm
Provide (12 mm Φ at 120 *c/c*)(15.48)

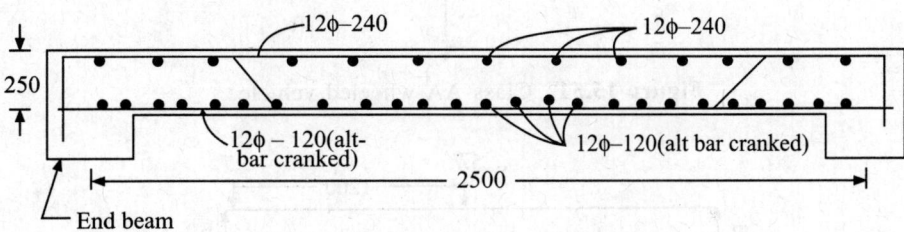

Figure 15.48 Deck slab (shorter span).

12. Beam at End Span (Figure 15.49 to 15.52)

The deck slab span is only 2.5 m while its width is 7.5 m plus footpath. The total width is not resting on the pier cap or the abutments. Two beams, one at each edge of the slab parallel to one width are provided. For each of the beams, two bearings spaced at 2.5 m from centre line of the road way are provided.

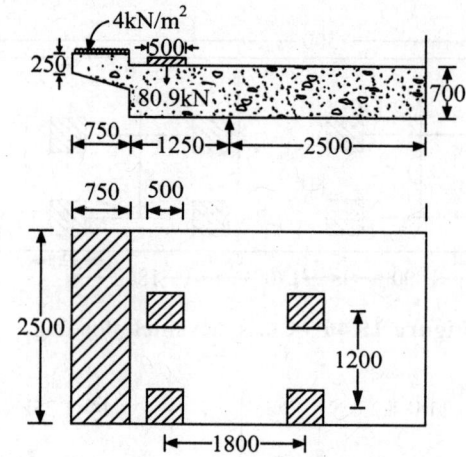

Figure 15.49 Class A vehicle loading on beam.

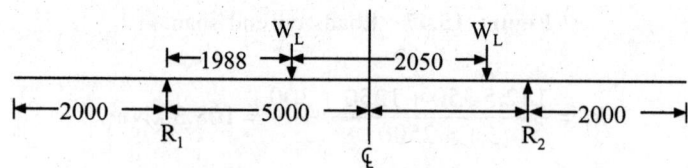

Figure 15.50 Class AA tracked vehicle.

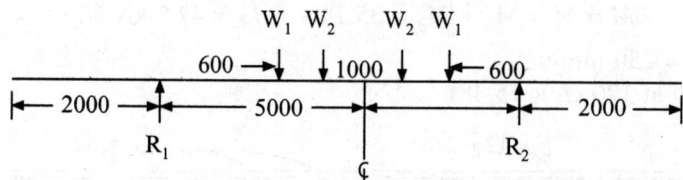

Figure 15.51 Class AA wheeled vehicle.

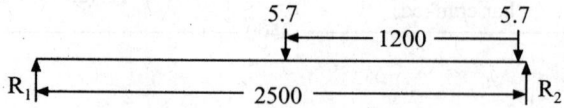

Figure 15.52 Class A vehicle.

12.1 Dead load

Slab weight = $(1/2)9(25)0.25(25/9) = 7.81$ kN/m

Beam weight = $25(0.7)0.4 = 7$ kN/m

$M_1 = $ B.M. from slab $= 7.81(9/2)2.5 - 7.81(4.5)^2/2 = 8.8$ kNm

$M_2 = $ BM from beam $= 7(3.5)(2.5) - 7(3.5)^2/2 = 18.4$ kNm

$M_3 = M_1 + M_2 = 27.2$ kNm

Max + Ve BM due to class A A tracked vehicle

$$W_L = \frac{35}{2} \times \frac{1.25 \times 2.5}{36} = 152 \text{ kN}$$

$$R_1 = \frac{3.0125 \times 152 + 0.9625 \times 152}{5.0} = 120.8 \text{ kN}$$

$$M_{max} = M_L = 120.8(1.988) = 240.1 \text{ kNm}$$

12.2 Class AA Wheeled Vehicle

$$W_1 = \left(37.5 + 37.5 \times \frac{1.3}{1.5}\right)1.25 = 71.25 \text{ kN}$$

$$W_2 = \left(62.5 + 62.5 \times \frac{1.3}{1.5}\right)1.25 = 118.75 \text{ kN}$$

$$R_1 = \frac{W_1 \times 0.85 + W_2 \times 1.45 + W_2 \times 3.05 + W_1 \times 3.65}{5.0} = 171 \text{ kN}$$

$$M_L = BM_{max} = 171(2.5 - 1.1/2) - 71.25 \times 0.6 = 290.7 \text{ kNm}$$

$$\text{Max + ve } BM = M = 290.7 + 27.2 = 318 \text{ kNm}$$

Max (–ve) BM

Class *AA* vehicle will not cause the – ve moment as the clearance required from the curb (*C*) is too vide.

12.3 Consider class A

$$L = 2.5 \text{ m}$$

$$\text{Impact factor} = \frac{4.5}{6.0 + L} = 0.53$$

$$R_1 = \frac{5.7(1.2 + 0.125) + 5.7 \times 0.125}{2.5} = 14.1 \text{ kN}$$

$$R_2 = 11.4 - 1.41 = 80.9 \text{ kN}$$

Max (–ve) B.M. = *due to live load on footpath + due to class A vehicle*

$$= 5 \times 0.75 \times (1.25 + 0.75/2) + 80.9 \times 0.85 \times 1.53 = 6.1 + 105.2$$

$$M_L(\text{–ve}) = -111.3 \text{ kNm}$$

Moment due to dead load

$$M_D(-ve) = -25[0.25/2)(2)^2/2 + 0.4 \times 0.6 \times \left(\frac{1.25}{2}\right)^2$$

$$+ 0.1 \times 1.5 \times 1.25 \times 1.8] = -28.75 \text{ kNm}$$

$$M(-ve) = -28.75 - 111.3 = -140.0 \text{ kNm}$$

Provide a doubly reinforced section (Figure 15.53)

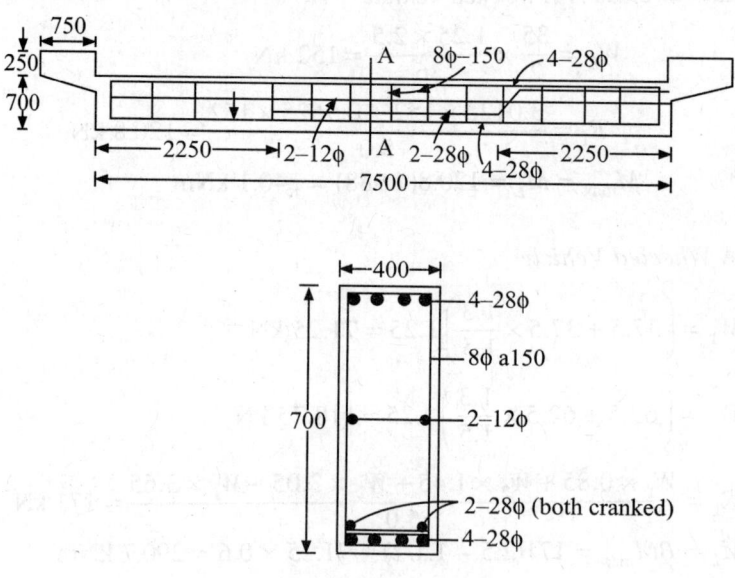

Figure 15.53   End beam.

13.   Abutment (Figure 15.52)

Two abutments, one at each end are to be concrete masonry and designed as gravity type. Figure 15.54 illustrates the assumed dimensions for the purpose of stability and bearing pressure calculations. The loads are indicated as W_1, W_2 etc. on the figure and used in Table 15.13.

13.1   *Stability and bearing pressure* (Figure 15.54)

Factored stabilized moment = $0.7 M_s$ = 10930 kNm
Factored overturning moment = $1.2 M_0$ = 8524 kNm

Stabilizing moment is greater than the overturning one so the abutment is safe against overturning. The bearing pressure under the abutment is computed first assuming a linear (trapezoidal) distribution as shown in Figure 15.55.

$$\frac{(p_1 + p_2)}{2} 10 \times b = W \text{ or}$$

TABLE 15.13 Vertical Loads and Moments About Toe

Load	Calculations	Value (kN)	Distance from toe	Moment about (kNm)
W_1	$0.4 \times 5.66 \times 24 \times 10$	= 544	2.55	1386
W_2	$0.6 \times 4.0 \times 24 \times 10$	= 576	2.05	1181
W_3	$(1/2)1.5 \times 3.4 \times 24 \times 10$	= 612	1.25	765
W_4	$(1/2)5.66 \times 2.4 \times 24 \times 10$	= 1630	3.55	5787
W_5	$0.25 \times 5.4 \times 24 \times 10$	= 324	2.70	875
W_6	$(1/2 \times 2.65 + 0.25)5.66 \times 18 \times 10$	= 1478	3.63	5363
W_7	$250/2$	= 125	2.05	256
Total		$W = 5289$		$M_s = 15613$

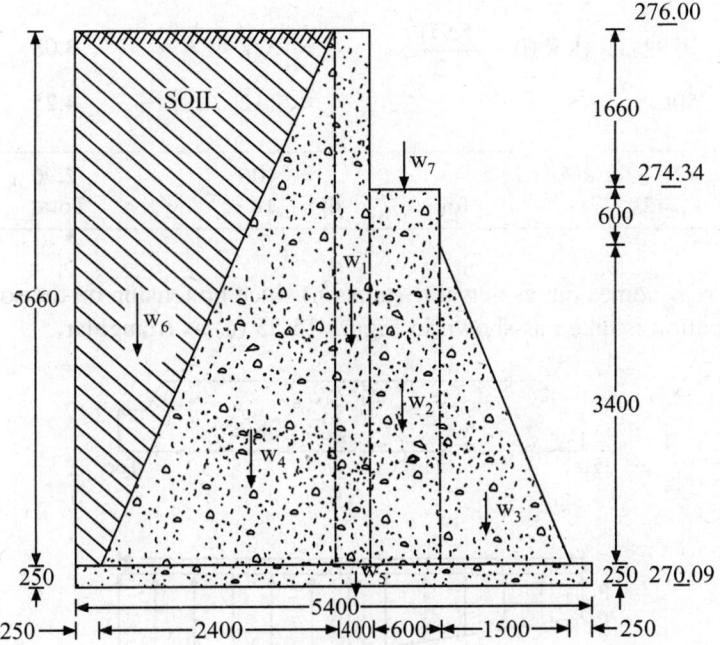

Figure 15.54 Ambutment notation.

$$p_1 + p_2 = 5289(2/10)(1/5.4)$$

$$p_1 + p_2 = 193.0 \text{ kN/m}^2 \tag{i}$$

$$(p_1 b/2)b/3 + (p_2/2)b(2b/3)10 = (M_s - M_0)$$

$$= p_1 + 2p_2 = \frac{(15613 - 7103) \times 6}{10(5.4)^2}$$

$$p_1 + 2p_2 = 175.0 \text{ kN/m}^2 \tag{ii}$$

TABLE 15.14 Horizontal Loads and Moments About Toe

Load	Calculations ($\alpha = 0.09$)		Value (kN)	Distance from toe	Moment (kNm)
αW_1	0.09×544	$=$	49	3.08	151
αW_2	0.09×576	$=$	52	2.25	117
αW_3	0.09×612	$=$	55	1.38	76
αW_4	0.09×1630	$=$	147	2.14	314
αW_5	0.09×195	$=$	29	0.125	4
Soil press (static)					
	$0.524 \times 18 \times 10 \dfrac{(5.91)^2}{2}$	$=$	1603	2.14	3430
Increment					
	$(0.682 - 0.527) \times 18 \times 10 \times \dfrac{(5.91)^2}{2}$	$=$	572	3.08	1762
W_7	$0.09 \times 250/2$	$=$	12	4.25	48
Live load					
	0.527×770 (Class AA)	$=$	406	2.96	1202
$k_{ae} = 0.692;$	$k_a = 0.527;$	for $\Phi = 18°$		Total	$M_0 = 7103$

From (i) and (ii) p_2 comes out as negative (tension) since no tension on the soil is allowed the pressure distribution is taken as shown in Figure 15.55 (c) as triangular.

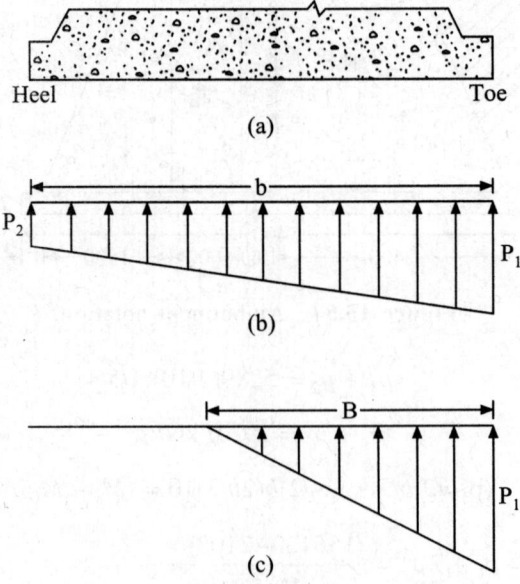

Figure 15.55 Pressure distribution.

$$\frac{pB\,10}{2} = W; \quad pB = 1058 \tag{iii}$$

$$(pB/2)(B/3)10 = (M_s - M_0)\ pB^2 = 5106 \tag{iv}$$

From (iii) and (iv) $B = 4.84$ m; $(p - p_1)$

$$p = 2W\,/\,(B)10 = \frac{2 \times 5160}{4.84 \times 10} = 213.2 \text{ kN/m}^2$$

Allowable bearing capacity:
Overburden Height = 274.0 − 270.0 = 4.0 m
Allowable bearing pressure = $1.25 \times 120 + 4 \times 18 = 222$ kN/m^2. Hence (safe)

13.2 *Checking stresses in the masonry, taking moments about A (see Figure 15.54)*

$$\left(\frac{p_1 + p_2}{2}\right) \times b \times 10 = W$$

$$(p_1 + p_2) = \frac{4837 \times 2}{10 \times 5.4}$$

TABLE 15.15   Vertical Loads

Load	Calculations Load type	Value (kN)	Eccentricity (m)	Moment (kNm)
W_1	Same as before	544	2.3	0151
W_2	"	576	1.8	6137
W_3	"	612	1.0	212
W_4	"	1630	3.3	5379
W_5	–	–	–	–
W_6	$(1/2)2.65 \times 5.66 \times 18 \times 10$	1350	3.38	4563
W_7	"	125	1.8	225
	Total $W = 4837$			$M_s = 13067$ kNm

$$p_1 + p_2 = 179.1 \tag{a}$$

$$(p_1 + p_2) \times \frac{10 \times b^2}{6} = (M_s - M_0)$$

$$(p_1 + 2p_2) = 7013 \times 6/(10 \times 5.4)^2) = 144.3 \tag{b}$$

from (a) and (b)

$$p_1 = 213.8 \text{ kN/m}^2$$

(safe for a masonry)

$$p_2 = 34.8 \text{ kN/m}^2$$

TABLE 15.16 Horizontal Loads

Load	Calculations Load type	Value (kN)	Eccentricity (m)	Moment (kNm)
αW_1	Same as before	49	2.83	139
αW_2	"	52	2.00	104
αW_3	"	55	1.13	62
αW_4	"	147	1.89	278
αW_5	–	–	–	–
Static				
	$0.51 \times 18 \times 10 \times \dfrac{(5.66)^2}{2}$	1470	1.89	2779
Increment				
	$(0.692 - 0.51) \times 18 \times 10 \times \dfrac{(5.66)^2}{2}$	524	2.83	1483
αW_7	$0.09 \times 250/2$	12	4.0	48
Live load				
	0.527×770	406	2.83	1150
			$M_0 =$	6054 kNm

Tables

TABLE 1 Values of Δ' (for full stress condition)

η	f_{ck}	c'_{tt}	c'_{ct}	c'_{ce}	c'_{te}	Δ'
			Post-tensioned			
0.80	35	.0286	.5000	.4000	.0286	1.0135
	40	.0250	.4806	.3889	.0250	1.0014
	45	.0222	.4611	.3778	.0222	0.9888
	50	.0200	.4417	.3667	.0200	0.9756
	53	.0189	.4300	.3600	.0189	0.9674
0.85	35	.0286	.5000	.4000	.0286	1.0690
	40	.0250	.4806	.3889	.0250	1.0569
	45	.0222	.4611	.3778	.0222	1.0441
	50	.0200	.4417	.3667	.0200	1.0306
	53	.0189	.4300	.3600	.0189	1.0222
			Pre-tensioned			
0.80	42	.0238	.5000	.4000	.0238	1.0114
	50	.0200	.4741	.3852	.0200	0.9952
	55	.0182	.4579	.3759	.0182	0.9847
	60	.0167	.4417	.3667	.0167	0.9737
	63	.0157	.4300	.3600	.0157	0 9655
0.85	42	.0238	.5000	.4000	.0238	1.0680
	50	.0200	.4741	.3852	.0200	1.0517
	55	.0182	.4579	.3759	.0182	1.0409
	60	.0167	4417	.3667	.0167	1.0295
	63	.0157	.4300	.3600	.0157	1.0210

TABLE 2 Sectional properties

c_t	c_b	c_a	c_i	c_{yb}	C_{yt}	ρ	Δ
		$c_f = 0.40$			$c_w = 0.12$		
0.08	.08	.212	.027	.603	.396	.128	1.523
0.08	.10	.218	.028	.590	.409	.132	1.442
0.08	.12	.224	.030	.578	.421	.134	1.372
0.10	.08	.230	.028	.627	.372	.125	1.682
0.10	.10	.236	.030	.614	.385	.129	1.593
0.10	.12	.241	.031	.602	.397	.131	1.517
0.12	.08	.248	.030	.645	.354	.121	1.823
0.12	.10	.253	.031	.633	.366	.125	1.728
0.12	.12	.259	.033	.622	377	.128	1.647
		$c_f = 0.40$			$c_w = 0.16$		
0.08	.08	.246	.029	.589	.410	.120	1.436
0.08	.10	.251	.030	.580	.419	.122	1.381
0.08	.12	.256	.031	.571	.428	.124	1.332
0.10	.08	.263	.031	.610	.389	.118	1.564
0.10	.10	.268	.032	.600	.399	.121	1.504
0.10	.12	.272	.033	.592	.407	.123	1.451
0.12	.08	.280	.032	.626	.373	.116	1.679
0.12	.10	.284	.033	.617	.382	.119	1.616
0.12	.12	.289	.035	.609	.390	.121	1.560
		$c_f = 0.40$			$c_w = 0.20$		
0.08	.08	.280	.031	.578	.421	.113	1.374
0.08	.10	.284	.032	.571	.428	.115	1.336
0.08	.12	.288	.033	.565	.434	.116	1.301
0.10	.08	.296	.033	.596	.403	.113	1.479
0.10	.10	.300	.034	.590	.410	.115	1.439
0.10	.12	.304	.035	.583	.416	.116	1.402
0.12	.08	.312	.034	.611	.388	.111	1.575
0.12	.10	.316	.035	.605	.394	.113	1.532
0.12	.12	.320	.036	.599	.401	.115	1.493

TABLE 2 *(Contd.)*

		$c_f = 0.50$			$c_w = 0.12$		
0.08	.08	.220	.029	.583	.416	.135	1.399
0.08	.10	.228	.031	.566	.433	.138	1.309
0.08	.12	.236	.033	.552	.447	.140	1.233
0.10	.08	.238	.031	.607	.392	.132	1.547
0.10	.10	.246	.033	.591	.408	.136	1.447
0.10	.12	.253	.035	.577	.422	.139	1.364
0.12	.08	.256	.032	.626	.373	.128	1.680
0.12	.10	.263	.035	.611	.388	.132	1.573
0.12	.12	.271	.036	.597	.402	.136	1.483
		$c_f = 0.50$			$c_w = 0.16$		
0.08	.08	.254	.032	.572	.427	.125	1.338
0.08	.10	.261	.033	.559	.440	.128	1.271
0.08	.12	.268	.034	.548	.451	.130	1.214
0.10	.08	.271	.033	.593	.406	.124	1.458
0.10	.10	.278	.035	.580	.419	.127	1.386
0.10	.12	.284	.036	.569	.430	.129	1.323
0.12	.08	.288	.035	.610	.389	.122	1.567
0.12	.10	.294	.037	.598	.401	.125	1.490
0.12	.12	.301	.038	.587	.412	.128	1.424
		$c_f = 0.50$			$c_w = 0.20$		
0.08	.08	.288	.034	.563	.436	.118	1.292
0.08	.10	.294	.035	.554	.445	.120	1.243
0.08	.12	.300	.036	.545	.454	.122	1.199
0.10	.08	.304	.035	.582	.417	.118	1.392
0.10	.10	.310	.037	.572	.427	.120	1.339
0.10	.12	.316	.038	.563	.436	.122	1.292
0.12	.08	.320	.037	.597	.402	.116	1.484
0.12	.10	.326	.038	.588	.411	.119	1.428
0.12	.12	.332	.040	.579	.420	.121	1.378

TABLE 2 *(Contd.)*

		$c_f = 0.60$			$c_w = 0.12$		
0.08	.08	.228	.032	.564	.435	.140	1.295
0.08	.10	.238	.034	.545	.454	.143	1.198
0.08	.12	.248	.035	.528	.471	.144	1.120
0.10	.08	.246	.034	.589	.410	.138	1.433
0.10	.10	.256	.036	.570	.429	.142	1.327
0.10	.12	.265	.038	.553	.446	.144	1.240
0.12	.08	.264	.035	.609	.390	.134	1.558
0.12	.10	.273	.038	.590	.409	.139	1.444
0.12	.12	.283	.040	.574	.425	.142	1.350
		$c_f = 0.60$			$c_w = 0.16$		
0.08	.08	.262	.034	.556	.443	.130	1.252
0.08	.10	.271	.036	.540	.459	.133	1.178
0.08	.12	.280	.037	.527	.472	.134	1.116
0.10	.08	.279	.036	.577	.422	.129	1.366
0.10	.10	.288	.038	.562	.437	.132	1.285
0.10	.12	.296	.039	.549	.450	.134	1.217
0.12	.08	.296	.037	.595	.404	.127	1.469
0.12	.10	.304	.039	.580	.419	.131	1.384
0.12	.12	.313	.041	.567	.432	.133	1.311
		$c_f = 0.60$			$c_w = 0.20$		
0.08	.08	.296	.036	.549	.450	.122	1.220
0.08	.10	.304	.037	.537	.462	.124	1.162
0.08	.12	.312	.039	.526	.473	.126	1.112
0.10	.08	.312	.038	.568	.431	.122	1.315
0.10	.10	.320	.040	.556	.443	.125	1 253
0.10	.12	.328	.041	.545	.454	.126	1.199
0.12	.08	.328	.039	.583	.416	.121	1.403
0.12	.10	.336	.041	.572	.427	.124	1.337
0 12	.12	.344	.043	.561	.438	.126	1.279

TABLE 2 *(Contd.)*

		$c_f = 0.70$			$c_w = 0.12$		
0.08	.08	.236	.034	.546	.453	.144	1.205
0.08	.10	.248	.036	.525	.474	.147	1.106
0.08	.12	.260	.038	.506	.493	.148	1.027
0.10	.08	.254	.036	.571	.428	.143	1.335
0.10	.10	.266	.039	.550	.449	.146	1.225
0.10	.12	.277	.041	.532	.467	.148	1.138
0.12	.08	.272	.038	.592	.407	.140	1.453
0.12	.10	.283	.044	.571	.428	.144	1.335
0.12	.12	.295	.043	.553	.446	.146	1.240

		$c_f = 0.70$			$c_w = 0.16$		
0.08	.08	.270	.036	.540	.459	.134	1.177
0.08	.10	.281	.038	.523	.476	.136	1.098
0.08	.12	.292	.040	.508	.491	.137	1.033
0.10	.08	.287	.038	.562	.437	.133	1.285
0.10	.10	.298	.040	.545	.454	.136	1.199
0.10	.12	.308	.042	.530	.469	.138	1.128
0.12	.08	.304	.040	.580	.419	.132	1.383
0.12	.10	.314	.042	.563	.436	.135	1.291
0.12	.12	.325	.044	.548	.451	.137	1.215

		$c_f = 0.70$			$c_w = 0.20$		
0.08	.08	.304	.038	.536	.463	.126	1.156
0.08	.10	.314	.040	.522	.477	.128	1.092
0.08	.12	.324	.041	.509	.490	.129	1.038
0.10	.08	.320	.040	.555	.445	.126	1.247
0.10	.10	.330	.042	.540	.459	.128	1.178
0.10	.12	.340	.044	.528	.471	.130	1.119
0.12	.08	.336	.042	.570	.429	.125	1.330
0.12	.10	.346	.044	.557	.442	.128	1.257
0.12	.12	.356	.046	.544	.455	.130	1.195

TABLE 2   *(Contd.)*

		$c_f = 0.80$			$c_w = 0.12$		
0.08	.08	.244	.036	.530	.469	.148	1.127
0.08	.10	.258	.038	.506	.493	.149	1.028
0.08	.12	.272	.040	.487	.512	.149	0.949
0.10	.08	.262	.038	.555	.444	.147	1.249
0.10	.10	.276	.041	.523	.467	.150	1.139
0.10	.12	.289	.043	.512	.487	.151	1.052
0.12	.08	.280	.040	.576	.432	.144	1.361
0.12	.10	.293	.043	.554	.445	.148	1.242
0.12	.12	.307	.046	.534	.465	.150	1.147

		$c_f = 0.80$			$c_w = 0.16$		
0.08	.08	.278	.038	.526	.473	.137	1.111
0.08	.10	.291	.040	.507	.492	.139	1.029
0.08	.12	.304	.042	.490	.509	.139	0.962
0.10	.08	.295	.040	.548	.451	.137	1.213
0.10	.10	.308	.043	.529	.470	.140	1.124
0.10	.12	.320	.045	.512	.487	.141	1.051
0.12	.08	.312	.042	.566	.433	.136	1.307
0.12	.10	.324	.045	.547	.452	.139	1.211
0.12	.12	.337	.047	.531	.468	.140	1.133

		$c_f = 0.80$			$c_w = 0.20$		
0.08	.08	.312	.040	.523	.476	.129	1.099
0.08	.10	.324	.042	.507	.492	.130	1.030
0.08	.12	.336	.044	.493	.506	.131	0.973
0.10	.08	.328	.042	.542	.457	.129	1.185
0.10	.10	.340	.044	.526	.473	.132	1.111
0.10	.12	.352	.046	.512	.487	.133	1.050
0.12	.08	.344	.044	.558	.441	.128	1.265
0.12	.10	.356	.046	.542	.457	.131	1.187
0.12	.12	.368	.049	.528	.471	.133	1.121

TABLE 2 *(Contd.)*

		$c_f = 0.90$			$c_w = 0.12$		
0.08	.08	.252	.038	.514	.485	.150	1.059
0.08	.10	.268	.040	.489	.510	.151	0.960
0.08	.12	.284	.042	.469	.530	.151	0.883
0.10	.08	.270	.040	.540	.459	.150	1.175
0.10	.10	.286	.043	.515	.484	.152	1.064
0.10	.12	.301	.046	.494	.505	.152	0.979
0.12	.08	.288	.042	.561	.438	.148	1.281
0.12	.10	.303	.046	.537	.462	.151	1.161
0.12	.12	.319	.048	.516	.483	.152	1.068

		$c_f = 0.90$			$c_w = 0.16$		
0.08	.08	.286	.040	.512	.487	.140	1.052
0.08	.10	.301	.042	.492	.507	.141	0.968
.0.08	.12	.316	.044	.474	.525	.141	0.901
0.10	.08	.303	.042	.534	.465	.140	1.149
0.10	.10	.318	.045	.514	.485	.142	1.058
0.10	.12	.332	.047	.496	.503	.143	0.984
0.12	.08	.320	.044	.553	.446	.139	1.239
0.12	.10	.334	.047	.533	.466	.142	1.141
0.12	.12	.349	.050	.515	.484	.143	1.062

		$c_f = 0.90$			$c_w = 0.20$		
0.08	.08	.320	.042	.511	.488	.131	1.047
0.08	.10	.334	.044	.493	.506	.133	0.975
0.08	.12	.348	.046	.478	.521	.133	0.917
0.10	.08	.336	.044	.530	.469	.132	1.129
0.10	.10	.350	.047	.512	.487	.134	1.052
0.10	.12	.364	.049	.497	.502	.135	0.989
0.12	.08	.352	.046	.546	.453	.132	1.206
0.12	.10	.366	.049	.529	.470	.134	1.124
0.12	.12	.380	.051	.513	.486	.135	1.057

TABLE 2 *(Contd.)*

		$c_f = 1.00$			$c_w = 0.12$		
0.08	.08	.260	.039	.500	.500	.152	1.000
0.08	.10	.278	.042	.474	.525	.153	0.901
0.08	.12	.296	.044	.452	.547	.151	0.826
0.10	.08	.278	.042	.525	.474	.153	1.109
0.10	.10	.296	.045	.500	.500	.154	1.000
0.10	.12	.313	.048	.478	.521	.154	0.916
0.12	.08	.296	.044	.547	.452	.151	1.210
0.12	.10	.313	.048	.521	.478	.154	1.091
0.12	.12	.331	.051	.500	.500	.154	1.000

		$c_f = 1.00$			$c_w = 0.16$		
0.08	.08	.294	.041	.500	.500	.142	1.000
0.08	.10	.311	.044	.477	.522	.143	0.915
0.08	.12	.328	.046	.459	.540	.142	0.848
0.10	.08	.311	.044	.522	.477	.143	1.092
0.10	.10	.328	.047	.500	.500	.144	1.000
0.10	.12	.344	.049	.480	.519	.144	0.926
0.12	.08	.328	.046	.540	.459	.142	1.178
0.12	.10	.344	.049	.519	.480	.144	1.079
0.12	.12	.361	.052	.500	.500	.145	1.000

		$c_f = 1.00$			$c_w = 0.20$		
0.08	.08	.328	.043	.500	.500	.133	1.000
0.08	.10	.344	.046	.480	.519	.134	0.926
0.08	.12	.360	.048	.464	.535	.134	0.867
0.10	.08	.344	.046	.519	.480	.134	1.079
0.10	.10	.360	.049	.500	.500	.136	1.000
0.10	.12	.376	.051	.483	.516	.137	0.935
0.12	.08	.360	.048	.535	.464	.134	1.153
0.12	.10	.376	.051	.516	.483	.137	1.068
0.12	.12	.392	.054	.500	.509	.137	1.000

TABLE 3 Fixed and moments due to prestressing force

Tendon Profile	$A \vdash\!\!\!\!\!\!-\!\!\!\!\!\!\dashv B$
(diagram: rectangle with horizontal tendon, eccentricity e, length L)	$M'_A = Pe$ $M_A = 0$ $M'_B = -Pe$ $M_B = 0$
(diagram: rectangle with sloping tendon from e_A to e_B, length L)	$M'_A = -Pe_A$ $M_A = 0$ $M'_B = -Pe_B$ $M_B = 0$
(diagram: rectangle with parabolic tendon, sag g, length L)	$\left.\begin{array}{l} M'_A \\ M_A \\ M'_B \\ M_B \end{array}\right\} = -\dfrac{2}{3}\,Pg$
(diagram: rectangle with V-shaped tendon, drop g at $\frac{1}{2}$)	$\left.\begin{array}{l} M'_A \\ M_A \\ M'_B \\ M_B \end{array}\right\} = -\dfrac{1}{2}\,Pg$
(diagram: rectangle with tendon dropping to g at βL)	$\left.\begin{array}{l} M'_A \\ M_A \end{array}\right\} = -Pg\beta$ $\left.\begin{array}{l} M'_B \\ M_B \end{array}\right\} = -Pg(1-\beta)$
(diagram: rectangle with parabolic tendon, eccentricities e_A, e_B, sag g at $\frac{1}{2}$)	$M'_A = -\dfrac{P}{3}(2g + 3e_A)$ $M_A = -\dfrac{2}{3}\,Pg$ $M'_B = -\dfrac{P}{3}(2g + 3e_B)$ $M_B = -\dfrac{2}{3}\,Pg$

TABLE 4 Membrane Forces and Displacements in Simply Supported Cylindrical Shells; Loads Uniformly Distributed Along the Length of the Barrel

(a) Uniform Transverse Load[a]

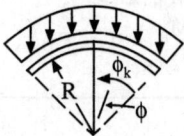 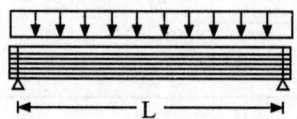

Longitudinal Force N_x –

$$qR\left(\frac{L}{R}\right)^2\left(\frac{x}{L}\right)\left(1-\frac{x}{L}\right)\text{Col.(1)}$$

Shearing Force $N_{x\phi}$ –

$$qR\left(\frac{L}{R}\right)\left(1-\frac{2x}{L}\right)\text{Col.(2)}$$

Transverse Force N_ϕ –
$qR \times$ Col. (3)

Vertical Displacement[b] w –

$$qR\frac{L^4}{R^3tE}\left\{2\left(\frac{R}{L}\right)^4+\left(\frac{x}{L}\right)^4-2\left(\frac{x}{L}\right)^3+\frac{x}{L}+6\left(\frac{R}{L}\right)^2\right.$$

$$\left.\left[\frac{x}{L}-\left(\frac{x}{L}\right)^2\right]\right\}\times\text{Col. (4)}$$

Horizontal Displacement[c] v

$$qR\frac{L^4}{R^3tE}\left\{\left(\frac{R}{L}\right)^4\text{Col. (5)}+\left[2\left(\frac{R}{L}\right)^4-2\left(\frac{x}{L}\right)^2+\left(\frac{x}{L}\right)^4+\frac{x}{L}\right.\right.$$

$$\left.\left.+6\left(\frac{R}{L}\right)^2\left[\frac{x}{L}-\left(\frac{x}{L}\right)^2\right]\right]\right\}\text{Col. (6)}$$

L = Span

R = radius

x = distance from support

[a]The use of column numbers in the formulas refers to the appropriate coffiecient in the column cited. [b]Downward direction is positive. [c]Inward direction is positive.

TABLE 4 Membrane Forces and Displacements in Simply Supported Cylindrical Shells; Loads Uniformly Distributed Along the Length of the Barrel

(b) *Dead Weight Load*[a]

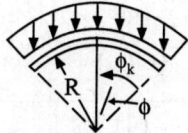 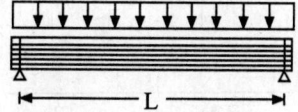

Longitudinal Force N_x –

$$q_g R \left(\frac{L}{R}\right)^2 \frac{x}{L}\left(L - \frac{x}{L}\right) \text{Col. (7)}$$

Shearing Force $N_{x\phi}$ –

$$q_g R \left(\frac{L}{R}\right)\left(1 - \frac{2x}{L}\right) \text{Col. (8)}$$

Transverse Force N_ϕ –
$$q_g R \times \text{Col. (9)}$$
Vertical Displacement[b] w –

$$q_g R \frac{L^4}{R^3 t E}\left\{ -\frac{1}{6}\left(\frac{x}{L}\right)^3 + \frac{1}{12}\left(\frac{x}{L}\right)^4 + \frac{1}{12}\left(\frac{x}{L}\right) + 2\left(\frac{R}{L}\right)^2\left[\frac{x}{L} - \left(\frac{x}{L}\right)^2\right] + \left(\frac{R}{L}\right)^2 \times \text{Col. (10)}\right\}$$

Horizontal Displacement[c] v –

$$q_g R \frac{L^4}{R^3 t E}\left(\frac{R}{L}\right)^4 \times \text{Col. (11)}$$

TABLE 4 Membrane Forces and Displacements in Simply Supported
Cylindrical Shells; Loads Uniformly Distributed
Along the Length of the Barrel (*Contd.*)

(*a*) *Uniform Transverse Load*

$\phi K - \phi$	N_x (C_{xm})	$N_{x\phi}$ (C_{sm})	N_ϕ $(C_{\phi m})$	w (C_{wm})	v (C_{vm1})	v (C_{vm2})
	(1)	(2)	(3)	(4)	(5)	(6)
0	−1.500	0	−1.0000	0.5000	0	0
5	−1.477	−0.1302	−0.9924	0.4944	0.0872	−0.0003
10	−1.409	−0.2565	−0.9698	0.4775	0.1736	−0.0026
15	−1.295	−0.3749	−0.9330	0.4506	0.2588	−0.0026
20	−1.150	−0.4820	−0.8830	0.4148	0.3420	−0.0200
25	−0.965	−0.5746	−0.8214	0.3722	0.4226	−0.0377
30	−0.750	−0.6503	−0.7500	0.3447	0.5000	−0.0625
35	−0.513	−0.7048	−0.6710	0.2748	0.5736	−0.0944
40	−0.261	−0.7385	−0.5868	0.2247	0.6428	−0.1328
45	0	−0.7500	−0.5000	0.1767	0.7071	−0.1768
50	0.261	−0.7385	−0.4132	0.1328	0.7660	−0.2247
55	0.513	−0.7048	−0.3290	0.0944	0.8191	−0.2749
60	0.750	−0.6503	−0.2500	0.0625	0.8660	−0.3248
65	0.965	−0.5746	−0.1786	0.0377	0.9063	−0.3722
70	1.150	−0.4820	−0.1170	0.0199	0.9397	−0.4149
75	1.295	−0.3749	−0.0669	0.0086	0.9659	−0.4506
80	1.409	−0.2565	−0.0301	0.0026	0.9848	−0.4776
85	1.477	−0.1302	−0.0076	0.0004	0.9962	−0.4943
90	1.500	0	0	0	1.0000	−0.5000

TABLE 4 Membrane Forces and Displacements in Simply Supported
Cylindrical Shells; Loads Uniformly Distributed
Along the Length of the Barrel (*Contd.*)

(*b*) *Dead Weight Load*

N_x (C_{xm}) (7)	$N_{x\phi}$ (C_{sm}) (8)	N_ϕ $(C_{\phi m})$ (9)	w (C_{wm}) (10)	v (C_{vm}) (11)	$\phi K - \phi$
−1.0000	0	−1.0000	1.0000	0	0
−0.9962	−0.0871	−0.9962	0.9924	0.0868	5
−o.9848	−0.1736	−0.9848	0.9698	0.1710	10
−0.9659	−0.2589	−0.9659	0.9330	0.2500	15
−0.9397	−0.3421	−0.9397	0.8830	0.3214	20
−0.9063	−0.4225	−0.9063	0.8214	0.3830	25
−0.8660	−0.5000	−0.8660	0.7500	0.4330	30
−0.8191	−0.5737	−0.8191	0.6710	0.4698	35
−0.7660	−0.6428	−0.7660	0.5868	0.4924	40
−0.7071	−0.7071	−0.7071	0.5000	0.5000	45
−0.6428	−0.7661	−0.6428	0.4132	0.4924	50
−0.5736	−0.8192	−0.5736	0.3290	0.4698	55
−0.5000	−0.8660	−0.5000	0.2500	0.4330	60
−0.4226	−0.9062	−0.4226	0.1786	0.3830	65
−0.3420	−0.9397	−0.3420	0.1170	0.3214	70
−0.2588	−0.9659	−0.2588	0.0669	0.2500	75
−0.1736	−0.9847	−0.1736	0.0301	0.1710	80
−0.0872	−0.9962	−0.0872	0.0076	0.0868	85
0	−1.0000	0	0	0	90

TABLE 5 Membrane Forces and Displacements in Simply Supported Cylindrical Shells; Loads Varying from Zero at the Ends to Maximum at the Middle

(a) Uniform Transverse Load

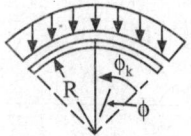 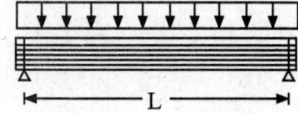

Longitudinal Force N_x –

$$qR\left[\left(\frac{L}{R}\right)^2 \text{Col. (1)}\right]\sin\frac{\pi x}{L}$$

Shearing Force $N_{x\phi}$ –

$$qR\left[\left(\frac{L}{R}\right)\text{Col. (2)}\right]\cos\frac{\pi x}{L}$$

Transverse Force N_ϕ –

$$qR\text{Col. (3)}\sin\frac{\pi x}{L}$$

Vertical Displacement w –

$$qR\frac{L^4}{R^3 tE}\left[\left(1+\frac{1}{2}\left(\frac{\pi R}{L}\right)^2 + \frac{1}{12}\left(\frac{\pi R}{L}\right)^4\right)\text{Col. (4)}\right]\sin\frac{\pi x}{L}$$

Horizontal Displacement v –

$$qR\frac{L^4}{R^3 tE}\left\{\left(\frac{R}{L}\right)^4 \text{Col. (5)} + \left[1+\frac{1}{2}\left(\frac{\pi R}{L}\right)^2 + \frac{1}{12}\left(\frac{\pi R}{L}\right)^4\right]\text{Col. (6)}\right\}\sin\frac{\pi x}{L}$$

TABLE 5 Membrane Forces and Displacements in Simply Supported Cylindrical Shells; Loads Varying from Zero at the Ends to Maximum at the Middle

(b) Dead Weight Load

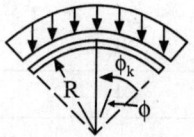 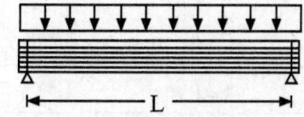

Longitudinal Force N_x –

$$q_g R \left[\left(\frac{L}{R} \right)^2 \text{Col. (7)} \right] \sin \frac{\pi x}{L}$$

Shearing Force $N_{x\phi}$ –

$$q_g R \left[\left(\frac{L}{R} \right) \text{Col. (8)} \right] \cos \frac{\pi x}{L}$$

Transverse Force N_ϕ –

$$q_g R \text{ Col. (9)} \sin \frac{\pi x}{L}$$

Vertical Displacement w –

$$q R \frac{L^4}{R^3 t E} \left[\left(\frac{2R}{\pi L} \right)^2 + \frac{2}{\pi^4} + \left(\frac{R}{L} \right)^4 \text{Col. (10)} \right] \sin \frac{\pi x}{L}$$

Horizontal Displacement v –

$$q R \frac{L^4}{R^3 t E} \left[\left(\frac{R}{L} \right)^4 \text{Col. (11)} \right] \sin \frac{\pi x}{L}$$

542

TABLE 5 Membrane Forces and Displacements in Simply Supported Cylindrical Shells; Loads Varying from Zero at the Ends to Maximum at the Middle (*Contd.*)

(*a*) *Uniform Transverse Load*

$\phi K - \phi$	N_x C_{xm}	$N_{x\phi}$ C_{sm}	N_ϕ $C_{\phi m}$	w C_{wm}	v C_{vm1}	v C_{vm2}
	(1)	(2)	(3)	(4)	(5)	(6)
0	−0.3040	0	−1.0000	0.12319	0	0
5	−0.2993	−0.0829	−0.9924	0.12180	0.0872	−0.00009
10	−0.2856	−0.1633	−0.9698	0.11766	0.1736	−0.00064
15	−0.2623	−0.2387	−0.9330	0.11102	0.2588	−0.00213
20	−0.2329	−0.3069	−0.8830	0.10222	0.3420	−0.00490
25	−0.1954	−0.3658	−0.8214	0.09170	0.4226	−0.00930
30	−0.1520	−0.4140	−0.7500	0.08001	0.5000	−0.01539
35	−0.1040	−0.4487	−0.6710	0.06771	0.5736	−0.02325
40	−0.0528	−0.4702	−0.5868	0.05537	0.6428	−0.03272
45	0	−0.4775	−0.5000	0.04355	0.7071	−0 04355
50	0.0528	−0.4702	−0.4132	0.03272	0.7660	−0.05537
55	0.1040	−0.4487	−0.3290	0.02325	0.8191	−0.06771
60	0.1520	−0.4140	−0.2500	0.01539	0.8660	−0.08001
65	0.1954	−0.3658	−0.1786	0.00930	0.9063	−0.09170
70	0.2329	−0.3069	−0.1170	0.00490	0.9397	−0.10222
75	0.2623	−0.2387	−0.0669	0.00213	0.9659	−0.11102
80	0.2856	−0.1633	−0.0301	0.00064	0.9848	−0.11766
85	0.2993	−0.0829	−0.0076	0.00009	0.9962	−0.12180
90	0.3040	0	0	0	1.0000	−0.12319

TABLE 5 Membrane Forces and Displacements in Simply Supported Cylindrical Shells; Loads Varying from Zero at the Ends to Maximum at the Middle (*Contd.*)

(b) Dead Weight Load

N_x (C_{xm}) (7)	$N_{x\phi}$ (C_{sm}) (8)	N_ϕ $(C_{\phi m})$ (9)	w (C_{wm}) (10)	v (C_{vm}) (11)	$\phi K - \phi$
−0.2026	0	−1.000	1.0000	0	0
−0.2019	−0.0555	−0.9962	0.9924	0.0868	5
−0.1996	−0.1105	−0.9848	0.9698	0.1710	10
−0.1957	−0.1648	−0.9659	0.9330	0.2500	15
−0.1904	−0.2178	−0.9397	0.8830	0.3214	20
−0.1837	−0.2690	−0.9063	0.8214	0.3830	25
−0.1754	−0.3183	−0.8660	0.7500	0.4330	30
−0.1660	−0.3652	−0.8191	0.6710	0.4698	35
−0.1552	−0.4092	−0.7660	0.5868	0.4924	40
−0.1433	−0.4502	−0.7071	0.5000	0.5000	45
−0.1302	−0.4877	−0.6428	0.4132	0.4924	50
−0.1162	−0.5215	−0.5736	0.3290	0.4698	55
−0.1013	−0.5513	−0.5000	0.2500	0.4330	60
−0.0856	−0.5769	−0.4226	0.1786	0.3830	65
−0.0693	−0.5982	−0.3420	0.1170	0.3214	70
−0.0524	−0.6149	−0.2588	0.0669	0.2500	75
−0.0351	−0.6269	−0 1736	0.0301	0 1710	80
−0.0177	−0.6342	−0.0872	0.0076	0.0868	85
0	−0.6366	0	0	0	90

TABLE 6 Symmetrical Edge Loads on Simply Supported Cylindrical Shell
(Force coefficients)

VERTICAL EDGE LOAD	HORIZONTAL EDGE LOAD

(a) *Basic Formulas and Loading Diagrams*

Longitudinal Force N_x	Longitudinal Force N_x
$V_L\left[\left(\dfrac{L}{R}\right)^2 \text{Col. (1)}\right]\sin\dfrac{\pi x}{L}$	$H_L\left[\left(\dfrac{L}{R}\right)^2 \text{Col. (5)}\right]\sin\dfrac{\pi x}{L}$
Shearing Force $N_{x\phi}$	Shearing Force $N_{x\phi}$
$V_L\left[\dfrac{L}{R}\text{Col. (2)}\right]\cos\dfrac{\pi x}{L}$	$H_L\left[\dfrac{L}{R}\text{Col. (6)}\right]\cos\dfrac{\pi x}{L}$
Transverse Force N_ϕ	Transverse Force N_ϕ
$V_L \text{ Col. (3)}\sin\dfrac{\pi x}{L}$	$H_L \text{ Col. (7)}\sin\dfrac{\pi x}{L}$
Transverse Moment M_ϕ	Transverse Moment M_ϕ
$V_L\left[R\text{ Col. (4)}\right]\sin\dfrac{\pi x}{L}$	$H_L\left[R\text{ Col. (8)}\right]\sin\dfrac{\pi x}{L}$

ϕ	N_x (1) C_{xV}	$N_{x\phi}$ (2) C_{sV}	N_ϕ (3) $C_{\phi V}$	M_ϕ (4) C_{MV}	N_x (5) C_{xH}	$N_{x\phi}$ (6) C_{sH}	N_ϕ (7) $C_{\phi H}$	M_ϕ (8) C_{NH}
			(b) $R/t = 100$ and $R/L = 0.1$					
$\phi K = 30°$								
30	−5.382	0	−3.471	−0.3278	−0.0013	0	+0.9997	+0.1310
20	−3.575	−2.619	−2.656	−0.2779	+0.0017	−0.0001	+0.9848	+0.1162
10	+1.809	−3.265	−0.775	−0.1512	+0.0034	+0.0017	+0.9399	+0.0722
0	+10.69	0	+0.500	0	−0.0151	0	+0.8660	0
$\phi K = 35°$								
35	−3.430	0	−2.959	−0.3703	−0.0061	0	+0.9986	+0.1754
30	−3.220	−0.921	−2.820	−0.3594	−0.0043	−0.0068	+0.9951	+0.1717
20	−1.538	−2.302	−1.812	−0.2781	+0.0060	+0.0136	+0.9663	+0.1425
10	+1.822	−2.301	−0.330	−0.1444	+0.0080	+0.0041	+0.9071	+0.0848
0	+6.845	0	+0.574	0	−0.0316	0	+0.8192	0

TABLE 6 Symmetrical Edge Loads on Simply Supported Cylindrical Shell (Force coefficients)

SHEAR EDGE LOAD	EDGE MOMENT LOAD

(a) Basic Formulas and Loading Diagrams

SHEAR EDGE LOAD	EDGE MOMENT LOAD
Longitudinal Force N_x	Longitudinal Force N_x
$S_L\left[\left(\dfrac{L}{R}\right)^2 \text{Col. (9)}\right]\sin\dfrac{\pi x}{L}$	$\dfrac{M_L}{R}\left[\left(\dfrac{L}{R}\right)^2 \text{Col. (13)}\right]\sin\dfrac{\pi x}{L}$
Shearing Force $N_{x\phi}$	Shearing Force $N_{x\phi}$
$S_L\left[\dfrac{L}{R}\text{Col. (10)}\right]\cos\dfrac{\pi x}{L}$	$\dfrac{M_L}{R}\left[\dfrac{L}{R}\text{Col. (14)}\right]\cos\dfrac{\pi x}{L}$
Transverse Force N_ϕ	Transverse Force N_ϕ
$S_L\,\text{Col. (11)}\sin\dfrac{\pi x}{L}$	$\dfrac{M_L}{R}\,\text{Col. (15)}\sin\dfrac{\pi x}{L}$
Transverse Moment M_ϕ	Transverse Moment M_ϕ
$S_L\left[R\,\text{Col. (12)}\right]\sin\dfrac{\pi x}{L}$	$M_L\,\text{Col. (16)}\sin\dfrac{\pi x}{L}$

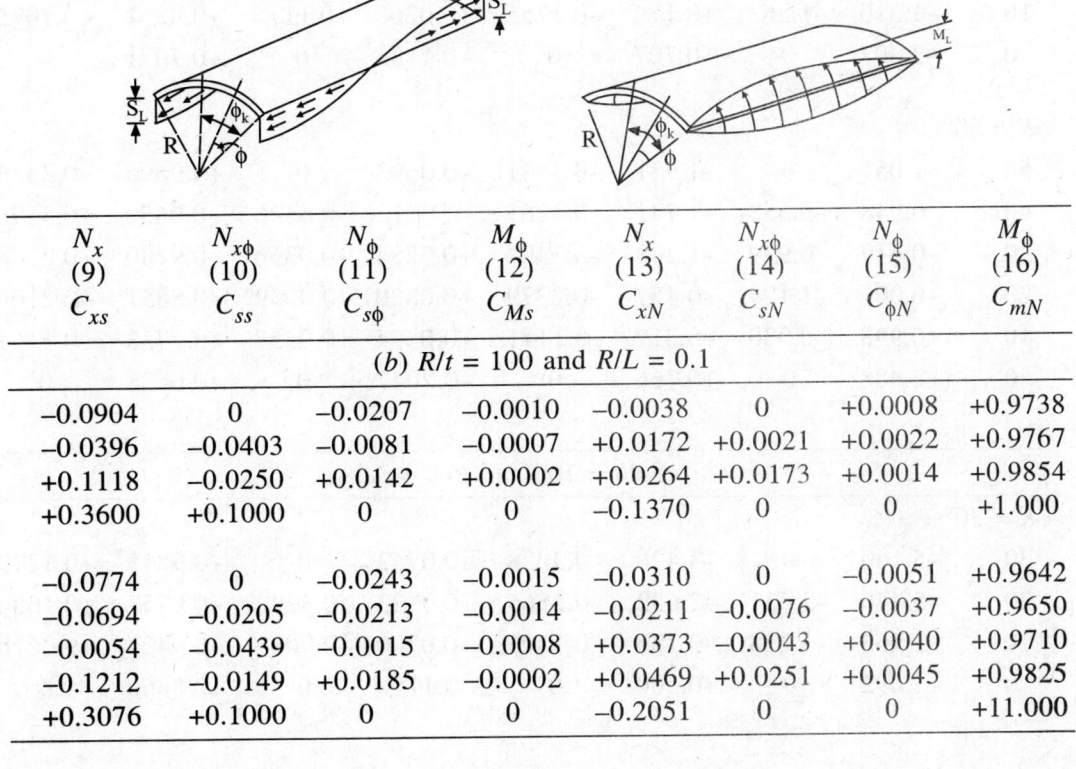

N_x (9) C_{xs}	$N_{x\phi}$ (10) C_{ss}	N_ϕ (11) $C_{s\phi}$	M_ϕ (12) C_{Ms}	N_x (13) C_{xN}	$N_{x\phi}$ (14) C_{sN}	N_ϕ (15) $C_{\phi N}$	M_ϕ (16) C_{mN}
(b) $R/t = 100$ and $R/L = 0.1$							
−0.0904	0	−0.0207	−0.0010	−0.0038	0	+0.0008	+0.9738
−0.0396	−0.0403	−0.0081	−0.0007	+0.0172	+0.0021	+0.0022	+0.9767
+0.1118	−0.0250	+0.0142	+0.0002	+0.0264	+0.0173	+0.0014	+0.9854
+0.3600	+0.1000	0	0	−0.1370	0	0	+1.000
−0.0774	0	−0.0243	−0.0015	−0.0310	0	−0.0051	+0.9642
−0.0694	−0.0205	−0.0213	−0.0014	−0.0211	−0.0076	−0.0037	+0.9650
−0.0054	−0.0439	−0.0015	−0.0008	+0.0373	−0.0043	+0.0040	+0.9710
−0.1212	−0.0149	+0.0185	−0.0002	+0.0469	+0.0251	+0.0045	+0.9825
+0.3076	+0.1000	0	0	−0.2051	0	0	+11.000

TABLE 6 (*Contd.*)

	VERTICAL EDGE LOAD				HORIZONTAL EDGE LOAD			
ϕ	N_x	$N_{x\phi}$	N_ϕ	M_ϕ	N_x	$N_{x\phi}$	N_ϕ	M_ϕ
	(1)	(2)	(3)	(4)	(5)	(6)	(7)	(8)
$\phi K = 40°$								
40	−2.210	0	−2.534	−0.4048	−0.0165	0	+0.9954	+0.2247
30	−1.806	−1.138	−2.168	−0.3683	−0.0049	−0.0068	+0.9836	+0.2041
20	−0.578	−1.830	−1.205	−0.2691	+0.0176	−0.0032	+0.9420	+0.1672
10	+1.516	−1.614	−0.045	−0.1352	+0.0156	+0.0081	+0.8685	+0.1047
0	+4.534.	0	+0.643	0	−0.0631	0	+0.7660	0
$\phi K = 45°$								
45	−1.602	0	−2.231	−0.4355	−0.0359	0	+0.9873	+2.2779
40	−1.546	−0.434	−2.163	−0.4274	−0.0309	−0.0094	+0.9849	+0.2743
30	−1.098	−1.180	−1.646	−0.3654	+0.0033	−0.0181	+0.9642	+0.2461
20	−0.166	−1.550	−0.763	−0.2562	+0.0412	−0.0050	+0.9138	+0.1899
10	+1.310	−1.263	+0.179	−0.1255	+0.0260	+0.1175	+0.8251	+0.1072
0	+3.407	0	+0.707	0	−0.1185	0	+0.7071	0
$\phi K = 50°$								
50	−1.051	0	−1.931	−0.4541	−0.0693	0	+0.9696	+0.3336
40	−0.948	−0.558	−1.741	−0.4265	−0.0213	−0.0326	+0.9650	+0.3198
30	−0.619	−0.999	−1.208	−0.3492	+0.0284	+0.0366	+0.9430	+0.2786
20	−0.001	−1.184	−0.451	−0.2379	+0.0820	−0.0039	+0.8851	+0.2104
10	+0.998	−0.930	+0.317	−0.1141	+0.0374	+0.0357	+0.7738	+0.1167
0	+2.488	0	+0.766	0	−0.2047	0	+0.6428	0
(*c*) $R/t = 100$ and $R/L = 0.2$								
$\phi K = 30°$								
30	−5.160	0	−3.420	−0.3028	−0.0742	0	+0.9815	+0.1219
20	−3.440	−2.531	−2.632	−0.2565	−0.0123	−0.3292	+0.9757	+0.1084
10	−1.763	−3.228	−0.787	−0.1395	+0.0700	−0.0087	+0.9427	+0.0678
0	+1.092	0	+0.500	0	−0.1140	0	+0.8660	0

SHEAR EDGE LOAD				EDGE MOMENT LOAD			
N_x	$N_{x\phi}$	N_ϕ	M_ϕ	N_x	$N_{x\phi}$	N_ϕ	M_ϕ
(9)	(10)	(11)	(12)	(13)	(14)	(15)	(16)
−0.0675	0	−0.0279	−0.0022	−0.0712	0	−0.0184	+0.9530
−0.0461	−0.0331	−0.0179	−0.0018	−0.0196	−0.0293	−0.0074	+0.9570
+0.0181	−0.0423	+0.0054	−0.0009	+0.0815	−0.0118	+0.0112	+0.9653
+0.1236	−0.0057	+0.0221	−0.0002	+0.0724	+0.0998	+0.0106	+0.9798
+0.2682	+0.1000	0	0	−0.3099	0	0	+1.000
−0.0596	0	−0.0315	−0.0031	−0.1280	0	−0.0437	+0.9392
−0.0559	−0.0160	−0.0292	−0.0030	−0.1102	−0.0325	−0.0385	+0.9402
−0.0258	−0.0398	−0.0123	−0.0022	+0.0128	−0.0642	−0.0047	+0.9469
+0.0338	−0.0389	+0.0120	−0.0010	+0.1523	−0.0158	+0.0277	+0.9595
+0.1222	+0.0025	+0.0250	−0.0002	+0.0981	+0.0680	+0.0212	+0.9771
+0.2378	+0.1000	0	0	−0.4613	0	0	+1.000
−0.0531	0	−0.0350	−0.0042	−0.2043	0	−0.0885	+0.9218
−0.0423	−0.0271	−0.0268	−0.0037	−0.1192	−0.0964	−0.0573	+0.9258
−0.0098	−0.0424	−0.0057	−0.0025	+0.0849	−0.1086	+0.0111	+0.9371
+0.0440	−0.0340	+0.0181	−0.0010	+0.2503	−0.0099	+0.0576	+0.9537
+0.1188	+0.0277	+0.0275	−0.0002	+0.1176	+0.1119	+0.0373	+0.9745
+0.2138	+0.1000	0	0	−0.6695	0	0	+1.000

(c) $R/t = 100$ and $R/L = 0.2$							
−0.1946	0	−0.0414	−0.0005	−0.549	0	−0.0813	+0.9009
0.0755	−0.0776	−0.0145	0	−0.073	−0.2142	−0.0153	+0.9124
0.2208	−0.0469	+0.0287	+0.0009	+0.557	−0.0429	+0.0568	+0.9456
−0.7119	+0.2000	0	0	−1.067	0	0	+1.000

TABLE 6 (*Contd.*)

ϕ	VERTICAL EDGE LOAD				HORIZONTAL EDGE LOAD			
	N_x	$N_{x\phi}$	N_ϕ	M_ϕ	N_x	N_ϕ	$N_{x\phi}$	M_ϕ
	(1)	(2)	(3)	(4)	(5)	(6)	(7)	(8)
$\phi K = 35°$								
35	−3.087	0	−2.871	−0.3347	−0.1552	0	+0.9603	+0.1591
30	−2.939	−0.833	−2.744	−0.3248	−0.1260	−0.0399	+0.9622	+0.1559
20	−1.639	−2.156	−1.807	−0.2514	+0.0561	−0.0634	+0.9649	+0.1301
10	+1.501	−2.296	−0.365	−0.1302	+0.1526	+0.0074	+0.9224	+0.0781
0	+7.434	0	+0.574	0	−0.3244	0	+0.8192	0
$\phi K = 40°$								
40	−1.764	0	−2.398	−0.3556	−0.3096	0	+0.9087	+0.1962
30	−1.640	−0.947	−2.105	−0.3239	−0.1336	−0.1389	+0.9351	+0.1845
20	−0.965	−1.697	−1.251	−0.2374	+0.2243	−0.1133	+0.9680	+0.1487
10	+1.111	−1.753	−0.109	−0.1188	+0.2753	+0.0557	+0.9101	+0.0870
0	+5.836	0	+0.643	0	−0.7210	0	+0.7660	0
$\phi K = 45°$								
45	−0.775	0	−1.934	−0.3619	−0.5679	0	+0.7941	+0.2293
40	−0.820	−0.217	−1.897	−0.3556	−0.4987	−0.1493	+0.8118	+0.2269
30	−1.062	−0.731	−1.584	−03068	−0.0210	−0.3078	+0.9229	+0.2071
20	−0.934	−1.317	−0.910	−0.2175	+0.5381	−0.1552	+1.008	+0.1641
10	+0.700	−1.478	+0.039	−0.1073	+0.4245	+0.1611	+0.9154	+3.0946
0	+5.344	0	+0.707	0	−1.392	0	+0.7071	0
$\phi K = 50°$								
50	+0.052	0	−1.437	−0.3493	−0.9469	0	+0.5813	+0.2514
40	−0.291	−0.028	−1.413	−0.3315	−0.5827	−0.4585	+0.7040	+0.2452
30	−1.016	−0.383	−1.246	−0.2786	+0.2885	−0.5526	+0.9656	+0.2228
20	−1.244	−1.059	−0.745	−0.1953	+0.1015	−0.1620	+1.112	+0.1764
10	+0.381	−1.427	+0.110	−0.0943	+0.5624	+0.3610	+0.9471	+0.1015
0	+5.670	0	+0.765	0	−2.398	0	+0.6428	0

SHEAR EDGE LOAD				EDGE MOMENT LOAD			
N_x (9)	$N_{x\phi}$ (10)	N_ϕ (11)	M_ϕ (12)	N_x (13)	$N_{x\phi}$ (14)	N_ϕ (15)	M_ϕ (16)
−0.1487	0	−0.0458	−0.0024	−0.912	0	−0.1914	+0.8637
−0.1334	−0.0394	−0.0401	−0.0022	−0.742	−0.2343	−0.1585	+0.8669
−0.0106	−0.0845	−0.0018	−0.0012	+0.327	−0.3747	−0.0315	+0.8914
+0.2378	−0.0280	+0.0369	−0.0001	+0.928	+0.0486	+0.1212	+0.9362
+0.6159	+0.2000	0	0	−2.045	0	0	+1.000
−0.1273	0	−0.0525	−0.0035	−1.421	0	−0.3885	+0.1856
−0.0880	−0.0626	−0.0335	−0.0029	−0.631	−0.6461	−0.2007	+0.8306
+0.0323	−0.0817	+0.0109	−0.0013	+1.015	−0.5396	+0.1574	+0.8707
+0.2408	−0.0111	+0.0434	−0.0007	+1.316	+0.2523	+0.2259	+0.9271
+0.5460	+0.2000	0	0	−3.452	0	0	+1.000
−0.1067	0	−0.0577	−0.0050	−2.074	0	−0.7365	+0.7509
−0.0988	−0.0287	−0.0535	−0.0048	−1.828	−0.5460	−0.6582	+0.7558
−0.0506	−0.0726	−0.0228	−0.0034	−0.116	−1.138	−0.1410	+0.7920
+0.0570	−0.0737	+0.0224	−0.0014	+1.958	−0.5987	+0.3956	+0.8510
+0.2353	+0.0028	+0.0482	0	+1.641	+0.5811	+0.3778	+0.9196
+0.4995	+0.2000	0	0	−5.299	0	0	+1.000
−0.0848	0	−0.0601	−0.0065	−2.814	0	−1.245	+0.6613
−0.0714	−0.0439	−0.0468	−0.0056	−1.762	−1.381	−0.8336	+0.6809
−0.0258	−0.0723	−0.0113	−0.0035	+0.797	−1.690	+0.0859	+0.7523
+0.0664	−0.0639	+0.0312	−0.0013	+3.038	−0.536	+0.7667	+0.8338
+0.2257	+0.0125	+0.0513	0	+1.811	+1.083	−0.5718	−0.9154
+0.4748	+0.2000	0	0	−7.473	0	0	+1.000

TABLE 6 (*Contd.*)

	VERTICAL EDGE LOAD				HORIZONTAL EDGE LOAD			
ϕ	N_x	$N_{x\phi}$	N_ϕ	M_ϕ	N_x	$N_{x\phi}$	N_ϕ	M_ϕ
	(1)	(2)	(3)	(4)	(5)	(6)	(7)	(8)
				(*d*) $R/t = 100$ and $R/L = 0.3$				
$\phi K = 30°$								
30	−4.709	0	−3.362	−0.2769	−0.2662	0	+0.9571	+0.1099
20	−3.568	−2.381	−2.632	−0.2347	−0.0094	−0.0961	+0.9752	+0.0982
10	+1.053	−3.290	−0.837	−0.1275	+0.2942	+0.0035	+0.9639	+0.0622
0	+12.43	0	+0.500	0	−0.6606	0	+0.8660	0
$\phi K = 35°$								
35	−2.156	0	−2.682	−0.2903	−0.6039	0	+0.8706	+0.1362
30	−2.203	−0.595	−2.591	−0.2821	−0.4823	−0.1544	+0.8892	+0.1338
20	−2.106	−1.817	−1.850	−0.2201	+0.2698	−0.2300	+0.9859	+0.1137
10	+0.513	−2.456	−0.482	−0.1148	+0.6242	+0.0755	+0.9800	+0.0698
0	+10.03	0	+0.574	0	−1.543	0	+0.8192	0
$\phi K = 40°$								
40	−0.216	0	−1.977	−0.2853	−1.176	0	+0.6733	+0.1553
30	−1.020	−0.275	−1.892	−0.2631	−0.4865	−0.5149	+0.8208	+0.1488
20	−2.210	−1.203	−1.430	−0.1982	+0.9141	−0.3958	+1.069	+0.1252
10	−0.295	−2.141	−0.338	−0.1018	+1.073	+0.2676	+1.040	+0.0762
0	+10.01	0	+0.643	0	−3.068	0	+0.7660	0
$\phi K = 45°$								
45	+1.400	0	−1.178	−0.2565	−1.970	0	+0.3055	+0.1591
40	+1.083	+0.354	−1.224	−0.2537	−1.728	−0.5178	+0.3766	+0.1589
30	−1.040	+0.432	−1.454	−0.2296	−0.0472	−1.061	+0.8385	+0.1550
20	−3.110	−0.788	−1.326	−0.1743	+1.941	−0.5051	+1.279	+0.1338
10	−0.946	−2.214	−0.330	−0.0899	+1.513	+0.6328	+1.158	+0.0821
0	+11.24	0	+0.707	0	−5.257	,0	+0.7071	0

	SHEAR EDGE LOAD				EDGE MOMENT LOAD		
N_x	$N_{x\phi}$	N_ϕ	M_ϕ	N_x	$N_{x\phi}$	N_ϕ	M_ϕ
(9)	(10)	(11)	(12)	(13)	(14)	(15)	(16)

(d) $R/t = 100$ and $R/L = 0.3$

N_x	$N_{x\phi}$	N_ϕ	M_ϕ	N_x	$N_{x\phi}$	N_ϕ	M_ϕ
−0.2618	0	−0.0596	−0.0024	−2.101	0	−0.3444	+0.7859
−01182	−0.1174	−0.0231	−0.0016	−0.108	−0.7669	−0.0802	+0.8119
+0.3251	−0.0750	+0.0421	−0.0003	+2.372	+0.0141	+0.1945	+0.8845
+1.102	+0.3000	0	0	−5.456	0	0	+1.000
−0.2140	0	−0.0685	−0.0036	−3.493	0	−0.7613	+0.7084
−0.1937	−0.0568	−0.0603	−0.0033	−2.808	−0.8945	−0.6139	+0.7161
−0.0244	−0.1248	−0.0046	−0.0018	+1.506	−1.358	+0.1098	+0.7726
+0.3456	−0.0471	+0.0535	−0.0002	+3.764	+0.4236	+0.4372	+0.8689
+0.9676	+0.3000	0	0	−9.403	0	0	+1.000
−0.1668	0	−0.0736	−0.0048	−5.193	0	−1.469	+0.6071
−0.1238	−0.0837	−0.0485	−0.0039	−2.222	−2.288	−0.7475	+0.6444
+0.0263	−0.1164	+0.0125	−0.0018	+3.993	−1.809	+0.5716	+0.7382
+0.3397	−0.0252	+0.0611	−0.0001	+5.004	+1.180	+0.7966	+0.8572
+0.8934	+0.3000	0	0	−14.38	0	0	+1.000
−0.1142	0	+0.0719	−0.0059	−6.844	0	−2.463	+0.4746
−0.1112	−0.0310	−0.0674	−0.0056	−6.032	−1.802	−2.202	+0.4871
−0.0776	−0.0851	−0.0328	−0.0040	−0.316	−3.743	−0.4779	+0.5768
+0.0382	−0.1013	+0.0241	−0.0016	+6.770	−1.867	+1.312	+0.7124
+0.3209	−0.0128	+0.0645	−0.0001	+5.696	+2.230	+1.229	+0.8510
+0.8743	+0.3000	0	0	−19.61	0	0	+1.000

TABLE 6 (*Contd.*)

φ	VERTICAL EDGE LOAD				HORIZONTAL EDGE LOAD			
	N_x	$N_{x\phi}$	N_ϕ	M_ϕ	N_x	$N_{x\phi}$	N_ϕ	M_ϕ
	(1)	(2)	(3)	(4)	(5)	(6)	(7)	(8)

$\phi K = 50°$

φ	(1)	(2)	(3)	(4)	(5)	(6)	(7)	(8)
50	+2.605	0	−0.326	−0.2300	−2.821	0	−0.2407	+0.1409
40	+1.276	+1.180	−0.676	−0 2032	−1.744	−1.346	+0.1577	+0.1474
30	−1.837	+1.062	−1.342	−0.1926	+0.8732	−1.624	+1.033	+0.1561
20	−4.099	−0.690	−1.432	−0.1527	+3.129	−0.4486	+1.623	+0.1422
10	−1.187	−2.497	−0.379	−0.0801	+1.766	+1.155	+1.315	+0.0885
0	+12.66	0	+0.766	0	−7.773	0	+0.6428	0

(*e*) $R/t = 100$ and $R/L = 0.4$

$\phi K = 30°$

φ	(1)	(2)	(3)	(4)	(5)	(6)	(7)	(8)
30	−3.508	0	−3.175	−0.2404	−0.736	0	+0.881	+0.0947
20	−3.500	−1.942	−2.592	−0.2048	−0.038	−0.2682	+0.958	+0.0856
10	−0.244	−3.281	−0.947	−2.1121	+0.804	−0.0008	+1.008	+0.0544
0	+15.23	0	+0.500	0	−1.772	0	+0.866	0

$\phi K = 35°$

φ	(1)	(2)	(3)	(4)	(5)	(6)	(7)	(8)
35	−0.217	0	−2.261	−0.2351	−1.518	0	+0.670	+0.1091
30	−0.656	−0.102	−2.241	−0.2293	−1.214	−0.3878	+0.723	+0.1078
20	−2.982	−1.086	−1.911	−0.1833	+0.680	−0.5777	+1.014	+0.0952
10	−1.512	−2.715	−0.712	−0.0982	+1.596	+0.1982	+1.092	+0.0612
0	+15.11	0	+0.574	0	−3.990	0	+0.819	0

$\phi K = 40°$

φ	(1)	(2)	(3)	(4)	(5)	(6)	(7)	(8)
40	+2.485	0	−1.221	−0.2044	−2.664	0	+0.250	+0.1088
30	+0.117	+0.913	−1.512	−0.1948	−1.123	−1.171	+0.606	+0.1093
20	−4.296	−0.278	−1.730	−0.1580	+2.058	−0.9139	+1.236	+0.1013
10	−2.814	−2.749	−0.749	−0.0862	+2.495	+0.6036	+1.272	+0.0665
0	+17.20	0	+0.643	0	−7.122	0	+0.766	0

SHEAR EDGE LOAD				EDGE MOMENT LOAD			
N_x (9)	$N_{x\phi}$ (10)	N_ϕ (11)	M_ϕ (12)	N_x (13)	$N_{x\phi}$ (14)	N_ϕ (15)	M_ϕ (16)
–0.0565	0	–0.0611	–0.0063	–7.871	0	–3.553	+0.3153
–0.0651	–0.0327	–0.0514	–0.0056	–4.994	–3.782	–2.394	+0.3754
–0.0643	–0.0700	–0.0208	–0.0036	+2.207	–4.666	+0.2204	+0.5257
+0.0214	–0.0879	+0.0276	–0.0013	+8.917	–1.422	+2.171	+0.6995
+0.3005	–0.0113	+0.0639	+0.0002	+5.588	+3.317	+1.625	+0.8508
+0.8992	+0.3000	0	0	–23.76	0	0	+1.000

(e) $R/t = 100$ and $R/L = 0.4$

N_x (9)	$N_{x\phi}$ (10)	N_ϕ (11)	M_ϕ (12)	N_x (13)	$N_{x\phi}$ (14)	N_ϕ (15)	M_ϕ (16)
–0.3318	0	–0.0772	–0.0028	–5.33	0	–0.879	+0.6558
–0.1568	–0.1503	–0.0304	–0.0028	–0.14	–1.920	–0.196	+0.6995
+0.4153	–0.0993	+0.0545	–0.0003	+6.27	+0.187	+0.502	+0.8169
+1.508	+0.4000	0	0	–15.27	0	0	+1.000
–0.2518	0	–0.0842	–0.0040	–8.48	0	–1.890	+0.5364
–0.2316	–0.0672	–0.0746	–0.0037	–6.82	–2.173	–1.574	+0.5500
–0.0480	–0.1540	–0.0073	–0.0019	+3.69	–3.300	+0.248	+0.6463
+0.4268	–0.0671	+0.0673	–0.0001	+9.37	+1.106	+1.064	+0.8000
+1.373	+0.4000	0	0	–23.99	0	0	+1.000
–0.1542	0	–0.0818	–0.0049	–11.15	0	–3.211	+0.3896
–0.1315	–0.0851	–0.0567	–0.0039	–4.83	–4.928	–0.164	+0.4541
+0.0023	–0.1351	+0.0099	–0.0017	+8.71	–3.904	+1.250	+0.6098
+0.4228	–0.0471	+0.0728	0	+11.30	+2.733	+1.794	+0.7918
+1.368	+0.4000	0	0	–33.39	0	0	+1.000

TABLE 6 (*Contd.*)

ϕ	VERTICAL EDGE LOAD				HORIZONTAL EDGE LOAD			
	N_x	$N_{x\phi}$	N_ϕ	M_ϕ	N_x	$N_{x\phi}$	N_ϕ	M_ϕ
	(1)	(2)	(3)	(4)	(5)	(6)	(7)	(8)
$\phi K = 45°$								
45	+4.329	0	−0.119	−0.1513	−3.817	0	−0.372	+0.0899
40	+3.663	+1.126	−0.277	−0.1526	−3.381	−1.005	−0.230	+0.0923
30	−0.918	+2.029	−1.249	−0.1564	−0.150	−2.079	+0.703	+0.1060
20	−6.034	−0.008	−1.889	−0.1377	+3.796	−1.015	+1.638	+0.1083
10	−3.355	−3.194	−0.854	−0.0780	+3.104	+1.260	+1.497	+0.0731
0	+19.64	0	+0.707	0	−10.78	0	+0.707	0
$\phi K = 50°$								
50	+4.973	0	+0.784	−0.0881	−4.477	0	−1.043	+0.0545
40	+2.829	+2.328	+0.079	+0.1028	−2.858	−2.155	−0.394	+0.0716
30	−2.433	+2.516	−1.399	−0.1279	+1.235	−2.673	+1.065	+0.1057
20	−6.881	−0.213	−2.118	−0.1244	+5.125	−0.8202	+2.120	+0.1184
10	−3.049	−3.540	−0.904	−0.0724	+3.212	+1.909	+1.707	−0.0810
0	+20.35	0	+0.766	0	−13.59	0	+0.643	0
(*f*) $R/t = 100$ and $R/L = 0.5$								
$\phi K = 30°$								
30	−1.757	0	−2.879	−0.2019	−1.420	0	+0.767	+0.0789
20	−3.508	−1.323	−2.524	−0.1738	−0.037	−0.5108	+0.932	+0.0727
10	−2.292	−3.378	−1.108	−0.0969	+1.616	+0.0381	+1.071	+0.0489
0	+20.22	0	+0.500	0	−3.767	0	+0.866	0
$\phi K = 35°$								
35	+2.471	0	−1.677	−0.1788	−2.776	0	+0.395	+0.0816
30	+1.493	+0.571	−1.758	−0.1758	−2.223	−0.7101	+0.495	+0.0816
20	−4.209	−0.060	−2.002	−0.1480	+1.246	−1.062	+1.056	+0.0775
10	−4.426	−3.090	−1.055	−0.0837	+2.986	+0.3737	+1.252	+0.0535
0	+22.54	0	+0.574	0	−7.555	0	+0.819	0

SHEAR EDGE LOAD				EDGE MOMENT LOAD			
N_x	$N_{x\phi}$	N_ϕ	M_ϕ	N_x	$N_{x\phi}$	N_ϕ	M_ϕ
(9)	(10)	(11)	(12)	(13)	(14)	(15)	(16)
−0.0653	0	−0.0657	−0.0051	−12.50	0	−4.647	+0.2210
−0.0723	−0.0186	−0.0630	−0.0049	−11.09	−3.299	−4.168	+0.2412
−0.1030	−0.0673	−0.0384	−0.0085	−0.92	−6.970	−0.971	+0.3846
−0.0344	−0.1142	+0.0158	−0.0013	+12.56	−3.637	+2.441	+0.5941
+0.3651	−0.0449	+0.0716	+0.0003	+11.51	+4.262	+2.363	+0.7936
+1.388	+0.4000	0	0	−39.59	0	0	+1.000
+0.0216	0	−0.0395	−0.0045	−11.45	0	−5.493	+0.0676
−0.0276	+0.0025	−0.0408	−0.0041	−7.65	−4.182	−3.773	+0.1539
−0.1165	−0.0382	−0.0313	−0.0029	+2.61	−3.745	+0.216	+0.3594
−0.0779	−0.1025	+0.0109	−0.0010	+13.82	−1.157	+3.391	+0.5981
+0.3428	−0.0541	+0.0667	+0.0003	+10.19	+5.359	+2.668	+0.7998
+1.459	+0.4000	0	0	−41.80	0	0	+1.000
			$(f)\ R/t = 100$ and $R/L = 0.5$				
−0.3849	0	−0.0912	−0.0030	−10.64	0	−1.779	+0.5185
−0.1963	−0.1773	−0.0360	−0.0020	−0.48	−3.877	−0.421	+0.5825
+0.4854	−0.1263	+0.0654	−0.0002	+12.53	+0.236	+0.998	+0.7480
+1.964	+0.5000	0	0	−30.35	0	0	+1.000
−0.2558	0	−0.0962	−0.0039	−14.99	0	−3.345	+0.3692
−0.2428	−0.0837	−0.0875	−0.0037	−12.12	−4.668	−2.786	+0.3888
−0.0879	−0.1706	−0.0227	−0.0018	+6.40	−5.907	+0.453	+0.5266
+0.4733	−0.0921	+0.0629	+0.0000	+17.12	+1.978	+1.931	+0.7345
+1.867	+0.5000	0	0	−44.56	0	0	+1.000

TABLE 6 (*Contd.*)

ϕ	VERTICAL EDGE LOAD				HORIZONTAL EDGE LOAD			
	N_x	$N_{x\phi}$	N_ϕ	M_ϕ	N_x	$N_{x\phi}$	N_ϕ	M_ϕ
	(1)	(2)	(3)	(4)	(5)	(6)	(7)	(8)
$\phi K = 40°$								
40	+5.472	0	−0.358	−0.1324	−4.282	0	−0.220	+0.0682
30	+1.404	+2.232	−1.069	−0.1354	−1.839	−1.889	+0.363	+0.0754
20	−6.661	+0.757	−2.062	−0.1254	+3.336	−1.488	+1.415	+0.0819
10	−5.826	−3.503	−1.214	−0.0751	+4.196	+1.017	+1.528	+0.0593
0	+26.12	0	+0.643	0	−12.17	0	+0.766	0
$\phi K = 45°$								
45	+6.655	0	+0.771	−0.0750	−4.823	0	−0.923	+0.0405
40	+5.749	+1.741	+0.524	−0.0793	−4.625	−1.375	−0.727	+0.0448
30	−0.652	+3.364	−1.053	−0.1039	−0.365	−2.902	+0.573	+0.0709
20	−8.443	+0.730	−2.362	−0.1125	+5.266	−1.497	+1.929	+0.0901
10	−5.758	−4.033	−1.326	−0.0708	+4.705	+1.781	+1.794	+0.0670
0	+27.76	0	+0.707	0	−16.12	0	+0.707	0
$\phi K = 50°$								
50	+6.074	0	+1.416	−0.0253	−5.122	0	−1.462	+0.0088
40	+3.738	+2.900	+0.542	−0.0469	−3.478	−2.507	+0.713	−0.0302
30	−2.427	+3.394	−1.371	−0.0900	+1.103	−3.275	+1.025	+0.0759
20	−8.536	+0.212	−2.512	−0.1065	+6.245	−1.164	+2.385	+0.1024
10	−4.760	−4.221	−1.261	−0.0672	+4.596	+2.387	+1.966	+0.0752
0	+26.68	0	+0.766	0	−18.51	0	+0.643	0
(g) $R/t = 100$ and $R/L = 0.6$								
$\phi K = 30°$								
30	+0.690	0	−2.477	−0.1646	−2.375	0	+0.609	+0.0636
20	−3.149	−2.438	−2.432	−0.1447	−0.079	−0.8580	+0.895	+0.0605
10	−5.088	−3.429	−1.334	−0.0837	+2.720	+0.0560	+1.160	+0.0431
0	+26.76	0	+0.500	0	−6.379	0	+0.866	0

	SHEAR EDGE LOAD				EDGE MOMENT LOAD		
N_x	$N_{x\phi}$	N_ϕ	M_ϕ	N_x	$N_{x\phi}$	N_ϕ	M_ϕ
(9)	(10)	(11)	(12)	(13)	(14)	(15)	(16)
−0.1143	0	−0.0775	−0.0034	−17.14	0	−5.060	+0.2033
−0.1408	−0.0683	−0.0585	−0.0034	−7.86	−7.678	−2.645	+0.2912
−0.0853	−01408	+0.0027	−0.0014	+13.14	−6.372	+1.904	+0.5001
+0.4218	−0.0802	+0.0780	+0.0003	+18.72	+4.129	+2.821	+0.7347
+1.908	+0.5000	0	0	−54.72	0	0	+1.000
+0.0153	0	−0.0477	−0.0038	−15.73	0	−6.128	+0.0523
−0.0054	+0.0023	−0.0480	−0.0037	−14.17	−4.167	−5.525	+0.0765
−0.1266	−0.0318	−0.0417	−0.0028	−2.11	−9.122	−1.418	+0.2487
−0.1395	−0.1180	+0.0012	−0.0011	+16.12	−5.253	−3200	+0.5015
+0.3807	−0.0882	+0.0726	+0.0003	+17.30	+5.658	+3.294	+0.7426
+2.002	+0.5000	0	0	−58.57	0	0	+1.000
+0.1028	0	−0.0139	−0.0027	−11.38	0	−6.094	−0.0520
+0.0147	+0.0398	−0.0269	−0.0027	−8.59	−5.742	−4.363	+0.0339
−0.1664	−0.0015	−0.0410	−0.0022	+0.92	−8.235	−0.093	+0.2533
−0.1868	−0.1144	−0.0079	−0.0010	+15.24	−3.756	+3.776	+0.5154
+0 3719	−0.1018	+0.0671	+0.0003	+14.96	+6.079	+3.305	+0.7497
+2.065	+0.5000	0	· 0	−56.51	0	0	+1.000
			(g) $R/t = 100$ and $R/L = 0.6$				
−0.4130	0	−0.1015	−0.0031	−17.24	0	−2.909	+2.909
−0.2349	−0.1448	−0.0424	−0.0010	−0.99	−6.327	−0.707	+0.4746
+0.5284	−0.1530	+0.0739	−0.0001	+16.89	+0.297	+1.631	+0.6844
+2.481	+0.6000	0	0	−50.48	0	0	+1.000

TABLE 6 (*Contd.*)

	VERTICAL EDGE LOAD				HORIZONTAL EDGE LOAD			
ϕ	N_x	$N_{x\phi}$	N_ϕ	M_ϕ	N_x	$N_{x\phi}$	N_ϕ	M_ϕ
	(1)	(2)	(3)	(4)	(5)	(6)	(7)	(8)
$\phi K = 35°$								
35	+5.530	0	+0.999	−0.1294	−4.187	0	+0.085	+0.0577
30	+3.964	+1.371	−1.194	−0.1291	−1.368	−1.072	+0.231	+0.0589
20	−5.521	+1.142	−2.096	−0.1187	+1.835	−1.625	+1.098	+0.0628
10	−7.862	−3.492	−1.452	−0.0731	+4.617	+0.5579	+1.438	+0.0477
0	+31.29	0	+0.574	0	−11.74	0	+0.891	0
$\phi K = 40°$								
40	+8.029	0	+0.412	−0.0781	−5.611	0	+0.632	+0.0378
30	+2.640	+3.389	−0.660	−0.0907	−2.518	−2.498	+0.140	+0.0500
20	−8.598	+1.754	−2.347	−0.1015	+4.328	−2.036	+1.563	+0.0643
10	−8.783	−4.118	−1.651	−0.0674	+5.859	+1.341	+1.421	+0.0538
0	+34.71	0	+0.643	0	−17.02	0	+0.766	0
$\phi K = 45°$								
45	+7.904	0	+1.346	−0.0293	−5.855	0	−1.258	+0.0115
40	+6.935	+2.078	+1.051	−0.0352	−5.259	−1.550	−1.038	+0.0166
30	−0.196	+4.195	−0.884	−0.0712	−0.727	−3.373	+0.459	+0.0490
20	−9.841	+1.352	−2.657	−0.0959	+6.025	−1.905	+9.094	+0.0775
10	−7.893	−4.563	−1.682	−0.0659	+6.150	+2.089	+2.020	+0.0623
0	+34.56	0	+0.707	0	−20.64	0	+0.707	0
$\phi K = 50°$								
50	+6.029	0	+1.647	+0.0038	−4.847	0	−1.571	−0.0114
40	+4.096	+2.955	+0.283	−0.0192	−3.656	−2.446	−0.851	+0.0100
30	−1.803	+3.754	−1.264	−0.0627	+0.458	−3.469	+0.919	+0.0574
20	−9.145	+0.659	−2.736	−0.0932	+6.530	−1.539	+2.469	+0.0897
10	−6.445	−4.528	−1.501	−0.0624	+6.010	+2.570	+2.141	+0.0696
0	+31.83	0	+0.766	0	−22.64	0	+0.643	0

SHEAR EDGE LOAD				EDGE MOMENT LOAD			
N_ϕ (9)	$N_{x\phi}$ (10)	N_ϕ (11)	M_ϕ (12)	N_x (13)	$N_{x\phi}$ (14)	N_ϕ (15)	M_ϕ (16)
−0.2245	0	−0.0937	−0.0036	−21.54	0	−4.900	+0.2286
−0.2258	−0.0617	−0.0849	−0.0034	−17.56	−5.539	−4.097	+0.2534
−0.1434	−0.1738	−0.0163	−0.0014	+8.78	−8.709	+0.609	+0.4257
+0.4825	−0.1215	+0.0821	+0.0005	+25.70	+2.778	+2.864	+0.6789
+2.453	+0.6000	0	0	−67.87	0	0	+1.000
−0.0378	0	−0.0653	−0.0034	−20.91	0	−6.314	+0.0741
−0.1275	−0.0385	−0.0561	−0.0020	−10.34	+9.508	−9.449	+0.1748
−0.1799	−0.1354	−0.0078	−0.0012	+16.02	−8.383	+2.316	+0.4147
+0.4136	−0.1192	+0.0787	+0.0004	+25.53	+5.071	+3.686	+0.6842
+2.558	+0.6000	0	0	−75.81	0	0	+1.000
+0.1009	0	−0.0267	−00026	−15.56	0	−6.665	−0.0364
+0.0669	−0.0245	−0.0302	−0.0025	−14.36	−4.156	−6.057	−0.0122
−0.1460	+0.0082	−0.0439	−0.0022	−3.96	−9.685	−1.817	+0.1633
−0.2510	−0.1174	−0.0148	−0.0010	+16.43	−6.537	+3.383	+0.4285
+0.3813	−0.1339	+0.0714	+0.0004	+22.52	+6.087	+3.873	+0.6947
+2.659	+0.6000	0	0	−74.68	0	0	+1.000
+0.1678	0	+0.0073	−0.0015	−8.26	0	−5.666.	−0.0950
+0.0523	+0.0699	−0.0151	−0.0017	−8.00	−3.483	−4.344	−0.0164
−0.2030	+0.0305	−0.0491	−0.0018	−1.93	−7.812	−0.623	+0.1835
−0.2862	−0.1225	−0.0245	−0.0010	+13.86	−5.044	+3.549	+0.4425
+0.3899	−0.1463	+0.0675	+0.0003	+20.05	+5.942	+3.658	+0.6986
+2.688	+0.6000	0	0	−70.37	0	0	+1.000

TABLE 6 (*Contd.*)

ϕ	VERTICAL EDGE LOAD				HORIZONTAL EDGE LOAD			
	N_x	$N_{x\phi}$	N_ϕ	M_ϕ	N_x	$N_{x\phi}$	N_ϕ	M_ϕ
	(1)	(2)	(3)	(4)	(5)	(6)	(7)	(8)

(*h*) $R/t = 200$ and $R/L = 0.1$

$\phi K = 30°$

30	−5.388	0	−3.482	−0.3278	−0.0129	0	+0.9979	+0.1310
20	−3.600	−2.628	−2.666	−0.2779	+0.0014	−0.0043	+0.9845	+0.1162
10	+1.787	−3.290	−0.781	−0.1511	−0.0164	+0.0020	+0.9410	+0.0722
0	+10.83	0	+0.500	0	−0.0447	0	+0.8660	0

$\phi K = 35°$

35	−3.373	0	−2.947	−0.3692	−0.0349	0	+0.9926	+0.1751
30	−3.175	−0.907	−2.810	−0.3584	−0.0272	−0.0089	+0.9901	+0.1715
20	−1.566	−2.282	−1.814	−0.2774	+0.0192	−0.0121	+0.9675	+0.1424
10	+1.763	−2.310	−0.337	−0.1441	+0.0378	+0.0075	+0.9107	+0.0832
0	+6.999	0	+0.574	0	−0.1053	0	+0.8192	0

$\phi K = 40°$

40	−2.193	0	−2.526	−0.4041	−0.0791	0	+0.9782	+0.2237
30	−1.834	−1.137	−2.168	−0.3677	−0.0303	−0.0341	+0.9742	+0.2095
20	−0.678	−1.866	−1.220	−0.2690	+0.0667	−0.0239	+0.9491	+0.1669
10	+1.483	−1.699	−0.048	−0.1354	+0.0718	+0.0226	+0.8777	+0.0965
0	+4.960	0	+0.643	0	−0.2243	0	+0.7660	0

$\phi K = 45°$

45	−1.415	0	−2.167	−0.4305	−0.1591	0	+0.9446	+0.2747
40	−1.384	−0.385	−2.106	−0.4226	−0.1386	−0.0418	+0.9469	+0.2713
30	−1.097	−1.081	−1.635	−0.3622	+0.0019	−0.0838	+0.9561	+0.2441
20	−0.350	−1.505	−0.798	−0.2550	+0.1623	−0.0353	+09371	+0.1892
10	+1.175	−1.323	+0.148	−0.1254	+0.1167	+0.0570	+0.8460	+0.1071
0	+3.887	0	+0.707	0	−0.4371	0	+0.7071	0

SHEAR EDGE LOAD				EDGE MOMENT LOAD			
N_x (9)	$N_{x\phi}$ (10)	N_ϕ (11)	M_ϕ (12)	N_x (13)	$N_{x\phi}$ (14)	N_ϕ (15)	M_ϕ (16)

(h) $R/t = 200$ and $R/L = 0.1$

N_x (9)	$N_{x\phi}$ (10)	N_ϕ (11)	M_ϕ (12)	N_x (13)	$N_{x\phi}$ (14)	N_ϕ (15)	M_ϕ (16)
−0.0905	0	−0.0207	−0.0009	−0.1005	0	−0.0164	+0.9730
−0.0397	−0.0403	−0.0080	−0.0006	+0.0123	−0.0332	−0.0024	+0.9761
+0.1117	−0.0251	+0.0142	−0.0002	+0.1340	+0.0177	+0.0099	+0.9852
+0.3604	+0.1000	0	0	−0.3807	0	0	+1.000
−0.0773	0	−0.0243	−0.0015	−0.2044	0	−0.0435	+0.9623
−0.0693	−0.0205	−0.0212	−0.0014	−0.1601	−0.0520	−0.0356	+0.9632
−0.0055	−0.0439	−0.0015	−0.0008	+0.1134	−0.0710	+0.0090	+0.9700
+0.1211	−0.0150	+0.0185	−0.0002	+0.2305	+0.0464	+0.0256	+0.9823
+0.3082	+0.1000	0	0	−0.6605	0	0	+1.000
−0.0670	0	−0.0278	−0.0022	−0.3607	0	−0.1000	+0.9476
−0.0458	−0.0329	−0.0178	−0.0018	−0.1407	−0.1563	−0.0487	+0.9521
+0.0177	−0.0425	+0.0054	−0.0009	+0.3060	−0.1102	+0.0428	+0.9635
+0.1231	−0.0058	+0.0220	−0.0002	+0.3419	+0.1068	+0.0540	+0.9795
+0.2696	+0.1000	0	0	−1.083	0	0	+1.000
−0.0585	0	−0.0311	−0.0031	−0.5815	0	−0.2039	+0.9270
−0.0549	−0.0127	−0.0288	−0.0030	−0.5076	−0.1526	−0.1817	+0.9284
−0.0259	−0.0392	−0.0122	−0.0021	+0.0014	−0.3085	−0.0344	+0.9392
+0.0326	−0.0387	+0.0118	−0.0010	+0.5987	−0.1323	+0.1135	+0.9566
+0.1213	+0.0021	+0.0248	−0.0002	+0.4491	+0.2141	+0.0996	+0.9768
+0.2407	+0.1000	0	0	−1.684	0	0	+1.000

TABLE 6 (*Contd.*)

	VERTICAL EDGE LOAD				HORIZONTAL EDGE LOAD			
ϕ	N_x	$N_{x\phi}$	N_ϕ	M_ϕ	N_x	$N_{x\phi}$	N_ϕ	M_ϕ
	(1)	(2)	(3)	(4)	(5)	(6)	(7)	(8)
$\phi K = 50°$								
50	−0.838	0	−1.831	−0.4457	−0.2911	0	+0.8743	+0.3250
40	−0.842	−0.461	−1.678	−0.4195	−0.1746	−0.1378	+0.9021	+0.3125
30	−0.759	−0.908	−1.225	−0.3455	+0.1012	−0.1615	+0.9516	+0.2745
20	−0.314	−1.227	−0.517	−0.2369	+0.3240	−00358	+0.9439	+0.2092
10	+0.907	−1.111	+0.281	−0.1142	+0.1645	+0.1245	+0.8204	+0.1167
0	+3.402	0	+0.766	0	−0.7874	0	+0.6428	0

(*i*) $R/t = 200$ and $R/L = 0.2$

$\phi K = 30°$								
30	−4.804	0	−3.394	−0.3054	−0.2444	0	+0.9630	+0.1216
20	−3.563	−2.414	−2.649	−0.2592	−0.0144	−0.0894	+0.9776	+0.1084
10	+1.170	−3.280	−0.836	−0.1412	+0.2619	−0.0028	+0.9629	+0.0679
0	+12.72	0	+0.500	0	−0.5594	0	+0.8660	0
$\phi K = 35°$								
35	−2.269	0	−2.720	−0.3304	−0.5645	0	+0.8832	+0.1560
30	−2.290	−0.624	−2.624	−0.3211	−0.4523	−0.1446	+0.9000	+0.1531
20	− 2.035	−1.854	−1.855	−0.2507	+0.2416	−0.2183	+0.9868	+0.1291
10	+0.648	−2.423	−0.474	−0.1313	+0.5770	+0.0604	+0.9767	+0.0783
0	+9.620	0	+0.574	0	−1.373	0	+0.8191	0
$\phi K = 40°$								
40	−0.324	0	−2.021	−0.3363	−1.133	0	+0.6921	+0.1847
30	−1.057	−0.319	−1.919	−0.3098	−0.4724	−0.4965	+0.8324	+0.1760
20	−2.100	−1.229	−1.425	−0.2331	+0.8653	−0.3869	+1.067	+0.1458
10	−0.178	−2.090	−0.327	−0.1202	+1.021	+0.2424	+1.036	+0.0873
0	+9.537	0	+0.643	0	−2.851	0	+0.7660	0

SHEAR EDGE LOAD				EDGE MOMENT LOAD			
N_x	$N_{x\phi}$	N_ϕ	M_ϕ	N_x	$N_{x\phi}$	N_ϕ	M_ϕ
(9)	(10)	(11)	(12)	(13)	(14)	(15)	(16)
−0.0506	0	+0.0340	−0.0041	−0.8725	0	−0.3803	+0.8948
−0.0408	−0.0260	−0.0262	−0.0036	−0.5294	−0.4145	−0 2512	+0.9030
−0.0107	−0.0411	−0.0176	−0.0024	+0.2947	−0.4901	+0.0344	+0.9238
+0.0369	−0.0337	+0.0174	−0.0010	+0.9856	−0.1131	+0.2362	+0.9497
+0.1175	+0.0087	+0.0271	−0.0001	+0.5268	+0.3830	+0.1664	+0.9741
+0.2199	+0.1000	0	0	−2.487	0	0	+1.000

(i) $R/t = 200$ and $R/L = 0.2$

N_x	$N_{x\phi}$	N_ϕ	M_ϕ	N_x	$N_{x\phi}$	N_ϕ	M_ϕ
−0.1770	0	−0.0407	−0.0017	−1.972	0	−0.3043	+0.8904
−0.0792	−0.0792	−0.0160	−0.0012	−0.154	−0.7298	−0.0618	+0.9051
+0.2192	−0.0502	+0.0279	−0.0003	+2.142	+0.0427	+0.1925	+0.9435
+0.7297	+0.2000	0	0	−4.639	0	0	+1.000
−0.1455	0	−0.0466	−0.0027	−3.344	0	−0.7035	+0.8373
−0.1313	−0.0386	−0.0410	−0.0325	−2.696	−0.8582	−0.5812	+0.8425
−0.0150	−0.0842	−0.0033	−0.0014	+1.373	−1.319	+0.1200	+0.8787
+0.2331	−0.0309	+0.0358	−0.0002	+3.519	+0.3350	+0.4303	+0.9335
+0.6376	+0.2000	0	0	−8.444	0	0	+1.000
−0.1144	0	−0.0503	−0.0037	−5.147	0	−1.422	+0.7549
−0.0838	−0.0572	−0.0331	−0.0030	−2.213	−2.270	−0.7127	+0.7838
+0.0202	−0.0785	+0.0084	−0.0015	+3.880	−1.816	+0.5816	+0.8519
+0.2295	−0.0158	+0.0411	−0.0002	+4.849	+1.085	+0.8003	+0.9255
+0.5858	+0.2000	0	0	−13.55	0	0	+1.000

TABLE 6 (*Contd.*)

	VERTICAL EDGE LOAD				HORIZONTAL EDGE LOAD			
ϕ	N_x	$N_{x\phi}$	N_ϕ	M_ϕ	N_x	$N_{x\phi}$	N_ϕ	M_ϕ
	(1)	(2)	(3)	(4)	(5)	(6)	(7)	(8)

$\phi K = 45°$

45	+1.358	0	−1.210	−0.3140	−1.967	0	+0.3169	+0.1973
40	+1.046	+0.344	−1.254	−0.3103	−1.725	−0.5171	+0.3877	+0.1967
30	−1.033	−0.270	−1.469	−0.2789	−0.050	−0.6113	+0.8459	+0.1884
20	−3.035	−0.788	−1.324	−0.2104	+1.916	−0.5110	+1.281	+0.1588
10	−0.885	−2.169	−0.327	−0.1087	+1.486	+0.6094	+1.159	+0.0955
0	+10.91	0	+0.707	0	−5.085	0	+0.7071	0

$\phi K = 50°$

50	+2.686	0	−0.312	−0.2588	−2.918	0	−0.2665	+0.1807
40	+1.320	+1.218	−0.671	−0.2562	−1.798	−1.391	+0.1445	+0.1856
30	−1.860	+1.105	−1.355	−0.2385	+0.904	−1.675	+1.046	+0.1889
20	−4.145	−0.669	−1.455	−0.1867	+3.198	−0.4671	+1.649	+0.1655
10	−1.190	−2.490	−0.395	−0.0979	−1.779	+1.160	+1.332	+0.0992
0	+12.58	0	+0.766	0	−7.777	0	+0.6428	0

(*j*) $R/t = 200$ and $R/L = 0.3$

$\phi K = 30°$

30	−2.875	0	−3.097	−0.2688	−1.018	0	+0.8433	+0.1065
20	−3.544	−1.727	−2.595	−0.2300	−0.025	−0.3675	+0.9558	+0.0961
10	−1.006	−3.351	−1.021	−0.1270	+1.142	+0.0251	+1.038	+0.0619
0	+17.18	0	+0.500	0	−2.618	0	+0.8660	0

$\phi K = 35°$

35	+1.157	0	−2.003	−0.2629	−3.540	0	+0.5377	+0.1233
30	+0.434	+0.250	−2.034	−0.2580	−1.763	−0.5643	+0.6157	+0.1220
20	−3.630	−0.571	−1.989	−0.2098	+1.001	−0.8379	+1.051	+0.1087
10	−2.937	−2.907	−0.903	−0.1149	+2.318	+0.2927	+1.184	+0.0698
0	+18.66	0	+0.574	0	−5.780	0	+0.8192	0

	SHEAR EDGE LOAD				EDGE MOMENT LOAD		
N_x (9)	$N_{x\phi}$ (10)	N_ϕ (11)	M_ϕ (12)	N_x (13)	$N_{x\phi}$ (14)	N_ϕ (15)	M_ϕ (16)
−0.0793	0	−0.0494	−0.0047	−7.076	0	−2.500	+0.6266
−0.0769	−0.0215	−0.0462	−0.0045	−6.231	−1.862	−2.231	+0.6378
−0.0516	−0.0642	−0.0223	−0.0032	−0.316	−2.536	−0.5827	+0.7167
+0.0289	−0.0681	+0.0165	−0.0014	+6.897	−1.941	+1.367	+0.8267
+0.2164	−0.0071	+0.0436	−0.0001	+5.708	+2.201	+1.273	+0.9200
+0.5724	+0.2000	0	0	−19.32	0	0	+1.000
−0.0395	0	−0.0419	−0.0052	−8.502	0	−3.770	+0.4461
−0.0445	−0.0226	−0.0351	−0.0046	−5.350	−4.060	−2.523	+0.5106
−0.0420	−0.0476	−0.0140	−0.0031	+2.433	−4.983	+0.2750	+0.6671
+0.0168	−0.0585	+0.0187	−0.0013	+9.459	−1.527	+2.336	+0.8314
+0.2016	−0.0062	+0.0428	−0.0001	+5.734	+3.392	+1.737	+0.9393
+0.5914	+0.2000	0	0	−24.30	0	0	+1.000
		(j) $R/t = 200$ and $R/L = 0.3$					
−0.2481	0	−0.0585	−0.0023	−8.01	0	−1.251	+0.7583
−0.1187	−0.0759	−0.0236	−0.0916	−0.32	−2.906	−0.242	+0.7953
+0.3098	−0.1126	+0.0402	−0.0003	+9.20	+0.158	+0.787	+0.8828
+1.138	+0.3000	0	0	−21.66	0	0	+1.000
−0.1777	0	−0.0616	−0.0033	−12.73	0	+2.708	+0.6320
−0.1650	−0.0475	−0.0548	−0.0031	−10.23	−3.260	−2.236	+0.6464
−0.0418	−0.1115	−0.0066	−0.0017	+5.57	−4.934	+0.474	+0.7427
+0.3091	−0.0522	+0.0487	−0.0002	+13.92	+1.650	+1.650	+0.8689
+1.058	+0.3000	0	0	−35.32	0	0	+1.000

TABLE 6 (*Contd.*)

	VERTICAL EDGE LOAD				HORIZONTAL EDGE LOAD			
ϕ	N_x (1)	$N_{x\phi}$ (2)	N_ϕ (3)	M_ϕ (4)	N_x (5)	$N_{x\phi}$ (6)	N_ϕ (7)	M_ϕ (8)
$\phi K = 40°$								
40	+4.491	0	−0.695	−0.2230	− 4.689	0	−0.0553	+0.1198
30	+0.927	+1.787	−1.263	−0.2170	−1.599	−1.680	+0.4612	+0.1225
20	−5.904	+0.374	−1.979	−0.1836	+2.992	−1.296	+1.378	+0.1168
10	−4.643	−3.248	−1.057	−0.1038	+3.576	+0.8937	+1.451	+0.0774
0	+22.55	0	+0.643	0	−10.29	0	+0.7660	0
$\phi K = 45°$								
45	+6.463	0	+0.608	−0.1511	−5.443	0	−0.8614	+0.0899
40	+5.541	+1.688	+0.369	−0.1543	−4.617	−1.379	−0.6652	+0.0937
30	−0.843	+3.190	−1.138	−0.1700	−0.203	−2.859	+0.6257	+0.1158
20	−8.160	+0.555	−2.324	−0.1610	+5.220	−1.401	+1.929	+0.1250
10	−5.055	−3.891	−1.245	−0.0955	+4.276	+1.728	+1.762	+0.0858
0	+25.47	0	+0.707	0	−14.77	0	+0.7071	0
$\phi K = 50°$								
50	+6.626	0	+1.510	−0.0742	−5.740	0	−1.603	+0.0441
40	+3.900	+3.127	+0.564	−0.0969	−3.696	−2.769	−0.7707	+0.0676
30	−2.878	+3.516	−1.463	−0.1393	+1.552	−3.453	+1.111	+0.1158
20	−8.823	+0.094	−2.598	+0.1467	+6.637	−1.072	+2.491	+0.1375
10	−4.273	−4.274	−1.262	−0.0891	+4.214	+2.483	+1.987	+0.0953
0	+25.53	0	+0.766	0	−17.75	0	+0.6428	0
			(*k*) $R/t = 200$ and $R/L = 0.4$					
$\phi K = 50°$								
0	+1.124	0	−2.479	−0.2211	−2.625	0	+0.5943	+0.0868
20	−3.463	−0.295	−2.481	−0.1937	−0.066	−0.9430	+0.9092	+0.0810
10	−5.528	−3.468	−1.407	−0.1117	+2.974	+0.0700	+1.194	+0.0553
0	+27.68	0	+0.500	0	−6.899	0	+0.8660	0

| | SHEAR EDGE LOAD | | | | EDGE MOMENT LOAD | | |
N_x (9)	$N_{x\phi}$ (10)	N_ϕ (11)	M_ϕ (12)	N_x (13)	$N_{x\phi}$ (14)	N_ϕ (15)	M_ϕ (16)
−0.0958	0	−0.0550	−0.0039	−16.80	0	−4.723	+0.4470
−0.0958	−0.0529	−0.0397	−0.0032	−7.30	−7.430	−2.380	+0.5253
−0.0298	−0.0936	+0.0040	−0.0016	+12.97	−5.929	+1.932	+0.6997
+0.2786	−0.0416	+0.0506	−0.0001	+16.75	+3.932	+2.673	+0.8621
+1.069	+0.3000	0	0	−48.57	0	0	+1.000
−0.0114	0	−0.0370	−0.0038	−17.92	0	−6.551	+0.2280
−0.0214	−0.0039	−0.0364	−0.0037	−15.90	−4.718	−5.866	+0.2542
−0.0772	−0.0304	−0.0273	−0.0028	−1.34	−9.994	−1.292	+0.4354
−0.0627	−0.0765	+0.0042	−0.0013	+17.97	−5.233	+3.575	+0.6795
+0.2474	−0.0456	+0.0471	−0.0000	+16.41	+6.050	+3.437	+0.8640
+1.134	+0.3000	0	0	−56.01	0	0	+1.000
+0.0544	0	−0.0135	−0.0031	−15.39	0	−7.343	+0.0368
+0.0026	+0.0201	−0.0201	−0.0029	−10.44	−7.529	−5.048	+0.1442
−0.1000	−0.0070	−0.0254	−0.0023	+3.25	−9.817	+0.304	+0.4060
−0.1008	−0.0715	−0.0025	−0.0011	+18.60	−3.576	+4.612	+0.6832
+0.2354	−0.0563	+0.0424	−0.0000	+14.04	+7.094	+3.666	+0.8700
+1.193	+0.3000	0	0	−56.30	0	0	+1.000
			(k) $R/t = 200$ and $R/L = 0.4$				
−0.2847	0	−0.0710	−0.0026	−20.19	0	−3.173	+0.5913
−0.1583	−0.1339	−0.0302	−0.0018	−0.87	−7.343	−0.619	+0.6619
+0.3622	−0.1029	+0.0493	−0.0003	+23.57	+0.435	+1.999	+0.8155
+1.634	+0.4000	0	0	−56.52	0	0	+1.000

TABLE 6 (*Contd.*)

ϕ	VERTICAL EDGE LOAD				HORIZONTAL EDGE LOAD			
	N_x	$N_{x\phi}$	N_ϕ	M_ϕ	N_x	$N_{x\phi}$	N_ϕ	M_ϕ
	(1)	(2)	(3)	(4)	(5)	(6)	(7)	(8)
$\phi_K = 35°$								
35	+6.873	0	−0.795	−0.1830	−4.932	0	−0.0417	+0.0837
30	+5.012	−1.712	−1.039	−0.1820	−3.951	−1.261	+0.1374	+0.0848
20	−6.203	+1.614	−2.206	−0.1647	+2.208	−1.889	+1.153	+0.0862
10	−9.087	−3.682	−1.633	−0.1001	+5.303	+0.6597	+1.536	+0.0621
0	+34.19	0	+0.574	0	−1336	0	+0.8192	0
$\phi_K = 40°$								
40	+11.34	0	+0.768	−0.1216	−7.637	0	−0.8602	+0.0620
30	+3.392	+4.247	−0.465	−0.1293	−2.992	−3.037	+0.0122	+0.0723
20	−10.20	+2.361	−2.620	−0.1409	+5.333	−2.419	+1.731	+0.0915
10	−10.34	−4.580	−1.955	−0.0931	+6.841	+1.629	+1.957	+0.0704
0	+39.09	0	+0.643	0	−19.78	0	+0.7660	0
$\phi_K = 45°$								
45	+9.787	0	+1.934	−0.0436	−7.229	0	−1.692	+0.0199
40	+8.573	+2.572	+1.567	−0.0513	−6.456	−1.911	−1.419	+0.0264
30	−0.227	+5.188	−0.825	−0.0982	−0.709	−4.101	+0.4128	+0.0678
20	−11.70	+1.769	−3.045	−0.1296	+7.382	−2.215	+2.378	+0.1027
10	−9.060	−5.153	−2.000	−0.0887	+7.017	+2.520	+2.255	+0.0800
0	+38.68	0	+0.707	0	−2.1.69	0	+0.7071	0
$\phi_K = 50°$								
50	+7.414	0	+2.207	+0.0042	−6.022	0	−2.055	−0.0134
40	+4.947	+3.616	+1.124	−0.0267	−4.389	−3.008	−1.161	+0.0151
30	−2.266	+4.540	−1.356	−0.0910	+0.8355	−4.146	+0.9927	+0.0776
20	−10.69	+0.849	−3.073	−0.1245	+7.857	−1.711	+2.811	+0.1179
10	−7.142	−5.081	−1.774	−0.0835	+6.637	+3.045	+2.378	+0.0892
0	+35.00	0	+0.766	0	−25.44	0	+0.6428	0

SHEAR EDGE LOAD				EDGE MOMENT LOAD			
N_x (9)	$N_{x\phi}$ (10)	N_ϕ (11)	M_ϕ (12)	N_x (13)	$N_{x\phi}$ (14)	N_ϕ (15)	M_ϕ (16)
−0.1462	0	−0.0635	−0.0032	−27.50	0	−5.935	+0.3911
−0.1482	−0.0403	−0.0577	−0.0030	−22.24	−7.055	−4.913	+0.4179
−0.0986	−0.1152	−0.0124	−0.0017	+11.73	−10.85	+1.001	+0.5937
+0.3204	−0.0822	+0.0539	−0.0001	+31.27	+3.583	+3.651	+0.8049
+1.639	+0.4000	0	0	−80.78	0	0	+1.000
+0.0319	0	−0.0448	−0.0032	−33.54	0	−8.301	+0.2027
−0.0781	−0.0171	−0.0365	−0.0026	−13.32	−12.80	−4.292	+0.2888
−0.1367	−0.0848	−0.0086	−0.0014	+21.73	−10.76	+3.315	+0.5642
+0.2620	−0.0845	+0.0492	−0.0000	+31.74	+6.881	+4.858	+0.8083
+1.744	+0.4000	0	0	−92.98	0	0	+1.000
+0.0957	0	−0.0099	−0.0065	−22.45	0	−8.938	−0.0169
+0.0691	+0.0238	−0.0133	−0.0043	−20.42	−5.970	−8.074	+0.0165
−0.0987	+0.0199	−0.0292	−0.0020	−4.08	−13.43	−2.116	+0.2506
−0.1953	−0.0728	−0.0161	−0.0012	+23.35	−8.260	+4.819	+0.5714
+0.2396	−0.0981	+0.0429	−0.0000	+27.40	+8.293	+5.115	+0.8168
+1.826	+0.4000	0	0	−91.16	0	0	+1.000
+0.1395	0	+0.0147	−0.0013	−13.14	0	−7.852	−0.1157
+0.0505	+0.0598	−0.0042	−0.0015	−11.30	−6.932	−5.801	−0.0061
−0.1446	+0.0347	−0.0352	−0.0017	−1.18	−10.95	−0.390	+0.2725
−0.2195	−0.0793	−0.0240	−0.0011	+19.84	−6.154	+5.079	+0.5892
+0.2513	−0.1076	+0.0402	+0.0001	+23.56	+7.944	+4.801	+0.8219
+1.810	+0.4000	0	0	−84.42	0	0	+1.000

TABLE 6 (*Contd.*)

	VERTICAL EDGE LOAD				HORIZONTAL EDGE LOAD			
ϕ	N_x	$N_{x\phi}$	N_ϕ	M_ϕ	N_x	$N_{x\phi}$	N_ϕ	M_ϕ
	(1)	(2)	(3)	(4)	(5)	(6)	(7)	(8)

(*l*) $R/t = 200$ and $R/L = 0.5$

$\phi K = 30°$

30	+6.908	0	−1.578	−0.1695	−4.932	0	+0.2325	+0.0656
20	−3.091	+1.791	−2.323	−0.1564	−0.320	−2.780	+0.8452	+0.0654
10	−12.18	−3.626	−1.972	−0.0957	+5.663	+0.128	+1.422	+0.0492
0	+43.35	0	+0.500	0	−13.29	0	+0.8660	0

$\phi_K = 35°$

35	+12.92	0	+0.520	−0.1106	−7.760	0	−0.6687	+0.0486
30	+10.14	+3.193	+0.037	−0.1151	−6.371	−1.952	−0.3803	+0.0518
20	−8.763	+4.022	−2.425	−0.1266	+3.367	−3.036	−1.253	+0.0670
10	−16.09	−4.484	−2.447	−0.0897	+8.692	+1.030	+1.926	+0.0564
0	+52.21	0	+0.574	0	−22.19	0	−0.8192	0

$\phi_K = 40°$

40	+13.93	0	+1.963	−0.0443	−8.898	0	−1.529	+0.0185
30	+5.361	+5.888	+0.146	−0.0720	−4.097	−3.897	−0.3259	+0.0398
20	−12.79	+3.938	−3.042	−0.1129	+6.622	−3.303	+1.951	+0.0737
10	−15.31	−5.476	−2.659	−0.0866	+9.721	+2.099	+2.350	+0.0645
0	+53.53	0	+0.643	0	+28.17	0	+0.7660	0

$\phi_K = 45°$

45	+10.22	0	+2.435	+0.0045	−7.196	0	−1.959	−0.0106
40	+9.385	+2.771	+2.035	−0.0052	−6.757	−1.967	−1.675	−0.0034
30	+0.957	+5.868	−0.565	−0.0616	−1.640	−4.438	+0.2286	+0.0431
20	−12.61	+2.686	−3.298	−0.1102	+7.723	−2.850	+2.505	+0.0881
10	−12.52	−5.653	−2.457	−0.0832	+9.498	+2.791	+2.554	+0.0750
0	+48.96	0	+0.707	0	−30.94	0	+0.7071	0

SHEAR EDGE LOAD				EDGE MOMENT LOAD			
N_x	$N_{x\phi}$	N_ϕ	M_ϕ	N_x	$N_{x\phi}$	N_ϕ	M_ϕ
(9)	(10)	(11)	(12)	(13)	(14)	(15)	(16)

(l) $R/t = 200$ and $R/L = 0.5$

–1.2715	0	–0.0759	–0.0026	–36.86	0	–5.858	+0.4164
–0.1974	–0.1375	–0.0355	–0.0018	– 2.00	–13.49	–1.174	+0.5258
+0.3600	–0.1314	+0.05366	–0.0002	+43.95	+0.71	+3.711	+0.7505
+2.253	+0.5000	0	0	–107.5	0	0	+1.000
+0.0566	0	–0.0528	–0.0027	–41.17	0	–9.181	+0.1851
–0.0847	–0.0181	–0.0505	–0.0025	–33.79	–10.62	–7.611	+0.2222
–0.1818	–0.0966	–0.0204	–0.0015	+16.53	–16.90	+1.429	+0.4668
+0.2700	–0.1200	+0.0516	–0.0001	+50.25	+5.26	+5.747	+0.7504
+2.374	+0.5000	0	0	–132.8	0	0	+1.000
+0.1093	0	–0.0175	–0.0020	–32.96	0	–10.30	–0.0072
–0.0475	+0.0298	–0.0294	–0.0019	–17.70	–15.29	–5.586	+0.1389
–0.2613	–0.0652	–0.0242	–0.0012	+24.15	–14.33	+3.905	+0.4653
+0.2136	–0.1319	+0.0433	–0.0000	+44.95	+8.13	+6.422	+0.7611
+2.509	+0.5000	0	0	–133.2	0	0	+1.000
+0.1830	0	+0.0132	–0.0011	–18.31	0	–8.981	–0.1089
+0.1446	+0.0466	+0.0066	–0.0012	–17.78	–4.98	–8.263	–0.0771
–0.1100	+0.0637	–0.0303	–0.0014	–8.90	–13.02	–2.879	+0.1523
–0.3155	–0.0676	–0.0341	–0.0011	+20.73	–10.69	+4.696	+0.1876
–0.2211	–0.1467	+0.0389	–0.0001	+38.31	+8.33	+6.068	+0.7681
+2.539	+0.5000	0	0	–123.1	0	0	+1.000

TABLE 6 (*Contd.*)

	VERTICAL EDGE LOAD				HORIZONTAL EDGE LOAD			
ϕ	N_x	$N_{x\phi}$	N_ϕ	M_ϕ	N_x	$N_{x\phi}$	N_ϕ	M_ϕ
	(1)	(2)	(3)	(4)	(5)	(6)	(7)	(8)
$\phi_K = 50°$								
50	+5.720	0	+2.151	+0.0279	−4.333	0	−1.931	−0.0295
40	+5.105	+3.070	+1.265	−0.0015	−4.431	−2.438	−1.238	−0.0037
30	−0.025	+4.652	−1.014	−00656	−1.252	−4.161	+0.6846	+0.0569
20	−10.40	+1.897	−3.112	−0.1093	+7.442	−2.639	+2.813	+0.1043
10	−12.65	−5.039	−2.643	−0.0852	+11.50	+3.115	+3.155	+0.0913
0	+42.10	0	+0.766	0	−31.89	0	+0.6428	0
(*m*) $R/t = 200$ and $R/L = 0.6$								
$\phi_K = 30°$								
30	+13.41	0	−0.550	−0.1215	−7.494	0	−0.177	+0.0459
20	−2.95	+4.174	−2.106	−0.1221	−0.317	−2.725	+2.756	+0.0512
10	−19.88	−3.751	−2.624	−0.0867	+8.756	+0.171	+1.682	+0.0443
0	+61.86	0	+0.500	0	−20.84	0	+0.866	0
$\phi_K = 35°$								
35	+17.48	0	+1.581	−0.0576	−9.794	0	−1.155	+0.0231
30	+13.76	+4.442	+0.942	−0.0648	−8.004	−2.520	−0.794	+0.0279
20	−10.37	+6.000	−2.570	−0.0989	+4.010	−3.973	+1.313	+0.0531
10	−22.33	−5.090	−3.136	−0.0827	+11.70	+1.270	+2.252	+0.0523
0	+68.85	0	+0.574	0	−30.32	0	+0.819	0
$\phi_K = 40°$								
40	+14.46	0	+2.530	−0.0052	−8.900	0	−1.788	−0.0029
30	+6.70	+6.504	+0.519	−0.0397	−4.840	−4.143	−0.527	+0.0213
20	−13.24	+4.992	−3.187	−0.0946	+6.608	−3.889	+2.003	+0.0630
10	−19.32	−5.810	−3.114	−0.0812	+12.10	+2.231	+2.601	+0.0616
0	+65.23	0	+0.643	0	−35.15	0	+0.766	0

SHEAR EDGE LOAD				EDGE MOMENT LOAD			
N_x (9)	$N_{x\phi}$ (10)	N_ϕ (11)	M_ϕ (12)	$N_{x\phi}$ (13)	N_x (14)	N_ϕ (15)	M_ϕ (16)
+0.1739	0	+0.0295	+0.0002	−4.66	0	−6.575	−0.1404
+0.0786	+0.0777	+0.0054	−0.0007	−9.77	−3.60	−5.629	−0.0518
−0.1578	+0.0593	−0.0387	−0.0013	−10.31	−9.96	−1.682	+0.1945
−0.2949	−0.0794	−0.0364	−0.0011	+16.18	−9.65	+4.675	+0.5157
+0.2141	−0.1221	+0.0366	−0.0000	+41.78	+9.30	+7.575	+0.8127
+2.441	+0.5000	0	0	−116.1	0	0	+1.000
(*m*) $R/t = 200$ and $R/L = 0.6$							
−0.2086	0	−0.0732	−0.0024	−53.95	0	−8.721	+0.2589
−0.2355	−0.1233	−0.0394	−0.0016	−4.01	−20.04	−1.828	+0.4042
+0.3019	−0.1613	+0.0532	−0.0001	+66.27	+0.71	+5.559	+0.6931
+3.003	+0.6000	0	0	−166.4	0	0	+1.000
+0.0614	0	−0.0363	−0.0020	−48.68	0	−11.31	+0.0433
+0.0019	+0.0113	−0.0381	−0.0020	−40.77	−12.62	−9.488	+0.0867
−0.2759	−0.0663	−0.0294	−0.0012	+17.11	−21.22	+1.522	+0.3710
+0.1839	−0.1613	+0.0454	−0.0000	+66.22	+5.83	+7.338	+0.7029
+3.204	+0.6000	0	0	−180.9	0	0	+1.000
+0.2096	0	+0.0030	−0.0012	−29.22	0	−10.53	−0.0850
−0.0157	+0.0719	−0.0224	−0.0013	−20.21	−14.50	−6.191	+0.0578
−0.3733	−0.0458	−0.0384	−0.0010	+19.37	−16.37	+3.594	+0.3909
+0.1569	−0.1797	+0.0373	−0.0001	+56.09	+7.56	+7.285	+0.7139
+3.297	+0.6000	0	0	−169.2	0	0	+1.000

TABLE 6 (*Contd.*)

	VERTICAL EDGE LOAD				HORIZONTAL EDGE LOAD			
ϕ	N_x	$N_{x\phi}$	N_ϕ	M_ϕ	N_x	$N_{x\phi}$	N_ϕ	M_ϕ
	(1)	(2)	(3)	(4)	(5)	(6)	(7)	(8)
$\phi K = 45°$								
45	+8.60	0	+2.397	+0.0217	−5.729	0	−1.858	−0.0205
40	+8.15	+2.319	+2.068	+0.0133	−5.622	−1.562	−1.636	−0.0140
30	+2.68	+5.666	−0.311	−0.0418	−2.980	−4.165	+0.041	+0.0293
20	−11.68	+3.500	−3.237	−0.0948	+6.764	−3.438	+2.416	+0.0759
10	−16.02	−5.641	−2.752	−0.0772	+12.12	+2.709	+2.752	+0.0695
0	+58.53	0	+0.707	0	−38.04	0	+0.707	0
$\phi_K = 50°$								
50	+3.09	0	+1.709	+0.0284	−1.816	0	−1.467	−0.0273
40	+4.42	+1.981	+1.179	+0.0063	−3.693	−1.374	−1.120	−0.0084
30	+2.57	+4.287	−0.638	−0.0476	−3.650	−3.722	+0.325	+0.0406
20	−9.53	+2.775	−2.920	−0.0908	+6.372	−3.425	+2.581	+0.0862
10	−14.34	−5.142	−2.383	−0.0706	+13.11	+2.955	+2.877	+0.0757
0	+53.46	0	+0.766	0	−41.80	0	+0.643	0

SHEAR EDGE LOAD				EDGE MOMENT LOAD			
N_x (9)	$N_{x\phi}$ (10)	N_ϕ (11)	M_ϕ (12)	N_x (13)	$N_{x\phi}$ (14)	N_ϕ (15)	M_ϕ (16)
+0.2347	0	+0.0286	−0.0004	−8.38	0	−7.554	−0.1241
+0.1924	+0.0605	+0.0199	−0.0005	−10.38	−2.49	−7.198	−0.0988
−0.1065	+0.0944	−0.0306	−0.0011	−15.36	−10.11	−3.686	+0.0939
−0.4030	−0.0596	−0.0476	−0.0011	+12.42	−13.03	+3.667	+0.4113
−0.1906	−0.1895	+0.0361	−0.0001	+50.54	+6.80	+6.668	+0.7160
+3.268	+0.6000	0	0	−158.3	0	0	+1.000
+0.1854	0	+0.0382	+0.0002	+7.18	0	−3.972	−0.1088
+0.1051	+0.0872	+0.0118	−0.0003	−5.77	+1.40	−4.634	−0.0572
−0.1481	+0.0826	−0.0419	−0.0012	−20.51	−6.83	−3.347	+0.1206
−0.3819	−0.0757	−0.0497	−0.0012	+7.78	−1.2.83	+3.041	+0.4149
+0.2199	−0.1883	+0.0384	−0.0001	+51.00	+5.85	+6.371	+0.7134
+3.218	+0.6000	0	0	−156.6	0	0	+1.000

TABLE 7 Symmetrical Edge Loads on Simply Supported Cylindrical Shell
(Displacement of Edge at $\phi = 0$)
(Deformation coefficients)

VERTICAL EDGE LOAD	HORIZONTAL EDGE LOAD
Vertical Displacement w (Downward Positive)	
$V_L \dfrac{L^4}{R^3 tE}(\text{Col.}(1))\sin\dfrac{\pi x}{L}$	$H_L \dfrac{L^4}{R^3 tE}(\text{Col.}(4))\sin\dfrac{\pi x}{L}$
Horizontal Displacement v (Inward Positive)	
$V_L \dfrac{L^4}{R^3 tE}(\text{Col.}(2))\sin\dfrac{\pi x}{L}$	$H_L \dfrac{L^4}{R^3 tE}(\text{Col.}(5))\sin\dfrac{\pi x}{L}$
Rotation θ	
$V_L \dfrac{R^2}{EI}(\text{Col.}(3))\sin\dfrac{\pi x}{L}$	$H_L \dfrac{R^2}{EI}(\text{Col.}(6))\sin\dfrac{\pi x}{L}$

ϕ_K	w (1)	v (2)	θ (3)	w (4)	v (5)	θ (6)
(a) R/t = 100						
						r/L = 0.1:
30	12.53	0.1445	0.1108	−0.1445	−0.0585	−0.04556
35	6.265	0.2508	0.1420	−0.2508	−0.1231	−0.07108
40	3.611	0.4009	0.1752	−0.4009	−0.2325	−0.1039
45	2.591	0.5987	0.2088	−0.5981	−0.4040	−0.1444
50	2.043	0.8434	0.2410	−0.8434	−0.6558	−0.1925
						R/L = 0.2:
30	17.38	2.031	0.0982	−2.031	−0.8742	−0.04288
35	12.87	3.538	0.1259	−3.538	−1.794	−0.06530
40	12.51	5.544	0.1524	−5.544	−3.282	−0.09256
45	13.47	7.956	0.1748	−7.956	−5.447	−0.1233
50	14.54	10.50	0.1899	−10.50	−8.270	−0.1546
						R/L = 0.3:
30	35.24	9.474	0.0891	−9.474	−4.010	−0.03866
35	37.31	15.65	0.1092	−15.65	−7.890	−0.05606
40	42.38	22.99	0.1245	−22.99	−13.59	−0.07595
45	46.06	30.12	0.1317	−30.12	−20.69	−0.09384
50	45.7	3511	0.1285	−35.11	−27.93	−0.1070

TABLE 7 Symmetrical Edge Loads on Simply Supported Cylindrical Shell
(Displacement of edge at $\phi = 0$)
(Deformation coefficients)

SHEAR EDGE LOAD	MOMENT EDGE LOAD
Vertical Displacement w (Downward Positive)	
$S_L \dfrac{L^4}{R^3 t E}\big(\text{Col.(7)}\big)\sin\dfrac{\pi x}{L}$	$M_L \dfrac{L^4}{R^3 t E}\big(\text{Col.(10)}\big)\sin\dfrac{\pi x}{L}$
Horizontal Displacement v (Inward Positive)	
$S_L \dfrac{L^4}{R^3 t E}\big(\text{Col.(8)}\big)\sin\dfrac{\pi x}{L}$	$M_L \dfrac{L^4}{R^3 t E}\big(\text{Col.(11)}\big)\sin\dfrac{\pi x}{L}$
Rotation θ	Rotation θ
$S_L \dfrac{R^2}{EI}\big(\text{Col.(9)}\big)\sin\dfrac{\pi x}{L}$	$M_L \dfrac{R}{EI}\big(\text{Col.(12)}\big)\sin\dfrac{\pi x}{L}$

w (7)	v (8)	θ (9)	w (10)	v (11)	θ (12)
			(a) $R/t = 100$		
0.3401	0.000480	0.000363	−1.330	−0.5469	−0.5144
0.2178	0.001006	0.000544	−1.704	−0.8531	−0.5964
0.1443	0.002009	0.000822	−2.102	−1.247	−0.6749
0.1084	0.003771	0.001223	−2.505	−1.734	−0.7542
0.0792	0.006516	0.001776	−2.892	−2.312	−0.8290
0.6050	0.00726	0.000353	−18.86	−8.233	−0.4893
0.4733	0.02065	0.000678	−24.18	−12.54	−0.5564
0.13715	0.04590	0.001145	−29.25	−16.77	−0.6159
0.3402	0.08862	0.001757	−33.56	−23.67	−0.6651
0.3610	0.1526	0.002478	−36.46	−29.70	−0.6986
1.187	0.06357	0.000536	−86.64	−37.58	−0.4499
0.9583	0.1474	0.000923	−106.2	−54.50	−0.4965
0.9561	0.2931	0.001413	−121.0	−73.83	−0.5288
1.073	0.5020	0.001927	−127.9	−91.21	−0.5434
1.209	0.7422	0.002334	−124.9	−104.0	−0.5402

TABLE 7 (*Contd.*)

	VERTICAL EDGE LOAD			HORIZONTAL EDGE LOAD		
ϕK	w (1)	v (2)	θ (3)	w (4)	v (5)	θ (6)
						$R/L = 0.4$:
30	75.06	26.10	0.07760	−26.10	−11.06	−0 03483
35	87.57	40.75	0.09046	−40.75	−20.25	−0 04716
40	96.68	55.00	0.09596	−55.00	−32.68	−0.05921
45	96.10	64.69	0.09256	−64.69	−44.95	−0.06746
50	85.15	67.32	0.08360	−67.32	−54.74	−0.07216
						$R/L = 0.5$:
30	142.2	54.14	0.06634	−54.14	−22.98	−0.02900
35	163.2	78.77	0.07239	−78.77	−39.97	−0.03811
40	167.7	97.43	0.07124	−97.43	−58.47	−0.04484
45	152.4	104.9	0.06505	−104.9	−74.31	−0.04883
50	127.5	103.9	0.05747	−103.9	−87.21	−0.05166
						$R/L = 0.6$:
30	238.5	88.47	0.05559	−88.47	−39.78	−0.02434
35	257.7	126.7	0 05714	−126.7	−64.74	−0.03049
40	245.4	1449	0.05366	_144.9	−88.10	−0.03396
45	212.3	149.5	0.04873	−149.5	−108.3	−0.03700
50	175.8	1437	0.04304	−147.9	−127.9	−0.03967
			(*b*) $R/t = 200$			
						$R/L = 0.1$:
30	13.59	0.5598	0.10563	−0.5598	−0.3112	−0.04555
35	7.732	0.9838	0.13813	−0.9838	−0.4916	−0.07100
40	5.711	1.576	0.1719	−1.576	−0 9269	−0.1036
45	5186	2.354	0.2050	−2.354	−1.462	−0.1434
50	5.265	3.292	0.2352	−3.292	−2.576	−0.1896

SHEAR EDGE LOAD			MOMENT EDGE LOAD		
w	v	θ	w	v	θ
(7)	(8)	(9)	(10)	(11)	(12)
1.940	0.2256	0.000632	−239.2	−104.0	−0.4040
1.924	0.5080	0.000994	−277.7	−145.2	−0.4320
2.189	0.9067	0.001384	−293.8	−181.4	−0.4417
2.500	1.372	0.001641	−284.1	−207.1	−0.4348
2.591	1.731	0.001732	−255.7	−220.8	−0.4228
3.218	0.5995	0.000644	−497.6	−216.6	−0.3602
3.588	1.202	0.000945	−542.9	−285.9	−0.3706
4.141	1.938	0.001160	−535.2	−335.5	−0.3668
4.418	2.566	0.001242	−487.9	−366.2	−0.3568
4.246	2.945	0.001199	−431.0	−387.5	−0.3493
5.112	1.218	0.000620	−864.6	−378.6	−0.3247
5.976	2.242	0.000834	−888.7	−474.2	−0.3185
6.629	3.250	0.000932	−834.6	−528.2	−0.3171
6.601	3.942	0.000918	−743.9	−575.4	−0.3042
6.678	4.324	0.000865	−669.4	−618.7	−0.3086
(b) $R/t = 200$					
0.3446	0.001423	0.000252	−5.070	−2.186	−0.5143
0.2228	0.003352	0.000438	−6.630	−3.408	−0.5958
0.1579	0.007139	0.000718	−8.250	−4.974	−0.6749
0.1237	0.01391	0.001117	−9.842	−6.883	−0.7503
0.1083	0.02506	0.001649	−11.29	−9.097	−0.8195

TABLE 7 *(Contd.)*

φK	VERTICAL EDGE LOAD			HORIZONTAL EDGE LOAD		
	w	v	θ	w	v	θ
	(1)	(2)	(3)	(4)	(5)	(6)
						R/L = 0.2:
30	32.02	8.267	0.0986	−8.267	−3.492	−0.04257
35	33.96	14.07	0.1245	−14.07	−7.090	−0.06423
40	39.51	21.38	0.1468	−21.38	−12.62	−0.08898
45	44.27	28.96	0.1600	−28.96	−19.84	−0.1132
50	45.16	34.63	0.1600	−34.63	−27.44	−0.1321
						R/L = 0.3:
30	110.9	37.12	0.08756	−37.12	−15.70	−0.03796
35	123.0	58.70	0.10342	−58.70	−29.63	−0.05150
40	136.4	78.62	0.10899	−78.62	−46.73	−0.06707
45	131.1	89.18	0.10274	−89.18	−62.16	−0.07481
50	111.5	89.27	0.09013	−89.27	−73.18	−0.07807
						R/L = 0.4:
30	247.6	98.82	0.07423	−98.82	−41.89	−0.03237
35	283.3	139.7	0.07945	−139.7	−70.98	−0.04180
40	273.8	161.7	0.07471	−161.7	− 97.64	−0.04717
45	232.0	163.1	0.06551	−163.1	−117.2	−0.04975
50	187.9	157.7	0.05768	−157.7	−135.5	−0.05270
						R/L = 0.5:
30	473.4	194 4	0.06058	−194.4	− 82.77	−0.02666
35	486.4	244.3	0.05905	−244.3	−125.5	−0.03167
40	424.2	256.6	0.05240	−256.6	−158.6	−0.03414
45	349.1	254.2	0.04653	−254.2	−189.1	−0.03671
50	295.1	260.4	0.04398	−260.4	−234.5	−0.04173

SHEAR EDGE LOAD			MOMENT EDGE LOAD		
w	v	θ	w	v	θ
(7)	(8)	(9)	(10)	(11)	(12)
0.7716	0.03561	0.000385	−75.70	−32.70	−0.4866
0.6124	0.08739	0.000700	−95.68	−49.33	−0.5497
0.6071	0.1815	0.001123	−112.7	−68.34	−0.5990
0.6944	0.3237	0.001601	−122.9	−86.91	−0.6273
0.8010	0.4950	0.002014	−122.8	−101.0	−0.6299
1.641	0.2500	0.000532	−340.4	−147.6	−0.4443
1.782	0.5520	0.000868	−402.2	−200.2	−0.4792
2.154	0.9824	0.001193	−423.7	−260.8	−0.4884
2.432	1.410	0.001376	−399.4	−290.9	−0.4749
2.437	1.695	0.001382	−350.4	−303.5	−0.4550
3.525	0.9471	0.000586	−912.9	−397.8	−0.3936
4.354	1.701	0.000837	−976.3	−513.6	−0.4011
4.977	2.518	0.000964	−918.0	−579.7	−0.3884
4.925	3.016	0.000945	−805.0	−611.3	−0.3732
4.456	3.239	0.000875	−708.8	−647.5	−0.3670
6.899	2.116	0.000570	−1817	−800.0	−0.3421
8.309	3.527	0.000704	−1772	−950.2	−0.3343
8.519	4.483	0.000707	−1572	−1024	−0.3210
7.792	4.924	0.000653	−1396	−1101	−0.3156
6.701	5.075	0.000616	−1319	−1252	−0.3226

TABLE 7 (*Contd.*)

ϕK	VERTICAL EDGE LOAD			HORIZONTAL EDGE LOAD		
	w	v	θ	w	v	θ
	(1)	(2)	(3)	(4)	(5)	(6)
						$R/L = 0.6$:
30	756.0	315.3	0.04838	−315.3	−135.1	−0.02155
35	706.1	360.6	0.04430	−360.0	−188.1	−0.02429
40	591.7	367.3	0 03930	−367.3	−232.8	−0.02637
45	498.8	374.8	0.03615	−374.8	−287.2	−002931
50	4319	389.3	0.03392	−389.3	−3S7.3	−0.03284

SHEAR EDGE LOAD			MOMENT EDGE LOAD		
w	v	θ	w	v	θ
(7)	(8)	(9)	(10)	(11)	(12)
11.82	3.979	0.000510	−3009	−1341	−0.2961
13.15	5.790	0.000555	−2756	−1511	−0.2847
12.48	6.714	0.000519	−2445	−1640	−0.2778
11.18	7.265	0.000486	−2249	−1823	−0.2782
10.21	7.983	0.000481	−2110	−2043	−0.2807

TABLE 8 Edge Loads on Simply Supported Cylindrical Shell

TANGENTIAL EDGE LOAD	RADIAL EDGE LOAD
Longitudinal Force N_x	Longitudinal Force N_x
$T_L\left(\text{Col. (1)}\right)\sin\dfrac{\pi x}{L}$	$R_L\dfrac{L}{t}\left(\text{Col. (5)}\right)\sin\dfrac{\pi x}{L}$
Shearing Force $N_{x\phi}$	Shearing Force $N_{x\phi}$
$T_L\left(\text{Col. (2)}\right)\cos\dfrac{\pi x}{L}$	$R_L\dfrac{L}{t}\left(\text{Col. (6)}\right)\cos\dfrac{\pi x}{L}$
Transverse Force N_ϕ	Transverse Force N_ϕ
$T_L\left(\text{Col. (3)}\right)\sin\dfrac{\pi x}{L}$	$R_L\dfrac{L}{t}\left(\text{Col. (7)}\right)\sin\dfrac{\pi x}{L}$
Transverse Moment M_ϕ	Transverse Moment M_ϕ
$T_L + t\left(\text{Col. (4)}\right)\sin\dfrac{\pi x}{L}$	$R_L L\left(\text{Col. (8)}\right)\sin\dfrac{\pi x}{L}$

$$s = R\phi$$

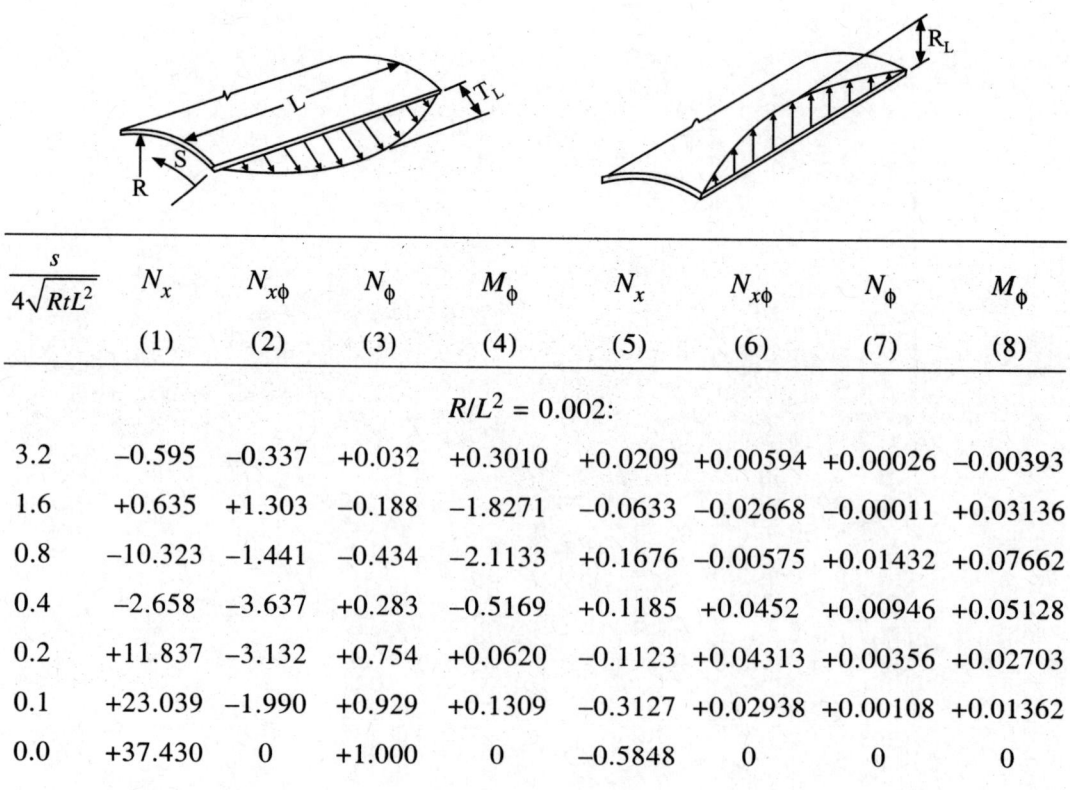

$\dfrac{s}{4\sqrt{RtL^2}}$	N_x	$N_{x\phi}$	N_ϕ	M_ϕ	N_x	$N_{x\phi}$	N_ϕ	M_ϕ
	(1)	(2)	(3)	(4)	(5)	(6)	(7)	(8)
$R/L^2 = 0.002$:								
3.2	−0.595	−0.337	+0.032	+0.3010	+0.0209	+0.00594	+0.00026	−0.00393
1.6	+0.635	+1.303	−0.188	−1.8271	−0.0633	−0.02668	−0.00011	+0.03136
0.8	−10.323	−1.441	−0.434	−2.1133	+0.1676	−0.00575	+0.01432	+0.07662
0.4	−2.658	−3.637	+0.283	−0.5169	+0.1185	+0.0452	+0.00946	+0.05128
0.2	+11.837	−3.132	+0.754	+0.0620	−0.1123	+0.04313	+0.00356	+0.02703
0.1	+23.039	−1.990	+0.929	+0.1309	−0.3127	+0.02938	+0.00108	+0.01362
0.0	+37.430	0	+1.000	0	−0.5848	0	0	0

TABLE 8 Edge Loads on Simply Supported Cylindrical Shell

SHEAR EDGE LOAD	MOMENT EDGE LOAD
Longitudinal Force N_x	Longitudinal Force N_x
$S_L\left(\text{Col. (9)}\right)\sin\dfrac{\pi x}{L}$	$M_L\dfrac{L}{t}\left(\text{Col. (13)}\right)\sin\dfrac{\pi x}{L}$
Shearing Force $N_{x\phi}$	Shearing Force $N_{x\phi}$
$S_L\left(\text{Col. (10)}\right)\cos\dfrac{\pi x}{L}$	$M_L\dfrac{L}{t}\left(\text{Col. (14)}\right)\cos\dfrac{\pi x}{L}$
Transverse Force N_ϕ	Transverse Force N_ϕ
$S_L\left(\text{Col. (11)}\right)\sin\dfrac{\pi x}{L}$	$M_L\dfrac{L}{t}\left(\text{Col. (15)}\right)\sin\dfrac{\pi x}{L}$
Transverse Moment M_ϕ	Transverse Moment M_ϕ
$S_L t\left(\text{Col. (12)}\right)\sin\dfrac{\pi x}{L}$	$M_L\left(\text{Col. (16)}\right)\sin\dfrac{\pi x}{L}$

$$s = R_\phi$$

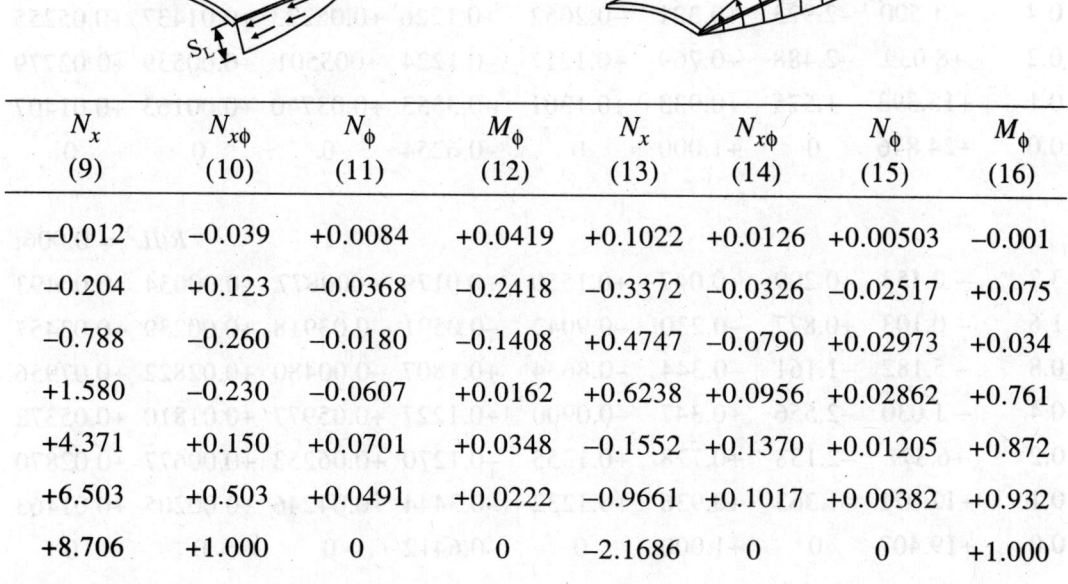

N_x (9)	$N_{x\phi}$ (10)	N_ϕ (11)	M_ϕ (12)	N_x (13)	$N_{x\phi}$ (14)	N_ϕ (15)	M_ϕ (16)
−0.012	−0.039	+0.0084	+0.0419	+0.1022	+0.0126	+0.00503	−0.001
−0.204	+0.123	−0.0368	−0.2418	−0.3372	−0.0326	−0.02517	+0.075
−0.788	−0.260	−0.0180	−0.1408	+0.4747	−0.0790	+0.02973	+0.034
+1.580	−0.230	−0.0607	+0.0162	+0.6238	+0.0956	+0.02862	+0.761
+4.371	+0.150	+0.0701	+0.0348	−0.1552	+0.1370	+0.01205	+0.872
+6.503	+0.503	+0.0491	+0.0222	−0.9661	+0.1017	+0.00382	+0.932
+8.706	+1.000	0	0	−2.1686	0	0	+1.000

TABLE 8 (*Contd.*)

$\dfrac{s}{4\sqrt{RtL^2}}$	TANGENTIAL EDGE LOAD				RADIAL EDGE LOAD			
	N_x	$N_{x\phi}$	N_ϕ	M_ϕ	N_x	$N_{x\phi}$	N_ϕ	M_ϕ
	(1)	(2)	(3)	(4)	(5)	(6)	(7)	(8)
								$Rt/L^2 = 0.003$:
3.2	−0.393	−0.288	+0.038	+0.2387	+0.0203	+0.00695	+0.00015	−0.00432
1.6	+0.316	+1.115	−0.204	−1.4220	−0.0629	−0.03104	+0.00041	+0.03275
0.8	−8.063	−1.331	−0.405	−1.5481	+0.1741	−0.00573	+0.01853	+0.07812
0.4	−1.920	−3.205	+0.305	−0.3150	+0.1214	+0.04716	+0.01212	+0.05225
0.2	+9.454	−2.742	+0.762	+0.1030	−0.1185	+0.04990	+0.00455	+0.02758
0.1	+18.230	−1.729	+0.931	+0.1323	−0.3269	+0.03395	+0.00138	+0.01394
0.0	+29.509	0	+1.000	0	−0.6104	0	0	0
								$Rt/L^2 = 0.004$:
3.2	− 0.278	−0.254	+0.042	+0.2008	+0.0196	+0.00771	+0.00000	−0.00459
1.6	+0.142	+0.991	−0.215	−1.1832	−0.16I8	−0.03434	+0.00103	+0.03355
0.8	− 6.733	−1.259	−0.381	−1.2254	+0 1776	−0.00550	+0.02214	+0.07856
0.4	− 1.500	−2.923	+0.321	−0.2052	+0.1226	+0.05223	+0.01437	+0.05255
0.2	+8.039	−2.488	+0.769	+0.1217	−0.1224	+005501	+0.00539	+0 02779
0.1	+15.393	−1.575	+0.933	+0.1301	−0.3353	+0.03740	+0.00163	+0.01407
0.0	+24.846	0	+1.000	0	−0.6254	0	0	0
								$Rt/L^2 = 0.006$:
3.2	− 0.153	−0.209	+0.047	+0.1550	+0.0179	+000877	−0.00034	−0.00493
1.6	− 0.103	+0.827	−0.230	−0.9042	−0.0591	−0.03918	+0.00239	+0.03457
0.8	− 5.182	−1.161	−0.344	−0.8634	+0.1807	−0.00480	+0.02822	+0.07956
0.4	− 1.030	−2.556	+0.347	−0,0900	+0.1227	+0.05977	+0.01810	+0.05372
0.2	+6.371	−2.158	+0.778	+0.1355	−0.1270	+0.06253	+0.00677	+0.02870
0.1	+12.071	−1.363	+0.936	+0.1232	−0.3444	+0.04246	+0.00205	+0.01463
0.0	+19.407	0	+1.000	0	−0.6412	0	0	0

SHEAR EDGE LOAD				EDGE MOMENT LOAD			
N_x	$N_{x\phi}$	N_ϕ	M_ϕ	N_x	$N_{x\phi}$	N_ϕ	M_ϕ
(9)	(10)	(11)	(12)	(13)	(14)	(15)	(16)
+0.000	−0.036	+0.0098	+0.0360	+0.0904	+0.0132	+0.00529	−0.002
−0.205	+0.113	−0.0421	−0.2075	−0.3010	−0.0517	−0.02790	+0.066
−0.665	−0.261	−0.0160	−0.1067	+0.4217	−0.0796	+0.03239	+0.493
+1.451	−0.219	+0.0697	+0.0260	+0.5681	+0.0942	+0.03157	+0.728
+3.919	+0.161	+0.0785	+0.0369	−0.1315	+0.1367	+0.01334	+0.852
+5.642	+0.510	+0.0546	+0.0227	−0.8691	+0.1019	+0.00424	+0.922
+7.750	+1.000	0	0	−1.9706	0	0	+1.000
+0.008	−0.033	+0.0108	+0.0320	+0.0817	+0.0134	+0.00539	−0.002
−0.205	+0.106	−0.0461	−0.1849	−0.2743	−0.0502	−0.02974	+0.060
−0.585	−0.262	−0.0136	−0.0851	+0.3819	−0.0794	+0.03386	+0.462
+1.366	−0.208	+0.0771	+0.0318	+0.5264	+0.0918	+0.03339	+0.702
+3.624	+0.169	+0.0851	+0.0378	−0.1135	+0.1348	+0.01418	+0.837
+5.200	+0.515	+0.0589	+0.0227	−0.7959	+0.1008	+0.00452	+0.913
+7.129	+1.000	0	0	−1.8212	0	0	+1.000
+0.017	−0.029	+0.0122	+0.0265	+0.0692	+0.0134	+0.00536	−0.002
−0.202	+0.094	−0.0522	−0.1554	−0.2359	−0.0467	−0.03215	+0.050
−0.481	−0.263	−0.0089	−0.0585	+0.3234	−0.0780	+0.03503	+0.417
+1.253	−0.197	+0.0890	+0.0380	+0.4651	+0.0860	+0.03538	+0.664
+3.240	+0.182	+0.0955	+0.0382	−0.0872	+0.1294	+0.01514	+0.814
+4.628	+0.523	+0.0655	+0.0224	−0.6882	+0.0975	+0.00484	+0.901
+6.328	+1.000	0	0	−1.6010	0	0	+1.000

TABLE 8 (*Contd.*)

$\dfrac{s}{4\sqrt{RtL^2}}$	TANGENTIAL EDGE LOAD				RADIAL EDGE LOAD			
	N_x (1)	$N_{x\phi}$ (2)	N_ϕ (3)	M_ϕ (4)	N_x (5)	$N_{x\phi}$ (6)	N_ϕ (7)	M_ϕ (8)
							Rt/L^2 = .008:	
3.2	−0.086	−0.178	+0.050	+0.1271	+0.0163	+0.00949	−0.00072	−0.00511
1.6	−0.131	+0.719	−0.240	−0.7409	−0.0562	−0.04265	+0.00383	+0.03459
0.8	−4.278	−1.094	−0.313	−0.6617	+0.1813	−0.00397	+0.03329	+0.07719
0.4	−6.770	−2.316	+0.367	−0.0315	+0.1217	+0.06531	+0.02116	+0.05188
0.2	+5.382	−1.945	+0.785	+0.1383	−0.1293	+0.06797	+0.00791	+0.02768
0.1	+10.120	−1.226	+0.938	+0.1163	−0.3483	+0.04611	+0.00239	+0.01411
0.0	+16.225	0	+1.000	0	−0.6479	0	0	0
							Rt/L^2 = 0.010:	
3.2	−0.045	−0.155	+0.052	+0.1079	+0.0148	+0.00997	−0.00111	−0 00521
1.6	−0.183	+0.640	−0.246	−0.6312	−0.0534	−0.04527	+0.00527	+0.03458
0.8	−3.671	−1.043	−0.288	−0.5322	+0.1807	−0.00313	+0.03767	+0.07597
0.4	−0.604	−2.141	+0.384	+0.0026	+0.1202	+0.06960	+0.02378	+0.05126
0.2	+4.712	−1.789	+0.791	+0.1370	−0.1305	+0.07215	+0.00887	+0.02748
0.1	+8.807	−1.127	+0.940	+0.1100	−0.3496	+0.04892	+0.00268	+0.01406
0.0	+14.089	0	+1.000	0	−0.6501	0	0	0
							Rt/L^2 = 0.015:	
3.2	+0.011	−0.115	+0.055	+0.0775	+0.0117	+0.01060	−0.00205	−0.00527
1.6	−0.243	+0.504	−0.255	−0.4648	−0.0474	−0.04966	+0.00882	+0.03405
0.8	−2.753	−0.952	−0.236	−0.3476	+0.1769	−0.00114	+0.04659	+0.07280
0.4	−0.371	−1.845	+0.416	+0.0440	+0.1159	+0.07720	+0.02902	+0.04970
0.2	+3.678	−1.529	+0.803	+0.1287	−0.1309	+0.07946	+0.01080	+0.02699
0.1	+6.801	−0.961	+0.943	+0.0969	−0.3472	+0.05383	+0.00326	+0.01393
0.0	+10.843	0	+1.000	0	−0.6465	0	0	0

SHEAR EDGE LOAD				MOMENT EDGE LOAD			
N_x (9)	$N_{x\phi}$ (10)	N_ϕ (11)	M_ϕ (12)	N_x (13)	$N_{x\phi}$ (14)	N_ϕ (15)	M_ϕ (16)
+0.022	−0.025	+0.0132	+0.0227	+0.0605	+0.0132	+0.00522	−0.002
−0.199	+0.084	−0.0567	−0.1360	−0.2087	−0.0432	−0.03372	+0.043
−0.414	−0.263	−0.0042	−0.0422	+0.2814	−0.0763	+0.03497	+0.385
+1.177	−0.187	+0.0988	+0.0110	+0.4211	+0.0803	+0.03618	+0.637
+2.990	+0.191	+0.1036	+0.0379	−0.0676	+0.1235	+0.01561	+0.796
+4.256	+0.530	+0.0707	+0.0218	−0.6093	+0.0937	+0.00501	+0.892
+5.810	+1.000	0	0	−1.4397	0	0	+1.000
+0.026	−0.022	+0.0138	+0.0199	+0.0539	+0.0128	+0.00503	−0.002
−0.196	+0.076	−0.0601	−0.1218	−0.1880	−0.0399	−0.03485	+0.038
−0.365	−0.262	+0.0004	−0.0312	+0.2487	−0.0746	+0.03426	+0.359
+1.121	−0.178	+0.1071	+0.0425	+0.3866	+0.0748	+0.03637	+0.614
+2.807	+0.199	+0.1105	+0.0372	−0.0524	+0.1178	+0.01581	+0.783
+3.986	+0.534	+0.0750	+0.0212	−0.5477	+0.0899	+0.00509	+0.884
+5.435	+1.000	0	0	−1.3134	0	0	+1.000
+0.030	−0.016	+0.0146	+0.0149	+0.0425	+0.0119	+0.00455	−0.002
−0.187	+0.059	−0.0657	−0.0980	−0.1518	−0.0325	−0.03674	+0.028
−0.284	−0.260	+0.0112	−0.0145	+0.1904	−0.0708	+0.03121	+0.313
+1.023	−0.160	+0.1242	+0.0435	+0.3249	+0.0626	+0.03557	+0.573
+2.498	+0.214	+0.1242	+0.0352	−0.0252	+0.1047	+0.01574	+0.757
+3.533	+0.544	+0.0836	+0.0197	−0.4365	+0.0811	+0.00511	+0.869
+4.811	+1.000	0	0	−1.0849	0	0	+1.000

TABLE 8 (*Contd.*)

	TANGENTIAL EDGE LOAD				RADIAL EDGE LOAD			
$\dfrac{s}{4\sqrt{RtL^2}}$	N_x	$N_{x\phi}$	N_ϕ	M_ϕ	N_x	$N_{x\phi}$	N_ϕ	M_ϕ
	(1)	(2)	(3)	(4)	(5)	(6)	(7)	(8)

$Rt/L^2 = 0.020$:

3.2	+0.033	−0.088	+0.054	+0.0594	+0.0091	+0.01073	−0.00290	−0.00518
1.6	−0.263	+0.415	−0.258	−0.3689	−0.0424	−0.05224	+0.01217	+0.03322
0.8	−2.225	−0.888	−0.195	−0.2500	+0.1720	+0.00061	+0.05358	+0.06986
0.4	−0.250	−1.652	+0.440	+0.0601	+0.1117	+0.08214	+0.03306	+0.04833
0.2	+3.068	−1.361	+0.812	+0.1192	−0.1296	+0.08417	+0.01230	+0.02659
0.1	+5.634	−0.854	+0.946	+0.0868	−0.3426	+0.05700	+0.00371	+0.01384
0.0	+8.964	0	+1.000	0	−0.6374	0	0	0

$Rt/L^2 = 0.025$:

3.2	+0.044	−0.069	+0.053	+0.0472	+0.0071	+0.01060	−0.00364	−0.00502
1.6	−0.268	+0.351	−0.257	−0.3054	−0.0382	−0.05380	+0.01529	+0.03229
0.8	−1.875	−0.839	−0.161	−0.1901	+0.1667	+0.00212	+0.05928	+0.06722
0.4	−0.177	−1.512	+0.460	+0.0668	+0.1078	+0.08554	+0.03631	+0.04747
0.2	+2.656	−1.240	+0.819	+0.1104	−0.1275	+0.08738	+0.01350	+0.02630
0.1	+4.853	−0.777	+0.948	+0.0787	−0.3362	+0.05918	+0.00407	+0.01380
0.0	+7.714	0	+1.000	0	−0.6263	0	0	0

$Rt/L^2 = 0.030$:

3.2	+0.050	−0.054	+0.051	+0.0383	+0.0054	+0.01031	−0.00428	−0.00482
1.6	−0.265	+0.301	−0.254	−0.2599	−0.0346	−0.05470	+0.01818	+0.03133
0.8	−1.625	−0.798	−0.131	−0.1497	+0.1616	+0.00344	+0.06403	+0.06485
0.4	−0.129	−1.403	+0.477	+0.0693	+0.1043	+0.08796	+0.03900	+0.04617
0.2	+2.355	−1.147	+0.825	+0.1028	−6.1249	+0.08966	+0.01449	+0.02607
0.1	+4.287	−0.719	+0.949	+0.0722	−0.3292	+0.06074	+0 00437	+0.01379
0.0	+6.811	0	+1.000	0	−0.6144	0	0	0

SHEAR EDGE LOAD				MOMENT EDGE LOAD			
N_x	$N_{x\phi}$	N_ϕ	M_ϕ	N_x	$N_{x\phi}$	N_ϕ	M_ϕ
(9)	(10)	(11)	(12)	(13)	(14)	(15)	(16)
+0.031	−0.011	+0.0147	+0.0116	+0.0352	+0.0111	+0.00412	−0.002
−0.178	+0.047	−0.0689	−0.0825	−0.1279	−0.0264	−0.03805	+0.022
−0.233	−0.257	+0.0212	−0.0052	+0.1510	−0.0676	+0.02741	+0.281
+0.957	−0.146	+0.1381	+0.0428	+0.2824	+0.0523	+0.03381	+0.544
+2.297	+0.225	+0.1350	+0.0331	−0.0068	+0.0934	+0.01527	+0.738
+3.241	+0.550	+0.0902	+0.0183	−0.3599	+0.0735	+0.00500	+0.859
+4.410	+1.000	0	0	−0.9265	0	0	+1.000
+0.031	−0.008	+0.0143	+0.0092	+0.0301	+0.0103	+0.00376	−0.001
−0.170	+0.036	−0.0705	−0.0715	−0.1107	−0.0213	−0.03912	+0.017
−0.197	−0.253	+0.0305	+0.0005	+0.1221	−0.0649	+0.02345	+0.257
+0.906	−0.135	+0.1499	+0.0416	+0.2507	+0.0436	+0.03181	+0.521
+2.150	+0.234	−0.1440	+0.0312	+0.0067	+0.0838	+0.01466	+0.723
+3.029	+0.556	+0.0958	+0.0170	−0.3028	+0.0670	+0.00484	+0.850
+4.123	+1.000	0	0	−0.8076	0	0	+1.000
+0.030	−0.005	+0.0137	+0.0073	+0.0263	+0.0096	+0.00345	−0.001
−0.163	+0.028	−0.0712	−0.0630	−0.0976	−0.0170	−0.04002	+0.013
−0.169	−0.250	+0.0392	+0.0044	+0.0998	−0.0627	+0.01962	+0.237
+0.866	−0.125	+0.1604	+0.0402	+0.2259	+0.0360	+0.02977	+0.502
+2.035	+0.242	+0.1518	+0.0294	+0.0171	+0.0754	+0.01401	+0.710
+2.866	+0.560	+0.1005	+0.0159	−0.2581	+0.0613	+0.00467	+0.843
+3.902	+1.000	0	0	−0.7141	0	0	+1.000

TABLE 8 (*Contd.*)

$\dfrac{s}{4\sqrt{RtL^2}}$	TANGENTIAL EDGE LOAD				RADIAL EDGE LOAD			
	N_x	$N_{x\phi}$	N_ϕ	M_ϕ	N_x	$N_{x\phi}$	N_ϕ	M_ϕ
	(1)	(2)	(3)	(4)	(5)	(6)	(7)	(8)

Rt/L = 0.040:

3.2	+0.053	−0.033	+0.045	+0.0263	+0.0028	+0.00947	−0.00526	−0.00439
1.6	−0.253	+0.227	−0.246	−0.1986	−0.0289	−0.05524	+0.02329	+0.02949
0.8	−1.286	−0.735	−0.081	−0.1002	+0.1516	+0.00551	+0.07145	+0.06091
0.4	−0.072	−1.241	+0.505	+0.0682	+0.0980	+0.09076	+0.04319	+0.04469
0.2	+1.930	−1.010	+0.834	+0.0895	−0.1195	+0.09230	+0.01605	+0.02586
0.1	+3.512	−0.632	+0.952	+0.0618	−0.3152	+0.06258	+0.00485	+0.01389
0.0	+5.579	0	+1.000	0	−0.5904	0	0	0

Rt/L² = 0.060:

3.2	+0.047	−0.009	+0.034	+0.0135	−0.0003	+0.00747	−0.00631	−0.00356
1.6	−0.222	+0.137	−0.222	−0.1313	−0.0213	−0.05387	+0.03126	+0.02620
0.8	− 0.910	−0.645	−0.005	−0.0531	+0.1343	+0.00814	+0.08038	+0.05495
0.4	−0.023	−1.035	+0.545	+0.0605	+0.0881	+0.09213	+0.04853	+0.04274
0.2	+1.456	−0.838	+0.848	+0.0708	−0.1084	+0.09368	−0.01807	+0.02586
0.1	+2.629	−0.524	+0.956	+0.0480	−0.2886	+0.06370	+0.00547	+0.01425
0.0	+4.187	+1.000	0	0	−0.5452	0	0	0

Rt/L² = 0.080:

3.2	+0.039	+0.002	+0.023	+0.0072	−0.0018	+0.00559	−0.00651	−0.00285
1.6	−0.195	+0.083	−0.196	−0.0951	−0.0166	−0.05117	+0.03683	+0.02346
0.8	−0.705	−0.581	+0.052	−0.0320	+0.1199	+0.00946	+0.08620	+0.05066
0.4	−0.005	−0.904	+0.574	+0.0523	+0.0801	+0.09073	+0.05150	+0.04162
0.2	+1.179	−0.730	+0.857	+0.0582	−0.0980	+0.09256	+0.01923	+0.02615
0.1	+2.130	−0.458	+0.959	+0.0391	−0.2648	+0.06315	+0.00583	+0.01471
0.0	+3.407	0	+1.000	0	−0.5051	0	0	0

	SHEAR EDGE LOAD				MOMENT EDGE LOAD			
N_x (9)	$N_{x\phi}$ (10)	N_ϕ (11)	M_ϕ (12)	N_x (13)	$N_{x\phi}$ (14)	N_ϕ (15)	M_ϕ (16)	
+0.028	+0.000	+0.0120	+0.0047	+0.0210	+0.0085	+0.00297	−0.000	3.2
−0.150	+0.014	−0.0704	−0.0507	−0.0787	−0.0100	−0.04169	+0.008	1.6
−0.131	−0.242	+0.0550	+0.0088	+0.0678	−0.0589	+0.01218	+0.208	0.8
+0.804	−0.109	+0.1781	+0.0372	+0.1887	+0.0238	+0.02556	+0.473	0.4
+1.866	+0.253	+0.1648	+0.0263	+0.0315	+0.0615	+0.01258	+0.690	0.2
+2.625	+0.567	+0.1085	+0.0140	−0.1925	+0.0517	+0.00426	+0.131	0.1
+3.580	+1.000	0	0	−0.5749	0	0	+1.000	0.0
+0.023	+0.005	+0.0081	+0.0017	+0.0151	+0.0070	+0.00225	−0.001	3.2
−0.130	−0.005	−0.0646	−0.0359	−0.0563	−0.0007	−0.04437	+0.001	1.6
−0.085	−0.230	+0.0822	+0.0122	+0.0302	−0.0533	−0.00023	+0.170	0.8
+0.718	−0.086	+0.2049	+0.0319	+0.1416	+0.0065	+0.01821	+0.431	0.4
+1.647	+0.269	+0.1850	+0.0216	+0.0474	+0.0415	+0.01001	+0.661	0.2
+2.321	+0.575	+0.1207	+0.0112	−0.1122	+0.0378	+0.00352	+0.814	0.8
+3.179	+1.000	0	0	−0.4003	0	0	+1.000	0.0
+0.018	+0.008	+0.0044	+0.0002	+0.0118	+00061	+0.00166	−0.000	3.2
−0.114	−0.018	−0.0564	−0.0271	−0.0433	+0.0050	−0.04634	−0.002	1.6
−0.060	−0.218	+0.1045	+0.0128	+0.0100	−0.0491	−0.00982	+0.146	0.8
+0.659	−0.071	+0.2273	+0.0277	+0.1128	−0.0047	+0.01226	+0.402	0.4
+1.506	+0.278	+0.2004	+0.0182	+0.0549	+0.0280	+0.00787	+0.639	0.2
+2.130	+0.580	+0.1301	+0.0093	+0.0654	+0.0283	+0.00289	+0.801	0.1
+2.931	+1.000	0	0	−0.2949	0	0	+1.000	0.0

TABLE 8 (*Contd.*)

$\dfrac{s}{4\sqrt{RtL^2}}$	TANGENTIAL EDGE LOAD				RADIAL EDGE LOAD			
	N_x	$N_{x\phi}$	N_ϕ	M_ϕ	N_x	$N_{x\phi}$	N_ϕ	M_ϕ
	(1)	(2)	(3)	(4)	(5)	(6)	(7)	(8)
$Rt/L^2 = 0.100$:								
3.2	+0.032	+0.008	+0.015	+0.0038	−0.0026	+0.00400	−0.00620	−0.00229
1.6	−0.171	+0.048	−0.171	−0.0728	−0.0135	−0.04807	+0.04061	+0.02117
0.8	−0.576	−0.533	+0.097	−0.1078	+0.0210	+0.00995	+0.08883	+0.04740
0.4	+0.002	−0.811	+0.595	+0.0454	+0.0747	+0.08805	+0.053O8	+0.04093
0.2	+0.997	−0.655	+0.865	+0.0490	−0.0886	+0.09028	+0.01989	+0.32656
0.1	+1.807	−0.411	+0.961	+0.0328	−0.2437	+0.06183	+0.00605	+0.01520
0.0	+2.903	0	+1.000	0	−0.4697	0	0	0

SHEAR EDGE LOAD				MOMENT EDGE LOAD			
N_x (9)	$N_{x\phi}$ (10)	N_ϕ (11)	M_ϕ (12)	N_x (13)	$N_{x\phi}$ (14)	N_ϕ (15)	M_ϕ (16)
+0.015	+0.009	+0.0014	−0.0006	+0.0097	+0.0056	+0.00110	−0.000
−0.102	−0.026	−0.0474	−0.0214	−0.0347	+0.0086	−0.04764	−0.004
−0.045	−0.209	+0.1231	+0.0126	−0.0027	−0.0456	−0.01715	+0.128
+0.614	−0.059	+0.2446	+0.0243	+0.0922	−0.0123	+0.00751	+0.380
+1.406	+0.585	+0.2129	+0.0156	+0.0578	+0.0183	+0.00611	+0.622
+1.995	+0.583	+0.1378	+0.0078	−0.0310	+0.0213	+0.00237	+0.790
+2.760	+1.000	0	0	−0.2250	0	0	+1.000

Objective Questions on Prestressed Concrete Structures

Prestressed concrete structure may be defined as the one in which the initial stresses are induced with tensioned steel such that the stresses caused by the working loads are partly or fully compensated by the initially induced stresses. Concrete is reasonably strong in compression but weak in tension, therefore prestresses are introduced in the concrete members through tensioned steel. Two methods of prestressing in practice are:

1. Pretensioning and
2. Post-tensioning

The wires are pretensioned against some external abutments and then the concrete is placed around the wires. As the concrete gets hardened, the tensioned wires develop bond with the concrete. The pretensioned force of the wires is transferred from the abutments to the members when the concrete has gained the desired strength. Such a construction is called pretensioned construction. Whereas in the case of post-tensioned construction, the concrete is allowed to harden leaving ducts for the cables or wires. The cables are tensioned and anchored on the hardened concrete at the appropriate time and locations. Each of the methods has its advantages and applicability.

In addition to the strains caused by the working loads, there are secondary strains caused in the concrete due to shrinkage and creep in concrete, creep in steel, relaxation in the cable force at the time of transferring the force to the member etc. Such strains cause a loss of prestress of the wires thus resulting into an effective prestresses less than the original tension.

The total loss of prestress is usually high enough to make the mild steel as a redundant tensioned steel. Hence high tensile steels of strength in the range of 1200 to 2000 MPa are normally used in the construction. Smaller diameter wires, or cables or wire ropes need to be used for practical convenience of the construction. They are manufactured and transported in coils so that appropriate lengths can be cut and used without wastage in the construction. Consequent to the use of high tensile steels in prestressed concrete the ordinary concrete having strength less than 30 MPa is not only uneconomical but reduces the effectiveness of the high tensile strength of the steel. Therefore high strength concrete need to be used in the prestressed concrete structures. Concrete having a minimum characteristic strength of 30 MPa is recommended in the construction.

Three methods are normally used in the design of prestressed concrete structures. They are:

1. Working stress design (WSD)
2. Ultimate strength design (USD)
3. Limit states design (LSD)

In the working stress design, the structure is analysed by the ?elastic structural analysis for a set of working loads. The stresses thus developed in the concrete and the steel are limited to a set of allowable stresses. These allowable stresses are obtained after applying the factors of safety to the ultimate strengths of the materials. In the ultimate strength design, the working loads are enhanced by multiplying them with load factors and the stress resultants are obtained by the elastic analysis. The members are proportioned such that the ultimate capacities of the sections are more than the stress resultants acting on them. In case of the limit states design, the structure is designed for different limit states. The main limit states are :

(a) Strength limit state
(b) Serviceability limit state
(c) Durability limit state

The working loads are multiplied by an appropriate set of partial safety factors, the resulting loads are called design loads. The collapse load on the structure is obtained by the plastic or limit analysis., The sections are proportioned so as 'to Withstand the stress resultants at the collapse stags. The structure is analysed by the elastic analysis in the service limit state, and the stresses or strains or cracks or deflections etc. thus developed are limited to the allowable values.

The design criteria-in each of the methods of the design vary and they are broadly listed here :

1. The stresses in the concrete and the steel in any load condition must be limited to the allowable stresses.
2. The prestressed concrete structures are designed as uncracked or cracked sections. Most of the structures are designed as uncracked sections, and limited tensile resistance of the concrete is accounted in the design.
3. Transfer of prestress to the concrete member or structure is considered as one of the

important load conditions in which the force in the cable is highest and the strength of the concrete is on the lower side.

4. The shear force is normally resisted by the concrete. The allowable shear stress in the concrete is governed by the tensile strength. Principal stress including the effect of the normal stress is computed for the determination of the shear capacity.

5. The concrete sections should also be designed to resist the anchorage forces such as bursting and spalling stresses generated at the anchor zones.

6. Bond strength of the wire and the bond length of pretensioned wires must be checked for safety.

7. Buckling of prestressed concrete members is likely to exist, therefore a check for buckling safety be made.

8. A check on deflection limits in the working load condition be made.

9. In case, the sections are proportioned by the ultimate strength design, then the tensile strength of the concrete is neglected.

10. Minimum percentage of steel be provided to withstand the shrinkage and temperature effects.

11. Minimum clear cover limitations to the cables or wires be adhered to so as to have an effective working of the wires..

12. Protection of the cables against corrosion must be ensured by proper grouting of the ducts in case of post-tensioned construction.

13. Anchorages of the cables be properly protected against damage and slippage.

OBJECTIVE QUESTIONS

10. GENERAL

10.1 External progressing of concrete structures can be obtained from:
    (a) Exterior ends of the beam   (b) Rigid abutments only
    (c) Rigid supports   (d) Internal wire ropes

10.2 External prestressing of the concrete structures can best be obtained from :
    (a) Flat jacks   (b) Mechanical jacks
    (c) Hydraulic jacks   (d) Anchor cones

10.3 External prestressing can best be applied to:
    (a) Statically indeterminate structures
    (b) Statically determinate structures
    (c) Continuous beams
    (d) None of the above

10.4 External prestressing of simply supported beams can be achieved:
    (a) Jacking up
    (b) Jacking against rigid abutments
    (c) Chemical treatment
    (d) Cables

10.5 Prestress in the externally prestressed concrete beams is:
    (a) Variable under working loads
    (b) Fixed unlike the internal prestress
    (c) Losses are very high
    (d) Sensitive to temperature, shrinkage etc.

10.6 Linear prestressing is used in:
    (a) Beams   (b) Columns only
    (c) Water tanks   (d) Pipes

10.7 Linear prestressing is best suited to:
    (a) Straight elements   (b) Overhead water tanks
    (c) Containment vessels   (d) None of the above set

10.8 Circular prestressing can be applied best to:
    (a) Circular columns   (b) Ring beam in buildings
    (c) Water tanks   (d) Beams

10.9 Circular prestressing is best suited to:
    (a) Circular beams in buildings   (b) Pipes
    (c) Railway sleepers   (d) Columns

10.10 Prestressed concrete electrical poles are prestressed efficiently with:
    (a) Axial prestress   (b) Eccentric prestress
    (c) Axial or eccentric prestress   (d) None of the above set

10.11 In case of prestressed concrete cantilever beams the steel cables are usually close to:
    (a) Bottom fibre   (b) Top fibre
    (c) Neutral axis   (d) Centroidal axis of the beam

10.12 In case of prestressed concrete simply supported beams, the steel cables are placed at:
    (a) Away from the centroid   (b) Close to the top fibre
    (c) Close to the bottom fibre   (d) Away from the neutral axis

10.13 The net effect of the high tensile wires used in prestressed concrete poles causes:
  (a) Axial force only
  (b) Bending and compression force
  (c) Bending only
  (d) Torsion, axial and bending forces

10.14 The prestressing wires in prestressed concrete electric poles are:
  (a) Concentric
  (b) Eccentric
  (c) Parabolic
  (d) Tapering

10.15 The tolerance in the overall length of prestressed concrete pole is about:
  (a) ±5 mm
  (b) +15 mm
  (c) One per cent
  (d) 0.5 per cent

10.16 The tolerance in the cross sectional size of prestressed concrete pole is about :
  (a) +1 mm
  (b) ±3 mm
  (c) +10 mm
  (d) One percent

10.17 Minimum strength (150 mm cube) of concrete used in pretensioned concrete poles is (in N/mm^2) :
  (a) 25
  (b) 30
  (c) 35
  (d) 40

10.18 Minimum strength of concrete (150 mm cube) used in post-tensioned concrete electric poles is (in N/mm^2) :
  (a) 25
  (b) 30
  (c) 35
  (d) 40

10.19 Minimum quantity of cement to be used in concrete for pretensioned concrete electric poles in kg m of concrete is:
  (a) 200
  (b) 280
  (c) 380
  (d) 480

10.20 Minimum quantity of cement to be used in concrete of post-tensioned prestressed concrete electric poles in kg/m^3 or concrete is:
  (a) 200
  (b) 260
  (c) 360
  (d) 460

10.21 Maximum quantity of cement to be used in prestressed concrete electric poles be limited to (in kg/m^3 of concrete):
  (a) 300
  (b) 420
  (c) 530
  (d) 650

10.22 Minimum depth of planting of prestressed concrete electric poles in the ground for small lengths is:
  (a) 0.8 to 1.2 m
  (b) 1.2 to 1.5 m
  (c) 1.5 to 2 m
  (d) 2 to 2.5 m

10.23 Cracking on the liquid face of a liquid storage prestressed concrete tank is:
  (a) Allowed in some cases
  (b) Not allowed at all
  (c) Allowed
  (d) Limited to 0.2 mm crack width

10.24 Tensile stresses in the concrete of liquid storage prestressed concrete tank is:
  (a) Allowed in some cases
  (b) Not allowed
  (c) Allowed
  (d) 1.5N/mm^2 is

10.25 Minimum clear cover in liquid storage tanks on the liquid face is:
  (a) 15 mm
  (b) 25 mm
  (c) 15 mm
  (d) 45 mm

10.26 Minimum strength of concrete (150 mm rubes at 28 days) used in post-tensioned prestressed concrete bridge girders is (in N/mm^2):
    (a) 25   (b) 30
    (c) 35   (d) 40

10.27 Minimum quantity of cement (in kg per m^3 of concrete) in post-tensioned concrete beam is about:
    (a) 260   (b) 360
    (c) 460   (d) 500

10.28 Maximum quantity of cement (in kg per m^3 of concrete) in post-tensioned concrete girders is about:
    (a) 240   (b) 340
    (c) 440   (d) 540

10.29 Minimum crushing strength (of 150 mm cube) concrete for which transfer prestress is permitted in post-tensioned concrete bridge girder is:
    (a) 15 N/mm^2 (b) 20 N/mm^2
    (c) 25 N/mm^2 (d) 50 per cent of design strength

10.30 Maximum permissible compressive stress at the time of handling including transfer load condition in post-tensioned prestressed concrete beam is:
    (a) 20 N/mm^2 or 50 per cent of cube strength
    (b) 18 N/mm^2 or 40 per cent of cube strength
    (c) 33 per cent of cube strength
    (d) One third more than that allowed at working load

10.31 Permissible tensile stress in concrete in temporary load condition in post-tensioned prestressed concrete bridge beams is limited to:
    (a) 1.0 N/mm^2 (b) 1.5 N/mm^2
    (c) Zero   (d) 10 per cent of that allowed in compression

10.32 Maximum compressive stress allowed in M35 concrete in post-tensioned prestressed concrete bridge girders under working load condition is :
    (a) 14 N/mm^2 (b) 18 N/mm^2
    (c) 0.3 times the cube strength   (d) 0.4 times the cube strength

10.33 Maximum permissible tensile stress in high tensile steel in post-tensioned prestressed concrete bridge girders under working load condition is :
    (a) 0.6 times the ultimate stress or 0 8 times the proof stress
    (b) 0.5 times the ultimate stress or 0.8 times proof stress
    (c) 0 7 times the ultimate stress or 0.9 times proof stress
    (d) None of the above

10.34 Maximum permissible tensile stress in steel in post-tensioned prestressed concrete bridge girder at anchorage after seating is equal to:
    (a) 0.6 times the ultimate stress   (b) 0.8 times the proof stress
    (c) 0.7 times the ultimate stress   (d) 0.9 times the proof stress

10.35 Permissible tension stress in steel at jacking in post-tensioned bridge beams is equal to:
    (a) 0.8 times proof stress   (b) 0.7 times ultimate stress
    (c) Proof stress   (d) 0.8 times ultimate stress

10.36 Long line prestressing in prestressed concrete is used in:
    (a) Post-tensioned construction   (b) Pretensioned construction
    (c) Cast in-situ construction   (d) In any construction

10.37 Long line prestressing in prestressed concrete is usually used in:
- (a) Precast construction
- (b) Cast in-situ construction
- (c) Long beams construction
- (d) None of the above set

10.38 Unit mould method of prestressing is used in:
- (a) Post-tensioned construction
- (b) Cast in-situ construction
- (c) Pretensioned construction
- (d) Cantilever construction

10.39 Unit mould method of prestressing is commonly used in:
- (a) Cantilever construction
- (b) Post-tension construction
- (c) Any type of construction
- (d) Precast construction

11. CONCEPTS AND DEFINITIONS

11.1 In prestressed concrete structures, prestressing of the concrete is done to compensate the stresses caused by:
- (a) Dead load
- (b) Working loads
- (c) Live loads
- (d) Dynamic loads

11.2 Net force due to prestressing on concrete in prestressed concrete beams is usually:
- (a) Tension
- (b) Bending and tension
- (c) Compression
- (d) Bending and compression

11.3 Net force caused by prestressing on concrete in prestressed concrete pipes is usually:
- (a) Tension
- (b) Bending and tension
- (c) Compression
- (d) Bending and compression

11.4 Net force due to prestressing on concrete in prestressed concrete columns is usually:
- (a) Tension
- (b) Bending and tension
- (c) Compression
- (d) Bending and compression

11.5 Plane section before applying prestressing on concrete beams .... after the application of the prestressing force:
- (a) Remains plane all the time
- (b) Does not remain plane
- (c) Remains plane only if the stress in steel is less than yield stress
- (d) Remains plane in the elastic zone

11.6 A straight prestressing cable in a beam introduces:
- (a) Compression
- (b) Compressive and tensile stresses
- (c) Bending and shear stresses
- (d) Shear and compressive stresses

11.7 Prestressing in concrete beams usually obtained by:
- (a) External prestressing
- (b) Chemical reaction
- (c) Internal prestressing
- (d) Steam curing

11.8 A tendon with two straight line segments and an intermediate kink introduces in a prestressed concrete beam:
- (a) Compression only
- (b) Bending and compression only
- (c) Compression, bending and shear only
- (d) Axial and shear forces only

11.9 A curved profile tendon in a prestressed concrete beam introduces:
    (a) Compression only
    (b) Bending and compression only
    (c) Compression, bending and shear only
    (d) Axial and shear forces only

11.10 A tendon anchored perpendicular to the plane of a concrete in prestressed beams introduces at the end section:
    (a) Compression only
    (b) Bending and compression only
    (c) Compression, bending and shear only
    (d) Axial and shear forces only

11.11 A tendon anchored at an inclination to the plane of concrete section in prestressed beams introduces at the section:
    (a) Compression only
    (b) Bending and compression only
    (c) Compression, bending and shear only
    (d) Axial and shear forces only

11.12 A tendon anchored at the top surface of a concrete beam introduces :
    (a) Compression only
    (b) Bending and compression only
    (c) Compression, bending and shear only
    (d) Axial and shear forces only

11.13 A tendon anchored at the centroid of a cross section and normal to the section in prestressed beam introduces:
    (a) Compression only
    (b) Bending and compression only
    (c) Compression, bending and shear only
    (d) Axial and shear forces only

11.14 A tendon anchored at the centroid of a section and inclined to it in prestressed beams introduces:
    (a) Compression only
    (b) Bending and compression only
    (c) Compression, bending and shear only
    (d) Axial and shear forces only

11.15 A parabolic tendon in prestressed beams causes an equivalent balancing:
    (a) Concentrated transverse force
    (b) Distributed force
    (c) Uniformly distributed force
    (d) Constant moment

11.16 A straight tendon with an eccentricity in prestressed beams causes an equivalent balancing:
    (a) Concentrated transverse force
    (b) Distributed force
    (c) Uniformly distributed force
    (d) Constant moment

11.17 A tendon with variable curved profile causes an equivalent balancing:
    (a) Concentrated transverse force
    (b) Distributed force
    (c) Uniformly distributed force
    (d) Constant moment

604

11.18 A devise which is used to impart prestress force to concrete from tendon in prestressed concrete is called:
(a) Cone
(b) Anchorage
(c) Wedge
(d) Anchor plate

11.19 The prestress at the time of transfer of tendon force to the concrete in prestressed concrete is called:
(a) Initial prestress
(b) Anchor prestress
(c) Final prestress
(d) Partial prestress

11.20 The prestress in a prestressed concrete beam at the time of working load is called:
(a) Initial prestress
(b) Anchor prestress
(c) Final prestress
(d) Partial prestress

11.21 A method of prestressing in which prestressing is done against hardened concrete is called:
(a) Pre-tensioning
(b) Post-tensioning
(c) Partial tensioning
(d) Prestressing

11.22 Decrease in stress at constant strain is called:
(a) Creep loss
(b) Relaxation
(c) Shrinkage
(d) Transfer stress

11.23 A method of prestressing concrete in which tensioning is done before placing concrete is called:
(a) Pre-tensioning
(b) Post-tensioning
(c) Partial tensioning
(d) Prestressing

11.24 Increase in strain at constant stress is called:
(a) Creep loss
(b) Relaxation
(c) Shrinkage
(d) Transfer stress

11.25 Strain in concrete independent of stress is called:
(a) Creep
(b) Relaxation
(c) Shrinkage
(d) Transfer stress

11.26 The stage at which the tendon force is released permanently on the concrete is called:
(a) Pre-tensioning
(b) Post-tensioning
(c) Anchoring
(d) Transfer

11.27 A prestressed concrete section which is likely to fail by yielding of steel is called:
(a) Balanced section
(b) Over-reinforced
(c) Under reinforced
(d) Yielded section

11.28 Fully prestressed concrete beams:
(a) Resists the full live load by prestress
(b) Resists all the working loads by prestress
(c) Resists part of the load by prestress
(d) None of the above

11.29 Fully prestressed concrete beams are called by the name because:
(a) No tension is permitted in the beams
(b) No cracking is permitted in the beams
(c) Working loads are completed resisted by the prestressing force
(d) Full prestressing is applied to start with

11.30 Fully prestressed concrete beams require:
(a) Nominal mild steel reinforcement for shrinkage & temperature
(b) No mild steel reinforcement is needed at, all
(c) Shear reinforcement nominal or otherwise
(d) Reinforcement for creep

11.31 The cables used in statically determinate prestressed concrete beams are usually:
    (a) Concordant cables   (b) Non-concordant cables
    (c) Balanced cables   (d) None of the above

11.32 Prestressing is most efficient when applied to:
    (a) Columns   (b) Struts
    (c) Beams   (d) Wall panels

11.33 Prestressing is most efficient when applied to:
    (a) Struts   (b) Ties
    (c) Pipes   (d) Panels

12. MATERIAL AND PERMISSIBLE STRESS

12.1 Ultimate tensile stress of high tensile wires .... with decrease in the diameter of the wire:
    (a) Decreases   (b) Increases
    (c) Constant   (d) No relation

12.2 Approximate percentage elongation of high tensile plain cold-drawn wires is about:
    (a) 0 to 2   (b) 2 to 4
    (c) 4 to 10   (d) 10 to 20

12.3 Approximate percentage elongation of high tensile steel as-drawn wires is about:
    (a) 0 to 2   (b) 2 to 4
    (c) 4 to 10   (d) 10 to 20

12.4 Approximate percentage elongation of high tensile steel bars is about:
    (a) 0 to 2   (b) 2 to 4
    (c) 4 to 10   (d) 10 to 20

12.5 Permissible relaxation of stress in cold drawn high tensile wires at 1000 h (in N/mm^2) is:
    (a) 10 to 50   (b) 10 to 100
    (c) 100 to 150   (d) 150 to 200

12.6 Approximate Young's modulus of high tensile steel is (in N/mm^2):
    (a) 2×10^5 (b) 2×10^6
    (c) 3×10^5 (d) 2×10^7

12.7 Approximate ratio of the permissible stress in the concrete at transfer to the crushing stress of 150 mm cube in prestressed concrete is about:
    (a) 0.30 to 040   (b) 0.35 to 0.45
    (c) 0.40 to 0.50   (d) 0.45 to 0.55

12.8 Approximate ratio of the permissible stress in concrete at working load to the crushing stress of 150 mm cube in prestressed concrete is:
    (a) 0.30 to 0.40   (b) 0.35 to 0.45
    (c) 0.40 to 0.50   (d) 0.45 to 0.55

12.9 The quality of concrete used in pre-tensioned construction is about:
    (a) M 30 to M 50   (b) M 35 to M 55
    (c) M 40 to M 60   (d) M 45 to M 70
    (M 30 means a 150 mm concrete cube at 28 days of water curing gives 30 N/mm^2 crushing strength)

12.10 The quality of concrete used in post-tensioned concrete is about:
    (a) M 30 to M 50   (b) M 35 to M 55
    (c) M 40 to M 60   (d) M 45 to M 70

12.11 Approximate ratio of permissible stress at transfer to the ultimate stress in high tensile steel wires used in prestressed concrete is:
(a) 0.40 to 0.55   (b) 0.55 to 0.65
(c) 0.65 to 0.75   (d) 0.70 to 0.80

12.12 Approximate ratio of permissible stress at working load to the ultimate stress in high tensile steel in prestressed concrete is about:
(a) 0.40 to 0.55   (b) 0.55 to 0.65
(c) 0.65 to 0.70   (d) 0.70 to 0.80

12.13 Approximate ratio of the permissible stress at the time of jacking to the ultimate stress of high tensile steel in prestressed concrete is about:
(a) 0.40 to 0.55   (b) 0 55 to 0.65
(c) 0 65 to 0.70   (d) 0.70 to 0.80

12.14 Approximate ratio of permissible bearing stress in concrete at anchor to the 150 mm cube crushing stress is about:
(a) 0.4 to 0 5   (b) 0.5 to 0.6
(c) 0.6 to 0.7   (d) 0.7 to 0.9

12.15 Maximum permissible tensile stress in concrete in post-tensioned prestressed concrete at working load (in N/mm^2) in building is about:
(a) 0   (b) 0 to 1
(c) 1 to 3   (d) 2.5 to 4.5

12.16 Maximum permissible tensile stress in concrete in post-tensioned prestressed concrete in bridges, at working load is (in N/mm^2):
(a) 0   (b) 0 to 1
(c) 1 to 2.5   (d)' 2.5 to 4.5

12.17 Maximum permissible tensile stress in prestressed concrete at transfer condition in bridges is about (in N/mm^2):
(a) 0   (b) 0 to 1
(c) 1 to 2   (d) 1.5 to 2.5

12.18 Maximum permissible tensile stress in pre-tensioned prestressed concrete in building is about (N/mm^2):
(a) 0   (b) 0 to 1
(c) 1 to 3   (d) 2.5 to 4.5

12.19 Maximum tensile stress allowed in prestressed concrete in water tanks is about (in N/mm^2):
(a) 0   (b) 0 to 1
(c). 1 to 2   (d) 1.5 to 2.5

12.20 Approximate ultimate tensile stress of high tensile steel bars (in N/mm^2) is:
(a) 900 to 1200   (b) 1200 to 1500
(c) 1500 to 2000   (d) 2000 to 2500

12.21 Approximate ultimate tensile stress of high tensile steel wires (in N/mm^2) is:
(a) 900 to 1200   (b) 1200 to 1500
(c) 1500 to 2000   (d) 2000 to 2500

12.22 Approximate ultimate tensile stress of high tensile cables (in N/mm^2) is:
(a) 900 to 1200   (b) 1200 to 1500
(c) 1500 to 2000   (d) 2000 to 2500

13. WORKING STRESS DESIGN

13.1 Working stress design assumes:
  (a) Elastic stress-strain relation
  (b) Linear stress-strain relation
  (c) Elastic and linear stress-strain relation
  (d) Homogeneous material

13.2 Working stress design assumes:
  (a) Stresses are always within the elastic limit
  (b) Stresses are likely to exceed the elastic limit in combined load condition
  (c) Stresses are always less than the ultimate stresses
  (d) None of the above set

13.3 Working stress design assumes:
  (a) Partial recovery of deflections caused during wind load condition
  (b) Full recovery of deflections caused in all types of load conditions
  (c) Full recovery of deflections caused in some load conditions
  (d) The deflection recovery is irrelevant

13.4 Stress caused by different loads can be super-imposed in:
  (a) Working stress design
  (b) Ultimate strength design
  (c) Limit state design
  (d) Plastic design

13.5 The superposition of stresses caused by different load conditions in working stress design is:
  (a) Allowed only in limited loads
  (b) Allowed in all loads
  (c) Allowed in all loads except earthquake
  (d) Allowed in loads limited to plastic loads

13.6 Critical load conditions for which a design is to be made in prestressed concrete structures are :
  (a) Different working loads
  (b) Transfer and live loads
  (c) Transfer and working loads
  (d) Dead and live loads

13.7 Plane sections of prestressed concrete structures before bending will remain plane even after bending is:
  (a) Only true in working stress design
  (b) True in working stress design of slender members only
  (c) Not true in the ultimate strength design of slender members
  (d) Only true in elastic design

13.8 Working stress design makes use of:
  (a) Elastic structural analysis
  (b) Plastic structural analysis
  (c) Elastic and also plastic analysis
  (d) Working stress analysis

13.9 Working loads may be defined as :
  (a) Maximum loads consisting of dead and live loads
  (b) A set of critical loads in service condition
  (c) A set of maximum loads for which no failure will occur
  (d) All loads that occur during service conditions

608

14. ULTIMATE STRENGTH DESIGN

14.1 The ratio of ultimate bending compressive strength of a prism to the 150 rim cube strength of concrete is:
(a) 0.50 to 0.60   (b) 0.55 to 0.65
(c) 0.65 to 0.70   (d) 0.70 to 0.85

14.2 The ratio of the ultimate compressive strength of prism to that of the 150 mm cube of concrete is:
(a) 0.50 to 0.60   (b) 0.55 to 0.65
(c) 0.65 to 0.70   (d) 0.70 to 0.85

14.3 Yield (or proof) strain used in high tensile steel in the computation of ultimate moment capacity of a prestressed concrete section is:
(a) 0.002 to 0.005   (b) 0.004 to 0.008
(c) 0.007 to 0.01   (d) 0.009 to 0.012

14.4 The total compressive force on concrete in a prestressed concrete balanced cross section at ultimate moment level is:
(a) Equal to the ultimate force in steel
(b) Equal to the ultimate force plus the prestressing force in steel
(c) Less than the ultimate force in steel
(d) Not connected to force in steel

14.5 The effect of prestressing force on the ultimate moment capacity of prestressed concrete balanced cross section is:
(a) Dominant   (b) Reasonable
(c) Hardly any   (d) Not connected

14.6 The ultimate moment capacity of a prestressed concrete balanced section is:
(a) Unaffected by the magnitude of initial prestress
(b) Proportional to initial prestress
(c) Affected by the nominal loss of prestress
(d) None of the above set

14.7 The ultimate moment capacity of a prestressed concrete cross section is directly proportional to:
(a) Strength of concrete
(b) Effective depth of steel
(c) Yield strain in steel
(d) Crushing strain in concrete

14.8 The ultimate moment capacity of a prestressed concrete balanced section is:
(a) Directly proportional to yield strain in steel
(b) Crushing strain in concrete
(c) Proportional to the square of the effective depth of steel
(d) None of the above set

14.9 The ultimate moment capacity of a prestressed concrete balanced section is directly proportional to:
(a) Yield strain in steel
(b) Crushing strain in concrete
(c) Effective depth of steel
(d) Neutral axis distance

14.10 The ultimate moment capacity of post-tensioned prestressed concrete beam with bonded wires when compared with the unbounded wires is:
(a) Same
(b) More
(c) Less
(d) Independent

14.11 Grouting of ducts of post-tensioned prestressed concrete beams causes:
(a) Enhanced ultimate moment capacity
(b) Does not influence the moment capacity
(c) Decreases the moment capacity
(d) Increases deflections

14.12 Approximate ratio of neutral axis to the effective depth of steel in balanced prestressed concrete sections is:
(a) 0.2 to 0.3
(b) 0.3 to 0.4
(c) 0.4 to 0.45
(d) 0.45 to 0 55

14.13 Approximate percentage of area of high tensile steel in pretensioned prestressed concrete balanced rectangular sections is:
(a) 0.2 to 0.5
(b) 0 5 to 1.0
(c) 1 .0 to 2.0
(d) 1.5 to 4.0

14.14 Approximate percentage of area of high tensile steel in post-tensioned prestressed concrete, balanced rectangular section is:
(a) 0.2 to 0.5
(b) 0.5 to 1.0
(c) 1.0 to 2.0
(d) 1.5 to 4.0

14.15 Approximate percentage of steel in prestressed concrete balanced T-beams is:
(a) 0.2 to 0.5
(b) 0.5 to 1.0
(c) 1.0 to 2.0
(d) 1.5 to 4.0

14.16 Approximate percentage of steel in prestressed concrete balanced I-beam is:
(a) 0.2 to 0.5
(b) 0.5 to 1.0
(c) 1.0 to 2.0
(d) 1.5 to 4.0

14.17 Ultimate moment capacity of bonded prestressed concrete balanced rectangular section is about:
(a) $0.185\, bd^2 f_{ck}$
(b) $0 135\, bd^2 f_{ck}$
(c) $0.185\, bd^3 f_{ck}$
(d) $0.135\, bd^3 f_{ck}$
(where b = width, d = effective depth, f_{ck} = characteristic cube strength)

14.18 Ultimate moment capacity of unbonded prestressed concrete balanced rectangular sections is about:
(a) $0 185\, bd^2 f_{ck}$
(b) $0.135\, bd^2 f_{ck}$
(c) $0.185\, bd^3 f_{ck}$
(d) $0.135\, bd^3 f_{ck}$

14.19 Load factors for dead and live loads in the ultimate strength design of prestressed concrete beams in buildings are:
(a) 1.5 and 2.2
(b) 1.8 and 2.0
(c) 1.2 and 2.5
(d) 1.2 and 3.0

14.20 Combined load factor in the ultimate strength design of prestressed concrete beams in buildings is:
(a) 1.2
(b) 1.4
(c) 1.8
(d) 2.3

14.21 Load factors for dead and live loads in the ultimate strength design of prestressed concrete beams in which the dead and live loads act in the opposite directions are:
(a) 1.2 and 2.5
(b) 0.9 and 2.5
(c) 1.5 and 2.5
(d) 1.2 and 3.0

610

14.22 Very low percentage of steel in prestressed concrete beams results into:
    (a) Primary compression failure
    (b) Very low deflections
    (c) Very large deflections
    (d) Secondary compression failure

14.23 Secondary compression failure in prestressed concrete can be due to:
    (a) Low strength of the concrete
    (b) High strength of steel
    (c) Low strength of steel
    (d) Large deflections

14.24 Secondary compression failure in prestressed concrete beams can be due to:
    (a) Low strength of concrete
    (b) High strength of steel
    (c) High losses in prestress
    (d) High percentage of steel

14.25 Primary compression failure in prestressed concrete beams can be due to:
    (a) Low strength of concrete
    (b) High strength of concrete
    (c) High strength of steel
    (d) High percentage of steel

14.26 The endurance limit is defined as:
    (a) Maximum stress a material can withstand under repeated loading
    (b) Maximum number of repetitions a material can withstand
    (c) The limit of time (in hours) a material can withstand load
    (d) None of the above set

15. TOLERANCES, COVERS, SPACING, ETC.

15.1 Percentage tolerance in the nominal diameter of high tensile Steel cold-wires is about:
    (a) −1 to +2   (b) −2 to +5
    (c) −3 to +3   (d) −2 to +5

15.2 Nominal diameters (in mm) of finished high tensile steel cold drawn produced in India are:
    (a) 5, 6, 7, 8, 10   (b) 4, 5, 6, 7, 8
    (c) 3, 4, 6, 7, 8   (d) 2, 3, 4, 5, 6

15.3 Nominal diameter (in mm) of finished as-drawn high tensile steel wires is about :
    (a) 2, 3. 4, 5   (b) 3, 4. 5, 6
    (c) 3.25, 4, 5   (d) 2.5, 3.5, 5, 7

15.4 Tolerance in the nominal diameter of high tensile as-drawn steel wire is about :
    (a) ±0.1 mm   (b) ±0.05 mm
    (c) ±0.03 mm   (d) ±0.01 mm

15.5 Allowable percentage variation in the measurement of concrete mix proportions is about:
    (a) ±1   (b) ±5
    (c) ±10   (d) ±15

15.6 Allowable tolerance in the length of prestressed concrete beams from 6 to 15m long is about:
    (a) ±;10 mm   (b) ±15 mm
    (c) ±20 mm   (d) ±25 mm

15.7 Allowable tolerance in the length of prestressed concrete beams front 15 to 30 m length are about:
  (a) ±10 mm
  (b) ±15 mm
  (c) ±20 mm
  (d) ±25 mm

15.8 Allowable tolerance in length of prestressed concrete beams of span more than 30 m is about:
  (a) ±10 mm
  (b) ±15 mm
  (c) ±20 mm
  (d) ±25 mm

15.9 Allowable tolerance in depth of prestressed concrete beams upto 1 m deep shall be:
  (a) 2 mm
  (b) 5 mm
  (c) 10 mm
  (d) 15 mm

15.10 Minimum cover in any pretensioned prestressed concrete protected from weather is about:
  (a) 15 mm
  (b) 25 mm
  (c) 35 mm
  (d) 40 mm

15.11 Minimum cover to cables in prestressed concrete construction is about:
  (a) 30 mm
  (b) Size of the cable
  (c) 20 mm or size of the cable (larger of the two)
  (d) 30 mm or size of the cable (larger of the two)

15.12 Minimum cover in any pretensioned concrete exposed to weather is about:
  (a) 15 mm
  (b) 25 mm
  (c) 35 mm
  (d) 45 mm

15.13 Minimum clear spacing of single wires in pretensioned concrete is about:
  (a) 1.33 times the size of the aggregate
  (b) 2 times the diameter of the wire
  (c) 3 times the diameter of the wire or that listed in (a) whichever is bigger
  (d) 1.5 times the size of the aggregate or that listed in (b) whichever is bigger

15.14 Minimum clear spacing of non-grouped cables in prestressed concrete is :
  (a) 40 mm
  (b) Diameter of the cable
  (c) Larger of that in (a) or (b)
  (d) Larger of : 30 mm or diameter of the cable

15.15 Minimum clear spacing of non-grouped cables in prestressed concrete is about:
  (a) 40 mm
  (b) Maximum size of the aggregate
  (c) Larger of those listed in (a) and (b)
  (d) Diameter of the cable or the size of the aggregate whichever is larger

15.16 Minimum clear spacing of strands in prestressed concrete is about :
  (a) 40 mm
  (b) Maximum size of the aggregate
  (c) Larger of those listed in (a) and (b)
  (d) Diameter of the cable or the size of the aggregate whichever is larger

15.17 Minimum number of cables which can be grouped in on horizontal layer is:
  (a) Two
  (b) Three
  (c) Four
  (d) Six

15.18 Maximum number of cables which can be grouped in one vertical layer is:
  (a) Two
  (b) Three
  (c) Four
  (d) One

612

15.19 Minimum clear horizontal spacing between grouped cables in prestressed concrete is :
(a) 30 mm (b) 50 mm
(c) 65 mm (d) 75 mm

15.20 Minimum clear vertical spacing between grouped cables is about:
(a) 30 mm (b) 50 mm
(c) 65 mm (d) 75 mm

15.21 In case of fire, the temperature at which a prestressed wire is likely to become useless is:
(a) 200°C (b) 300°C
(c) 400°C (d) 500°C

15.22 Minimum cover to steel in prestressed concrete structures for fire protection up to 1 hour is about:
(a) 25 mm (b) 40 mm
(c) 60 mm (d) 100 mm

15.23 Minimum cover to steel in prestressed concrete structures for fire protection up to 2 hours is about:
(a) 25 mm (b) 40 mm
(c) 60 mm (d) 100mm

15.24 Minimum cover to steel in prestressed concrete structures for fire protection up to 4 hours is about:
(a) 25 mm (b) 40 mm
(c) 60 mm (d) 100 mm

15.25 A light steel mesh is to be placed in the cover concrete of prestressed concrete members in case thickness of the cover exceeds:
(a) 40 mm (b) 60 mm
(c) 80 mm (d) 100 mm

15.26 Vermiculite concrete cover over prestressed concrete helps in:
(a) Water proofing (b) Strengthening
(c) Durability (d) Fire resistance

15.27 Vermiculite gypsum cover over prestressed concrete helps in:
(a) Water proofing (b) Strengthening
(c) Durability (d) Fire resistance

15.28 Sprayed asbestos over the surface of prestressed concrete helps in:
(a) Water proofing (b) Strengthening
(c) Durability (d) Fire resistance

15.29 25 mm thick vermiculite concrete over the surface of prestressed concrete can protect the member against fire a for period of (in minutes):
(a) 30 (b) 120
(c) 200 (d) 300

15.30 22 mm thick vermiculite gypsum coating on the surface of prestressed concrete protects the concrete against fire for a period of (in minutes):
(a) 30 (b) 120
(c) 200 (d) 300

15.31 20 mm thick asbestos coating over the surface of prestressed concrete predicts the member against fire for a period of (in minutes):
(a) 30 (b) 120
(c) 200 (d) 300

16. CONCRETING, CURING, GROUTING, FORMWORK, ETC.

16.1 Proportioning of concrete mix in prestressed concrete should be by:
(a) Volume
(b) Weight
(c) Bagwise
(d) Volume or weight

16.2 Water-cement ratio in the concrete of prestressed concrete is about:
(a) 0.24 to 0.40
(b) 0.33 to 0.45
(c) 0.40 to 0.55
(d) 0.50 to 0.65

16.3 Mixing of concrete in prestressed concrete is by:
(a) Batching plant only
(b) Hand mixing also
(c) Mechanical mixer
(d) Any method of concrete mixing

16.4 Minimum period of mixing of concrete in prestressed concrete construction should be:
(a) 20 minutes
(b) 10 minutes
(c) 5 minutes
(d) 2 minutes

16.5 Compaction of concrete in prestressed concrete is by:
(a) Mechanical vibration only
(b) Hand compaction also
(c) Pnumatic vibrations only
(d) Electronic vibrations

16.6 Minimum curing of concrete in prestressed concrete is:
(a) 7 days under water
(b) 7 days under moist gunny bags
(c) 3 days under water
(d) 10 days under moist hessian cover

16.7 Curing of concrete in prestressed concrete structures must be by:
(a) Submersion under water only
(b) Sprinkling of water
(c) Steam curing or water sprinkling
(d) Moist covering or water sprinkling or by steam curing

16.8 In cold working conditions, the temperature of the green concrete be maintained at least at:
(a) 20°C
(b) 15°C
(c) 7°C
(d) 2°C

16.9 In cold working conditions, the temperature of the concrete while mixing for prestressed concrete construction be maintained at least at:
(a) 20°C
(b) 15°C
(c) 10°C
(d) 5°C

16.10 Temperature of concrete during hardening in prestressed concrete structures should not be below:
(a) 20°C
(b) 15°C
(c) 8°C
(d) 2°C

16.11 The maximum temperature of concrete while placing in the prestressed concrete structures should not be more than:
(a) 28°C
(b) 35°C
(c) 48°C
(d) 58°C

16.12 The mix proportions of cement-mortar to be placed at the construction joints must be:
(a) 1:1
(b) 1:2
(c) 1:3
(d) Same as that of the concrete

16.13 The thickness of the cement mortar to be placed at the constriction joint is to be about:
(a) 6 mm
(b) 12 mm
(c) 18 mm
(d) 25 mm

16 14 The thickness of cement grout to be used at the butted joints of precast prestressed concrete members can be about:
(a) 6 mm
(b) 12 mm
(c) 18 mm
(d) 25 mm

614

16.15 The thickness of cement mortar to be used at the butted joints of precast prestressed concrete segments can be about:
    (a) 15 to 40 mm   (b) 40 to 80 mm
    (c) 0 to 10 mm   (d) Any thickness

16.16 The mix proportion of cement mortar to be used in joints in precast prestressed beams is about:
    (a) 1 : 1.5   (b) 1 : 3
    (c) 1 : 4   (d) 1 : 5

16.17 Grouting of the ducts in post-tensioned prestressed concrete causes:
    (a) Reduction in deflections   (b) Increase in deflections
    (c) Increase in corrosion   (d) Reduction in strength

16.18 Grouting of the ducts in post-tensioned prestressed concrete causes:
    (a) Increase in deflections   (b) Reduction in crack width
    (c) Increase in crack width   (d) Reduction in strength

16.19 Grouting of the ducts in post-tensioned prestressed concrete causes:
    (a) Reduction in prestress: (b) Increase in prestress
    (c) Increase in corrosion   (d) Reduction in corrosion

16.20 Grouting of the ducts in post-tensioned prestressed concrete causes:
    (a) Decrease in durability   (b) Decrease in prestress
    (c) Increase in prestress   (d) Increase in durability

16.21 In case of small ducts in post-tensioned concrete construction, the following pre-grouting process is recommended:
    (a) Clean the duct with pressure   (b) Clean the duct with water
    (c) No pre-grout cleaning is needed   (d) Clean the duct with sodium chloride

16.22 In case of large ducts in post-tensioned concrete construction, the following pre-grouting process is recommended:
    (a) Clean the duct with pressure
    (b) Clean the duct with water
    (c) No pre-grout cleaning is needed
    (d) Clean the duct with sodium chloride

16.23 The following grout mix can be used in grouting of ducts in post-tensioned prestressed concrete:
    (a) Cement grout only
    (b) Cement or cement-sand mortar grout
    (c) Cement mortar grout only
    (d) Cement with resin

16.24 The grouting of ducts in post-tensioned concrete construction be done under a pressure of about (N/mm^2):
    (a) 0.1   (b) 0.5
    (c) 1.0   (d) 1.5

16.25 Vertical sides of formwork in prestressed concrete construction can be stripped off after .... hours casting:
    (a) 24   (b) 48
    (c) 72   (d) 120

16.26 Formwork in prestressed concrete can be stripped:
    (a) 3 days after casting
    (b) 7 days after casting
    (c) When the strength of the concrete reaches a value to withstand the working.
    (d) When the strength of the concrete reaches that needed of transfer strength

16.27 Mould of prestressed concrete should be:
    (a) Damp proof   (b) Water tight
    (c) Sealed tight   (d) Oil free

16.28 The deformation of the moulds used in prestressed concrete construction must be limited to:
    (a) Span/200   (b) Span/500
    (c) Span/1000   (d) Span/1500

16.29 The interior surface of the moulds in prestressed concrete construction be :
    (a) Treated with grease   (b) Treated with heavy duty oil
    (c) Washed with water and oiled   (d) Cleaned and oiled

17. PRESTRESS, CABLES, SHEETING

17.1 Approximate percentage loss of prestress in prestressed concrete caused by creep in steel is about:
    (a) 0 to 5   (b) 4 to 8
    (c) 7 to 13   (d) 10 to 15

17.2 Approximate percentage loss of prestress in prestressed concrete caused by creep in concrete is about:
    (a) 0 to 5   (b) 4 to 8
    (c) 7 to 13   (d) 10 to 15

17.3 Approximate percentage loss of prestress in prestressed concrete caused by shrinkage of concrete is about:
    (a) 0 to 5   (b) 4 to 8
    (c) 7 to 13   (d) 10 to 15

17.4 Approximate percentage of loss of prestress in prestressed concrete caused by slippage of anchorages in long span beams is:
    (a) 0 to 5   (b) 4 to 8
    (c) 7 to 13   (d) 10 to 15

17.5 Total percentage loss of prestress in pretensioned concrete beams is about:
    (a) 0 to 15   (b) 15 to 25
    (c) 20 to 30   (d) 30 to 40

17.6 Total percentage loss of prestress in post-tensioned prestressed concrete beams:
    (a) 0 to 15   (b) 15 to 25
    (c) 20 to 30   (d) 30 to 40

17.7 Loss of prestress in steam cured prestressed concrete beams is .... when compared to those water cured:
    (a) More   (b) Less
    (c) Same   (d) Not connected

17.8 Percentage loss of prestress caused by friction in shallow curved tendons is about:
    (a) 0 to 5   (b) 4 to 8
    (c) 7 to 13   (d) 10 to 15

17.9 Percentage loss of prestress caused by friction in curved element like pipes etc. is about:
    (a) 0 to 5   (b) 4 to 8
    (c) 7 to 13   (d) 10 to 10

17.10 High tensile steel is used in prestressed concrete because:
    (a) To minimise losses   (b) High strength is essential
    (c) High durability   (d) Low yield point

17.11 Mild or high yield deformed bars cannot be used as prestressing bars in prestressed concrete because:
  (a) Loss of prestress is very high
  (b) Low percentage elongation
  (c) Loss of prestress is high when compared to yield stress
  (d) Not economical

17.12 High strength steel wires can be supplied in rolls because:
  (a) They are light
  (b) They can be rolled without residual stress
  (c) They cannot be exposed to weather
  (d) It is less expensive to transport

17.13 Stress corrosion is a phenomenon which results into:
  (a) Deteriorated failure   (b) Sudden failure
  (c) Stress concentrations   (d) Stress failure

17.14 Stress corrosion in steel used in prestressed concrete results in:
  (a) Splitting failure of the steel   (b) Ductile failure
  (c) Slow failure   (d) Excessive stresses

17.15 Stress corrosion in steel is common in:
  (a) Deformed bars   (b) Mild steel
  (c) High yield steel   (d) High tensile steel

17.16 The magnitude of prestress force during prestressing shall be measured by:
  (a) Pressure gauge   (b) Extension of wire
  (c) Extensometer   (d) Pressure gauge and extension of wire

17.17 The high tensile steel cables used in prestressed concrete can be stored in :
  (a) Open shed also   (b) Damp proof shed only
  (c) Water proof shed only   (d) Water and damp proof shed

17.18 The prestressed cables or wires should be cut by:
  (a) Power saw   (b) Mechanical cutting tool only
  (c) Flame cutting only   (d) Mechanical tool or flame cutting

17.19 Coefficient of friction between steel wires moving on smooth concrete in prestressed concrete beams is about:
  (a) 0.5 to 0.55   (b) 0.35 to 0.40
  (c) 0.2 to 0.35   (d) 0.10 to 0.25

17.20 Coefficient of friction between steel wires moving on steel Sued in a duct in prestressed concrete beam is about:
  (a) 0.5 to 0.55   (b) 0.35 to 0.40
  (c) 0.2 to 0.35   (d) 010 to 0.25

17.21 Coefficient of friction between steel wires moving on steel fixed in concrete of prestressed concrete is about:
  (a) 0.5 to 0.55   (b) 0.35 to 0.40
  (c) 0.2 to 0.35   (d) 0.10 to 0.25

17.22 Coefficient of friction between steel stranded cables moving on steel sheath in prestressed concrete is about:
  (a) 0.5 to 0.55   (b) 0.35 to 0.40
  (c) 0.2 to 0.35   (d) 0.15 to 0.25

17.23 Coefficient of friction between steel stranded cables moving on spacer plates in prestressed concrete is about:
    (a) 0.5 to 0.55   (b) 0.35 to 0.40
    (c) 0.2 to 0.35   (d) 0.15 to 0.25

17.24 The following sheathing be used to form ducts in post-tensioned concrete :
    (a) Metallic only   (b) PVC or rubber only
    (c) Metallic or rubber   (d) Steel

17.25 Ducts are required in the following type of prestressed concrete construction:
    (a) Post-tensioning   (b) Pre-tensioning
    (c) Partial prestressing   (d) In all types

17.26 Pretensioned prestressed concrete requires 'the following type of ducts for wires:
    (a) Metallic only   (b) Rubber or metallic
    (c) Metallic   (d) None at all

17.27 The sheating of ducts in prestressed concrete construction be:
    (a) Water tight   (b) Damp proof
    (c) Not necessarily water tight   (d) Air tight

17.28 Sheathing used in the ducts of prestressed concrete must have a minimum tensile strength of (N/mm^2)
    (a) 1   (b) 5
    (c) 10   (d) Nominal

17.29 The wires or cables used in prestressed concrete be:
    (a) Greased before use to prevent corrosion
    (b) Cleaned free from oil
    (c) Washed with water
    (d) Cleaned with alcohol against rusting

17.30 The wires used in prestressed concrete must be treated as listed below before grouting
    (a) Grease the wires
    (b) Wash the ducts and wires with water
    (c) No special attention is needed
    (d) Wash it with cement water

17.31 A concordant cable in prestressed concrete is:
    (a) One which causes no secondary moments
    (b) One which causes secondary moments
    (c) Cojncides with centroidal axis of the beams
    (d) Parallel to the beam axis

18. SHEAR

18.1 The net shear force on concrete in prestressed concrete members is more when compared with that in RCC because:
    (a) Pre-compression
    (b) Transverse component of the cable force
    (c) Anchorage force
    (d) The statement is wrong to start with

18.2 The net shear force on concrete in prestressed concrete member is less when compared with that in RCC because:
    (a) Pre-compression
    (b) Transverse component of the cable force
    (c) Anchorage force
    (d) The statement is wrong to start with

18.3 Shear force on an element depends on:
    (a) Distributed loads only
    (b) Concentrated, loads only 1
    (c) Bending moment variation
    (d) Torsion force

18.4 Shear force on an element is proportional to:
    (a) Rate of change of moment
    (b) Uniformly distributed load
    (c) Concentrated loads
    (d) Distributed and concentrated loads

18.5 Shear force on section of a member acts:
    (a) Along the axis of the member
    (b) Normal to the axis of the member
    (c) Parallel to the plane of the section
    (d) In the vertical direction

18.6 Shear force on an arch section acts:
    (a) Along the axis of the arch
    (b) Vertically upwards
    (c) Normal to the plane
    (d) Normal to the axis of the arch

18.7 Shear force on slender prestressed concrete members acts:
    (a) Normal to the axis
    (b) Parallel to the axis
    (c) Vertically upwards
    (d) None of the above set

18.8 Direct action of shear force on prestressed concrete member has the tendency of:
    (a) Bending of the axis
    (b) Relative sliding of the planes
    (c) Twisting of the axis
    (d) Extension of the axis

18.9 Shear force on a prestressed concrete member can cause:
    (a) Torsion
    (b) Shear stress only
    (c) Bending moment also
    (d) Axial force

18.10 Pure bending moment on a prestressed concrete member causes:
    (a) Normal stresses only
    (b) Shear stresses also
    (c) Axial and shear stresses
    (d) Any type of stresses

18.11 Shear stress caused by transverse shear force in prestressed concrete beams is given by:

    (a) $\dfrac{VQ}{Ib}$
    (b) $\dfrac{V}{bjd}$

    (c) $\dfrac{V_e Q}{Ib}$
    (d) $\dfrac{V_e}{bjd}$

    (Where V = shear force, V_e = effective shear force, b = width, I = moment of inertia, Q = first moment of area, jd = lever arm distance.)

18.12 Shear stress caused by transverse shear force in fully prestressed concrete beams can be calculated similar to that of:
    (a) Beam made of steel
    (b) PCC beam
    (c) RCC beam
    (d) Composite beam

19. ANCHORAGE. BOND

19.1 Anchorage plates in post-tensioned prestressed concrete are placed:
    (a) Perpendicular to the plane of the concrete
    (b) Perpendicular to the axis of the tendon
    (c) Inclined to the plane of the concrete
    (d) No special relation

19.2 The size of the anchor plate in post-tensioned prestressed concrete depends on:
    (a) Stress in the cable and bearing stress of the concrete
    (b) Prestressing force
    (c) Bearing capacity of the concrete
    (d) Prestressing force and bearing capacity of the concrete

19.3 The size of the anchor plate in prestressed concrete is:
   (a) Equal to the net area of cross section of the concrete
   (b) Equal to the gross area of cross-section of the concrete at anchorage
   (c) Connected with the strength of the concrete
   (d) Connected with the bursting stress in concrete

19.4 Tendons in post-tensioned prestressed concrete:
   (a) Must be anchored on steel anchor plates
   (b) Can be anchored through cones bearing directly on concrete
   (c) Can be anchored through special buttoning without steel plates
   (d) Anchored on special concrete ends

19.5 Maximum tensile stresses in concrete at the anchorage zone of the post-tensioned prestressed concrete are generated on planes (approximately) :
   (a) Parallel to the axis of the cable
   (b) Normal to the end planes of the beam
   (c) Normal to the axis of the cable
   (d) Parallel to the end planes of the beam

19.6 Bursting stress in concrete at the anchorage zone of post-tensioned prestressed concrete is generated on planes approximately:
   (a) Parallel to the axis of the cable
   (b) Normal to the end planes of the beam
   (c) Normal to the axis of the cable
   (d) Parallel to the' end planes of the beam

19.7 In prestressed concrete construction, the steel anchor plates are needed in:
   (a) Pretensioned construction
   (b) Post-tensioned construction
   (c) Pretensioned and post-tensioned construction
   (d) Some types of post-tensioned construction

19.8 Anchorage cones to be embedded in concrete for holding tendons in prestressed concrete construction are needed in:
   (a) Pretensioned construction
   (b) Post-tensioned construction
   (c) Pretensioned and post-tensioned construction
   (d) Some types of post-tenstioned construction

19.9 Tendons anchored exactly at the centroid of a cross-section in prestressed concrete:
   (a) Will cause bursting stresses
   (b) Will not cause bursting stresses
   (c) Will cause secondary moments
   (d) None of the above set

19.10 Bursting stresses in prestressed concrete are developed at:
   (a) Anchorage zone
   (b) Along the tendon length
   (c) The maximum bending moment zone
   (d) The time of transfer of prestress

19.11 Transmission zone in post-tensioned prestressed concrete may be defined as the zone in which:
   (a) The prestressing force is effective
   (b) The prestressing force is ineffective
   (c) Bursting stresses are developed
   (d) Anchor plate is fixed

19.12 The approximate length of transmission zone in post-tensioned prestressed concrete is:
    (a) Equal to depth of the beam
    (b) Equal to width of the beam
    (c) Smaller of the above two
    (d) Equal to the smaller size of the anchor plate

19.13 The secondary stresses caused at the anchor zone of post-tensioned prestressed concrete beams consists of:
    (a) Tensile stresses
    (b) Compressive stresses
    (c) Compressive and tensile stresses
    (d) Compressive, tensile and shear stresses

19.14 Reinforcement provided in the anchor zone in post-tensioned prestressed concrete consists of:
    (a) Stirrups   (b) Mesh or grid
    (c) Closed loops or spirals   (d) Cross and longitudinal wings

19.15 Spelling stresses are produced in the post-tensioned prestressed concrete at:
    (a) Maximum bending moment zone
    (b) Maximum shear zone
    (c) Anchorage zone
    (d) Bond zone

19.16 Spalling stresses are produced in the post-tensioned prestressed concrete because of:
    (a) High concentrated tendon force
    (b) Bursting force
    (c) Inadequate anchor block
    (d) Insufficient bond length

19.17 Prestressing force in post-tensioned concrete beams is transferred to concrete through:
    (a) Bond between the cables and concrete
    (b) Anchor plates or cones
    (c) Bond and anchor
    (d) External anchor

19.18 Bonding of wires in post-tensioned construction is done:
    (a) To transfer the tendon force to the concrete
    (b) To minimise anchorage slip
    (c) To prevent corrosion
    (d) None of the above

19.19 Anchor cones in post-tensioned prestressed concrete beams are designed:
    (a) Primarily to resist hoop tension
    (b) Primarily to resist hoop compression
    (c) For bearing compression
    (d) For compression

19.20 Anchor plates in post-tensioned prestressed concrete are subjected to:
    (a) Compression   (b) Hoop tension
    (c) Bending   (d) Torsion

19.21 Anchor cones in post-tensioned prestressed concrete are:
    (a) Usually left open at the end sections
    (b) Embedded
    (c) Covered later
    (d) Left as they are and greased regularly

19.22 Anchor plates in post-tensioned prestressed concrete are:
  (a) Usually left open at the end sections
  (b) Embedded
  (c) Covered later
  (d) Left as they are and greased regularly

19.23 Anchor plates are used in pretensioned prestressed concrete beams for:
  (a) To transfer prestress
  (b) To bond the wires
  (c) To protect the wires from slipping
  (d) There are no such plates at all

19.24 Prestressing force from wires is transferred to the concrete in pretensioned prestressed concrete through:
  (a) Direct bond between wire and the concrete
  (b) Bond between ducts and the wires
  (c) Anchor cones
  (d) Anchor plates etc.

19.25 Commonly used high tensile steel in pre-tensioned prestressed concrete is in the form of:
  (a) Cables   (b) Wires and strands
  (c) Strands   (d) Bars and wires

19.26 High tensile steel wires are usually used in pretensioned prestressed concrete construction because of:
  (a) Ease in workability and handling   (b) Bond strength considerations
  (c) Ease in prestressing   (d) High strength

19.27 Secondary stresses caused in the pretensioned prestressed concrete around the wires are called:
  (a) Bending stresses   (b) Spalling stresses
  (c) Bond stresses   (d) None of the above set

19.28 Bond stresses are produced around the wires in the prestressed concrete at:
  (a) Transmission zone only   (b) Maximum bending moment zone
  (c) Maximum shear zone   (d) Concentrated loads

19.29 Grouting is done in pretensioned prestressed concrete to:
  (a) Provide bond   (b) Prevent corrosion
  (c) Improve the ultimate strength   (d) No grouting is really done

19.31 The length of the wire in pretensioned prestressed concrete in which the bond stress reduces from maximum to zero can be called:
  (a) Bond length
  (b) Transfer length
  (c) Anchor length
  (d) Kern length

19.32 The bond stress along the length of the wire in pre-tensioned prestressed concrete:
  (a) Is almost constant
  (b) Varies proportional to the bending moment
  (c) Decreases from maximum to zero in a short length
  (d) None of the above

19.33 Transfer length in pretensioned prestressed concrete is approximately equal to:
  (a) 20 D   (b) 48 D
  (c) 100 D   (d) 200 D
      Where D = diameter of the wire

19.34 The external moment in the transmission zone of pretensioned prestressed concrete:
   (a) Must be almost zero
   (b) Can be any thing provided enough prestress and concrete are provided
   (c) It should be less than the moment causing the bending stresses less than those permissible
   (d) None of above set

19.35 The shear force in the transfer zone of pretensioned prestressed concrete:
   (a) Must be almost zero.
   (b) Must be less than that which causes permissible shear stress in concrete
   (c) Any thing provided enough prestress is provided
   (d) None of the above set

KEY SHEET: 10   Prestressed concrete structures-General

	1	2	3	4	5	6	7	8	9	10
0	c	a	c	b	d	a	a	c	b	a
10	b	c	a	a	b	b	d	c	c	c
20	c	b	b	c	c	c	b	d	b	a
30	d	a	a	c	c	b	a	c	d	

KEY SHEET: 11   Concepts and definitions

	1	2	3	4	5	6	7	8	9	10
0	b	d	c	c	a	c	c	c	c	b
10	c	c	a	d	c	d	b	b	a	c
20	b	b	a	a	c	d	c	b	c	c
30	a	c	c							

KEY SHEET: 12   Material and permissible stresses

	1	2	3	4	5	6	7	8	9	10
0	b	b	b	c	a	a	c	a	c	b
10	c	b	d	d	c	a	a	c	a	a
20	c	b								

KEY SHEET: 13 Working stress design

	1	2	3	4	5	6	7	8	9	10
0	c	a	b	a	b	c	b	a	b	

KEY SHEET: 14 Ultimate Strength design

	1	2	3	4	5	6	7	8	9	10
0	c	b	b	a	c	a	a	c	d	b
10	a	c	b	b	c	d	a	b	a	c
20	b	d	d	c	d	a				

KEY SHEET: 15 Tolerances, covers, spacing, etc.

	1	2	3	4	5	6	7	8	9	10
0	a	c	c	c	b	a	b	c	b	a
10	d	b	c	c	c	c	a	a	b	b
20	c	b	c	d	b	d	d	d	b	b
30	b									

KEY SHEET: 16 Concreting, curing, grouting, form work, etc.

	1	2	3	4	5	6	7	8	9	10
0	b	b	c	d	a	b	d	d	d	d
10	c	d	b	b	a	a	a	b	d	d
20	b	b	b	a	a	c	b	c	c	

KEY SHEET: 17 Prestress, cables, sheeting

	1	2	3	4	5	6	7	8	9	10
0	a	c	a	a	c	b	a	b	d	b
10	c	b	b	a	d	d	d	d	a	b
20	c	d	d	c	a	d	a	d	b	b
30	a									

KEY SHEET: 18   Shear

	1	2	3	4	5	6	7	8	9	10
0	d	b	c	a	c	d	a	b	c	a
10	c	a								

KEY SHEET: 19   Anchorage, Bond, etc.

	1	2	3	4	5	6	7	8	9	10
0	b	d	c	b	a	a	d	d	a	a
10	c	a	d	c	c	a	b	c	a	c
20	c	c	d	d	a	d	b	d	a	d
30	b	c	c	a	a					

(Time required for working the solution of each problem without asterisk is 15 minutes. Add extra five minutes for every asterisk marked)

Questions in Prestressed Concrete Structures

(SOLUTIONS ARE GIVEN AT THE END OF THIS SECTION)

1. A simply supported prestressed concrete beam of effective span 8 m is subjected to a uniformly distributed load of 50 kN/m including dead load. The cross section of the beam 300 by 800 mm and is prestressed with a parabolic cable having prestressed force of 2400 kN. The cable is anchored at the CG of the end section with a maximum sag of 200 mm at mid span. Determine the maximum stress caused in the beam during the working load condition.

2. A cantilever beam of span 10 m is subjected to a dead load of 10 kN/m and live load of 30 kN/m. The overall depth of the beam section is 800 mm. Design a cable profile by load balancing method and by balancing full dead load and half live load. Show that the cable profile is given by

$$e = \frac{12.5\, x^2}{P}$$

where e = profile of the cable taken origin at the mid height of the beam at free end

P = prestressing force.

x = distance from the origin at free end.

3. A prestressed concrete electric pole is subjected to a transverse wind load of 1800 N at 6 m above the ground level. Design a rectangular cross section for the pole at the ground level subject to the following allowable stresses in the concrete. Transfer load condition

$$\sigma_{ti} = 0$$

$$\sigma_{ci} = 20 \text{ MPa}$$

Stresses at the effective wind load condition

$$\sigma_{te} = 0$$

$$\sigma_{ce} = 18 \text{ MPa}$$

4. A oneway roof slab of 5 m span axially prestressed to carry a live load of 2 kN/m^2. Design the section with the following allowable stresses in the concrete section: Allowable stresses at transfer are:

$$\sigma_{ci} = 20 \text{ MPa}$$

$$\sigma_{ti} = 0 \text{ MPa}$$

Allowable stresses at effective working load

$$\sigma_{ce} = 16 \text{ MPa}$$

$$\sigma_{te} = -1 \text{ MPa } (= 1 \text{ MPa compression})$$

The loss of prestress from the transfer prestress be taken as 20 per cent.

5. Based on the allowable stress design of prestressed concrete beam section, show that the upper and lower bounds for the prestressing force are

$$P_i \leq \frac{M_g}{e} + (\sigma_{ci} + \sigma_{ti}) \frac{Z}{2e} \text{ and}$$

$$P_t \geq \frac{M_w}{e\eta} - (\sigma_{ce} + \sigma_{te}) \frac{Z}{2e}$$

in which M_g and M_w are the external bending moments on the cross section at transfer and working load conditions.

Z = modulus of the section

e = eccentricity

η = ratio of the effective prestress to that at transfer σ_{ci}, σ_{ti} and σ_{ce}, σ_{te} are the allowable compressive and tensile stress in concrete at transfer and working load. The section is symmetrical about both the axes of the section.

6. Show that the upper and lower bounds of prestressing force of a prestressed concrete section are given by

$$P_i \leq \frac{M_w}{e} (+ \sigma_{ci} + \sigma_{ti}) \frac{I}{eh} \text{ and}$$

$$\eta P_i \geq \frac{M_w}{e} - (\sigma_{ce} + \sigma_{te}) \frac{I}{eh}$$

in which

P_i = prestress force at transfer

ηP_i = prestress force at working load

e = eccentricityof the prestress force below the CG

M_g, M_w = the bending moments on the section at transfer and working
load considered positive if causing compression on the top fibre

I = moment of inertia

h = overall depth

σ_{ci}, σ_{ti}, σ_{ce} and σ_{te}

are the allowable stresses on concrete.

7. Show that a prestressed concrete sectional property for the condition of upper and lower
bounds of prestressings to be equal is given by

$$\frac{I}{h} = \frac{M_w - \eta M_g}{\eta(\sigma_{ci} + \sigma_{ti}) + (\sigma_{ce} + \sigma_{te})}$$

in which

I = moment of inertia of the section

h = overall depth of the section

M_w = BM at working load

M_g = BM at transfer (due to self weight)

σ_{ij} = allowable stress, the first subscript refers to compression or tension,
the second subscript refers to transfer (i) or working (e) load condition

η = ratio of the prestressing force at working load to that at transfer.

8. A simply supported beam of span 10 m is subjected to a live load of 20 kN/m. Design a
rectangular section for the following allowable stresses.
At transfer condition :

$$\sigma_{ti} = 0, \quad \sigma_{ci} = 17 \text{ MPa}, \quad \sigma_{si} = 1050 \text{ MPa}$$

At working load condition

$$\sigma_{te} = 0, \quad \sigma_{ce} = 14 \text{ MPa}, \quad \sigma_{se} = 1050 \text{ MPa}$$

Total loss of prestress = 25 per cent or $\eta = 0.75$.

9. A section shown in Figure 1 is used in a cantilever beam of span 3 m. It is subjected to
a live load of 2.8 kN/m. Design an appropriate prestressing force for the following allowable
stresses.

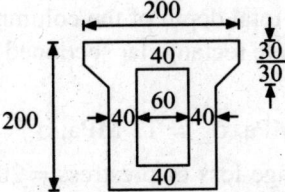

Figure 1 Cross section

$$\sigma_{ti} = 0, \sigma_{ci} = 18 \text{ MPa}, \sigma_{te} = 0 \text{ and } \sigma_{ce} = 15 \text{ MPa}$$

Capacity of 5 mm high tensile steel wire at transfer = 25 kN. The properties of the section are given here for convenience.

$$A = 22900 \text{ mm}^2, \ I = 97.24 \ (10^6) \text{ mm}^4$$

$$y_t = 93 \text{ mm}, \ y_b = 107 \text{ mm}.$$

10. A column is subjected to axial force of F and bending moment M. The column is axially prestressed. Show that the area of the column section is given by

$$A_c = \frac{F\left(1 - pm + \dfrac{Mh}{2Fr^2}\right)}{\sigma_{ce} - \eta p \sigma_{si}}$$

where $p = \dfrac{A_s}{A_c}$

   m = modular ratio

   h = depth of the section

   r = radius of gyration

   σ_{ce} = allowable compressive stress in concrete at working load

   σ_{si} = allowable tensile stress in steel at transfer

   η = ratio of effective prestress to that at transfer.

11. Design a circular cross-sectioned prestressed concrete column subjected to an axial force of 500 kN and bending moment of (10^7)Nmm. The allowable stress are:

$$\sigma_{si} = 1000 \text{ MPa}, \ \sigma_{ce} = 12 \text{ MPa}$$

$$\sigma_{ci} = 2 \text{ MPa},$$

   The modular ratio = m = 6

   Percentage of loss of prestress = 25

12. A rectangular cross-sectioned column is subjected to an axial compression of 800 kN and a bending moment of 400 kNm. The depth of the section is limited to 360 mm. Design a prestressed concrete column with the following allowable stresses:

$$\sigma_{ci} = 2 \text{ MPa}, \ \sigma_{ce} = 14 \text{ MPa}, \ \sigma_{si} = 1000 \text{ MPa}$$

   Modular ratio = 5; Percentage loss of prestress = 20.

13. A rectangular cross-sectioned column is subjected to an axial force of 1000 kN with eccentricity of 75 mm. The total depth of the column to be restricted to 450 mm. Design an axially prestressed concrete rectangular sectioned column for the allowable stresses as given below.

$$\sigma_{ci} = 1 \text{ MPa}, \ \sigma_{ce} = 12 \text{ MPa}, \ \sigma_{si} = 1000 \text{ MPa}$$

   Modular ratio = 6; Percentage loss of prestress = 20

14. A rectangular sectioned prestressed concrete column is subjected to an eccentric load with small eccentricity. Assume no tension is caused in the section. Show that the section of the column is given by

$$bh = \frac{F_u}{\sigma_c}\left(1 + \frac{6e}{h}\right) + \frac{\sigma_s A_p}{\sigma_c}$$

in which b = width, h = depth

$\quad\quad F_u$ = ultimate axial load

$\quad\quad e$ = eccentricity of the load

$\quad\quad \sigma_c$ = crushing stress in concrete

$\quad\quad A_p$ = area of pretensioned steel

$\quad\quad \sigma_s$ = net prestress in the steel

The value of σ_s is in the range of 260 MPa.

15. A prestressed concrete tie is subjected to axial tension of F and precompressed with a prestressing force of P_i at transfer condition. Show that the area of cross section of the concrete tie is given by

$$A_c = \frac{F(1 - mp)}{\eta p \sigma_{si} - \sigma_{ce}}$$

in which

$\quad\quad \sigma_{ce}$ = net compressive stress on the concrete section at working load

$\quad\quad m$ = modular ratio

$$p = \frac{A_p}{A_c}$$

$\quad\quad A_p$ = area of steel

$\quad\quad \eta$ = ratio of effective prestress to that at transfer.

16. A tie member is subjected to axial tension of 500 kN. The member is designed as a prestressed concrete tie with axially prestressed subject to the following allowable stresses.

Minimum compressive stress in concrete $\quad\quad\quad\quad$ = 3 MPa

Maximum compressive stress in concrete $\quad\quad\quad\quad$ = 20 MPa

Maximum allowable tensile stress in steel at working = 900 MPa

Loss of prestress = 20%, modular ratio $\quad\quad\quad\quad\quad$ = 5.

17. A cylindrical water tank has a diameter of 16 m and stores water up to a head of 6 m. Design a prestressed concrete cylindrical shell with minimum and maximum compressive stresses in the concrete as a 2 and 20 MPa. The maximum working allowable tension stress in high tensile steel is 900 MPa and that at transfer is 1100 MPa. The modular ratio is 5 and percentage loss of prestress is 20.

18. A penstock conduit of 2 m radius is carrying water at 200 m head of water. Design a prestressed conduit with a minimum of 3 MPa compressive stress at working load. The precompression in the concrete is restricted to 16 MPa. The loss of prestress can be taken as 20 per cent and modular ratio as 5. The allowable stress in steel wire at working load is 900 MPa.

19. Show that the ultimate moment capacity of under-reinforced rectangular sectioned prestressed concrete sections is

$$M_r = \gamma_0 A_p f_p \left(d - \frac{0.75 A_p f_p}{b f_{ck}} \right)$$

in which

γ_0 = material reduction factor

A_p = area of the tensioned steel

b = width of the section

d = depth of steel from the extreme compression fibre

f_p = characteristic stress in steel

f_{ck} = cube characteristic strength of concrete.

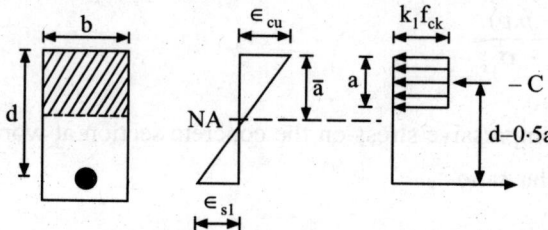

Figure 2 Idealised rectangular stress block on rectangular section.

Assume a rectangular stress block with ultimate strength of prism in bending as sixty seven per cent of that of the cube.

20. Show that the ultimate balanced moment capacity of prestressed concrete rectangular section is given by

$$M_r = 0.193 \, bd^2 f_{ck}$$

in which

the depth of rectangular stress block is assumed to be = 0.85 times the neutral axis distance.

Compressive strength of the prism is   = $0.67 f_{ck}$
The material reduction factor   = 0.9
f_{ck} = characteristic cube strength of the concrete
Yield strain in steel   = 0.008
Effective prestrain in steel   = 0.004
Crushing strain in concrete   = 0.0035.

21. Show that the percentage of high tensile steel in balanced prestressed concrete rectangular section in ultimate strength of the section is given by

$$\frac{100 \, A_p}{bd} = \frac{26.8 \, f_{ck}}{f_p}$$

where $\quad A_p$ = area of cross section of steel

$\qquad b$ = width of the section

$\qquad d$ = depth of CG of steel from extreme compression face

$\qquad f_{ck}$ = characteristic strength of concrete (cube)

$\qquad f_p$ = characteristic strength of steel

Yield strata in steel $\qquad\qquad\qquad\qquad$ = 0.008

Effective prestrain in steel $\qquad\qquad\qquad$ = 0.004

Crushing strain in concrete $\qquad\qquad\qquad$ = 0.0035

Use a rectangular compression stress block at ultimate moment in which its depth = 0.85 times the depth of the neutral axis and strength of the concrete prism in bending = 0.67 times the characteristic strength.

22. Design a rectangular sectioned pretensioned prestressed concrete beam to resist an ultimate bending moment of 600 kNm. Derive the ultimate moment capacity of the cross-section and area of steel are as

$$M_r = 0.193\ bd^2 f_{ck}$$

$$A_p = 0.268\ bd\ \frac{f_{ck}}{f_p}$$

where $\quad b$ = width of the section

$\qquad d$ = depth of CG of steel from extreme compression face

$\qquad f_{ck}$ = characteristic strength of concrete cube

$\qquad f_p$ = characteristic strength of steel

Yield strain in steel $\qquad\qquad\qquad\qquad$ = 0.008

Effective prestrain in steel $\qquad\qquad\qquad$ = 0.004

Crushing strain in concrete $\qquad\qquad\qquad$ = 0.0035

Also use the strength of concrete prism in bending equal to 0.67 times the characteristic strength and rectangular stress block at ultimate moment.

Use f_{ck} = 35 MPa, f_p = 1500 MPa and b = 300 mm.

23. Show that the ultimate moment capacity of flanged section of a prestressed concrete beam is

$$M_r = \gamma_0 t\ b_t\ k_p\ f_{ck} \left(d - \frac{t}{2} \right)$$

in which

$\qquad \gamma_0$ = material reduction coefficient

$\qquad t$ = thickness of the compression flange

$\qquad b_t$ = width of the compression flange

$\qquad d$ = depth of CG of steel from extreme compression fibre

$k_p f_{ck}$ = compressive strength of concrete prism in bending

and assume the depth of the neutral axis is more than the thickness of the flange and a rectangular compression block at ultimate moment.

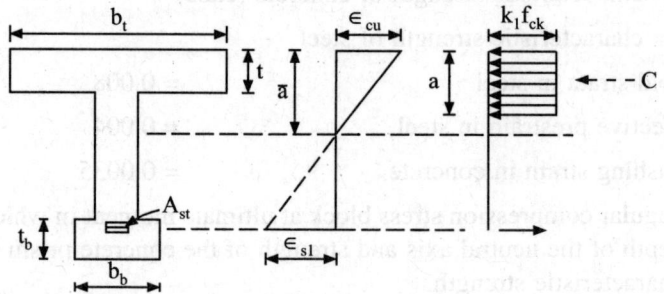

Figure 3   Idealised rectangular stess block on flanged section.

24. Show that the neutral axis will fall in the web portion in a prestressed concrete flanged section at the ultimate moment capacity if

    $d \geq 2.5\, t$

    in which d = depth of CG of steel from extreme compression fibre

    t = thickness of the compression flange provided the strains are:

Yield strain of steel	= 0.008
Effective prestrain in steel	= 0.004
Crushing strain in concrete	= 0.0035
Depth of rectangular stress block	= 0.85 depth of neutral axis.

25. Design a prestressed concrete I-beam section to resist an ultimate bending moment of 2400 kNm with top flange dimensions as 1250 mm width and 120 mm thickness. The characteristic strengths of steel and concrete are 1500 MPa and 35 MPa respectively. Design the web and bottom flange areas and the area of steel.

26. A two equal span continuous beam is provided with a straight prestressing cable having an eccentricity of e below the centroidal axis. The cross section of the beam is uniform and supported on three supports at same level as shown in the Figure 4. Determine the reactions caused by the prestressing cable alone and also compute secondary and net bending moments caused at the mid support.

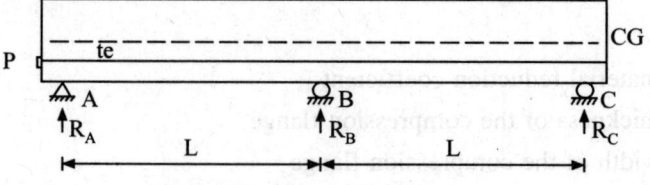

Figure 4   Continuous beam.

27. A prestressed concrete beam of uniform cross section is prestressed with a straight cable having an eccentricity of e below the centroidal axis. The beam is fixed at both ends even before the prestress is introduced. Show that the net bending moment at any section of the beam caused by the prestressed cable is zero.

28. A fixed ended prestressed concrete beam of uniform cross section is provided with a parabolic cable having a sag of g at mid span and zero eccentricity at the two end supports. The ends of the beam are fixed against rotation. Show that the net bending moment caused by the cable at the end section is given by $-2\,Pg/3$.

29. A two equal span continuous beam having a uniform cross section is provided with a prestressing cable having zero eccentricities at the end section and an eccentricity of e_B above the CG line at the mid support. The cable sags into a parabola in each of the spans as shown in Figure 5. The sag of the cable is g in each span. Show that the support secondary reaction at end support is equal to $R'_A = P(g - e_B)/L$.

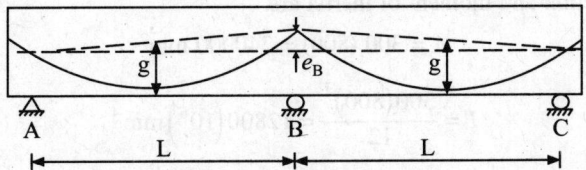

Figure 5   Continuous beam.

30. A two equal span continuous beam is uniform in cross section and has a prestressing cable as shown in Figure 5. Show that the cable can be a concordant cable if $g = e_B$ in which g = sag of the parabolic cable profile and e_B = eccentricity at mid support.

SOLUTIONS FOR QUESTIONS OF PRESTRESSED CONCRETE STRUCTURES

1.   Span = L = 8m
   Section :

$$b = 300 \text{ mm}$$
$$h = 800 \text{ mm}$$
$$P = 2400 \ kN$$
$$e = 200 \text{ mm}$$

   Load $w = 50 \ kN/m$
   Area of cross section and moment of inertia are

$$A = 300 \ (800) = 240000 \text{ mm}^2$$

$$I = \frac{300 (800)^3}{12} = 12800 (10^6) \text{ mm}^4$$

Maximum bending moment at mid span is

$$M = \frac{wL^2}{8} = \frac{50 \ (64)}{8} = 400 \ k\text{Nm}$$

The stresses caused by the prestressing force are:

$$-\frac{P}{A} = \frac{-2400 \ 000}{240 \ 000} = -10 \text{ N/mm}^2$$

$$\frac{P_e y}{I} = \pm \frac{2400 \ 000 (200)(400)}{12800 (10^6)} = \pm 12.5 \text{ N/mm}^2$$

The stress caused by the bending moment at mid span:

$$\frac{My}{I} = \pm \frac{400 \ (10^6)(400)}{12800 (10^6)} = \pm 12.5 \text{ N/mm}^2$$

The stress at the top fibre at mid span is

$$\sigma_t = -\frac{P}{A} + \frac{P_e y}{I} - \frac{My}{I}$$

$$= -10 + 15 - 12.5 = -7.5 \text{ N/mm}^2 \text{ (compression)}$$

The stress at the bottom fibre is

$$\sigma_b = -\frac{P}{A} - \frac{P_e y}{I} + \frac{My}{I}$$

$$= -10 - 15 + 12.5 = -12.5 \text{ N/mm}^2$$

2.   Span = L = 10 m

h = 800 mm

w_g = 10 kN/m

w_l = 30 kN/m

The balancing load is

$$w_b = 10 + \frac{30}{2} = 25 \ kN/m$$

Let the profile of the cable be given by a quadratic expression as the balancing load to be achieved is a UDL.

$$e = a + bx + cx^2 \tag{1}$$

where e = profile of the cable a, b, c are constants, and the origin of the coordinates be taken at the mid height of the beam section at free end.

The boundary conditions are:

At free end, that is at x = 0,

Shear force = 0 and

Bending moment = 0

which requires:

$$\text{Slope} = \frac{de}{dx} = 0 \ \text{and}$$

$$\text{eccentricity} = e = 0, \text{at } x = 0$$

This leads to $b = 0$ and

$a = 0$ then

The balancing force is set equal to w_b:

$$P\frac{d^2e}{dx^2} = 25 \ kN/m \tag{2}$$

Equations (1) and (2) along with $a = b = 0$ give

$$P(2c) = 25 \ \text{or} \ c = \frac{25}{2P} = \frac{12.5}{P}$$

The cable profile therefore is given by

$$e = cx^2 = \frac{12.5}{P}x^2$$

3.   Moment span = L = 6 m

Load w = 1800 N

Bending moment at the ground level of the pole.

$$M = wL = 6 \ (1800) = 10800 \ \text{Nm}$$

$$= 10.8 \ (10^6) \ \text{Nmm}$$

The moment is caused by the wind load, therefore the section must be designed to resist the moment acting from either direction. This means the prestressing force to be axial one and the section to be symmetrical. The allowable stress criteria are : (The units used are N and N/mm^2, mm)

At transfer condition :

$$\frac{P_i}{A} < \sigma_{ci} = 20 \text{ N/mm}^2 \qquad (2)$$

where A = area of the cross section at base
P_i = prestressing force at transfer.

The working load condition gives

$$\frac{P_e}{A} + \frac{M}{Z} \leq \sigma_{ce} = 18 \text{ N/mm}^2 \quad \text{and} \qquad (3)$$

$$\frac{P_e}{A} - \frac{M}{Z} \geq \sigma_{te} = 0 \qquad (4)$$

where Z = modulus of the section,

The Eq. (4) can be rearranged as

$$-\frac{P_e}{A} + \frac{M}{Z} \leq \sigma_{te} = 0 \qquad (5)$$

Eqs. (3) and (5) will yield

$$\frac{3M}{Z} \leq \sigma_{ce} + \sigma_{te} = 18 \quad \text{and} \qquad (6)$$

$$\frac{2P_e}{A} \leq \sigma_{ce} - \sigma_{te} = 18 \text{ N/mm}^2 \qquad (7)$$

or from Eq. (4)

$$P_e \geq \frac{MA}{Z} = \frac{6M}{h} \qquad (8)$$

$$Z \geq \frac{2M}{18} = \frac{2(10.8)(10^6)}{18} = 1.2(10^6) \text{ mm}^2 \qquad (9)$$

$$P_e \leq \frac{18A}{1} = 9A \text{ V} \qquad (10)$$

Let the section be rectangular one, then

$$Z = \frac{bh^2}{6} \geq 1.2(10^6) \text{ mm}^3 \qquad 1.2$$

Let b = 100 mm then

$$h^2 \geq \frac{1.2(10^6)(6)}{100} \text{ mm}^2 \quad \text{or}$$

$$h \geq 268 \text{ mm}$$

From Eq. (10)

$$P_e \leq 9A = 9(100) \ (268) = 24\ 1200 \text{ N} \qquad (11)$$

and from Eq. (9)

$$P_e \geq \frac{6M}{h} = \frac{6(10.8)(10^6)}{268} = 241791 \text{ N}$$

If $P_e = 0.8\, P_i$ (that is 20 per cent loss of prestressed) then from Eq. (2) we have

$$\frac{P_{ti}}{A} = \frac{P_e}{0.8A} \leq 20 \text{ or}$$

$$P_e \leq 16\, A \tag{12}$$

Eq. (11) will supersede over Eq. (12), therefore
Provide $P_e = 250\,000$ N

$$A_p = \frac{P_e}{\sigma_{se}} = \frac{250\,0000}{0.6(1500)} = 278 \text{ mm}^2$$

in which σ_{se} = allowable stress in high tensile steel at effective working load condition.
Minimum number of 5 mm wires required

$$= \frac{278}{19.6} = 15$$

Provide   8 numbers of 5 mm wires on each face of 100 by 268 mm section.

4.   Span $= L = 5\,m = 5000$ mm
Self weight $= w_g = $ A (density of the concrete)
Let the depth $= h$ mm; (normal working units are N, mm, or m)
The slab be designed for 1 m = 1000 lm width

$$w_g = \frac{1000h\,(1000)(24000)}{1000^3} = 24h \text{ N/m} \tag{1}$$

Bending moment due to self weight is

$$M_g = \frac{w_g L^2}{8} = \frac{24h(5^2)}{8} = 75h \text{ Nm}$$

$$= 75000\, h \text{ Nmm} \tag{2}$$

Bending moment due to live load is

$$M_1 = \frac{w_1 L^2}{8} = \frac{2000(25)}{8} = 6250 \text{ Nm}$$

$$= 6.25\,(10^6) \text{ Nmm} \tag{3}$$

$$M_w = M_g + M_l = 75000h + 6.25(10^6) \text{ Nmm} \tag{4}$$

The allowable stress criterion at working load gives :

$$\frac{P_e}{A} + \frac{M_w y}{I} \leq \sigma_{ce} \tag{5}$$

$$-\frac{P_e}{A} + \frac{M_w y}{I} \leq \sigma_{te} \tag{6}$$

In which eccentricity = 0 and P_e = effective prestressing force. For a rectangular section :

$$I = \frac{bh^2}{12}$$

$$y = \frac{h}{2}$$

Eqs. (5) and (6) give

$$\frac{2M_w y}{I} \leq \sigma_{ce} + \sigma_{te} \quad \text{or} \tag{7}$$

$$\frac{12M_w}{bh^2} \leq \sigma_{ce} + \sigma_{te} \quad \text{or}$$

$$h^2 \geq \frac{12M_w}{b(\sigma_{ce} + \sigma_{te})} \tag{8}$$

Given $\sigma_{ce} = 16$ MPa, $\sigma_{te} = -1$ MPa and $b = 1000$ mm.
Eqs. (4) and (8) give

$$h^2 \geq \frac{12\left(75000h + 6.25\left(10^6\right)\right)}{1000(15)} \quad \text{or}$$

$$\geq 60h + 5000 \quad \text{or}$$

$$h \geq \frac{60 \pm \sqrt{(3600 + 20000)}}{2} = 30 + 77 = 107 \text{ mm}$$

Eq. (6) gives

$$\frac{P_e}{A} \geq \frac{M_w y}{I} - \sigma_{te} \quad \text{or}$$

$$P_e \geq \left(\frac{6M_w}{bh^2} - \sigma_{ce}\right) bh = \frac{6M_w}{h} - bh\,\sigma_{te}$$

$$\geq \frac{6\left(75000h + 6.25\left(10^6\right)\right)}{h} - 1000\left(107(-1)\right)$$

$$\geq 6(75000) + \frac{6.25\left(10^6\right)(6)}{107} + 107\,000 = 907\,467 \text{ N}$$

with 20 per cent loss of prestress, the transfer prestress force should be

$$P_e = 0.8\,P_i \quad \text{or}$$

$$P_i = \frac{P_e}{0.8} = \frac{907\,467}{0.8} = 1134\,333 \text{ N}$$

Capacity of 5 mm wire at transfer condition

$$\frac{\pi}{4}(25)(0.7)(1500) = 206\,16 \text{ N}$$

(with allowable stress in steel at transfer = 0.7 (1500) N/mm²)
Number of wires needed

$$\frac{1134333}{20616} = 55$$

Provide 26 wires at top and 26 wires at bottom in 1000 mm width.

5.   The section is symmetric about both axes.
      Let
      P_i and P_e = prestressing force at transfer and working load respectively at the critical section
               e = eccentricity of the cable at the critical section
               Z = modulus of the section
      σ_{ti}, σ_{ci}, and σ_{te}, σ_{ce} are the allowable stresses in tension and compression at transfer and working load conditions.
      A and I = area and moment of inertia of the section.
      The allowable stress governing conditions at transfer load condition are:

$$-\frac{P_i}{A} + \frac{P_i s}{Z} + \frac{M_g}{Z} \le \sigma_{ti} \tag{1}$$

$$\frac{P_i}{A} + \frac{P_i e}{Z} - \frac{M_g}{Z} \le \sigma_{ei} \tag{2}$$

Similarly the allowable stress governing conditions at working load are:

$$\frac{P_e}{A} - \frac{P_e e}{Z} + \frac{M_w}{Z} \le \sigma_{ce} \tag{3}$$

$$-\frac{P_e}{A} - \frac{P_e e}{Z} + \frac{M_w}{Z} \le \sigma_{te} \tag{4}$$

Addition of in Eq. (1) and (2) gives

$$\frac{2P_i e}{Z} \le \left(\sigma_{ci} + \sigma_{ti}\right) + \frac{2M_g}{Z}$$

or $$P_i e \le \left(\sigma_{ct} + \sigma_{ti}\right)\frac{Z}{2} + M_g \tag{5}$$

Similarly addition of Ineq. (3) and (4) gives

$$P_e e \ge -\left(\sigma_{ce} + \sigma_{te}\right)\frac{Z}{2} + M_w \tag{6}$$

      Therefore the upper and lower bounds for the prestressing force from Ineqs. (5) and (6) with $P_e = \eta P_i$ are:

$$P_i \le \frac{M_g}{e} + \left(\sigma_{ci} + \sigma_{ti}\right)\frac{Z}{2e} \tag{7}$$

$$P_i \ge \frac{M_w}{\eta e} - \left(\sigma_{ce} + \sigma_{te}\right)\frac{Z}{2e\eta} \tag{8}$$

6.   The compressive and tensile stresses on a cross section at transfer condition to be less than the corresponding allowable stresses of the concrete results:

$$\frac{P_i}{A} + \frac{P_i e y_b}{I} - \frac{M_g y_b}{I} \le \sigma_{ct} \tag{1}$$

$$-\frac{P_i}{A} + \frac{P_i e y_t}{I} - \frac{M_g y_t}{I} \le \sigma_{ti} \tag{2}$$

      in which
          P_i = prestress force at transfer
          A = area of the section
          y_b, y_t = distance of the extreme bottom and top fibres from the *CG*

I = moment of inertia of the section

M_g = bending moment due to self weight compression causing on top fibre is positive

e = eccentricity of prestress below the *CG*.

Addition of Ineq. (1) and (2) gives

$$P_i e \frac{(y_b + y_t)}{I} \leq (\sigma_{ci} + \sigma_{ti}) + M_g \frac{(y_b + y_t)}{I}$$

It can be rearranged as

$$P_i \leq (\sigma_{ci} + \sigma_{ti}) \frac{I}{he} + \frac{M_g}{e} \tag{3}$$

in which $y_b + y_t = h$ = total depth of the section.

The actual compressive and tensile stresses on a prestressed concrete section at working load are to be less than the corresponding allowable stresses, this condition gives

$$\frac{P_e}{A} - \frac{P_e e y_t}{I} + \frac{M_w y_t}{I} \leq \sigma_{ce} \tag{4}$$

$$-\frac{P_e}{A} - \frac{P_e e y_b}{I} + \frac{M_w y_b}{I} \leq \sigma_{te} \tag{5}$$

The addition of the above two inequalities gives

$$P_e e \frac{h}{I} \geq M_w \frac{h}{I} - (\sigma_{cw} + \sigma_{tw})$$

or

$$P_e \geq \frac{M_w}{e} - (\sigma_{ce} + \sigma_{te}) \frac{1}{eh} \tag{6}$$

In eqs. (3) and (6) give the upper and tower bounds for prestressing force.

7. Vide solution 10.6 for the upper and lower bounds of the prestressing forces on a cross section, and they are

$$P_i \leq \frac{M_g}{e} + (\sigma_{cl} + \sigma_{ti}) \frac{I}{he} \tag{1}$$

$$P_i \geq \frac{M_w}{e\eta} - (\sigma_{te} + \sigma_{ce}) \frac{I}{eh\eta} \tag{2}$$

Equating the upper bound to the lower bound of in eq. (1) and (2), we have

$$\frac{M_w}{e\eta} - (\sigma_{te} + \sigma_{ce}) \frac{1}{eh\eta} = \frac{M_g}{e} + (\sigma_{ci} + \sigma_{ti}) \frac{I}{eh}$$

Rearrangement of the above equation gives

$$\frac{I}{h} \left(\eta(\sigma_{ci} + \sigma_{ti}) + \sigma_{te} + \sigma_{ce} \right) = M_w - \eta M_g$$

or

$$\frac{I}{h} = \frac{M_w - M_g}{\eta(\sigma_{ci} + \sigma_{ti}) + \sigma_{te} + \sigma_{ce}} \tag{3}$$

8. Simply supported span = L = 10 m

Live load $= w_1 = 20 \ kN/m$

Let the size of the beam for the purpose of dead load be

$$= \frac{L}{30} \times \frac{L}{20}$$

Self weight
$$= \frac{L^2}{600}(25) = 4\ kN/m$$

$$M_g = \frac{w_g L^2}{8} = \frac{4(100)}{8} = 50\ kNm$$

$$M_l = \frac{w_l L^2}{8} = \frac{20(100)}{8} = 250\ kNm$$

$$M_w = M_g + M_l = 300\ kNm$$

The upper and lower bounds of prestressing force

$$P_i \le \frac{M_g}{e} + (\sigma_{ci} + \sigma_{ti})\frac{I}{eh} \tag{1}$$

$$P_i \ge \frac{M_w}{e\eta} - (\sigma_{ce} + \sigma_{te})\frac{I}{eh\eta} \tag{2}$$

In the limit when the two bounds are equal to each other, we have

$$M_g + (\sigma_{ci} + \sigma_{ti})\frac{1}{eh} = \frac{M_w}{e\eta} - (\sigma_{ce} + \sigma_{te})\frac{1}{eh\eta}$$

or
$$\frac{I}{h} = \frac{M_w - M_g}{\eta(\sigma_{ci} + \sigma_{ti}) + \sigma_{ce} + \sigma_{te}}$$

$$= \frac{300(10^6) - 0.75(50)(10^6)}{0.75(17 + 0) + (14 + 0)} = 9.813(10^6)\ mm^3$$

but
$$I = \frac{bh^3}{12}$$

$$\frac{I}{h} = \frac{bh^2}{12}$$

Let (b is about $\frac{L}{30}$)

$$b = 350\ mm\ then$$

$$\frac{I}{h} = \frac{350h^2}{12} = 9.813(10^6)\ mm^3$$

$$h = \sqrt{\left(\frac{9.813(12)(10^6)}{350}\right)} = 580mm^3$$

Actual
$$M_g = (0.35)(.58)\frac{(25)(100)}{8} = 61\ kNm$$

$$I = \frac{350(580)^3}{12} = 5691\left(10^6\right) \text{mm}^4$$

$$\frac{I}{h} = 9.812\left(10^6\right) \text{mm}^3$$

Maximum or limiting $e = h/2 - \text{cover} = 280 - 60 = 220$ mm
The substitution of different quantities in Eqs. (1) and (2) gives

$$P_i = \frac{61(10^6)}{220} + \frac{17(9.812)(10^6)}{220} = 1.035(10^6) \text{ N}$$

$$= 1035 \text{ } kN$$

$$\sigma_{si} = 1050 \text{ MPa}$$

If σ_{si} is allowed at transfer the actual stress in the steel at working condition for 25 per cent losses is 787.5 N/mm² which is less than the allowable 900 N/mm². Therefore allowable stress at transfer governs. The area of prestress steel is

$$A_p = \frac{P_i}{\sigma_{si}} = \frac{1.035(10^6)}{1050} = 985 \text{ mm}^2$$

If 5 mm wire are used in the beam, then the number of wires are: (area of cross section of each wire = 19.6 mm²)

$$N = \frac{985}{19.6} = 51$$

Actual area of steel and prestress provided are:

$$A_p = 51 \, (19.6) = 999.6 \text{ mm}^2$$

$$P_i = A_p \sigma_{si} = 999.6(1050) = 1.0496(10^6) \text{ N}$$

Percentage of steel $= \dfrac{999.6(100)}{350(580)} = 0.492$

The actual stress in the section are computed and checked against the allowable values

$$A = 350 \, (580) \text{ mm}^2, \, y_b = y_t = \frac{h}{2} = 290 \text{ mm}$$

$$\frac{P_i}{A} = \frac{1.0496(10^6)}{350(580)} = 5.17 \text{ N/mm}^2$$

$$Z = \frac{I}{y_t} = \frac{I}{y_b} = \frac{5691(10^6)}{290} = 19.624(10^6) \text{ mm}^3$$

$$\frac{P_I e y_t}{I} = \frac{P_i e}{Z} = \frac{1.0496(10^6)(220)}{19.624(10^6)} = 11.77 \text{ N/mm}^2$$

$$\frac{M_g}{Z} = \frac{61(10^6)}{19.624(10^6)} = 3.11 \text{ N/mm}^2$$

$$\frac{M_e}{Z} = \frac{311(10^6)}{19.624(10^6)} = 15.85 \text{ N/mm}^2$$

Actual stresses at transfer condition are:

$$\frac{P_i}{A} + \frac{P_i e}{Z} - \frac{M_g}{Z} = 5.17 + 11.77 - 3.11 < \sigma_{ct} = 17$$

$$-\frac{P_i}{A} + \frac{P_i e}{Z} - \frac{M_g}{Z} = -5.17 + 11.77 - 3.11 > \sigma_{ti} \text{ (Not allowable)}$$

Actual stresses at working load condition are:

$$\frac{\eta P_i}{A} - \frac{\eta P_i e}{Z} + \frac{M_w}{Z} = 0.75 (5.17) - 0.75 (11.77) + 15.85$$

$$= 10.9 < \sigma_{ce} = 14$$

$$-\frac{\eta P_i}{A} - \frac{\eta P_i e}{Z} + \frac{M_w}{Z} = -0.75 (5.17) - 0.75 (11.77) + 15.85$$

$$= 3.145 > \sigma_{te} = 0 \text{ (Not allowable)}$$

The actual stresses in tension at transfer load and at working load are far higher than the allowable. It can be observed from the values that an increase in the prestressing increases the tension at transfer, therefore the section has to be increased with lowering of the eccentricity.

Let $\quad\quad\quad\quad b = 350$ mm, $h = 700$ mm

$$e = 160 \text{ mm}$$

$$P_i = 80 \ (19.6)(1050) = 1.6464 \ (10^6) \text{ N}$$

(using 80 nos. of 5 mm wires)

$$Z = \frac{700^2 (350)}{6} = 28.58(10^6) \text{ mm}^3$$

$$A = 700 \ (350) = 225000 \text{ mm}^2$$

$$= 0.225 \text{ m}^2$$

$$w_g = 0.225(25) = 5.5 \ k\text{N/m}$$

$$M_g = \frac{5.5(100)}{8} = 67.5 \ k\text{Nm}$$

$$= 67.5 \ (10^6) \text{ Nmm}$$

$$M_w = (250 + 67.5)(10^6) = 317.5(10^6) \text{ Nmm}$$

$$\frac{P_i}{A} = \frac{1.6464 \ (10^6)}{225000} = 7.32 \text{ N/mm}^2$$

$$\frac{P_i e}{Z} = \frac{1.6464\,(160)(10^6)}{28.58(10^6)} = 9.21 \text{ N/mm}^2$$

$$\frac{M_g}{Z} = \frac{67.5(10^6)}{28.58(10^6)} = 2.36 \text{ N/mm}^2$$

$$\frac{M_w}{Z} = \frac{317.5(10^6)}{28.58(10^6)} = 11.11 \text{ N/mm}^3$$

Actual stresses at transfer load condition are:

$$-\frac{P_i}{A} + \frac{P_i e}{Z} - \frac{M_g}{Z} = -7.32 + 9.22 - 2.36 = -0.46 < \sigma_{ti}$$

$$\frac{P_i}{A} + \frac{P_i e}{Z} - \frac{M_g}{Z} = 7.32 + 9.22 - 2.36 = 14.18 < \sigma_{ci}$$

Actual stresses at working load condition are:

$$\frac{\eta P_i}{A} - \frac{\eta P_i e}{Z} + \frac{M_w}{Z} = 0.75\,(7.32 - 9.22) + 11.11 = 9.685 < \sigma_{ce}$$

$$-\frac{\eta P_i}{A} - \frac{\eta P_i e}{Z} + \frac{M_w}{Z} = -0.75(7.32 - 9.22) + 11.11 = -1.3 < \sigma_{te}$$

All the stresses are less than the allowable and reasonably dose to the limiting values, hence, the design acceptable.

9. Cantilever span = 3m
   Live load = w_1 = 5 kN/m
   Self weight = w_g = 22900 (10^{-6}) 25000 = 550 N/m

$$\text{BM due to self weight} = M_g = \frac{w_g L^2}{2} = \frac{550\,(3)\,(3)}{2} = 2475 \text{ Nm}$$

$$= 2.5(10^6) \text{ Nmm}$$

The bending moment is causing tension on the top fibre in this beam and it is treated as positive for convenience.

$$\text{BM due to live load} = M_1 = \frac{w_1 L^2}{2}$$

$$= \frac{2.8(3)(3)}{2} = 12.6 \ k\text{Nm} = 12.6(10^6) \text{ Nmm}$$

$$M_w = M_g + M_l = (2.5 + 12.6)(10^6) = 15.1(10^6) \text{ Nmm}$$

The section properties and allowable stress are:

$$A = 22900 \text{ mm}^2,\ I = 97.24\ (10^6) \text{ mm}^4,\ y_t = 93 \text{ mm},\ y_b = 107 \text{ mm},$$

$$\sigma_{tl} = \sigma_{te} = 0,\ \sigma_{ci} = 18 \text{ MPa} : \sigma_{ce} = 15 \text{ MPa} = 15\text{N/mm}^2$$

The limiting prestressing forces are given by

$$P_i \le \frac{M_g}{e\eta} + (\sigma_{ti} + \sigma_{ci})\frac{I}{he} \tag{1}$$

$$P_i \ge \frac{M_w}{e\eta} - (\sigma_{ce} + \sigma_{te})\frac{I}{he\eta} \tag{2}$$

The distance of the top fibre from $CG = y_t = 93$ mm.

Assume $e = y_t - 50 = 43$ mm

Eqs. (1) and (2) give

$$P_i \le \frac{2.5(10^6)}{43} + \frac{(18)(97.24)(10^6)}{200(43)} = 0.26 (10^6) \text{ N}$$

$$\eta P_i \ge \frac{15.1(10^6)}{43} - \frac{15(97.24)(10^6)}{200(43)} = 0.19 (10^6) \text{ N}$$

$$P_i \ge \frac{0.19(10^6)}{\eta} = \frac{0.19(10^6)}{0.8} = 0.24 (10^6) \text{ N}$$

Maximum force in 5 mm wire at transfer = 25 kN

Number of 5 mm wtres = $\dfrac{P_i}{25000} \ge \dfrac{0.24 (10^6)}{25000}$

Provide 10 numbers of 5 mm wires

$$P_i = 24000 (10) = 240\ 000 \text{ N} = 0.24 (10^6) \text{ N}$$

$$P_e = 0.8(P_i) = 192\ 000 \text{ N}$$

Check for the actual stresses (stresses in N/mm²)

Stresses at transfer condition are:

Maximum compression which occurs at top fibre

$$\frac{P_i}{A} + \frac{P_i e y_i}{I} - \frac{M_g y_t}{I}$$

$$= \frac{0.25 (10^6)}{22900} + \frac{0.25 (10^6)(43)(93)}{97.24 (10^9)} - \frac{2.5 (10^6)(93)}{97.24 (10^6)}$$

$$= 10.9 + 10.3 - 2.4 = 10.9 + 7.9 = 18.8 \text{ N/mm}^2$$

Maximum tension which occurs at bottom fibre

$$-\frac{P_i}{A} + \frac{P_i e y_b}{I} - \frac{M_g y_b}{I}$$

$$= -10.9 + 7.9 \left(\frac{107}{93} \right) = -1.8$$

The stresses at working load are:

Maximum tension which occurs at top fibre is

$$= -\frac{P_e}{A} - \frac{P_e e y_t}{I} + \frac{M_w y_t}{I}$$

$$= -\frac{0.192 (10^6)}{22900} - \frac{0.2 (10^6)(43)(93)}{97.24 (10^6)} + \frac{15.1 (10^6)(93)}{97.24 (10^6)}$$

$$= -8.4 + (-8.6 + 15.1)\left(\frac{93}{97.24} \right) = -8.4 + 6.2 = -2.2$$

Maximum compressive stress which occurs at bottom fibre is:

$$\frac{P_e}{A} + (M_w - P_e e)\frac{y_b}{I} = 8.4 + 6.8\left(\frac{107}{97.24}\right) = 15.9$$

The actual stresses in compression are about six per cent more than the allowable values

10. F = external axial compressive force

M = external bending moment

The net axial strain in pretensioned steel

$$\epsilon_s = \epsilon_{sc} - \frac{F}{A_c E_c} \tag{1}$$

where ϵ_{se} = effective prestrain after losses.
The net axial prestressing force is

$$P = A_p \epsilon_s E_s$$

$$= A_p E_s \left(\epsilon_{se} - \frac{F}{A_c E_c}\right)$$

$$= P_e - \left(\frac{A_s E_s}{A_c E_c}\right)F$$

$$P = P_e - pnF$$

$$= \eta P_i - pmF \tag{2}$$

where p = ratio of steel = $\dfrac{A_p}{A_c}$

m = modular ratio = $\dfrac{E_s}{E_c}$

P_i = prestressing force at transfer.

The maximum compressive stress caused in the column under the combined action of P and M is

$$\sigma_{ce} = \frac{P}{A_c} + \frac{F}{A_c} + \frac{My}{I}$$

$$= \frac{\eta P_i - pmF}{A_c} + \frac{F}{A_c} + \frac{Mh}{2I}$$

$$= \eta A_p \sigma_{si} - \frac{(pm-1)F}{A_c} + \frac{Mh}{2I}$$

$$= \eta p \sigma_{si} + \frac{F}{A_c}(1 - pm) + \frac{Mh}{2I}$$

$$= \eta p \sigma_{si} + \frac{F}{A_c}\left(1 - pm + \frac{Mh}{2Fr^2}\right)$$

or

$$A_c = \frac{F(1 - pm + Mh/2Fr^2)}{\sigma_{cw} - \eta p \sigma_{st}} \tag{3}$$

11. At transfer load condition the compressive stress in concrete is

$$\sigma_{ci} = \frac{P_i}{A_c} = \frac{A_p \sigma_{si}}{A_c} = p\sigma_{si}$$

or

$$p = \frac{\sigma_{ci}}{\sigma_{si}} = \frac{2}{1000} = 0.002$$

$$F = 0.5\,(10^6)\ \text{N}$$

$$M = 10\,(10^6)\ \text{N}$$

$$\frac{M}{F} = 20\ \text{mm}$$

Let R = radius of the cross section, then the area of the section and radius of gyration are

$$A_c = \pi R^2$$

$$r^2 = \frac{I_c}{A_c} = \frac{R^2}{4}$$

or

$$r = \frac{R}{2}$$

The area of the cross section based on the maximum allowable compressive stress is

$$A_c = \frac{F\left(1 - pm + \dfrac{Mh}{2Fr^2}\right)}{\sigma_{ce} - \eta p\sigma_{si}}$$

or

$$\pi R^2 = \frac{0.5\,(10^6)\left(1 - 0.002(6) + \dfrac{20\,(2R)(4)}{2R^2}\right)}{12 - 0.75\,(0.002)\,(1000)}$$

$$= 5(10^5)\frac{(1 - 0.012 + 80/R)}{10.5}$$

$$\pi R^2 = (0.4705 + 38.1/R)10^5$$

$$\pi R^3 - 47050\,R - 3810000 = 0$$

The value of R by trial works out to be

$$R = 152.6\ \text{mm, then use } R = 153\ \text{mm}$$

$$A_c = \pi\,(153)^2 = 73542\ \text{mm}^2$$

$$A_p = pA_c = 0.002\,(73542) = 147.08\ \text{mm}^2$$

Provide 5 numbers of 7 mm wires.

12. Bending moment = $M = 40\,(10^6)\ \text{Nmm}$

Axial force = $F = 800\,000\ \text{N}$

Depth of the section = $h = 360\ \text{mm}$.

The axial stress in concrete at transfer load condition is

$$\sigma_{ci} = \frac{P_i}{A_c} - \frac{A_p \sigma_{si}}{A_c} = p\sigma_{si}$$

$$p = \frac{\sigma_{ci}}{\sigma_{si}} = \frac{2}{1000} = 0.002$$

The area of the cross section for columns is given by

$$A_c = \frac{F\left(1 - pm + \frac{MhA_c}{2FI}\right)}{\sigma_{ce} - \eta p\sigma_{si}} \tag{1}$$

$$\frac{2I}{A_c h} = \frac{2h^2}{12h} = \frac{h}{6} = \frac{360}{6} = 60 \tag{2}$$

$$e_0 = \frac{M}{F} = \frac{40\,(10^6)}{0.8\,(10^6)} = 50 \tag{3}$$

$$\frac{h}{e_0} = \frac{360}{50} > 6$$

in which e_0 = eccentricity of the external force

13.　　　$F = 1000\,kN = 10^6 N$

$e = 75$ mm

$h = 450$ mm

$m = 6$

$\eta = 0.8$

The stress in concrete at transfer is

$$\sigma_{ci} = \frac{P_i}{A_c} = \frac{A_p\sigma_{si}}{A_c} = p\sigma_{si}$$

or

$$p = \frac{\sigma_{si}}{\sigma_{si}} = \frac{1}{1000} = 0.001$$

$$\frac{e}{h} = \frac{75}{450} = \frac{1}{5}$$

There will be small tension in the section if prestress is not introduced. The column can be treated as under compressive stress throughout. The area of the section is given by

$$A_c = \frac{F\left(1 - pm + \frac{6e}{h}\right)}{\sigma_{ce} - \eta p\sigma_{si}}$$

$$= \frac{10^6(1 - 0.006 + 1.2)}{12 - 0.8(1)} = 0.196(10^6)\,\text{mm}^2$$

$$b = \frac{A_c}{h} = \frac{196000}{450} = 435\text{ mm}$$

$$A_p = pA_c = 196\,\text{mm}^2$$

Provide 435 by 450 mm section with 5 numbers of 7 mm wires.

14. Let P_e = axial effective prestressing force

P = net prestressing force at crushing of the concrete

ϵ_{se} = effective prestress strain = 0.004

ϵ_{cM} = crushing strain in concrete = 0.003

ϵ_{ce} = elastic compressive strain in concrete = 0.0003

Net strain in prestress steel at crushing of concrete is

$$\epsilon_s = \epsilon_{se} - (\epsilon_{cu} - \epsilon_{ce}) = 0.004 - 0.0027 = 0.0013$$

The net prestress is

$$P = \epsilon_c E_s A_p = 0.0013\ (2)(10^5)\ A_s$$

$$= 260\ A_p.\text{N (For } A_p \text{ in mm}^2)$$

Maximum compressive stress in the column section is

$$\sigma_c = \frac{F_u}{A_c} + \frac{P}{A_c} + \frac{M}{Z}$$

$$= \frac{F_u + P}{A_c} + \frac{6F_u e}{bh^2}$$

$$= \frac{F_u}{bh}\left(1 + \frac{6e}{h}\right) + \frac{P}{bh}$$

or

$$bh = \frac{F_u}{\sigma_c}\left(1 + \frac{6e}{h}\right) + \frac{P}{\sigma_c}$$

$$= \frac{F_u}{\sigma_c}\left(1 + \frac{6e}{h}\right) + \frac{\sigma_s A_p}{\sigma_c}$$

15. Let F = axial tension

P_i = axial prestressing force at transfer

P_e = effective axial prestress force

The compressive stress in a member axially precompressed is

$$\sigma_{cl} = \frac{P_i}{A_c} = \frac{A_p \sigma_{si}}{A_c} = p\sigma_{si} \tag{1}$$

where A_c = area of cross section of the concrete

σ_{si} = stress in steel at transfer

The net prestressing force under the action of an external tension of F is

$$P = P_i + \left(\frac{FE_s}{A_c E_c}\right) A_s$$

$$= P_i + pmF \tag{2}$$

The net compressive stress on the tie under working load condition is

$$\sigma_{cw} = \frac{P}{A_c} - \frac{F}{A_c} = \frac{A_p \sigma_{si}}{A_c} + \frac{pmF}{A_c} - \frac{F}{A_c}$$

$$= p\sigma_{si} - \frac{F}{A_c}(1 - mp) \tag{3}$$

or
$$A_c = \frac{F(1 - mp)}{p\sigma_{si} - \sigma_{cw}} \tag{4}$$

16. Axial tension = $F = 500\ kN$

Precompression on concrete = $\sigma_{ci} = 20$ N/mm^2

Net compressive stress at working load = $\sigma_{cw} = 3$ N/mm^2

The substitution of various quantities in the equation results into

$$A_c = \frac{500\ 000(1 - 5(0.02))}{0.8(0.02)(1000) - 3} = \frac{500\ 000(0.9)}{13} = 34615\ \text{mm}^2$$

Use 190 by 190 mm size section

A_c (provided) = 36 100 mm^2

The stress in steel at working load condition is

$$\sigma_{sw} = \sigma_{si} + \frac{EM}{A_c}$$

$$= 800 + \frac{500\ 000\ (5)}{36100} = 869\ \text{MPa}$$

which is less than the allowable value.

17. Diameter of the tank = $\quad D = 16$ m

Head of water = $\quad H = 6$ m

Maximum water pressure = $\ w = \gamma H = 10\ 000\ (6)$ N/m^2

Hoop tension = $\quad F = \dfrac{wD}{2} = \dfrac{60000(16)}{2} = 480\ 000$ N/m

The compressive stress at transfer (for lm width)

$$\sigma_{ci} = \frac{P_i}{A_c}\frac{A_p}{A_c}\ \sigma_{si} = p\sigma_{si}\ \text{ or}$$

$$p = \sigma_{ci}/\sigma_{si} = 20/1100.$$

The working stress in steel is

$$\sigma_{sw} = \sigma_{si} + \frac{Fm}{A_c}$$

The compressive stress on the wall at working load is

Allowable tension in steel at transfer = $\sigma_{si} = 1000$ MPa

Allowable tension in steel at working = $\sigma_{se} = 900$ MPa

$$\eta = 0.3 \text{ and } m = 5$$

The compressive stress at transfer is

$$\sigma_{ci} = \frac{P_i}{A_c} + \frac{A_p}{A_c}\sigma_{si} = p\sigma_{si}\ \text{ or} \tag{1}$$

$$p = \frac{\sigma_{ci}}{\sigma_{si}} = \frac{20}{1000} = 0.02$$

The stress in steel at working load condition is

$$\sigma_{sw} = \sigma_{se} + \frac{FE_s}{A_c E_c}$$

$$= \eta \sigma_{si} + \frac{Fm}{A_c}$$

The force in steel at working load is

$$P = A_p \sigma_{sw} = \eta A_p \sigma_{si} + Fmp$$

The compressive stress in concrete at working load is

$$\sigma_{ce} = \frac{P}{A_c} - \frac{F}{A_c}$$

$$= \eta \frac{A_p \sigma_{st}}{A_c} + \frac{Fmp}{A_c} - \frac{F}{A_c} \quad \text{or}$$

$$= \eta p \sigma_{si} - \frac{F}{A_c}(1 - mp)$$

$$A_c = \frac{F(1 - mp)}{\eta p \sigma_{si} - \sigma_{cw}}$$

$$\sigma_{ce} = \frac{P}{A_c} - \frac{F}{A_c} = \frac{A_p \sigma_{sw}}{A_c} - \frac{F}{A_c}$$

$$= p \left(\eta \sigma_{si} + \frac{Fm}{A_c} \right) - \frac{F}{A_c}$$

$$= \eta p \sigma_{si} - \frac{F}{A_c}(1 - mp) \quad \text{or} \tag{2}$$

$$A_c = \frac{F(1 - mp)}{\eta p \sigma_{si} - \sigma_{ce}} \tag{3}$$

The substitution of various quantities in the above gives

$$A_c = \frac{480\,000(1 - 5(20)/1100)}{0.8(20) - 2} = 31169 \text{ mm}^2$$

$$= bh \quad \text{or}$$

$$h = \frac{31169}{1000} = 31.169 \text{ mm}$$

The thickness is rather too small therefore provide a minimum thickness of 100 mm.

A_c (provided) = 100 (1000) = 100 000 mm²

The ratio of steel can be obtained from Eq. (2).

$$2 = 0.8p(1100) - \frac{480\,000}{100\,000}(1 - 5p)$$

$$= 880p - 4.8 + 24p = 904p - 4.8 \quad \text{or}$$

$$p = \frac{6.8}{904}$$

$$A_p = pA_c = \frac{6.8}{904}(100\,000) = 752 \text{ mm}^2/\text{m}$$

Provide 40 rounds of 5mm HT wire per metre.
The stress in steel at working load is

$$\sigma_{sw} = \eta\sigma_{si} + \frac{Fm}{A_c}$$

$$= 0.8\,(1100) + \frac{480\,000(5)}{100\,000} = 880 + 24 = 904 \text{ MPa}$$

Allowable stress is 900 MPa which slightly less than the actual stress 904 MPa.

18. Radius of the conduit = $R = 2$ m
Head of water = $w = 200$ m $= 200\,(10^4)$ N/m^2
Hoop tension = $F = wR = 4(10^6)$ N/m

Let $p = \dfrac{\sigma_{ci}}{\sigma_{si}} = \dfrac{16}{1100} = 0.0145$

The area of the concrete wall for 1000 mm depth is

$$A_c = \frac{F(1-mp)}{\eta\sigma_{ci} - \sigma_{cw}}$$

$$= \frac{4(10^6)(1 - 5(16)/1100)}{0.8(16) - 3} = 378478.66 \text{ mm}^2$$

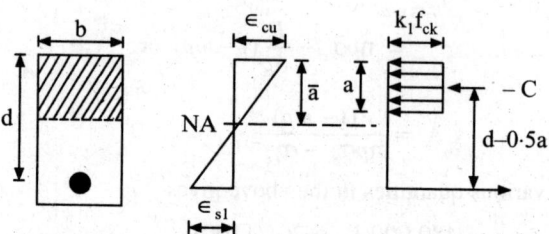

Figure 2   Idealised rectangular stress block.

Thickness of the wall is

$$t = \frac{A_c}{b} = \frac{378\,479}{1000} = 378 \text{ mm} \approx 380 \text{ mm}$$

Area of steel = $pA_c = \dfrac{16(378\,479)}{1100} = 5055 \text{ mm}^2$

Provide 258 rounds of 5 mm wire per metre.

19. Vide Figure 2 which illustrates the section, the strain across the depth and the rectangular stress block of compressive stress.
The equilibrium of forces on the section gives $(C = T)$

$$bak_1f_{ck} = A_pf_p \text{ or}$$ (1)

$$a = \frac{A_pf_p}{bk_1f_{ck}}$$

In which a = depth of the rectangular stress block.

   b = width of the section

The section is under-reinforced therefore, the stress in steel reaches the proof stress. The ultimate moment resistance of the section is

$$M_r = T\left(d - \frac{a}{2}\right)$$

$$= A_pf_p\left(d - \frac{A_pf_p}{22bk_1f_{ck}}\right)$$

If $k_1 = 0.67$, the above equation reduces to

$$M_r = A_pf_p\left(d - \frac{0.75A_pf_p}{bf_{ck}}\right)$$

20.   Vide Figure 2 for the rectangular stress block over the concrete cross section.

The neutral axis distance based on the simultaneous yielding of steel and crushing of concrete gives.

$$\frac{\bar{a}}{d - \bar{a}} = \frac{\epsilon_{cu}}{\epsilon_{yp} - \epsilon_{se}}$$

where ϵ_{yp} = yield strain in steel

   ϵ_{se} = effective prestrain in steel

   ϵ_{cu} = crushing strain in concrete

Let $\epsilon_{yp} = 0.008$, $\epsilon_{se} = 0.004$, $\epsilon_{cu} = 0.0035$, then the Eq. (2) gives

$$\frac{\bar{a}}{d - \bar{a}} = \frac{0.0035}{0.008 - 0.004} = \frac{35}{40}$$

$$\bar{a} = \frac{35d}{40 + 35} = \frac{7d}{15}$$

Let $a = 0.85\,\bar{a}$ while idealising the rectangular stress block, then

$$a = \frac{0.85(7)d}{15} = 0.4d$$ (2)

The ultimate moment capacity of the section (after applying a material reduction coefficient) is obtained by taking moment about the e steel is

$$M_r = \gamma_0(abk_1f_{ck})\left(d - \frac{a}{2}\right)$$ (3)

Substitution of Eq. (2) in Eq. (3) and $k_1 = 0.67$
and $\gamma_0 = 0.9$, we have

$$M_r = 0.9\,(0.4)\,(0.67)\,(0.8)\,bd^2f_{ck} = 0.193\,bd^2f_{ck}$$ (5)

21. Vide Figure 2 for the rectangular stress block and notations. The equilibrium of forces on the section gives

$$abk_1 f_{ck} = A_p f_p \tag{1}$$

where f_{ck} and f_p are the characteristic strength of concrete and steel.

a = depth of compression block, b = width of the section. The compatibility of strains in the steel and concrete for simultaneous occurrence of yielding in steel and crushing in concrete gives

$$\frac{\bar{a}}{d - \bar{a}} = \frac{\epsilon_{cu}}{\epsilon_{yp} - \epsilon_{se}} \tag{2}$$

in which

$\bar{a}$ = neutral axis distance

ϵ_{cu} = crushing strain in concrete taken equal to 0.0035

ϵ_{yp} = yield strain in steel taken equal to 0.008

ϵ_{se} = effective prestrain in steel (= 0.004)

d = depth of steel from the extreme compression fibre.

The Eq. (2) can be rearranged as

$$\bar{a} = \frac{\epsilon_{cu}\, d}{\epsilon_{yp} - \epsilon_{se} + \epsilon_{cu}} = \frac{0.0035d}{0.0075} = \frac{7d}{15} \tag{3}$$

The depth of the rectangular stress block is taken as

$$a = 0.85\, \bar{a} \; \frac{0.85(7)d}{15} = 0.4d \tag{4}$$

let $k_1 = 0.67$

Eqs. (1) and (4) give

$$A_p f_p = (0.4)(0.67)\, bdf_{ck} = 0.268\, bdf_{ck}$$

or

$$\frac{A_p}{bd} = \frac{0.268 f_{ck}}{f_{sk}} \tag{5}$$

For $f_{ck} = 35$ and $f_p = 1500$, the percentage of the reinforcement ratio is

$$\frac{100 A_p}{bd} = 0.268\left(\frac{35}{1500}\right)(100) = 0.625$$

22. Vide solutions 20 and 21 for the derivations.

$$A_p = 0.268bd\frac{f_{ck}}{f_p} \tag{1}$$

$$M_r = 0.193bd^2 f_{ck} \tag{2}$$

Given $f_{ck} = 35$ N/mm^2 and $f_p = 1500$ N/mm^2

Ultimate moment on the section is

$$M_u = 600\ kNm = 600(10^6)\ Nmm$$

Let

$$b = 300\ mm$$

then equating the ultimate moment to the resisting capacity, we have

$$0.193(300)\, d^2(35) = 600(10^6)$$

or

$$d = \sqrt{\left(\frac{600\,(10^6)}{0.193(300)(35)}\right)} = 545 \text{ mm}$$

$$A_p = 0.268bd\,\frac{f_{ck}}{f_p} = 0.268(300)\,\frac{(545)(35)}{1500} = 1022 \text{ mm}^2$$

Provide 27 numbers of 7 mm diameter wires, then area of steel provided
= 1039 mm².

Let the overall depth of the section be

$$d = d + 65 = 610 \text{ mm}.$$

23. Vide figure 3 in which the dimensions of a flanged sections are shown. The following assumptions are made:

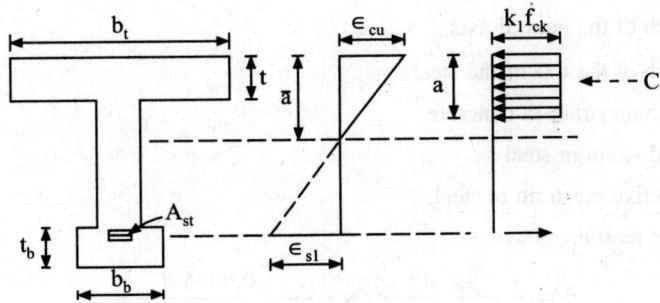

Figure 3

(1) Plane sections remain plane even after bending,

(2) Concrete cannot resist tension.

The depth of neutral axis is more than the thickness of the top flange. Strength of concrete prism in bending is = $k_1 f_{ck}$. The compression on the web is neglected in comparison to that on the flange. Total compressive force is:

$$C = tb_t\, k_1 f_{ck} \tag{1}$$

The total tension force is

$$T = A_p f_p \tag{2}$$

where t = thickness of the compression flange

b_t = width of the compression flange

f_{ck} = characteristic strength of concrete cube

f_p = characteristic strength of steel

A_p = area of cross section of steel.

The equilibrium of forces on the section gives

656

$$A_p f_p = t b_t k_p f_{ck}$$

or
$$A_p = t b_t k_p \frac{f_{ck}}{f_p} \tag{3}$$

The moment capacity of the section is obtained by taking the equilibrium of moments about the *CG* steel and it is

$$M_r = \gamma_0 C \left(d - \frac{t}{2} \right)$$

$$= \gamma_0 t b_t k_1 f_{ck} \left(d - \frac{t}{2} \right) \tag{4}$$

where γ_0 = material reduction factor.

24. Let the yielding of steel and crushing of concrete occurs simultaneously. Then the strain compatibility at the time of failure of the section is

$$\frac{\bar{a}}{d - \bar{a}} = \frac{\epsilon_{cu}}{\epsilon_{yp} - \epsilon_{sc}} \tag{1}$$

in which

$\bar{a}$ = depth of the neutral axis

d = depth of the *CG* of the steel

ϵ_{cu} = crushing strain in concrete

ϵ_{yp} = yield strain in steel

ϵ_{se} = effective prestrain in steel.

Eq. (1) can be rearranged as

$$\bar{a} = \frac{\epsilon_{cu} \, d}{\epsilon_{yp} - \epsilon_{sc} + \epsilon_{cu}} = \frac{0.0035 \, d}{0.008 - 0.004 + 0.0035}$$

$$= \frac{35 \, d}{75} \tag{2}$$

The depth of compression block is

$$a = 0.85\bar{a} = 0.85 \frac{(35) \, d}{75} = 0.4d$$

If $a = 0.4d \geq t$ then the neutral axis is in the web, this leads to
$d \geq 2.5 \, t$.

25. The ultimate moment on the section is

$$M_u = 2400 \ kNm = 2400 \ (10^6) \ Nmm.$$

The flange dimensions are:

$t = 120$ mm, $\qquad\qquad b_t = 1250$ mm

$f_p = 1500$ MPa, $\qquad\qquad f_{ck} = 35$ MPa

The material reduction factor $= \gamma_c = 0.9$

Vide solution 24 for the moment capacity and area of steel

$$M_r = \gamma_c b_t t \, k_1 f_{ck} \left(d - \frac{t}{2} \right) \text{ and} \qquad (1)$$

$$A_p = b_t t \, k_1 \frac{f_{ck}}{f_p} \qquad (2)$$

The Eq. (1) can be rearranged as

$$d = \frac{M_r}{b_t t k_1 f_{ck}} + 0.5t$$

Equating M_r to M_u and $k_1 = 0.67$, we have

$$d = \frac{2400(10^6)}{(0.9)(1259)(20)(0.67)(35)} + 60 = 818 \text{ mm}$$

$$A_p = \frac{1250(120)(0.67)(35)}{1500} = 2345 \text{ mm}^2$$

Provide 61 numbers of 7 mm diameter wires

$$A_p \text{ (provided)} = 2347 \text{ mm}^2$$

Select the overall depth as

$$h \simeq 1.15 \, d = 940 \text{ mm}$$

width of web = 150 mm

Also provide a bottom flange equal to half the area of the top flange to resist the precompression,

$$b_b = 625 \text{ mm}$$

$$t_b = 120 \text{ mm}$$

Ultimate tensile strength of the steel

$$T_u = A_p f_p = 2347 \,(1500) = 35\,20500 \text{ N}$$

Assuming the ultimate pretension is resisted by the bottom flange alone, the stress on the bottom flange above is

$$= \frac{T_u}{b_b t_b} = \frac{520500}{(625)(120)} = 47 \text{ N/mm}^2$$

and it is less than that caused by self weight.

26. The two span continuous beam is made statically determinate by releasing the central support B. The beam is now treated as a simply supported at A and C with bending moment equal to that caused by the eccentricity of the cable. The eccentricity of the cable is constant and it is equal to e for $x = 0$ to $x = 2L$. Therefore the bending moment due to the prestressing force

$$M_{ox} = -Pe \qquad (1)$$

where M_{ox} = BM due to P on the determinate form of the structure. The negative sign indicates the tension is caused on the top fibre.

P = prestressing force

e = eccentricity below the CG is positive.

The deflection at B caused by the constant bending moment is given by

$$v_{oB} = M_{ox}(2L) = -\frac{P_e L^2}{2EI}$$
(2)

The negative sign indicates the deflection is upwards.

Let R'_B = reaction of support B acting upwards is positive.

The reaction R'_B can be treated as a concentrated load acting at mid span of the beam AC and R'_A and R'_C are the reactions from the R'_B:

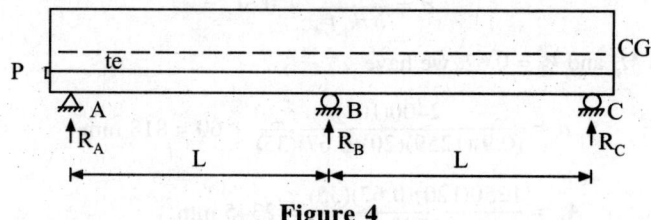

Figure 4

then each of them is equal and opposite to half of R'_B from symmetry The deflection caused by the concentrated upward reaction R'_B on the span AC is

$$v^1_B = -\frac{R^1 x(2L)^3}{48EI} = -\frac{B^1 x L^3}{6EI}$$
(3)

The negative sign is because the deflections are measured positive downward.

The compatibility condition at the support B is that the net deflection must be equal to zero; then

$$v_{oB} + v^1_B = 0$$
(4)

which gives from Eqns. (2) and (3)

$$-\frac{PeL^2}{2EI} - \frac{R^1_B L^2}{6EI} = 0 \text{ or}$$

$$R^1_B = -\frac{3Pe}{L}$$
(5)

This means, a downward reaction at B. This in turn gives rise to upward reactions at A and C and they are

$$R^1_A = R^1_C = -\frac{R^1_B}{2} = \frac{3Pe}{2L}$$
(6)

The secondary bending moment at B caused by the secondary reactions is

$$M^1_B = R^1_A L = \frac{3Pe}{2}$$
(7)

The net bending moment caused by the prestressing force at any distance x from A is

$$M_x = -Pe + R_A x = -Pe + \frac{3Pex}{2L} = Pe\left(\frac{3x}{2L} - 1\right)$$
(8)

The bending moment at mid support is

$$M_B = M_{oB} + M^1_B = -Pe + \frac{3Pe}{2} - \frac{Pe}{2}$$

27. An eccentric cable with constant eccentricity produces a bending moment of

$$M_{cx} = -Pe \qquad (2)$$

where the negative sign indicates bending moment causing tension on the top fibres. The slope of the beam caused at the support due to the moment M_{cx} if the beam is assumed to be simply supported is

$$\theta_{0A} = \frac{1}{2EI} \int_{0}^{L} M_{ox} dx = -\frac{PeL}{2EI} \qquad (1)$$

in which

θ_{0A} = slope of the beam at support A (left end support by the cable alone in case the beam is simply supported. The clockwise slope is treated as positive.)

Because the beam is symmetric, the fixed end moments can be assumed to be equal at the two end supports.

Let M^1_A = secondary fixed moment at the left end support causing compression at the top fibre.

The secondary reactions at both the supports must be equal to zero from symmetry. The slope of the beam at the left end support caused by the secondary moment is

$$\theta^1_A = \frac{M^1_A L}{2EI} \qquad (3)$$

The compatibility condition of the slope at the support gives

$$\theta_{0A} + \theta^1_A = 0$$

or

$$-\frac{PeL}{2EI} + \frac{M^1_A L}{2EI} = 0 \qquad (4)$$

The net moment at support is

$$M_A = M_{0A} + M^1_A = 0$$

Similarly the net moment caused by the prestressed cable at mid span is

$$M_C = -Pe + M^1_A = 0$$

28. A parabolic cable profile can be replaced by a uniformly distributed load, and the intensity of the load is

$$q = -\frac{8Pg}{L^2} \qquad (1)$$

The negative sign indicates an upward load. The fixed bending moment due to q is

$$M^1_A = -\frac{qL^2}{12} = +\frac{8PgL^2}{12L^2} = \frac{2Pg}{3} \qquad (2)$$

The net bending moment at support is that due to the eccentricity plus the secondary moment. Therefore

$$M_A = O + M^1_A = \frac{2Pg}{3}$$

The net bending moment at mid span is

$$M_c = -Pg + \frac{2Pg}{3} = \frac{Pg}{3}$$

29. The cable can be replaced by the equivalent balancing force. As the cable is anchored at the centroid at the end sections, there is no end moments. The cable is parabolic in each span with a sag equal to g. The balancing force from the cable is UDL equal to

$$w_b = \frac{-8Pg}{L^2} \tag{1}$$

The two span beams therefore develop a bending moment at mid support equal to

$$M_B = -\frac{w_b L^2}{8} \tag{2}$$

The load w_b is assumed positive (acting downwards) therefore the balancing load is indicated negative the bending moment at the middle support is also negative as the downward load causes tension at the top fibre near to the intermediate support.

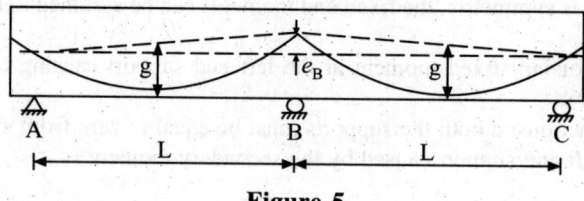

Figure 5

Eqs. (1) and (2) give

$$M_B = -\frac{w_b L^2}{8} = Pg \tag{3}$$

The final moment at B is independent of the eccentricity at B and it is a function of the sag. The secondary bending moment at B is

$$M^1{}_B = M_B - Pe_B = P(g - e_B) \tag{4}$$

The support secondary reaction at A is given by:

$$R^1{}_A L = M^1{}_B = P(g - e_B) \text{ or}$$

$$R^1{}_A = \frac{P}{L}(g - e_B) \tag{5}$$

and

$$R^1{}_B = -2R^1 A = -\frac{2P}{L}(g - e_B) \tag{6}$$

30. Vide solution 29 for secondary reactions

$$R^1{}_A = \frac{P}{L}(g - e_B)$$

For a concordant cable, the secondary reactions must be equal to zero.

So

$$R^1{}_A = 0 \text{ or}$$

$$(g - e_B) = 0$$

$$g = e_B.$$

References

1.1. Lin, T.Y., Load Balancing for Design and Analysis of Prestressed Concrete Structures, *J. of the ACI*, Vol. 60, No. 6, June 1963, pp. 719–42.

1.2. Indian Standard Code of Practice for Prestressed Concrete, IS: 1343–1960, New Delhi, (under revision in 1981).

1.3. Desai, P. and Krishnan, S., Equation for Stress Strain Curve of Concrete, *J. of the ACI*, Vol. 61, No. 3 March 1964.

1 4. Evans, R.H. and Kong, F.K., Estimation of Shrinkage of Concrete in Reinforced and Prestressed Concrete Design, *J. of Civil Engineering and Public Works Review*, Vol. 62, No. 5, May 1967, pp. 563–75.

1.5. Meyers, B.L., Branson, D.E. and Schumann, C.G., Prediction of Creep and Shrinkage Behaviour for Design from Short Terms Tests, *J. of PCI*, Vol. 17, No. 3, May–June 1972, pp. 29–45.

1.6. Muspratt, M.A., The effects of elevated temperature on concrete prestressed with alloy steel bars, *Indian Concrete Journal*, Dec. 1968, Vol. 42, p. 516.

1.7. Dayaratnam, P. and Rao, C.V.S.K., "Theory of Failure of Concrete by Distortion Energy", Research report CE 1–67, April 1967, Indian Institute of Technology, Kanpur.

1.8. Baikov, V.N., *Reinforced Concrete Constructions*, Rajkamal Prakashan, Delhi, 1963, p. 12.

1.9. Pattanayak, U.C. and Dayaratnam, P., Shear Flexure Capacity of Prestressed Concrete Beams, Proc. of the International Conference on Shear, Bend and Torsion of Reinforced and Prestressed Concrete, Coimbatore, January 1969.

1.10. Sheikh, M.A., de Paiva, H.A.R. and Neville, A.M., Calculation of Flexure Shear Strength of Prestressed Concrete Beam, *J. of PCI*, Vol. 13, No. 1, February 1968, pp. 68–85.

2.1. PCI Committee on Post-tensioning Tentative Specifications for Post-tensioning Materials, *J. of PCI*, Vol. 16, No. 1, Jan.–Feb. 1971, pp. 14–20.

2.2. Saglio, Puyo, A. and Picaut, J., Prestressing with High-tensile Steel Strips, *J. of PCI*, Vol. 16, No. 1, Jan.–Feb. 1971, pp. 48–60.

2.3. Schechter, E. and Boecker, H.C. Wedge, Anchorage System for Strand Post-tensioning, *J. of PCI*, Vol. 16, No. 4, July–Aug. 1971, pp, 49–63.

2.4. Branson, D.E. and Kripnarayanan, K.M., Loss of Prestress, Camber and Deflection of Non-composite and Composite Prestressed Concrete Structure, *J. of PCI*, Vol. 16, No. 5, Sept.–Oct. 1971, pp. 22–52.

2.5. Hilderbrand, J.F., Buttonheads for Tendon Wires of a Prestressed Concrete Reactor Vessel, *J. of PCI*, Vol. 16, No. 5, Sept.–Oct. 1971, pp. 78–99.

2.6. Glodowski, R.J. and Lorenzetti, *J.J.*, A Method for Predicting Prestress Losses in a Prestressed Concrete Structure, *J. of PCI*, Vol. 17, No. 2. Mar.–Apr. 1972, pp. 17–31.

2.7. Schupack Large, Post-tendoning Tendons, *J. of PCI*, Vol. 17, No. 3, May–June 1972, pp. 14–23.

2.8. Sinno, R. and Furr, H.L., Computer Programme for Predicting Prestress Loss and Camber, *J. of PCI*, Vol. 17, No. 5, Sept.–Oct. 1972, pp. 27–38.

2.9. Kourkene J., Post-tensioning Applications in Petroleum Industry, *J. of PCI,* Vol. 17. No. 5. Sept.–Oct. 1972, pp. 20–6.

2.10. Abeles, P.W. and Kung, R., Prestress Losses Due to the Effect of Shrinkage and Creep on Nontensioned Steel, *J. of ACI*, Proc. V. 70, No. 1, Jan. 1973, pp. 19–7.

2.11. Grouni, H.N., Prestressed Concrete—A Simplified Method for Loss Computation, *J. of ACI*, Proc. V. 70, No. 2, Feb. 1973, pp. 108–14.

2.12. Kidron, Z. Pretensioned Concrete Members with Varying Prestress, *J. of PCI*, Vol. 18, No. 4, July–Aug. 1973, pp. 64–76.

2.13. Shockey, D. Involvement—The Key to Successful Loss Prevention, *J, of PCI*, Vol. 18, No. 4, July–Aug. 1973, pp. 61–63.

2.14. Ghali, A. and Dileger, W. Rapid Accurate Evaluation of Prestress Losses, *J. of ACI*, Proc. V. 70, No. 11, Nov. 1973, pp. 759–763.

2.15. Hollington, M.R., An ABC of Prestressed Concrete published through the Cement and Concrete Association, London, *Indian Concrete Journal*, Nov. 1963.

2.16. Narasimhan, K.Y., A Prestressing Device, *Indian Concrete Journal*, May 1964, Vol. 38, p. 182.

2.17. Muspratt, M.A., Stress Losses in Concrete in Prestressed Members, *Indian Concrete Journal*, May 1971, Vol. 45, No. 3.

2.18. Martin, L.D. and Norman Scott L., Development of Prestressing Strand in Pretensioned Members, *J. of ACI*, August 1976, No. 8.

2.19. Whittakar, John B., Low Temp. Curing of Precast P:S.C. Products, JPCI, July–Aug. 1977, Vol. 22, No. 4.

2.20. Zia Paul, Kent, Preston H., Norman Scott L. and Workmen Edmin B., Estimating Prestress Losses, Concrete International, Design & Construction, *A.C.I.*, June 1979, Vol. 1, No. 6.

2.21. Johns Michael R., Post-compression Prestressing Technique, Concrete International, *A.C.I.*, March 1979, Vol. 1, No. 3.

3.1. Design Criteria for Prestressed Concrete Road Bridges, (Post-tensioned Concrete), IRC, New Delhi, 1969.

3.2. IS Code of Practice for Concrete Structures for the Storage of Liquids, Part III, Prestressed Concrete Structures, IS: 3370 (Part III), 1967.

3.3. Dayaratnam, P. and Patnaik, S.N., Feasibility of Full Stress Design, *J. of the American Institute of Aeronautics and Astronautics*, Vol. 7, No. 4, pp. 773–74, 1969.

3.4. IS Specifications for High Tensile Steel Bars Used in Prestressed Concrete, IS: 2090–1962, New Delhi.

3.5. IS Specifications for Plain Hand Drawn Steel Wire for Prestressed Concrete (Cold Drawn Stress Relieved Wire), IS: 1785–1966, New Delhi.

3.6. IS Specifications for Plain Hard Drawn Steel Wire for Prestressed Concrete, Part II as Drawn Wire, IS: 1785 (Part II)–1967, New Delhi.

3.7. IS Specifications for Prestressed Concrete Street Lighting Columns, IS: 2193–1962, New Delhi.

3.8. ACI–Building Code Requirements for Reinforced Concrete (ACI 313–71), *ACI*, Detroit.

3.9. IS Specifications for Prestressed Concrete Poles for Overhead Power, Traction and Telecommunication Lines, IS: 1678–1960, New Delhi.

3.10. Khachaturian, N., Service Load Design of Prestressed Concrete Beams–Part I, Non-composite Simple Beams, Bulletin of the Engineering Experimentation Station, University of Illinois.

3.11. Dayaratnam, P., Minimum Weight Design of Simply Supported Prestressed Concrete Beams, *Cement and Concrete*, Vol. 9, No. 2, July–September 1968.

3.12. PCI Committee on Allowable Stresses in Prestressed Concrete Design, Allowable Tensile Stresses for Prestressed Concrete, *J. of PCI*, Vol. 15, No. 1, Feb. 1970, pp. 37–42.

3.13. Inomata, S., Comparative Study on Behaviour of Prestressed and Reinforced Concrete Beams Subject to Loading Reversals, *J. of PCI*, Vol. 16, No. 1, Jan.–Feb. 1971, pp. 21–37.

3.14. Salmons. J.R. and Kagay, W.J., The Composite U-Beam Bridge Superstructure, *J. of PCI*, Vol. 16, No. 3, May–June 1971, pp. 20–2.

3.15. Salmons, JR. and Mokhtari, S., Structural Performance of the Composite U-Beam Bridge Superstructure, *J. of PCI*, Vol. 16, No. 4, July–Aug. 1971, pp. 21–3.

3.16. Libby, J.R., Long Span Prestressed Concrete Girders, *J. of PCI*, Vol. 16, No 4, July–August 1971, pp. 80–8.

3.17. Goble, C.G. and LaPay, W.S., Optimum Design of Prestressed Beams, *J. of ACI* Proc. V. 68, No. 9, Sept. 1971, pp. 712–18.

3.18. Lacey. G.C., Breen, J.E. and Burns, N.H., State of Art for Long Span Prestressed Concrete Bridges of Segmental Construction, *J. of PCI*, Vol. 16, No. 5, September–October 1971, pp. 53–7.

3.19. Huang, T. Estimating Stress for a Prestressed Concrete Member, *J. of PCI*, Vol. 17, No. 1, January–February 1972, pp. 29–34.

3.20. Naarnan, A., Computer Program for Selection and Design of Simple Span Prestressed Concrete Highway Girders. *J. of PCI*, Vol. 17, No. 1, January–February, 1962, pp. 73–81.

3.21. Fogarasi, G., General Method to Determine Optimum Shapes of Ribs and Stiffeners, *J. of PCI*, Vol. 17, No. 2, March–April 1972, pp. 76–86.

664

3.22. Hamoudi, A.A., Bierwelier, R.A. and Phang, M.K.S., Prestressed Concrete Beams with a Longitudinal Cavity, *J. of PCI*, Vol. 18, No. 1, January–February 1973, pp. 39–49.

3.23. Water Podolny, Jr, Cable-Stayed Bridges of Prestressed Concrete, *J. of PCI*, Vol. 18, No. 1, January–February 1973, pp. 68–79.

3.24. Jofriet, J.C., McNeice, G.M. and Csagoly, F., Finite Element Analysis of Prestressed Concrete Voided Bridge Decks, *J. of PCI*, Vol. 18, No. 3, May–June 1973, pp. 51–66.

3.25. Anderson, A.A., Stretched-out AASHO-PCI Beams Types III and IV for Longer Span Highway Bridges, *J. of PCI*, Vol. 18, No. 5, September–October 1973, pp. 32–49.

3.26. Lama, G.C., Designing Post-tensioned Beams Having Sections with Compression Steel, *J. of PCI*, Vol. 19, No. 2 March–April 1974, pp. 125–28.

3.27. Chung, H.W. and Gardner, N.J. Field Test of a Curved Prestressed Concrete Bridge and Comparison with Model Test, *J. of ACI*, Proc. V. 71, No. 5, May 1974, pp. 251–56.

3.28. IS Indented Wire for Prestressed Concrete, IS: 6003–1970, New Delhi.

3.29. IS-Uncoated Stress Relieved Strand for Prestressed Concrete, IS: 6006–1970, New Delhi.

3.30. St. John, AD. Computer Design of Prestressed Concrete, *J. of PCI*, Vol. 8, August 1963.

3.31. Nathan, N.D., Distribution of Loads in Precast Prestressed Concrete Members without Interior Diaphragms, *J. of PCI*, Vol. 8, October 1963.

3.32. Huang, L.Y., Angles, N.P., May, H.R. and Thornton, K.C., Design of Giant Post-tensioned Girders, *J. of ACI*, Vol. 64, August 1967, pp. 475–87.

3.33. Oladaps, I.O. Stability of Tensile Cracks in Prestressed Concrete Beams, *J. of ASCE*, Vol. 95, STL, January 1969, pp. 17–31.

3.34. Dayaratnam, P. and Patnaik, S.K., Behaviour of Prestress Concrete Beams Subjected to Variable Pulsating Loads, *J. of Bridge and Structural Engg.*, ING/IABSE, March 1974.

4.1. Khachaturian, N. Service Load Design of Prestressed Concrete Beams–Part II, Composite Simple Beams, Bulletin of the Engineering Experimentation station, University of Illinois.

4.2. Lu-Ln Wang, A Direct Method for Designing Composite Sections in Prestressed Concrete, *J. of PCI*, Vol. 10, No. 5, October 1965.

4.3. Pretzef, C.A., Unusual Application of Prestressed Waffle Slabs and Composite Beams, *J. of ACI*, Proc. Vol. 69, No. 12, Dec. 1972, pp. 765–69.

4.4. OGara, M.B. and Bezouska, T.J., Field Study of Curved Continuous Prestressed Bridge, *J. of PCI*, Vol. 8, No. 6, December 1963.

4.5. Evans, R.H. and Chung, H.W., Flexural Cracks in Composite Prestressed Light-weight Concrete Beams, *Civil Engineering and Public Works Review*, Vol. 63, January 1968.

4.6. Dudra, J. Design and Construction of Hudson Hope Bridge, *J. of PCI*, Vol. 11, No. 2, April 1966.

4.7. Gurfinkel, G., Design of Hold Downs in Prestressed Concrete Girders, *Indian Concrete Journal*, Vol. 41, No. 2, February 1967, p. 62.

4.8. Maver, J.L. Precast and Prestressed Concrete Bridges and Structures in Irrigation Projects, *J. of IE* (Australia), Vol. 39, No. 6, June 1967, pp. 67–74.

4.9. Narayanswamy, Torsional Resistance of Solid and Hollow Sections-Reinforced Concrete and Prestressed Concrete–*Cement and Concrete*, Vol. 4, No. 4, January–March 1964, pp. 13–42.

4.10. Development and Use of Prestressed Steel Flexural Members, Report by Subcommittee 3 on Prestressed Steel of Joint ASCE-AASHO Committee on Steel Flexural Members, C.E. Eckberg, Jr., Chmn., *J. of ASCE*, Vol. 94, No. ST9, September 1968, pp. 2033–60.

4.11. Connolly, W.H. Design of Prestressed Concrete Beams, F.W. Doge Corporation, 1960, New York.

4.12. Sarkar, S., Prasada Rao, Kapla, A.S. and Chhuanda, M.S., J.N. *Handbook for Prestressed Concrete* Bridges, Today and Tomorrow Printers & Publishers, Faridabad, 1969.

5.1. Leonhardt, F., *Prestressed Concrete Design and Construction*, Wilhelm Ernst and Sohn, Berlin, 1964, pp. 364–63.

5.2. Bailey, D.M. and Ferguson, M. Phil. Fixed End Moment Equations for Continuous Prestressed Concrete Beams, *J. of PCI*, Vol. 11, No. 1, February 1966, pp. 76–94.

5.3. Burns, N.H. Development of Continuity between Precast Prestressed Concrete Beams, *J. of PCI*, Vol. 11, No. 3, June 1966, pp. 23–36.

5.4. Libby, J.R. *Prestressed Concrete Design and Construction*, The Ronald Press Co., New York, 1961.

5.5. Turula, P. and. L-Freyermuth, C. Moment Influence Coefficients for Continuous Post-tensioned Structures, *J. of PCI*, Vol. 17, No. 1, January–February 1972, pp. 35–57.

5.6. Lin, T.Y. and Thornton, K. Secondary Moment and Moment Redistribution in Continuous Prestressed Concrete Beams, *J. of PCI*, Vol. 17, No. I, January–February 1972, pp. 8–20.

5.7. Bishara, A.G. and Mahmoud, M.H. Continuous Prestressed Concrete Girders with Keyed Scarf Connections, *J. of ACI*, Proc. V. 69, No. 9, Sep. 1972, pp. 569–77.

5.8. O'Neil, H.M., Computerized Design of Post-tensioned Continuous Beams and Flit Plates, *J. of PCI*, Vol. 18, No. 3, May–June 1973, pp. 42–50.

5.9. Dayaratnam, P. and Gowda, N.T.C., Prefabricated Portal Frames Subjected to Variable Repetitive Loads, *J. of the IABSE*, V. 34–11, Zurich, 1974, pp. 41–52.

5.10. Sherbourne, A.N. and Parameswar, H.C., Limit Analysis of Continuous Prestressed Beams. *J. of ASCE*, Vol. 94, No. STI, January 1968, pp. 19–40.

5.11. Mallick, S.K. and Sastry, M.K.L.N., Plastic Rotations in Computation of Redistribution in Continuous Prestressed Concrete Beams, *J. of ICJ*, Vol. 41, No. 2, February 1967, p. 83.

5.12. Koons, R.L. and Schlegel, G.J.A. Practical Approach to the Design of Continuous Structures in Prestressed Concrete, *J. of PCI*, Vol. 8, No. 4, August 1963.

5.13. Bailey, D.M. and Ferguson, P.M. Fixed End Moment Equation for Continuous Prestressed Concrete Beam, *J. of PCI*, Vol. II, No. 1, February 1966.

5.14. Sheikh, M.A., Rawdon De Paiva, H.A. and Neville, A.M., Strength and Behaviour of Two-span Continuous Preten-sioned Concrete Beams, *J. of ACl*, Vol. 65, No. 1, January 1965, p. 37.

5.15. Vernigora, E. and Marcil, J.R.M. Bridge Rehabilitation and Strengthening by Continuous Post-tensioning, *J. of PCI*, April 1969.

5.16. Frinkas, S. Continuity in Prestressed Concrete *J. of Boston Society of C.E.* January 1964.

5.17. Nasser George D., A Practical Approach to the Design of Continuous Prestressed Concrete Structures, *Indian Concrete Journal*, May 1968, Vol 42, No. 5.

5.18. Nasser George D., A Practical Approach to the Design of Continuous Prestressed Concrete Structures, *Indian Concrete Journal*, May 1968, Vol. 42.

5.19. Wawal, M.B. and Narasimhan R.K. Tests on Continuous Concrete Beams with Prestressed Reinforcement, *Indian Concrete Journal*, July–Aug. 1978, Vol. 52, No. 7 and 8.

6.1. Hoppe, Mass Production of Prestressed Concrete Auxiliary Members, *J. of PCI*, Vol. 8, No. 5, October 1963.

6.2. Elliston Jr., W.J., Connections between Precast Beam and Column Members, *J. of PCI*, Vol. 8, No. 4, August 1963.

6.3. LaFrangh, R.W. and Magura, D.D. Connections in Precast Concrete Structures–Column Base Plates, *J. of PCI*, Vol. 11, No 6, December 1966, pp. 18–39.

6.4. Gerwick Jr., B.C. Long Span Prestressed Concrete Bridges Utilising Precast Elements, *J. of PCI*, Vol. 9, No. 1, February 1964.

6.5. Sauter, F. Post-tensioned Multi-storey Rigid Frames for the New Bank Central in Costa Rico, *J. of PCI*, Vol, 8, No. 3, June 1963.

6.6. Riley, W.E. Post-tensioned Cast-in-place Multi-storey Building Frame, *J. of the ACI*, Vol. 63, No. 3, March 1966.

6.7. Report of PCI Committee, Recommended Practice for Driving Prestressed Concrete Piling, *J. of PCI*, Vol. II, No. 4, August 1966, pp. 18–27.

6.8. Weller, N.H.E. Prestressed Concrete Piles, *J. of PCI*, Vol. 7, No. 5, October 1962.

6.9. Sauter, S. Tall Precast Towers, *J. of PCI*, Vol. 6, No. 3, September 1961, pp. 72–74.

6.10. Swanson, H.V., Design of Prestressed Concrete Pressure Pipe, *J. of PCI*, Vol. 10, No. 4, August 1965, pp. 69–82.

6.11. Renz, C.F. and Melville, P.L., Experience with Prestressed Concrete Airfield Pavement in the United States, *J. of PCI*, Vol. 6, No. 1, March 1961, pp. 75–92.

6.12. Vendepitte, D., Prestressed Concrete Pavements–A Reviev–of European Practice, *J. of PCI*, Vol. 6. No. I, March 1960, pp. 60–74.

6.13. Mclntyre, J.P., Prestressed Concrete Airport Pavement, *J. of PCI*. Vol. 5, No. 2, June 1960, pp. 69–72.

6.14. Young, L.E., A Sliding Joint Prestressed Concrete Pavement, *J. of PCI*, Vol. 6, No. 3, September 1961, pp. 92–99.

6.15. Zetlin, L., A Novel Concept of Prestressed Concrete Pavement, *J. of PCI*. Vol. 6, No. 1, March 1961, pp. 93–102.

6.16. Hargett, E.R., Prestressed Concrete Panels for Pavement Construction, *J. of PCI*, Vol. 15, No. 1, Feb. 1970, pp. 37–42.

6.17. Lin, T.D., Tendon Geometry for Nuclear Containment Structures, *J. of PCI*, Vol. 15, No. 1, Feb. 1970, pp. 60–66.

6.18. Li, S. and Liu, T.C., Prestressed Concrete Piling–Contemporary Design Practice and Recommendations, *J, of ACI*, proc. V. 67, No. 3, March 1970, pp. 201–20.

6.19. Anderson, A.R. and Moustafa, S.E., Ultimate Strength of Prestressed Concrete Piles and Columns, *J. of ACI*, Proc. V. 67, No. 8, August 1970, pp. 620–35.

6.20. ACI Committee 344, Design and Construction of Circular Prestressed Concrete Structures, *J. of ACI*, Proc. V. 67, No. 9, Sept. 1970, pp. 657–72.

6.21. Benuska, K.L., Bertero, V.V. and Polivka, M., Self-stressed Concrete for Precast Building Units, *J. of PCI*, Vol. 16, No. 2, March–April 1971, pp. 72–84.

6.22. Chandwani, R. and Nathan, N.D., Precast Prestressed Sections Under Axial Load Bending, *J. of PCI*, Vol. 16, No. 3, May–June 1971, pp. 10–19.

6.23. Varkay, I., Load Bearing Wall Panels, *J. of PCI*, Vol.16, No. 4, July–August 1971, pp. 34–48.

6.24. Sargious, M. and Boecker, H. C., Economical Design of Prestressed Concrete Pavements, *J. of PCI*, Vol. 16, No. 4, July–August 1971, pp. 64–79.

6.25. Lalonde. Jr., W.S. Expanding a Precast Prestressed Concrete Garrage in New York City, *J. of ACI*, Proc. V. 68, No. 10, Oct. 1971, pp. 733–739.

6.26. Sargious, M. and Tadros, G., Step and Loads Effect on Stresses in Prestressed Concrete Short Brackets, *J. of ACI*, Proc. V. 68. No. 11, Nov. 1971 pp. 861–66.

6.27. Sargious, M. and King, S.K.. Design of Prestressed Concrete Airfield Pavements under Duel and Duel Tandem Wheel Loading, *J. of PCI*. Vol. 16. No. 6, November–December 1971, pp. 19–30.

6.28. Butts, E.A. The First Bison Building in the United States, *J. of PCI*, Vol. 16, No. 6, November–December 1971, pp. 31–39.

6.29. Pastoro J., P.R., Town Street Parking Garage; Design, Fabrication and Erection of Tree Columns, *J. of PCI*, Vol. 17, No. 1, January–February 1972, pp. 21–28.

6.30. Hughe, G. and Celis, R., Joint Free Experimental Prestressed Pavement, *J. of PCI*, Vol. 17, No. 1, January–February 1972, pp. 58–72.

6.31. Geer, E., Unusual Failure of Joint in Prestressed Slab, *J. of ACI*, Proc. V. 69, No. 2, Feb. 1972, pp. 93–95.

6.32. Pasko Jr., T.J. Prestressed Highway Pavement at Dulles Airport for Transpo 72, *J. of PCI*, Vol. 17, No. 2, March–April 1972, pp. 46–54.

6.33. Rodgers Jr. T.E., A Utility's Development and Use of Prestressed Concrete Poles, *J. of PCI*, Vol. 17, No. 3, May–June 1972, pp. 8–13.

6.34. Johnson, T. and Ghadiali, Z., Load Distribution Test on Precast Hollow Core Slabs with Openings, *J. of PCI*, VoL 17, No. 5, September–October 1972, pp. 9–19.

6.35. Li, S. and Ramakrishnan, V., Optimum Design of Prestressed Concrete Combined Bearing and Sheet Piles, *J. of PCI*, Vol. 17, No. 5, September–October 1972, pp. 39–48.

6.36. Mikhailov, V.V., Design of Slender Prestressed Concrete Columns Based on Stability Criteria, *J. of PCI*, Vol. 17, No. 5, September–October 1972, pp. 49–56.

6.37. Butts Jr.. E.A., Erection of the Versa Frame System, *J. of PCI*, Vol. 17, No 6, November–December 1972, pp. 26–36.

6.38. Nathan, N.D., Slenderness of Prestressed Concrete Beam Columns, *J. of PCI*, Vol. 17. No. 6, November–December 1972, pp. 45–57.

6.39. Scott, N.L., Performance of Precast Prestressed Hollow Core Slab with Composite Concrete Topping, *J. of PCI*, Vol. 18, No. 2, March–April 1973, pp. 64–77.

668

6.40. Dunbar, N.F., A Medley of Precast and Prestressed Concrete Systems, *J. of PCI*, Vol. 18. No. 2, March–April 1973, pp. 78–86.

6.41. PCI Feature, New Precast Prestressed System Saves Money in Hawaii Hotel, *J. of PCI*, Vol. 18, No. 3, May–June 1973, pp. 10–13.

6.42. PCI Construction Feature, Post-tensioning Plus Special Vibration System Aids Production of Giant Precast Prestressed Girders, *J. of PCI*, Vol. 18, No. 4, July–August 1973, pp. 33–36.

6.43. Fiorato, A.E., Geometric Variation in the Columns of a Precast Concrete Industrial Building, *J. of PCI*, Vol. 18, No. 4, July–August 1973, pp. 50–60.

6.44. Kelly, J.B. and Pike, K.J., Design and Production of Prestressed L-Shaped Bleacher Seat Units, *J. of PCI*, Vol. 18, No. 5, September–October 1973, pp. 73–84.

6.45. Christianson, J.V., Desigu Considerations for a Precast Prestressed Apartment Building Analysis of a Lateral Load Resisting Elements, *J. of PCI*, Vol. 18, No. 6, November–December 1973, pp. 54–71.

6.46. Holland, E.P. and Svab, L.E., Design Considerations for a Precast Prestressed Apartment Building-Design Problem, *J. of PCI*, Vol. 18, No. 6, November–December 1973, pp. 50–53.

6.47. Lamberson, E.A., Post-tensioned Structural Systems- Dallas-Ft Worth Airport, *J. of PCI*, Vol. 18, No. 6, November–December 1973, pp. 72–91.

6.48. Raths, C.H., Design Considerations for a Precast Prestressed Apartment Building-Design of Load Bearing Wall Panels, *J. of PCI*, vol. 19, No. 2, March–April 1974, pp. 68–92.

6.49. Barrett, M.H., Dunbar, N.F. and Gillaspie, D.D., Design Considerations for a Precast Prestressed Apartment Building-Design of Secondary Floor Members, *J. of PCI*, Vol. 19, No. 3, May–June 1974, pp. 32–52.

6.50. United States Gypsum Company, Floor and Ceiling Assembly Consisting of Precast Concrete Double Tee Units with a Wall-board Ceiling, *J. of PCI*, Vol. 19, No. 3, May–June 1974, pp. 53–73.

6.51. Fotinos, G.C., Design Considerations for a Precast Prestressed Apartment Building-Design of Prestressed Pile Foundations. *J. of PCI*, Vol. 19, No. 5, September–October 1974, pp. 40–53.

6.52. Bruce, Jr., R.N. and Hebert, D.C., Splicing of Precast Prestressed Concrete Piles Part–1-Review and Performance of Splices, *J. of PCI*, Vol. 19, No. 5, September–October 1974, pp. 70–97.

6.53. Shewab, F.C., Chow, P.Y. and Kulka, F., Prestressed Ring Beam Saves Time and Money on Domed Convention Hall, *J. of PCI*, Vol. 19, No. 5, September–October 1974, pp. 98–106.

6.54. Kaldjian Prestressed Bowstring Arch., *J. of ASCE*, ST, October 1961.

6.55. Osawa Strength of Prestressed Concrete Pavements, *J. of ASCE*, ST, December 1962.

6.56. Harrop. J., Analysis of the Standpipe Zone of Prestressed Concrete Pressure Vessels, *J. of PCI*, Vol. 14, June 1969.

6.57. Galezewski, S. A., Three Hinged Arch Storage Building in Precast Prestressed Concrete, *J. of PCI*, Vol. 14, August 1969.

6.58. Spencer, R.A. Stiffness of Damping of Nine Cyclically Loaded Prestressed Concrete Members, *J. of PCI*, June 1969.

6.59. Lin, T.Y. and Lakhwarn, T.R., Ultimate Strength of Eccentrically Loaded Partially Prestressed Columns, *J. of PCI*, Vol. 11. June 1966.

6.60. Brerweiler, R.A., Prestressed Concrete Pipe Beams, *J. of PCI*, Vol. 11, April 1966.

6.61. Deszkiewicz, S. and Parzniewski, Z., Simplified Erection Method for Shell Structures, *J. of ACI*, Vol. 64, June 196,7 pp. 295–300.

6.62. Rucker, W.H., Alternate Design Proves Economy of Precast, Prestressed Design for Pier Shed, *J. of Civil Engineering*. May 1964, pp. 42–46.

6.63. Corum, James M., White, Richard N. and Smith, Jack E., Mortar Models of Prestressed Concrete Reactor Vessels, *J. of ASCE*, Vol. 95, No. ST2, February 1969, pp. 229–248.

6.64. Aroni, Samuel, Slender Prestressed Concrete Columns, *J. of ASCE*, Vol. 94, No. ST4, April 1968, pp. 875–904.

6.65. Dayaratnam, P. and Gowda, N.T.C., Behaviour of Precast Portal Frames Under Pulsating Loads, Bulletin of lABSE, Vol. 34–II, 1974.

6.66. Rao, Ram Mohan, The Use of Prestressing Techniques in the Construction of Dams, *Indian Concrete Journal*, Vol. 38, Aug. 1964.

6.67. Thamharkar, A.S., Prasad Rao, A.S. and Kapla M.S., Standardization of Precast Prestressed Concrete Bridge Beams, *Indian Concrete journal*, Vol. 42, No. 6, June 1968.

6.68. Goyal S.C. and Shri Prakash, Prestressed Concrete Pavements, *Indian Concrete Journal*, Vol. 42, Sept. 1968.

6.69. Pliskin, L. Prestressed Concrete Reservoirs, *Indian Concrete Journal*, Vol. 43, No. 12, Dec. 1969.

70. Thomas, P.K. Gross Sectional Design of Prestressed Concrete Bridges, *Indian Concrete Journal* Vol. 47, No. 11, Nov. 1973.

6.71. Raina, V.K., Transverse Analysis of Some Typical Prestressed Concrete Box-section Bridge decks–1, *Indian Concrete Journal*, Vol. 48, No. 8, August 1974.

6.72. Raina, V.K. Transverse Analysis of Some Typical Prestressed Concrete Box-Section Bridge decks–2, *Indian Concrete Journal*, Vol. 48, No. 10, Oct. 1974.

6.73. RainavV.K., Transverse Analysis of Some Typical Prestressed Concrete Box-section Bridge Decks–3, Indian Concrete Journal, Vol. 48, No. 11, Nov. 1974.

6.74. Shah, N.K. and Natarajan V.K.,The Prestressing in Underground Structures, *Indian Concrete Journal*, Vol. 50, No. 5, May 1976.

6.75. Shah Navin, Precast Prestressed Concrete Components for a Large Span Industrial Building, *Indian Concrete Journal*, Vol. 50, No. 9, Sept. 1976.

7.1. Scordelies, A.C., Branson, D.E. and Sozen, M.A. Deflection of Prestressed Concrete Members, *J. of ACI*, Vol. 60, No. 12, December 1963, pp. 1697–1728.

7.2. Potyondy, J.G. and Nawy, G.. Deflection Behaviour of Spirally Confined Prestressed Concrete Flanged Beams, *J. of PCI*, Vol. 16, No. 3, May–June 1971, pp. 44–59.

7.3. Chetty, S.M.K., Prasad Rao, A.S. and Bhargava, R.N., Selection and Design of Prestressed Concrete Sections for Flexure–1, *Indian Concrete Journal*, Vol. 38, August 1964.

670

7.4. Swamy Narayan, Some Aspects of the Diagonal Cracking Failure of Prestre5sed Hollow Beams without Web Rein-forcement, *Indian Concrete Journal*, Vol. 38, Sept. 1964.

7.5. Chetty, S.M.K., Prasada Rao, A.S. and Bhargava R.N., Selection and Design of Prestressed Concrete Sections for Flexure–2, *Indian Concrete Journal*, Vol. 38, Sept. 1964.

7.6. Chetty, S.M.K., Bhargava, R.N. and Prasada Rao, A.S., Selection and Design of Prestressed Concrete Selection for Flexure–3, *Indian Concrete Journal*, Vol. 3, Nov. 1964.

7.7. Chetty, S.M.K. and Bhargava, R.N., Selection and Design of Prestressed Concrete Sections for Flexure–4., *Indian Concrete Journal*, Vol. 39, Jan. 1965.

7.8. Chetty, S.M.K. and Bhargava, R.N., Selection and Design of Prestressed Concrete Sections for Flexure–5, *Indian Concrete Journal*, Vol. 39, Feb. 1965.

7.9. Chetty, S.M.K. Bhargava, R.N. Selection and Design of Prestressed Concrete Sections for Flexure–6, *Indian Concrete Journal*, Vol. 39, June 1965.

7.10. Chetty, S.M.K. and Bhargava, R.N., Selection and Design of Prestressed Concrete Sections for Flexure–7, *Indian Concrete Journal*, Vol. 39, Sept. 1965.

7.11. Evans, R.H. and Chung H.W., Long-term Deflection of. Prestressed Composite Beams without in-situ Lightweight Concrete, *Indian Concrete Journal*, Vol. 41, Dec. 1967.

7.12. Pandit, G.S. and Sharma, A.K., Design of Prestressed Concrete Beams Subjected to Combined Bending and Torsion, *Indian Concrete Journal*, Vol. 45, No.5, May, 1971.

7.13. Ajimani, J.L., Long-term Deflections in Pre-tensioned Beams, *Indian Concrete Journal*, Vol. 46, No. 2, Feb. 1972.

7.14. Denis McGee and Paul Zia, P.S.C. under Torsion, Shear and Bending, *J. of ACI.*, Proceeding V. 73, No. 1, Jan. 1976.

7.15. Krishna Raju, N. Basavarajaiah, B. S. and Govinda D., Flexural Behaviour of Prestressed Concrete Beams with Confined Concrete in the Compression Zone, *Indian Concrete Journal*, Vol. 50, No. 6, June, 1976.

7.16. ACI Committee 443, P.S.C. Bridge Design, *J. of ACI*. No. 11, Nov. 1976.

7.17. Sengupfa, B. and Som P. Failure of Pretensioned Members due to Web-distress under Combined Shear and Flexure, *Indian Concrete Journal*, Vol. 51. No. 3, March 1977.

7.18. Denis Mitchell, and Collins Michael P. Influence of Pre-tressing on Torsional Response of Concrete Beams, *J. of PCI*, Vol. 23, No. 3, May–June 1978.

7.19. Donald R. Buettner and James R. Libtjy, Camber, Requirements for Pretensioned Concrete International, Design and Construction, *A.C.I.*, Vol. 1, No. 2, Feb. 1979.

7.20. Campbell, T.I., Barrington de V. Batchelor and Larp Chitnuyanandh, Web Crushing in Concrete Girders with Prestressing Ducts in the Web, *PCI.*, Vol. 24, No. 5, Sept./Oct. 1979.

7.21. Edward G. Nawy and Jin Y. Chiang. Serviceability Behaviour of Post-tensioned Beam, *J. of PCI.*, Vol. 25, No. 1, Jan.–Feb. 1980.

7.22. Sozen, M.A., Research on Shear Strength of Prestressed Beams, *J. of PCI*, Vol. 5, No. 2, June 1960, pp. 40–49.

7.23. McGregor, J.F., Sozen, M.A. and Siess, CP., Effect of Draped Reinforcement on Behaviour of Prestressed Concrete Beams, *J. of ACI,*, Vol. 57, No. 6, December 1960.

7.24. Venuti, W.J., Diaphragm Shear Connectors between Flanges of Prestressed Concrete T-Beams, *J. of PCI*, Vol. 15, No. 1, Feb. 1970, pp. 67–79.

7.25. Sargious, M.A. Principal Stresses at the Intermediate Support of Prestressed Concrete Continuous Beams, *J. of ACI*, Proc. V. 67, No. 10, Oct. 1970, pp. 828–35.

7.26. Bennett, E.W. and Balasooriya, B.M.A., Shear Strength of Prestressed Beams with Thin Webs Falling in Inclined Compression, *J. of ACI*, Proc. V. 68, No. 3, March 1971, pp. 204–212.

7.27. Price, K.M. and Edwards, A.D., Fatigue Strength in Shear Stress of Prestressed Concrete I-Beams, *J. of ACI*, Proc. V. 68. No. 4, April 1971, pp. 282–92.

7.28. Blakeby, R.W.C. and Paik, R., Seismic Resistance of Prestressed Concrete Beams-Column Assemblies, *J. of ACI*, Proc. V. 58, No. 9, Sep. 1971, pp. 677–92.

7.29. Benette, E.W., Abdul-Ahad, H.Y. and Neville, A.M., Shear Strength of Reinforced and Prestressed Concrete Beams Subject to Moving Loads, *J. of PCI*, Vol. 17, No. 6, November–December 1972, pp. 58–69.

7.30. Arthur, P.D., Bhatt, P, and Duncan, W., Experimental and Analytical Studies on the Shear Failure of Pretensioned I-Beams Under Distributed Loading, *J. of PCI* Vol. 18, No. 1, January–February 1973, pp. 50–67.

7.31. Werner, M.P. andDiiger, W.H., Shear Design of Prestressed Concrete Stepped Beams, *J. of PCI*, Vol. 18, No. 4, July–August 1973, pp. 37–49.

7.32. McMullen, A.E. and Woodhead, R., Experimental Study of Prestressed Concrete under Combined Torsion, Bending and shear, *J. of PCI*, Vol. 18, No. 5, September–October !973, pp. 85–100.

7.33. Woodhead, H.R. and McMullen, A.E., Design of Prestressed Concrete Beams Subjected to Torsion, *J. of ACI*, Proc. V. 70, No. 11, Nov. 1973, pp. 745–48.

7.34. Rangan B.V. and Hall A.S., Strength of Rectangular Prestressed Concrete Beams in Combined Torsion, Bending and Shear, *J. of ACI*, Proc, V. 70, No. 4, April 1973, pp. 270–78.

7.35. Zia, P. and McGee, WD. Torsion Design of Prestressed Concrete, *J. of PCI*, Vol. 19, No. 2, March–April 1974, pp. 46–65.

7.36. Jacques, F.J., Graphs for Direct Shear Design of Prestressed Concrete Members, *J. of PCI*, Vol. 14, August 1969.

7.37. Kar, J.N. Diagonal Cracking in Prestressed ConcreXt Beams, *J. of ASCE*, Vol. 94, No. STI, January 1968, pp. 83–109.

7.38. Cowan, H.J. and Lyalin, I.M., Reinforced and Prestressed Concrete in Torsion.

8.1. Gurfinkel, G. and Khachaturian, N., Ultimate Design of Prestressed Concrete Beams, *Research Bulletin*, No. 478 Engineering Experimentation Station, University of Illinois.

8.2. BurrBennet Jr., W., Limit Design with Prestressed CoDcrete, *J. of PCI*, Vol. 8, No. 2, April 1963.

8.3. Bennett, E.W., Cook, N. and Naughton, L.P., Deformation of Continuous Prestressed Concrete Beams and its Effect on LUtimate Load, *Proc. Inst, of Civil Engineers (Loud.)*, Vol. 37, May 1967, pp. 57–74.

8.4. Raina, V.K., Ultimate Load Carrying Capacity of Continuous Prestressed Concrete Beams, *The Construction Engineer*, Vol. 31, No. 7, July 1967, pp. 77–86.

8.5. Gausel, E., Ultimate Strength of Prestressed I-Beams Under Combined Torsion, Bending and Shear, *J. of ACI*, Proc. V. 67, No. 9, September 1970, pp. 675–77.

8.6. Jacques, F.J., Study on Long Span Prestressed Concrete Bridge Girders. *J. of PCI*, Vol. 16, No. 2, March–April 1971, pp. 24–42.

8.7. Nawy, E.G. and Potyondy, J.G., Flexural Cracking Behaviour of Pretensioned Prestressed Concrete I- and T-Beams, *J. of ACI*, Proc. V. 68, No. 5, May 1971, pp. 355–60.

8.8. Li, S. and Ramakrishnan, V., Optimum Prestress, Analysis and Ultimate Strength Design of Prestressed Concrete Sheet Piles, *J. of PCI*, Vol. 16, No. 3, May–June 1971, pp. 60–74.

8.9. Gerber, L.L. and Burns, N.H., Ultimate Strength Tests of Post-tensioned Flat Plates, *J. of PCI*, Vol. 16. No. 6, November–December 1971, pp. 40–58.

8.10. Saeed, M.A. and Kennedy, J.B., Critical Sections for Flexure in Prestressed Girders with Inclined Top Flanges, *J. of PCI*, Vol. 16, No. 6, November–December 1971, pp. 59–69.

8.11. Corley, W.G., Carpenter, J.E., Russel, H.G., Hanson, N. W., Cardenas, A.E., Helgason, T., Hanson, J.M. and Hog-nestad, E., Design Ultimate Load Test of 1/10 Scale Micro-concrete Model of a New Potomac River Crossing, 1–266, *J. of PCI*, Vol. 16, November–December 1971, pp. 70–84.

8.12. Rarrrakrishnan, V. and Carlo Jr., C, Limit Design of Prestressed Concrete Bridges, *J. of PCI*, Vol. 17, No. 3, May–June 1972, pp. 61–74.

8.13. Bennett, E.W. and Veerasubramanian, N., Behaviour of N on-rectangular Beams with Limited Prestress after Flexural Cracking, *J. of ACI*, Proc. V. 69, No. 9, September 1972, pp. 533–42.

8.14. Chung, H.W., Unbonded Tendons in Post-tensioned Concrete Beams Under Repeated Loading, *J. of ACI*, Proc. V. 70, No. 12, December 1973, pp. 814–16.

8.15. Burdette, E.G and Goodpasture, D.W. Test to Failure of a Prestressed Concrete Bridge, *J. of PCI*, Vol. 19, No. 3, May–June 1974, pp. 92–103.

8.16. Freyermuth, C.L., New Load Factors for Prestressed Concrete Highway Bridges, *J. of PCI*, Vol. 18, No. 3, May–June 1973, pp. 14–34.

9.1. Abeles, P.W. and Czuprynski, L., Partial Prestressing: Its history, research, application and future development, *Annales Travanx Publics de Belgigue* (Brussells), No. 2, April 1966, pp. 131–39.

9.2. Abeles, P.W., Saving Reinforcement by Prestressing, *Concrete and Constructional Engineering*, V. 35, 1944, pp. 328–32.

9.3. Abeles, P.W. Static and Fatigue Tests on Partially Prestressed Concrete Construction, *J.* of ACI, Vol. 26, No. 4, 1954, pp. 361–76.

9.4. Abeles, P.W., The Condition of Partially Prestressed Concrete Structures after 3–7 Years Use, IABSE, Sixth Congress, Final Report, Lisbon, 1956, pp. 625–30.

9.5. Practical Recommendations for the Design and Construction of Prestressed Concrete Structures, *FIP-CEB Joint Committee Report*, Provisional Edition, June 1966, Cement and Concrete Association, London.

9.6. Abeles, P.W., Design of Partially Prestressed Concrete Beams, *J. of ACI*, October 1967, Vol.64, No. 10, pp. 669–77.

9.7. Thurlimann, B., Partially Prestressed Members, IABSE, Eighth Congress 1968, New York.

9.8. Abeles. P.W., The Practical Application of Partial Prestress-ing'Research on Cracking and Deflection under Static, Sustained and Fatigue Loading, IABSE, 8th Congress, 1968.

9.9. Abeles, P.W. and Gill, V.L. The Behaviour of Partially Prestressed Beams Containing Bonded, Nontensioned Strands and Curved, Nontensioned Tendons, *IABSE*, 8th Congress, 1968, New York.

9.10. Visvesvaraya, H.C. and Raghavendra, N., Partial Prestress-ing in Concrete Structures, *Proceedings of the Seminar on Problems of Prestressing, ING IABSE*, January–February 1970, pp. I1–I–II–20.

9.11. Shaikh, A.F. and Branson, D.E., Non-tensioned Steel in Prestressed Concrete Beams, *J. of PCI*, Vol. 15, No. 1, Feb. 1970, pp. 14–36.

9.12. Magura, D.D. and Hognested, E, Tests on Partially Prestressed Concrete Girders, *J. of ASCE*, ST 1, Feb. 1966.

9.13. Abdul Rahman, P.M., Natarajan. P.R., Madhava Rao, A.G. and Zacharia George Precast Partially Prestressed Beams in Structural Frames for a Laboratory Building complex, *Indian Concrete Journal*, Vol. 47, No. 4, Apr. 1973.

9.14. Limit State Behaviour of Pretensioned I-Beams with Limited Prestress, *Indian Concrete Journal*, Vol. 49, No. 7, July 1975.

9.15. Lyngberg Bent S., Shear Resistance of Partially Prestressed Reinforced Concrete I-Beams, *J. of ACI*, No. 4, Proceedings Vol. 73 April 1976.

9.16. Naamen. Antine E., Siriaksorn Amnuayporer, Serviceability Based Design of Partially Prestressed Beam-Part I, Analytical Formulation, *PCI*, March–April 1979, Vol. 24, No. 2. 9.17. Park Robert, Thompson, KerinJ. Ductility of Prestressed and Partially Prestressed Concrete Beam Sections, March–April 1980, Vol. 25, No. 2.

10.1. Dayaratnam, P., Design of Two-way Slab as an Orthotropic Plate, *Indian Concrete Journal*, Vol. 38, No. 3, March 1964, p. 89.

10.2. Timosheako, S. and Woinowsky-Krieger, S., *Theory of Plates and Shells*, McGraw-Hill, 1959, p. 120.

10.3. Zielinski, Z. A., Prestressed Slabs and Shell Made of Prefabricated Components, *J. of PCI*, Vol. 9, No. 4, August 1964, pp. 69–80.

10.4. Felix, Adler., Precast Plate Units for Roof Construction, *Proc. of Inst, of Civil Engineers (Lond.)*, Vol. 23, November 1962, pp. 321–26.

10.5. Chen-Hwa Wang. Direct Design Method for Prestressed Concrete Slab, *J. of PCI*, Vol. 13, No. 3, June 1968, pp. 62–72.

10.6. Omsted, H., Notes'on Post-Tensioned Lift Slab, *J. of PCI*, Vol. 10, No. 6. December 1965, pp. 84–90.

10.7. Rozvany, G.I.N, and Hampson, A.J.K., Optimum Design of Prestressed Plates, *J. of ACI*, Vol. 60, No. 8, August 1963, pp. 1065–82.

10.8. Timoshenko, S. and Woinowsky-Krieger, S., *Theory of Plates and Shells*, McGraw-Hill, 1959, pp. 245–58.

10.9. Indian Standard Code for Design of Plain and Reinforced Concrete, IS: 456–1964, New Delhi.

10.10. Lague, D J., Load Distribution Test on Precast Prestressed Hollow-core Slab Construction, *J. of PCI*, Vol. 16, No 6, November–December 1971, pp. 10–18.

10.11. ACI-ASCE Committee 423, Tentative Recommendations for Prestressed Concrete Flat.Plates, *J. of ACI*, Proc. Vol. 71, No. 2, Feb. 1974, pp. 61–71

10.12. Smith, S.W. and Burns, N.H., Post-tensioned Flat Plate to Column Connection Behaviour, *J. of PCI*, Vol. 19, No. 3, May–June 1974, pp. 74–91.

10.13. Sacther, K., A Structural Membrane Theory Applied to the Design of Prestressed Concrete Flat Slab, *J. of PCI*, Vol. 8, No. 5, October 1963.

10.14. Korb, J.L., Construction of Post-Tensioned Roof Panels, *J. of ACI*, Vol. 64, No. 7, August 1967, pp– 488–91.

10.15. Raju N. Krishpa, The Design of Prestressed Concrete Grid floors, *Indian Concrete Journal*, Jan. 1965, Vol. 39.

10.16. Abdul Rahman, P.M., Sreenath. G.H., Precast Prestressed Concrete Slab Incorporating Structural Hollow Clay Blocks in Floors and Roofs, *I.C.J.*, May 1975, Vol. 49, No. 5.

10.17. Nawy. Edward G. Chakraborti Pinaki, Deflection of P.S.C. Flat Plates *J. of PCI*, March–April 1976, Vol. 21, No. 2.

10.18. Naaman. Antoine E., Minimum Cost vs. Minimum Weight of PSC Slabs, *J. of Structural Div., A.S.C.E.* Vol. 102, No. ST&, July 1976.

11.1. Flugge, W., *Stresses in Shells*, Springer-Verlog, Berlin, 1962.

11.2. Ramaswamy, G.S. *Design and Construction of Concrete Shell Roofs*, McGraw-Hill, New York, 1968.

11.3. Billington, D.P.. *Thin Shell Concrete Structures*, McGraw-Hill, New York, 1965.

11.4. Gibson, J.E., *The Design of Cylindrical Shell Roofs*, Van Nostrand, Princeton, 1961.

11.5. Novozhiloy, V.V., *Thin Shell Theory*, P. Noordhoff Ltd., Grouingen, The Netherlands, 1964.

11.6. Hass, A.M. *Design of Thin Concrete Shells*, John Wiley & Sons. Inc., New York, 1962.

11.7. Powell, GH., Comparison of Simplified Theories for Folded Plates, *J. of Struct. Div. of ASCE*, Vol. 91, No. ST6, December 1965, pp. 1–33.

11.8. Brough, J.C. and Stephens, B.H., Long Span Prestressed Concrete Folded Plate Roofs, *J. of Struct. Div. of ASCE*, October 1960, Paper No. 2630.

11.9. Power, J.O. Folded Plate Concrete Roof in Post-tensioned in Plate, *J. of PCI*, Vol. 5, No. 1, February 1963, pp. 29–34.

11.10. Ramaswamy, G.S, Tamahankar M.G. and Naithani, K.C., Precast Pretensioned Folded Plate Roof, *Indian Concrete Journal*, January 1966, 6–12.

11.11. Edwards, H.H., Facts on Precast Prestressed Folded Plate Slabs, *Modern Concrete*, Vol. 24, No. 2, June 1960, pp. 42–46.

11.12. Edwards, H.H., Precast Prestressed Folded Plate Slabs, *J. of ACI*, Vol. 57, April 1956, pp. 1313–22.

11.13. Timoshenko, S. and Woinowisky-Kriegeri S., *Theory of Plates and Shells*, McGraw-Hill, 1959, pp. 471–87.

11.14. Timoshenko, S. and Woinbwsky-Krieger, S., *Theory of Plates and Shells*, McGraw-Hill, 1959, p. 436.

11.15. ASCE Manual on Design of Cylindrical Concrete Shell Roofs, *Manual 31*, ASCE, New York.

11.16. Lin, T.Y. and Kulka, F", Concrete Shells Prestressed for Load Balancing, *Proc.of World Conference on Shell Structures*, San Francisco, 1962, pp. 423–30.

11.17. Agarwal, S.K., Optimal Design of Prestressed Cylindrical Shells, *Ph. D. Thesis.*, I.I.T. Kanpur, 1973.

11.18. Lundgren, H., *Cylindrical Shells*, Vol. I, The Danish Technical Press, Copenhagen, 1959.

11.19. Chinn, J. Cylindrical Shell Analysis Simplified by Beam Theory, *J. of ACI*, Vol. 55, May 1959.

11.20. Panne, A.L. and Conner, H.W., *J. of ACI*, Vol. 55, Part 2, Dec. 1959.

11.21. Finsterwalder, U., Die Theorie der Kreiszylindrischen Schai-engewolbe System Zeiss Dywidag Und ihre Anwendung auf die Grossmarkthalle in Budapest, Vol. 1, *IABSE*, Zurich, 1932.

11.22. Aas-Jakobsen, A., Einzellarten auf Krieszylinderschalen, Bauingenieur, Vol. 22, 1941.

11.23. Firnkas, S., A Prestressed Folded Plate, Hanger for Allegheny Airlines-Design and Construction, *J. of PCI*, Vol. 16, No. 2, March-April 1971, pp. 43–62.

11.24. Lin, T.Y., Kulka, F. and Kam Lo, Gaint Prestressed HP Shell for Ponce Coliseum, *J. of PCI*, Vol. 18, No. 5, September–October 1973, pp. 69–72.

11.25. Graham, D.E., Low Slump Concrete Results in Cost Savings for Prestressed Containment Structures, *J. of ACI*, Proc. Vol. 71, No. 3, March, 1974, pp. 126–29.

11.26. Dayaratnam, P. and Indubushen, P., Tie Bar Effect in Prestressed Concrete Folded Plate Models, *J. of Institution of Engineers* (India), Vol. 55, May 1975, pp. 186–88.

11.27. IS Code of Practice for Construction of Reinforced Con-crete Shell Roof, IS: 2204–1962, New Delhi.

11.28. Martin, I., Full Scale Load Test of a Prestressed Folded Plate Unit, *J. of ACI*, Proc. Vol. 68, No. 12, Dec. 1971, pp. 937–44

11.29. IS–Criteria for Design of Reinforced Concrete Shell Structures and Folded Plates, IS: 2210–1962, New Delhi.

11.30. Mirza, J.F., Analysis of Post-tensioned Hyperbolic Paraboloid Shell Roof, *J. of PCI*, Vol. 12, No. 1, February 1967, pp. 18–36.

11.31. Lin, T.Y. and I lay a, R., Tests of a Prestressed Concrete Hyperbolic Paraboloid Shell, *J. of PCI*, Vol. 3, June 1958, pp. 70–78.

11.32. Dayaratnam, P., Segmental Precast Prestressed Concrete Folded Plates, *J. of lASS*, Vol. 49, 1972, pp. 27–32.

11.33. Rao.G.S., Kalara, M.L., The Influence of Prestressing on Shell Stresses, *Indian Concrete Journal*, Dec. 1964. Vol. 38.

11.34. Tamhankar M.G., The Analysis and Design of Prestressed Concrete Cylindrical Shells, *Indian Concrete Journal*, Aug. 1966, Vol. 140.

11.35. Kalara, M. L., The Influence of Prestressed on Stresses in Single Shells. *Indian Concrete Journal*, Jan. 1967, Vol.

11.36. Abdul Rahman, P.M., Deshmukh, R.S., and Jain S.S. Precast Prestressed Hyperbolid Shells for Industrial Roofs, *Indian Concrete Journal*, Nov. 1968, Vol. 42.

11.37. Krishna, Prem and Kohli A.K. Design of a Doubly Curved Suspended Cable Roof with Prestressed Concrete Beams, *Indian Concrete Journal* Nov. 1968, Vol. 42.

676

11.38. Sur; S., Precast Prestressed Conoidal Northlight Roof for an Industrial Building, *Indian Concrete Journal*, Feb. 1976, Vol. 50, No. 2.

11.39. ParkG. Arthur and JohnC, Soriscner Prestressed Cylindrical Shells after Cracking., *J. of the Structural Div., ASCE*, Vol. 103, No. ST 9, Sept. 1977.

11.40. Nori Vasudev V., Precast-pretensioned Hyper Shell Units for Factory Roofs, *Indian Concrete Journal*, Jan. 1978, Vol. 52, No. 1.

11.41. Stefanus F. Van Zyland, Scordelis, Phenander C, Analysis of Curved, Prestressed, Segmental Bridges, *J. of the Structural Div. ASCE*. Vol. 105, No. ST 11, Nov. 1979.

12.1. Marshall, W.T. A Theory for End Zone Stresses in Pretensioned Concrete Beams, *J. of PCI*, Vol. 11, No. 2, April 1966, pp. 45–51.

12.2. Jenney, J.R., Report of Stress Transfer Length Studies on 270k Prestressing Strand, *J. of PCI*, Vol. 8, No. 1, February 1963.

12.3. Linger, D.A. and Bhonsle, S.R., An Investigation of Transfer Length in Pretensioned Concrete Using Photo-elasticity, *J. of PCI*, Vol. 8, No. 4, August 1963.

12.4. Nawy, E.G. and Salek, F. Moment-Rotation Relationships of Non-Bonded Post-Tensioned I- and T-Beams, *J. of PCI*, Vol. 13, No. 4, August 1968, pp. 40–55.

12.5. Bondy, K.B., Realistic Requirements for Unbonded Post-Tensioning Tendons, *J. of PCI*, Vol. 15, No. 1, Feb. 1970, pp. 50–9.

12.6. Sbhupack, M., Grouting Tests on Large Post-tensiohhig Tendons for Secondary Nuclear Containment Structures, *J. of PCI*, Vol. 16, No. 2, March–April 1971, pp. 85–97.

12.7. PCI Commitee on Post-tensioning. Recommended Practice for Grouting of Post-tensioned Prestressed Concrete, *J. of PCI*, Vol. 17, No. 6, November–December 1972, pp. 18–25.

12.8. Schupack, M. and Johnston, D.W., Bond Development Length Tests of a Grouted 54 Strand Post-tensioning Tendon, *J. of ACI*, Proc. Vol. 71, No. 10, October 1974, pp. 522–25.

12.9. Linger, D.A. and Bhonsle, S.R., An Investigation of Transfer Length in Pretensioned Concrete Using Photoelasttcity, *J. of PCI*, Vol. 8, August 1963.

12.10. Kaar, P.H., LaFraugh, R.W. and Mass, M.A. Influence of Concrete Strength on Strand Transfer Length, *J. of PCI*, Vol. 8, No. 5, October 1963.

12.11. Marshall, W.T., Theory for End Zone Stresses in Pretensioned Concrete Beams, *J. of PCI*, Vol. 11, No. 2, April 1966.

12.12. Ganguli S., End-zone cracking in Pretensioned Prestressed Concrete Beams, *Indian Concrete Journal*, March 1965, Vol. 39.

12.13. Ganguli S., Transmission Length in Pretensioned Prestressed Concrete, *Indian Concrete Journal*, Jan. 1966, Vol. 40.

12.14. Krishnamurthy D. A Method of Determining the Tensile Stresses in the End Zones of Pre-tensioned Beams, *Indian Concrete Journal*, July 1971 Vol. 45, No. 7,

12.15. Krishnamurthy D., Design of End Zone Reinforcement to Control Horizontal Cracking in Pretensioned Concrete Members at Transfer–1, *Indian Concrete Journal*, Sept. 1973, Vol. 47. No. 9.

12.16. Krishnamurthy D., Design of End Zone Reinforcement to Control Horizontal Cracking in Pretensioned Concrete Members at Transfer–2, Indian Concrete Journal, Oct. 1973, Vol. 47, No. 10.

12.17. Kalyanasundaram P., Krishnamoorthy C.S. and Srinivas Rao P., Hnd-Zone Stresses in Pretensioned Prestressed Concrete Beams, *Indian Concrete Journal*, Oct. 1976, Vol. 50, No. 10.

12.18. Salon.ms John R., McGrate Tirnthy E. Bond. Characteristics of Untcnsioned Prestressing Strand, *J. of PCI*, Jan.–Feb. 1977. Vol. 22, No. 1.

12.19. Sriryyas Rao P., Kalyanasundaram P and Sharieff. Fazllul-lah Md., Transmission Length of Ribbed Bars in Pretensioned Concrete, *Indian Concrete Journal*, May 1977, Vol. 51, No. 5.

13.1. Guyon, Y. *Prestressed Concrete*, Vol. I, John Wiley and Sens, New York. 1953.

13.2. Leonhardt, F., Prestressed Concrete Design and Construction, Wilhelm Ernst and John Berlin, 1964, p. 271.

13.3. Som, Pijush Kanti and Ghosh, Kalyanmay, Anchor Z one Stresses in Prestressed Concrete Beams, *J. of ASCE*, Vol. 90, No. ST 4, August 1964, pp. 49–62.

13.4. Eberhardt, A. and Veltrop, J.A, Prestressed Anchorage for Large Tainter Gate, *J. of ASCE*, Vol. 90, No. ST 6, December 1964, pp. 123–48.

13.5. Middendorf, K.H. Practical Aspects of End Zone Bearing of Post-Tensioning Tendons, *J. of PCI*, Vol. 8, No. 4, August 1963.

13.6. Dowrick, D.J., Anchorage Zone Reinforcement for Post-Tensioned Concrete, *Civil Engineering and Public Works Review*, Vol. 62, January 1967, pp. 51–54.

13.7. Chandrashekhar K., Abraham Jacob K and Sunduraraju Iyengar K.T., Design of Anchorage Zones in Post-Tensioned Prestressed Concrete Beams, *Indian Concrete Journal*, June 1977. Vol. 51, No. 6, p. 176.

14.1. Searl, E.N., Fire Rating and Building Codes, *J. of PCI*, Vol. 3, December 1958, pp. 17–22.

14.2. PAC Fire Test Centre Works on Prestressed Concrete, *J. of PCI*, Vol. 3, March 1959, pp. 61–62.

14.3. Woods, H., Research onFire Resistance of Prestressed Concrete, *J. of ASCE*, ST, November, 1960.

14.4. Troxell, G.E., *Fire Resistance Concrete, Proc. of Western Conference on Prestressed Concrete Buildings*, Gordon and Breach, New York, 1962, pp. 205–21.

14.5. Hill, A.W. and Aston, L.A., The Fire Resistance of Prestressed Concrete, *World Conf.* on Prestressed Concrete, San Fransisco, 1957,

14.6. Ashton, L.A. and Bate, S.C.C., The Fire Resistance of Prestressed Concrete Beams, *Proc. of Institution of Civil Engineers*, London, 1960, pp. 15–38.

14.7. Gerwick Jr., B.C., Ocean Structures of Prestressed Concrete, *J. of PCI*, Vol. 16, No. 2, March–April 1971, pp. 10–23.

14.8. Industry Report (PCI Editor), Accoustical Properties of Prestressed Concrete, *J. of PCI*, Vol. 16, No. 2, March–April 1971, pp. 63–71.

14.9. Industry Report (PCI Editor), Thermal Properties of Precast Concrete, *J. of PCI*, Vol. 16 No. 3, May–June 1971, pp. 33–43.

14.10. PCI Committee on Architectural Precast Concrete Guide Specifications. Guide Specifications for Architectural Precast Concrete, *J. of PCI*. Vol. 16, No. 4, July–August 1971. pp. 10–20.

14.11. Abrams, M.S.. and Gustaferro, A.H., Fire Endurance of Prestressed Concrete Units Coated with Spray-Applied Insulation, *J. of PCI*, Vol. 17, No. 1, January–February 1972, pp. 82–103.

14.12. Elliot, A.L., Hindsight and Foresight on the Performance of Prestressed Concrete Bridges in the Sam Fernando Earthquake, *J. of PCI*, Vol. 17, No. 2, March–April 1972, pp. 8–16.

14.13. LaGue, D.J., Thermal Movements in Prestressed Hollow Core Slabs, *J. of PCI*, Vol. 17, No. 2, March–April 1972, pp. 32–45.

14.14. Calhoun, W.D. and Hemsley, W.T., A Technique to Gain Extra Capacity in Double Tees, *J. of PCI*, Vol. 17, No. 3, May–June 1972, pp. 46–60.

14.15. PCI Committee on Fire Resistance Ratings. Fire Endurance of Prestressed Concrete Double Tee Wall Assemblies, *J. of PCI*, Vol. 17, No. 4, July–August 1972, pp. 19–28.

14.16. Parme, A.L., American Practice in Seismic Region, *J. of PCI*, Voi. 17, No. 4, July–August 1972, pp. 31–44.

14.17. Mast, R.I–, Seismic Design oi~24-Storey Building with Precast Elements. *J. of PCI*, Vol. 17, No. 4, July–August 1972, pp. 45–59.

14.18. Walocha, H.J., Parking Structure Solutions in a Seismic Zone, *J. of PCI*, Vol. 17, No. 4, July–August 1972. pp. 60–4.

14.19. Pond, W.F., Performance of Bridges During San Fernando Earthquake, *J. of PCI*, Vol. 17, No. 14, July–August 1972, pp. 65–75.

14.20. Inomata, S., Japanese Practice in Seismic Design of Prestressed Bridges, *J. of PCI*, Vol. 17, No. 4, July–August 1972, pp. 76–85.

14.21. Leonhardt, E., Improving the Seismic Safety of Prestressed Concrete Bridges, *J. of PCI*, Vol. 17, No. 6, November–December 1972, pp. 37–44.

14.22. Gustaferro, A.H., Fire Resistance of Post-tensioned Structures, *J. of PCI*, Vol. 18, No. 2, March–April 1973, pp. 38–63.

14.23. Lin, T. Y. and Kulka. F., Construction of Rio Colorado Bridge, *J. of PCI*, Vol. 18, No. 6, November–December 1973, pp. 92–101.

14.24. Gustaferro, A.H., Design of Prestressed Concrete for Fire Resistance, *J. of PCI*, Vol. 18, No– 6, November December 1973, pp. 102–116.

14.25. Kar, A.K., Pre'stressed Applications in Distressed Structures, *J. of PCI*, Vol. 19, No. 2, March–April 1974, pp. 93–97.

14.26. Hugenschmidt, F., Epoxy Adhesives in Precast Prestressed Concrete Construction, *J. of PCI*, Vol. 19, No. 2. March–April 1974, pp. 122–24.

14.27. Gentilini, B. and Gentilini, L., Giant Precast Prestressed Girder Bridge Erected Using Special Launching Technique, *J. of PCI*, Vol. 19, No. 3, May–June 1974, pp. 18–29.

14.28. Leonhardt, F., To New Frontiers for Prestressed Concrete Design and Construction, *J. of PCI*, Vol. 19, No. 5, September–October 1974, pp. 54–69,

14.29. Prestressed Concrete Works in India, *ING of IABSE*, 1973, New Delhi.

14.30. Warner, R.F. and Hulsbos, CL., Probable Fatigue Life of Prestressed Concrete Beams, *J. of PCI*, Vol. 11, April 1966.

14.31. Warner, R.F. and Hulsbos, C.L., Fatigue Properties of Prestressed Strand, *J. of PCI*, Vol. 11, February 1966.

14.32. Liu, Y. and Zuk, W., Thermoelastic Effects in Prestressed Flexural Members, *J. of PCI*, Vol. 8, June 1963.

14.33. Sbarounis, J.A. and Amrhein, J.E., Behavior of Prestressed Concrete Structures During the Alaskan Earthquake, *J. of PCI*, April 1965.

14.34. PCI Seismic Committee. Principle of Design and Construction of Earthquakes Resistant Prestressed Concrete Structures, *J. of PCI*, Vol. 11, June 1966.

14.35. Lin, T.Y., Earthquake Resistant Structures, *J. of ASCE*, ST, October 1965.

14.36. Johnson, D., Erection of Precast Concrete. *J, of PCI*, Vol. 14, April 1969.

14.37. Bryan, R.H., Aluminium Rolling Mill Framed in Precast Construction, *J. of PCI*, Vol. 14, August 1969.

14.38. Dayaratnam, P., Lateral Buckling of Prestressed Concrete Girders, *Proc. of the 9th Congress of* STBM, Kanpur, 1964.

14.39. Wilby, C.H., *Elastic Stability of Post-tensioned Prestressed Concrete* Members, Edward Arnold Ltd., 1964.

14.40. Feld, J. Difficulties and Incidents in Prestressed Concrete, *J of Boston Society of Civil Engineers*, January 1964.

14.41. Jatna B.L., The Free Cantilever Method of Constructing Prestressed Concrete Bridges, *Indian Concrete Journal*, May 1963.

14.42. Editorial. Research on the Fire of Prestressed Concrete, *Indian Concrete Journal*, Aug. 1963.

14.43. Tamhankar M.G., Rao Prasad A.S, Kapta M.S., Standardisation of Precast Prestressed Concrete Bridge Beams, *Indian Concrete Journal*, June 1968, Vol. 42.

14.44. Muspratt, M.A., The Effects of Elevated Temperatures on Concrete-Prestressed with Alloy Steel Bars, *Indian Concrete Journal*, Dec. 1968 Vol. 42, No. 12,

14.45. Alfred, Yee A. Prestressed Concrete for Building, *J. of Prestressed Concrete Institute*, Sept.–Oct. 1976, Vol. 21, No. 5.

14.46. Rangaswamy N.S. and Rajagopalan K.S., Corrosion of High Tensile Steel in Prestressed Concrete 1-Electrochemical Studies, *Indian Concrete Journal*, Oct. 1977, Vol. 51, No. 10.

14.47. Rangaswamy N.S., Chandrasekharan S. and Rajagopalan K.S., Corrosion of High Strength Steel in Prestressed Concrete: 2-Corrosion Studies in High-strength Post-tensioned Concrete, *Indian Concrete Journal*, No. 1977, Vol. 51, No. 11.

14.48. Sawko F., *Developments in Prestressed Concrete-2*, Applied Science Publisher Ltd., 1978.

14.49. Mirza J.F. Tawpik M.E., End Cracking in PSC Members During Detensioning, *PCI*, Mar.–Apr. 1978 Vol. 23, No. 2.

14.50. Ramasay B., The Incremental Lavaching Method in Prestressed Concrete Bridge Construction, *Concrete International*, Feb 1980, No. 2.

Index